T0211810

Graduate Texts in Mathematics 52

Editorial Board

S. Axler   F.W. Gehring   K.A. Ribet

Graduate Texts in Mathematics

most recent titles in the GTM series

Robin Hartshorne

Algebraic Geometry

Springer

Robin Hartshorne
Department of Mathematics
University of California
Berkeley, California 94720
USA

Mathematics Subject Classification (2000): 13-xx, 14A10, 14A15, 14Fxx, 14Hxx, 14Jxx

Library of Congress Cataloging-in-Publication Data
Hartshorne, Robin.
    Algebraic geometry.

    (Graduate texts in mathematics: 52)
    Bibliography: p.
    Includes index.
    1. Geometry, Algebraic.   I. Title   II. Series.
QA564.H25   516′.35   77-1177

ISBN 978-1-4419-2807-8   ISBN 978-1-4757-3849-0 (eBook)   Printed on acid-free paper.
DOI 10.1007/978-1-4757-3849-0

(ASC/SBA)

15 14

springeronline.com

For Edie, Jonathan, and Benjamin

Preface

This book provides an introduction to abstract algebraic geometry using the methods of schemes and cohomology. The main objects of study are algebraic varieties in an affine or projective space over an algebraically closed field; these are introduced in Chapter I, to establish a number of basic concepts and examples. Then the methods of schemes and cohomology are developed in Chapters II and III, with emphasis on applications rather than excessive generality. The last two chapters of the book (IV and V) use these methods to study topics in the classical theory of algebraic curves and surfaces.

The prerequisites for this approach to algebraic geometry are results from commutative algebra, which are stated as needed, and some elementary topology. No complex analysis or differential geometry is necessary. There are more than four hundred exercises throughout the book, offering specific examples as well as more specialized topics not treated in the main text. Three appendices present brief accounts of some areas of current research.

This book can be used as a textbook for an introductory course in algebraic geometry, following a basic graduate course in algebra. I recently taught this material in a five-quarter sequence at Berkeley, with roughly one chapter per quarter. Or one can use Chapter I alone for a short course. A third possibility worth considering is to study Chapter I, and then proceed directly to Chapter IV, picking up only a few definitions from Chapters II and III, and *assuming* the statement of the Riemann–Roch theorem for curves. This leads to interesting material quickly, and may provide better motivation for tackling Chapters II and III later.

The material covered in this book should provide adequate preparation for reading more advanced works such as Grothendieck [EGA], [SGA], Hartshorne [5], Mumford [2], [5], or Shafarevich [1].

Preface

Acknowledgements

In writing this book, I have attempted to present what is essential for a basic course in algebraic geometry. I wanted to make accessible to the nonspecialist an area of mathematics whose results up to now have been widely scattered, and linked only by unpublished "folklore." While I have reorganized the material and rewritten proofs, the book is mostly a synthesis of what I have learned from my teachers, my colleagues, and my students. They have helped in ways too numerous to recount. I owe especial thanks to Oscar Zariski, J.-P. Serre, David Mumford, and Arthur Ogus for their support and encouragement.

Aside from the "classical" material, whose origins need a historian to trace, my greatest intellectual debt is to A. Grothendieck, whose treatise [EGA] is the authoritative reference for schemes and cohomology. His results appear without specific attribution throughout Chapters II and III. Otherwise I have tried to acknowledge sources whenever I was aware of them.

In the course of writing this book, I have circulated preliminary versions of the manuscript to many people, and have received valuable comments from them. To all of these people my thanks, and in particular to J.-P. Serre, H. Matsumura, and Joe Lipman for their careful reading and detailed suggestions.

I have taught courses at Harvard and Berkeley based on this material, and I thank my students for their attention and their stimulating questions.

I thank Richard Bassein, who combined his talents as mathematician and artist to produce the illustrations for this book.

A few words cannot adequately express the thanks I owe to my wife, Edie Churchill Hartshorne. While I was engrossed in writing, she created a warm home for me and our sons Jonathan and Benjamin, and through her constant support and friendship provided an enriched human context for my life.

For financial support during the preparation of this book, I thank the Research Institute for Mathematical Sciences of Kyoto University, the National Science Foundation, and the University of California at Berkeley.

August 29, 1977
Berkeley, California   ROBIN HARTSHORNE

viii

Contents

Contents

Introduction

The author of an introductory book on algebraic geometry has the difficult task of providing geometrical insight and examples, while at the same time developing the modern technical language of the subject. For in algebraic geometry, a great gap appears to separate the intuitive ideas which form the point of departure from the technical methods used in current research.

The first question is that of language. Algebraic geometry has developed in waves, each with its own language and point of view. The late nineteenth century saw the function-theoretic approach of Riemann, the more geometric approach of Brill and Noether, and the purely algebraic approach of Kronecker, Dedekind, and Weber. The Italian school followed with Castelnuovo, Enriques, and Severi, culminating in the classification of algebraic surfaces. Then came the twentieth-century "American" school of Chow, Weil, and Zariski, which gave firm algebraic foundations to the Italian intuition. Most recently, Serre and Grothendieck initiated the French school, which has rewritten the foundations of algebraic geometry in terms of schemes and cohomology, and which has an impressive record of solving old problems with new techniques. Each of these schools has introduced new concepts and methods. In writing an introductory book, is it better to use the older language which is closer to the geometric intuition, or to start at once with the technical language of current research?

The second question is a conceptual one. Modern mathematics tends to obliterate history: each new school rewrites the foundations of its subject in its own language, which makes for fine logic but poor pedagogy. Of what use is it to know the definition of a scheme if one does not realize that a ring of integers in an algebraic number field, an algebraic curve, and a compact Riemann surface are all examples of a "regular scheme of dimension one''? How then can the author of an introductory book indicate the inputs to algebraic geometry coming from number theory, commutative algebra, and complex analysis, and also introduce the reader to the main objects of study, which are algebraic varieties in affine or projective space, while at the same time developing the modern language of schemes and cohomology? What choice of topics will convey the meaning of algebraic geometry, and still serve as a firm foundation for further study and research?

My own bias is somewhat on the side of classical geometry. I believe that the most important problems in algebraic geometry are those arising from old-fashioned varieties in affine or projective spaces. They provide the geometric intuition which motivates all further developments. In this book, I begin with a chapter on varieties, to establish many examples and basic ideas in their simplest form, uncluttered with technical details. Only after that do I develop systematically the language of schemes, coherent sheaves, and cohomology, in Chapters II and III. These chapters form the technical heart of the book. In them I attempt to set forth the most important results, but without striving for the utmost generality. Thus, for example, the cohomology theory is developed only for quasi-coherent sheaves on noetherian schemes, since this is simpler and sufficient for most applications; the theorem of ''coherence of direct image sheaves'' is proved only for projective morphisms, and not for arbitrary proper morphisms. For the same reasons I do not include the more abstract notions of representable functors, algebraic spaces, étale cohomology, sites, and topoi.

The fourth and fifth chapters treat classical material, namely nonsingular projective curves and surfaces, but they use techniques of schemes and cohomology. I hope these applications will justify the effort needed to absorb all the technical apparatus in the two previous chapters.

As the basic language and logical foundation of algebraic geometry, I have chosen to use commutative algebra. It has the advantage of being precise. Also, by working over a base field of arbitrary characteristic, which is necessary in any case for applications to number theory, one gains new insight into the classical case of base field $\mathbf{C}$. Some years ago, when Zariski began to prepare a volume on algebraic geometry, he had to develop the necessary algebra as he went. The task grew to such proportions that he produced a book on commutative algebra only. Now we are fortunate in having a number of excellent books on commutative algebra: Atiyah–Macdonald [1], Bourbaki [1], Matsumura [2], Nagata [7], and Zariski–Samuel [1]. My policy is to quote purely algebraic results as needed, with references to the literature for proof. A list of the results used appears at the end of the book.

Originally I had planned a whole series of appendices—short expository accounts of some current research topics, to form a bridge between the main text of this book and the research literature. Because of limited time and space only three survive. I can only express my regret at not including the others, and refer the reader instead to the Arcata volume (Hartshorne, ed. [1]) for a series of articles by experts in their fields, intended for the nonspecialist. Also, for the historical development of algebraic geometry let me refer to Dieudonné [1]. Since there was not space to explore the relation of algebraic geometry to neighboring fields as much as I would have liked, let me refer to the survey article of Cassels [1] for connections with number theory, and to Shafarevich [2, Part III] for connections with complex manifolds and topology.

Because I believe strongly in active learning, there are a great many exercises in this book. Some contain important results not treated in the main text. Others contain specific examples to illustrate general phenomena. I believe that the study of particular examples is inseparable from the development of general theories. The serious student should attempt as many as possible of these exercises, but should not expect to solve them immediately. Many will require a real creative effort to understand. An asterisk denotes a more difficult exercise. Two asterisks denote an unsolved problem.

See (I, §8) for a further introduction to algebraic geometry and this book.

Terminology

For the most part, the terminology of this book agrees with generally accepted usage, but there are a few exceptions worth noting. A *variety* is always irreducible and is always over an algebraically closed field. In Chapter I all varieties are quasi-projective. In (Ch. II, §4) the definition is expanded to include *abstract varieties,* which are integral separated schemes of finite type over an algebraically closed field. The words *curve, surface,* and *3-fold* are used to mean varieties of dimension 1, 2, and 3 respectively. But in Chapter IV, the word *curve* is used only for a nonsingular projective curve; whereas in Chapter V a *curve* is any effective divisor on a nonsingular projective surface. A *surface* in Chapter V is always a nonsingular projective surface.

A *scheme* is what used to be called a prescheme in the first edition of [EGA], but is called scheme in the new edition of [EGA, Ch. I].

The definitions of a *projective morphism* and a *very ample invertible sheaf* in this book are not equivalent to those in [EGA]—see (II, §4, 5). They are technically simpler, but have the disadvantage of not being local on the base.

The word *nonsingular* applies only to varieties; for more general schemes, the words *regular* and *smooth* are used.

Results from algebra

I assume the reader is familiar with basic results about rings, ideals, modules, noetherian rings, and integral dependence, and is willing to accept or look up other results, belonging properly to commutative algebra or homological algebra, which will be stated as needed, with references to the literature. These results will be marked with an A: e.g., Theorem 3.9A, to distinguish them from results proved in the text.

The basic conventions are these: All rings are commutative with identity element 1. All homomorphisms of rings take 1 to 1. In an integral domain or a field, $0 \neq 1$. A *prime ideal* (respectively, *maximal ideal*) is an ideal $\mathfrak{p}$ in a ring A such that the quotient ring $A/\mathfrak{p}$ is an integral domain (respectively, a field). Thus the ring itself is not considered to be a prime ideal or a maximal ideal.

A *multiplicative system* in a ring A is a subset S, containing 1, and closed under multiplication. The *localization* $S^{-1}A$ is defined to be the ring formed by equivalence classes of fractions a/s, $a \in A$, $s \in S$, where a/s and a'/s' are said to be *equivalent* if there is an $s'' \in S$ such that $s''(s'a - sa') = 0$ (see e.g. Atiyah–Macdonald [1, Ch. 3]). Two special cases which are used constantly are the following. If $\mathfrak{p}$ is a prime ideal in A, then $S = A - \mathfrak{p}$ is a multiplicative system, and the corresponding localization is denoted by $A_\mathfrak{p}$. If f is an element of A, then $S = \{1\} \cup \{f^n \mid n \geq 1\}$ is a multiplicative system, and the corresponding localization is denoted by A_f. (Note for example that if f is nilpotent, then A_f is the zero ring.)

References

Bibliographical references are given by author, with a number in square brackets to indicate which work, e.g. Serre, [3, p. 75]. Cross references to theorems, propositions, lemmas within the same chapter are given by number in parentheses, e.g. (3.5). Reference to an exercise is given by (Ex. 3.5). References to results in another chapter are preceded by the chapter number, e.g. (II, 3.5), or (II, Ex. 3.5).

CHAPTER I
Varieties

Our purpose in this chapter is to give an introduction to algebraic geometry with as little machinery as possible. We work over a fixed algebraically closed field k. We define the main objects of study, which are algebraic varieties in affine or projective space. We introduce some of the most important concepts, such as dimension, regular functions, rational maps, nonsingular varieties, and the degree of a projective variety. And most important, we give lots of specific examples, in the form of exercises at the end of each section. The examples have been selected to illustrate many interesting and important phenomena, beyond those mentioned in the text. The person who studies these examples carefully will not only have a good understanding of the basic concepts of algebraic geometry, but he will also have the background to appreciate some of the more abstract developments of modern algebraic geometry, and he will have a resource against which to check his intuition. We will continually refer back to this library of examples in the rest of the book.

The last section of this chapter is a kind of second introduction to the book. It contains a discussion of the "classification problem," which has motivated much of the development of algebraic geometry. It also contains a discussion of the degree of generality in which one should develop the foundations of algebraic geometry, and as such provides motivation for the theory of schemes.

1 Affine Varieties

Let k be a fixed algebraically closed field. We define *affine n-space* over k, denoted $\mathbf{A}^n_k$ or simply $\mathbf{A}^n$, to be the set of all n-tuples of elements of k. An element $P \in \mathbf{A}^n$ will be called a *point*, and if $P = (a_1, \ldots, a_n)$ with $a_i \in k$, then the a_i will be called the *coordinates* of P.

Let $A = k[x_1, \ldots, x_n]$ be the polynomial ring in n variables over k. We will interpret the elements of A as functions from the affine n-space to k, by defining $f(P) = f(a_1, \ldots, a_n)$, where $f \in A$ and $P \in \mathbf{A}^n$. Thus if $f \in A$ is a polynomial, we can talk about the set of *zeros* of f, namely $Z(f) = \{P \in \mathbf{A}^n | f(P) = 0\}$. More generally, if T is any subset of A, we define the *zero set* of T to be the common zeros of all the elements of T, namely

$$Z(T) = \{P \in \mathbf{A}^n | f(P) = 0 \text{ for all } f \in T\}.$$

Clearly if $\mathfrak{a}$ is the ideal of A generated by T, then $Z(T) = Z(\mathfrak{a})$. Furthermore, since A is a noetherian ring, any ideal $\mathfrak{a}$ has a finite set of generators $f_1, \ldots, f_r$. Thus $Z(T)$ can be expressed as the common zeros of the finite set of polynomials $f_1, \ldots, f_r$.

Definition. A subset Y of $\mathbf{A}^n$ is an *algebraic set* if there exists a subset $T \subseteq A$ such that $Y = Z(T)$.

Proposition 1.1. *The union of two algebraic sets is an algebraic set. The intersection of any family of algebraic sets is an algebraic set. The empty set and the whole space are algebraic sets.*

PROOF. If $Y_1 = Z(T_1)$ and $Y_2 = Z(T_2)$, then $Y_1 \cup Y_2 = Z(T_1 T_2)$, where $T_1 T_2$ denotes the set of all products of an element of T_1 by an element of T_2. Indeed, if $P \in Y_1 \cup Y_2$, then either $P \in Y_1$ or $P \in Y_2$, so P is a zero of every polynomial in $T_1 T_2$. Conversely, if $P \in Z(T_1 T_2)$, and $P \notin Y_1$ say, then there is an $f \in T_1$ such that $f(P) \neq 0$. Now for any $g \in T_2, (fg)(P) = 0$ implies that $g(P) = 0$, so that $P \in Y_2$.

If $Y_\alpha = Z(T_\alpha)$ is any family of algebraic sets, then $\bigcap Y_\alpha = Z(\bigcup T_\alpha)$, so $\bigcap Y_\alpha$ is also an algebraic set. Finally, the empty set $\varnothing = Z(1)$, and the whole space $\mathbf{A}^n = Z(0)$.

Definition. We define the *Zariski topology* on $\mathbf{A}^n$ by taking the open subsets to be the complements of the algebraic sets. This is a topology, because according to the proposition, the intersection of two open sets is open, and the union of any family of open sets is open. Furthermore, the empty set and the whole space are both open.

Example 1.1.1. Let us consider the Zariski topology on the affine line $\mathbf{A}^1$. Every ideal in $A = k[x]$ is principal, so every algebraic set is the set of zeros of a single polynomial. Since k is algebraically closed, every nonzero polynomial $f(x)$ can be written $f(x) = c(x - a_1) \cdots (x - a_n)$ with $c, a_1, \ldots, a_n \in k$. Then $Z(f) = \{a_1, \ldots, a_n\}$. Thus the algebraic sets in $\mathbf{A}^1$ are just the finite subsets (including the empty set) and the whole space (corresponding to $f = 0$). Thus the open sets are the empty set and the complements of finite subsets. Notice in particular that this topology is not Hausdorff.

Definition. A nonempty subset Y of a topological space X is *irreducible* if it cannot be expressed as the union $Y = Y_1 \cup Y_2$ of two proper subsets, each one of which is closed in Y. The empty set is not considered to be irreducible.

Example 1.1.2. $\mathbf{A}^1$ is irreducible, because its only proper closed subsets are finite, yet it is infinite (because k is algebraically closed, hence infinite).

Example 1.1.3. Any nonempty open subset of an irreducible space is irreducible and dense.

Example 1.1.4. If Y is an irreducible subset of X, then its closure $\bar{Y}$ in X is also irreducible.

Definition. An *affine algebraic variety* (or simply *affine variety*) is an irreducible closed subset of $\mathbf{A}^n$ (with the induced topology). An open subset of an affine variety is a *quasi-affine variety*.

These affine and quasi-affine varieties are our first objects of study. But before we can go further, in fact before we can even give any interesting examples, we need to explore the relationship between subsets of $\mathbf{A}^n$ and ideals in A more deeply. So for any subset $Y \subseteq \mathbf{A}^n$, let us define the *ideal of* Y in A by

$$I(Y) = \{f \in A | f(P) = 0 \text{ for all } P \in Y\}.$$

Now we have a function Z which maps subsets of A to algebraic sets, and a function I which maps subsets of $\mathbf{A}^n$ to ideals. Their properties are summarized in the following proposition.

Proposition 1.2.
(a) *If $T_1 \subseteq T_2$ are subsets of A, then $Z(T_1) \supseteq Z(T_2)$.*
(b) *If $Y_1 \subseteq Y_2$ are subsets of $\mathbf{A}^n$, then $I(Y_1) \supseteq I(Y_2)$.*
(c) *For any two subsets Y_1, Y_2 of $\mathbf{A}^n$, we have $I(Y_1 \cup Y_2) = I(Y_1) \cap I(Y_2)$.*
(d) *For any ideal $\mathfrak{a} \subseteq A$, $I(Z(\mathfrak{a})) = \sqrt{\mathfrak{a}}$, the radical of $\mathfrak{a}$.*
(e) *For any subset $Y \subseteq \mathbf{A}^n$, $Z(I(Y)) = \bar{Y}$, the closure of Y.*

PROOF. (a), (b) and (c) are obvious. (d) is a direct consequence of Hilbert's Nullstellensatz, stated below, since the radical of $\mathfrak{a}$ is defined as

$$\sqrt{\mathfrak{a}} = \{f \in A | f^r \in \mathfrak{a} \text{ for some } r > 0\}.$$

To prove (e), we note that $Y \subseteq Z(I(Y))$, which is a closed set, so clearly $\bar{Y} \subseteq Z(I(Y))$. On the other hand, let W be any closed set containing Y. Then $W = Z(\mathfrak{a})$ for some ideal $\mathfrak{a}$. So $Z(\mathfrak{a}) \supseteq Y$, and by (b), $IZ(\mathfrak{a}) \subseteq I(Y)$. But certainly $\mathfrak{a} \subseteq IZ(\mathfrak{a})$, so by (a) we have $W = Z(\mathfrak{a}) \supseteq ZI(Y)$. Thus $ZI(Y) = \bar{Y}$.

Theorem 1.3A (Hilbert's Nullstellensatz). *Let* k *be an algebraically closed field, let* $\mathfrak{a}$ *be an ideal in* $A = k[x_1, \ldots, x_n]$, *and let* $f \in A$ *be a polynomial which vanishes at all points of* $Z(\mathfrak{a})$. *Then* $f^r \in \mathfrak{a}$ *for some integer* $r > 0$.

PROOF. Lang [2, p. 256] or Atiyah–Macdonald [1, p. 85] or Zariski–Samuel [1, vol. 2, p. 164].

Corollary 1.4. *There is a one-to-one inclusion-reversing correspondence between algebraic sets in* $\mathbf{A}^n$ *and radical ideals (i.e., ideals which are equal to their own radical) in* A, *given by* $Y \mapsto I(Y)$ *and* $\mathfrak{a} \mapsto Z(\mathfrak{a})$. *Furthermore, an algebraic set is irreducible if and only if its ideal is a prime ideal.*

PROOF. Only the last part is new. If Y is irreducible, we show that $I(Y)$ is prime. Indeed, if $fg \in I(Y)$, then $Y \subseteq Z(fg) = Z(f) \cup Z(g)$. Thus $Y = (Y \cap Z(f)) \cup (Y \cap Z(g))$, both being closed subsets of Y. Since Y is irreducible, we have either $Y = Y \cap Z(f)$, in which case $Y \subseteq Z(f)$, or $Y \subseteq Z(g)$. Hence either $f \in I(Y)$ or $g \in I(Y)$.

Conversely, let $\mathfrak{p}$ be a prime ideal, and suppose that $Z(\mathfrak{p}) = Y_1 \cup Y_2$. Then $\mathfrak{p} = I(Y_1) \cap I(Y_2)$, so either $\mathfrak{p} = I(Y_1)$ or $\mathfrak{p} = I(Y_2)$. Thus $Z(\mathfrak{p}) = Y_1$ or Y_2, hence it is irreducible.

Example 1.4.1. $\mathbf{A}^n$ is irreducible, since it corresponds to the zero ideal in A, which is prime.

Example 1.4.2. Let f be an irreducible polynomial in $A = k[x, y]$. Then f generates a prime ideal in A, since A is a unique factorization domain, so the zero set $Y = Z(f)$ is irreducible. We call it the *affine curve* defined by the equation $f(x, y) = 0$. If f has degree d, we say that Y is a curve of *degree d*.

Example 1.4.3. More generally, if f is an irreducible polynomial in $A = k[x_1, \ldots, x_n]$, we obtain an affine variety $Y = Z(f)$, which is called a *surface* if $n = 3$, or a *hypersurface* if $n > 3$.

Example 1.4.4. A maximal ideal $\mathfrak{m}$ of $A = k[x_1, \ldots, x_n]$ corresponds to a minimal irreducible closed subset of $\mathbf{A}^n$, which must be a point, say $P = (a_1, \ldots, a_n)$. This shows that every maximal ideal of A is of the form $\mathfrak{m} = (x_1 - a_1, \ldots, x_n - a_n)$, for some $a_1, \ldots, a_n \in k$.

Example 1.4.5. If k is not algebraically closed, these results do not hold. For example, if $k = \mathbf{R}$, the curve $x^2 + y^2 + 1 = 0$ in $\mathbf{A}^2_{\mathbf{R}}$ has no points. So (1.2d) is false. See also (Ex. 1.12).

Definition. If $Y \subseteq \mathbf{A}^n$ is an affine algebraic set, we define the *affine coordinate ring* $A(Y)$ of Y, to be $A/I(Y)$.

Remark 1.4.6. If Y is an affine variety, then $A(Y)$ is an integral domain. Furthermore, $A(Y)$ is a finitely generated k-algebra. Conversely, any finitely generated k-algebra B which is a domain is the affine coordinate ring of some affine variety. Indeed, write B as the quotient of a polynomial ring $A = k[x_1, \ldots, x_n]$ by an ideal $\mathfrak{a}$, and let $Y = Z(\mathfrak{a})$.

Next we will study the topology of our varieties. To do so we introduce an important class of topological spaces which includes all varieties.

Definition. A topological space X is called *noetherian* if it satisfies the *descending chain condition* for closed subsets: for any sequence $Y_1 \supseteq Y_2 \supseteq \ldots$ of closed subsets, there is an integer r such that $Y_r = Y_{r+1} = \ldots$.

Example 1.4.7. $\mathbf{A}^n$ is a noetherian topological space. Indeed, if $Y_1 \supseteq Y_2 \supseteq \ldots$ is a descending chain of closed subsets, then $I(Y_1) \subseteq I(Y_2) \subseteq \ldots$ is an ascending chain of ideals in $A = k[x_1, \ldots, x_n]$. Since A is a noetherian ring, this chain of ideals is eventually stationary. But for each i, $Y_i = Z(I(Y_i))$, so the chain Y_i is also stationary.

Proposition 1.5. *In a noetherian topological space X, every nonempty closed subset Y can be expressed as a finite union $Y = Y_1 \cup \ldots \cup Y_r$ of irreducible closed subsets Y_i. If we require that $Y_i \not\supseteq Y_j$ for $i \neq j$, then the Y_i are uniquely determined. They are called the irreducible components of Y.*

PROOF. First we show the existence of such a representation of Y. Let $\mathfrak{S}$ be the set of nonempty closed subsets of X which *cannot* be written as a finite union of irreducible closed subsets. If $\mathfrak{S}$ is nonempty, then since X is noetherian, it must contain a minimal element, say Y. Then Y is not irreducible, by construction of $\mathfrak{S}$. Thus we can write $Y = Y' \cup Y''$, where Y' and Y'' are proper closed subsets of Y. By minimality of Y, each of Y' and Y'' can be expressed as a finite union of closed irreducible subsets, hence Y also, which is a contradiction. We conclude that every closed set Y can be written as a union $Y = Y_1 \cup \ldots \cup Y_r$ of irreducible subsets. By throwing away a few if necessary, we may assume $Y_i \not\supseteq Y_j$ for $i \neq j$.

Now suppose $Y = Y_1' \cup \ldots \cup Y_s'$ is another such representation. Then $Y_1' \subseteq Y = Y_1 \cup \ldots \cup Y_r$, so $Y_1' = \bigcup (Y_1' \cap Y_i)$. But Y_1' is irreducible, so $Y_1' \subseteq Y_i$ for some i, say $i = 1$. Similarly, $Y_1 \subseteq Y_j'$ for some j. Then $Y_1' \subseteq Y_j'$, so $j = 1$, and we find that $Y_1 = Y_1'$. Now let $Z = (Y - Y_1)^-$. Then $Z = Y_2 \cup \ldots \cup Y_r$ and also $Z = Y_2' \cup \ldots \cup Y_s'$. So proceeding by induction on r, we obtain the uniqueness of the Y_i.

Corollary 1.6. *Every algebraic set in $\mathbf{A}^n$ can be expressed uniquely as a union of varieties, no one containing another.*

Definition. If X is a topological space, we define the *dimension* of X (denoted $\dim X$) to be the supremum of all integers n such that there exists a chain $Z_0 \subset Z_1 \subset \ldots \subset Z_n$ of distinct irreducible closed subsets of X. We define the *dimension* of an affine or quasi-affine variety to be its dimension as a topological space.

Example 1.6.1. The dimension of $\mathbf{A}^1$ is 1. Indeed, the only irreducible closed subsets of $\mathbf{A}^1$ are the whole space and single points.

Definition. In a ring A, the *height* of a prime ideal $\mathfrak{p}$ is the supremum of all integers n such that there exists a chain $\mathfrak{p}_0 \subset \mathfrak{p}_1 \subset \ldots \subset \mathfrak{p}_n = \mathfrak{p}$ of distinct prime ideals. We define the *dimension* (or *Krull dimension*) of A to be the supremum of the heights of all prime ideals.

Proposition 1.7. *If Y is an affine algebraic set, then the dimension of Y is equal to the dimension of its affine coordinate ring $A(Y)$.*

PROOF. If Y is an affine algebraic set in $\mathbf{A}^n$, then the closed irreducible subsets of Y correspond to prime ideals of $A = k[x_1, \ldots, x_n]$ containing $I(Y)$. These in turn correspond to prime ideals of $A(Y)$. Hence dim Y is the length of the longest chain of prime ideals in $A(Y)$, which is its dimension.

This proposition allows us to apply results from the dimension theory of noetherian rings to algebraic geometry.

Theorem 1.8A. *Let k be a field, and let B be an integral domain which is a finitely generated k-algebra. Then:*
    (a) *the dimension of B is equal to the transcendence degree of the quotient field $K(B)$ of B over k;*
    (b) *For any prime ideal $\mathfrak{p}$ in B, we have*

$$\text{height } \mathfrak{p} + \dim B/\mathfrak{p} = \dim B.$$

PROOF. Matsumura [2, Ch. 5, §14] or, in the case k is algebraically closed, Atiyah–Macdonald [1, Ch. 11]

Proposition 1.9. *The dimension of $\mathbf{A}^n$ is n.*

PROOF. According to (1.7) this says that the dimension of the polynomial ring $k[x_1, \ldots, x_n]$ is n, which follows from part (a) of the theorem.

Proposition 1.10. *If Y is a quasi-affine variety, then* dim $Y = $ dim $\bar{Y}$.

PROOF. If $Z_0 \subset Z_1 \subset \ldots \subset Z_n$ is a sequence of distinct closed irreducible subsets of Y, then $\bar{Z}_0 \subset \bar{Z}_1 \subset \ldots \subset \bar{Z}_n$ is a sequence of distinct closed irreducible subsets of $\bar{Y}$ (1.1.4), so we have dim $Y \leqslant$ dim $\bar{Y}$. In particular, dim Y is finite, so we can choose a maximal such chain $Z_0 \subset \ldots \subset Z_n$, with $n = $ dim Y. In that case Z_0 must be a point P, and the chain $P = \bar{Z}_0 \subset \ldots \subset \bar{Z}_n$ will also be maximal (1.1.3). Now P corresponds to a maximal ideal $\mathfrak{m}$ of the affine coordinate ring $A(\bar{Y})$ of $\bar{Y}$. The $\bar{Z}_i$ correspond to prime ideals contained in $\mathfrak{m}$, so height $\mathfrak{m} = n$. On the other hand, since P is a point in affine space, $A(\bar{Y})/\mathfrak{m} \cong k$ (1.4.4). Hence by (1.8Ab) we find that $n = $ dim $A(\bar{Y}) = $ dim $\bar{Y}$. Thus dim $Y = $ dim $\bar{Y}$.

Theorem 1.11A (Krull's Hauptidealsatz). *Let A be a noetherian ring, and let $f \in A$ be an element which is neither a zero divisor nor a unit. Then every minimal prime ideal $\mathfrak{p}$ containing f has height 1.*

PROOF. Atiyah–Macdonald [1, p. 122].

Proposition 1.12A. *A noetherian integral domain A is a unique factorization domain if and only if every prime ideal of height 1 is principal.*

PROOF. Matsumura [2, p. 141], or Bourbaki [1, Ch. 7, §3].

Proposition 1.13. *A variety Y in $\mathbf{A}^n$ has dimension $n - 1$ if and only if it is the zero set $Z(f)$ of a single nonconstant irreducible polynomial in $A = k[x_1, \ldots, x_n]$.*

PROOF. If f is an irreducible polynomial, we have already seen that $Z(f)$ is a variety. Its ideal is the prime ideal $\mathfrak{p} = (f)$. By (1.11A), $\mathfrak{p}$ has height 1, so by (1.8A), $Z(f)$ has dimension $n - 1$. Conversely, a variety of dimension $n - 1$ corresponds to a prime ideal of height 1. Now the polynomial ring A is a unique factorization domain, so by (1.12A), $\mathfrak{p}$ is principal, necessarily generated by an irreducible polynomial f. Hence $Y = Z(f)$.

Remark 1.13.1. A prime ideal of height 2 in a polynomial ring cannot necessarily be generated by two elements (Ex. 1.11).

EXERCISES

1.1. (a) Let Y be the plane curve $y = x^2$ (i.e., Y is the zero set of the polynomial $f = y - x^2$). Show that $A(Y)$ is isomorphic to a polynomial ring in one variable over k.

   (b) Let Z be the plane curve $xy = 1$. Show that $A(Z)$ is not isomorphic to a polynomial ring in one variable over k.

   *(c) Let f be any irreducible quadratic polynomial in $k[x, y]$, and let W be the conic defined by f. Show that $A(W)$ is isomorphic to $A(Y)$ or $A(Z)$. Which one is it when?

1.2. *The Twisted Cubic Curve.* Let $Y \subseteq \mathbf{A}^3$ be the set $Y = \{(t, t^2, t^3) | t \in k\}$. Show that Y is an affine variety of dimension 1. Find generators for the ideal $I(Y)$. Show that $A(Y)$ is isomorphic to a polynomial ring in one variable over k. We say that Y is given by the *parametric representation* $x = t, y = t^2, z = t^3$.

1.3. Let Y be the algebraic set in $\mathbf{A}^3$ defined by the two polynomials $x^2 - yz$ and $xz - x$. Show that Y is a union of three irreducible components. Describe them and find their prime ideals.

1.4. If we identify $\mathbf{A}^2$ with $\mathbf{A}^1 \times \mathbf{A}^1$ in the natural way, show that the Zariski topology on $\mathbf{A}^2$ is not the product topology of the Zariski topologies on the two copies of $\mathbf{A}^1$.

1.5. Show that a k-algebra B is isomorphic to the affine coordinate ring of some algebraic set in A^n, for some n, if and only if B is a finitely generated k-algebra with no nilpotent elements.

1.6. Any nonempty open subset of an irreducible topological space is dense and irreducible. If Y is a subset of a topological space X, which is irreducible in its induced topology, then the closure $\overline{Y}$ is also irreducible.

1.7. (a) Show that the following conditions are equivalent for a topological space X: (i) X is noetherian; (ii) every nonempty family of closed subsets has a minimal element; (iii) X satisfies the ascending chain condition for open subsets; (iv) every nonempty family of open subsets has a maximal element.

(b) A noetherian topological space is *quasi-compact*, i.e., every open cover has a finite subcover.

(c) Any subset of a noetherian topological space is noetherian in its induced topology.

(d) A noetherian space which is also Hausdorff must be a finite set with the discrete topology.

1.8. Let Y be an affine variety of dimension r in A^n. Let H be a hypersurface in A^n, and assume that $Y \nsubseteq H$. Then every irreducible component of $Y \cap H$ has dimension $r - 1$. (See (7.1) for a generalization.)

1.9. Let $a \subseteq A = k[x_1, \ldots, x_n]$ be an ideal which can be generated by r elements. Then every irreducible component of $Z(a)$ has dimension $\geqslant n - r$.

1.10. (a) If Y is any subset of a topological space X, then $\dim Y \leqslant \dim X$.

(b) If X is a topological space which is covered by a family of open subsets $\{U_i\}$, then $\dim X = \sup \dim U_i$.

(c) Give an example of a topological space X and a dense open subset U with $\dim U < \dim X$.

(d) If Y is a closed subset of an irreducible finite-dimensional topological space X, and if $\dim Y = \dim X$, then $Y = X$.

(e) Give an example of a noetherian topological space of infinite dimension.

*__1.11.__ Let $Y \subseteq A^3$ be the curve given parametrically by $x = t^3$, $y = t^4$, $z = t^5$. Show that $I(Y)$ is a prime ideal of height 2 in $k[x, y, z]$ which cannot be generated by 2 elements. We say Y is *not a local complete intersection*—cf. (Ex. 2.17).

1.12. Give an example of an irreducible polynomial $f \in R[x, y]$, whose zero set $Z(f)$ in A_R^2 is not irreducible (cf. 1.4.2).

2 Projective Varieties

To define projective varieties, we proceed in a manner analogous to the definition of affine varieties, except that we work in projective space.

Let k be our fixed algebraically closed field. We defined *projective n-space* over k, denoted P_k^n, or simply P^n, to be the set of equivalence classes of $(n + 1)$-tuples $(a_0, \ldots, a_n)$ of elements of k, not all zero, under the equivalence relation given by $(a_0, \ldots, a_n) \sim (\lambda a_0, \ldots, \lambda a_n)$ for all $\lambda \in k$, $\lambda \neq 0$. Another way of saying this is that P^n as a set is the quotient of the set

$\mathbf{A}^{n+1} - \{(0, \ldots, 0)\}$ under the equivalence relation which identifies points lying on the same line through the origin.

An element of $\mathbf{P}^n$ is called a point. If P is a point, then any $(n + 1)$-tuple $(a_0, \ldots, a_n)$ in the equivalence class P is called a *set of homogeneous coordinates for P*.

Let S be the polynomial ring $k[x_0, \ldots, x_n]$. We want to regard S as a graded ring, so we recall briefly the notion of a graded ring.

A *graded ring* is a ring S, together with a decomposition $S = \bigoplus_{d \geq 0} S_d$ of S into a direct sum of abelian groups S_d, such that for any $d, e \geq 0$, $S_d \cdot S_e \subseteq S_{d+e}$. An element of S_d is called a *homogeneous element* of *degree* d. Thus any element of S can be written uniquely as a (finite) sum of homogeneous elements. An ideal $\mathfrak{a} \subseteq S$ is a *homogeneous ideal* if $\mathfrak{a} = \bigoplus_{d \geq 0} (\mathfrak{a} \cap S_d)$. We will need a few basic facts about homogeneous ideals (see, for example, Matsumura [2, §10] or Zariski–Samuel [1, vol. 2, Ch. VII, §2]). An ideal is homogeneous if and only if it can be generated by homogeneous elements. The sum, product, intersection, and radical of homogeneous ideals are homogeneous. To test whether a homogeneous ideal is prime, it is sufficient to show for any two *homogeneous* elements f, g, that $fg \in \mathfrak{a}$ implies $f \in \mathfrak{a}$ or $g \in \mathfrak{a}$.

We make the polynomial ring $S = k[x_0, \ldots, x_n]$ into a graded ring by taking S_d to be the set of all linear combinations of monomials of total weight d in $x_0, \ldots, x_n$. If $f \in S$ is a polynomial, we cannot use it to define a function on $\mathbf{P}^n$, because of the nonuniqueness of the homogeneous coordinates. However, if f is a homogeneous polynomial of degree d, then $f(\lambda a_0, \ldots, \lambda a_n) = \lambda^d f(a_0, \ldots, a_n)$, so that the property of f being zero or not depends only on the equivalence class of $(a_0, \ldots, a_n)$. Thus f gives a function from $\mathbf{P}^n$ to $\{0, 1\}$ by $f(P) = 0$ if $f(a_0, \ldots, a_n) = 0$, and $f(P) = 1$ if $f(a_0, \ldots, a_n) \neq 0$.

Thus we can talk about the *zeros* of a homogeneous polynomial, namely $Z(f) = \{P \in \mathbf{P}^n | f(P) = 0\}$. If T is any set of homogeneous elements of S, we define the *zero set* of T to be

$$Z(T) = \{P \in \mathbf{P}^n | f(P) = 0 \text{ for all } f \in T\}.$$

If $\mathfrak{a}$ is a homogeneous ideal of S, we define $Z(\mathfrak{a}) = Z(T)$, where T is the set of all homogeneous elements in $\mathfrak{a}$. Since S is a noetherian ring, any set of homogeneous elements T has a finite subset $f_1, \ldots, f_r$ such that $Z(T) = Z(f_1, \ldots, f_r)$.

Definition. A subset Y of $\mathbf{P}^n$ is an *algebraic set* if there exists a set T of homogeneous elements of S such that $Y = Z(T)$.

Proposition 2.1. *The union of two algebraic sets is an algebraic set. The intersection of any family of algebraic sets is an algebraic set. The empty set and the whole space are algebraic sets.*

PROOF. Left to reader (it is similar to the proof of (1.1) above).

Definition. We define the *Zariski topology* on $\mathbf{P}^n$ by taking the open sets to be the complements of algebraic sets.

Once we have a topological space, the notions of irreducible subset and the dimension of a subset, which were defined in §1, will apply.

Definition. A *projective algebraic variety* (or simply *projective variety*) is an irreducible algebraic set in $\mathbf{P}^n$, with the induced topology. An open subset of a projective variety is a *quasi-projective variety*. The *dimension* of a projective or quasi-projective variety is its dimension as a topological space.

If Y is any subset of $\mathbf{P}^n$, we define the *homogeneous ideal* of Y in S, denoted $I(Y)$, to be the ideal generated by $\{f \in S \,|\, f$ is homogeneous and $f(P) = 0$ for all $P \in Y\}$. If Y is an algebraic set, we define the *homogeneous coordinate ring* of Y to be $S(Y) = S/I(Y)$. We refer to (Ex. 2.1–2.7) below for various properties of algebraic sets in projective space and their homogeneous ideals.

Our next objective is to show that projective n-space has an open covering by affine n-spaces, and hence that every projective (respectively, quasi-projective) variety has an open covering by affine (respectively, quasi-affine) varieties. First we introduce some notation.

If $f \in S$ is a linear homogeneous polynomial, then the zero set of f is called a *hyperplane*. In particular we denote the zero set of x_i by H_i, for $i = 0, \ldots, n$. Let U_i be the open set $\mathbf{P}^n - H_i$. Then $\mathbf{P}^n$ is covered by the open sets U_i, because if $P = (a_0, \ldots, a_n)$ is a point, then at least one $a_i \neq 0$, hence $P \in U_i$. We define a mapping $\varphi_i : U_i \to \mathbf{A}^n$ as follows: if $P = (a_0, \ldots, a_n) \in U_i$, then $\varphi_i(P) = Q$, where Q is the point with affine coordinates

$$\left(\frac{a_0}{a_i}, \ldots, \frac{a_n}{a_i} \right),$$

with a_i/a_i omitted. Note that φ_i is well-defined since the ratios a_j/a_i are independent of the choice of homogeneous coordinates.

Proposition 2.2. *The map φ_i is a homeomorphism of U_i with its induced topology to $\mathbf{A}^n$ with its Zariski topology.*

PROOF. φ_i is clearly bijective, so it will be sufficient to show that the closed sets of U_i are identified with the closed sets of $\mathbf{A}^n$ by φ_i. We may assume $i = 0$, and we write simply U for U_0 and $\varphi : U \to \mathbf{A}^n$ for φ_0.

Let $A = k[y_1, \ldots, y_n]$. We define a map α from the set S^h of homogeneous elements of S to A, and a map β from A to S^h. Given $f \in S^h$, we set $\alpha(f) = f(1, y_1, \ldots, y_n)$. On the other hand, given $g \in A$ of degree e, then

$x_0^e g(x_1/x_0, \ldots, x_n/x_0)$ is a homogeneous polynomial of degree e in the x_i, which we call $\beta(g)$.

Now let $Y \subseteq U$ be a closed subset. Let $\overline{Y}$ be its closure in $\mathbf{P}^n$. This is an algebraic set, so $\overline{Y} = Z(T)$ for some subset $T \subseteq S^h$. Let $T' = \alpha(T)$. Then straightforward checking shows that $\varphi(Y) = Z(T')$. Conversely, let W be a closed subset of $\mathbf{A}^n$. Then $W = Z(T')$ for some subset T' of A, and one checks easily that $\varphi^{-1}(W) = Z(\beta(T')) \cap U$. Thus φ and φ^{-1} are both closed maps, so φ is a homeomorphism.

Corollary 2.3. *If Y is a projective (respectively, quasi-projective) variety, then Y is covered by the open sets $Y \cap U_i, i = 0, \ldots, n$, which are homeomorphic to affine (respectively, quasi-affine) varieties via the mapping φ_i defined above.*

EXERCISES

2.1. Prove the "homogeneous Nullstellensatz," which says if $\mathfrak{a} \subseteq S$ is a homogeneous ideal, and if $f \in S$ is a homogeneous polynomial with deg $f > 0$, such that $f(P) = 0$ for all $P \in Z(\mathfrak{a})$ in $\mathbf{P}^n$, then $f^q \in \mathfrak{a}$ for some $q > 0$. [*Hint*: Interpret the problem in terms of the affine $(n + 1)$-space whose affine coordinate ring is S, and use the usual Nullstellensatz, (1.3A).]

2.2. For a homogeneous ideal $\mathfrak{a} \subseteq S$, show that the following conditions are equivalent:

  (i) $Z(\mathfrak{a}) = \varnothing$ (the empty set);
  (ii) $\sqrt{\mathfrak{a}} = $ either S or the ideal $S_+ = \bigoplus_{d>0} S_d$;
  (iii) $\mathfrak{a} \supseteq S_d$ for some $d > 0$.

2.3. (a) If $T_1 \subseteq T_2$ are subsets of S^h, then $Z(T_1) \supseteq Z(T_2)$.
  (b) If $Y_1 \subseteq Y_2$ are subsets of $\mathbf{P}^n$, then $I(Y_1) \supseteq I(Y_2)$.
  (c) For any two subsets Y_1, Y_2 of $\mathbf{P}^n$, $I(Y_1 \cup Y_2) = I(Y_1) \cap I(Y_2)$.
  (d) If $\mathfrak{a} \subseteq S$ is a homogeneous ideal with $Z(\mathfrak{a}) \neq \varnothing$, then $I(Z(\mathfrak{a})) = \sqrt{\mathfrak{a}}$.
  (e) For any subset $Y \subseteq \mathbf{P}^n$, $Z(I(Y)) = \overline{Y}$.

2.4. (a) There is a 1-1 inclusion-reversing correspondence between algebraic sets in $\mathbf{P}^n$, and homogeneous radical ideals of S not equal to S_+, given by $Y \mapsto I(Y)$ and $\mathfrak{a} \mapsto Z(\mathfrak{a})$. *Note*: Since S_+ does not occur in this correspondence, it is sometimes called the *irrelevant* maximal ideal of S.
  (b) An algebraic set $Y \subseteq \mathbf{P}^n$ is irreducible if and only if $I(Y)$ is a prime ideal.
  (c) Show that $\mathbf{P}^n$ itself is irreducible.

2.5. (a) $\mathbf{P}^n$ is a noetherian topological space.
  (b) Every algebraic set in $\mathbf{P}^n$ can be written uniquely as a finite union of irreducible algebraic sets, no one containing another. These are called its *irreducible components*.

2.6. If Y is a projective variety with homogeneous coordinate ring $S(Y)$, show that dim $S(Y) = $ dim $Y + 1$. [*Hint*: Let $\varphi_i: U_i \to \mathbf{A}^n$ be the homeomorphism of (2.2), let Y_i be the affine variety $\varphi_i(Y \cap U_i)$, and let $A(Y_i)$ be its affine coordinate ring.

Show that $A(Y_i)$ can be identified with the subring of elements of degree 0 of the localized ring $S(Y)_{x_i}$. Then show that $S(Y)_{x_i} \cong A(Y_i)[x_i, x_i^{-1}]$. Now use (1.7), (1.8A), and (Ex 1.10), and look at transcendence degrees. Conclude also that dim Y = dim Y_i whenever Y_i is nonempty.]

2.7. (a) dim $\mathbf{P}^n = n$.
    (b) If $Y \subseteq \mathbf{P}^n$ is a quasi-projective variety, then dim Y = dim $\bar{Y}$.
        [*Hint*: Use (Ex. 2.6) to reduce to (1.10).]

2.8. A projective variety $Y \subseteq \mathbf{P}^n$ has dimension $n - 1$ if and only if it is the zero set of a single irreducible homogeneous polynomial f of positive degree. Y is called a *hypersurface* in $\mathbf{P}^n$.

2.9. *Projective Closure of an Affine Variety.* If $Y \subseteq \mathbf{A}^n$ is an affine variety, we identify $\mathbf{A}^n$ with an open set $U_0 \subseteq \mathbf{P}^n$ by the homeomorphism φ_0. Then we can speak of $\bar{Y}$, the closure of Y in $\mathbf{P}^n$, which is called the *projective closure* of Y.
    (a) Show that $I(\bar{Y})$ is the ideal generated by $\beta(I(Y))$, using the notation of the proof of (2.2).
    (b) Let $Y \subseteq \mathbf{A}^3$ be the twisted cubic of (Ex. 1.2). Its projective closure $\bar{Y} \subseteq \mathbf{P}^3$ is called the *twisted cubic curve* in $\mathbf{P}^3$. Find generators for $I(Y)$ and $I(\bar{Y})$, and use this example to show that if $f_1, \ldots, f_r$ generate $I(Y)$, then $\beta(f_1), \ldots, \beta(f_r)$ do *not* necessarily generate $I(\bar{Y})$.

2.10. *The Cone Over a Projective Variety* (Fig. 1). Let $Y \subseteq \mathbf{P}^n$ be a nonempty algebraic set, and let $\theta: \mathbf{A}^{n+1} - \{(0, \ldots, 0)\} \rightarrow \mathbf{P}^n$ be the map which sends the point with affine coordinates $(a_0, \ldots, a_n)$ to the point with homogeneous coordinates $(a_0, \ldots, a_n)$. We define the *affine cone* over Y to be

$$C(Y) = \theta^{-1}(Y) \cup \{(0, \ldots, 0)\}.$$

    (a) Show that $C(Y)$ is an algebraic set in $\mathbf{A}^{n+1}$, whose ideal is equal to $I(Y)$, considered as an ordinary ideal in $k[x_0, \ldots, x_n]$.
    (b) $C(Y)$ is irreducible if and only if Y is.
    (c) dim $C(Y)$ = dim $Y + 1$.
Sometimes we consider the projective closure $\overline{C(Y)}$ of $C(Y)$ in $\mathbf{P}^{n+1}$. This is called the *projective cone* over Y.

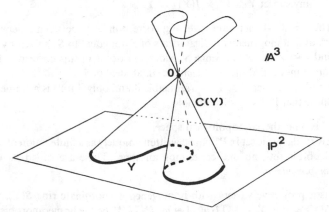

Figure 1.   The cone over a curve in $\mathbf{P}^2$.

2.11. *Linear Varieties in* $\mathbf{P}^n$. A hypersurface defined by a linear polynomial is called a *hyperplane*.
   (a) Show that the following two conditions are equivalent for a variety Y in $\mathbf{P}^n$:
      (i) $I(Y)$ can be generated by linear polynomials.
      (ii) Y can be written as an intersection of hyperplanes.
      In this case we say that Y is a *linear variety* in $\mathbf{P}^n$.
   (b) If Y is a linear variety of dimension r in $\mathbf{P}^n$, show that $I(Y)$ is minimally generated by $n - r$ linear polynomials.
   (c) Let Y, Z be linear varieties in $\mathbf{P}^n$, with dim $Y = r$, dim $Z = s$. If $r + s - n \geqslant 0$, then $Y \cap Z \neq \varnothing$. Furthermore, if $Y \cap Z \neq \varnothing$, then $Y \cap Z$ is a linear variety of dimension $\geqslant r + s - n$. (Think of $\mathbf{A}^{n+1}$ as a vector space over k, and work with its subspaces.)

2.12. *The d-Uple Embedding.* For given $n, d > 0$, let $M_0, M_1, \ldots, M_N$ be all the monomials of degree d in the $n + 1$ variables $x_0, \ldots, x_n$, where $N = \binom{n+d}{n} - 1$. We define a mapping $\rho_d : \mathbf{P}^n \to \mathbf{P}^N$ by sending the point $P = (a_0, \ldots, a_n)$ to the point $\rho_d(P) = (M_0(a), \ldots, M_N(a))$ obtained by substituting the a_i in the monomials M_j. This is called the *d-uple embedding* of $\mathbf{P}^n$ in $\mathbf{P}^N$. For example, if $n = 1, d = 2$, then $N = 2$, and the image Y of the 2-uple embedding of $\mathbf{P}^1$ in $\mathbf{P}^2$ is a conic.
   (a) Let $\theta : k[y_0, \ldots, y_N] \to k[x_0, \ldots, x_n]$ be the homomorphism defined by sending y_i to M_i, and let $\mathfrak{a}$ be the kernel of θ. Then $\mathfrak{a}$ is a homogeneous prime ideal, and so $Z(\mathfrak{a})$ is a projective variety in $\mathbf{P}^N$.
   (b) Show that the image of ρ_d is exactly $Z(\mathfrak{a})$. (One inclusion is easy. The other will require some calculation.)
   (c) Now show that ρ_d is a homeomorphism of $\mathbf{P}^n$ onto the projective variety $Z(\mathfrak{a})$.
   (d) Show that the twisted cubic curve in $\mathbf{P}^3$ (Ex. 2.9) is equal to the 3-uple embedding of $\mathbf{P}^1$ in $\mathbf{P}^3$, for suitable choice of coordinates.

2.13. Let Y be the image of the 2-uple embedding of $\mathbf{P}^2$ in $\mathbf{P}^5$. This is the *Veronese surface*. If $Z \subseteq Y$ is a closed curve (a *curve* is a variety of dimension 1), show that there exists a hypersurface $V \subseteq \mathbf{P}^5$ such that $V \cap Y = Z$.

2.14. *The Segre Embedding.* Let $\psi : \mathbf{P}^r \times \mathbf{P}^s \to \mathbf{P}^N$ be the map defined by sending the ordered pair $(a_0, \ldots, a_r) \times (b_0, \ldots, b_s)$ to $(\ldots, a_i b_j, \ldots)$ in lexicographic order, where $N = rs + r + s$. Note that ψ is well-defined and injective. It is called the *Segre embedding*. Show that the image of ψ is a *subvariety* of $\mathbf{P}^N$. [*Hint*: Let the homogeneous coordinates of $\mathbf{P}^N$ be $\{z_{ij} | i = 0, \ldots, r, j = 0, \ldots, s\}$, and let $\mathfrak{a}$ be the kernel of the homomorphism $k[\{z_{ij}\}] \to k[x_0, \ldots, x_r, y_0, \ldots, y_s]$ which sends z_{ij} to $x_i y_j$. Then show that Im $\psi = Z(\mathfrak{a})$.]

2.15. *The Quadric Surface in* $\mathbf{P}^3$ (Fig. 2). Consider the surface Q (a *surface* is a variety of dimension 2) in $\mathbf{P}^3$ defined by the equation $xy - zw = 0$.
   (a) Show that Q is equal to the Segre embedding of $\mathbf{P}^1 \times \mathbf{P}^1$ in $\mathbf{P}^3$, for suitable choice of coordinates.
   (b) Show that Q contains two families of lines (a *line* is a linear variety of dimension 1) $\{L_t\}, \{M_t\}$, each parametrized by $t \in \mathbf{P}^1$, with the properties that if $L_t \neq L_u$, then $L_t \cap L_u = \varnothing$; if $M_t \neq M_u$, $M_t \cap M_u = \varnothing$, and for all t, u, $L_t \cap M_u =$ one point.
   (c) Show that Q contains other curves besides these lines, and deduce that the Zariski topology on Q is not homeomorphic via ψ to the product topology on $\mathbf{P}^1 \times \mathbf{P}^1$ (where each $\mathbf{P}^1$ has its Zariski topology).

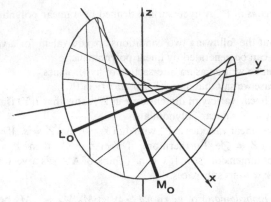

Figure 2.   The quadric surface in $\mathbf{P}^3$.

2.16. (a) The intersection of two varieties need not be a variety. For example, let Q_1 and Q_2 be the quadric surfaces in $\mathbf{P}^3$ given by the equations $x^2 - yw = 0$ and $xy - zw = 0$, respectively. Show that $Q_1 \cap Q_2$ is the union of a twisted cubic curve and a line.

   (b) Even if the intersection of two varieties is a variety, the ideal of the intersection may not be the sum of the ideals. For example, let C be the conic in $\mathbf{P}^2$ given by the equation $x^2 - yz = 0$. Let L be the line given by $y = 0$. Show that $C \cap L$ consists of one point P, but that $I(C) + I(L) \neq I(P)$.

2.17. *Complete intersections.* A variety Y of dimension r in $\mathbf{P}^n$ is a *(strict) complete intersection* if $I(Y)$ can be generated by $n - r$ elements. Y is a *set-theoretic complete intersection* if Y can be written as the intersection of $n - r$ hypersurfaces.

   (a) Let Y be a variety in $\mathbf{P}^n$, let $Y = Z(\mathfrak{a})$; and suppose that $\mathfrak{a}$ can be generated by q elements. Then show that dim $Y \geq n - q$.

   (b) Show that a strict complete intersection is a set-theoretic complete intersection.

   *(c) The converse of (b) is false. For example let Y be the twisted cubic curve in $\mathbf{P}^3$ (Ex. 2.9). Show that $I(Y)$ cannot be generated by two elements. On the other hand, find hypersurfaces H_1, H_2 of degrees 2,3 respectively, such that $Y = H_1 \cap H_2$.

   **(d) It is an unsolved problem whether every closed irreducible curve in $\mathbf{P}^3$ is a set-theoretic intersection of two surfaces. See Hartshorne [1] and Hartshorne [5, III, §5] for commentary.

3 Morphisms

So far we have defined affine and projective varieties, but we have not discussed what mappings are allowed between them. We have not even said when two are isomorphic. In this section we will discuss the regular functions on a variety, and then define a morphism of varieties. Thus we will have a good category in which to work.

Let Y be a quasi-affine variety in $\mathbf{A}^n$. We will consider functions f from Y to k.

Definition. A function $f : Y \rightarrow k$ is *regular at a point* $P \in Y$ if there is an open neighborhood U with $P \in U \subseteq Y$, and polynomials $g, h \in A = k[x_1, \ldots, x_n]$, such that h is nowhere zero on U, and $f = g/h$ on U. (Here of course we interpret the polynomials as functions on $\mathbf{A}^n$, hence on Y.) We say that f is *regular on* Y if it is regular at every point of Y.

Lemma 3.1. *A regular function is continuous, when k is identified with* $\mathbf{A}^1_k$ *in its Zariski topology.*

PROOF. It is enough to show that f^{-1} of a closed set is closed. A closed set of $\mathbf{A}^1_k$ is a finite set of points, so it is sufficient to show that $f^{-1}(a) = \{P \in Y \mid f(P) = a\}$ is closed for any $a \in k$. This can be checked locally: a subset Z of a topological space Y is closed if and only if Y can be covered by open subsets U such that $Z \cap U$ is closed in U for each U. So let U be an open set on which f can be represented as g/h, with $g, h \in A$, and h nowhere 0 on U. Then $f^{-1}(a) \cap U = \{P \in U \mid g(P)/h(P) = a\}$. But $g(P)/h(P) = a$ if and only if $(g - ah)(P) = 0$. So $f^{-1}(a) \cap U = Z(g - ah) \cap U$ which is closed. Hence $f^{-1}(a)$ is closed in Y.

Now let us consider a quasi-projective variety $Y \subseteq \mathbf{P}^n$.

Definition. A function $f : Y \rightarrow k$ is *regular at a point* $P \in Y$ if there is an open neighborhood U with $P \in U \subseteq Y$, and homogeneous polynomials $g, h \in S = k[x_0, \ldots, x_n]$, of the same degree, such that h is nowhere zero on U, and $f = g/h$ on U. (Note that in this case, even though g and h are not functions on $\mathbf{P}^n$, their quotient is a well-defined function whenever $h \neq 0$, since they are homogeneous of the same degree.) We say that f is *regular on* Y if it is regular at every point.

Remark 3.1.1. As in the quasi-affine case, a regular function is necessarily continuous (proof left to reader). An important consequence of this is the fact that if f and g are regular functions on a variety X, and if $f = g$ on some nonempty open subset $U \subseteq X$, then $f = g$ everywhere. Indeed, the set of points where $f - g = 0$ is closed and dense, hence equal to X.

Now we can define the category of varieties.

Definition. Let k be a fixed algebraically closed field. A *variety over* k (or simply *variety*) is any affine, quasi-affine, projective, or quasi-projective variety as defined above. If X, Y are two varieties, a *morphism* $\varphi : X \rightarrow Y$

is a continuous map such that for every open set $V \subseteq Y$, and for every regular function $f : V \to k$, the function $f \circ \varphi : \varphi^{-1}(V) \to k$ is regular.

Clearly the composition of two morphisms is a morphism, so we have a category. In particular, we have the notion of isomorphism: an *isomorphism* $\varphi : X \to Y$ of two varieties is a morphism which admits an inverse morphism $\psi : Y \to X$ with $\psi \circ \varphi = \mathrm{id}_X$ and $\varphi \circ \psi = \mathrm{id}_Y$. Note that an isomorphism is necessarily bijective and bicontinuous, but a bijective bicontinuous morphism need not be an isomorphism (Ex. 3.2).

Now we introduce some rings of functions associated with any variety.

Definition. Let Y be a variety. We denote by $\mathcal{O}(Y)$ the ring of all regular functions on Y. If P is a point of Y, we define the *local ring of P on Y*, $\mathcal{O}_{P,Y}$ (or simply $\mathcal{O}_P$) to be the ring of germs of regular functions on Y near P. In other words, an element of $\mathcal{O}_P$ is a pair $\langle U, f \rangle$ where U is an open subset of Y containing P, and f is a regular function on U, and where we identify two such pairs $\langle U, f \rangle$ and $\langle V, g \rangle$ if $f = g$ on $U \cap V$. (Use (3.1.1) to verify that this is an equivalence relation!)

Note that $\mathcal{O}_P$ is indeed a local ring: its maximal ideal $\mathfrak{m}$ is the set of germs of regular functions which vanish at P. For if $f(P) \neq 0$, then $1/f$ is regular in some neighborhood of P. The residue field $\mathcal{O}_P/\mathfrak{m}$ is isomorphic to k.

Definition. If Y is a variety, we define the *function field $K(Y)$ of Y* as follows: an element of $K(Y)$ is an equivalence class of pairs $\langle U, f \rangle$ where U is a nonempty open subset of Y, f is a regular function on U, and where we identify two pairs $\langle U, f \rangle$ and $\langle V, g \rangle$ if $f = g$ on $U \cap V$. The elements of $K(Y)$ are called *rational functions* on Y.

Note that $K(Y)$ is in fact a field. Since Y is irreducible, any two nonempty open sets have a nonempty intersection. Hence we can define addition and multiplication in $K(Y)$, making it a ring. Then if $\langle U, f \rangle \in K(Y)$ with $f \neq 0$, we can restrict f to the open set $V = U - U \cap Z(f)$ where it never vanishes, so that $1/f$ is regular on V, hence $\langle V, 1/f \rangle$ is an inverse for $\langle U, f \rangle$.

Now we have defined, for any variety Y, the ring of global functions $\mathcal{O}(Y)$, the local ring $\mathcal{O}_P$ at a point of Y, and the function field $K(Y)$. By restricting functions we obtain natural maps $\mathcal{O}(Y) \to \mathcal{O}_P \to K(Y)$ which in fact are injective by (3.1.1). Hence we will usually treat $\mathcal{O}(Y)$ and $\mathcal{O}_P$ as subrings of $K(Y)$.

If we replace Y by an isomorphic variety, then the corresponding rings are isomorphic. Thus we can say that $\mathcal{O}(Y)$, $\mathcal{O}_P$, and $K(Y)$ are *invariants* of the variety Y (and the point P) up to isomorphism.

Our next task is to relate $\mathcal{O}(Y)$, $\mathcal{O}_P$, and $K(Y)$ to the affine coordinate ring $A(Y)$ of an affine variety, and the homogeneous coordinate ring $S(Y)$

of a projective variety, which were introduced earlier. We will find that for an affine variety Y, $A(Y) = \mathcal{O}(Y)$, so it is an invariant up to isomorphism. However, for a projective variety Y, $S(Y)$ is not an invariant: it depends on the embedding of Y in projective space (Ex. 3.9).

Theorem 3.2. *Let $Y \subseteq \mathbf{A}^n$ be an affine variety with affine coordinate ring $A(Y)$. Then:*

(a) $\mathcal{O}(Y) \cong A(Y)$;

(b) *for each point $P \in Y$, let $\mathfrak{m}_P \subseteq A(Y)$ be the ideal of functions vanishing at P. Then $P \mapsto \mathfrak{m}_P$ gives a 1-1 correspondence between the points of Y and the maximal ideals of $A(Y)$;*

(c) *for each P, $\mathcal{O}_P \cong A(Y)_{\mathfrak{m}_P}$, and $\dim \mathcal{O}_P = \dim Y$;*

(d) *$K(Y)$ is isomorphic to the quotient field of $A(Y)$, and hence $K(Y)$ is a finitely generated extension field of k, of transcendence degree $= \dim Y$.*

PROOF. We will proceed in several steps. First we define a map $\alpha : A(Y) \to \mathcal{O}(Y)$. Every polynomial $f \in A = k[x_1, \ldots, x_n]$ defines a regular function on $\mathbf{A}^n$ and hence on Y. Thus we have a homomorphism $A \to \mathcal{O}(Y)$. Its kernel is just $I(Y)$, so we obtain an injective homomorphism $\alpha : A(Y) \to \mathcal{O}(Y)$.

From (1.4) we know there is a 1-1 correspondence between points of Y (which are the minimal algebraic subsets of Y) and maximal ideals of A containing $I(Y)$. Passing to the quotient by $I(Y)$, these correspond to the maximal ideals of $A(Y)$. Furthermore, using α to identify elements of $A(Y)$ with regular functions on Y, the maximal ideal corresponding to P is just $\mathfrak{m}_P = \{f \in A(Y) | f(P) = 0\}$. This proves (b).

For each P there is a natural map $A(Y)_{\mathfrak{m}_P} \to \mathcal{O}_P$. It is injective because α is injective, and it is surjective by definition of a regular function! This shows that $\mathcal{O}_P \cong A(Y)_{\mathfrak{m}_P}$. Now $\dim \mathcal{O}_P =$ height $\mathfrak{m}_P$. Since $A(Y)/\mathfrak{m}_P \cong k$, we conclude from (1.7) and (1.8A) that $\dim \mathcal{O}_P = \dim Y$.

From (c) it follows that the quotient field of $A(Y)$ is isomorphic to the quotient field of $\mathcal{O}_P$ for every P, and this is equal to $K(Y)$, because every rational function is actually in some $\mathcal{O}_P$. Now $A(Y)$ is a finitely generated k-algebra, so $K(Y)$ is a finitely generated field extension of k. Furthermore, the transcendence degree of $K(Y)/k$ is equal to $\dim Y$ by (1.7) and (1.8A). This proves (d).

To prove (a) we note that $\mathcal{O}(Y) \subseteq \bigcap_{P \in Y} \mathcal{O}_P$, where all our rings are regarded as subrings of $K(Y)$.

Using (b) and (c) we have

$$A(Y) \subseteq \mathcal{O}(Y) \subseteq \bigcap_{\mathfrak{m}} A(Y)_{\mathfrak{m}},$$

where $\mathfrak{m}$ runs over all the maximal ideals of $A(Y)$. The equality now follows from the simple algebraic fact that if B is an integral domain, then B is equal to the intersection (inside its quotient field) of its localizations at all maximal ideals.

Propositi⌐ 1 3.3. *Let $U_i \subseteq \mathbf{P}^n$ be the open set defined by the equation $x_i \neq 0$. Then tne mapping $\varphi_i : U_i \to \mathbf{A}^n$ of (2.2) above is an isomorphism of varieties.*

PROOF. We have already shown that it is a homeomorphism, so we need only check that the regular functions are the same on any open set. On U_i the regular functions are locally quotients of homogeneous polynomials in $x_0, \ldots, x_n$ of the same degree. On $\mathbf{A}^n$ the regular functions are locally quotients of polynomials in $y_1, \ldots, y_n$. One can check easily that these two concepts are identified by the maps α and β of the proof of (2.2).

Before stating the next result, we introduce some notation. If S is a graded ring, and $\mathfrak{p}$ a homogeneous prime ideal in S, then we denote by $S_{(\mathfrak{p})}$ the subring of elements of degree 0 in the localization of S with respect to the multiplicative subset T consisting of the homogeneous elements of S not in $\mathfrak{p}$. Note that $T^{-1}S$ has a natural grading given by $\deg(f/g) = \deg f - \deg g$ for f homogeneous in S and $g \in T$. $S_{(\mathfrak{p})}$ is a local ring, with maximal ideal $(\mathfrak{p} \cdot T^{-1}S) \cap S_{(\mathfrak{p})}$. In particular, if S is a·domain, then for $\mathfrak{p} = (0)$ we obtain a field $S_{((0))}$. Similarly, if $f \in S$ is a homogeneous element, we denote by $S_{(f)}$ the subring of elements of degree 0 in the localized ring S_f.

Theorem 3.4. *Let $Y \subseteq \mathbf{P}^n$ be a projective variety with homogeneous coordinate ring $S(Y)$. Then:*
   (a) $\mathcal{O}(Y) = k$;
   (b) *for any point $P \in Y$, let $\mathfrak{m}_P \subseteq S(Y)$ be the ideal generated by the set of homogeneous $f \in S(Y)$ such that $f(P) = 0$. Then $\mathcal{O}_P = S(Y)_{(\mathfrak{m}_P)}$;*
   (c) $K(Y) \cong S(Y)_{((0))}$.

PROOF. To begin with, let $U_i \subseteq \mathbf{P}^n$ be the open set $x_i \neq 0$, and let $Y_i = Y \cap U_i$. Then U_i is isomorphic to $\mathbf{A}^n$ by the isomorphism φ_i of (3.3), so we can consider Y_i as an affine variety. There is a natural isomorphism φ_i^* of the affine coordinate ring $A(Y_i)$ with the localization $S(Y)_{(x_i)}$ of the homogeneous coordinate ring of Y. We first make an isomorphism of $k[y_1, \ldots, y_n]$ with $k[x_0, \ldots, x_n]_{(x_i)}$ by sending $f(y_1, \ldots, y_n)$ to $f(x_0/x_i, \ldots, x_n/x_i)$, leaving out x_i/x_i, as in the proof of (2.2). This isomorphism sends $I(Y_i)$ to $I(Y)S_{(x_i)}$ (cf. Ex. 2.6), so passing to the quotient, we obtain the desired isomorphism $\varphi_i^* : A(Y_i) \cong S(Y)_{(x_i)}$.

Now to prove (b), let $P \in Y$ be any point, and choose i so that $P \in Y_i$. Then by (3.2), $\mathcal{O}_P \cong A(Y_i)_{\mathfrak{m}_P'}$, where $\mathfrak{m}_P'$ is the maximal ideal of $A(Y_i)$ corresponding to P. One checks easily that $\varphi_i^*(\mathfrak{m}_P') = \mathfrak{m}_P \cdot S(Y)_{(x_i)}$. Now $x_i \notin \mathfrak{m}_P$, and localization is transitive, so we find that $A(Y_i)_{\mathfrak{m}_P'} \cong S(Y)_{(\mathfrak{m}_P)}$, which proves (b).

To prove (c), we use (3.2) again to see that $K(Y)$, which is equal to $K(Y_i)$, is the quotient field of $A(Y_i)$. But by φ_i^*, this is isomorphic to $S(Y)_{((0))}$.

To prove (a), let $f \in \mathcal{O}(Y)$ be a global regular function. Then for each i, f is regular on Y_i, so by (3.2), $f \in A(Y_i)$. But we have just seen that $A(Y_i) \cong S(Y)_{(x_i)}$, so we conclude that f can be written as $g_i/x_i^{N_i}$ where $g_i \in S(Y)$ is homogeneous of degree N_i. Thinking of $\mathcal{O}(Y)$, $K(Y)$ and $S(Y)$ all as sub-rings of the quotient field L of $S(Y)$, this means that $x_i^{N_i} f \in S(Y)_{N_i}$, for each i. Now choose $N \geqslant \sum N_i$. Then $S(Y)_N$ is spanned as a k-vector space by monomials of degree N in $x_0, \ldots, x_n$, and in any such monomial, at least one x_i occurs to a power $\geqslant N_i$. Thus we have $S(Y)_N \cdot f \subseteq S(Y)_N$. Iterating, we have $S(Y)_N \cdot f^q \subseteq S(Y)_N$ for all $q > 0$. In particular, $x_0^N f^q \in S(Y)$ for all $q > 0$. This shows that the subring $S(Y)[f]$ of L is contained in $x_0^{-N} S(Y)$, which is a finitely generated $S(Y)$-module. Since $S(Y)$ is a noetherian ring, $S(Y)[f]$ is a finitely generated $S(Y)$-module, and therefore f is *integral* over $S(Y)$ (see, e.g., Atiyah–Macdonald [1, p. 59]). This means that there are elements $a_1, \ldots, a_m \in S(Y)$ such that

$$f^m + a_1 f^{m-1} + \cdots + a_m = 0.$$

Since f has degree 0, we can replace the a_i by their homogeneous components of degree 0, and still have a valid equation. But $S(Y)_0 = k$, so the $a_i \in k$, and f is algebraic over k. But k is algebraically closed, so $f \in k$, which completes the proof.

Our next result shows that if X and Y are affine varieties, then X is iso-morphic to Y if and only if $A(X)$ is isomorphic to $A(Y)$ as a k-algebra. Actually the proof gives more, so we state the stronger result.

Proposition 3.5. *Let X be any variety and let Y be an affine variety. Then there is a natural bijective mapping of sets*

$$\alpha: \operatorname{Hom}(X, Y) \xrightarrow{\sim} \operatorname{Hom}(A(Y), \mathcal{O}(X))$$

where the left Hom *means morphisms of varieties, and the right* Hom *means homomorphisms of k-algebras.*

PROOF. Given a morphism $\varphi: X \to Y$, φ carries regular functions on Y to regular functions on X. Hence φ induces a map $\mathcal{O}(Y)$ to $\mathcal{O}(X)$, which is clearly a homomorphism of k-algebras. But we have seen (3.2) that $\mathcal{O}(Y) \cong A(Y)$, so we get a homomorphism $A(Y) \to \mathcal{O}(X)$. This defines α.

Conversely, suppose given a homomorphism $h: A(Y) \to \mathcal{O}(X)$ of k-algebras. Suppose that Y is a closed subset of $\mathbf{A}^n$, so that $A(Y) = k[x_1, \ldots, x_n]/I(Y)$. Let $\bar{x}_i$ be the image of x_i in $A(Y)$, and consider the elements $\xi_i = h(\bar{x}_i) \in \mathcal{O}(X)$. These are global functions on X, so we can use them to define a mapping $\psi: X \to \mathbf{A}^n$ by $\psi(P) = (\xi_1(P), \ldots, \xi_n(P))$ for $P \in X$.

We show next that the image of ψ is contained in Y. Since $Y = Z(I(Y))$, it is sufficient to show that for any $P \in X$ and any $f \in I(Y)$, $f(\psi(P)) = 0$. But

$$f(\psi(P)) = f(\xi_1(P), \ldots, \xi_n(P)).$$

Now f is a polynomial, and h is a homomorphism of k-algebras, so we have

$$f(\xi_1(P), \ldots, \xi_n(P)) = h(f(\bar{x}_1, \ldots, \bar{x}_n))(P) = 0$$

19

since $f \in I(Y)$. So ψ defines a map from X to Y, which induces the given homomorphism h.

To complete the proof, we must show that ψ is a morphism. This is a consequence of the following lemma.

Lemma 3.6. *Let X be any variety, and let $Y \subseteq \mathbf{A}^n$ be an affine variety. A map of sets $\psi : X \to Y$ is a morphism if and only if $x_i \circ \psi$ is a regular function on X for each i, where $x_1, \ldots, x_n$ are the coordinate functions on $\mathbf{A}^n$.*

PROOF. If ψ is a morphism, the $x_i \circ \psi$ must be regular functions, by definition of a morphism. Conversely, suppose the $x_i \circ \psi$ are regular. Then for any polynomial $f = f(x_1, \ldots, x_n)$, $f \circ \psi$ is also regular on X. Since the closed sets of Y are defined by the vanishing of polynomial functions, and since regular functions are continuous, we see that ψ^{-1} takes closed sets to closed sets, so ψ is continuous. Finally, since regular functions on open subsets of Y are locally quotients of polynomials, $g \circ \psi$ is regular for any regular function g on any open subset of Y. Hence ψ is a morphism.

Corollary 3.7. *If X, Y are two affine varieties, then X and Y are isomorphic if and only if $A(X)$ and $A(Y)$ are isomorphic as k-algebras.*

PROOF. Immediate from the proposition.

In the language of categories, we can express the above result as follows:

Corollary 3.8. *The functor $X \mapsto A(X)$ induces an arrow-reversing equivalence of categories between the category of affine varieties over k and the category of finitely generated integral domains over k.*

We include here an algebraic result which will be used in the exercises.

Theorem 3.9A (Finiteness of Integral Closure). *Let A be an integral domain which is a finitely generated algebra over a field k. Let K be the quotient field of A, and let L be a finite algebraic extension of K. Then the integral closure A' of A in L is a finitely generated A-module, and is also a finitely generated k-algebra.*

PROOF. Zariski–Samuel [1, vol. 1, Ch. V., Thm. 9, p. 267.]

EXERCISES

3.1. (a) Show that any conic in $\mathbf{A}^2$ is isomorphic either to $\mathbf{A}^1$ or $\mathbf{A}^1 - \{0\}$ (cf. Ex. 1.1).
 (b) Show that $\mathbf{A}^1$ is *not* isomorphic to any proper open subset of itself. (This result is generalized by (Ex. 6.7) below.)
 (c) Any conic in $\mathbf{P}^2$ is isomorphic to $\mathbf{P}^1$.
 (d) We will see later (Ex. 4.8) that any two curves are homeomorphic. But show now that $\mathbf{A}^2$ is not even homeomorphic to $\mathbf{P}^2$.

(e) If an affine variety is isomorphic to a projective variety, then it consists of only one point.

3.2. A morphism whose underlying map on the topological spaces is a homeomorphism need not be an isomorphism.
   (a) For example, let $\varphi: \mathbf{A}^1 \to \mathbf{A}^2$ be defined by $t \mapsto (t^2, t^3)$. Show that φ defines a bijective bicontinuous morphism of $\mathbf{A}^1$ onto the curve $y^2 = x^3$, but that φ is not an isomorphism.
   (b) For another example, let the characteristic of the base field k be $p > 0$, and define a map $\varphi: \mathbf{A}^1 \to \mathbf{A}^1$ by $t \mapsto t^p$. Show that φ is bijective and bicontinuous but not an isomorphism. This is called the *Frobenius morphism*.

3.3. (a) Let $\varphi: X \to Y$ be a morphism. Then for each $P \in X$, φ induces a homomorphism of local rings $\varphi_P^*: \mathcal{O}_{\varphi(P),Y} \to \mathcal{O}_{P,X}$.
   (b) Show that a morphism φ is an isomorphism if and only if φ is a homeomorphism, and the induced map φ_P^* on local rings is an isomorphism, for all $P \in X$.
   (c) Show that if $\varphi(X)$ is dense in Y, then the map φ_P^* is *injective* for all $P \in X$.

3.4. Show that the d-uple embedding of $\mathbf{P}^n$ (Ex. 2.12) is an isomorphism onto its image.

3.5. By abuse of language, we will say that a variety "is affine" if it is isomorphic to an affine variety. If $H \subseteq \mathbf{P}^n$ is any hypersurface, show that $\mathbf{P}^n - H$ is affine. [*Hint*: Let H have degree d. Then consider the d-uple embedding of $\mathbf{P}^n$ in $\mathbf{P}^N$ and use the fact that $\mathbf{P}^N$ minus a hyperplane is affine.]

3.6. There are quasi-affine varieties which are not affine. For example, show that $X = \mathbf{A}^2 - \{(0,0)\}$ is not affine. [*Hint*: Show that $\mathcal{O}(X) \cong k[x,y]$ and use (3.5). See (III, Ex. 4.3) for another proof.]

3.7. (a) Show that any two curves in $\mathbf{P}^2$ have a nonempty intersection.
   (b) More generally, show that if $Y \subseteq \mathbf{P}^n$ is a projective variety of dimension ≥ 1, and if H is a hypersurface, then $Y \cap H \neq \varnothing$. [*Hint*: Use (Ex. 3.5) and (Ex. 3.1e). See (7.2) for a generalization.]

3.8. Let H_i and H_j be the hyperplanes in $\mathbf{P}^n$ defined by $x_i = 0$ and $x_j = 0$, with $i \neq j$. Show that any regular function on $\mathbf{P}^n - (H_i \cap H_j)$ is constant. (This gives an alternate proof of (3.4a) in the case $Y = \mathbf{P}^n$.)

3.9. The homogeneous coordinate ring of a projective variety is not invariant under isomorphism. For example, let $X = \mathbf{P}^1$, and let Y be the 2-uple embedding of $\mathbf{P}^1$ in $\mathbf{P}^2$. Then $X \cong Y$ (Ex. 3.4). But show that $S(X) \not\cong S(Y)$.

3.10. *Subvarieties.* A subset of a topological space is *locally closed* if it is an open subset of its closure, or, equivalently, if it is the intersection of an open set with a closed set.

   If X is a quasi-affine or quasi-projective variety and Y is an irreducible locally closed subset, then Y is also a quasi-affine (respectively, quasi-projective) variety, by virtue of being a locally closed subset of the same affine or projective space. We call this the *induced structure* on Y, and we call Y a *subvariety* of X.

   Now let $\varphi: X \to Y$ be a morphism, let $X' \subseteq X$ and $Y' \subseteq Y$ be irreducible locally closed subsets such that $\varphi(X') \subseteq Y'$. Show that $\varphi|_{X'}: X' \to Y'$ is a morphism.

3.11. Let X be any variety and let $P \in X$. Show there is a 1-1 correspondence between the prime ideals of the local ring $\mathcal{O}_P$ and the closed subvarieties of X containing P.

3.12. If P is a point on a variety X, then dim $\mathcal{O}_P$ = dim X. [*Hint:* Reduce to the affine case and use (3.2c).]

3.13. *The Local Ring of a Subvariety.* Let $Y \subseteq X$ be a subvariety. Let $\mathcal{O}_{Y,X}$ be the set of equivalence classes $\langle U,f \rangle$ where $U \subseteq X$ is open, $U \cap Y \neq \emptyset$, and f is a regular function on U. We say $\langle U,f \rangle$ is equivalent to $\langle V,g \rangle$, if $f = g$ on $U \cap V$. Show that $\mathcal{O}_{Y,X}$ is a local ring, with residue field $K(Y)$ and dimension $=$ dim X − dim Y. It is the *local ring* of Y on X. Note if $Y = P$ is a point we get $\mathcal{O}_P$, and if $Y = X$ we get $K(X)$. Note also that if Y is not a point, then $K(Y)$ is not algebraically closed, so in this way we get local rings whose residue fields are not algebraically closed.

3.14. *Projection from a Point.* Let $\mathbf{P}^n$ be a hyperplane in $\mathbf{P}^{n+1}$ and let $P \in \mathbf{P}^{n+1} - \mathbf{P}^n$. Define a mapping $\varphi: \mathbf{P}^{n+1} - \{P\} \to \mathbf{P}^n$ by $\varphi(Q) =$ the intersection of the unique line containing P and Q with $\mathbf{P}^n$.
 (a) Show that φ is a morphism.
 (b) Let $Y \subseteq \mathbf{P}^3$ be the twisted cubic curve which is the image of the 3-uple embedding of $\mathbf{P}^1$ (Ex. 2.12). If t,u are the homogeneous coordinates on $\mathbf{P}^1$, we say that Y is the curve given *parametrically* by $(x,y,z,w) = (t^3, t^2u, tu^2, u^3)$. Let $P = (0,0,1,0)$, and let $\mathbf{P}^2$ be the hyperplane $z = 0$. Show that the projection of Y from P is a cuspidal cubic curve in the plane, and find its equation.

3.15. *Products of Affine Varieties.* Let $X \subseteq \mathbf{A}^n$ and $Y \subseteq \mathbf{A}^m$ be affine varieties.
 (a) Show that $X \times Y \subseteq \mathbf{A}^{n+m}$ with its induced topology is irreducible. [*Hint:* Suppose that $X \times Y$ is a union of two closed subsets $Z_1 \cup Z_2$. Let $X_i = \{x \in X | x \times Y \subseteq Z_i\}$, $i = 1,2$. Show that $X = X_1 \cup X_2$ and X_1, X_2 are closed. Then $X = X_1$ or X_2 so $X \times Y = Z_1$ or Z_2.] The affine variety $X \times Y$ is called the *product* of X and Y. Note that its topology is in general not equal to the product topology (Ex. 1.4).
 (b) Show that $A(X \times Y) \cong A(X) \otimes_k A(Y)$.
 (c) Show that $X \times Y$ is a product in the category of varieties, i.e., show (i) the projections $X \times Y \to X$ and $X \times Y \to Y$ are morphisms, and (ii) given a variety Z, and the morphisms $Z \to X$, $Z \to Y$, there is a unique morphism $Z \to X \times Y$ making a commutative diagram

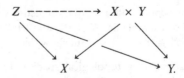

 (d) Show that dim $X \times Y$ = dim X + dim Y.

3.16. *Products of Quasi-Projective Varieties.* Use the Segre embedding (Ex. 2.14) to identify $\mathbf{P}^n \times \mathbf{P}^m$ with its image and hence give it a structure of projective variety. Now for any two quasi-projective varieties $X \subseteq \mathbf{P}^n$ and $Y \subseteq \mathbf{P}^m$, consider $X \times Y \subseteq \mathbf{P}^n \times \mathbf{P}^m$.
 (a) Show that $X \times Y$ is a quasi-projective variety.
 (b) If X,Y are both projective, show that $X \times Y$ is projective.
 *(c) Show that $X \times Y$ is a product in the category of varieties.

3.17. *Normal Varieties.* A variety Y is *normal at a point* $P \in Y$ if $\mathcal{O}_P$ is an integrally closed ring. Y is *normal* if it is normal at every point.
(a) Show that every conic in $\mathbf{P}^2$ is normal.
(b) Show that the quadric surfaces Q_1, Q_2 in $\mathbf{P}^3$ given by equations $Q_1 : xy = zw$; $Q_2 : xy = z^2$ are normal (cf. (II. Ex. 6.4) for the latter.)
(c) Show that the cuspidal cubic $y^2 = x^3$ in $\mathbf{A}^2$ is not normal.
(d) If Y is affine, then Y is normal $\Leftrightarrow A(Y)$ is integrally closed.
(e) Let Y be an affine variety. Show that there is a normal affine variety $\tilde{Y}$, and a morphism $\pi : \tilde{Y} \to Y$, with the property that whenever Z is a normal variety, and $\varphi : Z \to Y$ is a *dominant* morphism (i.e., $\varphi(Z)$ is dense in Y), then there is a unique morphism $\theta : Z \to \tilde{Y}$ such that $\varphi = \pi \circ \theta$. $\tilde{Y}$ is called the *normalization* of Y. You will need (3.9A) above.

3.18. *Projectively Normal Varieties.* A projective variety $Y \subseteq \mathbf{P}^n$ is *projectively normal* (with respect to the given embedding) if its homogeneous coordinate ring $S(Y)$ is integrally closed.
(a) If Y is projectively normal, then Y is normal.
(b) There are normal varieties in projective space which are not projectively normal. For example, let Y be the twisted quartic curve in $\mathbf{P}^3$ given parametrically by $(x,y,z,w) = (t^4, t^3u, tu^3, u^4)$. Then Y is normal but not projectively normal. See (III, Ex. 5.6) for more examples.
(c) Show that the twisted quartic curve Y above is isomorphic to $\mathbf{P}^1$, which is projectively normal. Thus projective normality depends on the embedding.

3.19. *Automorphisms of $\mathbf{A}^n$.* Let $\varphi : \mathbf{A}^n \to \mathbf{A}^n$ be a morphism of $\mathbf{A}^n$ to $\mathbf{A}^n$ given by n polynomials $f_1, \ldots, f_n$ of n variables $x_1, \ldots, x_n$. Let $J = \det|\partial f_i / \partial x_j|$ be the *Jacobian* polynomial of φ.
(a) If φ is an isomorphism (in which case we call φ an *automorphism* of $\mathbf{A}^n$) show that J is a nonzero constant polynomial.
**(b) The converse of (a) is an unsolved problem, even for $n = 2$. See, for example, Vitushkin [1].

3.20. Let Y be a variety of dimension ≥ 2, and let $P \in Y$ be a normal point. Let f be a regular function on $Y - P$.
(a) Show that f extends to a regular function on Y.
(b) Show this would be false for dim $Y = 1$.
See (III, Ex. 3.5) for generalization.

3.21. *Group Varieties.* A group variety consists of a variety Y together with a morphism $\mu : Y \times Y \to Y$, such that the set of points of Y with the operation given by μ is a group, and such that the inverse map $y \to y^{-1}$ is also a morphism of $Y \to Y$.
(a) The *additive group* $\mathbf{G}_a$ is given by the variety $\mathbf{A}^1$ and the morphism $\mu : \mathbf{A}^2 \to \mathbf{A}^1$ defined by $\mu(a,b) = a + b$. Show it is a group variety.
(b) The *multiplicative group* $\mathbf{G}_m$ is given by the variety $\mathbf{A}^1 - \{(0)\}$ and the morphism $\mu(a,b) = ab$. Show it is a group variety.
(c) If G is a group variety, and X is any variety, show that the set $\mathrm{Hom}(X,G)$ has a natural group structure.
(d) For any variety X, show that $\mathrm{Hom}(X,\mathbf{G}_a)$ is isomorphic to $\mathcal{O}(X)$ as a group under addition.
(e) For any variety X, show that $\mathrm{Hom}(X,\mathbf{G}_m)$ is isomorphic to the group of units in $\mathcal{O}(X)$, under multiplication.

4 Rational Maps

In this section we introduce the notions of rational map and birational equivalence, which are important for the classification of varieties. A rational map is a morphism which is only defined on some open subset. Since an open subset of a variety is dense, this already carries a lot of information. In this respect algebraic geometry is more "rigid" than differential geometry or topology. In particular, the concept of birational equivalence is unique to algebraic geometry.

Lemma 4.1. *Let X and Y be varieties, let φ and ψ be two morphisms from X to Y, and suppose there is a nonempty open subset $U \subseteq X$ such that $\varphi|_U = \psi|_U$. Then $\varphi = \psi$.*

PROOF. We may assume that $Y \subseteq \mathbf{P}^n$ for some n. Then by composing with the inclusion morphism $Y \to \mathbf{P}^n$, we reduce to the case $Y = \mathbf{P}^n$. We consider the product $\mathbf{P}^n \times \mathbf{P}^n$, which has a structure of projective variety given by its Segre embedding (Ex. 3.16). The morphisms φ and ψ determine a map $\varphi \times \psi : X \to \mathbf{P}^n \times \mathbf{P}^n$, which in fact is a morphism (Ex. 3.16c). Let $\Delta = \{P \times P | P \in \mathbf{P}^n\}$ be the *diagonal* subset of $\mathbf{P}^n \times \mathbf{P}^n$. It is defined by the equations $\{x_i y_j = x_j y_i | i,j = 0,1,\dots,n\}$ and so is a closed subset of $\mathbf{P}^n \times \mathbf{P}^n$. By hypothesis $\varphi \times \psi(U) \subseteq \Delta$. But U is dense in X, and Δ is closed, so $\varphi \times \psi(X) \subseteq \Delta$. This says that $\varphi = \psi$.

Definition. Let X,Y be varieties. A *rational map* $\varphi : X \to Y$ is an equivalence class of pairs $\langle U, \varphi_U \rangle$ where U is a nonempty open subset of X, φ_U is a morphism of U to Y, and where $\langle U, \varphi_U \rangle$ and $\langle V, \varphi_V \rangle$ are equivalent if φ_U and φ_V agree on $U \cap V$. The rational map φ is *dominant* if for some (and hence every) pair $\langle U, \varphi_U \rangle$, the image of φ_U is dense in Y.

Note that the lemma implies that the relation on pairs $\langle U, \varphi_U \rangle$ just described is an equivalence relation. Note also that a rational map $\varphi : X \to Y$ is *not* in general a map of the set X to Y. Clearly one can compose dominant rational maps, so we can consider the category of varieties and dominant rational maps. An "isomorphism" in this category is called a birational map:

Definition. A *birational map* $\varphi : X \to Y$ is a rational map which admits an inverse, namely a rational map $\psi : Y \to X$ such that $\psi \circ \varphi = \mathrm{id}_X$ and $\varphi \circ \psi = \mathrm{id}_Y$ as rational maps. If there is a birational map from X to Y, we say that X and Y are *birationally equivalent*, or simply *birational*.

The main result of this section is that the category of varieties and dominant rational maps is equivalent to the category of finitely generated field extensions of k, with the arrows reversed. Before giving this result, we need a couple of lemmas which show that on any variety, the open affine subsets form a base for the topology. We say loosely that a variety is *affine* if it is isomorphic to an affine variety.

Lemma 4.2. *Let Y be a hypersurface in $\mathbf{A}^n$ given by the equation $f(x_1,\ldots,x_n) = 0$. Then $\mathbf{A}^n - Y$ is isomorphic to the hypersurface H in $\mathbf{A}^{n+1}$ given by $x_{n+1}f = 1$. In particular, $\mathbf{A}^n - Y$ is affine, and its affine ring is $k[x_1,\ldots,x_n]_f$.*

PROOF. For $P = (a_1,\ldots,a_{n+1}) \in H$, let $\varphi(P) = (a_1,\ldots,a_n)$. Then clearly φ is a morphism from H to $\mathbf{A}^n$, corresponding to the homomorphism of rings $A \to A_f$, where $A = k[x_1,\ldots,x_n]$. It is also clear that φ gives a bijective mapping of H onto its image, which is $\mathbf{A}^n - Y$. To show that φ is an isomorphism, it is sufficient to show that φ^{-1} is a morphism. But $\varphi^{-1}(a_1,\ldots,a_n) = (a_1,\ldots,a_n,1/f(a_1,\ldots,a_n))$, so the fact that φ^{-1} is a morphism on $\mathbf{A}^n - Y$ follows from (3.6).

Proposition 4.3. *On any variety Y, there is a base for the topology consisting of open affine subsets.*

PROOF. We must show for any point $P \in Y$ and any open set U containing P, that there exists an open affine set V with $P \in V \subseteq U$. First, since U is also a variety, we may assume $U = Y$. Secondly, since any variety is covered by quasi-affine varieties (2.3), we may assume that Y is quasi-affine in $\mathbf{A}^n$. Let $Z = \bar{Y} - Y$, which is a closed set in $\mathbf{A}^n$, and let $\mathfrak{a} \subseteq A = k[x_1,\ldots,x_n]$ be the ideal of Z. Then, since Z is closed, and $P \notin Z$, we can find a polynomial $f \in \mathfrak{a}$ such that $f(P) \neq 0$. Let H be the hypersurface $f = 0$ in $\mathbf{A}^n$. Then $Z \subseteq H$ but $P \notin H$. Thus $P \in Y - Y \cap H$, which is an open subset of Y. Furthermore, $Y - Y \cap H$ is a closed subset of $\mathbf{A}^n - H$, which is affine by (4.2), hence $Y - Y \cap H$ is affine. This is the required affine neighborhood of P.

Now we come to the main result of this section. Let $\varphi: X \to Y$ be a dominant rational map, represented by $\langle U, \varphi_U \rangle$. Let $f \in K(Y)$ be a rational function, represented by $\langle V, f \rangle$ where V is an open set in Y, and f is a regular function on V. Since $\varphi_U(U)$ is dense in Y, $\varphi_U^{-1}(V)$ is a nonempty open subset of X, so $f \circ \varphi_U$ is a regular function on $\varphi_U^{-1}(V)$. This gives us a rational function on X, and in this manner we have defined a homomorphism of k-algebras from $K(Y)$ to $K(X)$.

Theorem 4.4. *For any two varieties X and Y, the above construction gives a bijection between*

(i) *the set of dominant rational maps from X to Y, and*
(ii) *the set of k-algebra homomorphisms from $K(Y)$ to $K(X)$.*

Furthermore, this correspondence gives an arrow-reversing equivalence of categories of the category of varieties and dominant rational maps with the category of finitely generated field extensions of k.

PROOF. We will construct an inverse to the mapping given by the construction above. Let $\theta: K(Y) \to K(X)$ be a homomorphism of k-algebras. We wish to define a rational map from X to Y. By (4.3), Y is covered by affine varieties, so we may assume Y is affine. Let $A(Y)$ be its affine coordinate ring, and let $y_1, \ldots, y_n$ be generators for $A(Y)$ as a k-algebra. Then $\theta(y_1), \ldots, \theta(y_n)$ are rational functions on X. We can find an open set $U \subseteq X$ such that the functions $\theta(y_i)$ are all regular on U. Then θ defines an injective homomorphism of k-algebras $A(Y) \to \mathcal{O}(U)$. By (3.5) this corresponds to a morphism $\varphi: U \to Y$, which gives us a dominant rational map from X to Y. It is easy to see that this gives a map of sets (ii) $\to$ (i) which is inverse to the one defined above.

To see that we have an equivalence of categories as stated, we need only check that for any variety Y, $K(Y)$ is finitely generated over k, and conversely, if K/k is a finitely generated field extension, then $K = K(Y)$ for some Y. If Y is a variety, then $K(Y) = K(U)$ for any open affine subset, so we may assume Y affine. Then by (3.2d), $K(Y)$ is a finitely generated field extension of k. On the other hand, let K be a finitely generated field extension of k. Let $y_1, \ldots, y_n \in K$ be a set of generators, and let B be the sub-k-algebra of K generated by $y_1, \ldots, y_n$. Then B is a quotient of the polynomial ring $A = k[x_1, \ldots, x_n]$, so $B \cong A(Y)$ for some variety Y in $\mathbf{A}^n$. Then $K \cong K(Y)$ so we are done.

Corollary 4.5. *For any two varieties X, Y the following conditions are equivalent:*

  (i) *X and Y are birationally equivalent;*
 (ii) *there are open subsets $U \subseteq X$ and $V \subseteq Y$ with U isomorphic to V,*
(iii) *$K(X) \cong K(Y)$ as k-algebras.*

PROOF.
  (i) $\Rightarrow$ (ii). Let $\varphi: X \to Y$ and $\psi: Y \to X$ be rational maps which are inverse to each other. Let φ be represented by $\langle U, \varphi \rangle$ and let ψ be represented by $\langle V, \psi \rangle$. Then $\psi \circ \varphi$ is represented by $\langle \varphi^{-1}(V), \psi \circ \varphi \rangle$, and since $\psi \circ \varphi = \mathrm{id}_X$ as a rational map, $\psi \circ \varphi$ is the identity on $\varphi^{-1}(V)$. Similarly $\varphi \circ \psi$ is the identity on $\psi^{-1}(U)$. We now take $\varphi^{-1}(\psi^{-1}(U))$ as our open set in X, and $\psi^{-1}(\varphi^{-1}(V))$ as our open set in Y. It follows from the construction that these two open sets are isomorphic via φ and ψ.
  (ii) $\Rightarrow$ (iii) follows from the definition of function field.
  (iii) $\Rightarrow$ (i) follows from the theorem.

As an illustration of the notion of birational correspondence, we will use some algebraic results on field extensions to show that every variety is birational to a hypersurface. We assume familiarity with the notion of separable algebraic field extensions, and the notions of transcendence base and transcendence degree for infinite field extensions (see, e.g., Zariski–Samuel [1, Ch. II]).

Theorem 4.6A (Theorem of the Primitive Element). *Let L be a finite separable extension field of a field K. Then there is an element $\alpha \in L$ which generates L as an extension field of K. Furthermore, if $\beta_1, \ldots, \beta_n$ is any set of generators of L over K, and if K is infinite, then α can be taken to be a linear combination $\alpha = c_1\beta_1 + \ldots + c_n\beta_n$ of the β_i with coefficients $c_i \in K$.*

PROOF. Zariski–Samuel [1, Ch. II, Theorem 19, p. 84]. The second statement follows from the proof given there.

Definition. A field extension K/k is *separably generated* if there is a transcendence base $\{x_i\}$ for K/k such that K is a separable algebraic extension of $k(\{x_i\})$. Such a transcendence base is called a *separating transcendence base.*

Theorem 4.7A. *If a field extension K/k is finitely generated and separably generated, then any set of generators contains a subset which is a separating transcendence base.*

PROOF. Zariski–Samuel [1, Ch. II, Theorem 30, p. 104].

Theorem 4.8A. *If k is a perfect field (hence in particular if k is algebraically closed), any finitely generated field extension K/k is separably generated.*

PROOF. Zariski–Samuel [1, Ch. II, Theorem 31, p. 105], or Matsumura [2, Ch. 10, Corollary, p. 194].

Proposition 4.9. *Any variety X of dimension r is birational to a hypersurface Y in $\mathbf{P}^{r+1}$.*

PROOF. The function field K of X is a finitely generated extension field of k. By (4.8A), K is separably generated over k. Hence we can find a transcendence base $x_1, \ldots, x_r \in K$ such that K is a finite separable extension of $k(x_1, \ldots, x_r)$. Then by (4.6A) we can find one further element $y \in K$ such that $K = k(x_1, \ldots, x_r, y)$. Now y is algebraic over $k(x_1, \ldots, x_r)$, so it satisfies a polynomial equation with coefficients which are rational functions in $x_1, \ldots, x_r$. Clearing denominators, we get an irreducible polynomial $f(x_1, \ldots, x_r, y) = 0$. This defines a hypersurface in $\mathbf{A}^{r+1}$ with function field K, which, according to (4.5), is birational to X. Its projective closure (Ex. 2.9) is the required hypersurface $Y \subseteq \mathbf{P}^{r+1}$.

27

Blowing Up

As another example of a birational map, we will now construct the blowing-up of a variety at a point. This important construction is the main tool in the resolution of singularities of an algebraic variety.

First we will construct the blowing-up of $\mathbf{A}^n$ at the point $O = (0, \ldots, 0)$. Consider the product $\mathbf{A}^n \times \mathbf{P}^{n-1}$, which is a quasi-projective variety (Ex. 3.16). If $x_1, \ldots, x_n$ are the affine coordinates of $\mathbf{A}^n$, and if $y_1, \ldots, y_n$ are the homogeneous coordinates of $\mathbf{P}^{n-1}$ (observe the unusual notation!), then the closed subsets of $\mathbf{A}^n \times \mathbf{P}^{n-1}$ are defined by polynomials in the x_i, y_j, which are homogeneous with respect to the y_j.

We now define the *blowing-up of $\mathbf{A}^n$ at the point O* to be the closed subset X of $\mathbf{A}^n \times \mathbf{P}^{n-1}$ defined by the equations $\{x_i y_j = x_j y_i \mid i,j = 1, \ldots, n\}$.

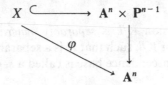

We have a natural morphism $\varphi : X \to \mathbf{A}^n$ obtained by restricting the projection map of $\mathbf{A}^n \times \mathbf{P}^{n-1}$ onto the first factor. We will now study the properties of X.

(1) If $P \in \mathbf{A}^n$, $P \neq O$, then $\varphi^{-1}(P)$ consists of a single point. In fact, φ gives an isomorphism of $X - \varphi^{-1}(O)$ onto $\mathbf{A}^n - O$. Indeed, let $P = (a_1, \ldots, a_n)$, with some $a_i \neq 0$. Now if $P \times (y_1, \ldots, y_n) \in \varphi^{-1}(P)$, then for each j, $y_j = (a_j/a_i)y_i$, so $(y_1, \ldots, y_n)$ is uniquely determined as a point in $\mathbf{P}^{n-1}$. In fact, setting $y_i = a_i$, we can take $(y_1, \ldots, y_n) = (a_1, \ldots, a_n)$. Thus $\varphi^{-1}(P)$ consists of a single point. Furthermore, for $P \in \mathbf{A}^n - O$, setting $\psi(P) = (a_1, \ldots, a_n) \times (a_1, \ldots, a_n)$ defines an inverse morphism to φ, showing $X - \varphi^{-1}(O)$ is isomorphic to $\mathbf{A}^n - O$.

(2) $\varphi^{-1}(O) \cong \mathbf{P}^{n-1}$. Indeed, $\varphi^{-1}(O)$ consists of all points $O \times Q$, with $Q = (y_1, \ldots, y_n) \in \mathbf{P}^{n-1}$, subject to no restriction.

(3) The points of $\varphi^{-1}(O)$ are in 1-1 correspondence with the set of lines through O in $\mathbf{A}^n$. Indeed, a line L through O in $\mathbf{A}^n$ can be given by parametric equations $x_i = a_i t$, $i = 1, \ldots, n$, where $a_i \in k$ are not all zero, and $t \in \mathbf{A}^1$. Now consider the line $L' = \varphi^{-1}(L - O)$ in $X - \varphi^{-1}(O)$. It is given parametrically by $x_i = a_i t$, $y_i = a_i t$, with $t \in \mathbf{A}^1 - 0$. But the y_i are homogeneous coordinates in $\mathbf{P}^{n-1}$, so we can equally well describe L' be the equations $x_i = a_i t, y_i = a_i$, for $t \in \mathbf{A}^1 - 0$. These equations make sense also for $t = 0$, and give the closure $\bar{L}'$ of L' in X. Now $\bar{L}'$ meets $\varphi^{-1}(O)$ in the point $Q = (a_1, \ldots, a_n) \in \mathbf{P}^{n-1}$, so we see that sending L to Q gives a 1-1 correspondence between lines through O in $\mathbf{A}^n$ and points of $\varphi^{-1}(O)$.

(4) X is irreducible. Indeed, X is the union of $X - \varphi^{-1}(O)$ and $\varphi^{-1}(O)$. The first piece is isomorphic to $\mathbf{A}^n - O$, hence irreducible. On the other hand, we have just seen that every point of $\varphi^{-1}(O)$ is in the closure of some

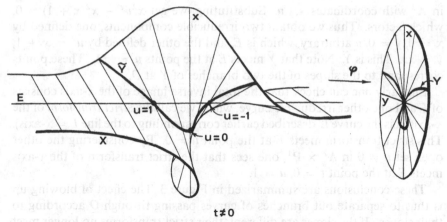

Figure 3. Blowing up.

subset (the line L') of $X - \varphi^{-1}(O)$. Hence $X - \varphi^{-1}(O)$ is dense in X, and X is irreducible.

Definition. If Y is a closed subvariety of $\mathbf{A}^n$ passing through O, we define the *blowing-up of Y at the point O* to be $\tilde{Y} = (\varphi^{-1}(Y - O))^-$, where $\varphi : X \to \mathbf{A}^n$ is the blowing-up of $\mathbf{A}^n$ at the point O described above. We denote also by $\varphi : \tilde{Y} \to Y$ the morphism obtained by restricting $\varphi : X \to \mathbf{A}^n$ to $\tilde{Y}$. To blow up at any other point P of $\mathbf{A}^n$, make a linear change of coordinates sending P to O.

Note that φ induces an isomorphism of $\tilde{Y} - \varphi^{-1}(O)$ to $Y - O$, so that φ is a birational morphism of $\tilde{Y}$ to Y. Note also that this definition apparently depends on the embedding of Y in $\mathbf{A}^n$, but in fact, we will see later that blowing-up is intrinsic (II, 7.15.1).

The effect of blowing up a point of Y is to "pull apart" Y near O according to the different directions of lines through O. We will illustrate this with an example.

Example 4.9.1. Let Y be the plane cubic curve given by the equation $y^2 = x^2(x + 1)$. We will blow up Y at O (Fig. 3). Let t,u be homogeneous coordinates for $\mathbf{P}^1$. Then X, the blowing-up of $\mathbf{A}^2$ at O, is defined by the equation $xu = ty$ inside $\mathbf{A}^2 \times \mathbf{P}^1$. It looks like $\mathbf{A}^2$, except that the point O has been replaced by a $\mathbf{P}^1$ corresponding to the slopes of lines through O. We will call this $\mathbf{P}^1$ the *exceptional curve*, and denote it by E.

We obtain the total inverse image of Y in X by considering the equations $y^2 = x^2(x + 1)$ and $xu = ty$ in $\mathbf{A}^2 \times \mathbf{P}^1$. Now $\mathbf{P}^1$ is covered by the open sets $t \neq 0$ and $u \neq 0$, which we consider separately. If $t \neq 0$, we can set $t = 1$, and use u as an affine parameter. Then we have the equations

$$y^2 = x^2(x + 1)$$

$$y = xu$$

in A^3 with coordinates x, y, u. Substituting, we get $x^2u^2 - x^2(x + 1) = 0$, which factors. Thus we obtain two irreducible components, one defined by $x = 0, y = 0, u$ arbitrary, which is E, and the other defined by $u^2 = x + 1$, $y = xu$. This is $\tilde{Y}$. Note that $\tilde{Y}$ meets E at the points $u = \pm 1$. These points correspond to the slopes of the two branches of Y at O.

Similarly one can check that the total inverse image of the x-axis consists of E and one other irreducible curve, which we call the *strict transform* of the x-axis (it is the curve $\bar{L}'$ described earlier corresponding to the line $L = x$-axis). This strict transform meets E at the point $u = 0$. By considering the other open set $u \neq 0$ in $A^2 \times P^1$, one sees that the strict transform of the y-axis meets E at the point $t = 0, u = 1$.

These conclusions are summarized in Figure 3. The effect of blowing up is thus to separate out branches of curves passing through O according to their slopes. If the slopes are different, their strict transforms no longer meet in X. Instead, they meet E at points corresponding to the different slopes.

EXERCISES

4.1. If f and g are regular functions on open subsets U and V of a variety X, and if $f = g$ on $U \cap V$, show that the function which is f on U and g on V is a regular function on $U \cup V$. Conclude that if f is a *rational* function on X, then there is a largest open subset U of X on which f is represented by a regular function. We say that f *is defined* at the points of U.

4.2. Same problem for rational maps. If φ is a rational map of X to Y, show there is a largest open set on which φ is represented by a morphism. We say the rational map *is defined* at the points of that open set.

4.3. (a) Let f be the rational function on P^2 given by $f = x_1/x_0$. Find the set of points where f is defined and describe the corresponding regular function.
   (b) Now think of this function as a rational map from P^2 to A^1. Embed A^1 in P^1, and let $\varphi: P^2 \to P^1$ be the resulting rational map. Find the set of points where φ is defined, and describe the corresponding morphism.

4.4. A variety Y is *rational* if it is birationally equivalent to P^n for some n (or, equivalently by (4.5), if $K(Y)$ is a pure transcendental extension of k).
   (a) Any conic in P^2 is a rational curve.
   (b) The cuspidal cubic $y^2 = x^3$ is a rational curve.
   (c) Let Y be the nodal cubic curve $y^2z = x^2(x + z)$ in P^2. Show that the projection φ from the point $P = (0,0,1)$ to the line $z = 0$ (Ex. 3.14) induces a birational map from Y to P^1. Thus Y is a rational curve.

4.5. Show that the quadric surface $Q: xy = zw$ in P^3 is birational to P^2, but not isomorphic to P^2 (cf. Ex. 2.15).

4.6. *Plane Cremona Transformations.* A birational map of P^2 into itself is called a *plane Cremona transformation*. We give an example, called a *quadratic transformation*. It is the rational map $\varphi: P^2 \to P^2$ given by $(a_0, a_1, a_2) \to (a_1a_2, a_0a_2, a_0a_1)$ when no two of a_0, a_1, a_2 are 0.

    (a) Show that φ is birational, and is its own inverse.

    (b) Find open sets $U, V \subseteq \mathbf{P}^2$ such that $\varphi : U \to V$ is an isomorphism.

    (c) Find the open sets where φ and φ^{-1} are defined, and describe the corresponding morphisms. See also (V, 4.2.3).

4.7. Let X and Y be two varieties. Suppose there are points $P \in X$ and $Q \in Y$ such that the local rings $\mathcal{O}_{P,X}$ and $\mathcal{O}_{Q,Y}$ are isomorphic as k-algebras. Then show that there are open sets $P \in U \subseteq X$ and $Q \in V \subseteq Y$ and an isomorphism of U to V which sends P to Q.

4.8. (a) Show that any variety of positive dimension over k has the same cardinality as k. [*Hints:* Do $\mathbf{A}^n$ and $\mathbf{P}^n$ first. Then for any X, use induction on the dimension n. Use (4.9) to make X birational to a hypersurface $H \subseteq \mathbf{P}^{n+1}$. Use (Ex. 3.7) to show that the projection of H to $\mathbf{P}^n$ from a point not on H is finite-to-one and surjective.]

    (b) Deduce that any two *curves* over k are homeomorphic (cf. Ex. 3.1).

4.9. Let X be a projective variety of dimension r in $\mathbf{P}^n$. with $n \geq r + 2$. Show that for suitable choice of $P \notin X$, and a linear $\mathbf{P}^{n-1} \subseteq \mathbf{P}^n$, the projection from P to $\mathbf{P}^{n-1}$ (Ex. 3.14) induces a *birational* morphism of X onto its image $X' \subseteq \mathbf{P}^{n-1}$. You will need to use (4.6A). (4.7A), and (4.8A). This shows in particular that the birational map of (4.9) can be obtained by a finite number of such projections.

4.10. Let Y be the cuspidal cubic curve $y^2 = x^3$ in $\mathbf{A}^2$. Blow up the point $O = (0,0)$, let E be the exceptional curve, and let $\tilde{Y}$ be the strict transform of Y. Show that E meets $\tilde{Y}$ in one point, and that $\tilde{Y} \cong \mathbf{A}^1$. In this case the morphism $\varphi : \tilde{Y} \to Y$ is bijective and bicontinuous, but it is not an isomorphism.

5 Nonsingular Varieties

The notion of nonsingular variety in algebraic geometry corresponds to the notion of manifold in topology. Over the complex numbers, for example, the nonsingular varieties are those which in the "usual" topology are complex manifolds. Accordingly, the most natural (and historically first) definition of nonsingularity uses the derivatives of the functions defining the variety:

Definition. Let $Y \subseteq \mathbf{A}^n$ be an affine variety, and let $f_1, \ldots, f_t \in A = k[x_1, \ldots, x_n]$ be a set of generators for the ideal of Y. Y is *nonsingular at a point* $P \in Y$ if the rank of the matrix $\|(\partial f_i / \partial x_j)(P)\|$ is $n - r$, where r is the dimension of Y. Y is *nonsingular* if it is nonsingular at every point.

    A few comments are in order. In the first place, the notion of partial derivative of a polynomial with respect to one of its variables makes sense over any field. One just applies the usual rules for differentiation. Thus no limiting process is needed. But funny things can happen in characteristic $p > 0$. For example, if $f(x) = x^p$, then $df/dx = px^{p-1} = 0$, since $p = 0$ in k. In any case, if $f \in A$ is a polynomial, then for each i, $\partial f / \partial x_i$ is a polynomial.

The matrix $\|(\partial f_i/\partial x_j)(P)\|$ is called the *Jacobian matrix* at P. One can show easily that this definition of nonsingularity is independent of the set of generators of the ideal of Y chosen.

One drawback of our definition is that it apparently depends on the embedding of Y in affine space. However, it was shown in a fundamental paper, Zariski [1], that nonsingularity could be described intrinsically in terms of the local rings. In our case the result is this.

Definition. Let A be a noetherian local ring with maximal ideal $\mathfrak{m}$ and residue field $k = A/\mathfrak{m}$. A is a *regular local ring* if $\dim_k \mathfrak{m}/\mathfrak{m}^2 = \dim A$.

Theorem 5.1. *Let $Y \subseteq \mathbf{A}^n$ be an affine variety. Let $P \in Y$ be a point. Then Y is nonsingular at P if and only if the local ring $\mathcal{O}_{P,Y}$ is a regular local ring.*

PROOF. Let P be the point $(a_1, \ldots, a_n)$ in $\mathbf{A}^n$, and let $\mathfrak{a}_P = (x_1 - a_1, \ldots, x_n - a_n)$ be the corresponding maximal ideal in $A = k[x_1, \ldots, x_n]$. We define a linear map $\theta: A \to k^n$ by

$$\theta(f) = \left\langle \frac{\partial f}{\partial x_1}(P), \ldots, \frac{\partial f}{\partial x_n}(P) \right\rangle$$

for any $f \in A$. Now it is clear that $\theta(x_i - a_i)$ for $i = 1, \ldots, n$ form a basis of k^n, and that $\theta(\mathfrak{a}_P^2) = 0$. Thus θ induces an isomorphism $\theta': \mathfrak{a}_P/\mathfrak{a}_P^2 \to k^n$.

Now let $\mathfrak{b}$ be the ideal of Y in A, and let $f_1, \ldots, f_t$ be a set of generators of $\mathfrak{b}$. Then the rank of the Jacobian matrix $J = \|(\partial f_i/\partial x_j)(P)\|$ is just the dimension of $\theta(\mathfrak{b})$ as a subspace of k^n. Using the isomorphism θ', this is the same as the dimension of the subspace $(\mathfrak{b} + \mathfrak{a}_P^2)/\mathfrak{a}_P^2$ of $\mathfrak{a}_P/\mathfrak{a}_P^2$. On the other hand, the local ring $\mathcal{O}_P$ of P on Y is obtained from A by dividing by $\mathfrak{b}$ and localizing at the maximal ideal $\mathfrak{a}_P$. Thus if $\mathfrak{m}$ is the maximal ideal of $\mathcal{O}_P$, we have

$$\mathfrak{m}/\mathfrak{m}^2 \cong \mathfrak{a}_P/(\mathfrak{b} + \mathfrak{a}_P^2).$$

Counting dimensions of vector spaces, we have $\dim \mathfrak{m}/\mathfrak{m}^2 + \operatorname{rank} J = n$.

Now let $\dim Y = r$. Then $\mathcal{O}_P$ is a local ring of dimension r (3.2), so $\mathcal{O}_P$ is regular if and only if $\dim_k \mathfrak{m}/\mathfrak{m}^2 = r$. But this is equivalent to $\operatorname{rank} J = n - r$, which says that P is a nonsingular point of Y.

Note. Later we will give another characterization of nonsingular points in terms of the sheaf of differential forms on Y (II, 8.15).

Now that we know the concept of nonsingularity is intrinsic, we can extend the definition to arbitrary varieties.

Definition. Let Y be any variety. Y is *nonsingular* at a point $P \in Y$ if the local ring $\mathcal{O}_{P,Y}$ is a regular local ring. Y is *nonsingular* if it is nonsingular at every point. Y is *singular* if it is not nonsingular.

Our next objective is to show that most points of a variety are nonsingular. We need an algebraic preliminary.

Proposition 5.2A. *If A is a noetherian local ring with maximal ideal* m *and residue field k, then* $\dim_k \mathfrak{m}/\mathfrak{m}^2 \geqslant \dim A$.

PROOF. Atiyah–Macdonald [1, Cor. 11.15, p. 121] or Matsumura [2, p. 78].

Theorem 5.3. *Let Y be a variety. Then the set* Sing *Y of singular points of Y is a proper closed subset of Y.*

PROOF. (See also II, 8.16.) First we show Sing Y is a closed subset. It is sufficient to show for some open covering $Y = \bigcup Y_i$ of Y, that Sing Y_i is closed for each i. Hence by (4.3) we may assume that Y is affine. By (5.2) and the proof of (5.1) we know that the rank of the Jacobian matrix is always $\leqslant n - r$. Hence the set of singular points is the set of points where the rank is $< n - r$. Thus Sing Y is the algebraic set defined by the ideal generated by $I(Y)$ together with all determinants of $(n - r) \times (n - r)$ submatrices of the matrix $\|\partial f_i/\partial x_j\|$. Hence Sing Y is closed.

To show that Sing Y is a proper subset of Y, we first apply (4.9) to get Y birational to a hypersurface in $\mathbf{P}^n$. Since birational varieties have isomorphic open subsets, we reduce to the case of a hypersurface. It is enough to consider any open affine subset of Y, so we may assume that Y is a hypersurface in $\mathbf{A}^n$, defined by a single irreducible polynomial $f(x_1, \dots, x_n) = 0$.

Now Sing Y is the set of points $P \in Y$ such that $(\partial f/\partial x_i)(P) = 0$ for $i = 1, \dots, n$. If Sing $Y = Y$, then the functions $\partial f/\partial x_i$ are zero on Y, and hence $\partial f/\partial x_i \in I(Y)$ for each i. But $I(Y)$ is the principal ideal generated by f, and $\deg(\partial f/\partial x_i) \leqslant \deg f - 1$ for each i, so we must have $\partial f/\partial x_i = 0$ for each i.

In characteristic 0 this is already impossible, because if x_i occurs in f, then $\partial f/\partial x_i \neq 0$. So we must have char $k = p > 0$, and then the fact that $\partial f/\partial x_i = 0$ implies that f is actually a polynomial in x_i^p. This is true for each i, so by taking pth roots of the coefficients (possible since k is algebraically closed), we get a polynomial $g(x_1, \dots, x_n)$ such that $f = g^p$. But this contradicts the hypothesis that f was irreducible, so we conclude that Sing $Y < Y$.

Completion

For the local analysis of singularities we will now describe the technique of completion. Let A be a local ring with maximal ideal $\mathfrak{m}$. The powers of $\mathfrak{m}$ define a topology on A, called the $\mathfrak{m}$-*adic topology*. By completing with respect to this topology, one defines the *completion* of A, denoted $\hat{A}$. Alternatively, one can define $\hat{A}$ as the inverse limit $\varprojlim A/\mathfrak{m}^n$. See Atiyah–Macdonald [1, Ch. 10], Matsumura [2, Ch. 9], or Zariski–Samuel [1, vol. 2, Ch. VIII] for general information on completions.

The significance of completion in algebraic geometry is that by passing to the completion $\hat{\mathcal{O}}_P$ of the local ring of a point P on a variety X, one can study the very local behavior of X near P. We have seen (Ex. 4.7) that if points $P \in X$ and $Q \in Y$ have isomorphic local rings, then already P and Q have isomorphic neighborhoods, so in particular X and Y are birational. Thus the ordinary local ring $\mathcal{O}_P$ carries information about almost all of X. However, the completion $\hat{\mathcal{O}}_P$, as we will see, carries much more local information, closer to our intuition of what "local" means in topology or differential geometry.

We will recall some of the algebraic properties of completion and then give some examples.

Theorem 5.4A. *Let A be a noetherian local ring with maximal ideal $\mathfrak{m}$, and let $\hat{A}$ be its completion.*

(a) *$\hat{A}$ is a local ring, with maximal ideal $\hat{\mathfrak{m}} = \mathfrak{m}\hat{A}$, and there is a natural injective homomorphism $A \to \hat{A}$.*

(b) *If M is a finitely generated A-module, its completion $\hat{M}$ with respect to its $\mathfrak{m}$-adic topology is isomorphic to $M \otimes_A \hat{A}$.*

(c) *$\dim A = \dim \hat{A}$.*

(d) *A is regular if and only if $\hat{A}$ is regular.*

PROOF. See Atiyah–Macdonald [1, Ch. 10, 11] or Zariski–Samuel [1, vol. 2, Ch. VIII].

Theorem 5.5A (Cohen Structure Theorem). *If A is a complete regular local ring of dimension n containing some field, then $A \cong k[[x_1, \ldots, x_n]]$, the ring of formal power series over the residue field k of A.*

PROOF. Matsumura [2, Cor. 2, p. 206] or Zariski–Samuel [1, vol. 2, Cor., p. 307].

Definition. We say two points $P \in X$ and $Q \in Y$ are *analytically isomorphic* if there is an isomorphism $\hat{\mathcal{O}}_P \cong \hat{\mathcal{O}}_Q$ as k-algebras.

Example 5.6.1. If $P \in X$ and $Q \in Y$ are analytically isomorphic, then $\dim X = \dim Y$. This follows from (5.4A) and the fact that any local ring of a point on a variety has the same dimension as the variety (Ex. 3.12).

Example 5.6.2. If $P \in X$ and $Q \in Y$ are nonsingular points on varieties of the same dimension, then P and Q are analytically isomorphic. This follows from (5.4A) and (5.5A). This example is the algebraic analogue of the fact that any two manifolds (topological, differentiable, or complex) of the same dimension are locally isomorphic.

Example 5.6.3. Let X be the plane nodal cubic curve given by the equation $y^2 = x^2(x + 1)$. Let Y be the algebraic set in $\mathbf{A}^2$ defined by the equation $xy = 0$. We will show that the point $O = (0,0)$ on X is analytically isomorphic to the point O on Y. (Since we haven't yet developed the general theory of local rings of points on reducible algebraic sets, we use an ad hoc definition $\mathcal{O}_{O,Y} = (k[x,y]/(xy))_{(x,y)}$. Thus $\hat{\mathcal{O}}_{O,Y} \cong k[[x,y]]/(xy)$.) This example corresponds to the geometric fact that near O, X looks like two lines crossing.

To prove this result, we consider the completion $\hat{\mathcal{O}}_{O,X}$ which is isomorphic to $k[[x,y]]/(y^2 - x^2 - x^3)$. The key point is that the leading form of the equation, namely $y^2 - x^2$, factors into two distinct factors $y + x$ and $y - x$ (we assume char $k \neq 2$). I claim there are formal power series

$$g = y + x + g_2 + g_3 + \ldots$$
$$h = y - x + h_2 + h_3 + \ldots$$

in $k[[x,y]]$, where g_i, h_i are homogeneous of degree i, such that $y^2 - x^2 - x^3 = gh$. We construct g and h step by step. To determine g_2 and h_2, we need to have

$$(y - x)g_2 + (y + x)h_2 = -x^3.$$

This is possible, because $y - x$ and $y + x$ generate the maximal ideal of $k[[x,y]]$. To determine g_3 and h_3, we need

$$(y - x)g_3 + (y + x)h_3 = -g_2 h_2$$

which is again possible, and so on.

Thus $\hat{\mathcal{O}}_{O,X} = k[[x,y]]/(gh)$. Since g and h begin with linearly independent linear terms, there is an automorphism of $k[[x,y]]$ sending g and h to x and y, respectively. This shows that $\hat{\mathcal{O}}_{O,X} \cong k[[x,y]]/(xy)$ as required.

Note in this example that $\mathcal{O}_{O,X}$ is an integral domain, but its completion is not.

We state here an algebraic result which will be used in (Ex. 5.15) below.

Theorem 5.7A (Elimination Theory). *Let $f_1, \ldots, f_r$ be homogeneous polynomials in $x_0, \ldots, x_n$, having indeterminate coefficients a_{ij}. Then there is a set $g_1, \ldots, g_t$ of polynomials in the a_{ij}, with integer coefficients, which are homogeneous in the coefficients of each f_i separately, with the following property: for any field k, and for any set of special values of the $a_{ij} \in k$, a necessary and sufficient condition for the f_i to have a common zero different from $(0,0,\ldots,0)$ is that the a_{ij} are a common zero of the polynomials g_j.*

PROOF. Van der Waerden [1, vol. II, §80, p. 8].

EXERCISES

5.1. Locate the singular points and sketch the following curves in $\mathbf{A}^2$ (assume char $k \neq 2$). Which is which in Figure 4?
   (a) $x^2 = x^4 + y^4$;
   (b) $xy = x^6 + y^6$;
   (c) $x^3 = y^2 + x^4 + y^4$;
   (d) $x^2 y + xy^2 = x^4 + y^4$.

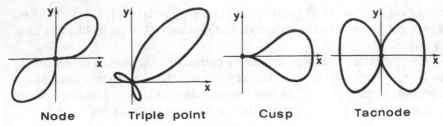

Node **Triple point** **Cusp** **Tacnode**

Figure 4. Singularities of plane curves.

5.2. Locate the singular points and describe the singularities of the following surfaces in A^3 (assume char $k \neq 2$). Which is which in Figure 5?
(a) $xy^2 = z^2$;
(b) $x^2 + y^2 = z^2$;
(c) $xy + x^3 + y^3 = 0$.

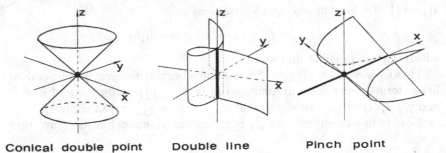

Conical double point **Double line** **Pinch point**

Figure 5. Surface singularities.

5.3. *Multiplicities.* Let $Y \subseteq A^2$ be a curve defined by the equation $f(x,y) = 0$. Let $P = (a,b)$ be a point of A^2. Make a linear change of coordinates so that P becomes the point $(0,0)$. Then write f as a sum $f = f_0 + f_1 + \ldots + f_d$, where f_i is a homogeneous polynomial of degree i in x and y. Then we define the *multiplicity* of P on Y, denoted $\mu_P(Y)$, to be the least r such that $f_r \neq 0$. (Note that $P \in Y \Leftrightarrow \mu_P(Y) > 0$.) The linear factors of f_r are called the *tangent directions* at P.
(a) Show that $\mu_P(Y) = 1 \Leftrightarrow P$ is a nonsingular point of Y.
(b) Find the multiplicity of each of the singular points in (Ex. 5.1) above.

5.4. *Intersection Multiplicity.* If $Y, Z \subseteq A^2$ are two distinct curves, given by equations $f = 0$, $g = 0$, and if $P \in Y \cap Z$, we define the *intersection multiplicity* $(Y \cdot Z)_P$ of Y and Z at P to be the length of the $\mathcal{O}_P$-module $\mathcal{O}_P/(f,g)$.
(a) Show that $(Y \cdot Z)_P$ is finite, and $(Y \cdot Z)_P \geq \mu_P(Y) \cdot \mu_P(Z)$.
(b) If $P \in Y$, show that for almost all lines L through P (i.e., all but a finite number), $(L \cdot Y)_P = \mu_P(Y)$.
(c) If Y is a curve of degree d in P^2, and if L is a line in P^2, $L \neq Y$, show that $(L \cdot Y) = d$. Here we define $(L \cdot Y) = \sum (L \cdot Y)_P$ taken over all points $P \in L \cap Y$, where $(L \cdot Y)_P$ is defined using a suitable affine cover of P^2.

5.5. For every degree $d > 0$, and every $p = 0$ or a prime number, give the equation of a nonsingular curve of degree d in $\mathbf{P}^2$ over a field k of characteristic p.

5.6. *Blowing Up Curve Singularities.*
(a) Let Y be the cusp or node of (Ex. 5.1). Show that the curve $\tilde{Y}$ obtained by blowing up Y at $O = (0,0)$ is nonsingular (cf. (4.9.1) and (Ex. 4.10)).
(b) We define a *node* (also called *ordinary double point*) to be a double point (i.e., a point of multiplicity 2) of a plane curve with distinct tangent directions (Ex. 5.3). If P is a node on a plane curve Y, show that $\varphi^{-1}(P)$ consists of two distinct nonsingular points on the blown-up curve $\tilde{Y}$. We say that "blowing up P resolves the singularity at P".
(c) Let $P \in Y$ be the tacnode of (Ex. 5.1). If $\varphi: \tilde{Y} \to Y$ is the blowing-up at P, show that $\varphi^{-1}(P)$ is a node. Using (b) we see that the tacnode can be resolved by two successive blowings-up.
(d) Let Y be the plane curve $y^3 = x^5$, which has a "higher order cusp" at O. Show that O is a triple point; that blowing up O gives rise to a double point (what kind?) and that one further blowing up resolves the singularity.
Note: We will see later (V, 3.8) that any singular point of a plane curve can be resolved by a finite sequence of successive blowings-up.

5.7. Let $Y \subseteq \mathbf{P}^2$ be a nonsingular plane curve of degree > 1, defined by the equation $f(x,y,z) = 0$. Let $X \subseteq \mathbf{A}^3$ be the affine variety defined by f (this is the cone over Y; see (Ex. 2.10)). Let P be the point $(0,0,0)$, which is the *vertex* of the cone. Let $\varphi: \tilde{X} \to X$ be the blowing-up of X at P.
(a) Show that X has just one singular point, namely P.
(b) Show that $\tilde{X}$ is nonsingular (cover it with open affines).
(c) Show that $\varphi^{-1}(P)$ is isomorphic to Y.

5.8. Let $Y \subseteq \mathbf{P}^n$ be a projective variety of dimension r. Let $f_1, \ldots, f_t \in S = k[x_0, \ldots, x_n]$ be homogeneous polynomials which generate the ideal of Y. Let $P \in Y$ be a point, with homogeneous coordinates $P = (a_0, \ldots, a_n)$. Show that P is nonsingular on Y if and only if the rank of the matrix $\|(\partial f_i/\partial x_j)(a_0, \ldots, a_n)\|$ is $n - r$. [*Hint:* (a) Show that this rank is independent of the homogeneous coordinates chosen for P; (b) pass to an open affine $U_i \subseteq \mathbf{P}^n$ containing P and use the affine Jacobian matrix; (c) you will need Euler's lemma, which says that if f is a homogeneous polynomial of degree d, then $\sum x_i(\partial f/\partial x_i) = d \cdot f$.]

5.9. Let $f \in k[x,y,z]$ be a homogeneous polynomial, let $Y = Z(f) \subseteq \mathbf{P}^2$ be the algebraic set defined by f, and suppose that for every $P \in Y$, at least one of $(\partial f/\partial x)(P), (\partial f/\partial y)(P), (\partial f/\partial z)(P)$ is nonzero. Show that f is irreducible (and hence that Y is a nonsingular variety). [*Hint:* Use (Ex. 3.7).]

5.10. For a point P on a variety X, let $\mathfrak{m}$ be the maximal ideal of the local ring $\mathcal{O}_P$. We define the *Zariski tangent space* $T_P(X)$ of X at P to be the dual k-vector space of $\mathfrak{m}/\mathfrak{m}^2$.
(a) For any point $P \in X$, $\dim T_P(X) \geq \dim X$, with equality if and only if P is nonsingular.
(b) For any morphism $\varphi: X \to Y$, there is a natural induced k-linear map $T_P(\varphi): T_P(X) \to T_{\varphi(P)}(Y)$.
(c) If φ is the vertical projection of the parabola $x = y^2$ onto the x-axis, show that the induced map $T_0(\varphi)$ of tangent spaces at the origin is the zero map.

5.11. *The Elliptic Quartic Curve in* $\mathbf{P}^3$. Let Y be the algebraic set in $\mathbf{P}^3$ defined by the equations $x^2 - xz - yw = 0$ and $yz - xw - zw = 0$. Let P be the point $(x,y,z,w) = (0,0,0,1)$, and let φ denote the projection from P to the plane $w = 0$. Show that φ induces an isomorphism of $Y - P$ with the plane cubic curve $y^2 z - x^3 + xz^2 = 0$ minus the point $(1,0,-1)$. Then show that Y is an irreducible nonsingular curve. It is called the *elliptic quartic curve* in $\mathbf{P}^3$. Since it is defined by two equations it is another example of a complete intersection (Ex. 2.17).

5.12. *Quadric Hypersurfaces.* Assume char $k \neq 2$, and let f be a homogeneous polynomial of degree 2 in $x_0, \dots, x_n$.
   (a) Show that after a suitable linear change of variables, f can be brought into the form $f = x_0^2 + \dots + x_r^2$ for some $0 \leqslant r \leqslant n$.
   (b) Show that f is irreducible if and only if $r \geqslant 2$.
   (c) Assume $r \geqslant 2$, and let Q be the quadric hypersurface in $\mathbf{P}^n$ defined by f. Show that the singular locus $Z = \operatorname{Sing} Q$ of Q is a *linear* variety (Ex. 2.11) of dimension $n - r - 1$. In particular, Q is nonsingular if and only if $r = n$.
   (d) In case $r < n$, show that Q is a cone with axis Z over a nonsingular quadric hypersurface $Q' \subseteq \mathbf{P}^r$. (This notion of cone generalizes the one defined in (Ex. 2.10). If Y is a closed subset of $\mathbf{P}^r$, and if Z is a linear subspace of dimension $n - r - 1$ in $\mathbf{P}^n$, we embed $\mathbf{P}^r$ in $\mathbf{P}^n$ so that $\mathbf{P}^r \cap Z = \varnothing$, and define the *cone over Y with axis Z* to be the union of all lines joining a point of Y to a point of Z.)

5.13. It is a fact that any regular local ring is an integrally closed domain (Matsumura [2, Th. 36, p. 121]). Thus we see from (5.3) that any variety has a nonempty open subset of normal points (Ex. 3.17). In this exercise, show directly (without using (5.3)) that the set of nonnormal points of a variety is a proper closed subset (you will need the finiteness of integral closure: see (3.9A)).

5.14. *Analytically Isomorphic Singularities.*
   (a) If $P \in Y$ and $Q \in Z$ are analytically isomorphic plane curve singularities, show that the multiplicities $\mu_P(Y)$ and $\mu_Q(Z)$ are the same (Ex. 5.3).
   (b) Generalize the example in the text (5.6.3) to show that if $f = f_r + f_{r+1} + \dots \in k[[x,y]]$, and if the leading form f_r of f factors as $f_r = g_s h_t$, where g_s, h_t are homogeneous of degrees s and t respectively, and have no common linear factor, then there are formal power series

$$g = g_s + g_{s+1} + \dots$$
$$h = h_t + h_{t+1} + \dots$$

in $k[[x,y]]$ such that $f = gh$.
   (c) Let Y be defined by the equation $f(x,y) = 0$ in $\mathbf{A}^2$, and let $P = (0,0)$ be a point of multiplicity r on Y, so that when f is expanded as a polynomial in x and y, we have $f = f_r + $ higher terms. We say that P is an *ordinary r-fold point* if f_r is a product of r *distinct* linear factors. Show that any two ordinary double points are analytically isomorphic. Ditto for ordinary triple points. But show that there is a one-parameter family of mutually nonisomorphic ordinary 4-fold points.
   *(d) Assume char $k \neq 2$. Show that any double point of a plane curve is analytically isomorphic to the singularity at $(0,0)$ of the curve $y^2 = x^r$, for a uniquely

determined $r \geqslant 2$. If $r = 2$ it is a node (Ex. 5.6). If $r = 3$ we call it a *cusp*; if $r = 4$ a *tacnode*. See (V, 3.9.5) for further discussion.

5.15. *Families of Plane Curves.* A homogeneous polynomial f of degree d in three variables x,y,z has $\binom{d+2}{2}$ coefficients. Let these coefficients represent a point in $\mathbf{P}^N$, where $N = \binom{d+2}{2} - 1 = \frac{1}{2}d(d + 3)$.

(a) Show that this gives a correspondence between points of $\mathbf{P}^N$ and algebraic sets in $\mathbf{P}^2$ which can be defined by an equation of degree d. The correspondence is 1-1 except in some cases where f has a multiple factor.

(b) Show under this correspondence that the (irreducible) nonsingular curves of degree d correspond 1-1 to the points of a nonempty Zariski-open subset of $\mathbf{P}^N$. [*Hints:* (1) Use elimination theory (5.7A) applied to the homogeneous polynomials $\partial f/\partial x_0, \ldots, \partial f/\partial x_n$; (2) use the previous (Ex. 5.5, 5.8, 5.9) above.]

6 Nonsingular Curves

In considering the problem of classification of algebraic varieties, we can formulate several subproblems, based on the idea that a nonsingular projective variety is the best kind: (a) classify varieties up to birational equivalence; (b) within each birational equivalence class, find a nonsingular projective variety; (c) classify the nonsingular projective varieties in a given birational equivalence class.

In general, all three problems are very difficult. However, in the case of curves, the situation is much simpler. In this section we will answer problems (b) and (c) by showing that in each birational equivalence class, there is a unique nonsingular projective curve. We will also give an example to show that not all curves are birationally equivalent to each other (Ex. 6.2). Thus for a given finitely generated extension field K of k of transcendence degree 1 (which we will call a *function field of dimension* 1) we can talk about *the* nonsingular projective curve C_K with function field equal to K. We will see also that if K_1, K_2 are two function fields of dimension 1, then any k-homomorphism $K_2 \to K_1$ is represented by a morphism of C_{K_1} to C_{K_2}.

We will begin our study in an oblique manner by defining the notion of an "abstract nonsingular curve" associated with a given function field. It will not be clear a priori that this is a variety. However, we will see in retrospect that we have defined nothing new.

First we have to recall some basic facts about valuation rings and Dedekind domains.

Definition. Let K be a field and let G be a totally ordered abelian group. A *valuation* of K with values in G is a map $v: K - \{0\} \to G$ such that for all $x, y \in K$, $x, y \neq 0$, we have:

(1) $v(xy) = v(x) + v(y)$;

(2) $v(x + y) \geqslant \min(v(x), v(y))$.

If v is a valuation, then the set $R = \{x \in K | v(x) \geqslant 0\} \cup \{0\}$ is a subring of K, which we call the *valuation ring* of v. The subset $\mathfrak{m} = \{x \in K | v(x) > 0\} \cup \{0\}$ is an ideal in R, and $R, \mathfrak{m}$ is a local ring. A *valuation ring* is an integral domain which is the valuation ring of some valuation of its quotient field. If R is a valuation ring with quotient field K, we say that R is a *valuation ring of K*. If k is a subfield of K such that $v(x) = 0$ for all $x \in k - \{0\}$, then we say v is a *valuation of K/k*, and R is a *valuation ring of K/k*. (Note that valuation rings are not in general noetherian!)

Definition. If A, B are local rings contained in a field K, we say that B *dominates* A if $A \subseteq B$ and $\mathfrak{m}_B \cap A = \mathfrak{m}_A$.

Theorem 6.1A. *Let K be a field. A local ring R contained in K is a valuation ring of K if and only if it is a maximal element of the set of local rings contained in K, with respect to the relation of domination. Every local ring contained in K is dominated by some valuation ring of K.*

PROOF. Bourbaki [2, Ch. VI, §1, 3] or Atiyah–Macdonald [1, Ch. 5, p. 65, and exercises, p. 72].

Definition. A valuation v is *discrete* if its value group G is the integers. The corresponding valuation ring is called a *discrete valuation ring*.

Theorem 6.2A. *Let A be a noetherian local domain of dimension one, with maximal ideal $\mathfrak{m}$. Then the following conditions are equivalent:*

(i) *A is a discrete valuation ring;*
(ii) *A is integrally closed;*
(iii) *A is a regular local ring;*
(iv) *$\mathfrak{m}$ is a principal ideal.*

PROOF. Atiyah–Macdonald [1, Prop. 9.2, p. 94].

Definition. A *Dedekind domain* is an integrally closed noetherian domain of dimension one.

Because integral closure is a local property (Atiyah–Macdonald [1, Prop. 5.13, p. 63]), every localization of a Dedekind domain at a nonzero prime ideal is a discrete valuation ring.

Theorem 6.3A. *The integral closure of a Dedekind domain in a finite extension field of its quotient field is again a Dedekind domain.*

PROOF. Zariski–Samuel [1, vol. 1, Th. 19, p. 281].

We now turn to the case of a function field K of dimension 1 over k, where k is our fixed algebraically closed base field. We wish to establish a connection between non-singular curves with function field K and the set of discrete valuation rings of K/k. If P is a point on a nonsingular curve Y, then by (5.1) the local ring $\mathcal{O}_P$ is a regular local ring of dimension one, and so by (6.2A) it is a discrete valuation ring. Its quotient field is the function field K of Y, and since $k \subseteq \mathcal{O}_P$, it is a valuation ring of K/k. Thus the local rings of Y define a subset of the set C_K of all discrete valuation rings of K/k. This motivates the definition of an abstract nonsingular curve below. But first we need a few more preliminaries.

Lemma 6.4. *Let Y be a quasi-projective variety, let $P,Q \in Y$, and suppose that $\mathcal{O}_Q \subseteq \mathcal{O}_P$ as subrings of $K(Y)$. Then $P = Q$.*

PROOF. Embed Y in $\mathbf{P}^n$ for some n. Replacing Y by its closure, we may assume Y is projective. After a suitable linear change of coordinates in $\mathbf{P}^n$, we may assume that neither P nor Q is in the hyperplane H_0 defined by $x_0 = 0$. Thus $P,Q \in Y \cap (\mathbf{P}^n - H_0)$ which is affine, so we may assume that Y is an affine variety.

Let A be the affine ring of Y. Then there are maximal ideals $\mathfrak{m},\mathfrak{n} \subseteq A$ such that $\mathcal{O}_P = A_\mathfrak{m}$ and $\mathcal{O}_Q = A_\mathfrak{n}$. If $\mathcal{O}_Q \subseteq \mathcal{O}_P$, we must have $\mathfrak{m} \subseteq \mathfrak{n}$. But $\mathfrak{m}$ is a maximal ideal, so $\mathfrak{m} = \mathfrak{n}$, hence $P = Q$, by (3.2b).

Lemma 6.5. *Let K be a function field of dimension one over k, and let $x \in K$. Then $\{R \in C_K | x \notin R\}$ is a finite set.*

PROOF. If R is a valuation ring, then $x \notin R$ if and only if $1/x \in \mathfrak{m}_R$. So letting $y = 1/x$, we have to show that if $y \in K$, $y \neq 0$, then $\{R \in C_K | y \in \mathfrak{m}_R\}$ is a finite set. If $y \in k$, there are no such R, so let us assume $y \notin k$.

We consider the subring $k[y]$ of K generated by y. Since k is algebraically closed, y is transcendental over k, hence $k[y]$ is a polynomial ring. Furthermore, since K is finitely generated and of transcendence degree 1 over k, K is a finite field extension of $k(y)$. Now let B be the integral closure of $k[y]$ in K. Then by (6.3A), B is a Dedekind domain, and it is also a finitely generated k-algebra (3.9A).

Now if y is contained in a discrete valuation ring R of K/k, then $k[y] \subseteq R$, and since R is integrally closed in K, we have $B \subseteq R$. Let $\mathfrak{n} = \mathfrak{m}_R \cap B$. Then $\mathfrak{n}$ is a maximal ideal of B, and B is dominated by R. But $B_\mathfrak{n}$ is also a discrete valuation ring of K/k, hence $B_\mathfrak{n} = R$ by the maximality of valuation rings (6.1A).

If furthermore $y \in \mathfrak{m}_R$, then $y \in \mathfrak{n}$. Now B is the affine coordinate ring of some affine variety Y (1.4.6). Since B is a Dedekind domain, Y has dimension one and is nonsingular. To say that $y \in \mathfrak{n}$ says that y, as a regular function on Y, vanishes at the point of Y corresponding to $\mathfrak{n}$. But $y \neq 0$, so it vanishes only at a finite set of points; these are in 1-1 correspondence with the maximal ideals of B by (3.2), and $R = B_\mathfrak{n}$ is determined by the maximal ideal $\mathfrak{n}$. Hence we conclude that $y \in \mathfrak{m}_R$ for only finitely many $R \in C_K$, as required.

Corollary 6.6. *Any discrete valuation ring of K/k is isomorphic to the local ring of a point on some nonsingular affine curve.*

PROOF. Given R, let $y \in R - k$. Then the construction used in the proof of (6.5) gives such a curve.

We now come to the definition of an abstract nonsingular curve. Let K be a function field of dimension 1 over k (i.e., a finitely generated extension field of transcendence degree 1). Let C_K be the set of all discrete valuation rings of K/k. We will sometimes call the elements of C_K *points*, and write $P \in C_K$, where P stands for the valuation ring R_P. Note that the set C_K is infinite, because it contains all the local rings of any nonsingular curve with function field K; those local rings are all distinct (6.4), and there are infinitely many of them (Ex. 4.8). We make C_K into a topological space by taking the closed sets to be the finite subsets and the whole space. If $U \subseteq C_K$ is an open subset of C_K, we define the ring of *regular functions* on U to be $\mathcal{O}(U) = \bigcap_{P \in U} R_P$. An element $f \in \mathcal{O}(U)$ defines a function from U to k by taking $f(P)$ to be the residue of f modulo the maximal ideal of R_P. (Note by (6.6) that for any $R \in C_K$, the residue field of R is k.) If two elements $f, g \in \mathcal{O}(U)$ define the same function, then $f - g \in \mathfrak{m}_P$ for infinitely many $P \in C_K$, so by (6.5) and its proof, $f = g$. Thus we can identify the elements of $\mathcal{O}(U)$ with functions from U to k. Note also by (6.5) that any $f \in K$ is a regular function on some open set U. Thus the function field of C_K, defined as in §3, is just K.

Definition. An *abstract nonsingular curve* is an open subset $U \subseteq C_K$, where K is a function field of dimension 1 over k, with the induced topology, and the induced notion of regular functions on its open subsets.

Note that it is not clear a priori that such an abstract curve is a variety. So we will enlarge the category of varieties by adjoining the abstract curves:

Definition. A *morphism* $\varphi : X \to Y$ between abstract nonsingular curves or varieties is a continuous mapping such that for every open set $V \subseteq Y$, and every regular function $f : V \to k$, $f \circ \varphi$ is a regular function on $\varphi^{-1}(V)$.

Now that we have apparently enlarged our category, our task will be to show that every nonsingular quasi-projective curve is isomorphic to an abstract nonsingular curve, and conversely. In particular, we will show that C_K itself is isomorphic to a nonsingular projective curve.

Proposition 6.7. *Every nonsingular quasi-projective curve Y is isomorphic to an abstract nonsingular curve.*

PROOF. Let K be the function field of Y. Then each local ring $\mathcal{O}_P$ of a point $P \in Y$ is a discrete valuation ring of K/k, by (5.1) and (6.2A). Furthermore, by (6.4), distinct points give rise to distinct subrings of K. So let $U \subseteq C_K$ be the set of local rings of Y, and let $\varphi : Y \to U$ be the bijective map defined by $\varphi(P) = \mathcal{O}_P$.

First, we need to show that U is an open subset of C_K. Because open sets are complements of finite sets, it is sufficient to show that U contains a nonempty open set. Thus, by (4.3), we may assume Y is affine, with affine ring A. Then A is a finitely generated k-algebra, and by (3.2), K is the quotient field of A, and U is the set of localizations of A at its maximal ideals. Since these local rings are all discrete valuation rings, U consists in fact of all discrete valuation rings of K/k containing A. Now let $x_1, \ldots, x_n$ be a set of generators of A over k. Then $A \subseteq R_P$ if and only if $x_1, \ldots, x_n \in R_P$. Thus $U = \bigcap U_i$, where $U_i = \{P \in C_K | x_i \in R_P\}$. But by (6.5), $\{P \in C_K | x_i \notin R_P\}$ is a finite set. Therefore each U_i and hence also U is open.

So we have shown that the U defined above is an abstract nonsingular curve. To show that φ is an isomorphism, we need only check that the regular functions on any open set are the same. But this follows from the definition of the regular functions on U and the fact that for any open set $V \subseteq Y$, $\mathcal{O}(V) = \bigcap_{P \in V} \mathcal{O}_{P,Y}$.

Now we need a result about extensions of morphisms from curves to projective varieties, which is interesting in its own right.

Proposition 6.8. *Let X be an abstract nonsingular curve, let $P \in X$, let Y be a projective variety, and let $\varphi : X - P \to Y$ be a morphism. Then there exists a unique morphism $\bar{\varphi} : X \to Y$ extending φ.*

PROOF. Embed Y as a closed subset of $\mathbf{P}^n$ for some n. Then it will be sufficient to show that φ extends to a morphism of X into $\mathbf{P}^n$, because if it does, the image is necessarily contained in Y. Thus we reduce to the case $Y = \mathbf{P}^n$.

Let $\mathbf{P}^n$ have homogeneous coordinates $x_0, \ldots, x_n$, and let U be the open set where $x_0, \ldots, x_n$ are all nonzero. By using induction on n, we may assume that $\varphi(X - P) \cap U \neq \varnothing$. Because if $\varphi(X - P) \cap U = \varnothing$, then $\varphi(X - P) \subseteq \mathbf{P}^n - U$. But $\mathbf{P}^n - U$ is the union of the hyperplanes H_i defined by $x_i = 0$. Since $\varphi(X - P)$ is irreducible, it must be contained in H_i for some i. Now $H_i \cong \mathbf{P}^{n-1}$, so the result would follow by induction. So we will assume that $\varphi(X - P) \cap U \neq \varnothing$.

For each i, j, x_i/x_j is a regular function on U. Pulling it back by φ, we obtain a regular function f_{ij} on an open subset of X, which we view as a rational function on X, i.e., $f_{ij} \in K$, where K is the function field of X.

Let v be the valuation of K associated with the valuation ring R_P. Let $r_i = v(f_{i0}), i = 0, 1, \ldots, n, r_i \in \mathbf{Z}$. Then since $x_i/x_j = (x_i/x_0)/(x_j/x_0)$, we have

$$v(f_{ij}) = r_i - r_j \qquad i, j = 0, \ldots, n.$$

43

Choose k such that r_k is minimal among $r_0, \ldots, r_n$. Then $v(f_{ik}) \geqslant 0$ for all i, hence $f_{0k}, \ldots, f_{nk} \in R_P$. Now define $\bar{\varphi}(P) = (f_{0k}(P), \ldots, f_{nk}(P))$, and $\bar{\varphi}(Q) = \varphi(Q)$ for $Q \neq P$. I claim that $\bar{\varphi}$ is a morphism of X to $\mathbf{P}^n$ which extends φ, and that $\bar{\varphi}$ is unique. The uniqueness is clear by construction (it also follows from (4.1)). To show that $\bar{\varphi}$ is a morphism, it will be sufficient to show that regular functions in a neighborhood of $\bar{\varphi}(P)$ pull back to regular functions on X. Let $U_k \subseteq \mathbf{P}^n$ be the open set where $x_k \neq 0$. Then $\bar{\varphi}(P) \in U_k$, since $f_{kk}(P) = 1$. Now U_k is affine, with affine coordinate ring equal to

$$k[x_0/x_k, \ldots, x_n/x_k].$$

These functions pull back to $f_{0k}, \ldots, f_{nk}$ which are regular at P by construction. It follows immediately that for any smaller neighborhood $\bar{\varphi}(P) \in V \subseteq U_k$, regular functions on V pull back to regular functions on X. Hence $\bar{\varphi}$ is a morphism, which completes the proof.

Now we come to our main result.

Theorem 6.9. *Let K be a function field of dimension 1 over k. Then the abstract nonsingular curve C_K defined above is isomorphic to a nonsingular projective curve.*

PROOF. The idea of the proof is this: we first cover $C = C_K$ with open subsets U_i which are isomorphic to nonsingular affine curves. Let Y_i be the projective closure of this affine curve. Then we use (6.8) to define a morphism $\varphi_i : C \to Y_i$. Next, we consider the product mapping $\varphi : C \to \prod Y_i$, and let Y be the closure of the image of C. Then Y is a projective curve, and we show that φ is an isomorphism of C onto Y.

To begin with, let $P \in C$ be any point. Then by (6.6) there is a nonsingular affine curve V and a point $Q \in V$ with $R_P \cong \mathcal{O}_Q$. It follows that the function field of V is K, and then by (6.7), V is isomorphic to an open subset of C. Thus we have shown that every point $P \in C$ has an open neighborhood which is isomorphic to an affine variety.

Since C is quasi-compact, we can cover it with a finite number of open subsets U_i, each of which is isomorphic to an affine variety V_i. Embed $V_i \subseteq \mathbf{A}^{n_i}$, think of $\mathbf{A}^{n_i}$ as an open subset of $\mathbf{P}^{n_i}$, and let Y_i be the closure of V_i in $\mathbf{P}^{n_i}$. Then Y_i is a projective variety, and we have a morphism $\varphi_i : U_i \to Y_i$ which is an isomorphism of U_i onto its image.

By (6.8) applied to the finite set of points $C - U_i$, we can find a morphism $\bar{\varphi}_i : C \to Y_i$ extending φ_i. Let $\prod Y_i$ be the product of the projective varieties Y_i (Ex. 3.16). Then $\prod Y_i$ is also a projective variety. Let $\varphi : C \to \prod Y_i$ be the "diagonal" map $\varphi(P) = \prod \bar{\varphi}_i(P)$, and let Y be the closure of the image of φ. Then Y is a projective variety, and $\varphi : C \to Y$ is a morphism whose image is dense in Y. (It follows that Y is a curve.)

Now we must show that φ is an isomorphism. For any point $P \in C$, we have $P \in U_i$ for some i. There is a commutative diagram

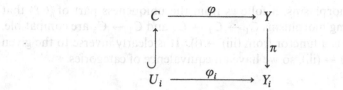

of dominant morphisms, where π is the projection map onto the ith factor. Thus we have inclusions of local rings

$$\mathcal{O}_{\varphi_i(P), Y_i} \hookrightarrow \mathcal{O}_{\varphi(P), Y} \hookrightarrow \mathcal{O}_{P, C}$$

by (Ex. 3.3). The two outside ones are isomorphic, so the middle one is also. Thus we see that for any $P \in C$, the map $\varphi_P^* : \mathcal{O}_{\varphi(P), Y} \to \mathcal{O}_{P, C}$ is an isomorphism.

Next, let Q be any point of Y. Then $\mathcal{O}_Q$ is dominated by some discrete valuation ring R of K/k (take for example a localization of the integral closure of $\mathcal{O}_Q$ at a maximal ideal). But $R = R_P$ for some $P \in C$, and $\mathcal{O}_{\varphi(P)} \cong R$, so by (6.4) we must have $Q = \varphi(P)$. This shows that φ is surjective. But φ is clearly injective, because distinct points of C correspond to distinct subrings of K.

Thus φ is a bijective morphism of C to Y, and for every $P \in C$, φ_P^* is an isomorphism, so by (Ex. 3.3b), φ is an isomorphism.

Corollary 6.10. *Every abstract nonsingular curve is isomorphic to a quasi-projective curve. Every nonsingular quasi-projective curve is isomorphic to an open subset of a nonsingular projective curve.*

Corollary 6.11. *Every curve is birationally equivalent to a nonsingular projective curve.*

PROOF. Indeed, if Y is any curve, with function field K, then Y is birationally equivalent to C_K which is nonsingular and projective.

Corollary 6.12. *The following three categories are equivalent:*

(i) *nonsingular projective curves, and dominant morphisms;*
(ii) *quasi-projective curves, and dominant rational maps;*
(iii) *function fields of dimension 1 over k, and k-homomorphisms.*

PROOF. We have an obvious functor from (i) to (ii). We have the functor $Y \to K(Y)$ from (ii) to (iii), which induces an equivalence of categories by (4.4). To complete the cycle, we need a functor from (iii) to (i).

To a function field K, associate the curve C_K, which by the theorem is a projective nonsingular curve. If $K_2 \to K_1$ is a homomorphism, then by (ii) $\simeq$ (iii), it induces a rational map of the corresponding curves. This can be represented by a morphism $\varphi : U \to C_{K_2}$, where $U \subseteq C_{K_1}$ is an open subset. By (6.8) φ extends to a morphism $\bar{\varphi} : C_{K_1} \to C_{K_2}$. If $K_3 \to K_2 \to K_1$

are two homomorphisms, it follows from the uniqueness part of (6.8) that the corresponding morphisms $C_1 \to C_2 \to C_3$ and $C_1 \to C_3$ are compatible. Hence $K \mapsto C_K$ is a functor from (iii) $\to$ (i). It is clearly inverse to the given functor (i) $\to$ (ii) $\to$ (iii), so we have an equivalence of categories.

EXERCISES

6.1. Recall that a curve is *rational* if it is birationally equivalent to $\mathbf{P}^1$ (Ex. 4.4). Let Y be a nonsingular rational curve which is not isomorphic to $\mathbf{P}^1$.
  (a) Show that Y is isomorphic to an open subset of $\mathbf{A}^1$.
  (b) Show that Y is affine.
  (c) Show that $A(Y)$ is a unique factorization domain.

6.2. *An Elliptic Curve.* Let Y be the curve $y^2 = x^3 - x$ in $\mathbf{A}^2$, and assume that the characteristic of the base field k is $\neq 2$. In this exercise we will show that Y is not a rational curve, and hence $K(Y)$ is not a pure transcendental extension of k.
  (a) Show that Y is nonsingular, and deduce that $A = A(Y) \simeq k[x,y]/(y^2 - x^3 + x)$ is an integrally closed domain.
  (b) Let $k[x]$ be the subring of $K = K(Y)$ generated by the image of x in A. Show that $k[x]$ is a polynomial ring, and that A is the integral closure of $k[x]$ in K.
  (c) Show that there is an automorphism $\sigma: A \to A$ which sends y to $-y$ and leaves x fixed. For any $a \in A$, define the *norm* of a to be $N(a) = a \cdot \sigma(a)$. Show that $N(a) \in k[x]$, $N(1) = 1$, and $N(ab) = N(a) \cdot N(b)$ for any $a,b \in A$.
  (d) Using the norm, show that the units in A are precisely the nonzero elements of k. Show that x and y are irreducible elements of A. Show that A is *not* a unique factorization domain.
  (e) Prove that Y is not a rational curve (Ex. 6.1). See (II, 8.20.3) and (III, Ex. 5.3) for other proofs of this important result.

6.3. Show by example that the result of (6.8) is false if either (a) dim $X \geqslant 2$, or (b) Y is not projective.

6.4. Let Y be a nonsingular projective curve. Show that every nonconstant rational function f on Y defines a surjective morphism $\varphi: Y \to \mathbf{P}^1$, and that for every $P \in \mathbf{P}^1$, $\varphi^{-1}(P)$ is a finite set of points.

6.5. Let X be a nonsingular projective curve. Suppose that X is a (locally closed) subvariety of a variety Y (Ex. 3.10). Show that X is in fact a closed subset of Y. See (II, Ex. 4.4) for generalization.

6.6. *Automorphisms of* $\mathbf{P}^1$. Think of $\mathbf{P}^1$ as $\mathbf{A}^1 \cup \{\infty\}$. Then we define a *fractional linear transformation* of $\mathbf{P}^1$ by sending $x \mapsto (ax + b)/(cx + d)$, for $a,b,c,d \in k$, $ad - bc \neq 0$.
  (a) Show that a fractional linear transformation induces an *automorphism* of $\mathbf{P}^1$ (i.e., an isomorphism of $\mathbf{P}^1$ with itself). We denote the group of all these fractional linear transformations by PGL(1).
  (b) Let Aut $\mathbf{P}^1$ denote the group of all automorphisms of $\mathbf{P}^1$. Show that Aut $\mathbf{P}^1 \simeq$ Aut $k(x)$, the group of k-automorphisms of the field $k(x)$.
  (c) Now show that every automorphism of $k(x)$ is a fractional linear transformation, and deduce that PGL(1) $\to$ Aut $\mathbf{P}^1$ is an isomorphism.

Note: We will see later (II, 7.1.1) that a similar result holds for $\mathbf{P}^n$: every automorphism is given by a linear transformation of the homogeneous coordinates.

6.7. Let $P_1, \ldots, P_r, Q_1, \ldots, Q_s$ be distinct points of $\mathbf{A}^1$. If $\mathbf{A}^1 - \{P_1, \ldots, P_r\}$ is isomorphic to $\mathbf{A}^1 - \{Q_1, \ldots, Q_s\}$, show that $r = s$. Is the converse true? Cf. (Ex. 3.1).

7   Intersections in Projective Space

The purpose of this section is to study the intersection of varieties in a projective space. If Y, Z are varieties in $\mathbf{P}^n$, what can one say about $Y \cap Z$? We have already seen (Ex. 2.16) that $Y \cap Z$ need not be a variety. But it is an algebraic set, and we can ask first about the dimensions of its irreducible components. We take our cue from the theory of vector spaces: if U, V are subspaces of dimensions r, s of a vector space W of dimension n, then $U \cap V$ is a subspace of dimension $\geqslant r + s - n$. Furthermore, if U and V are in sufficiently general position, the dimension of $U \cap V$ is equal to $r + s - n$ (provided $r + s - n \geqslant 0$). This result on vector spaces immediately implies the analogous result for linear subspaces of $\mathbf{P}^n$ (Ex. 2.11). Our first result in this section will be to prove that if Y, Z are subvarieties of dimensions r, s of $\mathbf{P}^n$, then every irreducible component of $Y \cap Z$ has dimension $\geqslant r + s - n$. Furthermore, if $r + s - n \geqslant 0$, then $Y \cap Z$ is nonempty.

Knowing something about the dimension of $Y \cap Z$, we can ask for more precise information. Suppose for example that $r + s = n$, and that $Y \cap Z$ is a finite set of points. Then we can ask, how many points are there? Let us look at a special case. If Y is a curve of degree d in $\mathbf{P}^2$, and if Z is a line in $\mathbf{P}^2$, then $Y \cap Z$ consists of at most d points, and the number comes to d exactly if we count them with appropriate multiplicities (Ex. 5.4). This result generalizes to the well-known theorem of Bézout, which says that if Y, Z are plane curves of degrees d, e, with $Y \neq Z$, then $Y \cap Z$ consists of de points, counted with multiplicities. We will prove Bézout's theorem later in this section (7.8).

The ideal generalization of Bézout's theorem to $\mathbf{P}^n$ would be this. First, define the degree of any projective variety. Let Y, Z be varieties of dimensions r, s, and of degrees d, e in $\mathbf{P}^n$. Assume that Y and Z are in a sufficiently general position so that all irreducible components of $Y \cap Z$ have dimension $= r + s - n$, and assume that $r + s - n \geqslant 0$. For each irreducible component W of $Y \cap Z$, define the intersection multiplicity $i(Y, Z; W)$ of Y and Z along W. Then we should have

$$\sum i(Y, Z; W) \cdot \deg W = de,$$

where the sum is taken over all irreducible components of $Y \cap Z$.

The hardest part of this generalization is the correct definition of the intersection multiplicity. (And, by the way, historically it took many attempts before a satisfactory treatment was given by Severi [3] geometrically and by Chevalley [1] and Weil [1] algebraically). We will define the intersection multiplicity only in the case where Z is a hypersurface. See Appendix A for the general case.

Our main task in this section will be the definition of the degree of a variety Y of dimension r in $\mathbf{P}^n$. Classically, the degree of Y is defined as the number of points of intersection of Y with a sufficiently general linear space L of dimension $n - r$. However, this definition is difficult to use. Cutting Y successively with r sufficiently general hyperplanes, one can find a linear space L of dimension $n - r$ which meets Y in a finite number of points (Ex. 1.8). But the number of intersection points may depend on L, and it is hard to make precise the notion "sufficiently general."

Therefore we will give a purely algebraic definition of degree, using the Hilbert polynomial of a projective variety. This definition is less geometrically motivated, but it has the advantage of being precise. In an exercise we show that it agrees with the classical definition in a special case (Ex. 7.4).

Proposition 7.1 (Affine Dimension Theorem). *Let Y, Z be varieties of dimensions r, s in $\mathbf{A}^n$. Then every irreducible component W of $Y \cap Z$ has dimension $\geq r + s - n$.*

PROOF. We proceed in several steps. First, suppose that Z is a hypersurface, defined by an equation $f = 0$. If $Y \subseteq Z$, there is nothing to prove. If $Y \not\subseteq Z$, we must show that each irreducible component W of $Y \cap Z$ has dimension $r - 1$. Let $A(Y)$ be the affine coordinate ring of Y. Then the irreducible components of $Y \cap Z$ correspond to the minimal prime ideals $\mathfrak{p}$ of the principal ideal (f) in $A(Y)$. Now by Krull's Hauptidealsatz (1.11A), each such $\mathfrak{p}$ has height one, so by the dimension theorem (1.8A), $A(Y)/\mathfrak{p}$ has dimension $r - 1$. By (1.7) this shows that each irreducible component W has dimension $r - 1$.

Now for the general case. We consider the product $Y \times Z \subseteq \mathbf{A}^{2n}$, which is a variety of dimension $r + s$ (Ex. 3.15). Let Δ be the diagonal $\{P \times P \mid P \in \mathbf{A}^n\} \subseteq \mathbf{A}^{2n}$. Then $\mathbf{A}^n$ is isomorphic to Δ by the map $P \to P \times P$, and under this isomorphism, $Y \cap Z$ corresponds to $(Y \times Z) \cap \Delta$. Since Δ has dimension n, and since $r + s - n = (r + s) + n - 2n$, we reduce to proving the result for the two varieties $Y \times Z$ and Δ in $\mathbf{A}^{2n}$. Now Δ is an intersection of exactly n hypersurfaces, namely, $x_1 - y_1 = 0, \ldots, x_n - y_n = 0$, where $x_1, \ldots, x_n, y_1, \ldots, y_n$ are the coordinates of $\mathbf{A}^{2n}$. Now applying the special case above n times, we have the result.

Theorem 7.2 (Projective Dimension Theorem). *Let Y, Z be varieties of dimensions r, s in $\mathbf{P}^n$. Then every irreducible component of $Y \cap Z$ has dimension $\geq r + s - n$. Furthermore, if $r + s - n \geq 0$, then $Y \cap Z$ is nonempty.*

PROOF. The first statement follows from the previous result, since $\mathbf{P}^n$ is covered by affine n-spaces. For the second result, let $C(Y)$ and $C(Z)$ be the cones over Y, Z in $\mathbf{A}^{n+1}$ (Ex. 2.10). Then $C(Y), C(Z)$ have dimensions $r + 1$, $s + 1$, respectively. Furthermore, $C(Y) \cap C(Z) \neq \varnothing$, because both contain the origin $P = (0, \ldots, 0)$. By the affine dimension theorem, $C(Y) \cap C(Z)$ has dimension $\geq (r + 1) + (s + 1) - (n + 1) = r + s - n + 1 > 0$. Hence $C(Y) \cap C(Z)$ contains some point $Q \neq P$, and so $Y \cap Z \neq \varnothing$.

Next, we come to the definition of the Hilbert polynomial of a projective variety. The idea is to associate to each projective variety $Y \subseteq \mathbf{P}_k^n$ a polynomial $P_Y \in \mathbf{Q}[z]$ from which we can obtain various numerical invariants of Y. We will define P_Y starting from the homogeneous coordinate ring $S(Y)$. In fact, more generally, we will define a Hilbert polynomial for any graded S-module, where $S = k[x_0, \ldots, x_n]$. Although the next few results are almost pure algebra, we include their proofs, for lack of a suitable reference.

Definition. A *numerical polynomial* is a polynomial $P(z) \in \mathbf{Q}[z]$ such that $P(n) \in \mathbf{Z}$ for all $n \gg 0$, $n \in \mathbf{Z}$.

Proposition 7.3.
(a) *If $P \in \mathbf{Q}[z]$ is a numerical polynomial, then there are integers $c_0, c_1, \ldots, c_r$ such that*

$$P(z) = c_0 \binom{z}{r} + c_1 \binom{z}{r-1} + \ldots + c_r,$$

where

$$\binom{z}{r} = \frac{1}{r!} z(z - 1) \cdots (z - r + 1)$$

is the binomial coefficient function. In particular $P(n) \in \mathbf{Z}$ for all $n \in \mathbf{Z}$.
(b) *If $f: \mathbf{Z} \to \mathbf{Z}$ is any function, and if there exists a numerical polynomial $Q(z)$ such that the difference function $\Delta f = f(n + 1) - f(n)$ is equal to $Q(n)$ for all $n \gg 0$, then there exists a numerical polynomial $P(z)$ such that $f(n) = P(n)$ for all $n \gg 0$.*

PROOF.
(a) By induction on the degree of P, the case of degree 0 being obvious. Since $\binom{z}{r} = z^r/r! + \ldots$, we can express any polynomial $P \in \mathbf{Q}[z]$ of degree r in the above form, with $c_0, \ldots, c_r \in \mathbf{Q}$. For any polynomial P we define the *difference polynomial* ΔP by $\Delta P(z) = P(z + 1) - P(z)$. Since $\Delta \binom{z}{r} = \binom{z}{r-1}$,

$$\Delta P = c_0 \binom{z}{r-1} + c_1 \binom{z}{r-2} + \ldots + c_{r-1}.$$

By induction, $c_0, \ldots, c_{r-1} \in \mathbf{Z}$. But then $c_r \in \mathbf{Z}$ since $P(n) \in \mathbf{Z}$ for $n \gg 0$.
(b) Write

$$Q = c_0 \binom{z}{r} + \ldots + c_r$$

with $c_0, \ldots, c_r \in \mathbf{Z}$. Let

$$P = c_0 \binom{z}{r+1} + \ldots + c_r \binom{z}{1}.$$

Then $\Delta P = Q$, so $\Delta(f - P)(n) = 0$ for all $n \gg 0$, so $(f - P)(n) = $ constant c_{r+1} for all $n \gg 0$, so

$$f(n) = P(n) + c_{r+1}$$

for all $n \gg 0$, as required.

Next, we need some preparations about graded modules. Let S be a graded ring (cf. §2). A *graded S-module* is an S-module M, together with a decomposition $M = \bigoplus_{d \in \mathbf{Z}} M_d$, such that $S_d \cdot M_e \subseteq M_{d+e}$. For any graded S-module M, and for any $l \in \mathbf{Z}$, we define the *twisted* module $M(l)$ by shifting l places to the left, i.e., $M(l)_d = M_{d+l}$. If M is a graded S-module, we define the *annihilator* of M, Ann $M = \{s \in S | s \cdot M = 0\}$. This is a homogeneous ideal in S.

The next result is the analogue for graded modules of a well-known result for modules of finite type over a noetherian ring (Bourbaki [1, Ch. IV, §1, no. 4] or Matsumura [2, p. 51]). Again, we include the proof for lack of an adequate reference.

Proposition 7.4. *Let M be a finitely generated graded module over a noetherian graded ring S. Then there exists a filtration $0 = M^0 \subseteq M^1 \subseteq \ldots \subseteq M^r = M$ by graded submodules, such that for each i, $M^i/M^{i-1} \simeq (S/\mathfrak{p}_i)(l_i)$, where $\mathfrak{p}_i$ is a homogeneous prime ideal of S, and $l_i \in \mathbf{Z}$. The filtration is not unique, but for any such filtration we do have:*

   (a) *if $\mathfrak{p}$ is a homogeneous prime ideal of S, then $\mathfrak{p} \supseteq$ Ann $M \Leftrightarrow \mathfrak{p} \supseteq \mathfrak{p}_i$ for some i. In particular, the minimal elements of the set $\{\mathfrak{p}_1, \ldots, \mathfrak{p}_r\}$ are just the minimal primes of M, i.e., the primes which are minimal containing Ann M;*

   (b) *for each minimal prime of M, the number of times which $\mathfrak{p}$ occurs in the set $\{\mathfrak{p}_1, \ldots, \mathfrak{p}_r\}$ is equal to the length of $M_\mathfrak{p}$ over the local ring $S_\mathfrak{p}$ (and hence is independent of the filtration).*

PROOF. For the existence of the filtration, we consider the set of graded submodules of M which admit such a filtration. Clearly, the zero module does, so the set is nonempty. M is a noetherian module, so there is a maximal such submodule $M' \subseteq M$. Now consider $M'' = M/M'$. If $M'' = 0$, we are done. If not, we consider the set of ideals $\mathfrak{I} = \{I_m = \text{Ann}(m) | m \in M'' \text{ is a homogeneous element}, m \neq 0\}$. Each I_m is a homogeneous ideal, and $I_m \neq S$. Since S is a noetherian ring, we can find an element $m \in M''$, $m \neq 0$, such that I_m is a maximal element of the set $\mathfrak{I}$. I claim that I_m is a prime ideal. Let $a, b \in S$. Suppose that $ab \in I_m$, but $b \notin I_m$. We wish to show $a \in I_m$. By splitting into homogeneous components, we may assume that a, b are homogeneous elements. Now consider the element $bm \in M''$. Since $b \notin I_m$,

$bm \neq 0$. We have $I_m \subseteq I_{bm}$, so by maximality of I_m, $I_m = I_{bm}$. But $ab \in I_m$, so $abm = 0$, so $a \in I_{bm} = I_m$ as required. Thus I_m is a homogeneous prime ideal of S. Call it $\mathfrak{p}$. Let m have degree l. Then the module $N \subseteq M''$ generated by m is isomorphic to $(S/\mathfrak{p})(-l)$. Let $N' \subseteq M$ be the inverse image of N in M. Then $M' \subseteq N'$, and $N'/M' \simeq (S/\mathfrak{p})(-l)$. So N' also has a filtration of the type required. This contradicts the maximality of M'. We conclude that M' was equal to M, which proves the existence of the filtration.

Now suppose given such a filtration of M. Then it is clear that $\mathfrak{p} \supseteq$ Ann $M \Leftrightarrow \mathfrak{p} \supseteq \mathrm{Ann}(M^i/M^{i-1})$ for some i. But $\mathrm{Ann}((S/\mathfrak{p}_i)(l)) = \mathfrak{p}_i$ so this proves (a).

To prove (b) we localize at a minimal prime $\mathfrak{p}$. Since $\mathfrak{p}$ is minimal in the set $\{\mathfrak{p}_1, \ldots, \mathfrak{p}_r\}$, after localization, we will have $M_\mathfrak{p}^i = M_\mathfrak{p}^{i-1}$ except in the cases where $\mathfrak{p}_i = \mathfrak{p}$. And in those cases $M_\mathfrak{p}^i/M_\mathfrak{p}^{i-1} \simeq (S/\mathfrak{p})_\mathfrak{p} = k(\mathfrak{p})$, the quotient field of $S/\mathfrak{p}$ (we forget the grading). This shows that $M_\mathfrak{p}$ is an $S_\mathfrak{p}$-module of finite length equal to the number of times $\mathfrak{p}$ occurs in the set $\{\mathfrak{p}_1, \ldots, \mathfrak{p}_r\}$.

Definition. If $\mathfrak{p}$ is a minimal prime of a graded S-module M, we define the *multiplicity* of M at $\mathfrak{p}$, denoted $\mu_\mathfrak{p}(M)$, to be the length of $M_\mathfrak{p}$ over $S_\mathfrak{p}$.

Now we can define the Hilbert polynomial of a graded module M over the polynomial ring $S = k[x_0, \ldots, x_n]$. First, we define the *Hilbert function* φ_M of M, given by

$$\varphi_M(l) = \dim_k M_l$$

for each $l \in \mathbf{Z}$.

Theorem 7.5. (Hilbert–Serre). *Let M be a finitely generated graded $S = k[x_0, \ldots, x_n]$-module. Then there is a unique polynomial $P_M(z) \in \mathbf{Q}[z]$ such that $\varphi_M(l) = P_M(l)$ for all $l \gg 0$. Furthermore, $\deg P_M(z) = \dim Z(\mathrm{Ann}\ M)$, where Z denotes the zero set in $\mathbf{P}^n$ of a homogeneous ideal (cf. §2).*

PROOF. If $0 \to M' \to M \to M'' \to 0$ is a short exact sequence, then $\varphi_M = \varphi_{M'} + \varphi_{M''}$, and $Z(\mathrm{Ann}\ M) = Z(\mathrm{Ann}\ M') \cup Z(\mathrm{Ann}\ M'')$, so if the theorem is true for M' and M'', it is also true for M. By (7.4), M has a filtration with quotients of the form $(S/\mathfrak{p})(l)$ where $\mathfrak{p}$ is a homogeneous prime ideal, and $l \in \mathbf{Z}$. So we reduce to $M \simeq (S/\mathfrak{p})(l)$. The shift l corresponds to a change of variables $z \mapsto z + l$, so it is sufficient to consider the case $M = S/\mathfrak{p}$. If $\mathfrak{p} = (x_0, \ldots, x_n)$, then $\varphi_M(l) = 0$ for $l > 0$, so $P_M = 0$ is the corresponding polynomial, and $\deg P_M = \dim Z(\mathfrak{p})$, where we make the convention that the zero polynomial has degree -1, and the empty set has dimension -1.

If $\mathfrak{p} \neq (x_0, \ldots, x_n)$, choose $x_i \notin \mathfrak{p}$, and consider the exact sequence $0 \to M \xrightarrow{x_i} M \to M'' \to 0$, where $M'' = M/x_iM$. Then $\varphi_{M''}(l) = \varphi_M(l) - \varphi_M(l-1) = (\Delta\varphi_M)(l-1)$. On the other hand, $Z(\mathrm{Ann}\ M'') = Z(\mathfrak{p}) \cap H$, where H is the hyperplane $x_i = 0$, and $Z(\mathfrak{p}) \not\subseteq H$ by choice of x_i, so by (7.2), $\dim Z(\mathrm{Ann}\ M'') = \dim Z(\mathfrak{p}) - 1$. Now using induction on $\dim Z(\mathrm{Ann}\ M)$, we may assume that $\varphi_{M''}$ is a polynomial function, corresponding to a polynomial $P_{M''}$ of degree $= \dim Z(\operatorname{Ann} M'')$. Now, by (7.3), it follows that φ_M is a polynomial function, corresponding to a polynomial of degree $= \dim Z(\mathfrak{p})$. The uniqueness of P_M is clear.

Definition. The polynomial P_M of the theorem is the *Hilbert polynomial* of M.

Definition. If $Y \subseteq \mathbf{P}^n$ is an algebraic set of dimension r, we define the *Hilbert polynomial* of Y to be the Hilbert polynomial P_Y of its homogeneous coordinate ring $S(Y)$. (By the theorem, it is a polynomial of degree r.) We define the *degree* of Y to be r! times the leading coefficient of P_Y.

Proposition 7.6.
   (a) *If $Y \subseteq \mathbf{P}^n$, $Y \neq \varnothing$, then the degree of Y is a positive integer.*
   (b) *Let $Y = Y_1 \cup Y_2$, where Y_1 and Y_2 have the same dimension r, and where $\dim(Y_1 \cap Y_2) < r$. Then $\deg Y = \deg Y_1 + \deg Y_2$.*
   (c) $\deg \mathbf{P}^n = 1$.
   (d) *If $H \subseteq \mathbf{P}^n$ is a hypersurface whose ideal is generated by a homogeneous polynomial of degree d, then $\deg H = d$. (In other words, this definition of degree is consistent with the degree of a hypersurface as defined earlier (1.4.2).)*

PROOF.
   (a) Since $Y \neq \varnothing$, P_Y is a nonzero polynomial of degree $r = \dim Y$. By (7.3a), $\deg Y = c_0$, which is an integer. It is a positive integer because for $l \gg 0$, $P_Y(l) = \varphi_{S/I}(l) \geqslant 0$.
   (b) Let I_1, I_2 be the ideals of Y_1 and Y_2. Then $I = I_1 \cap I_2$ is the ideal of Y. We have an exact sequence

$$0 \to S/I \to S/I_1 \oplus S/I_2 \to S/(I_1 + I_2) \to 0.$$

Now $Z(I_1 + I_2) = Y_1 \cap Y_2$, which has smaller dimension. Hence $P_{S/(I_1+I_2)}$ has degree $<r$. So the leading coefficient of $P_{S/I}$ is the sum of the leading coefficients of P_{S/I_1} and P_{S/I_2}.
   (c) We calculate the Hilbert polynomial of $\mathbf{P}^n$. It is the polynomial P_S, where $S = k[x_0, \ldots, x_n]$. For $l > 0$, $\varphi_S(l) = \binom{l+n}{n}$, so $P_S = \binom{z+n}{n}$. In particular, its leading coefficient is $1/n!$, so $\deg \mathbf{P}^n = 1$.
   (d) If $f \in S$ is homogeneous of degree d, then we have an exact sequence of graded S-modules

$$0 \to S(-d) \xrightarrow{f} S \to S/(f) \to 0.$$

Hence

$$\varphi_{S/(f)}(l) = \varphi_S(l) - \varphi_S(l - d).$$

Therefore we can find the Hilbert polynomial of H, as

$$P_H(z) = \binom{z+n}{n} - \binom{z-d+n}{n} = \frac{d}{(n-1)!} z^{n-1} + \ldots.$$

Thus $\deg H = d$.

Now we come to our main result about the intersection of a projective variety with a hypersurface, which is a partial generalization of Bézout's theorem to higher projective spaces. Let $Y \subseteq \mathbf{P}^n$ be a projective variety of dimension r. Let H be a hypersurface not containing Y. Then, by (7.2), $Y \cap H = Z_1 \cup \ldots \cup Z_s$, where Z_j are varieties of dimension $r - 1$. Let $\mathfrak{p}_j$ be the homogeneous prime ideal of Z_j. We define the *intersection multiplicity* of Y and H along Z_j to be $i(Y,H; Z_j) = \mu_{\mathfrak{p}_j}(S/(I_Y + I_H))$. Here I_Y, I_H are the homogeneous ideals of Y and H. The module $M = S/(I_Y + I_H)$ has annihilator $I_Y + I_H$, and $Z(I_Y + I_H) = Y \cap H$, so $\mathfrak{p}_j$ is a minimal prime of M, and μ is the multiplicity introduced above.

Theorem 7.7. *Let Y be a variety of dimension ≥ 1 in $\mathbf{P}^n$, and let H be a hypersurface not containing Y. Let $Z_1, \ldots, Z_s$ be the irreducible components of $Y \cap H$. Then*

$$\sum_{j=1}^{s} i(Y,H; Z_j) \cdot \deg Z_j = (\deg Y)(\deg H).$$

PROOF. Let H be defined by the homogeneous polynomial f of degree d. We consider the exact sequence of graded S-modules

$$0 \to (S/I_Y)(-d) \xrightarrow{f} S/I_Y \to M \to 0,$$

where $M = S/(I_Y + I_H)$. Taking Hilbert polynomials, we find that

$$P_M(z) = P_Y(z) - P_Y(z - d).$$

Our result comes from comparing the leading coefficients of both sides of this equation. Let Y have dimension r and degree e. Then $P_Y(z) = (e/r!)z^r + \ldots$ so on the right we have

$$(e/r!)z^r + \ldots - [(e/r!)(z - d)^r + \ldots] = (de/(r - 1)!)z^{r-1} + \ldots .$$

Now consider the module M. By (7.4), M has a filtration $0 = M^0 \subseteq M^1 \subseteq \ldots \subseteq M^q = M$, whose quotients M^i/M^{i-1} are of the form $(S/\mathfrak{q}_i)(l_i)$. Hence $P_M = \sum_{i=1}^{q} P_i$, where P_i is the Hilbert polynomial of $(S/\mathfrak{q}_i)(l_i)$. If $Z(\mathfrak{q}_i)$ is a projective variety of dimension r_i and degree f_i, then

$$P_i = (f_i/r_i!)z^{r_i} + \ldots .$$

Note that the shift l_i does not affect the leading coefficient of P_i. Since we are interested only in the leading coefficient of P_i, we can ignore those P_i of degree $< r - 1$. We are left with those P_i, where $\mathfrak{q}_i$ is a minimal prime of M, namely, one of the primes $\mathfrak{p}_1, \ldots, \mathfrak{p}_s$ corresponding to the Z_j. Each one of these occurs $\mu_{\mathfrak{p}_j}(M)$ times, so the leading coefficient of P_M is

$$\left(\sum_{j=1}^{s} i(Y,H; Z_j) \cdot \deg Z_j \right) / (r - 1)!$$

Comparing with the above, we have our result.

Corollary 7.8 (Bézout's Theorem). *Let Y,Z be distinct curves in* $\mathbf{P}^2$, *having degrees d,e. Let* $Y \cap Z = \{P_1, \dots, P_s\}$. *Then*

$$\sum i(Y,Z; P_j) = de.$$

PROOF. We have only to observe that a point has Hilbert polynomial 1, hence degree 1. See (V, 1.4.2) for another proof.

Remark 7.8.1. Our definition of intersection multiplicity in terms of the homogeneous coordinate ring is different from the local definition given earlier (Ex. 5.4). However, it is easy to show that they coincide in the case of intersections of plane curves.

Remark 7.8.2. The proof of (7.8) extends easily to the case where Y and Z are "reducible curves," i.e., algebraic sets of dimension 1 in $\mathbf{P}^2$, provided they have no irreducible component in common.

EXERCISES

7.1. (a) Find the degree of the d-uple embedding of $\mathbf{P}^n$ in $\mathbf{P}^N$ (Ex. 2.12). [Answer: d^n]
   (b) Find the degree of the Segre embedding of $\mathbf{P}^r \times \mathbf{P}^s$ in $\mathbf{P}^N$ (Ex. 2.14). [Answer: $\binom{r+s}{r}$]

7.2. Let Y be a variety of dimension r in $\mathbf{P}^n$, with Hilbert polynomial P_Y. We define the *arithmetic genus* of Y to be $p_a(Y) = (-1)^r(P_Y(0) - 1)$. This is an important invariant which (as we will see later in (III, Ex. 5.3)) is independent of the projective embedding of Y.
   (a) Show that $p_a(\mathbf{P}^n) = 0$.
   (b) If Y is a plane curve of degree d, show that $p_a(Y) = \frac{1}{2}(d-1)(d-2)$.
   (c) More generally, if H is a hypersurface of degree d in $\mathbf{P}^n$, then $p_a(H) = \binom{d-1}{n}$.
   (d) If Y is a complete intersection (Ex. 2.17) of surfaces of degrees a,b in $\mathbf{P}^3$, then $p_a(Y) = \frac{1}{2}ab(a+b-4)+1$.
   (e) Let $Y^r \subseteq \mathbf{P}^n$, $Z^s \subseteq \mathbf{P}^m$ be projective varieties, and embed $Y \times Z \subseteq \mathbf{P}^n \times \mathbf{P}^m \to \mathbf{P}^N$ by the Segre embedding. Show that

$$p_a(Y \times Z) = p_a(Y)p_a(Z) + (-1)^s p_a(Y) + (-1)^r p_a(Z).$$

7.3. *The Dual Curve.* Let $Y \subseteq \mathbf{P}^2$ be a curve. We regard the set of lines in $\mathbf{P}^2$ as another projective space, $(\mathbf{P}^2)^*$, by taking (a_0,a_1,a_2) as homogeneous coordinates of the line $L : a_0 x_0 + a_1 x_1 + a_2 x_2 = 0$. For each nonsingular point $P \in Y$, show that there is a unique line $T_P(Y)$ whose intersection multiplicity with Y at P is >1. This is the *tangent line* to Y at P. Show that the mapping $P \mapsto T_P(Y)$ defines a *morphism* of Reg Y (the set of nonsingular points of Y) into $(\mathbf{P}^2)^*$. The closure of the image of this morphism is called the dual curve $Y^* \subseteq (\mathbf{P}^2)^*$ of Y.

7.4. Given a curve Y of degree d in $\mathbf{P}^2$, show that there is a nonempty open subset U of $(\mathbf{P}^2)^*$ in its Zariski topology such that for each $L \in U, L$ meets Y in exactly d points. [*Hint:* Show that the set of lines in $(\mathbf{P}^2)^*$ which are either tangent to Y or pass through a singular point of Y is contained in a proper closed subset.] This result shows that we could have defined the degree of Y to be the number d such that almost all lines in $\mathbf{P}^2$ meet Y in d points, where "almost all" refers to a nonempty open set of the set of lines, when this set is identified with the dual projective space $(\mathbf{P}^2)^*$.

7.5. (a) Show that an irreducible curve Y of degree $d > 1$ in $\mathbf{P}^2$ cannot have a point of multiplicity $\geqslant d$ (Ex. 5.3).

   (b) If Y is an irreducible curve of degree $d > 1$ having a point of multiplicity $d - 1$, then Y is a rational curve (Ex. 6.1).

7.6. *Linear Varieties.* Show that an algebraic set Y of pure dimension r (i.e., every irreducible component of Y has dimension r) has degree 1 if and only if Y is a linear variety (Ex. 2.11). [*Hint:* First, use (7.7) and treat the case dim $Y = 1$. Then do the general case by cutting with a hyperplane and using induction.]

7.7. Let Y be a variety of dimension r and degree $d > 1$ in $\mathbf{P}^n$. Let $P \in Y$ be a nonsingular point. Define X to be the closure of the union of all lines PQ, where $Q \in Y$, $Q \neq P$.

   (a) Show that X is a variety of dimension $r + 1$.

   (b) Show that deg $X < d$. [*Hint:* Use induction on dim Y.]

7.8. Let $Y^r \subseteq \mathbf{P}^n$ be a variety of degree 2. Show that Y is contained in a linear subspace L of dimension $r + 1$ in $\mathbf{P}^n$. Thus Y is isomorphic to a quadric hypersurface in $\mathbf{P}^{r+1}$ (Ex. 5.12).

8   What Is Algebraic Geometry?

Now that we have met some algebraic varieties, and have encountered some of the main concepts about them, it is appropriate to ask, what is this subject all about? What are the important problems in the field, and where is it going?

To define algebraic geometry, we could say that it is the study of the solutions of systems of polynomial equations in an affine or projective n-space. In other words, it is the study of algebraic varieties.

In any branch of mathematics, there are usually guiding problems, which are so difficult that one never expects to solve them completely, yet which provide stimulus for a great amount of work, and which serve as yardsticks for measuring progress in the field. In algebraic geometry such a problem is the classification problem. In its strongest form, the problem is to classify all algebraic varieties up to isomorphism. We can divide the problem into parts. The first part is to classify varieties up to birational equivalence. As we have seen, this is equivalent to the question of classifying function fields (finitely generated extension fields) over k up to isomorphism. The second part is to identify a good subset of a birational equivalence class, such as the nonsingular projective varieties, and classify them up to isomorphism. The third part is to study how far an arbitrary variety is from one of the good ones considered above. In particular, we want to know (a) how much do you have to add to a nonprojective variety to get a projective variety, and (b) what is the structure of singularities, and how can they be resolved to give a nonsingular variety?

Typically, the answer to any classification problem in algebraic geometry consists of a discrete part and a continuous part. So we can rephrase the problem as follows: define numerical invariants and continuous invariants of algebraic varieties, which allow one to distinguish among nonisomorphic varieties. Another special feature of the classification problem is that often when there is a continuous family of nonisomorphic objects, the parameter space can itself be given a structure of algebraic variety. This is a very powerful method, because then all the techniques of the subject can be applied to the study of the parameter space as well as to the original varieties.

Let us illustrate these ideas by describing what is known about the classification of algebraic curves (over a fixed algebraically closed field k). First, the birational classification. There is an invariant called the *genus* of a curve, which is a birational invariant, and which takes on all nonnegative values $g \geq 0$. For $g = 0$ there is exactly one birational equivalence class, namely, that of the rational curves (i.e., those curves which are birationally equivalent to $\mathbf{P}^1$). For each $g > 0$ there is a continuous family of birational equivalence classes, which can be parametrized by an irreducible algebraic variety $\mathfrak{M}_g$, called the *variety of moduli* of curves of genus g, which has dimension 1 if $g = 1$, and dimension $3g - 3$ if $g \geq 2$. Curves with $g = 1$ are called *elliptic* curves. Thus for curves, the birational classification question is answered by giving the genus, which is a discrete invariant, and a point on the variety of moduli, which is a continuous invariant. See Chapter IV for more details.

The second question for curves, namely, to describe all nonsingular projective curves in a given birational equivalence class, has a simple answer, as we have seen, since there is exactly one.

For the third question, we know that any curve can be completed to a projective curve by adding a finite number of points, so there is not much more to say there. As for the classification of singularities of curves, see (V, 3.9.4).

While we are discussing the classification problem, I would like to describe another special case where a satisfactory answer is known, namely, the classification of nonsingular projective surfaces within a given birational equivalence class. In this case one knows that (1) every birational equivalence class of surfaces has a nonsingular projective surface in it, (2) the set of nonsingular projective surfaces with a given function field K/k is a partially ordered set under the relation given by the existence of a birational morphism, (3) any birational morphism $f: X \to Y$ can be factored into a finite number of steps, each of which is a blowing-up of a point, and (4) unless K is *rational* (i.e., equal to $K(\mathbf{P}^2)$) or *ruled* (i.e., K is the function field of a product $\mathbf{P}^1 \times C$, where C is a curve), there is a unique minimal element of this partially ordered set, which is called the *minimal model* of the function field K. (In the rational and ruled cases, there are infinitely many minimal elements, and their structure is also well-known.) The theory of minimal models is a very beautiful branch of the theory of surfaces. The results were known to the Italians, but were first proved in all characteristics by Zariski [5], [6]. See Chapter V for more details.

From these remarks it should be clear that the classification problem is a very fruitful problem to keep in mind while studying algebraic geometry. This leads us to the next question: how does one go about defining invariants of an algebraic variety? So far, we have defined the dimension, and for projective varieties we have defined the Hilbert polynomial, and hence the degree and the arithmetic genus p_a. Of course the dimension is a birational invariant. But the degree and the Hilbert polynomial depend on the embedding in projective space, so they are not even invariants under isomorphism of varieties. Now it happens that the arithmetic genus is an invariant under isomorphism (III, Ex. 5.3), and is even a *birational* invariant in most cases (curves, surfaces, nonsingular varieties in characteristic 0; see (V, 5.6.1)), but this is not at all apparent from our definition.

To go further, we must study the intrinsic geometry on a variety, which we have not done at all yet. So, for example, we will study *divisors* on a variety X. A divisor is an element of the free abelian group generated by the subvarieties of codimension one. We will define *linear equivalence* of divisors, and then we can form the group of divisors modulo linear equivalence, called the *Picard group* of X. This is an intrinsic invariant of X. Another very important notion is that of a differential form on a variety X. Using differential forms, one can give an intrinsic definition of the tangent bundle and cotangent bundle on an algebraic variety. Then one can carry over many constructions from differential geometry to define numerical invariants. For example, one can define the genus of a curve as the dimension of the vector space of global differential forms on the nonsingular projective model. From this definition it is clear that it is a birational invariant. See (II, §6,7,8).

Perhaps the most important modern technique for defining numerical invariants is by cohomology. There are many cohomology theories, but we will be principally concerned in this book with the cohomology of coherent sheaves, which was introduced by Serre [3]. Cohomology is an extremely powerful and versatile tool. Not only can it be used to define numerical invariants (for example, the genus of a curve X can be defined as dim $H^1(X, \mathcal{O}_X)$), but it can be used to prove many important results which do not apparently have any connection with cohomology, such as "Zariski's main theorem," which has to do with the structure of birational transformations. To set up a cohomology theory requires a lot of work, but I believe it is well worth the effort. We will devote a whole chapter to cohomology later in the book (Chapter III). Cohomology is also a useful vehicle for understanding and expressing important results such as the Riemann–Roch theorem. This theorem was known classically for curves and surfaces, but it was by using cohomology that Hirzebruch [1] and Grothendieck (see Borel and Serre

[1]) were able to clarify and generalize it to varieties of any dimension (Appendix A).

Now that we have seen a little bit of what algebraic geometry is about, we should discuss the degree of generality in which to develop the foundations of the subject. In this chapter we have worked over an algebraically closed field, because that is the simplest case. But there are good reasons for allowing fields which are not algebraically closed. One reason is that the local ring of a subvariety on a variety has a residue field which is not algebraically closed (Ex. 3.13), and at times it is desirable to give a unified treatment of properties which hold along a subvariety and properties which hold at a point. Another strong reason for allowing non-algebraically closed fields is that many problems in algebraic geometry are motivated by number theory, and in number theory one is primarily concerned with solutions of equations over finite fields or number fields. For example, Fermat's problem is equivalent to the question, does the curve $x^n + y^n = z^n$ in $\mathbf{P}^2$ for $n \geqslant 3$ have any points rational over $\mathbf{Q}$ (i.e., points whose coordinates are in $\mathbf{Q}$), with $x, y, z \neq 0$.

The need to work over arbitrary ground fields was recognized by Zariski and Weil. In fact, perhaps one of the principal contributions of Weil's "Foundations" [1] was to provide a systematic framework for studying varieties over arbitrary fields, and the various phenomena which occur with change of ground field. Nagata [2] went further by developing the foundations of algebraic geometry over Dedekind domains.

Another direction in which we need to expand our foundations is to define some kind of abstract variety which does not a priori have an embedding in an affine or projective space. This is especially necessary in problems such as the construction of a variety of moduli, because there one may be able to make the construction locally, without knowing anything about a global embedding. In §6 we gave a definition of an abstract curve. In higher dimensions that method does not work, because there is no unique nonsingular model of a given function field. However, we can define an abstract variety by starting from the observation that any variety has an open covering by affine varieties. Thus one can define an abstract variety as a topological space X, with an open cover U_i, plus for each U_i a structure of affine variety, such that on each intersection $U_i \cap U_j$ the induced variety structures are isomorphic. It turns out that this generalization of the notion of variety is not illusory, because in dimension $\geqslant 2$ there are abstract varieties which are not isomorphic to any quasi-projective variety (II, 4.10.2).

There is a third direction in which it is useful to expand our notion of algebraic variety. In this chapter we have defined a variety as an irreducible algebraic set in affine or projective space. But it is often convenient to allow reducible algebraic sets, or even algebraic sets with multiple components. For example, this is suggested by what we have seen of intersection theory in §7, since the intersection of two varieties may be reducible, and the sum of the ideals of the two varieties may not be the ideal of the intersection. So one might be tempted to define a "generalized projective variety" in $\mathbf{P}^n$ to be an ordered pair $\langle V,I \rangle$, where V is an algebraic set in $\mathbf{P}^n$, and $I \subseteq S = k[x_0, \ldots, x_n]$ is any ideal such that $V = Z(I)$. This is not in fact what we will do, but it gives the general idea.

All three generalizations of the notion of variety suggested above are contained in Grothendieck's definition of a scheme. He starts from the observation that an affine variety corresponds to a finitely generated integral domain over a field (3.8). But why restrict one's attention to such a special class of rings? So for any commutative ring A, he defines a topological space Spec A, and a sheaf of rings on Spec A, which generalizes the ring of regular functions on an affine variety, and he calls this an affine scheme. An arbitrary scheme is then defined by glueing together affine schemes, thus generalizing the notion of abstract variety we suggested above.

One caution about working in extreme generality. There are many advantages to developing a theory in the most general context possible. In the case of algebraic geometry there is no doubt that the introduction of schemes has revolutionized the subject and has made possible tremendous advances. On the other hand, the person who works with schemes has to carry a considerable load of technical baggage with him: sheaves, abelian categories, cohomology, spectral sequences, and so forth. Another more serious difficulty is that some things which are always true for varieties may no longer be true. For example, an affine scheme need not have finite dimension, even if its ring is noetherian. So our intuition must be supported by a good knowledge of commutative algebra.

In this book we will develop the foundations of algebraic geometry using the language of schemes, starting with the next chapter.

CHAPTER II

Schemes

This chapter and the next form the technical heart of this book. In this chapter we develop the basic theory of schemes, following Grothendieck [EGA]. Sections 1 to 5 are fundamental. They contain a review of sheaf theory (necessary even to define a scheme), then the basic definitions of schemes, morphisms, and coherent sheaves. This is the language that we use for the rest of the book.

Then in Sections 6, 7, 8, we treat some topics which could have been done in the language of varieties, but which are already more convenient to discuss using schemes. For example, the notion of Cartier divisor, and of an invertible sheaf, which belong to the new language, greatly clarify the discussion of Weil divisors and linear systems, which belong to the old language. Then in §8, the systematic use of nonclosed scheme points gives much more flexibility in the discussion of sheaves of differentials and nonsingular varieties, improving the treatment of (I, §5).

In §9 we give the definition of a formal scheme, which did not have an analogue in the theory of varieties. It was invented by Grothendieck as a good way of dealing with Zariski's theory of "holomorphic functions," which Zariski regarded as an analogue in abstract algebraic geometry of the holomorphic functions in a neighborhood of a subvariety in the classical case.

1 Sheaves

The concept of a sheaf provides a systematic way of keeping track of local algebraic data on a topological space. For example, the regular functions on open subsets of a variety, introduced in Chapter I, form a sheaf, as we will see shortly. Sheaves are essential in the study of schemes. In fact, we cannot

60

even define a scheme without using sheaves. So we begin this chapter with sheaves. For additional information, see the book of Godement [1].

Definition. Let X be a topological space. A *presheaf* $\mathscr{F}$ of abelian groups on X consists of the data

   (a) for every open subset $U \subseteq X$, an abelian group $\mathscr{F}(U)$, and
   (b) for every inclusion $V \subseteq U$ of open subsets of X, a morphism of abelian groups $\rho_{UV} : \mathscr{F}(U) \to \mathscr{F}(V)$,

subject to the conditions

   (0) $\mathscr{F}(\varnothing) = 0$, where $\varnothing$ is the empty set,
   (1) ρ_{UU} is the identity map $\mathscr{F}(U) \to \mathscr{F}(U)$, and
   (2) if $W \subseteq V \subseteq U$ are three open subsets, then $\rho_{UW} = \rho_{VW} \circ \rho_{UV}$.

The reader who likes the language of categories may rephrase this definition as follows. For any topological space X, we define a category $\mathfrak{Top}(X)$, whose objects are the open subsets of X, and where the only morphisms are the inclusion maps. Thus $\mathrm{Hom}(V,U)$ is empty if $V \nsubseteq U$, and $\mathrm{Hom}(V,U)$ has just one element if $V \subseteq U$. Now a presheaf is just a contravariant functor from the category $\mathfrak{Top}(X)$ to the category $\mathfrak{Ab}$ of abelian groups.

We define a presheaf of rings, a presheaf of sets, or a presheaf with values in any fixed category $\mathfrak{C}$, by replacing the words "abelian group" in the definition by "ring", "set", or "object of $\mathfrak{C}$" respectively. We will stick to the case of abelian groups in this section, and let the reader make the necessary modifications for the case of rings, sets, etc.

As a matter of terminology, if $\mathscr{F}$ is a presheaf on X, we refer to $\mathscr{F}(U)$ as the *sections* of the presheaf $\mathscr{F}$ over the open set U, and we sometimes use the notation $\Gamma(U, \mathscr{F})$ to denote the group $\mathscr{F}(U)$. We call the maps ρ_{UV} *restriction* maps, and we sometimes write $s|_V$ instead of $\rho_{UV}(s)$, if $s \in \mathscr{F}(U)$.

A sheaf is roughly speaking a presheaf whose sections are determined by local data. To be precise, we give the following definition.

Definition. A presheaf $\mathscr{F}$ on a topological space X is a *sheaf* if it satisfies the following supplementary conditions:

   (3) if U is an open set, if $\{V_i\}$ is an open covering of U, and if $s \in \mathscr{F}(U)$ is an element such that $s|_{V_i} = 0$ for all i, then $s = 0$;
   (4) if U is an open set, if $\{V_i\}$ is an open covering of U, and if we have elements $s_i \in \mathscr{F}(V_i)$ for each i, with the property that for each i,j, $s_i|_{V_i \cap V_j} = s_j|_{V_i \cap V_j}$, then there is an element $s \in \mathscr{F}(U)$ such that $s|_{V_i} = s_i$ for each i. (Note condition (3) implies that s is unique.)

Note. According to our definition, a sheaf is a presheaf satisfying certain extra conditions. This is equivalent to the definition found in some other

61

books, of a sheaf as a topological space over X with certain properties (Ex. 1.13).

Example 1.0.1. Let X be a variety over the field k. For each open set $U \subseteq X$, let $\mathcal{O}(U)$ be the ring of regular functions from U to k, and for each $V \subseteq U$, let $\rho_{UV} : \mathcal{O}(U) \to \mathcal{O}(V)$ be the restriction map (in the usual sense). Then $\mathcal{O}$ is a sheaf of rings on X. It is clear that it is a presheaf of rings. To verify the conditions (3) and (4), we note that a function which is 0 locally is 0, and a function which is regular locally is regular, because of the definition of regular function (I, §3). We call $\mathcal{O}$ the *sheaf of regular functions* on X.

Example 1.0.2. In the same way, one can define the sheaf of continuous real-valued functions on any topological space, or the sheaf of differentiable functions on a differentiable manifold, or the sheaf of holomorphic functions on a complex manifold.

Example 1.0.3. Let X be a topological space, and A an abelian group. We define the *constant sheaf* $\mathscr{A}$ on X determined by A as follows. Give A the discrete topology, and for any open set $U \subseteq X$, let $\mathscr{A}(U)$ be the group of all continuous maps of U into A. Then with the usual restriction maps, we obtain a sheaf $\mathscr{A}$. Note that for every connected open set U, $\mathscr{A}(U) \cong A$, whence the name "constant sheaf." If U is an open set whose connected components are open (which is always true on a locally connected topological space), then $\mathscr{A}(U)$ is a direct product of copies of A, one for each connected component of U.

Definition. If $\mathscr{F}$ is a presheaf on X, and if P is a point of X, we define the stalk $\mathscr{F}_P$ of $\mathscr{F}$ at P to be the direct limit of the groups $\mathscr{F}(U)$ for all open sets U containing P, via the restriction maps ρ.

Thus an element of $\mathscr{F}_P$ is represented by a pair $\langle U, s \rangle$, where U is an open neighborhood of P, and s is an element of $\mathscr{F}(U)$. Two such pairs $\langle U, s \rangle$ and $\langle V, t \rangle$ define the same element of $\mathscr{F}_P$ if and only if there is an open neighborhood W of P with $W \subseteq U \cap V$, such that $s|_W = t|_W$. Thus we may speak of elements of the stalk $\mathscr{F}_P$ as *germs* of sections of $\mathscr{F}$ at the point P. In the case of a variety X and its sheaf of regular functions $\mathcal{O}$, the stalk $\mathcal{O}_P$ at a point P is just the local ring of P on X, which was defined in (I, §3).

Definition. If $\mathscr{F}$ and $\mathscr{G}$ are presheaves on X, a *morphism* $\varphi : \mathscr{F} \to \mathscr{G}$ consists of a morphism of abelian groups $\varphi(U) : \mathscr{F}(U) \to \mathscr{G}(U)$ for each open set U, such that whenever $V \subseteq U$ is an inclusion, the diagram

$$
\begin{array}{ccc}
\mathscr{F}(U) & \xrightarrow{\;\varphi(U)\;} & \mathscr{G}(U) \\
\Big\downarrow{\scriptstyle \rho_{UV}} & & \Big\downarrow{\scriptstyle \rho'_{UV}} \\
\mathscr{F}(V) & \xrightarrow{\;\varphi(V)\;} & \mathscr{G}(V)
\end{array}
$$

is commutative, where ρ and ρ' are the restriction maps in $\mathcal{F}$ and $\mathcal{G}$. If $\mathcal{F}$ and $\mathcal{G}$ are sheaves on X, we use the same definition for a morphism of sheaves. An *isomorphism* is a morphism which has a two-sided inverse.

Note that a morphism $\varphi:\mathcal{F} \to \mathcal{G}$ of presheaves on X induces a morphism $\varphi_P:\mathcal{F}_P \to \mathcal{G}_P$ on the stalks, for any point $P \in X$. The following proposition (which would be false for presheaves) illustrates the local nature of a sheaf.

Proposition 1.1. *Let* $\varphi:\mathcal{F} \to \mathcal{G}$ *be a morphism of sheaves on a topological space* X. *Then* φ *is an isomorphism if and only if the induced map on the stalk* $\varphi_P:\mathcal{F}_P \to \mathcal{G}_P$ *is an isomorphism for every* $P \in X$.

PROOF. If φ is an isomorphism it is clear that each φ_P is an isomorphism. Conversely, assume φ_P is an isomorphism for all $P \in X$. To show that φ is an isomorphism, it will be sufficient to show that $\varphi(U):\mathcal{F}(U) \to \mathcal{G}(U)$ is an isomorphism for all U, because then we can define an inverse morphism ψ by $\psi(U) = \varphi(U)^{-1}$ for each U. First we show $\varphi(U)$ is injective. Let $s \in \mathcal{F}(U)$, and suppose $\varphi(s) \in \mathcal{G}(U)$ is 0. Then for every point $P \in U$, the image $\varphi(s)_P$ of $\varphi(s)$ in the stalk $\mathcal{G}_P$ is 0. Since φ_P is injective for each P, we deduce that $s_P = 0$ in $\mathcal{F}_P$ for each $P \in U$. To say that $s_P = 0$ means that s and 0 have the same image in $\mathcal{F}_P$, which means that there is an open neighborhood W_P of P, with $W_P \subseteq U$, such that $s|_{W_P} = 0$. Now U is covered by the neighborhoods W_P of all its points, so by the sheaf property (3), s is 0 on U. Thus $\varphi(U)$ is injective.

Next, we show that $\varphi(U)$ is surjective. Suppose we have a section $t \in \mathcal{G}(U)$. For each $P \in U$, let $t_P \in \mathcal{G}_P$ be its germ at P. Since φ_P is surjective, we can find $s_P \in \mathcal{F}_P$ such that $\varphi_P(s_P) = t_P$. Let s_P be represented by a section $s(P)$ on a neighborhood V_P of P. Then $\varphi(s(P))$ and $t|_{V_P}$ are two elements of $\mathcal{G}(V_P)$, whose germs at P are the same. Hence, replacing V_P by a smaller neighborhood of P if necessary, we may assume that $\varphi(s(P)) = t|_{V_P}$ in $\mathcal{G}(V_P)$. Now U is covered by the open sets V_P, and on each V_P we have a section $s(P) \in \mathcal{F}(V_P)$. If P,Q are two points, then $s(P)|_{V_P \cap V_Q}$ and $s(Q)|_{V_P \cap V_Q}$ are two sections of $\mathcal{F}(V_P \cap V_Q)$, which are both sent by φ to $t|_{V_P \cap V_Q}$. Hence by the injectivity of φ proved above, they are equal. Then by the sheaf property (4), there is a section $s \in \mathcal{F}(U)$ such that $s|_{V_P} = s(P)$ for each P. Finally, we have to check that $\varphi(s) = t$. Indeed, $\varphi(s)$, t are two sections of $\mathcal{G}(U)$, and for each P, $\varphi(s)|_{V_P} = t|_{V_P}$, hence by the sheaf property (3) applied to $\varphi(s) - t$, we conclude that $\varphi(s) = t$.

Our next task is to define kernels, cokernels and images of morphisms of sheaves.

Definition. Let $\varphi:\mathcal{F} \to \mathcal{G}$ be a morphism of presheaves. We define the *presheaf kernel* of φ, *presheaf cokernel* of φ, and *presheaf image* of φ to be the presheaves given by $U \mapsto \ker(\varphi(U))$, $U \mapsto \operatorname{coker}(\varphi(U))$, and $U \mapsto \operatorname{im}(\varphi(U))$ respectively.

Note that if $\varphi:\mathcal{F} \to \mathcal{G}$ is a morphism of sheaves, then the presheaf kernel of φ is a sheaf, but the presheaf cokernel and presheaf image of φ are in general not sheaves. This leads us to the notion of a sheaf associated to a presheaf.

Proposition-Definition 1.2. *Given a presheaf $\mathcal{F}$, there is a sheaf $\mathcal{F}^+$ and a morphism $\theta:\mathcal{F} \to \mathcal{F}^+$, with the property that for any sheaf $\mathcal{G}$, and any morphism $\varphi:\mathcal{F} \to \mathcal{G}$, there is a unique morphism $\psi:\mathcal{F}^+ \to \mathcal{G}$ such that $\varphi = \psi \circ \theta$. Furthermore the pair $(\mathcal{F}^+,\theta)$ is unique up to unique isomorphism. $\mathcal{F}^+$ is called the* sheaf associated *to the presheaf $\mathcal{F}$.*

PROOF. We construct the sheaf $\mathcal{F}^+$ as follows. For any open set U, let $\mathcal{F}^+(U)$ be the set of functions s from U to the union $\bigcup_{P \in U} \mathcal{F}_P$ of the stalks of $\mathcal{F}$ over points of U, such that

(1) for each $P \in U$, $s(P) \in \mathcal{F}_P$, and
(2) for each $P \in U$, there is a neighborhood V of P, contained in U, and an element $t \in \mathcal{F}(V)$, such that for all $Q \in V$, the germ t_Q of t at Q is equal to $s(Q)$.

Now one can verify immediately (!) that $\mathcal{F}^+$ with the natural restriction maps is a sheaf, that there is a natural morphism $\theta:\mathcal{F} \to \mathcal{F}^+$, and that it has the universal property described. The uniqueness of $\mathcal{F}^+$ is a formal consequence of the universal property. Note that for any point P, $\mathcal{F}_P = \mathcal{F}^+_P$. Note also that if $\mathcal{F}$ itself was a sheaf, then $\mathcal{F}^+$ is isomorphic to $\mathcal{F}$ via θ.

Definition. A *subsheaf* of a sheaf $\mathcal{F}$ is a sheaf $\mathcal{F}'$ such that for every open set $U \subseteq X$, $\mathcal{F}'(U)$ is a subgroup of $\mathcal{F}(U)$, and the restriction maps of the sheaf $\mathcal{F}'$ are induced by those of $\mathcal{F}$. It follows that for any point P, the stalk $\mathcal{F}'_P$ is a subgroup of $\mathcal{F}_P$.

If $\varphi:\mathcal{F} \to \mathcal{G}$ is a morphism of sheaves, we define the *kernel* of φ, denoted ker φ, to be the presheaf kernel of φ (which is a sheaf). Thus ker φ is a subsheaf of $\mathcal{F}$.

We say that a morphism of sheaves $\varphi:\mathcal{F} \to \mathcal{G}$ is *injective* if ker $\varphi = 0$. Thus φ is injective if and only if the induced map $\varphi(U):\mathcal{F}(U) \to \mathcal{G}(U)$ is injective for every open set of X.

If $\varphi:\mathcal{F} \to \mathcal{G}$ is a morphism of sheaves, we define the *image* of φ, denoted im φ, to be the sheaf associated to the presheaf image of φ. By the universal property of the sheaf associated to a presheaf, there is a natural map im $\varphi \to \mathcal{G}$. In fact this map is injective (see Ex. 1.4), and thus im φ can be identified with a subsheaf of $\mathcal{G}$.

We say that a morphism $\varphi:\mathcal{F} \to \mathcal{G}$ of sheaves is *surjective* if im $\varphi = \mathcal{G}$.

We say that a sequence $\dots \to \mathcal{F}^{i-1} \xrightarrow{\varphi^{i-1}} \mathcal{F}^i \xrightarrow{\varphi^i} \mathcal{F}^{i+1} \to \dots$ of sheaves and morphisms is *exact* if at each stage ker $\varphi^i = $ im φ^{i-1}. Thus a sequence

$0 \to \mathscr{F} \xrightarrow{\varphi} \mathscr{G}$ is exact if and only if φ is injective, and $\mathscr{F} \xrightarrow{\varphi} \mathscr{G} \to 0$ is exact if and only if φ is surjective.

Now let $\mathscr{F}'$ be a subsheaf of a sheaf $\mathscr{F}$. We define the *quotient sheaf* $\mathscr{F}/\mathscr{F}'$ to be the sheaf associated to the presheaf $U \to \mathscr{F}(U)/\mathscr{F}'(U)$. It follows that for any point P, the stalk $(\mathscr{F}/\mathscr{F}')_P$ is the quotient $\mathscr{F}_P/\mathscr{F}'_P$.

If $\varphi:\mathscr{F} \to \mathscr{G}$ is a morphism of sheaves, we define the *cokernel* of φ, denoted coker φ, to be the sheaf associated to the presheaf cokernel of φ.

Caution 1.2.1. We saw that a morphism $\varphi:\mathscr{F} \to \mathscr{G}$ of sheaves is injective if and only if the map on sections $\varphi(U):\mathscr{F}(U) \to \mathscr{G}(U)$ is injective for each U. The corresponding statement for surjective morphisms is not true: if $\varphi:\mathscr{F} \to \mathscr{G}$ is surjective, the maps $\varphi(U):\mathscr{F}(U) \to \mathscr{G}(U)$ on sections need not be surjective. However, we can say that φ is surjective if and only if the maps $\varphi_P:\mathscr{F}_P \to \mathscr{G}_P$ on stalks are surjective for each P. More generally, a sequence of sheaves and morphisms is exact if and only if it is exact on stalks (Ex. 1.2). This again illustrates the local nature of sheaves.

So far we have talked only about sheaves on a single topological space. Now we define some operations on sheaves, associated with a continuous map from one topological space to another.

Definition. Let $f:X \to Y$ be a continuous map of topological spaces. For any sheaf $\mathscr{F}$ on X, we define the *direct image* sheaf $f_*\mathscr{F}$ on Y by $(f_*\mathscr{F})(V) = \mathscr{F}(f^{-1}(V))$ for any open set $V \subseteq Y$. For any sheaf $\mathscr{G}$ on Y, we define the *inverse image* sheaf $f^{-1}\mathscr{G}$ on X to be the sheaf associated to the presheaf $U \mapsto \lim_{V \supseteq f(U)} \mathscr{G}(V)$, where U is any open set in X, and the limit is taken over all open sets V of Y containing $f(U)$. Do not confuse $f^{-1}\mathscr{G}$ with the sheaf $f^*\mathscr{G}$ which will be defined later for a morphism of ringed spaces (§5).

Note that f_* is a functor from the category $\mathfrak{Ab}(X)$ of sheaves on X to the category $\mathfrak{Ab}(Y)$ of sheaves on Y. Similarly, f^{-1} is a functor from $\mathfrak{Ab}(Y)$ to $\mathfrak{Ab}(X)$.

Definition. If Z is a subset of X, regarded as a topological subspace with the induced topology, if $i:Z \to X$ is the inclusion map, and if $\mathscr{F}$ is a sheaf on X, then we call $i^{-1}\mathscr{F}$ the *restriction* of $\mathscr{F}$ to Z, and we often denote it by $\mathscr{F}|_Z$. Note that the stalk of $\mathscr{F}|_Z$ at any point $P \in Z$ is just $\mathscr{F}_P$.

EXERCISES

1.1. Let A be an abelian group, and define the *constant presheaf* associated to A on the topological space X to be the presheaf $U \mapsto A$ for all $U \neq \emptyset$, with restriction maps the identity. Show that the constant sheaf $\mathscr{A}$ defined in the text is the sheaf associated to this presheaf.

1.2. (a) For any morphism of sheaves $\varphi:\mathscr{F}\to\mathscr{G}$, show that for each point P, $(\ker\varphi)_P = \ker(\varphi_P)$ and $(\operatorname{im}\varphi)_P = \operatorname{im}(\varphi_P)$.

(b) Show that φ is injective (respectively, surjective) if and only if the induced map on the stalks φ_P is injective (respectively, surjective) for all P.

(c) Show that a sequence $\ldots\mathscr{F}^{i-1}\xrightarrow{\varphi^{i-1}}\mathscr{F}^{i}\xrightarrow{\varphi^{i}}\mathscr{F}^{i+1}\to\ldots$ of sheaves and morphisms is exact if and only if for each $P\in X$ the corresponding sequence of stalks is exact as a sequence of abelian groups.

1.3. (a) Let $\varphi:\mathscr{F}\to\mathscr{G}$ be a morphism of sheaves on X. Show that φ is surjective if and only if the following condition holds: for every open set $U\subseteq X$, and for every $s\in\mathscr{G}(U)$, there is a covering $\{U_i\}$ of U, and there are elements $t_i\in\mathscr{F}(U_i)$, such that $\varphi(t_i)=s|_{U_i}$ for all i.

(b) Give an example of a surjective morphism of sheaves $\varphi:\mathscr{F}\to\mathscr{G}$, and an open set U such that $\varphi(U):\mathscr{F}(U)\to\mathscr{G}(U)$ is not surjective.

1.4. (a) Let $\varphi:\mathscr{F}\to\mathscr{G}$ be a morphism of presheaves such that $\varphi(U):\mathscr{F}(U)\to\mathscr{G}(U)$ is injective for each U. Show that the induced map $\varphi^{+}:\mathscr{F}^{+}\to\mathscr{G}^{+}$ of associated sheaves is injective.

(b) Use part (a) to show that if $\varphi:\mathscr{F}\to\mathscr{G}$ is a morphism of sheaves, then $\operatorname{im}\varphi$ can be naturally identified with a subsheaf of $\mathscr{G}$, as mentioned in the text.

1.5. Show that a morphism of sheaves is an isomorphism if and only if it is both injective and surjective.

1.6. (a) Let $\mathscr{F}'$ be a subsheaf of a sheaf $\mathscr{F}$. Show that the natural map of $\mathscr{F}$ to the quotient sheaf $\mathscr{F}/\mathscr{F}'$ is surjective, and has kernel $\mathscr{F}'$. Thus there is an exact sequence

$$0\to\mathscr{F}'\to\mathscr{F}\to\mathscr{F}/\mathscr{F}'\to 0.$$

(b) Conversely, if $0\to\mathscr{F}'\to\mathscr{F}\to\mathscr{F}''\to 0$ is an exact sequence, show that $\mathscr{F}'$ is isomorphic to a subsheaf of $\mathscr{F}$, and that $\mathscr{F}''$ is isomorphic to the quotient of $\mathscr{F}$ by this subsheaf.

1.7. Let $\varphi:\mathscr{F}\to\mathscr{G}$ be a morphism of sheaves.

(a) Show that $\operatorname{im}\varphi\cong\mathscr{F}/\ker\varphi$.

(b) Show that $\operatorname{coker}\varphi\cong\mathscr{G}/\operatorname{im}\varphi$.

1.8. For any open subset $U\subseteq X$, show that the functor $\Gamma(U,\cdot)$ from sheaves on X to abelian groups is a left exact functor, i.e., if $0\to\mathscr{F}'\to\mathscr{F}\to\mathscr{F}''$ is an exact sequence of sheaves, then $0\to\Gamma(U,\mathscr{F}')\to\Gamma(U,\mathscr{F})\to\Gamma(U,\mathscr{F}'')$ is an exact sequence of groups. The functor $\Gamma(U,\cdot)$ need not be exact; see (Ex. 1.21) below.

1.9. *Direct Sum.* Let $\mathscr{F}$ and $\mathscr{G}$ be sheaves on X. Show that the presheaf $U\mapsto\mathscr{F}(U)\oplus\mathscr{G}(U)$ is a sheaf. It is called the *direct sum* of $\mathscr{F}$ and $\mathscr{G}$, and is denoted by $\mathscr{F}\oplus\mathscr{G}$. Show that it plays the role of direct sum and of direct product in the category of sheaves of abelian groups on X.

1.10. *Direct Limit.* Let $\{\mathscr{F}_i\}$ be a direct system of sheaves and morphisms on X. We define the *direct limit* of the system $\{\mathscr{F}_i\}$, denoted $\varinjlim\mathscr{F}_i$, to be the sheaf associated to the presheaf $U\mapsto\varinjlim\mathscr{F}_i(U)$. Show that this is a direct limit in the category of sheaves on X, i.e., that it has the following universal property: given a sheaf $\mathscr{G}$, and a collection of morphisms $\mathscr{F}_i\to\mathscr{G}$, compatible with the maps of the direct system, then there exists a unique map $\varinjlim \mathscr{F}_i \to \mathscr{G}$ such that for each i, the original map $\mathscr{F}_i \to \mathscr{G}$ is obtained by composing the maps $\mathscr{F}_i \to \varinjlim \mathscr{F}_i \to \mathscr{G}$.

1.11. Let $\{\mathscr{F}_i\}$ be a direct system of sheaves on a noetherian topological space X. In this case show that the presheaf $U \mapsto \varinjlim \mathscr{F}_i(U)$ is already a sheaf. In particular, $\Gamma(X, \varinjlim \mathscr{F}_i) = \varinjlim \Gamma(X, \mathscr{F}_i)$.

1.12. *Inverse Limit.* Let $\{\mathscr{F}_i\}$ be an inverse system of sheaves on X. Show that the presheaf $U \mapsto \varprojlim \mathscr{F}_i(U)$ is a sheaf. It is called the *inverse limit* of the system $\{\mathscr{F}_i\}$, and is denoted by $\varprojlim \mathscr{F}_i$. Show that it has the universal property of an inverse limit in the category of sheaves.

1.13. *Espace Étalé of a Presheaf.* (This exercise is included only to establish the connection between our definition of a sheaf and another definition often found in the literature. See for example Godement [1, Ch. II, §1.2].) Given a presheaf $\mathscr{F}$ on X, we define a topological space Spé($\mathscr{F}$), called the *espace étalé* of $\mathscr{F}$, as follows. As a set, Spé($\mathscr{F}$) $= \bigcup_{P \in X} \mathscr{F}_P$. We define a projection map π:Spé($\mathscr{F}$)$\to X$ by sending $s \in \mathscr{F}_P$ to P. For each open set $U \subseteq X$ and each section $s \in \mathscr{F}(U)$, we obtain a map $\bar{s} : U \to$ Spé($\mathscr{F}$) by sending $P \mapsto s_P$, its germ at P. This map has the property that $\pi \circ \bar{s} = \mathrm{id}_U$, in other words, it is a "section" of π over U. We now make Spé($\mathscr{F}$) into a topological space by giving it the strongest topology such that all the maps $\bar{s} : U \to$ Spé($\mathscr{F}$) for all U, and all $s \in \mathscr{F}(U)$, are continuous. Now show that the sheaf $\mathscr{F}^+$ associated to $\mathscr{F}$ can be described as follows: for any open set $U \subseteq X$, $\mathscr{F}^+(U)$ is the set of *continuous* sections of Spé($\mathscr{F}$) over U. In particular, the original presheaf $\mathscr{F}$ was a sheaf if and only if for each U, $\mathscr{F}(U)$ is equal to the set of all continuous sections of Spé($\mathscr{F}$) over U.

1.14. *Support.* Let $\mathscr{F}$ be a sheaf on X, and let $s \in \mathscr{F}(U)$ be a section over an open set U. The *support* of s, denoted Supp s, is defined to be $\{P \in U \mid s_P \neq 0\}$, where s_P denotes the germ of s in the stalk $\mathscr{F}_P$. Show that Supp s is a closed subset of U. We define the *support* of $\mathscr{F}$, Supp $\mathscr{F}$, to be $\{P \in X \mid \mathscr{F}_P \neq 0\}$. It need not be a closed subset.

1.15. *Sheaf Hom.* Let $\mathscr{F}$, $\mathscr{G}$ be sheaves of abelian groups on X. For any open set $U \subseteq X$, show that the set $\mathrm{Hom}(\mathscr{F}|_U, \mathscr{G}|_U)$ of morphisms of the restricted sheaves has a natural structure of abelian group. Show that the presheaf $U \mapsto \mathrm{Hom}(\mathscr{F}|_U, \mathscr{G}|_U)$ is a sheaf. It is called the *sheaf of local morphisms* of $\mathscr{F}$ into $\mathscr{G}$, "sheaf hom" for short, and is denoted $\mathscr{H}om(\mathscr{F}, \mathscr{G})$.

1.16. *Flasque Sheaves.* A sheaf $\mathscr{F}$ on a topological space X is *flasque* if for every inclusion $V \subseteq U$ of open sets, the restriction map $\mathscr{F}(U) \to \mathscr{F}(V)$ is surjective.
   (a) Show that a constant sheaf on an irreducible topological space is flasque. See (I, §1) for irreducible topological spaces.
   (b) If $0 \to \mathscr{F}' \to \mathscr{F} \to \mathscr{F}'' \to 0$ is an exact sequence of sheaves, and if $\mathscr{F}'$ is flasque, then for any open set U, the sequence $0 \to \mathscr{F}'(U) \to \mathscr{F}(U) \to \mathscr{F}''(U) \to 0$ of abelian groups is also exact.
   (c) If $0 \to \mathscr{F}' \to \mathscr{F} \to \mathscr{F}'' \to 0$ is an exact sequence of sheaves, and if $\mathscr{F}'$ and $\mathscr{F}$ are flasque, then $\mathscr{F}''$ is flasque.
   (d) If $f : X \to Y$ is a continuous map, and if $\mathscr{F}$ is a flasque sheaf on X, then $f_* \mathscr{F}$ is a flasque sheaf on Y.
   (e) Let $\mathscr{F}$ be any sheaf on X. We define a new sheaf $\mathscr{G}$, called the sheaf of *discontinuous sections* of $\mathscr{F}$ as follows. For each open set $U \subseteq X$, $\mathscr{G}(U)$ is the set of

maps $s: U \to \bigcup_{P \in U} \mathscr{F}_P$ such that for each $P \in U$, $s(P) \in \mathscr{F}_P$. Show that $\mathscr{G}$ is a flasque sheaf, and that there is a natural injective morphism of $\mathscr{F}$ to $\mathscr{G}$.

1.17. *Skyscraper Sheaves.* Let X be a topological space, let P be a point, and let A be an abelian group. Define a sheaf $i_P(A)$ on X as follows: $i_P(A)(U) = A$ if $P \in U$, 0 otherwise. Verify that the stalk of $i_P(A)$ is A at every point $Q \in \{P\}^-$, and 0 elsewhere, where $\{P\}^-$ denotes the closure of the set consisting of the point P. Hence the name "skyscraper sheaf." Show that this sheaf could also be described as $i_*(A)$, where A denotes the constant sheaf A on the closed subspace $\{P\}^-$, and $i: \{P\}^- \to X$ is the inclusion.

1.18. *Adjoint Property of f^{-1}.* Let $f: X \to Y$ be a continuous map of topological spaces. Show that for any sheaf $\mathscr{F}$ on X there is a natural map $f^{-1} f_* \mathscr{F} \to \mathscr{F}$, and for any sheaf $\mathscr{G}$ on Y there is a natural map $\mathscr{G} \to f_* f^{-1} \mathscr{G}$. Use these maps to show that there is a natural bijection of sets, for any sheaves $\mathscr{F}$ on X and $\mathscr{G}$ on Y,

$$\text{Hom}_X(f^{-1}\mathscr{G}, \mathscr{F}) = \text{Hom}_Y(\mathscr{G}, f_* \mathscr{F}).$$

Hence we say that f^{-1} is a *left adjoint* of f_*, and that f_* is a *right adjoint* of f^{-1}.

1.19. *Extending a Sheaf by Zero.* Let X be a topological space, let Z be a closed subset, let $i: Z \to X$ be the inclusion, let $U = X - Z$ be the complementary open subset, and let $j: U \to X$ be its inclusion.
  (a) Let $\mathscr{F}$ be a sheaf on Z. Show that the stalk $(i_* \mathscr{F})_P$ of the direct image sheaf on X is $\mathscr{F}_P$ if $P \in Z$, 0 if $P \notin Z$. Hence we call $i_* \mathscr{F}$ the sheaf obtained by extending $\mathscr{F}$ by zero outside Z. By abuse of notation we will sometimes write $\mathscr{F}$ instead of $i_* \mathscr{F}$, and say "consider $\mathscr{F}$ as a sheaf on X," when we mean "consider $i_* \mathscr{F}$."
  (b) Now let $\mathscr{F}$ be a sheaf on U. Let $j_!(\mathscr{F})$ be the sheaf on X associated to the presheaf $V \mapsto \mathscr{F}(V)$ if $V \subseteq U$, $V \mapsto 0$ otherwise. Show that the stalk $(j_!(\mathscr{F}))_P$ is equal to $\mathscr{F}_P$ if $P \in U$, 0 if $P \notin U$, and show that $j_! \mathscr{F}$ is the only sheaf on X which has this property, and whose restriction to U is $\mathscr{F}$. We call $j_! \mathscr{F}$ the sheaf obtained by *extending $\mathscr{F}$ by zero* outside U.
  (c) Now let $\mathscr{F}$ be a sheaf on X. Show that there is an exact sequence of sheaves on X,

$$0 \to j_!(\mathscr{F}|_U) \to \mathscr{F} \to i_*(\mathscr{F}|_Z) \to 0.$$

1.20. *Subsheaf with Supports.* Let Z be a closed subset of X, and let $\mathscr{F}$ be a sheaf on X. We define $\Gamma_Z(X, \mathscr{F})$ to be the subgroup of $\Gamma(X, \mathscr{F})$ consisting of all sections whose support (Ex. 1.14) is contained in Z.
  (a) Show that the presheaf $V \mapsto \Gamma_{Z \cap V}(V, \mathscr{F}|_V)$ is a sheaf. It is called the subsheaf of $\mathscr{F}$ with supports in Z, and is denoted by $\mathscr{H}_Z^0(\mathscr{F})$.
  (b) Let $U = X - Z$, and let $j: U \to X$ be the inclusion. Show there is an exact sequence of sheaves on X

$$0 \to \mathscr{H}_Z^0(\mathscr{F}) \to \mathscr{F} \to j_*(\mathscr{F}|_U).$$

Furthermore, if $\mathscr{F}$ is flasque, the map $\mathscr{F} \to j_*(\mathscr{F}|_U)$ is surjective.

1.21. *Some Examples of Sheaves on Varieties.* Let X be a variety over an algebraically closed field k, as in Ch. I. Let $\mathcal{O}_X$ be the sheaf of regular functions on X (1.0.1).
  (a) Let Y be a closed subset of X. For each open set $U \subseteq X$, let $\mathscr{I}_Y(U)$ be the ideal in the ring $\mathcal{O}_X(U)$ consisting of those regular functions which vanish

at all points of $Y \cap U$. Show that the presheaf $U \mapsto \mathscr{I}_Y(U)$ is a sheaf. It is
called the *sheaf of ideals* $\mathscr{I}_Y$ of Y, and it is a subsheaf of the sheaf of rings $\mathscr{O}_X$.
(b) If Y is a subvariety, then the quotient sheaf $\mathscr{O}_X/\mathscr{I}_Y$ is isomorphic to $i_*(\mathscr{O}_Y)$,
where $i: Y \to X$ is the inclusion, and $\mathscr{O}_Y$ is the sheaf of regular functions on Y.
(c) Now let $X = \mathbf{P}^1$, and let Y be the union of two distinct points $P, Q \in X$. Then
there is an exact sequence of sheaves on X, where $\mathscr{F} = i_*\mathscr{O}_P \oplus i_*\mathscr{O}_Q$,

$$0 \to \mathscr{I}_Y \to \mathscr{O}_X \to \mathscr{F} \to 0.$$

Show however that the induced map on global sections $\Gamma(X,\mathscr{O}_X) \to \Gamma(X,\mathscr{F})$
is not surjective. This shows that the global section functor $\Gamma(X,\cdot)$ is not exact
(cf. (Ex. 1.8) which shows that it is left exact).
(d) Again let $X = \mathbf{P}^1$, and let $\mathscr{O}$ be the sheaf of regular functions. Let $\mathscr{K}$ be the
constant sheaf on X associated to the function field K of X. Show that there
is a natural injection $\mathscr{O} \to \mathscr{K}$. Show that the quotient sheaf $\mathscr{K}/\mathscr{O}$ is isomorphic
to the direct sum of sheaves $\sum_{P \in X} i_P(I_P)$, where I_P is the group $K/\mathscr{O}_P$, and
$i_P(I_P)$ denotes the skyscraper sheaf (Ex. 1.17) given by I_P at the point P.
(e) Finally show that in the case of (d) the sequence

$$0 \to \Gamma(X,\mathscr{O}) \to \Gamma(X,\mathscr{K}) \to \Gamma(X,\mathscr{K}/\mathscr{O}) \to 0$$

is exact. (This is an analogue of what is called the "first Cousin problem" in
several complex variables. See Gunning and Rossi [1, p. 248].)

1.22. *Glueing Sheaves.* Let X be a topological space, let $\mathfrak{U} = \{U_i\}$ be an open cover of
X, and suppose we are given for each i a sheaf $\mathscr{F}_i$ on U_i, and for each i,j an iso-
morphism $\varphi_{ij}: \mathscr{F}_i|_{U_i \cap U_j} \xrightarrow{\sim} \mathscr{F}_j|_{U_i \cap U_j}$ such that (1) for each i, $\varphi_{ii} = \mathrm{id}$, and (2) for
each i,j,k, $\varphi_{ik} = \varphi_{jk} \circ \varphi_{ij}$ on $U_i \cap U_j \cap U_k$. Then there exists a unique sheaf
$\mathscr{F}$ on X, together with isomorphisms $\psi_i: \mathscr{F}|_{U_i} \xrightarrow{\sim} \mathscr{F}_i$ such that for each i,j, $\psi_j = \varphi_{ij} \circ \psi_i$ on $U_i \cap U_j$. We say loosely that $\mathscr{F}$ is obtained by *glueing* the sheaves $\mathscr{F}_i$
via the isomorphisms φ_{ij}.

2 Schemes

In this section we will define the notion of a scheme. First we define affine
schemes: to any ring A (recall our conventions about rings made in the
Introduction!) we associate a topological space together with a sheaf of
rings on it, called Spec A. This construction parallels the construction of
affine varieties (I, §1) except that the points of Spec A correspond to all prime
ideals of A, not just the maximal ones. Then we define an arbitrary scheme
to be something which locally looks like an affine scheme. This definition
has no parallel in Chapter I. An important class of schemes is given by the
construction of the scheme Proj S associated to any graded ring S. This
construction parallels the construction of projective varieties in (I, §2).
Finally, we will show that the varieties of Chapter I, after a slight modification,
can be regarded as schemes. Thus the category of schemes is an enlargement
of the category of varieties.

Now we will construct the space Spec A associated to a ring A. As a set, we define Spec A to be the set of all prime ideals of A. If $\mathfrak{a}$ is any ideal of A, we define the subset $V(\mathfrak{a}) \subseteq$ Spec A to be the set of all prime ideals which contain $\mathfrak{a}$.

Lemma 2.1.

(a) *If $\mathfrak{a}$ and $\mathfrak{b}$ are two ideals of A, then $V(\mathfrak{ab}) = V(\mathfrak{a}) \cup V(\mathfrak{b})$.*

(b) *If $\{\mathfrak{a}_i\}$ is any set of ideals of A, then $V(\sum \mathfrak{a}_i) = \bigcap V(\mathfrak{a}_i)$.*

(c) *If $\mathfrak{a}$ and $\mathfrak{b}$ are two ideals, $V(\mathfrak{a}) \subseteq V(\mathfrak{b})$ if and only if $\sqrt{\mathfrak{a}} \supseteq \sqrt{\mathfrak{b}}$.*

PROOF.

(a) Certainly if $\mathfrak{p} \supseteq \mathfrak{a}$ or $\mathfrak{p} \supseteq \mathfrak{b}$, then $\mathfrak{p} \supseteq \mathfrak{ab}$. Conversely, if $\mathfrak{p} \supseteq \mathfrak{ab}$, and if $\mathfrak{p} \not\supseteq \mathfrak{b}$ for example, then there is a $b \in \mathfrak{b}$ such that $b \notin \mathfrak{p}$. Now for any $a \in \mathfrak{a}$, $ab \in \mathfrak{p}$, so we must have $a \in \mathfrak{p}$ since $\mathfrak{p}$ is a prime ideal. Thus $\mathfrak{p} \supseteq \mathfrak{a}$.

(b) $\mathfrak{p}$ contains $\sum \mathfrak{a}_i$ if and only if $\mathfrak{p}$ contains each $\mathfrak{a}_i$, simply because $\sum \mathfrak{a}_i$ is the smallest ideal containing all of the ideals $\mathfrak{a}_i$.

(c) The radical of $\mathfrak{a}$ is the intersection of the set of all prime ideals containing $\mathfrak{a}$. So $\sqrt{\mathfrak{a}} \supseteq \sqrt{\mathfrak{b}}$ if and only if $V(\mathfrak{a}) \subseteq V(\mathfrak{b})$.

Now we define a topology on Spec A by taking the subsets of the form $V(\mathfrak{a})$ to be the closed subsets. Note that $V(A) = \varnothing$; $V((0)) = $ Spec A; and the lemma shows that finite unions and arbitrary intersections of sets of the form $V(\mathfrak{a})$ are again of that form. Hence they do form the set of closed sets for a topology on Spec A.

Next we will define a sheaf of rings $\mathcal{O}$ on Spec A. For each prime ideal $\mathfrak{p} \subseteq A$, let $A_\mathfrak{p}$ be the localization of A at $\mathfrak{p}$. For an open set $U \subseteq$ Spec A, we define $\mathcal{O}(U)$ to be the set of functions $s: U \to \coprod_{\mathfrak{p} \in U} A_\mathfrak{p}$, such that $s(\mathfrak{p}) \in A_\mathfrak{p}$ for each $\mathfrak{p}$, and such that s is locally a quotient of elements of A: to be precise, we require that for each $\mathfrak{p} \in U$, there is a neighborhood V of $\mathfrak{p}$, contained in U, and elements $a, f \in A$, such that for each $\mathfrak{q} \in V$, $f \notin \mathfrak{q}$, and $s(\mathfrak{q}) = a/f$ in $A_\mathfrak{q}$. (Note the similarity with the definition of the regular functions on a variety. The difference is that we consider functions into the various local rings, instead of to a field.)

Now it is clear that sums and products of such functions are again such, and that the element 1 which gives 1 in each $A_\mathfrak{p}$ is an identity. Thus $\mathcal{O}(U)$ is a commutative ring with identity. If $V \subseteq U$ are two open sets, the natural restriction map $\mathcal{O}(U) \to \mathcal{O}(V)$ is a homomorphism of rings. It is then clear that $\mathcal{O}$ is a presheaf. Finally, it is clear from the local nature of the definition that $\mathcal{O}$ is a sheaf.

Definition. Let A be a ring. The *spectrum* of A is the pair consisting of the topological space Spec A together with the sheaf of rings $\mathcal{O}$ defined above.

Let us establish some basic properties of the sheaf $\mathcal{O}$ on Spec A. For any element $f \in A$, we denote by $D(f)$ the open complement of $V((f))$. Note that open sets of the form $D(f)$ form a base for the topology of Spec A. Indeed, if $V(\mathfrak{a})$ is a closed set, and $\mathfrak{p} \notin V(\mathfrak{a})$, then $\mathfrak{p} \not\supseteq \mathfrak{a}$, so there is an $f \in \mathfrak{a}$, $f \notin \mathfrak{p}$. Then $\mathfrak{p} \in D(f)$ and $D(f) \cap V(\mathfrak{a}) = \varnothing$.

Proposition 2.2. *Let A be a ring, and* (Spec A, $\mathcal{O}$) *its spectrum.*

(a) *For any $\mathfrak{p} \in$ Spec A, the stalk $\mathcal{O}_\mathfrak{p}$ of the sheaf $\mathcal{O}$ is isomorphic to the local ring $A_\mathfrak{p}$.*

(b) *For any element $f \in A$, the ring $\mathcal{O}(D(f))$ is isomorphic to the localized ring A_f.*

(c) *In particular, $\Gamma(\text{Spec } A, \mathcal{O}) \cong A$.*

PROOF.

(a) First we define a homomorphism from $\mathcal{O}_\mathfrak{p}$ to $A_\mathfrak{p}$ by sending any local section s in a neighborhood of $\mathfrak{p}$ to its value $s(\mathfrak{p}) \in A_\mathfrak{p}$. This gives a well-defined homomorphism φ from $\mathcal{O}_\mathfrak{p}$ to $A_\mathfrak{p}$. The map φ is surjective, because any element of $A_\mathfrak{p}$ can be represented as a quotient a/f, with $a, f \in A$, $f \notin \mathfrak{p}$. Then $D(f)$ will be an open neighborhood of $\mathfrak{p}$, and a/f defines a section of $\mathcal{O}$ over $D(f)$ whose value at $\mathfrak{p}$ is the given element. To show that φ is injective, let U be a neighborhood of $\mathfrak{p}$, and let $s, t \in \mathcal{O}(U)$ be elements having the same value $s(\mathfrak{p}) = t(\mathfrak{p})$ at $\mathfrak{p}$. By shrinking U if necessary, we may assume that $s = a/f$, and $t = b/g$ on U, where $a, b, f, g \in A$, and $f, g \notin \mathfrak{p}$. Since a/f and b/g have the same image in $A_\mathfrak{p}$, it follows from the definition of localization that there is an $h \notin \mathfrak{p}$ such that $h(ga - fb) = 0$ in A. Therefore $a/f = b/g$ in every local ring $A_\mathfrak{q}$ such that $f, g, h \notin \mathfrak{q}$. But the set of such $\mathfrak{q}$ is the open set $D(f) \cap D(g) \cap D(h)$, which contains $\mathfrak{p}$. Hence $s = t$ in a whole neighborhood of $\mathfrak{p}$, so they have the same stalk at $\mathfrak{p}$. So φ is an isomorphism, which proves (a).

(b) and (c). Note that (c) is the special case of (b) when $f = 1$, and $D(f)$ is the whole space. So it is sufficient to prove (b). We define a homomorphism $\psi : A_f \to \mathcal{O}(D(f))$ by sending a/f^n to the section $s \in \mathcal{O}(D(f))$ which assigns to each $\mathfrak{p}$ the image of a/f^n in $A_\mathfrak{p}$.

First we show ψ is injective. If $\psi(a/f^n) = \psi(b/f^m)$, then for every $\mathfrak{p} \in D(f)$, a/f^n and b/f^m have the same image in $A_\mathfrak{p}$. Hence there is an element $h \notin \mathfrak{p}$ such that $h(f^m a - f^n b) = 0$ in A. Let $\mathfrak{a}$ be the annihilator of $f^m a - f^n b$. Then $h \in \mathfrak{a}$, and $h \notin \mathfrak{p}$, so $\mathfrak{a} \not\subseteq \mathfrak{p}$. This holds for any $\mathfrak{p} \in D(f)$, so we conclude that $V(\mathfrak{a}) \cap D(f) = \varnothing$. Therefore $f \in \sqrt{\mathfrak{a}}$, so some power $f^l \in \mathfrak{a}$, so $f^l(f^m a - f^n b) = 0$, which shows that $a/f^n = b/f^m$ in A_f. Hence ψ is injective.

The hard part is to show that ψ is surjective. So let $s \in \mathcal{O}(D(f))$. Then by definition of $\mathcal{O}$, we can cover $D(f)$ with open sets V_i, on which s is represented by a quotient a_i/g_i, with $g_i \notin \mathfrak{p}$ for all $\mathfrak{p} \in V_i$, in other words, $V_i \subseteq D(g_i)$. Now the open sets of the form $D(h)$ form a base for the topology, so we may assume that $V_i = D(h_i)$ for some h_i. Since $D(h_i) \subseteq D(g_i)$, we have $V((h_i)) \supseteq V((g_i))$, hence by (2.1c), $\sqrt{(h_i)} \subseteq \sqrt{(g_i)}$, and in particular, $h_i^n \in (g_i)$ for some n. So $h_i^n = cg_i$, so $a_i/g_i = ca_i/h_i^n$. Replacing h_i by h_i^n (since $D(h_i) = D(h_i^n)$) and a_i by ca_i, we may assume that $D(f)$ is covered by the open subsets $D(h_i)$, and that s is represented by a_i/h_i on $D(h_i)$.

Next we observe that $D(f)$ can be covered by a finite number of the $D(h_i)$. Indeed, $D(f) \subseteq \bigcup D(h_i)$ if and only if $V((f)) \supseteq \bigcap V((h_i)) = V(\sum(h_i))$. By (2.1c) again, this is equivalent to saying $f \in \sqrt{\sum(h_i)}$, or $f^n \in \sum(h_i)$ for some n. This means that f^n can be expressed as a *finite* sum $f^n = \sum b_i h_i$, $b_i \in A$. Hence a finite subset of the h_i will do. So from now on we fix a finite set $h_1, \ldots, h_r$ such that $D(f) \subseteq D(h_1) \cup \ldots \cup D(h_r)$.

For the next step, note that on $D(h_i) \cap D(h_j) = D(h_i h_j)$ we have two elements of $A_{h_i h_j}$, namely a_i/h_i and a_j/h_j both of which represent s. Hence, according to the injectivity of ψ proved above, applied to $D(h_i h_j)$, we must have $a_i/h_i = a_j/h_j$ in $A_{h_i h_j}$. Hence for some n,

$$(h_i h_j)^n (h_j a_i - h_i a_j) = 0.$$

Since there are only finitely many indices involved, we may pick n so large that it works for all i,j at once. Rewrite this equation as

$$h_j^{n+1}(h_i^n a_i) - h_i^{n+1}(h_j^n a_j) = 0.$$

Then replace each h_i by h_i^{n+1}, and a_i by $h_i^n a_i$. Then we still have s represented on $D(h_i)$ by a_i/h_i, and furthermore, we have $h_j a_i = h_i a_j$ for all i,j.

Now write $f^n = \sum b_i h_i$ as above, which is possible for some n since the $D(h_i)$ cover $D(f)$. Let $a = \sum b_i a_i$. Then for each j we have

$$h_j a = \sum_i b_i a_i h_j = \sum_i b_i h_i a_j = f^n a_j.$$

This says that $a/f^n = a_j/h_j$ on $D(h_j)$. So $\psi(a/f^n) = s$ everywhere, which shows that ψ is surjective, hence an isomorphism.

To each ring A we have now associated its spectrum (Spec $A, \mathcal{O}$). We would like to say that this correspondence is functorial. For that we need a suitable category of spaces with sheaves of rings on them. The appropriate notion is the category of locally ringed spaces.

Definition. A *ringed space* is a pair $(X, \mathcal{O}_X)$ consisting of a topological space X and a sheaf of rings $\mathcal{O}_X$ on X. A *morphism* of ringed spaces from $(X, \mathcal{O}_X)$ to $(Y, \mathcal{O}_Y)$ is a pair $(f, f^\#)$ of a continuous map $f: X \to Y$ and a map $f^\#: \mathcal{O}_Y \to f_* \mathcal{O}_X$ of sheaves of rings on Y. The ringed space $(X, \mathcal{O}_X)$ is a *locally ringed space* if for each point $P \in X$, the stalk $\mathcal{O}_{X,P}$ is a local ring. A *morphism* of locally ringed spaces is a morphism $(f, f^\#)$ of ringed spaces, such that for each point $P \in X$, the induced map (see below) of local rings $f_P^\#: \mathcal{O}_{Y,f(P)} \to \mathcal{O}_{X,P}$ is a *local homomorphism* of local rings. We explain this last condition. First of all, given a point $P \in X$, the morphism of sheaves $f^\#: \mathcal{O}_Y \to f_* \mathcal{O}_X$ induces a homomorphism of rings $\mathcal{O}_Y(V) \to \mathcal{O}_X(f^{-1}V)$, for every open set V in Y. As V ranges over all open neighborhoods of $f(P)$, $f^{-1}(V)$ ranges over a subset of the neighborhoods of P.

Taking direct limits, we obtain a map

$$\mathcal{O}_{Y,f(P)} = \varinjlim_V \mathcal{O}_Y(V) \to \varinjlim_V \mathcal{O}_X(f^{-1}V),$$

and the latter limit maps to the stalk $\mathcal{O}_{X,P}$. Thus we have an induced homomorphism $f_P^{\#}:\mathcal{O}_{Y,f(P)} \to \mathcal{O}_{X,P}$. We require that this be a local homomorphism: If A and B are local rings with maximal ideals $\mathfrak{m}_A$ and $\mathfrak{m}_B$ respectively, a homomorphism $\varphi:A \to B$ is called a *local homomorphism* if $\varphi^{-1}(\mathfrak{m}_B) = \mathfrak{m}_A$.

An *isomorphism* of locally ringed spaces is a morphism with a two-sided inverse. Thus a morphism $(f,f^{\#})$ is an isomorphism if and only if f is a homeomorphism of the underlying topological spaces, and $f^{\#}$ is an isomorphism of sheaves.

Proposition 2.3.

(a) *If A is a ring, then* (Spec A, $\mathcal{O}$) *is a locally ringed space.*

(b) *If $\varphi:A \to B$ is a homomorphism of rings, then φ induces a natural morphism of locally ringed spaces*

$$(f,f^{\#}):(\text{Spec } B, \mathcal{O}_{\text{Spec } B}) \to (\text{Spec } A, \mathcal{O}_{\text{Spec } A}).$$

(c) *If A and B are rings, then any morphism of locally ringed spaces from* Spec B *to* Spec A *is induced by a homomorphism of rings $\varphi:A \to B$ as in* (b).

PROOF.

(a) This follows from (2.2a).

(b) Given a homomorphism $\varphi:A \to B$, we define a map $f:\text{Spec } B \to \text{Spec } A$ by $f(\mathfrak{p}) = \varphi^{-1}(\mathfrak{p})$ for any $\mathfrak{p} \in \text{Spec } B$. If $\mathfrak{a}$ is an ideal of A, then it is immediate that $f^{-1}(V(\mathfrak{a})) = V(\varphi(\mathfrak{a}))$, so f is continuous. For each $\mathfrak{p} \in \text{Spec } B$, we can localize φ to obtain a local homomorphism of local rings $\varphi_{\mathfrak{p}}:A_{\varphi^{-1}(\mathfrak{p})} \to B_{\mathfrak{p}}$. Now for any open set $V \subseteq \text{Spec } A$ we obtain a homomorphism of rings $f^{\#}:\mathcal{O}_{\text{Spec } A}(V) \to \mathcal{O}_{\text{Spec } B}(f^{-1}(V))$ by the definition of $\mathcal{O}$, composing with the maps f and $\varphi_{\mathfrak{p}}$. This gives the morphism of sheaves $f^{\#}:\mathcal{O}_{\text{Spec } A} \to f_*(\mathcal{O}_{\text{Spec } B})$. The induced maps $f^{\#}$ on the stalks are just the local homomorphisms $\varphi_{\mathfrak{p}}$, so $(f,f^{\#})$ is a morphism of locally ringed spaces.

(c) Conversely, suppose given a morphism of locally ringed spaces $(f,f^{\#})$ from Spec B to Spec A. Taking global sections, $f^{\#}$ induces a homomorphism of rings $\varphi:\Gamma(\text{Spec } A, \mathcal{O}_{\text{Spec } A}) \to \Gamma(\text{Spec } B, \mathcal{O}_{\text{Spec } B})$. By (2.2c), these rings are A and B, respectively, so we have a homomorphism $\varphi:A \to B$. For any $\mathfrak{p} \in \text{Spec } B$, we have an induced local homomorphism on the stalks, $\mathcal{O}_{\text{Spec } A,f(\mathfrak{p})} \to \mathcal{O}_{\text{Spec } B,\mathfrak{p}}$ or $A_{f(\mathfrak{p})} \to B_{\mathfrak{p}}$, which must be compatible with the map φ on global sections and the localization homomorphisms. In other words, we have a commutative diagram

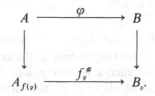

73

Since $f^\#$ is a local homomorphism, it follows that $\varphi^{-1}(\mathfrak{p}) = f(\mathfrak{p})$, which shows that f coincides with the map Spec $B \to$ Spec A induced by φ. Now it is immediate that $f^\#$ also is induced by φ, so that the morphism $(f, f^\#)$ of locally ringed spaces does indeed come from the homomorphism of rings φ.

Caution 2.3.0. Statement (c) of the proposition would be false, if in the definition of a morphism of locally ringed spaces, we did not insist that the induced maps on the stalks be *local* homomorphisms of local rings (see (2.3.2) below).

Now we come to the definition of a scheme.

Definition. An *affine scheme* is a locally ringed space $(X, \mathcal{O}_X)$ which is isomorphic (as a locally ringed space) to the spectrum of some ring. A *scheme* is a locally ringed space $(X, \mathcal{O}_X)$ in which every point has an open neighborhood U such that the topological space U, together with the restricted sheaf $\mathcal{O}_X|_U$, is an affine scheme. We call X the *underlying topological space* of the scheme $(X, \mathcal{O}_X)$, and $\mathcal{O}_X$ its *structure sheaf*. By abuse of notation we will often write simply X for the scheme $(X, \mathcal{O}_X)$. If we wish to refer to the underlying topological space without its scheme structure, we write sp(X), read "space of X." A *morphism* of schemes is a morphism as locally ringed spaces. An *isomorphism* is a morphism with a two-sided inverse.

Example 2.3.1. If k is a field, Spec k is an affine scheme whose topological space consists of one point, and whose structure sheaf consists of the field k.

Example 2.3.2. If R is a discrete valuation ring, then $T =$ Spec R is an affine scheme whose topological space consists of two points. One point t_0 is closed, with local ring R; the other point t_1 is open and dense, with local ring equal to K, the quotient field of R. The inclusion map $R \to K$ corresponds to the morphism Spec $K \to T$ which sends the unique point of Spec K to t_1. There is another morphism of ringed spaces Spec $K \to T$ which sends the unique point of Spec K to t_0, and uses the inclusion $R \to K$ to define the associated map $f^\#$ on structure sheaves. This morphism is *not* induced by any homomorphism $R \to K$ as in (2.3b,c), since it is not a morphism of *locally ringed spaces*.

Example 2.3.3. If k is a field, we define the *affine line* over k, $\mathbf{A}_k^1$, to be Spec $k[x]$. It has a point ξ, corresponding to the zero ideal, whose closure is the whole space. This is called a *generic point*. The other points, which correspond to the maximal ideals in $k[x]$, are all closed points. They are in one-to-one correspondence with the nonconstant monic irreducible polynomials in x. In particular, if k is algebraically closed, the closed points of $\mathbf{A}_k^1$ are in one-to-one correspondence with elements of k.

Example 2.3.4. Let k be an algebraically closed field, and consider the *affine plane* over k, defined as $\mathbf{A}_k^2 = \operatorname{Spec} k[x,y]$ (Fig. 6). The closed points of $\mathbf{A}_k^2$ are in one-to-one correspondence with ordered pairs of elements of k. Furthermore, the set of all closed points of $\mathbf{A}_k^2$, with the induced topology, is homeomorphic to the *variety* called $\mathbf{A}^2$ in Chapter I. In addition to the closed points, there is a *generic point* ξ, corresponding to the zero ideal of $k[x,y]$, whose closure is the whole space. Also, for each irreducible polynomial $f(x,y)$, there is a point η whose closure consists of η together with all closed points (a,b) for which $f(a,b) = 0$. We say that η is a *generic point* of the curve $f(x,y) = 0$.

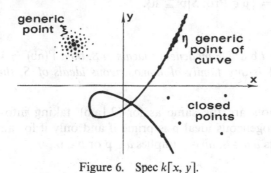

Figure 6. Spec $k[x, y]$.

Example 2.3.5. Let X_1 and X_2 be schemes, let $U_1 \subseteq X_1$ and $U_2 \subseteq X_2$ be open subsets, and let $\varphi:(U_1,\mathcal{O}_{X_1}|_{U_1}) \to (U_2,\mathcal{O}_{X_2}|_{U_2})$ be an isomorphism of locally ringed spaces. Then we can define a scheme X, obtained by *glueing* X_1 and X_2 along U_1 and U_2 via the isomorphism φ. The topological space of X is the quotient of the disjoint union $X_1 \cup X_2$ by the equivalence relation $x_1 \sim \varphi(x_1)$ for each $x_1 \in U_1$, with the quotient topology. Thus there are maps $i_1:X_1 \to X$ and $i_2:X_2 \to X$, and a subset $V \subseteq X$ is open if and only if $i_1^{-1}(V)$ is open in X_1 and $i_2^{-1}(V)$ is open in X_2. The structure sheaf $\mathcal{O}_X$ is defined as follows: for any open set $V \subseteq X$,

$$\mathcal{O}_X(V) = \{\langle s_1,s_2\rangle | s_1 \in \mathcal{O}_{X_1}(i_1^{-1}(V)) \text{ and } s_2 \in \mathcal{O}_{X_2}(i_2^{-1}(V)) \text{ and}$$
$$\varphi(s_1|_{i_1^{-1}(V) \cap U_1}) = s_2|_{i_2^{-1}(V) \cap U_2}\}.$$

Now it is clear that $\mathcal{O}_X$ is a sheaf, and that $(X,\mathcal{O}_X)$ is a locally ringed space. Furthermore, since X_1 and X_2 are schemes, it is clear that every point of X has a neighborhood which is affine, hence X is a scheme.

Example 2.3.6. As an example of glueing, let k be a field, let $X_1 = X_2 = \mathbf{A}_k^1$, let $U_1 = U_2 = \mathbf{A}_k^1 - \{P\}$, where P is the point corresponding to the maximal ideal (x), and let $\varphi: U_1 \to U_2$ be the identity map. Let X be obtained by glueing X_1 and X_2 along U_1 and U_2 via φ. We get an "affine line with the point P doubled."

————————:————————

This is an example of a scheme which is not an affine scheme (!). It is also an example of a nonseparated scheme, as we will see later (4.0.1).

Next we will define an important class of schemes, constructed from graded rings, which are analogous to projective varieties.

Let S be a graded ring. See (I, §2) for our conventions about graded rings. We denote by S_+ the ideal $\bigoplus_{d>0} S_d$.

We define the *set* Proj S to be the set of all homogeneous prime ideals $\mathfrak{p}$, which do not contain all of S_+. If $\mathfrak{a}$ is a homogeneous ideal of S, we define the subset $V(\mathfrak{a}) = \{\mathfrak{p} \in \text{Proj } S \,|\, \mathfrak{p} \supseteq \mathfrak{a}\}$.

Lemma 2.4.

    (a) *If $\mathfrak{a}$ and $\mathfrak{b}$ are homogeneous ideals in S, then $V(\mathfrak{ab}) = V(\mathfrak{a}) \cup V(\mathfrak{b})$.*

    (b) *If $\{\mathfrak{a}_i\}$ is any family of homogeneous ideals of S, then $V(\sum \mathfrak{a}_i) = \bigcap V(\mathfrak{a}_i)$.*

PROOF. The proofs are the same as for (2.1a,b), taking into account the fact that a homogeneous ideal $\mathfrak{p}$ is prime if and only if for any two *homogeneous* elements $a,b \in S$, $ab \in \mathfrak{p}$ implies $a \in \mathfrak{p}$ or $b \in \mathfrak{p}$.

Because of the lemma we can define a topology on Proj S by taking the closed subsets to be the subsets of the form $V(\mathfrak{a})$.

Next we will define a sheaf of rings $\mathcal{O}$ on Proj S. For each $\mathfrak{p} \in \text{Proj } S$, we consider the ring $S_{(\mathfrak{p})}$ of elements of degree zero in the localized ring $T^{-1}S$, where T is the multiplicative system consisting of all *homogeneous* elements of S which are not in $\mathfrak{p}$. For any open subset $U \subseteq \text{Proj } S$, we define $\mathcal{O}(U)$ to be the set of functions $s: U \to \coprod S_{(\mathfrak{p})}$ such that for each $\mathfrak{p} \in U$, $s(\mathfrak{p}) \in S_{(\mathfrak{p})}$, and such that s is locally a quotient of elements of S: for each $\mathfrak{p} \in U$, there exists a neighborhood V of $\mathfrak{p}$ in U, and homogeneous elements a,f in S, of the same degree, such that for all $\mathfrak{q} \in V$, $f \notin \mathfrak{q}$, and $s(\mathfrak{q}) = a/f$ in $S_{(\mathfrak{q})}$. Now it is clear that $\mathcal{O}$ is a presheaf of rings, with the natural restrictions, and it is also clear from the local nature of the definition that $\mathcal{O}$ is a sheaf.

Definition. If S is any graded ring, we define (Proj $S, \mathcal{O}$) to be the topological space together with the sheaf of rings constructed above.

Proposition 2.5. *Let S be a graded ring.*

    (a) *For any $\mathfrak{p} \in \text{Proj } S$, the stalk $\mathcal{O}_\mathfrak{p}$ is isomorphic to the local ring $S_{(\mathfrak{p})}$.*

    (b) *For any homogeneous $f \in S_+$, let $D_+(f) = \{\mathfrak{p} \in \text{Proj } S \,|\, f \notin \mathfrak{p}\}$.*

Then $D_+(f)$ is open in Proj S. *Furthermore, these open sets cover* Proj S, *and for each such open set, we have an isomorphism of locally ringed spaces*

$$(D_+(f), \mathcal{O}|_{D_+(f)}) \cong \operatorname{Spec} S_{(f)},$$

where $S_{(f)}$ is the subring of elements of degree 0 in the localized ring S_f.

(c) Proj S *is a scheme.*

PROOF. Note first that (a) says that Proj S is a locally ringed space, and (b) tells us it is covered by open affine schemes, so (c) is a consequence of (a) and (b).

The proof of (a) is practically identical to the proof of (2.2a) above, so is left to the reader.

To prove (b), first note that $D_+(f) = \operatorname{Proj} S - V((f))$, so it is open. Since the elements of Proj S are those homogeneous prime ideals $\mathfrak{p}$ of S which do not contain all of S_+, it follows that the open sets $D_+(f)$ for homogeneous $f \in S_+$ cover Proj S. Now fix a homogeneous $f \in S_+$. We will define an isomorphism $(\varphi, \varphi^\#)$ of locally ringed spaces from $D_+(f)$ to $\operatorname{Spec} S_{(f)}$. There is a natural homomorphism of rings $S \to S_f$, and $S_{(f)}$ is a subring of S_f. For any homogeneous ideal $\mathfrak{a} \subseteq S$, let $\varphi(\mathfrak{a}) = (\mathfrak{a}S_f) \cap S_{(f)}$. In particular, if $\mathfrak{p} \in D_+(f)$, then $\varphi(\mathfrak{p}) \in \operatorname{Spec} S_{(f)}$, so this gives the map φ as sets. The properties of localization show that φ is bijective as a map from $D_+(f)$ to $\operatorname{Spec} S_{(f)}$. Furthermore, if $\mathfrak{a}$ is a homogeneous ideal of S, then $\mathfrak{p} \supseteq \mathfrak{a}$ if and only if $\varphi(\mathfrak{p}) \supseteq \varphi(\mathfrak{a})$. Hence φ is a homeomorphism. Note also if $\mathfrak{p} \in D_+(f)$, then the local rings $S_{(\mathfrak{p})}$ and $(S_{(f)})_{\varphi(\mathfrak{p})}$ are naturally isomorphic. These isomorphisms and the homeomorphism φ induce a natural map of sheaves $\varphi^\# : \mathcal{O}_{\operatorname{Spec} S_{(f)}} \to \varphi_*(\mathcal{O}_{\operatorname{Proj} S}|_{D_+(f)})$ which one recognizes immediately to be an isomorphism. Hence $(\varphi, \varphi^\#)$ is an isomorphism of locally ringed spaces, as required.

Example 2.5.1. If A is a ring, we define *projective n-space* over A to be the scheme $\mathbf{P}_A^n = \operatorname{Proj} A[x_0, \ldots, x_n]$. In particular, if A is an algebraically closed field k, then $\mathbf{P}_k^n$ is a scheme whose subspace of closed points is naturally homeomorphic to the *variety* called projective n-space—see (Ex. 2.14d) below.

Next we will show that the notion of scheme does in fact generalize the notion of variety. It is not quite true that a variety is a scheme. As we have already seen in the examples above, the underlying topological space of a scheme such as $\mathbf{A}_k^1$ or $\mathbf{A}_k^2$ has more points than the corresponding variety. However, we will show that there is a natural way of adding generic points (Ex. 2.9) for every irreducible subset of a variety so that the variety becomes a scheme.

To state our result, we need a definition.

Definition. Let S be a fixed scheme. A *scheme over* S is a scheme X, together
with a morphism $X \to S$. If X and Y are schemes over S, a morphism
of X to Y as schemes over S, (also called an S-morphism) is a morphism
$f : X \to Y$ which is compatible with the given morphisms to S. We denote
by $\mathfrak{Sch}(S)$ the category of schemes over S. If A is a ring, then by abuse of
notation we write $\mathfrak{Sch}(A)$ for the category of schemes over Spec A.

Proposition 2.6. *Let k be an algebraically closed field. There is a natural
fully faithful functor $t : \mathfrak{Var}(k) \to \mathfrak{Sch}(k)$ from the category of varieties over
k to schemes over k. For any variety V, its topological space is homeo-
morphic to the set of closed points of $\mathrm{sp}(t(V))$, and its sheaf of regular
functions is obtained by restricting the structure sheaf of $t(V)$ via this
homeomorphism.*

PROOF. To begin with, let X be any topological space, and let $t(X)$ be the
set of (nonempty) irreducible closed subsets of X. If Y is a closed subset of
X, then $t(Y) \subseteq t(X)$. Furthermore, $t(Y_1 \cup Y_2) = t(Y_1) \cup t(Y_2)$ and $t(\bigcap Y_i) = \bigcap t(Y_i)$. So we can define a topology on $t(X)$ by taking as closed sets the
subsets of the form $t(Y)$, where Y is a closed subset of X. If $f : X_1 \to X_2$ is a
continuous map, then we obtain a map $t(f) : t(X_1) \to t(X_2)$ by sending an
irreducible closed subset to the closure of its image. Thus t is a functor on
topological spaces. Furthermore, one can define a continuous map $\alpha : X \to t(X)$ by $\alpha(P) = \{P\}^-$. Note that α induces a *bijection* between the set of
open subsets of X and the set of open subsets of $t(X)$.

Now let k be an algebraically closed field. Let V be a variety over
k, and let $\mathcal{O}_V$ be its sheaf of regular functions (1.0.1). We will show that
$(t(V), \alpha_*(\mathcal{O}_V))$ is a scheme over k. Since any variety can be covered by open
affine subvarieties (I, 4.3), it will be sufficient to show that if V is affine,
then $(t(V), \alpha_*(\mathcal{O}_V))$ is a scheme. So let V be an affine variety with affine
coordinate ring A. We define a morphism of locally ringed spaces

$$\beta : (V, \mathcal{O}_V) \to X = \mathrm{Spec}\ A$$

as follows. For each point $P \in V$, let $\beta(P) = \mathfrak{m}_P$, the ideal of A consisting
of all regular functions which vanish at P. Then by (I, 3.2b), β is a bijection
of V onto the set of closed points of X. It is easy to see that β is a homeomorphism onto its image. Now for any open set $U \subseteq X$, we will define a
homomorphism of rings $\mathcal{O}_X(U) \to \beta_*(\mathcal{O}_V)(U) = \mathcal{O}_V(\beta^{-1}U)$. Given a section
$s \in \mathcal{O}_X(U)$, and given a point $P \in \beta^{-1}(U)$, we define $s(P)$ by taking the image
of s in the stalk $\mathcal{O}_{X, \beta(P)}$, which is isomorphic to the local ring $A_{\mathfrak{m}_P}$, and then
passing to the quotient ring $A_{\mathfrak{m}_P}/\mathfrak{m}_P$ which is isomorphic to the field k. Thus
s gives a function from $\beta^{-1}(U)$ to k. It is easy to see that this is a regular
function, and that this map gives an isomorphism $\mathcal{O}_X(U) \cong \mathcal{O}_V(\beta^{-1}U)$.
Finally, since the prime ideals of A are in 1-1 correspondence with the irreducible closed subsets of V (see (I, 1.4) and proof), these remarks show that
$(X, \mathcal{O}_X)$ is isomorphic to $(t(V), \alpha_* \mathcal{O}_V)$, so the latter is indeed an affine scheme.

To give a morphism of $(t(V), \alpha_* \mathcal{O}_V)$ to Spec k, we have only to give a homomorphism of rings $k \to \Gamma(t(V), \alpha_* \mathcal{O}_V) = \Gamma(V, \mathcal{O}_V)$. We send $\lambda \in k$ to the constant function λ on V. Thus $t(V)$ becomes a scheme over k. Finally, if V and W are two varieties, then one can check (Ex. 2.15) that the natural map

$$\text{Hom}_{\mathfrak{Var}(k)}(V, W) \to \text{Hom}_{\mathfrak{Sch}(k)}(t(V), t(W))$$

is bijective. This shows that the functor $t : \mathfrak{Var}(k) \to \mathfrak{Sch}(k)$ is fully faithful. In particular it implies that $t(V)$ is isomorphic to $t(W)$ if and only if V is isomorphic to W.

It is clear from the construction that $\alpha : V \to t(V)$ induces a homeomorphism from V onto the set of closed points of $t(V)$, with the induced topology.

Note. We will see later (4.10) what the image of the functor t is.

EXERCISES

2.1. Let A be a ring, let $X = \text{Spec } A$, let $f \in A$ and let $D(f) \subseteq X$ be the open complement of $V((f))$. Show that the locally ringed space $(D(f), \mathcal{O}_X|_{D(f)})$ is isomorphic to Spec A_f.

2.2. Let $(X, \mathcal{O}_X)$ be a scheme, and let $U \subseteq X$ be any open subset. Show that $(U, \mathcal{O}_X|_U)$ is a scheme. We call this the *induced scheme structure* on the open set U, and we refer to $(U, \mathcal{O}_X|_U)$ as an *open subscheme* of X.

2.3. *Reduced Schemes.* A scheme $(X, \mathcal{O}_X)$ is *reduced* if for every open set $U \subseteq X$, the ring $\mathcal{O}_X(U)$ has no nilpotent elements.
   (a) Show that $(X, \mathcal{O}_X)$ is reduced if and only if for every $P \in X$, the local ring $\mathcal{O}_{X,P}$ has no nilpotent elements.
   (b) Let $(X, \mathcal{O}_X)$ be a scheme. Let $(\mathcal{O}_X)_{\text{red}}$ be the sheaf associated to the presheaf $U \mapsto \mathcal{O}_X(U)_{\text{red}}$, where for any ring A, we denote by A_{red} the quotient of A by its ideal of nilpotent elements. Show that $(X, (\mathcal{O}_X)_{\text{red}})$ is a scheme. We call it the *reduced scheme* associated to X, and denote it by X_{red}. Show that there is a morphism of schemes $X_{\text{red}} \to X$, which is a homeomorphism on the underlying topological spaces.
   (c) Let $f : X \to Y$ be a morphism of schemes, and assume that X is reduced. Show that there is a unique morphism $g : X \to Y_{\text{red}}$ such that f is obtained by composing g with the natural map $Y_{\text{red}} \to Y$.

2.4. Let A be a ring and let $(X, \mathcal{O}_X)$ be a scheme. Given a morphism $f : X \to \text{Spec } A$, we have an associated map on sheaves $f^\# : \mathcal{O}_{\text{Spec } A} \to f_* \mathcal{O}_X$. Taking global sections we obtain a homomorphism $A \to \Gamma(X, \mathcal{O}_X)$. Thus there is a natural map

$$\alpha : \text{Hom}_{\mathfrak{Sch}}(X, \text{Spec } A) \to \text{Hom}_{\mathfrak{Rings}}(A, \Gamma(X, \mathcal{O}_X)).$$

Show that α is bijective (cf. (I, 3.5) for an analogous statement about varieties).

2.5. Describe Spec $\mathbf{Z}$, and show that it is a final object for the category of schemes, i.e., each scheme X admits a unique morphism to Spec $\mathbf{Z}$.

2.6. Describe the spectrum of the zero ring, and show that it is an initial object for the category of schemes. (According to our conventions, all ring homomorphisms must take 1 to 1. Since $0 = 1$ in the zero ring, we see that each ring R admits a unique homomorphism to the zero ring, but that there is no homomorphism from the zero ring to R unless $0 = 1$ in R.)

2.7. Let X be a scheme. For any $x \in X$, let $\mathcal{O}_x$ be the local ring at x, and $\mathfrak{m}_x$ its maximal ideal. We define the *residue field* of x on X to be the field $k(x) = \mathcal{O}_x/\mathfrak{m}_x$. Now let K be any field. Show that to give a morphism of Spec K to X it is equivalent to give a point $x \in X$ and an inclusion map $k(x) \to K$.

2.8. Let X be a scheme. For any point $x \in X$, we define the *Zariski tangent space* T_x to X at x to be the dual of the $k(x)$-vector space $\mathfrak{m}_x/\mathfrak{m}_x^2$. Now assume that X is a scheme over a field k, and let $k[\varepsilon]/\varepsilon^2$ be the *ring of dual numbers* over k. Show that to give a k-morphism of Spec $k[\varepsilon]/\varepsilon^2$ to X is equivalent to giving a point $x \in X$, *rational over k* (i.e., such that $k(x) = k$), and an element of T_x.

2.9. If X is a topological space, and Z an irreducible closed subset of X, a *generic point* for Z is a point ζ such that $Z = \{\zeta\}^-$. If X is a scheme, show that every (nonempty) irreducible closed subset has a unique generic point.

2.10. Describe Spec $\mathbf{R}[x]$. How does its topological space compare to the set $\mathbf{R}$? To $\mathbf{C}$?

2.11. Let $k = \mathbf{F}_p$ be the finite field with p elements. Describe Spec $k[x]$. What are the residue fields of its points? How many points are there with a given residue field?

2.12. *Glueing Lemma.* Generalize the glueing procedure described in the text (2.3.5) as follows. Let $\{X_i\}$ be a family of schemes (possible infinite). For each $i \neq j$, suppose given an open subset $U_{ij} \subseteq X_i$, and let it have the induced scheme structure (Ex. 2.2). Suppose also given for each $i \neq j$ an isomorphism of schemes $\varphi_{ij}: U_{ij} \to U_{ji}$ such that (1) for each i,j, $\varphi_{ji} = \varphi_{ij}^{-1}$, and (2) for each i,j,k, $\varphi_{ij}(U_{ij} \cap U_{ik}) = U_{ji} \cap U_{jk}$, and $\varphi_{ik} = \varphi_{jk} \circ \varphi_{ij}$ on $U_{ij} \cap U_{ik}$. Then show that there is a scheme X, together with morphisms $\psi_i: X_i \to X$ for each i, such that (1) ψ_i is an isomorphism of X_i onto an open subscheme of X, (2) the $\psi_i(X_i)$ cover X, (3) $\psi_i(U_{ij}) = \psi_i(X_i) \cap \psi_j(X_j)$ and (4) $\psi_i = \psi_j \circ \varphi_{ij}$ on U_{ij}. We say that X is obtained by *glueing* the schemes X_i along the isomorphisms φ_{ij}. An interesting special case is when the family X_i is arbitrary, but the U_{ij} and φ_{ij} are all empty. Then the scheme X is called the *disjoint union* of the X_i, and is denoted $\coprod X_i$.

2.13. A topological space is *quasi-compact* if every open cover has a finite subcover.
   (a) Show that a topological space is noetherian (I, §1) if and only if every open subset is quasi-compact.
   (b) If X is an affine scheme, show that sp(X) is quasi-compact, but not in general noetherian. We say a scheme X is *quasi-compact* if sp(X) is.
   (c) If A is a noetherian ring, show that sp(Spec A) is a noetherian topological space.
   (d) Give an example to show that sp(Spec A) can be noetherian even when A is not.

2.14. (a) Let S be a graded ring. Show that Proj $S = \varnothing$ if and only if every element of S_+ is nilpotent.
   (b) Let $\varphi: S \to T$ be a graded homomorphism of graded rings (preserving degrees). Let $U = \{\mathfrak{p} \in \text{Proj } T \mid \mathfrak{p} \not\supseteq \varphi(S_+)\}$. Show that U is an open subset of Proj T, and show that φ determines a natural morphism $f: U \to \text{Proj } S$.

(c) The morphism f can be an isomorphism even when φ is not. For example, suppose that $\varphi_d : S_d \to T_d$ is an isomorphism for all $d \geqslant d_0$, where d_0 is an integer. Then show that $U = \operatorname{Proj} T$ and the morphism $f : \operatorname{Proj} T \to \operatorname{Proj} S$ is an isomorphism.

(d) Let V be a projective variety with homogeneous coordinate ring S (I, §2). Show that $t(V) \cong \operatorname{Proj} S$.

2.15. (a) Let V be a variety over the algebraically closed field k. Show that a point $P \in t(V)$ is a closed point if and only if its residue field is k.

(b) If $f : X \to Y$ is a morphism of schemes over k, and if $P \in X$ is a point with residue field k, then $f(P) \in Y$ also has residue field k.

(c) Now show that if V, W are any two varieties over k, then the natural map

$$\operatorname{Hom}_{\mathfrak{Var}}(V, W) \to \operatorname{Hom}_{\mathfrak{Sch}/k}(t(V), t(W))$$

is bijective. (Injectivity is easy. The hard part is to show it is surjective.)

2.16. Let X be a scheme, let $f \in \Gamma(X, \mathcal{O}_X)$, and define X_f to be the subset of points $x \in X$ such that the stalk f_x of f at x is not contained in the maximal ideal $\mathfrak{m}_x$ of the local ring $\mathcal{O}_x$.

(a) If $U = \operatorname{Spec} B$ is an open *affine* subscheme of X, and if $\bar{f} \in B = \Gamma(U, \mathcal{O}_X|_U)$ is the restriction of f, show that $U \cap X_f = D(\bar{f})$. Conclude that X_f is an open subset of X.

(b) Assume that X is quasi-compact. Let $A = \Gamma(X, \mathcal{O}_X)$, and let $a \in A$ be an element whose restriction to X_f is 0. Show that for some $n > 0$, $f^n a = 0$. [*Hint:* Use an open affine cover of X.]

(c) Now assume that X has a finite cover by open affines U_i such that each intersection $U_i \cap U_j$ is quasi-compact. (This hypothesis is satisfied, for example, if $\operatorname{sp}(X)$ is noetherian.) Let $b \in \Gamma(X_f, \mathcal{O}_{X_f})$. Show that for some $n > 0$, $f^n b$ is the restriction of an element of A.

(d) With the hypothesis of (c), conclude that $\Gamma(X_f, \mathcal{O}_{X_f}) \cong A_f$.

2.17. *A Criterion for Affineness.*

(a) Let $f : X \to Y$ be a morphism of schemes, and suppose that Y can be covered by open subsets U_i, such that for each i, the induced map $f^{-1}(U_i) \to U_i$ is an isomorphism. Then f is an isomorphism.

(b) A scheme X is affine if and only if there is a finite set of elements $f_1, \ldots, f_r \in A = \Gamma(X, \mathcal{O}_X)$, such that the open subsets X_{f_i} are affine, and $f_1, \ldots, f_r$ generate the unit ideal in A. [*Hint:* Use (Ex. 2.4) and (Ex. 2.16d) above.]

2.18. In this exercise, we compare some properties of a ring homomorphism to the induced morphism of the spectra of the rings.

(a) Let A be a ring, $X = \operatorname{Spec} A$, and $f \in A$. Show that f is nilpotent if and only if $D(f)$ is empty.

(b) Let $\varphi : A \to B$ be a homomorphism of rings, and let $f : Y = \operatorname{Spec} B \to X = \operatorname{Spec} A$ be the induced morphism of affine schemes. Show that φ is injective if and only if the map of sheaves $f^\# : \mathcal{O}_X \to f_* \mathcal{O}_Y$ is injective. Show furthermore in that case f is *dominant*, i.e., $f(Y)$ is dense in X.

(c) With the same notation, show that if φ is surjective, then f is a homeomorphism of Y onto a closed subset of X, and $f^\# : \mathcal{O}_X \to f_* \mathcal{O}_Y$ is surjective.

    (d) Prove the converse to (c), namely, if $f: Y \to X$ is a homeomorphism onto a closed subset, and $f^\#: \mathcal{O}_X \to f_* \mathcal{O}_Y$ is surjective, then φ is surjective. [*Hint*: Consider $X' = \operatorname{Spec}(A/\ker \varphi)$ and use (b) and (c).]

2.19. Let A be a ring. Show that the following conditions are equivalent:

    (i) Spec A is disconnected;
    (ii) there exist nonzero elements $e_1, e_2 \in A$ such that $e_1 e_2 = 0$, $e_1^2 = e_1$, $e_2^2 = e_2$, $e_1 + e_2 = 1$ (these elements are called *orthogonal idempotents*);
    (iii) A is isomorphic to a direct product $A_1 \times A_2$ of two nonzero rings.

3 First Properties of Schemes

In this section we will give some of the first properties of schemes. In particular we will discuss open and closed subschemes, and products of schemes. In the exercises we introduce the notion of constructible subsets, and study the dimension of the fibres of a morphism.

Definition. A scheme is *connected* if its topological space is connected. A scheme is *irreducible* if its topological space is irreducible.

Definition. A scheme X is *reduced* if for every open set U, the ring $\mathcal{O}_X(U)$ has no nilpotent elements. Equivalently (Ex. 2.3), X is reduced if and only if the local rings $\mathcal{O}_P$, for all $P \in X$, have no nilpotent elements.

Definition. A scheme X is *integral* if for every open set $U \subseteq X$, the ring $\mathcal{O}_X(U)$ is an integral domain.

Example 3.0.1. If $X = \operatorname{Spec} A$ is an affine scheme, then X is irreducible if and only if the nilradical nil A of A is prime; X is reduced if and only if nil $A = 0$; and X is integral if and only if A is an integral domain.

Proposition 3.1. *A scheme is integral if and only if it is both reduced and irreducible.*

PROOF. Clearly an integral scheme is reduced. If X is not irreducible, then one can find two nonempty disjoint open subsets U_1 and U_2. Then $\mathcal{O}(U_1 \cup U_2) = \mathcal{O}(U_1) \times \mathcal{O}(U_2)$ which is not an integral domain. Thus integral implies irreducible.

    Conversely, suppose that X is reduced and irreducible. Let $U \subseteq X$ be an open subset, and suppose that there are elements $f, g \in \mathcal{O}(U)$ with $fg = 0$. Let $Y = \{x \in U \,|\, f_x \in \mathfrak{m}_x\}$, and let $Z = \{x \in U \,|\, g_x \in \mathfrak{m}_x\}$. Then Y and Z are closed subsets (Ex. 2.16a), and $Y \cup Z = U$. But X is irreducible, so U is irreducible, so one of Y or Z is equal to U, say $Y = U$. But then the restriction of f to any open affine subset of U will be nilpotent (Ex. 2.18a), hence zero, so f is zero. This shows that X is integral.

Definition. A scheme X is *locally noetherian* if it can be covered by open affine subsets Spec A_i, where each A_i is a noetherian ring. X is *noetherian* if it is locally noetherian and quasi-compact. Equivalently, X is noetherian if it can be covered by a finite number of open affine subsets Spec A_i, with each A_i a noetherian ring.

Caution 3.1.1. If X is a noetherian scheme, then $\mathrm{sp}(X)$ is a noetherian topological space, but not conversely (Ex. 2.13) and (Ex. 3.17).

Note that in this definition we do not require that every open affine subset be the spectrum of a noetherian ring. So while it is obvious from the definition that the spectrum of a noetherian ring is a noetherian scheme, the converse is not obvious. It is a question of showing that the noetherian property is a "local property". We will often encounter similar situations later in defining properties of a scheme or of a morphism of schemes, so we will give a careful statement and proof of the local nature of the noetherian property, to illustrate this type of situation.

Proposition 3.2. *A scheme X is locally noetherian if and only if for every open affine subset $U =$ Spec A, A is a noetherian ring. In particular, an affine scheme $X =$ Spec A is a noetherian scheme if and only if the ring A is a noetherian ring.*

PROOF. The "if" part follows from the definition, so we have to show if X is locally noetherian, and if $U =$ Spec A is an open affine subset, then A is a noetherian ring. First note that if B is a noetherian ring, so is any localization B_f. The open subsets $D(f) \cong$ Spec B_f form a base for the topology of Spec B. Hence on a locally noetherian scheme X there is a base for the topology consisting of the spectra of noetherian rings. In particular, our open set U can be covered by spectra of noetherian rings.

So we have reduced to proving the following statement: let $X =$ Spec A be an affine scheme, which can be covered by open subsets which are spectra of noetherian rings. Then A is noetherian. Let $U =$ Spec B be an open subset of X, with B noetherian. Then for some $f \in A$, $D(f) \subseteq U$. Let $\bar{f}$ be the image of f in B. Then $A_f \cong B_{\bar{f}}$, hence A_f is noetherian. So we can cover X by open subsets $D(f) \cong$ Spec A_f with A_f noetherian. Since X is quasi-compact, a finite number will do.

So now we have reduced to a purely algebraic problem: A is a ring, $f_1, \ldots, f_r$ are a finite number of elements of A, which generate the unit ideal, and each localization A_{f_i} is noetherian. We have to show A is noetherian. First we establish a lemma. Let $\mathfrak{a} \subseteq A$ be an ideal, and let $\varphi_i : A \to A_{f_i}$ be the localization map, $i = 1, \ldots, r$. Then

$$\mathfrak{a} = \bigcap \varphi_i^{-1}(\varphi_i(\mathfrak{a}) \cdot A_{f_i}).$$

The inclusion $\subseteq$ is obvious. Conversely, given an element $b \in A$ contained in this intersection, we can write $\varphi_i(b) = a_i/f_i^{n_i}$ in A_{f_i} for each i, where $a_i \in \mathfrak{a}$, and $n_i > 0$. Increasing the n_i if necessary, we can make them all equal to a fixed n. This means that in A we have

$$f_i^{m_i}(f_i^n b - a_i) = 0$$

for some m_i. And as before, we can make all the $m_i = m$. Thus $f_i^{m+n}b \in \mathfrak{a}$ for each i. Since $f_1, \ldots, f_r$ generate the unit ideal, the same is true of their Nth powers for any N. Take $N = n + m$. Then we have $1 = \sum c_i f_i^N$ for suitable $c_i \in A$. Hence

$$b = \sum c_i f_i^N b \in \mathfrak{a}$$

as required.

Now we can easily show that A is noetherian. Let $\mathfrak{a}_1 \subseteq \mathfrak{a}_2 \subseteq \ldots$ be an ascending chain of ideals in A. Then for each i,

$$\varphi_i(\mathfrak{a}_1) \cdot A_{f_i} \subseteq \varphi_i(\mathfrak{a}_2) \cdot A_{f_i} \subseteq \ldots$$

is an ascending chain of ideals in A_{f_i}, which must become stationary because A_{f_i} is noetherian. There are only finitely many A_{f_i}, so from the lemma we conclude that the original chain is eventually stationary, and hence A is noetherian.

Definition. A morphism $f : X \to Y$ of schemes is *locally of finite type* if there exists a covering of Y by open affine subsets $V_i = \operatorname{Spec} B_i$, such that for each i, $f^{-1}(V_i)$ can be covered by open affine subsets $U_{ij} = \operatorname{Spec} A_{ij}$, where each A_{ij} is a finitely generated B_i-algebra. The morphism f is *of finite type* if in addition each $f^{-1}(V_i)$ can be covered by a finite number of the U_{ij}.

Definition. A morphism $f : X \to Y$ is a *finite* morphism if there exists a covering of Y by open affine subsets $V_i = \operatorname{Spec} B_i$, such that for each i, $f^{-1}(V_i)$ is affine, equal to $\operatorname{Spec} A_i$, where A_i is a B_i-algebra which is a finitely generated B_i-module.

Note in each of these definitions that a property of a morphism $f : X \to Y$ is defined by the existence of an open affine cover of Y with certain properties. In fact in each case it is equivalent to require the given property for every open affine subset of Y (Ex. 3.1–3.4).

Example 3.2.1. If V is a variety over an algebraically closed field k, then the associated scheme $t(V)$ (see (2.6)) is an integral noetherian scheme of finite type over k. Indeed, V can be covered by a finite number of open affine subvarieties (I, 4.3), so $t(V)$ can be covered by a finite number of open affines of the form $\operatorname{Spec} A_i$, where each A_i is an integral domain which is a finitely generated k-algebra and hence noetherian.

Example 3.2.2. If P is a point of a variety V, with local ring $\mathcal{O}_P$, then Spec $\mathcal{O}_P$ is an integral noetherian scheme, which is not in general of finite type over k.

Next we come to open and closed subschemes.

Definition. An *open subscheme* of a scheme X is a scheme U, whose topological space is an open subset of X, and whose structure sheaf $\mathcal{O}_U$ is isomorphic to the restriction $\mathcal{O}_X|_U$ of the structure sheaf of X. An *open immersion* is a morphism $f: X \to Y$ which induces an isomorphism of X with an open subscheme of Y.

Note that every open subset of a scheme carries a unique structure of open subscheme (Ex. 2.2).

Definition. A *closed immersion* is a morphism $f: Y \to X$ of schemes such that f induces a homeomorphism of $\mathrm{sp}(Y)$ onto a closed subset of $\mathrm{sp}(X)$, and furthermore the induced map $f^{\#}: \mathcal{O}_X \to f_*\mathcal{O}_Y$ of sheaves on X is surjective. A *closed subscheme* of a scheme X is an equivalence class of closed immersions, where we say $f: Y \to X$ and $f': Y' \to X$ are equivalent if there is an isomorphism $i: Y' \to Y$ such that $f' = f \circ i$.

Example 3.2.3. Let A be a ring, and let $\mathfrak{a}$ be an ideal of A. Let $X = $ Spec A and let $Y = $ Spec $A/\mathfrak{a}$. Then the ring homomorphism $A \to A/\mathfrak{a}$ induces a morphism of schemes $f: Y \to X$ which is a closed immersion. The map f is a homeomorphism of Y onto the closed subset $V(\mathfrak{a})$ of X, and the map of structure sheaves $\mathcal{O}_X \to f_*\mathcal{O}_Y$ is surjective because it is surjective on the stalks, which are localizations of A and $A/\mathfrak{a}$, respectively (Ex. 2.18).

Thus for any ideal $\mathfrak{a} \subseteq A$ we obtain a structure of closed subscheme on the closed set $V(\mathfrak{a}) \subseteq X$. In particular, every closed subset Y of X has many closed subscheme structures, corresponding to all the ideals $\mathfrak{a}$ for which $V(\mathfrak{a}) = Y$. In fact, every closed subscheme structure on a closed subset Y of an affine scheme X arises from an ideal in this way (Ex. 3.11b) or (5.10).

Example 3.2.4. For some more specific examples, let $A = k[x, y]$, where k is a field. Then Spec $A = \mathbf{A}_k^2$ is the affine plane over k. The ideal $\mathfrak{a} = (xy)$ gives a reducible subscheme, consisting of the union of the x and y axes. The ideal $\mathfrak{a} = (x^2)$ gives a subscheme structure with nilpotents on the y-axis. The ideal $\mathfrak{a} = (x^2, xy)$ gives another subscheme structure on the y-axis, this one having nilpotents only in the local ring at the origin. We say the origin is an *embedded point* for this subscheme.

Example 3.2.5. Let V be an affine variety over the field k, and let W be a closed subvariety. Then W corresponds to a prime ideal $\mathfrak{p}$ in the affine coordinate ring A of V (I, §1). Let $X = t(V)$ and $Y = t(W)$ be the associated schemes. Then $X = $ Spec A and Y is the closed subscheme defined by $\mathfrak{p}$. For each $n \geq 1$ let Y_n be the closed subscheme of X corresponding to the

ideal p^n. Then $Y_1 = Y$, but for each $n > 1$, Y_n is a nonreduced scheme structure on the closed set Y, which does not correspond to any subvariety of V.
We call Y_n the nth *infinitesimal neighborhood* of Y in X. The schemes Y_n reflect
properties of the embedding of Y in X. Later (§9) we will study the "formal
completion" of Y in X, which is roughly the limit of the schemes Y_n as $n \to \infty$.

Example 3.2.6. Let X be a scheme, and let Y be a closed subset. In general Y
will have many possible closed subscheme structures. However, there is one
which is "smaller" than any other, called the *reduced induced closed subscheme
structure*, which we now describe.

First let $X = \text{Spec } A$ be an affine scheme, and let Y be a closed subset.
Let $\mathfrak{a} \subseteq A$ be the ideal obtained by intersecting all the prime ideals in Y. This
is the largest ideal for which $V(\mathfrak{a}) = Y$. Then we take the reduced induced
structure on Y to be the one defined by $\mathfrak{a}$.

Now let X be any scheme, and let Y be a closed subset. For each open
affine subset $U_i \subseteq X$, consider the closed subset $Y_i = Y \cap U_i$ of U_i, and give
it the reduced induced structure just defined for affines (which may depend
on U_i). I claim that for any i,j, the restrictions to $Y_i \cap Y_j$ of the two structure
sheaves just defined on Y_i and Y_j are isomorphic, and furthermore, that the
three such isomorphisms on $Y_i \cap Y_j \cap Y_k$ are compatible for all i,j,k. One
reduces easily to showing that if $U = \text{Spec } A$ is an open affine, and if $f \in A$,
and if $V = D(f) = \text{Spec } A_f$, then the reduced induced structure on $Y \cap U$
obtained from A when restricted to $Y \cap V$ agrees with the one obtained
from A_f. This corresponds to the algebraic fact that if $\mathfrak{a}$ is the intersection
of those prime ideals of A which are in Y, then $\mathfrak{a}A_f$ is the intersection of those
prime ideals of A_f which are in $Y \cap D(f)$.

So now we can glue the sheaves defined on the Y_i to obtain a sheaf on Y
(Ex. 1.22), which gives us the desired reduced induced subscheme structure
on Y. See (Ex. 3.11) below for a universal property of the reduced induced
subscheme structure.

Definition. The *dimension* of a scheme X, denoted dim X, is its dimension as a
topological space (I, §1). If Z is an irreducible closed subset of X, then the
codimension of Z in X, denoted codim(Z,X) is the supremum of integers n
such that there exists a chain

$$Z = Z_0 < Z_1 < \ldots < Z_n$$

of distinct closed irreducible subsets of X, beginning with Z. If Y is any
closed subset of X, we define

$$\text{codim}(Y,X) = \inf_{Z \subseteq Y} \text{codim}(Z,X)$$

where the infimum is taken over all closed irreducible subsets of Y.

Example 3.2.7. If $X = \text{Spec } A$ is an affine scheme, then the dimension of X
is the same as the Krull dimension of A (I, §1).

Caution 3.2.8. Be careful in applying the concepts of dimension and codimension to arbitrary schemes. Our intuition is derived from working with schemes of finite type over a field, where these notions are well-behaved. For example, if X is an affine integral scheme of finite type over a field k, and if $Y \subseteq X$ is any closed irreducible subset, then (I, 1.8A) implies that $\dim Y + \text{codim}(Y,X) = \dim X$. But on arbitrary (even noetherian) schemes, funny things can happen. See (Ex. 3.20–3.22), and also Nagata [7], and Grothendieck [EGA IV, §5].

Definition. Let S be a scheme, and let X, Y be schemes over S, i.e., schemes with morphisms to S. We define the *fibred product* of X and Y over S, denoted $X \times_S Y$, to be a scheme, together with morphisms $p_1: X \times_S Y \to X$ and $p_2: X \times_S Y \to Y$, which make a commutative diagram with the given morphisms $X \to S$ and $Y \to S$, such that given any scheme Z over S, and given morphisms $f: Z \to X$ and $g: Z \to Y$ which make a commutative diagram with the given morphisms $X \to S$ and $Y \to S$, then there exists a unique morphism $\theta: Z \to X \times_S Y$ such that $f = p_1 \circ \theta$, and $g = p_2 \circ \theta$. The morphisms p_1 and p_2 are called the *projection morphisms* of the fibred product onto its factors.

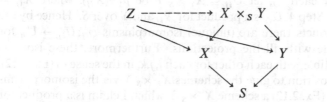

If X and Y are schemes given without reference to any base scheme S, we take $S = \text{Spec } \mathbf{Z}$ (Ex. 2.5) and define the *product* of X and Y, denoted $X \times Y$, to be $X \times_{\text{Spec } \mathbf{Z}} Y$.

Theorem 3.3. *For any two schemes X and Y over a scheme S, the fibred product $X \times_S Y$ exists, and is unique up to unique isomorphism.*

PROOF. The idea is first to construct products for affine schemes and then glue. We proceed in seven steps.

Step 1. Let $X = \text{Spec } A$, $Y = \text{Spec } B$, $S = \text{Spec } R$ all be affine. Then A and B are R-algebras, and I claim that $\text{Spec}(A \otimes_R B)$ is a product for X and Y over S. Indeed, for any scheme Z, to give a morphism of Z to $\text{Spec}(A \otimes_R B)$ is the same as to give a homomorphism of the ring $A \otimes_R B$ into the ring $\Gamma(Z, \mathcal{O}_Z)$, by (Ex. 2.4). But to give a homomorphism of $A \otimes_R B$ into any ring is the same as to give homomorphisms of A and B into that ring, inducing the same homomorphism on R. Applying (Ex. 2.4) again, we see that to give a morphism of Z into $\text{Spec}(A \otimes_R B)$ is the same as giving morphisms of Z into X and into Y, which give rise to the same morphism of Z into S. Thus $\text{Spec}(A \otimes_R B)$ is the desired product.

Step 2. It follows immediately from the universal property of the product that it is unique up to unique isomorphism, if it exists. We will need this uniqueness for those products already constructed, as we go along.

Step 3. Glueing morphisms. We have already seen how to glue sheaves (Ex. 1.22) and how to glue schemes (Ex. 2.12). Now we glue morphisms. If X and Y are schemes, then to give a morphism f from X to Y, it is equivalent to give an open cover $\{U_i\}$ of X, together with morphisms $f_i: U_i \to Y$, where U_i has the induced open subscheme structure, such that the restrictions of f_i and f_j to $U_i \cap U_j$ are the same, for each i,j. The proof is straightforward.

Step 4. If X,Y are schemes over a scheme S, if $U \subseteq X$ is an open subset, and if the product $X \times_S Y$ exists, then $p_1^{-1}(U) \subseteq X \times_S Y$ is a product for U and Y over S. Indeed, given a scheme Z, and morphisms $f:Z \to U$ and $g:Z \to Y$, f determines a map of Z to X by composing with the inclusion $U \subseteq X$. Hence there is a map $\theta:Z \to X \times_S Y$ compatible with f,g and the projections. But since $f(Z) \subseteq U$, we have $\theta(Z) \subseteq p_1^{-1}(U)$. So θ can be regarded as a morphism $Z \to p_1^{-1}(U)$. It is clearly unique, so $p_1^{-1}(U)$ is a product $U \times_S Y$.

Step 5. Suppose given X,Y schemes over S, suppose $\{X_i\}$ is an open covering of X, and suppose that for each i, $X_i \times_S Y$ exists. Then $X \times_S Y$ exists. Indeed, for each i,j, let $U_{ij} \subseteq X_i \times_S Y$ be $p_1^{-1}(X_{ij})$, where $X_{ij} = X_i \cap X_j$. Then by Step 4, U_{ij} is a product for X_{ij} and Y over S. Hence by the uniqueness of products there are (unique) isomorphisms $\varphi_{ij}: U_{ij} \to U_{ji}$ for each i,j compatible with all the projections. Furthermore, these isomorphisms are compatible with each other for each i,j,k, in the sense of (Ex. 2.12). Thus we are in a position to glue the schemes $X_i \times_S Y$ via the isomorphisms φ_{ij}. We obtain by (Ex. 2.12) a scheme $X \times_S Y$ which I claim is a product for X and Y over S. The projection morphisms p_1 and p_2 are defined by glueing the projections from the pieces $X_i \times_S Y$ (Step 3). Given a scheme Z and morphisms $f:Z \to X$, $g:Z \to Y$, let $Z_i = f^{-1}(X_i)$. Then we get maps $\theta_i:Z_i \to X_i \times_S Y$, hence by composition with the inclusions $X_i \times_S Y \subseteq X \times_S Y$ we get maps $\theta_i:Z_i \to X \times_S Y$. One verifies that these maps agree on $Z_i \cap Z_j$, so we can glue the morphisms (Step 3) to obtain a morphism $\theta:Z \to X \times_S Y$, compatible with the projections and f and g. The uniqueness of θ can be checked locally.

Step 6. We know from Step 1 that if X,Y, S are all affine, then $X \times_S Y$ exists. Thus using Step 5 we conclude that for any X, but Y, S affine, the product exists. Using Step 5 again, with X and Y interchanged, we find that the product exists for any X and any Y over an affine S.

Step 7. Given arbitrary X,Y, S, let $q:X \to S$ and $r:Y \to S$ be the given morphisms. Let S_i be an open affine cover of S. Let $X_i = q^{-1}(S_i)$ and let $Y_i = r^{-1}(S_i)$. Then by Step 6, $X_i \times_{S_i} Y_i$ exists. Note that this same scheme is a product for X_i and Y over S. Indeed, given morphisms $f:Z \to X_i$ and $g:Z \to Y$ over S, the image of g must land inside Y_i. Thus $X_i \times_S Y$ exists for each i, and one more application of Step 5 gives us $X \times_S Y$. This completes the proof.

Perhaps this is a good place to make some general remarks on the importance and uses of fibred products. To begin with, we can define the fibres of a morphism.

Definition. Let $f: X \to Y$ be a morphism of schemes, and let $y \in Y$ be a point. Let $k(y)$ be the residue field of y, and let Spec $k(y) \to Y$ be the natural morphism (Ex. 2.7). Then we define the *fibre* of the morphism f over the point y to be the scheme

$$X_y = X \times_Y \text{Spec } k(y).$$

The fibre X_y is a scheme over $k(y)$, and one can show that its underlying topological space is homeomorphic to the subset $f^{-1}(y)$ of X (Ex. 3.10).

The notion of the fibre of a morphism allows us to regard a morphism as a family of schemes (namely its fibres) parametrized by the points of the image scheme. Conversely, this notion of family is a good way of making sense of the idea of a family of schemes varying algebraically. For example, given a scheme X_0 over a field k, we define a *family of deformations* of X_0 to be a morphism $f: X \to Y$ with Y connected, together with a point $y_0 \in Y$, such that $k(y_0) = k$, and $X_{y_0} \cong X_0$. The other fibres X_y of f are called *deformations* of X_0.

An interesting kind of family arises when we have a scheme X over Spec **Z**. In this case, taking the fibre over the generic point gives a scheme $X_\mathbf{Q}$ over **Q**, while taking the fibre over a closed point, corresponding to a prime number p, gives a scheme X_p over the finite field $\mathbf{F}_p$. We say that X_p arises by *reduction* mod p of the scheme X.

Another important application of fibred products is to the notion of base extension. Let S be a fixed scheme which we think of as a *base scheme*, meaning that we are interested in the category of schemes over S. For example, think of $S = \text{Spec } k$, where k is a field. If S' is another base scheme, and if $S' \to S$ is a morphism, then for any scheme X over S, we let $X' = X \times_S S'$, which will be a scheme over S'. We say that X' is obtained from X by making a *base extension* $S' \to S$. For example, think of $S' = \text{Spec } k'$ where k' is an extension field of k. Note, by the way, that base extension is a transitive operation: if $S'' \to S' \to S$ are two morphisms, then $(X \times_S S') \times_{S'} S'' \cong X \times_S S''$.

This ties in with a general philosophy, emphasized by Grothendieck in his "Eléments de Géométrie Algébrique" ([EGA]), that one should try to develop all concepts of algebraic geometry in a relative context. Instead of always working over a fixed base field, and considering properties of one variety at a time, one should consider a morphism of schemes $f: X \to S$, and study properties of the morphism. It then becomes important to study the behavior of properties of f under base extension, and in particular, to relate properties of f to properties of the fibres of f. For example, if $f: X \to S$

is a morphism of finite type, and if $S' \to S$ is any base extension, then $f': X' \to S'$ is also a morphism of finite type, where $X' = X \times_S S'$. Hence we say the property of a morphism f being of finite type is *stable under base extension*. On the other hand, if for example $f: X \to S$ is a morphism of integral schemes, the fibres of f may be neither irreducible nor reduced. So the property of a scheme being integral is not stable under base extension.

Example 3.3.1. Let k be an algebraically closed field, let

$$X = \operatorname{Spec} k[x,y,t]/(ty - x^2),$$

let $Y = \operatorname{Spec} k[t]$, and let $f: X \to Y$ be the morphism determined by the natural homomorphism $k[t] \to k[x,y,t]/(ty - x^2)$. Then X and Y are integral schemes of finite type over k, and f is a surjective morphism. We identify the closed points of Y with elements of k. For $a \in k$, $a \neq 0$, the fibre X_a is the plane curve $ay = x^2$ in $\mathbf{A}_k^2$, which is an irreducible, reduced curve. But for $a = 0$, the fibre X_0 is the nonreduced scheme given by $x^2 = 0$ in $\mathbf{A}^2$. Thus we have a family (Fig. 7) in which most members are irreducible curves, but one is nonreduced. This shows how nonreduced schemes occur naturally even if one is primarily interested in varieties. We can say that the nonreduced scheme $x^2 = 0$ in $\mathbf{A}^2$ is a deformation of the irreducible parabola $ay = x^2$ as $a \to 0$.

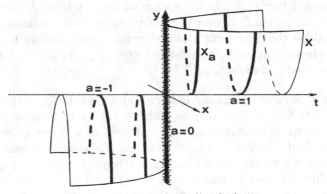

Figure 7. An algebraic family of schemes.

Example 3.3.2. Similarly, if $X = \operatorname{Spec} k[x,y,t]/(xy - t)$, we get a family whose general member X_a is an irreducible hyperbola $xy = a$, when $a \neq 0$, but whose special member X_0 is the reducible scheme $xy = 0$ consisting of two lines.

EXERCISES

3.1. Show that a morphism $f: X \to Y$ is locally of finite type if and only if for *every* open affine subset $V = \operatorname{Spec} B$ of Y, $f^{-1}(V)$ can be covered by open affine subsets $U_j = \operatorname{Spec} A_j$, where each A_j is a finitely generated B-algebra.

3.2. A morphism $f : X \to Y$ of schemes is *quasi-compact* if there is a cover of Y by open affines V_i such that $f^{-1}(V_i)$ is quasi-compact for each i. Show that f is quasi-compact if and only if for *every* open affine subset $V \subseteq Y$, $f^{-1}(V)$ is quasi-compact.

3.3. (a) Show that a morphism $f : X \to Y$ is of finite type if and only if it is locally of finite type and quasi-compact.

(b) Conclude from this that f is of finite type if and only if for *every* open affine subset $V = \operatorname{Spec} B$ of Y, $f^{-1}(V)$ can be covered by a finite number of open affines $U_j = \operatorname{Spec} A_j$, where each A_j is a finitely generated B-algebra.

(c) Show also if f is of finite type, then for *every* open affine subset $V = \operatorname{Spec} B \subseteq Y$, and for *every* open affine subset $U = \operatorname{Spec} A \subseteq f^{-1}(V)$, A is a finitely generated B-algebra.

3.4. Show that a morphism $f : X \to Y$ is finite if and only if for *every* open affine subset $V = \operatorname{Spec} B$ of Y, $f^{-1}(V)$ is affine, equal to $\operatorname{Spec} A$, where A is a finite B-module.

3.5. A morphism $f : X \to Y$ is *quasi-finite* if for every point $y \in Y$, $f^{-1}(y)$ is a finite set.

(a) Show that a finite morphism is quasi-finite.

(b) Show that a finite morphism is *closed*, i.e., the image of any closed subset is closed.

(c) Show by example that a surjective, finite-type, quasi-finite morphism need not be finite.

3.6. Let X be an integral scheme. Show that the local ring $\mathcal{O}_\xi$ of the generic point ξ of X is a field. It is called the *function field* of X, and is denoted by $K(X)$. Show also that if $U = \operatorname{Spec} A$ is any open affine subset of X, then $K(X)$ is isomorphic to the quotient field of A.

3.7. A morphism $f : X \to Y$, with Y irreducible, is *generically finite* if $f^{-1}(\eta)$ is a finite set, where η is the generic point of Y. A morphism $f : X \to Y$ is *dominant* if $f(X)$ is dense in Y. Now let $f : X \to Y$ be a dominant, generically finite morphism of finite type of integral schemes. Show that there is an open dense subset $U \subseteq Y$ such that the induced morphism $f^{-1}(U) \to U$ is finite. [*Hint*: First show that the function field of X is a finite field extension of the function field of Y.]

3.8. *Normalization.* A scheme is *normal* if all of its local rings are integrally closed domains. Let X be an integral scheme. For each open affine subset $U = \operatorname{Spec} A$ of X, let $\tilde{A}$ be the integral closure of A in its quotient field, and let $\tilde{U} = \operatorname{Spec} \tilde{A}$. Show that one can glue the schemes $\tilde{U}$ to obtain a normal integral scheme $\tilde{X}$, called the *normalization* of X. Show also that there is a morphism $\tilde{X} \to X$, having the following universal property: for every normal integral scheme Z, and for every dominant morphism $f : Z \to X$, f factors uniquely through $\tilde{X}$. If X is of finite type over a field k, then the morphism $\tilde{X} \to X$ is a finite morphism. This generalizes (I, Ex. 3.17).

3.9. *The Topological Space of a Product.* Recall that in the category of varieties, the Zariski topology on the product of two varieties is not equal to the product topology (I, Ex. 1.4). Now we see that in the category of schemes, the underlying point set of a product of schemes is not even the product set.

(a) Let k be a field, and let $\mathbf{A}_k^1 = \operatorname{Spec} k[x]$ be the affine line over k. Show that $\mathbf{A}_k^1 \times_{\operatorname{Spec} k} \mathbf{A}_k^1 \cong \mathbf{A}_k^2$, and show that the underlying point set of the product is not the product of the underlying point sets of the factors (even if k is algebraically closed).

(b) Let k be a field, let s and t be indeterminates over k. Then Spec $k(s)$, Spec $k(t)$, and Spec k are all one-point spaces. Describe the product scheme Spec $k(s) \times_{\text{Spec } k}$ Spec $k(t)$.

3.10. *Fibres of a Morphism.*

(a) If $f: X \to Y$ is a morphism, and $y \in Y$ a point, show that $\text{sp}(X_y)$ is homeomorphic to $f^{-1}(y)$ with the induced topology.

(b) Let $X = \text{Spec } k[s,t]/(s - t^2)$, let $Y = \text{Spec } k[s]$, and let $f: X \to Y$ be the morphism defined by sending $s \to s$. If $y \in Y$ is the point $a \in k$ with $a \neq 0$, show that the fibre X_y consists of two points, with residue field k. If $y \in Y$ corresponds to $0 \in k$, show that the fibre X_y is a nonreduced one-point scheme. If η is the generic point of Y, show that X_η is a one-point scheme, whose residue field is an extension of degree two of the residue field of η. (Assume k algebraically closed.)

3.11. *Closed Subschemes.*

(a) Closed immersions are stable under base extension: if $f: Y \to X$ is a closed immersion, and if $X' \to X$ is any morphism, then $f': Y \times_X X' \to X'$ is also a closed immersion.

*(b) If Y is a closed subscheme of an affine scheme $X = \text{Spec } A$, then Y is also affine, and in fact Y is the closed subscheme determined by a suitable ideal $\mathfrak{a} \subseteq A$ as the image of the closed immersion Spec $A/\mathfrak{a} \to \text{Spec } A$. [*Hints:* First show that Y can be covered by a finite number of open affine subsets of the form $D(f_i) \cap Y$, with $f_i \in A$. By adding some more f_i with $D(f_i) \cap Y = \varnothing$, if necessary, show that we may assume that the $D(f_i)$ cover X. Next show that $f_1, \ldots, f_r$ generate the unit ideal of A. Then use (Ex. 2.17b) to show that Y is affine, and (Ex. 2.18d) to show that Y comes from an ideal $\mathfrak{a} \subseteq A$.] *Note:* We will give another proof of this result using sheaves of ideals later (5.10).

(c) Let Y be a closed subset of a scheme X, and give Y the reduced induced subscheme structure. If Y' is any other closed subscheme of X with the same underlying topological space, show that the closed immersion $Y \to X$ factors through Y'. We express this property by saying that the reduced induced structure is the smallest subscheme structure on a closed subset.

(d) Let $f: Z \to X$ be a morphism. Then there is a unique closed subscheme Y of X with the following property: the morphism f factors through Y, and if Y' is any other closed subscheme of X through which f factors, then $Y \to X$ factors through Y' also. We call Y the *scheme-theoretic image* of f. If Z is a reduced scheme, then Y is just the reduced induced structure on the closure of the image $f(Z)$.

3.12. *Closed Subschemes of* Proj S.

(a) Let $\varphi: S \to T$ be a surjective homomorphism of graded rings, preserving degrees. Show that the open set U of (Ex. 2.14) is equal to Proj T, and the morphism $f: \text{Proj } T \to \text{Proj } S$ is a closed immersion.

(b) If $I \subseteq S$ is a homogeneous ideal, take $T = S/I$ and let Y be the closed subscheme of $X = \text{Proj } S$ defined as image of the closed immersion Proj $S/I \to X$. Show that different homogeneous ideals can give rise to the same closed subscheme. For example, let d_0 be an integer, and let $I' = \bigoplus_{d \geq d_0} I_d$. Show that I and I' determine the same closed subscheme.

We will see later (5.16) that every closed subscheme of X comes from a homogeneous ideal I of S (at least in the case where S is a polynomial ring over S_0).

3.13. *Properties of Morphisms of Finite Type.*
  (a) A closed immersion is a morphism of finite type.
  (b) A quasi-compact open immersion (Ex. 3.2) is of finite type.
  (c) A composition of two morphisms of finite type is of finite type.
  (d) Morphisms of finite type are stable under base extension.
  (e) If X and Y are schemes of finite type over S, then $X \times_S Y$ is of finite type over S.
  (f) If $X \xrightarrow{f} Y \xrightarrow{g} Z$ are two morphisms, and if f is quasi-compact, and $g \circ f$ is of finite type, then f is of finite type.
  (g) If $f : X \to Y$ is a morphism of finite type, and if Y is noetherian, then X is noetherian.

3.14. If X is a scheme of finite type over a field, show that the closed points of X are dense. Give an example to show that this is not true for arbitrary schemes.

3.15. Let X be a scheme of finite type over a field k (not necessarily algebraically closed).
  (a) Show that the following three conditions are equivalent (in which case we say that X is *geometrically irreducible*).
     (i) $X \times_k \bar{k}$ is irreducible, where $\bar{k}$ denotes the algebraic closure of k. (By abuse of notation, we write $X \times_k \bar{k}$ to denote $X \times_{\operatorname{Spec} k} \operatorname{Spec} \bar{k}$.)
     (ii) $X \times_k k_s$ is irreducible, where k_s denotes the separable closure of k.
     (iii) $X \times_k K$ is irreducible for every extension field K of k.
  (b) Show that the following three conditions are equivalent (in which case we say X is *geometrically reduced*).
     (i) $X \times_k \bar{k}$ is reduced.
     (ii) $X \times_k k_p$ is reduced, where k_p denotes the perfect closure of k.
     (iii) $X \times_k K$ is reduced for all extension fields K of k.
  (c) We say that X is *geometrically integral* if $X \times_k \bar{k}$ is integral. Give examples of integral schemes which are neither geometrically irreducible nor geometrically reduced.

3.16. *Noetherian Induction.* Let X be a noetherian topological space, and let $\mathscr{P}$ be a property of closed subsets of X. Assume that for any closed subset Y of X, if $\mathscr{P}$ holds for every proper closed subset of Y, then $\mathscr{P}$ holds for Y. (In particular, $\mathscr{P}$ must hold for the empty set.) Then $\mathscr{P}$ holds for X.

3.17. *Zariski Spaces.* A topological space X is a *Zariski space* if it is noetherian and every (nonempty) closed irreducible subset has a unique generic point (Ex. 2.9).
   For example, let R be a discrete valuation ring, and let $T = \operatorname{sp}(\operatorname{Spec} R)$. Then T consists of two points $t_0 = $ the maximal ideal, $t_1 = $ the zero ideal. The open subsets are $\varnothing$, $\{t_1\}$, and T. This is an irreducible Zariski space with generic point t_1.
  (a) Show that if X is a noetherian scheme, then $\operatorname{sp}(X)$ is a Zariski space.
  (b) Show that any minimal nonempty closed subset of a Zariski space consists of one point. We call these *closed points*.
  (c) Show that a Zariski space X satisfies the axiom T_0: given any two distinct points of X, there is an open set containing one but not the other.
  (d) If X is an irreducible Zariski space, then its generic point is contained in every nonempty open subset of X.
  (e) If x_0, x_1 are points of a topological space X, and if $x_0 \in \{x_1\}^-$, then we say that x_1 *specializes* to x_0, written $x_1 \rightsquigarrow x_0$. We also say x_0 is a *specialization*

of x_1, or that x_1 is a *generization* of x_0. Now let X be a Zariski space. Show that the minimal points, for the partial ordering determined by $x_1 > x_0$ if $x_1 \rightsquigarrow x_0$, are the closed points, and the maximal points are the generic points of the irreducible components of X. Show also that a closed subset contains every specialization of any of its points. (We say closed subsets are *stable under specialization*.) Similarly, open subsets are *stable under generization*.

(f) Let t be the functor on topological spaces introduced in the proof of (2.6). If X is a noetherian topological space, show that $t(X)$ is a Zariski space. Furthermore X itself is a Zariski space if and only if the map $\alpha : X \to t(X)$ is a homeomorphism.

3.18. *Constructible Sets.* Let X be a Zariski topological space. A *constructible subset* of X is a subset which belongs to the smallest family $\mathfrak{F}$ of subsets such that (1) every open subset is in $\mathfrak{F}$, (2) a finite intersection of elements of $\mathfrak{F}$ is in $\mathfrak{F}$, and (3) the complement of an element of $\mathfrak{F}$ is in $\mathfrak{F}$.

(a) A subset of X is *locally closed* if it is the intersection of an open subset with a closed subset. Show that a subset of X is constructible if and only if it can be written as a finite disjoint union of locally closed subsets.

(b) Show that a constructible subset of an irreducible Zariski space X is dense if and only if it contains the generic point. Furthermore, in that case it contains a nonempty open subset.

(c) A subset S of X is closed if and only if it is constructible and stable under specialization. Similarly, a subset T of X is open if and only if it is constructible and stable under generization.

(d) If $f : X \to Y$ is a continuous map of Zariski spaces, then the inverse image of any constructible subset of Y is a constructible subset of X.

3.19. The real importance of the notion of constructible subsets derives from the following theorem of Chevalley—see Cartan and Chevalley [1, exposé 7] and see also Matsumura [2, Ch. 2, §6]: let $f : X \to Y$ be a morphism of finite type of noetherian schemes. Then the image of any constructible subset of X is a constructible subset of Y. In particular, $f(X)$, which need not be either open or closed, is a constructible subset of Y. Prove this theorem in the following steps.

(a) Reduce to showing that $f(X)$ itself is constructible, in the case where X and Y are affine, integral noetherian schemes, and f is a dominant morphism.

*(b) In that case, show that $f(X)$ contains a nonempty open subset of Y by using the following result from commutative algebra: let $A \subseteq B$ be an inclusion of noetherian integral domains, such that B is a finitely generated A-algebra. Then given a nonzero element $b \in B$, there is a nonzero element $a \in A$ with the following property: if $\varphi : A \to K$ is any homomorphism of A to an algebraically closed field K, such that $\varphi(a) \neq 0$, then φ extends to a homomorphism φ' of B into K, such that $\varphi'(b) \neq 0$. [*Hint:* Prove this algebraic result by induction on the number of generators of B over A. For the case of one generator, prove the result directly. In the application, take $b = 1$.]

(c) Now use noetherian induction on Y to complete the proof.

(d) Give some examples of morphisms $f : X \to Y$ of varieties over an algebraically closed field k, to show that $f(X)$ need not be either open or closed.

3.20. *Dimension.* Let X be an integral scheme of finite type over a field k (not necessarily algebraically closed). Use appropriate results from (I, §1) to prove the following.

(a) For any closed point $P \in X$, $\dim X = \dim \mathcal{O}_P$, where for rings, we always mean the Krull dimension.

(b) Let $K(X)$ be the function field of X (Ex. 3.6). Then $\dim X = \text{tr.d. } K(X)/k$.

(c) If Y is a closed subset of X, then $\text{codim}(Y,X) = \inf\{\dim \mathcal{O}_{P,X} | P \in Y\}$.

(d) If Y is a closed subset of X, then $\dim Y + \text{codim}(Y,X) = \dim X$.

(e) If U is a nonempty open subset of X, then $\dim U = \dim X$.

(f) If $k \subseteq k'$ is a field extension, then every irreducible component of $X' = X \times_k k'$ has dimension $= \dim X$.

3.21. Let R be a discrete valuation ring containing its residue field k. Let $X = \text{Spec } R[t]$ be the affine line over $\text{Spec } R$. Show that statements (a), (d), (e) of (Ex. 3.20) are false for X.

*3.22. *Dimension of the Fibres of a Morphism.* Let $f: X \to Y$ be a dominant morphism of integral schemes of finite type over a field k.

(a) Let Y' be a closed irreducible subset of Y, whose generic point η' is contained in $f(X)$. Let Z be any irreducible component of $f^{-1}(Y')$, such that $\eta' \in f(Z)$, and show that $\text{codim}(Z,X) \leqslant \text{codim}(Y',Y)$.

(b) Let $e = \dim X - \dim Y$ be the *relative dimension* of X over Y. For any point $y \in f(X)$, show that every irreducible component of the fibre X_y has dimension $\geqslant e$. [*Hint:* Let $Y' = \{y\}^-$, and use (a) and (Ex. 3.20b).]

(c) Show that there is a dense open subset $U \subseteq X$, such that for any $y \in f(U)$, $\dim U_y = e$. [*Hint:* First reduce to the case where X and Y are affine, say $X = \text{Spec } A$ and $Y = \text{Spec } B$. Then A is a finitely generated B-algebra. Take $t_1, \ldots, t_e \in A$ which form a transcendence base of $K(X)$ over $K(Y)$, and let $X_1 = \text{Spec } B[t_1, \ldots, t_e]$. Then X_1 is isomorphic to affine e-space over Y, and the morphism $X \to X_1$ is generically finite. Now use (Ex. 3.7) above.]

(d) Going back to our original morphism $f: X \to Y$, for any integer h, let E_h be the set of points $x \in X$ such that, letting $y = f(x)$, there is an irreducible component Z of the fibre X_y, containing x, and having $\dim Z \geqslant h$. Show that (1) $E_e = X$ (use (b) above); (2) if $h > e$, then E_h is not dense in X (use (c) above); and (3) E_h is closed, for all h (use induction on $\dim X$).

(e) Prove the following theorem of Chevalley—see Cartan and Chevalley [1, exposé 8]. For each integer h, let C_h be the set of points $y \in Y$ such that $\dim X_y = h$. Then the subsets C_h are constructible, and C_e contains an open dense subset of Y.

3.23. If V, W are two varieties over an algebraically closed field k, and if $V \times W$ is their product, as defined in (I, Ex. 3.15, 3.16), and if t is the functor of (2.6), then $t(V \times W) = t(V) \times_{\text{Spec } k} t(W)$.

4 Separated and Proper Morphisms

We now come to two properties of schemes, or rather of morphisms between schemes, which correspond to well-known properties of ordinary topological spaces. Separatedness corresponds to the Hausdorff axiom for a topological space. Properness corresponds to the usual notion of properness, namely that the inverse image of a compact subset is compact. However, the usual definitions are not suitable in abstract algebraic geometry, because the Zariski topology is never Hausdorff, and the underlying topological space of a scheme does not accurately reflect all of its properties. So instead we will use definitions which reflect the functorial behavior of the morphism within the category of schemes. For schemes of finite type over $\mathbf{C}$, one can show that these notions, defined abstractly, are in fact the same as the usual notions if we consider those schemes as complex analytic spaces in the ordinary topology (Appendix B).

In this section we will define separated and proper morphisms. We will give criteria for a morphism to be separated or proper using valuation rings. Then we will show that projective space over any scheme is proper.

Definition. Let $f: X \to Y$ be a morphism of schemes. The *diagonal morphism* is the unique morphism $\Delta: X \to X \times_Y X$ whose composition with both projection maps $p_1, p_2: X \times_Y X \to X$ is the identity map of $X \to X$. We say that the morphism f is *separated* if the diagonal morphism Δ is a closed immersion. In that case we also say X is *separated* over Y. A scheme X is *separated* if it is separated over Spec $\mathbf{Z}$.

Example 4.0.1. Let k be a field, and let X be the affine line with the origin doubled (2.3.6). Then X is not separated over k. Indeed, $X \times_k X$ is the affine plane with doubled axes and four origins. The image of Δ is the usual diagonal, with two of those origins. This is not closed, because all four origins are in the closure of $\Delta(X)$.

Example 4.0.2. We will see later (4.10) that if V is any variety over an algebraically closed field k, then the associated scheme $t(V)$ is separated over k.

Proposition 4.1. *If $f: X \to Y$ is any morphism of affine schemes, then f is separated.*

PROOF. Let $X = \operatorname{Spec} A$, $Y = \operatorname{Spec} B$. Then A is a B-algebra, and $X \times_Y X$ is also affine, given by Spec $A \otimes_B A$. The diagonal morphism Δ comes from the *diagonal homomorphism* $A \otimes_B A \to A$ defined by $a \otimes a' \to aa'$. This is a surjective homomorphism of rings, hence Δ is a closed immersion.

Corollary 4.2. *An arbitrary morphism $f: X \to Y$ is separated if and only if the image of the diagonal morphism is a closed subset of $X \times_Y X$.*

PROOF. One implication is obvious, so we have only to prove that if $\Delta(X)$ is a closed subset, then $\Delta: X \to X \times_Y X$ is a closed immersion. In other words, we have to check that $\Delta: X \to \Delta(X)$ is a homeomorphism, and that the morphism of sheaves $\mathcal{O}_{X \times_Y X} \to \Delta_* \mathcal{O}_X$ is surjective. Let $p_1: X \times_Y X \to X$ be the first projection. Since $p_1 \circ \Delta = \operatorname{id}_X$, it follows immediately that Δ gives a homeomorphism onto $\Delta(X)$. To see that the map of sheaves $\mathcal{O}_{X \times_Y X} \to \Delta_* \mathcal{O}_X$ is surjective is a local question. For any point $P \in X$, let U be an open affine neighborhood of P which is small enough so that $f(U)$ is contained in an open affine subset V of Y. Then $U \times_V U$ is an open affine neighborhood of $\Delta(P)$, and by the proposition, $\Delta : U \to U \times_V U$ is a closed immersion. So our map of sheaves is surjective in a neighborhood of P, which completes the proof.

Next we will discuss the valuative criterion of separatedness. The rough idea is that in order for a scheme X to be separated, it should not contain any subscheme which looks like a curve with a doubled point, as in the example above. Another way of saying this is that if C is a curve, and P a point of C, then given any morphism of $C - P$ into X, it should admit at most one extension to a morphism of all of C into X. (Compare (I, 6.8) where we showed that a projective variety has this property.)

In practice, this rough idea has to be modified. The question is local, so we replace the curve by its local ring at P, which is a discrete valuation ring. Then since our schemes may be quite general, we must consider arbitrary (not necessarily discrete) valuation rings. Finally, we make the criterion relative over the image scheme Y of a morphism.

See (I, §6) for the definition and basic properties of valuation rings.

Theorem 4.3 (Valuative Criterion of Separatedness). *Let $f : X \to Y$ be a morphism of schemes, and assume that X is noetherian. Then f is separated if and only if the following condition holds. For any field K, and for any valuation ring R with quotient field K, let $T = \operatorname{Spec} R$, let $U = \operatorname{Spec} K$, and let $i : U \to T$ be the morphism induced by the inclusion $R \subseteq K$. Given a morphism of T to Y, and given a morphism of U to X which makes a commutative diagram*

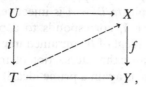

there is at most one morphism of T to X making the whole diagram commutative.

We will need two lemmas.

Lemma 4.4. *Let R be a valuation ring of a field K. Let $T = \operatorname{Spec} R$ and let $U = \operatorname{Spec} K$. To give a morphism of U to a scheme X is equivalent to giving a point $x_1 \in X$ and an inclusion of fields $k(x_1) \subseteq K$. To give a morphism of T to X is equivalent to giving two points x_0, x_1 in X, with x_0 a specialization (see Ex. 3.17e) of x_1, and an inclusion of fields $k(x_1) \subseteq K$,* such that R dominates the local ring $\mathcal{O}$ of x_0 on the subscheme $Z = \{x_1\}^-$ of X with its reduced induced structure.

PROOF. U is a one-point scheme, with structure sheaf K. To give a local homomorphism $\mathcal{O}_{x_1,X} \to K$ is the same as giving an inclusion of $k(x_1) \subseteq K$, so the first part is obvious. For the second part, let $t_0 = \mathfrak{m}_R$ be the closed point of T, and let $t_1 = (0)$ be the generic point of T. Given a morphism of T to X, let x_0 and x_1 be the images of t_0 and t_1. Since T is reduced, the morphism $T \to X$ factors through Z (Ex. 3.11). Furthermore, $k(x_1)$ is the function field of Z. So we have a local homomorphism of $\mathcal{O} = \mathcal{O}_{x_0,Z}$ to R compatible with the inclusion $k(x_1) \subseteq K$. In other words R dominates $\mathcal{O}$.

Conversely, given the data consisting of x_0, x_1, and the inclusion $k(x_1) \subseteq K$ such that R dominates $\mathcal{O}$, the inclusion $\mathcal{O} \to R$ gives a morphism $T \to \operatorname{Spec} \mathcal{O}$, which composed with the natural map $\operatorname{Spec} \mathcal{O} \to X$ gives the desired morphism $T \to X$.

Lemma 4.5. Let $f: X \to Y$ be a quasi-compact morphism of schemes (see Ex. 3.2). Then the subset $f(X)$ of Y is closed if and only if it is stable under specialization (Ex. 3.17e).

PROOF. One implication is obvious, so we have only to show that if $f(X)$ is stable under specialization, then it is closed. Clearly we may assume that X and Y are both reduced, and that $f(X)^- = Y$ (replace Y by the reduced induced structure on $f(X)^-$). So let $y \in Y$ be a point. We wish to show that $y \in f(X)$. Now we can replace Y by an affine neighborhood of y, and so assume that Y is affine. Then since f is quasi-compact, X will be a finite union of open affines X_i. We know that $y \in f(X)^-$. Hence $y \in f(X_i)^-$ for some i. Let $Y_i = f(X_i)^-$ with the reduced induced structure. Then Y_i is also affine, and we will consider the dominant morphism $X_i \to Y_i$ of reduced affine schemes. Let $X_i = \operatorname{Spec} A$ and $Y_i = \operatorname{Spec} B$. Then the corresponding ring homomorphism $B \to A$ is injective, because the morphism is dominant. The point $y \in Y_i$ corresponds to a prime ideal $\mathfrak{p} \subseteq B$. Let $\mathfrak{p}' \subseteq \mathfrak{p}$ be a minimal prime ideal of B contained in $\mathfrak{p}$. (Minimal prime ideals exist, by Zorn's lemma, because the intersection of any family of prime ideals, totally ordered by inclusion, is again a prime ideal!) Then $\mathfrak{p}'$ corresponds to a point y' of Y_i which specializes to y. I claim $y' \in f(X_i)$. Indeed, let us localize A and B at $\mathfrak{p}'$. Localization is an exact functor, so $B_{\mathfrak{p}'} \subseteq A \otimes B_{\mathfrak{p}'}$. Now $B_{\mathfrak{p}'}$ is a field. Let $\mathfrak{q}_0'$ be any prime ideal of $A \otimes B_{\mathfrak{p}'}$. Then $\mathfrak{q}_0' \cap B_{\mathfrak{p}'} = (0)$. Let $\mathfrak{q}' \subseteq A$ be the inverse image of $\mathfrak{q}_0'$ under the localization map $A \to A \otimes B_{\mathfrak{p}'}$. Then $\mathfrak{q}' \cap B = \mathfrak{p}'$. So $\mathfrak{q}'$ corresponds to a point $x' \in X_i$ with $f(x') = y'$. Now go back to the morphism $f: X \to Y$. We have $x' \in X$, $f(x') = y'$, so $y' \in f(X)$. But $f(X)$ is stable under specialization by hypothesis, and $y' \rightsquigarrow y$, so $y \in f(X)$, which is what we wanted to prove.

PROOF OF THEOREM 4.3. First suppose f is separated, and suppose given a diagram as above where there are two morphisms h, h' of T to X making the whole diagram commutative.

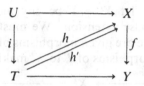

Then we obtain a morphism $h'': T \to X \times_Y X$. Since the restrictions of h and h' to U are the same, the generic point t_1 of T has image in the diagonal $\Delta(X)$. Since $\Delta(X)$ is closed, the image of t_0 is also in the diagonal. Therefore h and h' both send the points t_0, t_1 to the same points x_0, x_1 of X. Since the inclusions of $k(x_1) \subseteq K$ induced by h and h' are also the same, it follows from (4.4) that h and h' are equal.

Conversely, let us suppose the condition of the theorem satisfied. To show that f is separated, it is sufficient by (4.2) to show that $\Delta(X)$ is a closed subset of $X \times_Y X$. And since we have assumed that X is noetherian, the morphism Δ is quasi-compact, so by (4.5) it will be sufficient to show that $\Delta(X)$ is stable under specialization. So let $\xi_1 \in \Delta(X)$ be a point, and let $\xi_1 \leadsto \xi_0$ be a specialization. Let $K = k(\xi_1)$ and let $\mathcal{O}$ be the local ring of ξ_0 on the subscheme $\{\xi_1\}^-$ with its reduced induced structure. Then $\mathcal{O}$ is a local ring contained in K, so by (I, 6.1A) there is a valuation ring R of K which dominates $\mathcal{O}$. Now by (4.4) we obtain a morphism of $T = \operatorname{Spec} R$ to $X \times_Y X$ sending t_0 and t_1 to ξ_0 and ξ_1. Composing with the projections p_1, p_2 gives two morphisms of T to X, which give the same morphism to Y, and whose restrictions to $U = \operatorname{Spec} K$ are the same, since $\xi_1 \in \Delta(X)$. So by the condition, these two morphisms of T to X must be the same. Therefore the morphism $T \to X \times_Y X$ factors through the diagonal morphism $\Delta: X \to X \times_Y X$, and so $\xi_0 \in \Delta(X)$. This completes the proof. Note in the last step it would not be sufficient to know only that $p_1(\xi_0) = p_2(\xi_0)$. For in general if $\xi \in X \times_Y X$ then $p_1(\xi) = p_2(\xi)$ does *not* imply $\xi \in \Delta(X)$.

Corollary 4.6. *Assume that all schemes are noetherian in the following statements.*

(a) *Open and closed immersions are separated.*

(b) *A composition of two separated morphisms is separated.*

(c) *Separated morphisms are stable under base extension.*

(d) *If $f: X \to Y$ and $f': X' \to Y'$ are separated morphisms of schemes over a base scheme S, then the product morphism $f \times f': X \times_S X' \to Y \times_S Y'$ is also separated.*

(e) *If $f: X \to Y$ and $g: Y \to Z$ are two morphisms and if $g \circ f$ is separated, then f is separated.*

(f) *A morphism $f: X \to Y$ is separated if and only if Y can be covered by open subsets V_i such that $f^{-1}(V_i) \to V_i$ is separated for each i.*

PROOF. These statements all follow immediately from the condition of the theorem. We will give the proof of (c) to illustrate the method. Let $f: X \to Y$

be a separated morphism, let $Y' \to Y$ be any morphism, and let $X' = X \times_Y Y'$ be obtained by base extension. We must show that $f': X' \to Y'$ is separated. So suppose we are given morphisms of T to Y' and U to X' as in the theorem, and two morphisms of T to X' making the diagram

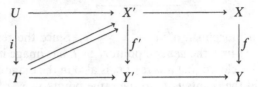

commutative. Composing with the map $X' \to X$, we obtain two morphisms of T to X. Since f is separated, these are the same. But X' is the fibred product of X and Y' over Y, so by the universal property of the fibred product, the two maps of T to X' are the same. Hence f' is separated.

Note on Noetherian Hypotheses. You have probably noticed that in order to apply the theorem, it is not necessary to assume that all the schemes mentioned in the corollary are noetherian. In fact, even in the theorem itself, you can get by with assuming something less than X noetherian (see Grothendieck [EGA I, new ed., 5.5.4]). My feeling is that if a noetherian hypothesis will make statements and proofs substantially simpler, then I will make that hypothesis, even though it may not be necessary. My justification for this attitude is that most of the motivation and examples in algebraic geometry come from schemes of finite type over a field, and constructions made from them, and practically all the schemes encountered in this way are noetherian. This attitude will prevail in Chapter III, where noetherian hypotheses are built into the very foundations of our treatment of cohomology. The reader who wishes to avoid noetherian hypotheses is advised to read [EGA], especially [EGA IV, §8].

Definition. A morphism $f: X \to Y$ is *proper* if it is separated, of finite type, and universally closed. Here we say that a morphism is *closed* if the image of any closed subset is closed. A morphism $f: X \to Y$ is *universally closed* if it is closed, and for any morphism $Y' \to Y$, the corresponding morphism $f': X' \to Y'$ obtained by base extension is also closed.

Example 4.6.1. Let k be a field and let X be the affine line over k. Then X is separated and of finite type over k, but it is not proper over k. Indeed, take the base extension $X \to k$. The map $X \times_k X \to X$ we obtain is the projection map of the affine plane onto the affine line. This is not a closed map. For example, the hyperbola given by the equation $xy = 1$ is a closed subset of the plane, but its image under projection consists of the affine line minus the origin, which is not closed.

Of course it is clear that what is missing in this example is the point at infinity on the hyperbola. This suggests that the *projective* line would be proper over k. In fact, we will see later (4.9) that any projective variety over a field is proper.

Theorem 4.7 (Valuative Criterion of Properness). *Let $f:X \to Y$ be a morphism of finite type, with X noetherian. Then f is proper if and only if for every valuation ring R and for every morphism of U to X and T to Y forming a commutative diagram*

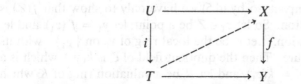

(using the notation of (4.3)), there exists a unique morphism $T \to X$ making the whole diagram commutative.

PROOF. First assume that f is proper. Then by definition f is separated, so the uniqueness of the morphism $T \to X$ will follow from (4.3), once we know it exists. For the existence, we consider the base extension $T \to Y$, and let $X_T = X \times_Y T$. We get a map $U \to X_T$ from the given maps $U \to X$ and $U \to T$.

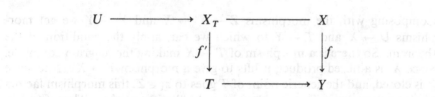

Let $\xi_1 \in X_T$ be the image of the unique point t_1 of U. Let $Z = \{\xi_1\}^-$. Then Z is a closed subset of X_T. Since f is proper, it is universally closed, so the morphism $f':X_T \to T$ must be closed, so $f'(Z)$ is a closed subset of T. But $f'(\xi_1) = t_1$, which is the generic point of T, so in fact $f'(Z) = T$. Hence there is a point $\xi_0 \in Z$ with $f'(\xi_0) = t_0$. So we get a local homomorphism of local rings $R \to \mathcal{O}_{\xi_0,Z}$ corresponding to the morphism f'. Now the function field of Z is $k(\xi_1)$, which is contained in K, by construction of ξ_1. By (I, 6.1A), R is maximal for the relation of domination between local subrings of K. Hence R is isomorphic to $\mathcal{O}_{\xi_0,Z}$, and in particular R dominates it. Hence by (4.4) we obtain a morphism of T to X_T sending t_0,t_1 to ξ_0,ξ_1. Composing with the map $X_T \to X$ gives the desired morphism of T to X.

Conversely, suppose the condition of the theorem holds. To show f is proper, we have only to show that it is universally closed, since it is of finite type by hypothesis, and it is separated by (4.3). So let $Y' \to Y$ be any morphism, and let $f':X' \to Y'$ be the morphism obtained from f by base extension. Let Z be a closed subset of X', and give it the reduced induced structure.

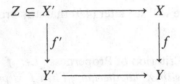

We need to show that $f'(Z)$ is closed in Y'. Since f is of finite type, so is f' and so is the restriction of f' to Z (Ex. 3.13). In particular, the morphism $f':Z \to Y'$ is quasi-compact, so by (4.5) we have only to show that $f(Z)$ is stable under specialization. So let $z_1 \in Z$ be a point, let $y_1 = f'(z_1)$, and let $y_1 \leadsto y_0$ be a specialization. Let $\mathcal{O}$ be the local ring of y_0 on $\{y_1\}^-$ with its reduced induced structure. Then the quotient field of $\mathcal{O}$ is $k(y_1)$, which is a subfield of $k(z_1)$. Let $K = k(z_1)$, and let R be a valuation ring of K which dominates $\mathcal{O}$ (which exists by (I, 6.1A)).

From this data, by (4.4) we obtain morphisms $U \to Z$ and $T \to Y'$ forming a commutative diagram

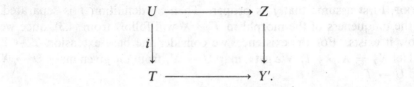

Composing with the morphisms $Z \to X' \to X$ and $Y' \to Y$, we get morphisms $U \to X$ and $T \to Y$ to which we can apply the condition of the theorem. So there is a morphism of $T \to X$ making the diagram commute. Since X' is a fibred product, it lifts to give a morphism $T \to X'$. And since Z is closed, and the generic point of T goes to $z_1 \in Z$, this morphism factors to give a morphism $T \to Z$. Now let z_0 be the image of t_0. Then $f'(z_0) = y_0$, so $y_0 \in f'(Z)$. This completes the proof.

Corollary 4.8. *In the following statements, we take all schemes to be noetherian.*
    (a) *A closed immersion is proper.*
    (b) *A composition of proper morphisms is proper.*
    (c) *Proper morphisms are stable under base extension.*
    (d) *Products of proper morphisms are proper as in (4.6d).*
    (e) *If $f:X \to Y$ and $g:Y \to Z$ are two morphisms, if $g \circ f$ is proper, and if g is separated, then f is proper.*
    (f) *Properness is local on the base as in (4.6f).*

PROOF. These results follow immediately from the condition of the theorem, taking into account (Ex. 3.13) which deals with the finite type property, and (4.6). We will give the proof of (e) to illustrate the method. Assume $g \circ f$ is proper and g is separated. Then f is of finite type by (Ex. 3.13). (We have assumed that X is noetherian, so f is automatically quasi-compact.) Also f is separated by (4.6). So we have to show that given a valuation ring R, and morphisms $U \to X$ and $T \to Y$ making a commutative diagram,

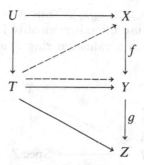

then there exists a morphism of T to X making the diagram commutative.

Let $T \to Z$ be the composed map. Then since $g \circ f$ is proper, there is a map of T to X commuting with the map of $T \to Z$. By composing with f, we get a second map of T to Y. But now since g is separated, the two maps of T to Y are the same, so we are done.

Our next objective is to define projective morphisms and to show that any projective morphism is proper. Recall that in Section 2 we defined projective n-space $\mathbf{P}_A^n$ over any ring A to be $\operatorname{Proj} A[x_0, \ldots, x_n]$. Note that if $A \to B$ is a homomorphism of rings, and $\operatorname{Spec} B \to \operatorname{Spec} A$ is the corresponding morphism of affine schemes, then $\mathbf{P}_B^n \cong \mathbf{P}_A^n \times_{\operatorname{Spec} A} \operatorname{Spec} B$. In particular, for any ring A, we have $\mathbf{P}_A^n \cong \mathbf{P}_{\mathbf{Z}}^n \times_{\operatorname{Spec} \mathbf{Z}} \operatorname{Spec} A$. This motivates the following definition for any scheme Y.

Definition. If Y is any scheme, we define *projective n-space* over Y, denoted $\mathbf{P}_Y^n$, to be $\mathbf{P}_{\mathbf{Z}}^n \times_{\operatorname{Spec} \mathbf{Z}} Y$. A morphism $f : X \to Y$ of schemes is *projective* if it factors into a closed immersion $i : X \to \mathbf{P}_Y^n$ for some n, followed by the projection $\mathbf{P}_Y^n \to Y$. A morphism $f : X \to Y$ is *quasi-projective* if it factors into an open immersion $j : X \to X'$ followed by a projective morphism $g : X' \to Y$. (This definition of projective morphism is slightly different from the one in Grothendieck [EGA II, 5.5]. The two definitions are equivalent in case Y itself is quasi-projective over an affine scheme.)

Example 4.8.1. Let A be a ring, let S be a graded ring with $S_0 = A$, which is finitely generated as an A-algebra by S_1. Then the natural map $\operatorname{Proj} S \to \operatorname{Spec} A$ is a projective morphism. Indeed, by hypothesis S is a quotient of a polynomial ring $S' = A[x_0, \ldots, x_n]$. The surjective homomorphism of graded rings $S' \to S$ gives rise to a closed immersion $\operatorname{Proj} S \to \operatorname{Proj} S' = \mathbf{P}_A^n$, which shows that $\operatorname{Proj} S$ is projective over A (Ex. 3.12).

Theorem 4.9. *A projective morphism of noetherian schemes is proper. A quasi-projective morphism of noetherian schemes is of finite type and separated.*

PROOF. Taking into account the results of (Ex. 3.13) and (4.6) and (4.8), it will be sufficient to show that $X = \mathbf{P}_{\mathbf{Z}}^n$ is proper over $\operatorname{Spec} \mathbf{Z}$. Recall by (2.5) that X is a union of open affine subsets $V_i = D_+(x_i)$, and that V_i is

103

isomorphic to Spec $\mathbf{Z}[x_0/x_i, \ldots, x_n/x_i]$. Thus X is of finite type. To show that X is proper, we will use the criterion of (4.7) and imitate the proof of (I, 6.8). So suppose given a valuation ring R and morphisms $U \to X$, $T \to$ Spec $\mathbf{Z}$ as shown:

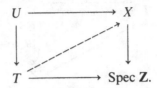

Let $\xi_1 \in X$ be the image of the unique point of U. Using induction on n, we may assume that ξ_1 is not contained in any of the hyperplanes $X - V_i$, which are each isomorphic to $\mathbf{P}^{n-1}$. In other words, we may assume that $\xi_1 \in \bigcap V_i$, and hence all of the functions x_i/x_j are invertible elements of the local ring $\mathcal{O}_{\xi_1}$.

We have an inclusion $k(\xi_1) \subseteq K$ given by the morphism $U \to X$. Let $f_{ij} \in K$ be the image of x_i/x_j. Then the f_{ij} are nonzero elements of K, and $f_{ik} = f_{ij} \cdot f_{jk}$ for all i,j,k. Let $v \colon K \to G$ be the valuation associated to the valuation ring R. Let $g_i = v(f_{i0})$ for $i = 0, \ldots, n$. Choose k such that g_k is minimal among the set $\{g_0, \ldots, g_n\}$, for the ordering of G. Then for each i we have

$$v(f_{ik}) = g_i - g_k \geqslant 0,$$

hence $f_{ik} \in R$ for $i = 0, \ldots, n$. Then we can define a homomorphism

$$\varphi \colon \mathbf{Z}[x_0/x_k, \ldots, x_n/x_k] \to R$$

by sending x_i/x_k to f_{ik}. It is compatible with the given field inclusion $k(\xi_1) \subseteq K$. This homomorphism φ gives a morphism $T \to V_k$, and hence a morphism of T to X which is the one required. The uniqueness of this morphism follows from the construction and the way the V_i patch together.

Proposition 4.10. *Let k be an algebraically closed field. The image of the functor $t \colon \mathfrak{Bar}(k) \to \mathfrak{Sch}(k)$ of (2.6) is exactly the set of quasi-projective integral schemes over k. The image of the set of projective varieties is the set of projective integral schemes. In particular, for any variety V, $t(V)$ is an integral, separated scheme of finite type over k.*

PROOF. We have already seen in Section 3 that for any variety V, the associated scheme $t(V)$ is integral and of finite type over k. Since varieties were defined as locally closed subsets of projective space (I, §3), it is clear that $t(V)$ is also quasi-projective.

For the converse, it will be sufficient to show that any projective integral scheme Y over k is in the image of t. Let Y be a closed subscheme of $\mathbf{P}_k^n$, and let V be the set of closed points of Y. Then V is a closed subset of the variety $\mathbf{P}^n$. Since V is dense in Y (Ex. 3.14) we see that V is irreducible, so V is a projective variety, and we see also that $t(V)$ and Y have the same underlying topological space. But they are both reduced closed subschemes of $\mathbf{P}_k^n$, so they are isomorphic (Ex. 3.11).

Definition. An *abstract variety* is an integral separated scheme of finite type over an algebraically closed field k. If it is proper over k, we will also say it is *complete*.

Remark 4.10.1. From now on we will use the word "variety" to mean "abstract variety" in the sense just defined. We will identify the varieties of Chapter I with their associated schemes, and refer to them as quasi-projective varieties. We will use the words "curve," "surface," "three-fold," etc., to mean an abstract variety of dimension 1, 2, 3, etc.

Remark 4.10.2. The concept of an abstract variety was invented by Weil [1]. He needed it to provide a purely algebraic construction of the Jacobian variety of a curve, which at first appeared only as an abstract variety (Weil [2]). Then Chow [3] gave a different construction of the Jacobian variety showing that it was in fact a projective variety. Later Weil [6] himself showed that all abelian varieties were projective.

Meanwhile Nagata [1] found an example of a complete abstract non-projective variety, showing that in fact the new class of abstract varieties is larger than the class of projective varieties.

We can sum up the present state of knowledge of this subject as follows.

(a) Every complete curve is projective (III, Ex. 5.8).
(b) Every nonsingular complete surface is projective (Zariski [5]). See also Hartshorne [5, II.4.2].
(c) There exist singular nonprojective complete surfaces (Nagata [3]). See also (Ex. 7.13) and (III, Ex. 5.9).
(d) There exist nonsingular complete nonprojective three-folds (Nagata [4], Hironaka [2], and (Appendix B)).
(e) Every variety can be embedded as an open dense subset of a complete variety (Nagata [6]).

The following algebraic result will be used in (Ex. 4.6).

Theorem 4.11A. *If A is a subring of a field K, then the integral closure of A in K is the intersection of all valuation rings of K which contain A.*

PROOF. Bourbaki [1, Ch. VI, §1, no. 3, Thm. 3, p. 92].

EXERCISES

4.1. Show that a finite morphism is proper.

4.2. Let S be a scheme, let X be a reduced scheme over S, and let Y be a separated scheme over S. Let f and g be two S-morphisms of X to Y which agree on an open dense subset of X. Show that $f = g$. Give examples to show that this result fails if either (a) X is nonreduced, or (b) Y is nonseparated. [*Hint*: Consider the map $h:X \to Y \times_S Y$ obtained from f and g.]

4.3. Let X be a separated scheme over an affine scheme S. Let U and V be open affine subsets of X. Then $U \cap V$ is also affine. Give an example to show that this fails if X is not separated.

4.4. Let $f:X \to Y$ be a morphism of separated schemes of finite type over a noetherian scheme S. Let Z be a closed subscheme of X which is proper over S. Show that $f(Z)$ is closed in Y, and that $f(Z)$ with its image subscheme structure (Ex. 3.11d) is proper over S. We refer to this result by saying that "the image of a proper scheme is proper." [*Hint*: Factor f into the graph morphism $\Gamma_f:X \to X \times_S Y$ followed by the second projection p_2, and show that Γ_f is a closed immersion.]

4.5. Let X be an integral scheme of finite type over a field k, having function field K. We say that a valuation of K/k (see I, §6) has *center* x on X if its valuation ring R dominates the local ring $\mathcal{O}_{x,X}$.
   (a) If X is separated over k, then the center of any valuation of K/k on X (if it exists) is unique.
   (b) If X is proper over k, then every valuation of K/k has a unique center on X.
  *(c) Prove the converses of (a) and (b). [*Hint*: While parts (a) and (b) follow quite easily from (4.3) and (4.7), their converses will require some comparison of valuations in different fields.]
   (d) If X is proper over k, and if k is algebraically closed, show that $\Gamma(X,\mathcal{O}_X) = k$. This result generalizes (I, 3.4a). [*Hint*: Let $a \in \Gamma(X,\mathcal{O}_X)$, with $a \notin k$. Show that there is a valuation ring R of K/k with $a^{-1} \in \mathfrak{m}_R$. Then use (b) to get a contradiction.]
   Note. If X is a variety over k, the criterion of (b) is sometimes taken as the definition of a complete variety.

4.6. Let $f:X \to Y$ be a proper morphism of affine varieties over k. Then f is a finite morphism. [*Hint*: Use (4.11A).]

4.7. *Schemes Over* $\mathbf{R}$. For any scheme X_0 over $\mathbf{R}$, let $X = X_0 \times_{\mathbf{R}} \mathbf{C}$. Let $\alpha:\mathbf{C} \to \mathbf{C}$ be complex conjugation, and let $\sigma:X \to X$ be the automorphism obtained by keeping X_0 fixed and applying α to $\mathbf{C}$. Then X is a scheme over $\mathbf{C}$, and σ is a *semi-linear* automorphism, in the sense that we have a commutative diagram

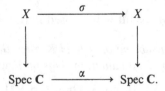

Since $\sigma^2 = \mathrm{id}$, we call σ an *involution*.
   (a) Now let X be a separated scheme of finite type over $\mathbf{C}$, let σ be a semilinear involution on X, and assume that for any two points $x_1,x_2 \in X$, there is an open affine subset containing both of them. (This last condition is satisfied for example if X is quasi-projective.) Show that there is a unique separated scheme X_0 of finite type over $\mathbf{R}$, such that $X_0 \times_{\mathbf{R}} \mathbf{C} \cong X$, and such that this isomorphism identifies the given involution of X with the one on $X_0 \times_{\mathbf{R}} \mathbf{C}$ described above.

For the following statements, X_0 will denote a separated scheme of finite type over $\mathbf{R}$, and X,σ will denote the corresponding scheme with involution over $\mathbf{C}$.

(b) Show that X_0 is affine if and only if X is.

(c) If X_0, Y_0 are two such schemes over $\mathbf{R}$, then to give a morphism $f_0 : X_0 \to Y_0$ is equivalent to giving a morphism $f : X \to Y$ which commutes with the involutions, i.e., $f \circ \sigma_X = \sigma_Y \circ f$.

(d) If $X \cong \mathbf{A}_\mathbf{C}^1$, then $X_0 \cong \mathbf{A}_\mathbf{R}^1$.

(e) If $X \cong \mathbf{P}_\mathbf{C}^1$, then either $X_0 \cong \mathbf{P}_\mathbf{R}^1$, or X_0 is isomorphic to the conic in $\mathbf{P}_\mathbf{R}^2$ given by the homogeneous equation $x_0^2 + x_1^2 + x_2^2 = 0$.

4.8. Let $\mathscr{P}$ be a property of morphisms of schemes such that:

(a) a closed immersion has $\mathscr{P}$;

(b) a composition of two morphisms having $\mathscr{P}$ has $\mathscr{P}$;

(c) $\mathscr{P}$ is stable under base extension.

Then show that:

(d) a product of morphisms having $\mathscr{P}$ has $\mathscr{P}$;

(e) if $f : X \to Y$ and $g : Y \to Z$ are two morphisms, and if $g \circ f$ has $\mathscr{P}$ and g is separated, then f has $\mathscr{P}$;

(f) If $f : X \to Y$ has $\mathscr{P}$, then $f_{red} : X_{red} \to Y_{red}$ has $\mathscr{P}$.

[*Hint:* For (e), consider the graph morphism $\Gamma_f : X \to X \times_Z Y$ and note that it is obtained by base extension from the diagonal morphism $\Delta : Y \to Y \times_Z Y$.]

4.9. Show that a composition of projective morphisms is projective. [*Hint:* Use the Segre embedding defined in (I, Ex. 2.14) and show that it gives a closed immersion $\mathbf{P}^r \times \mathbf{P}^s \to \mathbf{P}^{rs+r+s}$.] Conclude that projective morphisms have properties (a)–(f) of (Ex. 4.8) above.

*4.10. *Chow's Lemma.* This result says that proper morphisms are fairly close to projective morphisms. Let X be proper over a noetherian scheme S. Then there is a scheme X' and a morphism $g : X' \to X$ such that X' is projective over S, and there is an open dense subset $U \subseteq X$ such that g induces an isomorphism of $g^{-1}(U)$ to U. Prove this result in the following steps.

(a) Reduce to the case X irreducible.

(b) Show that X can be covered by a finite number of open subsets $U_i, i = 1, \ldots, n$, each of which is quasi-projective over S. Let $U_i \to P_i$ be an open immersion of U_i into a scheme P_i which is projective over S.

(c) Let $U = \bigcap U_i$, and consider the map

$$ f : U \to X \times_S P_1 \times_S \cdots \times_S P_n $$

deduced from the given maps $U \to X$ and $U \to P_i$. Let X' be the closed image subscheme structure (Ex. 3.11d) $f(U)^-$. Let $g : X' \to X$ be the projection onto the first factor, and let $h : X' \to P = P_1 \times_S \ldots \times_S P_n$ be the projection onto the product of the remaining factors. Show that h is a closed immersion, hence X' is projective over S.

(d) Show that $g^{-1}(U) \to U$ is an isomorphism, thus completing the proof.

4.11. If you are willing to do some harder commutative algebra, and stick to noetherian schemes, then we can express the valuative criteria of separatedness and properness using only *discrete* valuation rings.

(a) If $\mathcal{O},\mathfrak{m}$ is a noetherian local domain with quotient field K, and if L is a finitely generated field extension of K, then there exists a discrete valuation ring R of

L dominating $\mathcal{O}$. Prove this in the following steps. By taking a polynomial ring over $\mathcal{O}$, reduce to the case where L is a *finite* extension field of K. Then show that for a suitable choice of generators $x_1, \ldots, x_n$ of $\mathfrak{m}$, the ideal $\mathfrak{a} = (x_1)$ in $\mathcal{O}' = \mathcal{O}[x_2/x_1, \ldots, x_n/x_1]$ is not equal to the unit ideal. Then let $\mathfrak{p}$ be a minimal prime ideal of $\mathfrak{a}$, and let $\mathcal{O}'_{\mathfrak{p}}$ be the localization of $\mathcal{O}'$ at $\mathfrak{p}$. This is a noetherian local domain of dimension 1 dominating $\mathcal{O}$. Let $\tilde{\mathcal{O}}'_{\mathfrak{p}}$ be the integral closure of $\mathcal{O}'_{\mathfrak{p}}$ in L. Use the theorem of Krull–Akizuki (see Nagata [7, p. 115]) to show that $\tilde{\mathcal{O}}'_{\mathfrak{p}}$ is noetherian of dimension 1. Finally, take R to be a localization of $\tilde{\mathcal{O}}'_{\mathfrak{p}}$ at one of its maximal ideals.

(b) Let $f: X \to Y$ be a morphism of finite type of noetherian schemes. Show that f is separated (respectively, proper) if and only if the criterion of (4.3) (respectively, (4.7)) holds for all *discrete* valuation rings.

4.12. *Examples of Valuation Rings.* Let k be an algebraically closed field.

(a) If K is a function field of dimension 1 over k (I, §6), then every valuation ring of K/k (except for K itself) is discrete. Thus the set of all of them is just the abstract nonsingular curve C_K of (I, §6).

(b) If K/k is a function field of dimension two, there are several different kinds of valuations. Suppose that X is a complete nonsingular surface with function field K.

   (1) If Y is an irreducible curve on X, with generic point x_1, then the local ring $R = \mathcal{O}_{x_1, X}$ is a discrete valuation ring of K/k with center at the (nonclosed) point x_1 on X.

   (2) If $f: X' \to X$ is a birational morphism, and if Y' is an irreducible curve in X' whose image in X is a single closed point x_0, then the local ring R of the generic point of Y' on X' is a discrete valuation ring of K/k with center at the closed point x_0 on X.

   (3) Let $x_0 \in X$ be a closed point. Let $f: X_1 \to X$ be the blowing-up of x_0 (I, §4) and let $E_1 = f^{-1}(x_0)$ be the exceptional curve. Choose a closed point $x_1 \in E_1$, let $f_2: X_2 \to X_1$ be the blowing-up of x_1, and let $E_2 = f_2^{-1}(x_1)$ be the exceptional curve. Repeat. In this manner we obtain a sequence of varieties X_i with closed points x_i chosen on them, and for each i, the local ring $\mathcal{O}_{x_{i+1}, X_{i+1}}$ dominates $\mathcal{O}_{x_i, X_i}$. Let $R_0 = \bigcup_{i=0}^{\infty} \mathcal{O}_{x_i, X_i}$. Then R_0 is a local ring, so it is dominated by some valuation ring R of K/k by (I, 6.1A). Show that R is a valuation ring of K/k, and that it has center x_0 on X. When is R a discrete valuation ring?

Note. We will see later (V, Ex. 5.6) that in fact the R_0 of (3) is already a valuation ring itself, so $R_0 = R$. Furthermore, every valuation ring of K/k (except for K itself) is one of the three kinds just described.

5 Sheaves of Modules

So far we have discussed schemes and morphisms between them without mentioning any sheaves other than the structure sheaves. We can increase the flexibility of our technique enormously by considering sheaves of modules on a given scheme. Especially important are quasi-coherent and coherent sheaves, which play the role of modules (respectively, finitely generated modules) over a ring.

In this section we will develop the basic properties of quasi-coherent and coherent sheaves. In particular we will introduce the important "twisting sheaf" $\mathcal{O}(1)$ of Serre on a projective scheme.

We will start by defining sheaves of modules on a ringed space.

Definitions. Let $(X,\mathcal{O}_X)$ be a ringed space (see §2). A *sheaf of $\mathcal{O}_X$-modules* (or simply an *$\mathcal{O}_X$-module*) is a sheaf $\mathscr{F}$ on X, such that for each open set $U \subseteq X$, the group $\mathscr{F}(U)$ is an $\mathcal{O}_X(U)$-module, and for each inclusion of open sets $V \subseteq U$, the restriction homomorphism $\mathscr{F}(U) \to \mathscr{F}(V)$ is compatible with the module structures via the ring homomorphism $\mathcal{O}_X(U) \to \mathcal{O}_X(V)$. A *morphism $\mathscr{F} \to \mathscr{G}$* of sheaves of $\mathcal{O}_X$-modules is a morphism of sheaves, such that for each open set $U \subseteq X$, the map $\mathscr{F}(U) \to \mathscr{G}(U)$ is a homomorphism of $\mathcal{O}_X(U)$-modules.

Note that the kernel, cokernel, and image of a morphism of $\mathcal{O}_X$-modules is again an $\mathcal{O}_X$-module. If $\mathscr{F}'$ is a subsheaf of $\mathcal{O}_X$-modules of an $\mathcal{O}_X$-module $\mathscr{F}$, then the quotient sheaf $\mathscr{F}/\mathscr{F}'$ is an $\mathcal{O}_X$-module. Any direct sum, direct product, direct limit, or inverse limit of $\mathcal{O}_X$-modules is an $\mathcal{O}_X$-module. If $\mathscr{F}$ and $\mathscr{G}$ are two $\mathcal{O}_X$-modules, we denote the group of morphisms from $\mathscr{F}$ to $\mathscr{G}$ by $\mathrm{Hom}_{\mathcal{O}_X}(\mathscr{F},\mathscr{G})$, or sometimes $\mathrm{Hom}_X(\mathscr{F},\mathscr{G})$ or $\mathrm{Hom}(\mathscr{F},\mathscr{G})$ if no confusion can arise. A sequence of $\mathcal{O}_X$-modules and morphisms is *exact* if it is exact as a sequence of sheaves of abelian groups.

If U is an open subset of X, and if $\mathscr{F}$ is an $\mathcal{O}_X$-module, then $\mathscr{F}|_U$ is an $\mathcal{O}_X|_U$-module. If $\mathscr{F}$ and $\mathscr{G}$ are two $\mathcal{O}_X$-modules, the presheaf

$$U \mapsto \mathrm{Hom}_{\mathcal{O}_X|_U}(\mathscr{F}|_U,\mathscr{G}|_U)$$

is a sheaf, which we call the *sheaf $\mathscr{H}om$* (Ex. 1.15), and denote by $\mathscr{H}om_{\mathcal{O}_X}(\mathscr{F},\mathscr{G})$. It is also an $\mathcal{O}_X$-module.

We define the *tensor product $\mathscr{F} \otimes_{\mathcal{O}_X} \mathscr{G}$* of two $\mathcal{O}_X$-modules to be the sheaf associated to the presheaf $U \mapsto \mathscr{F}(U) \otimes_{\mathcal{O}_X(U)} \mathscr{G}(U)$. We will often write simply $\mathscr{F} \otimes \mathscr{G}$, with $\mathcal{O}_X$ understood.

An $\mathcal{O}_X$-module $\mathscr{F}$ is *free* if it is isomorphic to a direct sum of copies of $\mathcal{O}_X$. It is *locally free* if X can be covered by open sets U for which $\mathscr{F}|_U$ is a free $\mathcal{O}_X|_U$-module. In that case the *rank* of $\mathscr{F}$ on such an open set is the number of copies of the structure sheaf needed (finite or infinite). If X is connected, the rank of a locally free sheaf is the same everywhere. A locally free sheaf of rank 1 is also called an *invertible sheaf*.

A *sheaf of ideals* on X is a sheaf of modules $\mathscr{I}$ which is a subsheaf of $\mathcal{O}_X$. In other words, for every open set U, $\mathscr{I}(U)$ is an ideal in $\mathcal{O}_X(U)$.

Let $f:(X,\mathcal{O}_X) \to (Y,\mathcal{O}_Y)$ be a morphism of ringed spaces (see §2). If $\mathscr{F}$ is an $\mathcal{O}_X$-module, then $f_*\mathscr{F}$ is an $f_*\mathcal{O}_X$-module. Since we have the morphism $f^\#:\mathcal{O}_Y \to f_*\mathcal{O}_X$ of sheaves of rings on Y, this gives $f_*\mathscr{F}$ a natural structure of $\mathcal{O}_Y$-module. We call it the *direct image* of $\mathscr{F}$ by the morphism f.

Now let $\mathscr{G}$ be a sheaf of $\mathcal{O}_Y$-modules. Then $f^{-1}\mathscr{G}$ is an $f^{-1}\mathcal{O}_Y$-module. Because of the adjoint property of f^{-1} (Ex. 1.18) we have a morphism

$f^{-1}\mathcal{O}_Y \to \mathcal{O}_X$ of sheaves of rings on X. We define $f^*\mathcal{G}$ to be the tensor product

$$f^{-1}\mathcal{G} \otimes_{f^{-1}\mathcal{O}_Y} \mathcal{O}_X.$$

Thus $f^*\mathcal{G}$ is an $\mathcal{O}_X$-module. We call it the *inverse image* of $\mathcal{G}$ by the morphism f.

As in (Ex. 1.18) one can show that f_* and f^* are adjoint functors between the category of $\mathcal{O}_X$-modules and the category of $\mathcal{O}_Y$-modules. To be precise, for any $\mathcal{O}_X$-module $\mathcal{F}$ and any $\mathcal{O}_Y$-module $\mathcal{G}$, there is a natural isomorphism of groups

$$\mathrm{Hom}_{\mathcal{O}_X}(f^*\mathcal{G},\mathcal{F}) \cong \mathrm{Hom}_{\mathcal{O}_Y}(\mathcal{G},f_*\mathcal{F}).$$

Now that we have the general notion of a sheaf of modules on a ringed space, we specialize to the case of schemes. We start by defining the sheaf of modules $\tilde{M}$ on Spec A associated to a module M over a ring A.

Definition. Let A be a ring and let M be an A-module. We define the *sheaf associated* to M on Spec A, denoted by $\tilde{M}$, as follows. For each prime ideal $\mathfrak{p} \subseteq A$, let $M_{\mathfrak{p}}$ be the localization of M at $\mathfrak{p}$. For any open set $U \subseteq$ Spec A we define the group $\tilde{M}(U)$ to be the set of functions $s: U \to \coprod_{\mathfrak{p} \in U} M_{\mathfrak{p}}$ such that for each $\mathfrak{p} \in U$, $s(\mathfrak{p}) \in M_{\mathfrak{p}}$, and such that s is locally a fraction m/f with $m \in M$ and $f \in A$. To be precise, we require that for each $\mathfrak{p} \in U$, there is a neighborhood V of $\mathfrak{p}$ in U, and there are elements $m \in M$ and $f \in A$, such that for each $\mathfrak{q} \in V$, $f \notin \mathfrak{q}$, and $s(\mathfrak{q}) = m/f$ in $M_{\mathfrak{q}}$. We make $\tilde{M}$ into a sheaf by using the obvious restriction maps.

Proposition 5.1. *Let A be a ring, let M be an A-module, and let $\tilde{M}$ be the sheaf on $X = $ Spec A associated to M. Then:*

(a) *$\tilde{M}$ is an $\mathcal{O}_X$-module;*

(b) *for each $\mathfrak{p} \in X$, the stalk $(\tilde{M})_{\mathfrak{p}}$ of the sheaf $\tilde{M}$ at $\mathfrak{p}$ is isomorphic to the localized module $M_{\mathfrak{p}}$;*

(c) *for any $f \in A$, the A_f-module $\tilde{M}(D(f))$ is isomorphic to the localized module M_f;*

(d) *in particular, $\Gamma(X,\tilde{M}) = M$.*

PROOF. Recalling the construction of the structure sheaf $\mathcal{O}_X$ from §2, it is clear that $\tilde{M}$ is an $\mathcal{O}_X$-module. The proofs of (b), (c), (d) are identical to the proofs of (a), (b), (c) of (2.2), replacing A by M at appropriate places.

Proposition 5.2. *Let A be a ring and let $X = $ Spec A. Also let $A \to B$ be a ring homomorphism, and let $f: $ Spec $B \to$ Spec A be the corresponding morphism of spectra. Then:*

(a) *the map $M \to \tilde{M}$ gives an exact, fully faithful functor from the category of A-modules to the category of $\mathcal{O}_X$-modules;*

(b) *if M and N are two A-modules, then $(M \otimes_A N)^\sim \cong \tilde{M} \otimes_{\mathcal{O}_X} \tilde{N}$;*

(c) *if $\{M_i\}$ is any family of A-modules, then $(\bigoplus M_i)^\sim \cong \bigoplus \tilde{M}_i$;*

(d) *for any B-module N we have* $f_*(\tilde{N}) \cong (_A N)^\sim$, *where* $_A N$ *means N considered as an A-module*;

(e) *for any A-module M we have* $f^*(\tilde{M}) \cong (M \otimes_A B)^\sim$.

PROOF. The map $M \to \tilde{M}$ is clearly functorial. It is exact, because localization is exact, and exactness of sheaves can be measured at the stalks (use (Ex. 1.2) and (5.1b)). It commutes with direct sum and tensor product, because these commute with localization. To say it is fully faithful means that for any A-modules M and N, we have $\operatorname{Hom}_A(M,N) = \operatorname{Hom}_{\mathcal{O}_X}(\tilde{M},\tilde{N})$. The functor $\sim$ gives a natural map $\operatorname{Hom}_A(M,N) \to \operatorname{Hom}_{\mathcal{O}_X}(\tilde{M},\tilde{N})$. Applying Γ and using (5.1d) gives a map the other way. These two maps are clearly inverse to each other, hence isomorphisms. The last statements about f_* and f^* follow directly from the definitions.

These sheaves of the form $\tilde{M}$ on affine schemes are our models for quasi-coherent sheaves. A quasi-coherent sheaf on a scheme X will be an $\mathcal{O}_X$-module which is locally of the form $\tilde{M}$. In the next few lemmas and propositions, we will show that this is a local property, and we will establish some facts about quasi-coherent and coherent sheaves.

Definition. Let $(X,\mathcal{O}_X)$ be a scheme. A sheaf of $\mathcal{O}_X$-modules $\mathcal{F}$ is *quasi-coherent* if X can be covered by open affine subsets $U_i = \operatorname{Spec} A_i$, such that for each i there is an A_i-module M_i with $\mathcal{F}|_{U_i} \cong \tilde{M}_i$. We say that $\mathcal{F}$ is *coherent* if furthermore each M_i can be taken to be a finitely generated A_i-module.

Although we have just defined the notion of quasi-coherent and coherent sheaves on an arbitrary scheme, we will normally not mention coherent sheaves unless the scheme is noetherian. This is because the notion of coherence is not at all well-behaved on a nonnoetherian scheme.

Example 5.2.1. On any scheme X, the structure sheaf $\mathcal{O}_X$ is quasi-coherent (and in fact coherent).

Example 5.2.2. If $X = \operatorname{Spec} A$ is an affine scheme, if $Y \subseteq X$ is the closed subscheme defined by an ideal $\mathfrak{a} \subseteq A$ (3.2.3), and if $i: Y \to X$ is the inclusion morphism, then $i_*\mathcal{O}_Y$ is a quasi-coherent (in fact coherent) $\mathcal{O}_X$-module. Indeed, it is isomorphic to $(A/\mathfrak{a})^\sim$.

Example 5.2.3. If U is an open subscheme of a scheme X, with inclusion map $j: U \to X$, then the sheaf $j_!(\mathcal{O}_U)$ obtained by extending $\mathcal{O}_U$ by zero outside of U (Ex. 1.19), is an $\mathcal{O}_X$-module, but it is not in general quasi-coherent. For example, suppose X is integral, and $V = \operatorname{Spec} A$ is any open affine subset of X, not contained in U. Then $j_!(\mathcal{O}_U)|_V$ has no global sections over V, and yet it is not the zero sheaf. Hence it cannot be of the form $\tilde{M}$ for any A-module M.

Example 5.2.4. If Y is a closed subscheme of a scheme X, then the sheaf $\mathcal{O}_X|_Y$ is not in general quasi-coherent on Y. In fact, it is not even an $\mathcal{O}_Y$-module in general.

Example 5.2.5. Let X be an integral noetherian scheme, and let $\mathcal{K}$ be the constant sheaf with group K equal to the function field of X (Ex. 3.6). Then $\mathcal{K}$ is a quasi-coherent $\mathcal{O}_X$-module, but it is not coherent unless X is reduced to a point.

Lemma 5.3. *Let $X = \operatorname{Spec} A$ be an affine scheme, let $f \in A$, let $D(f) \subseteq X$ be the corresponding open set, and let $\mathcal{F}$ be a quasi-coherent sheaf on X.*
    (a) If $s \in \Gamma(X, \mathcal{F})$ is a global section of $\mathcal{F}$ whose restriction to $D(f)$ is 0, then for some $n > 0$, $f^n s = 0$.
    (b) Given a section $t \in \mathcal{F}(D(f))$ of $\mathcal{F}$ over the open set $D(f)$, then for some $n > 0$, $f^n t$ extends to a global section of $\mathcal{F}$ over X.

PROOF. First we note that since $\mathcal{F}$ is quasi-coherent, X can be covered by open affine subsets of the form $V = \operatorname{Spec} B$, such that $\mathcal{F}|_V \cong \tilde{M}$ for some B-module M. Now the open sets of the form $D(g)$ form a base for the topology of X (see §2), so we can cover V by open sets of the form $D(g)$, for various $g \in A$. An inclusion $D(g) \subseteq V$ corresponds to a ring homomorphism $B \to A_g$ by (2.3). Hence $\mathcal{F}|_{D(g)} \cong (M \otimes_B A_g)^\sim$ by (5.2). Thus we have shown that if $\mathcal{F}$ is quasi-coherent on X, then X can be covered by open sets of the form $D(g_i)$ where for each i, $\mathcal{F}|_{D(g_i)} \cong \tilde{M}_i$ for some module M_i over the ring A_{g_i}. Since X is quasi-compact, a finite number of these open sets will do.

(a) Now suppose given $s \in \Gamma(X, \mathcal{F})$ with $s|_{D(f)} = 0$. For each i, s restricts to give a section s_i of $\mathcal{F}$ over $D(g_i)$, in other words, an element $s_i \in M_i$ (using (5.1d)). Now $D(f) \cap D(g_i) = D(fg_i)$, so $\mathcal{F}|_{D(fg_i)} = (M_i)_f^\sim$ using (5.1c). Thus the image of s_i in $(M_i)_f$ is zero, so by the definition of localization, $f^n s_i = 0$ for some n. This n may depend on i, but since there are only finitely many i, we can pick n large enough to work for them all. Then since the $D(g_i)$ cover X, we have $f^n s = 0$.

(b) Given an element $t \in \mathcal{F}(D(f))$, we restrict it for each i to get an element t of $\mathcal{F}(D(fg_i)) = (M_i)_f$. Then by the definition of localization, for some $n > 0$ there is an element $t_i \in M_i = \mathcal{F}(D(g_i))$ which restricts to $f^n t$ on $D(fg_i)$. The integer n may depend on i, but again we take one large enough to work for all i. Now on the intersection $D(g_i) \cap D(g_j) = D(g_i g_j)$ we have two sections t_i and t_j of $\mathcal{F}$, which agree on $D(fg_i g_j)$ where they are both equal to $f^n t$. Hence by part (a) above, there is an integer $m > 0$ such that $f^m(t_i - t_j) = 0$ on $D(g_i g_j)$. This m depends on i and j, but we take one m large enough for all. Now the local sections $f^m t_i$ of $\mathcal{F}$ on $D(g_i)$ glue together to give a global section s of $\mathcal{F}$, whose restriction to $D(f)$ is $f^{n+m} t$.

Proposition 5.4. *Let X be a scheme. Then an $\mathcal{O}_X$-module $\mathcal{F}$ is quasi-coherent if and only if for every open affine subset $U = \text{Spec } A$ of X, there is an A-module M such that $\mathcal{F}|_U \cong \tilde{M}$. If X is noetherian, then $\mathcal{F}$ is coherent if and only if the same is true, with the extra condition that M be a finitely generated A-module.*

PROOF. Let $\mathcal{F}$ be quasi-coherent on X, and let $U = \text{Spec } A$ be an open affine. As in the proof of the lemma, there is a base for the topology consisting of open affines for which the restriction of $\mathcal{F}$ is the sheaf associated to a module. It follows that $\mathcal{F}|_U$ is quasi-coherent, so we can reduce to the case X affine $= \text{Spec } A$. Let $M = \Gamma(X,\mathcal{F})$. Then in any case there is a natural map $\alpha : \tilde{M} \to \mathcal{F}$ (Ex. 5.3). Since $\mathcal{F}$ is quasi-coherent, X can be covered by open sets $D(g_i)$ with $\mathcal{F}|_{D(g_i)} \cong \tilde{M}_i$ for some A_{g_i}-module M_i. Now the lemma, applied to the open set $D(g_i)$, tells us exactly that $\mathcal{F}(D(g_i)) \cong M_{g_i}$, so $M_i = M_{g_i}$. It follows that the map α, restricted to $D(g_i)$, is an isomorphism. The $D(g_i)$ cover X, so α is an isomorphism.

Now suppose that X is noetherian, and $\mathcal{F}$ coherent. Then, using the above notation, we have the additional information that each M_{g_i} is a finitely generated A_{g_i}-module, and we want to prove that M is finitely generated. Since the rings A and A_{g_i} are noetherian, the modules M_{g_i} are noetherian, and we have to prove that M is noetherian. For this we just use the proof of (3.2) with A replaced by M in appropriate places.

Corollary 5.5. *Let A be a ring and let $X = \text{Spec } A$. The functor $M \mapsto \tilde{M}$ gives an equivalence of categories between the category of A-modules and the category of quasi-coherent $\mathcal{O}_X$-modules. Its inverse is the functor $\mathcal{F} \mapsto \Gamma(X,\mathcal{F})$. If A is noetherian, the same functor also gives an equivalence of categories between the category of finitely generated A-modules and the category of coherent $\mathcal{O}_X$-modules.*

PROOF. The only new information here is that $\mathcal{F}$ is quasi-coherent on X if and only if it is of the form $\tilde{M}$, and in that case $M = \Gamma(X,\mathcal{F})$. This follows from (5.4).

Proposition 5.6. *Let X be an affine scheme, let $0 \to \mathcal{F}' \to \mathcal{F} \to \mathcal{F}'' \to 0$ be an exact sequence of $\mathcal{O}_X$-modules, and assume that $\mathcal{F}'$ is quasi-coherent. Then the sequence*

$$0 \to \Gamma(X,\mathcal{F}') \to \Gamma(X,\mathcal{F}) \to \Gamma(X,\mathcal{F}'') \to 0$$

is exact.

PROOF. We know already that Γ is a left-exact functor (Ex. 1.8) so we have only to show that the last map is surjective. Let $s \in \Gamma(X,\mathcal{F}'')$ be a global section of $\mathcal{F}''$. Since the map of sheaves $\mathcal{F} \to \mathcal{F}''$ is surjective, for any $x \in X$ there is an open neighborhood $D(f)$ of x, such that $s|_{D(f)}$ lifts to a section $t \in \mathcal{F}(D(f))$ (Ex. 1.3). I claim that for some $n > 0$, $f^n s$ lifts to a global section of $\mathcal{F}$. Indeed, we can cover X with a finite number of open

113

sets $D(g_i)$, such that for each i, $s|_{D(g_i)}$ lifts to a section $t_i \in \mathscr{F}(D(g_i))$. On $D(f) \cap D(g_i) = D(fg_i)$, we have two sections $t, t_i \in \mathscr{F}(D(fg_i))$ both lifting s. Therefore $t - t_i \in \mathscr{F}'(D(fg_i))$. Since $\mathscr{F}'$ is quasi-coherent, by (5.3b) for some $n > 0$, $f^n(t - t_i)$ extends to a section $u_i \in \mathscr{F}'(D(g_i))$. As usual, we pick one n to work for all i. Let $t'_i = f^n t_i + u_i$. Then t'_i is a lifting of $f^n s$ on $D(g_i)$, and furthermore t'_i and $f^n t$ agree on $D(fg_i)$. Now on $D(g_i g_j)$ we have two sections t'_i and t'_j of $\mathscr{F}$, both of which lift $f^n s$, so $t'_i - t'_j \in \mathscr{F}'(D(g_i g_j))$. Furthermore, t'_i and t'_j are equal on $D(fg_i g_j)$, so by (5.3a) we have $f^m(t'_i - t'_j) = 0$ for some $m > 0$, which we may take independent of i and j. Now the sections $f^m t'_i$ of $\mathscr{F}$ glue to give a global section t'' of $\mathscr{F}$ over X, which lifts $f^{n+m} s$. This proves the claim.

Now cover X by a finite number of open sets $D(f_i)$, $i = 1, \ldots, r$, such that $s|_{D(f_i)}$ lifts to a section of $\mathscr{F}$ over $D(f_i)$ for each i. Then by the claim, we can find an integer n (one for all i) and global sections $t_i \in \Gamma(X, \mathscr{F})$ such that t_i is a lifting of $f_i^n s$. Now the open sets $D(f_i)$ cover X, so the ideal $(f_1^n, \ldots, f_r^n)$ is the unit ideal of A, and we can write $1 = \sum_{i=1}^r a_i f_i^n$, with $a_i \in A$. Let $t = \sum a_i t_i$. Then t is a global section of $\mathscr{F}$ whose image in $\Gamma(X, \mathscr{F}'')$ is $\sum a_i f_i^n s = s$. This completes the proof.

Remark 5.6.1. When we have developed the techniques of cohomology, we will see that this proposition is an immediate consequence of the fact that $H^1(X, \mathscr{F}') = 0$ for any quasi-coherent sheaf $\mathscr{F}'$ on an affine scheme X (III, 3.5).

Proposition 5.7. *Let X be a scheme. The kernel, cokernel, and image of any morphism of quasi-coherent sheaves are quasi-coherent. Any extension of quasi-coherent sheaves is quasi-coherent. If X is noetherian, the same is true for coherent sheaves.*

PROOF. The question is local, so we may assume X is affine. The statement about kernels, cokernels and images follows from the fact that the functor $M \mapsto \tilde{M}$ is exact and fully faithful from A-modules to quasi-coherent sheaves (5.2a and 5.5). The only nontrivial part is to show that an extension of quasi-coherent sheaves is quasi-coherent. So let $0 \to \mathscr{F}' \to \mathscr{F} \to \mathscr{F}'' \to 0$ be an exact sequence of $\mathcal{O}_X$-modules, with $\mathscr{F}'$ and $\mathscr{F}''$ quasi-coherent. By (5.6), the corresponding sequence of global sections over X is exact, say $0 \to M' \to M \to M'' \to 0$. Applying the functor $\sim$, we get an exact commutative diagram

$$0 \to \tilde{M}' \to \tilde{M} \to \tilde{M}'' \to 0$$
$$\downarrow \qquad \downarrow \qquad \downarrow$$
$$0 \to \mathscr{F}' \to \mathscr{F} \to \mathscr{F}'' \to 0.$$

The two outside arrows are isomorphisms, since $\mathscr{F}'$ and $\mathscr{F}''$ are quasi-coherent. So by the 5-lemma, the middle one is also, showing that $\mathscr{F}$ is quasi-coherent.

In the noetherian case, if $\mathscr{F}'$ and $\mathscr{F}''$ are coherent, then M' and M'' are finitely generated, so M is also finitely generated, and hence $\mathscr{F}$ is coherent.

Proposition 5.8. *Let $f: X \to Y$ be a morphism of schemes.*

(a) *If $\mathscr{G}$ is a quasi-coherent sheaf of $\mathcal{O}_Y$-modules, then $f^*\mathscr{G}$ is a quasi-coherent sheaf of $\mathcal{O}_X$-modules.*

(b) *If X and Y are noetherian, and if $\mathscr{G}$ is coherent, then $f^*\mathscr{G}$ is coherent.*

(c) *Assume that either X is noetherian, or f is quasi-compact (Ex. 3.2) and separated. Then if $\mathscr{F}$ is a quasi-coherent sheaf of $\mathcal{O}_X$-modules, $f_*\mathscr{F}$ is a quasi-coherent sheaf of $\mathcal{O}_Y$-modules.*

PROOF.

(a) The question is local on both X and Y, so we can assume X and Y both affine. In this case the result follows from (5.5) and (5.2e).

(b) In the noetherian case, the same proof works for coherent sheaves.

(c) Here the question is local on Y only, so we may assume that Y is affine. Then X is quasi-compact (under either hypothesis) so we can cover X with a finite number of open affine subsets U_i. In the separated case, $U_i \cap U_j$ is again affine (Ex. 4.3). Call it U_{ijk}. In the noetherian case, $U_i \cap U_j$ is at least quasi-compact, so we can cover it with a finite number of open affine subsets U_{ijk}. Now for any open subset V of Y, giving a section s of $\mathscr{F}$ over $f^{-1}V$ is the same thing as giving a collection of sections s_i of $\mathscr{F}$ over $(f^{-1}V) \cap U_i$ whose restrictions to the open subsets $f^{-1}(V) \cap U_{ijk}$ are all equal. This is just the sheaf property (§1). Therefore, there is an exact sequence of sheaves on Y,

$$0 \to f_*\mathscr{F} \to \bigoplus_i f_*(\mathscr{F}|_{U_i}) \to \bigoplus_{i,j,k} f_*(\mathscr{F}|_{U_{ijk}}),$$

where by abuse of notation we denote also by f the induced morphisms $U_i \to Y$ and $U_{ijk} \to Y$. Now $f_*(\mathscr{F}|_{U_i})$ and $f_*(\mathscr{F}|_{U_{ijk}})$ are quasi-coherent by (5.2d). Thus $f_*\mathscr{F}$ is quasi-coherent by (5.7).

Caution 5.8.1. If X and Y are noetherian, it is *not* true in general that f_* of a coherent sheaf is coherent (Ex. 5.5). However, it is true if f is a finite morphism (Ex. 5.5) or a projective morphism (5.20) or (III, 8.8), or more generally, a proper morphism: see Grothendieck [EGA III, 3.2.1].

As a first application of these concepts, we will discuss the sheaf of ideals of a closed subscheme.

Definition. Let Y be a closed subscheme of a scheme X, and let $i: Y \to X$ be the inclusion morphism. We define the *ideal sheaf* of Y, denoted $\mathscr{I}_Y$, to be the kernel of the morphism $i^{\#}: \mathcal{O}_X \to i_*\mathcal{O}_Y$.

Proposition 5.9. *Let X be a scheme. For any closed subscheme Y of X, the corresponding ideal sheaf $\mathscr{I}_Y$ is a quasi-coherent sheaf of ideals on X. If X is noetherian, it is coherent. Conversely, any quasi-coherent sheaf of ideals on X is the ideal sheaf of a uniquely determined closed subscheme of X.*

PROOF. If Y is a closed subscheme of X, then the inclusion morphism $i: Y \to X$ is quasi-compact (obvious) and separated (4.6), so by (5.8), $i_*\mathscr{O}_Y$ is quasi-coherent on X. Hence $\mathscr{I}_Y$, being the kernel of a morphism of quasi-coherent sheaves, is also quasi-coherent. If X is noetherian, then for any open affine subset $U = \operatorname{Spec} A$ of X, the ring A is noetherian, so the ideal $I = \Gamma(U, \mathscr{I}_Y|_U)$, is finitely generated, so $\mathscr{I}_Y$ is coherent.

Conversely, given a scheme X and a quasi-coherent sheaf of ideals $\mathscr{I}$, let Y be the support of the quotient sheaf $\mathscr{O}_X/\mathscr{I}$. Then Y is a subspace of X, and $(Y, \mathscr{O}_X/\mathscr{I})$ is the unique closed subscheme of X with ideal sheaf $\mathscr{I}$. The unicity is clear, so we have only to check that $(Y, \mathscr{O}_X/\mathscr{I})$ is a closed subscheme. This is a local question, so we may assume $X = \operatorname{Spec} A$ is affine. Since $\mathscr{I}$ is quasi-coherent, $\mathscr{I} = \tilde{\mathfrak{a}}$ for some ideal $\mathfrak{a} \subseteq A$. Then $(Y, \mathscr{O}_X/\mathscr{I})$ is just the closed subscheme of X determined by the ideal $\mathfrak{a}$ (3.2.3).

Corollary 5.10. *If $X = \operatorname{Spec} A$ is an affine scheme, there is a 1-1 correspondence between ideals $\mathfrak{a}$ in A and closed subschemes Y of X, given by $\mathfrak{a} \mapsto$ image of $\operatorname{Spec} A/\mathfrak{a}$ in X (3.2.3). In particular, every closed subscheme of an affine scheme is affine.*

PROOF. By (5.5) the quasi-coherent sheaves of ideals on X are in 1-1 correspondence with the ideals of A.

Our next concern is to study quasi-coherent sheaves on the Proj of a graded ring. As in the case of Spec, there is a connection between modules over the ring and sheaves on the space, but it is more complicated.

Definition. Let S be a graded ring and let M be a graded S-module. (See (I, §7) for generalities on graded modules.) We define the *sheaf associated to M* on Proj S, denoted by $\tilde{M}$, as follows. For each $\mathfrak{p} \in \operatorname{Proj} S$, let $M_{(\mathfrak{p})}$ be the group of elements of degree 0 in the localization $T^{-1}M$, where T is the multiplicative system of homogeneous elements of S not in $\mathfrak{p}$ (cf. definition of Proj in §2). For any open subset $U \subseteq \operatorname{Proj} S$ we define $\tilde{M}(U)$ to be the set of functions s from U to $\coprod_{\mathfrak{p} \in U} M_{(\mathfrak{p})}$ which are locally fractions. This means that for every $\mathfrak{p} \in U$, there is a neighborhood V of $\mathfrak{p}$ in U, and homogeneous elements $m \in M$ and $f \in S$ of the same degree, such that for every $\mathfrak{q} \in V$, we have $f \notin \mathfrak{q}$, and $s(\mathfrak{q}) = m/f$ in $M_{(\mathfrak{q})}$. We make $\tilde{M}$ into a sheaf with the obvious restriction maps.

Proposition 5.11. *Let S be a graded ring, and M a graded S-module. Let $X = \operatorname{Proj} S$.*

(a) *For any* $\mathfrak{p} \in X$, *the stalk* $(\tilde{M})_\mathfrak{p} = M_{(\mathfrak{p})}$.

(b) *For any homogeneous* $f \in S_+$, *we have* $\tilde{M}|_{D_+(f)} \cong (M_{(f)})^\sim$ *via the isomorphism of* $D_+(f)$ *with* Spec $S_{(f)}$ *(see (2.5b)), where* $M_{(f)}$ *denotes the group of elements of degree 0 in the localized module* M_f.

(c) $\tilde{M}$ *is a quasi-coherent* $\mathcal{O}_X$*-module. If* S *is noetherian and* M *is finitely generated, then* $\tilde{M}$ *is coherent.*

PROOF. For (a) and (b), just repeat the proof of (2.5), with M in place of S. Then (c) follows from (b).

Definition. Let S be a graded ring, and let $X = $ Proj S. For any $n \in \mathbf{Z}$, we define the sheaf $\mathcal{O}_X(n)$ to be $S(n)^\sim$. We call $\mathcal{O}_X(1)$ the *twisting sheaf* of Serre. For any sheaf of $\mathcal{O}_X$-modules, $\mathscr{F}$, we denote by $\mathscr{F}(n)$ the *twisted sheaf* $\mathscr{F} \otimes_{\mathcal{O}_X} \mathcal{O}_X(n)$.

Proposition 5.12. *Let* S *be a graded ring and let* $X = $ Proj S. *Assume that* S *is generated by* S_1 *as an* S_0*-algebra.*

(a) *The sheaf* $\mathcal{O}_X(n)$ *is an invertible sheaf on* X.

(b) *For any graded* S*-module* M, $\tilde{M}(n) \cong (M(n))^\sim$. *In particular,* $\mathcal{O}_X(n) \otimes \mathcal{O}_X(m) \cong \mathcal{O}_X(n + m)$.

(c) *Let* T *be another graded ring, generated by* T_1 *as a* T_0*-algebra, let* $\varphi : S \to T$ *be a homomorphism preserving degrees, and let* $U \subseteq Y = $ Proj T *and* $f : U \to X$ *be the morphism determined by* φ *(Ex. 2.14). Then* $f^*(\mathcal{O}_X(n)) \cong \mathcal{O}_Y(n)|_U$ *and* $f_*(\mathcal{O}_Y(n)|_U) \cong (f_* \mathcal{O}_U)(n)$.

PROOF.

(a) Recall that invertible means locally free of rank 1. Let $f \in S_1$, and consider the restriction $\mathcal{O}_X(n)|_{D_+(f)}$. By the previous proposition this is isomorphic to $S(n)^\sim_{(f)}$ on Spec $S_{(f)}$. We will show that this restriction is free of rank 1. Indeed, $S(n)_{(f)}$ is a free $S_{(f)}$-module of rank 1. For $S_{(f)}$ is the group of elements of degree 0 in S_f, and $S(n)_{(f)}$ is the group of elements of degree n in S_f. We obtain an isomorphism of one to the other by sending s to $f^n s$. This makes sense, for any $n \in \mathbf{Z}$, because f is invertible in S_f. Now since S is generated by S_1 as an S_0-algebra, X is covered by the open sets $D_+(f)$ for $f \in S_1$. Hence $\mathcal{O}(n)$ is invertible.

(b) This follows from the fact that $(M \otimes_S N)^\sim \cong \tilde{M} \otimes_{\mathcal{O}_X} \tilde{N}$ for any two graded S-modules M and N, when S is generated by S_1. Indeed, for any $f \in S_1$ we have $(M \otimes_S N)_{(f)} = M_{(f)} \otimes_{S_{(f)}} N_{(f)}$.

(c) More generally, for any graded S-module M, $f^*(\tilde{M}) \cong (M \otimes_S T)^\sim|_U$ and for any graded T-module N, $f_*(\tilde{N}|_U) \cong (_S N)^\sim$. Furthermore, the sheaf $\tilde{T}$ on X is just $f_*(\mathcal{O}_U)$. The proofs are straightforward (cf. (5.2) for the affine case).

The twisting operation allows us to define a graded S-module associated to any sheaf of modules on $X = $ Proj S.

Definition. Let S be a graded ring, let $X = \text{Proj } S$, and let $\mathscr{F}$ be a sheaf of $\mathcal{O}_X$-modules. We define the *graded S-module associated* to $\mathscr{F}$ as a group, to be $\Gamma_*(\mathscr{F}) = \bigoplus_{n \in \mathbf{Z}} \Gamma(X, \mathscr{F}(n))$. We give it a structure of graded S-module as follows. If $s \in S_d$, then s determines in a natural way a global section $s \in \Gamma(X, \mathcal{O}_X(d))$. Then for any $t \in \Gamma(X, \mathscr{F}(n))$ we define the product $s \cdot t$ in $\Gamma(X, \mathscr{F}(n + d))$ by taking the tensor product $s \otimes t$ and using the natural map $\mathscr{F}(n) \otimes \mathcal{O}_X(d) \cong \mathscr{F}(n + d)$.

Proposition 5.13. *Let A be a ring, let $S = A[x_0, \ldots, x_r]$, $r \geq 1$, and let $X = \text{Proj } S$. (This is just projective r-space over A.) Then $\Gamma_*(\mathcal{O}_X) \cong S$.*

PROOF. We cover X with the open sets $D_+(x_i)$. Then to give a section $t \in \Gamma(X, \mathcal{O}_X(n))$ is the same as giving sections $t_i \in \mathcal{O}_X(n)(D_+(x_i))$ for each i, which agree on the intersections $D_+(x_i x_j)$. Now t_i is just a homogeneous element of degree n in the localization S_{x_i}, and its restriction to $D_+(x_i x_j)$ is just the image of that element in $S_{x_i x_j}$. Summing over all n, we see that $\Gamma_*(\mathcal{O}_X)$ can be identified with the set of $(r + 1)$-tuples $(t_0, \ldots, t_r)$ where for each i, $t_i \in S_{x_i}$, and for each i,j, the images of t_i and t_j in $S_{x_i x_j}$ are the same.

Now the x_i are not zero divisors in S, so the localization maps $S \to S_{x_i}$ and $S_{x_i} \to S_{x_i x_j}$ are all injective, and these rings are all subrings of $S' = S_{x_0 \cdots x_r}$. Hence $\Gamma_*(\mathcal{O}_X)$ is the intersection $\bigcap S_{x_i}$ taken inside S'. Now any homogeneous element of S' can be written uniquely as a product $x_0^{i_0} \cdots x_r^{i_r} f(x_0, \ldots, x_r)$, where the $i_j \in \mathbf{Z}$, and f is a homogeneous polynomial not divisible by any x_i. This element will be in S_{x_i} if and only if $i_j \geq 0$ for $j \neq i$. It follows that the intersection of all the S_{x_i} (in fact the intersection of any two of them) is exactly S.

Caution 5.13.1. If S is a graded ring which is not a polynomial ring, then it is not true in general that $\Gamma_*(\mathcal{O}_X) = S$ (Ex. 5.14).

Lemma 5.14. *Let X be a scheme, let $\mathscr{L}$ be an invertible sheaf on X, let $f \in \Gamma(X, \mathscr{L})$, let X_f be the open set of points $x \in X$ where $f_x \notin \mathfrak{m}_x \mathscr{L}_x$, and let $\mathscr{F}$ be a quasi-coherent sheaf on X.*

(a) Suppose that X is quasi-compact, and let $s \in \Gamma(X, \mathscr{F})$ be a global section of $\mathscr{F}$ whose restriction to X_f is 0. Then for some $n > 0$, we have $f^n s = 0$, where $f^n s$ is considered as a global section of $\mathscr{F} \otimes \mathscr{L}^{\otimes n}$.

(b) Suppose furthermore that X has a finite covering by open affine subsets U_i, such that $\mathscr{L}|_{U_i}$ is free for each i, and such that $U_i \cap U_j$ is quasi-compact for each i,j. Given a section $t \in \Gamma(X_f, \mathscr{F})$, then for some $n > 0$, the section $f^n t \in \Gamma(X_f, \mathscr{F} \otimes \mathscr{L}^{\otimes n})$ extends to a global section of $\mathscr{F} \otimes \mathscr{L}^{\otimes n}$.

PROOF. This lemma is a direct generalization of (5.3), with an extra twist due to the presence of the invertible sheaf $\mathscr{L}$. It also generalizes (Ex. 2.16). To prove (a), we first cover X with a finite number (possible since X is quasi-compact) of open affines $U = \text{Spec } A$ such that $\mathscr{L}|_U$ is free. Let $\psi : \mathscr{L}|_U \cong \mathcal{O}_U$ be an isomorphism expressing the freeness of $\mathscr{L}|_U$. Since $\mathscr{F}$ is quasi-coherent, by (5.4) there is an A-module M with $\mathscr{F}|_U \cong \tilde{M}$. Our section $s \in \Gamma(X,\mathscr{F})$ restricts to give an element $s \in M$. On the other hand, our section $f \in \Gamma(X,\mathscr{L})$ restricts to give a section of $\mathscr{L}|_U$, which in turn gives rise to an element $g = \psi(f) \in A$. Clearly $X_f \cap U = D(g)$. Now $s|_{X_f}$ is zero, so $g^n s = 0$ in M for some $n > 0$, just as in the proof of (5.3). Using the isomorphism

$$\mathrm{id} \times \psi^{\otimes n} : \mathscr{F} \otimes \mathscr{L}^n|_U \cong \mathscr{F}|_U,$$

we conclude that $f^n s \in \Gamma(U, \mathscr{F} \otimes \mathscr{L}^n)$ is zero. This statement is intrinsic (i.e., independent of ψ). So now we do this for each open set of the covering, pick one n large enough to work for all the sets of the covering, and we find $f^n s = 0$ on X.

To prove (b), we proceed as in the proof of (5.3), keeping track of the twist due to $\mathscr{L}$ as above. The hypothesis $U_i \cap U_j$ quasi-compact is used to be able to apply part (a) there.

Remark 5.14.1. The hypotheses on X made in the statements (a) and (b) above are satisfied either if X is noetherian (in which case every open set is quasi-compact) or if X is quasi-compact and separated (in which case the intersection of two open affine subsets is again affine, hence quasi-compact).

Proposition 5.15. *Let S be a graded ring, which is finitely generated by S_1 as an S_0-algebra. Let $X = \mathrm{Proj}\, S$, and let $\mathscr{F}$ be a quasi-coherent sheaf on X. Then there is a natural isomorphism $\beta : \Gamma_*(\mathscr{F})^\sim \to \mathscr{F}$.*

PROOF. First let us define the morphism β for any $\mathscr{O}_X$-module $\mathscr{F}$. Let $f \in S_1$. Since $\Gamma_*(\mathscr{F})^\sim$ is quasi-coherent in any case, to define β, it is enough to give the image of a section of $\Gamma_*(\mathscr{F})^\sim$ over $D_+(f)$ (see Ex. 5.3). Such a section is represented by a fraction m/f^d, where $m \in \Gamma(X, \mathscr{F}(d))$, for some $d \geqslant 0$. We can think of f^{-d} as a section of $\mathscr{O}_X(-d)$, defined over $D_+(f)$. Taking their tensor product, we obtain $m \otimes f^{-d}$ as a section of $\mathscr{F}$ over $D_+(f)$. This defines β.

Now let $\mathscr{F}$ be quasi-coherent. To show that β is an isomorphism we have to identify the module $\Gamma_*(\mathscr{F})_{(f)}$ with the sections of $\mathscr{F}$ over $D_+(f)$. We apply (5.14), considering f as a global section of the invertible sheaf $\mathscr{L} = \mathscr{O}(1)$. Since we have assumed that S is finitely generated by S_1 as an S_0-algebra, we can find finitely many elements $f_0, \ldots, f_r \in S_1$ such that X is covered by the open affine subsets $D_+(f_i)$. The intersections $D_+(f_i) \cap D_+(f_j)$ are also affine, and $\mathscr{L}|_{D_+(f_i)}$ is free for each i, so the hypotheses of (5.14) are satisfied. The conclusion of (5.14) tells us that $\mathscr{F}(D_+(f)) \cong \Gamma_*(\mathscr{F})_{(f)}$, which is just what we wanted.

Corollary 5.16. *Let A be a ring.*

(a) *If Y is a closed subscheme of $\mathbf{P}^r_A$, then there is a homogeneous ideal $I \subseteq S = A[x_0, \ldots, x_r]$ such that Y is the closed subscheme determined by I* (Ex. 3.12).

(b) *A scheme Y over* Spec *A is projective if and only if it is isomorphic to* Proj *S for some graded ring S, where* $S_0 = A$, *and S is finitely generated by* S_1 *as an* S_0-*algebra.*

PROOF.

(a) Let $\mathscr{I}_Y$ be the ideal sheaf of Y on $X = \mathbf{P}_A^r$. Now $\mathscr{I}_Y$ is a subsheaf of $\mathcal{O}_X$; the twisting functor is exact; the global section functor Γ is left exact; hence $\Gamma_*(\mathscr{I}_Y)$ is a submodule of $\Gamma_*(\mathcal{O}_X)$. But by (5.13), $\Gamma_*(\mathcal{O}_X) = S$. Hence $\Gamma_*(\mathscr{I}_Y)$ is a homogeneous ideal of S, which we will call I. Now I determines a closed subscheme of X (Ex. 3.12), whose sheaf of ideals will be $\tilde{I}$. Since $\mathscr{I}_Y$ is quasi-coherent by (5.9), we have $\mathscr{I}_Y \cong \tilde{I}$ by (5.15), and hence Y is the subscheme determined by I. In fact, $\Gamma_*(\mathscr{I}_Y)$ is the largest ideal in S defining Y (Ex. 5.10).

(b) Recall that by definition Y is projective over Spec A if it is isomorphic to a closed subscheme of $\mathbf{P}_A^r$ for some r (§4). By part (a), any such Y is isomorphic to Proj S/I, and we can take I to be contained in $S_+ = \bigoplus_{d>0} S_d$ (Ex. 3.12), so that $(S/I)_0 = A$. Conversely, any such graded ring S is a quotient of a polynomial ring, so Proj S is projective.

Definition. For any scheme Y, we define the *twisting sheaf* $\mathcal{O}(1)$ on $\mathbf{P}_Y^r$ to be $g^*(\mathcal{O}(1))$, where $g: \mathbf{P}_Y^r \to \mathbf{P}_Z^r$ is the natural map (recall that $\mathbf{P}_Y^r$ was defined as $\mathbf{P}_Z^r \times_Z Y$).

Note that if $Y = $ Spec A, this is the same as the $\mathcal{O}(1)$ already defined on $\mathbf{P}_A^r = $ Proj $A[x_0, \ldots, x_r]$, by (5.12c).

Definition. If X is any scheme over Y, an invertible sheaf $\mathscr{L}$ on X is *very ample* relative to Y, if there is an immersion $i: X \to \mathbf{P}_Y^r$ for some r, such that $i^*(\mathcal{O}(1)) \cong \mathscr{L}$. We say that a morphism $i: X \to Z$ is an *immersion* if it gives an isomorphism of X with an open subscheme of a closed subscheme of Z. (This definition of very ample differs slightly from the one in Grothendieck [EGA II, 4.4.2].)

Remark 5.16.1. Let Y be a noetherian scheme. Then a scheme X over Y is projective if and only if it is proper, and there exists a very ample sheaf on X relative to Y. Indeed, if X is projective over Y, then X is proper by (4.9). On the other hand, there is a closed immersion $i: X \to \mathbf{P}_Y^r$ for some r, so $i^*\mathcal{O}(1)$ is a very ample invertible sheaf on X. Conversely, if X is proper over Y, and $\mathscr{L}$ is a very ample invertible sheaf, then $\mathscr{L} \cong i^*(\mathcal{O}(1))$ for some immersion $i: X \to \mathbf{P}_Y^r$. But by (Ex. 4.4) the image of X is closed, so in fact i is a closed immersion, so X is projective over Y.

Note however that there may be several nonisomorphic very ample sheaves on a projective scheme X over Y. The sheaf $\mathscr{L}$ depends on the embedding of X into $\mathbf{P}_Y^r$ (Ex. 5.12). If $Y = $ Spec A, and if $X = $ Proj S, where S is a graded ring as in (5.16b), then the sheaf $\mathcal{O}(1)$ on X defined earlier is a very ample sheaf on X. However, there may be nonisomorphic graded rings having the same Proj and the same very ample sheaf $\mathcal{O}(1)$ (Ex. 2.14).

We end this section with some special results about sheaves on a projective scheme over a noetherian ring.

Definition. Let X be a scheme, and let $\mathscr{F}$ be a sheaf of $\mathcal{O}_X$-modules. We say that $\mathscr{F}$ is *generated by global sections* if there is a family of global sections $\{s_i\}_{i \in I}$, $s_i \in \Gamma(X, \mathscr{F})$, such that for each $x \in X$, the images of s_i in the stalk $\mathscr{F}_x$ generate that stalk as an $\mathcal{O}_x$-module.

Note that $\mathscr{F}$ is generated by global sections if and only if $\mathscr{F}$ can be written as a quotient of a free sheaf. Indeed, the generating sections $\{s_i\}_{i \in I}$ define a surjective morphism of sheaves $\bigoplus_{i \in I} \mathcal{O}_X \to \mathscr{F}$, and conversely.

Example 5.16.2. Any quasi-coherent sheaf on an affine scheme is generated by global sections. Indeed, if $\mathscr{F} = \tilde{M}$ on Spec A, any set of generators for M as an A-module will do.

Example 5.16.3. Let $X = \text{Proj } S$, where S is a graded ring which is generated by S_1 as an S_0-algebra. Then the elements of S_1 give global sections of $\mathcal{O}_X(1)$ which generate it.

Theorem 5.17 (Serre). *Let X be a projective scheme over a noetherian ring A, let $\mathcal{O}(1)$ be a very ample invertible sheaf on X, and let $\mathscr{F}$ be a coherent $\mathcal{O}_X$-module. Then there is an integer n_0 such that for all $n \geq n_0$, the sheaf $\mathscr{F}(n)$ can be generated by a finite number of global sections.*

PROOF. Let $i: X \to \mathbf{P}_A^r$ be a closed immersion of X into a projective space over A, such that $i^*(\mathcal{O}(1)) = \mathcal{O}_X(1)$. Then $i_*\mathscr{F}$ is coherent on $\mathbf{P}_A^r$ (Ex. 5.5), and $i_*(\mathscr{F}(n)) = (i_*\mathscr{F})(n)$ (5.12) or (Ex. 5.1d), and $\mathscr{F}(n)$ is generated by global sections if and only if $i_*(\mathscr{F}(n))$ is (in fact, their global sections are the same), so we reduce to the case $X = \mathbf{P}_A^r = \text{Proj } A[x_0, \dots, x_r]$.

Now cover X with the open sets $D_+(x_i)$, $i = 0, \dots, r$. Since $\mathscr{F}$ is coherent, for each i there is a finitely generated module M_i over $B_i = A[x_0/x_i, \dots, x_n/x_i]$ such that $\mathscr{F}|_{D_+(x_i)} \cong \tilde{M}_i$. For each i, take a finite number of elements $s_{ij} \in M_i$ which generate this module. By (5.14) there is an integer n such that $x_i^n s_{ij}$ extends to a global section t_{ij} of $\mathscr{F}(n)$. As usual, we take one n to work for all i, j. Now $\mathscr{F}(n)$ corresponds to a B_i-module M_i' on $D_+(x_i)$, and the map $x_i^n: \mathscr{F} \to \mathscr{F}(n)$ induces an *isomorphism* of M_i to M_i'. So the sections $x_i^n s_{ij}$ generate M_i', and hence the global sections $t_{ij} \in \Gamma(X, \mathscr{F}(n))$ generate the sheaf $\mathscr{F}(n)$ everywhere.

Corollary 5.18. *Let X be projective over a noetherian ring A. Then any coherent sheaf $\mathscr{F}$ on X can be written as a quotient of a sheaf $\mathscr{E}$, where $\mathscr{E}$ is a finite direct sum of twisted structure sheaves $\mathcal{O}(n_i)$ for various integers n_i.*

PROOF. Let $\mathscr{F}(n)$ be generated by a finite number of global sections. Then we have a surjection $\bigoplus_{i=1}^{N} \mathcal{O}_X \to \mathscr{F}(n) \to 0$. Tensoring with $\mathcal{O}_X(-n)$ we obtain a surjection $\bigoplus_{i=1}^{N} \mathcal{O}_X(-n) \to \mathscr{F} \to 0$ as required.

Theorem 5.19. *Let k be a field, let A be a finitely generated k-algebra, let X be a projective scheme over A, and let $\mathscr{F}$ be a coherent $\mathcal{O}_X$-module. Then $\Gamma(X,\mathscr{F})$ is a finitely generated A-module. In particular, if $A = k$, $\Gamma(X,\mathscr{F})$ is a finite-dimensional k-vector space.*

PROOF. First we write $X = \operatorname{Proj} S$, where S is a graded ring with $S_0 = A$ which is finitely generated by S_1 as an S_0-algebra (5.16b). Let M be the graded S-module $\Gamma_*(\mathscr{F})$. Then by (5.15) we have $\tilde{M} \cong \mathscr{F}$. On the other hand, by (5.17), for n sufficiently large, $\mathscr{F}(n)$ is generated by a finite number of global sections in $\Gamma(X,\mathscr{F}(n))$. Let M' be the submodule of M generated by these sections. Then M' is a finitely generated S-module. Furthermore, the inclusion $M' \hookrightarrow M$ induces an inclusion of sheaves $\tilde{M}' \hookrightarrow \tilde{M} = \mathscr{F}$. Twisting by n we have an inclusion $\tilde{M}'(n) \hookrightarrow \mathscr{F}(n)$ which is actually an isomorphism, because $\mathscr{F}(n)$ is generated by global sections in M'. Twisting by $-n$ we find that $\tilde{M}' \cong \mathscr{F}$. Thus $\mathscr{F}$ is the sheaf associated to a finitely generated S-module, and so we have reduced to showing that if M is a finitely generated S-module, then $\Gamma(X,\tilde{M})$ is a finitely generated A-module.

Now by (I, 7.4), there is a finite filtration

$$0 = M^0 \subseteq M^1 \subseteq \ldots \subseteq M^r = M$$

of M by graded submodules, where for each i, $M^i/M^{i-1} \cong (S/\mathfrak{p}_i)(n_i)$ for some homogeneous prime ideal $\mathfrak{p}_i \subseteq S$, and some integer n_i. This filtration gives a filtration of $\tilde{M}$, and the short exact sequences

$$0 \to \tilde{M}^{i-1} \to \tilde{M}^i \to \tilde{M}^i/\tilde{M}^{i-1} \to 0$$

give rise to left-exact sequences

$$0 \to \Gamma(X,\tilde{M}^{i-1}) \to \Gamma(X,\tilde{M}^i) \to \Gamma(X,\tilde{M}^i/\tilde{M}^{i-1}).$$

Thus to show that $\Gamma(X,\tilde{M})$ is finitely generated over A, it will be sufficient to show that $\Gamma(X,(S/\mathfrak{p})^{\sim}(n))$ is finitely generated, for each $\mathfrak{p}$ and n. Thus we have reduced to the following special case: Let S be a graded integral domain, finitely generated by S_1 as an S_0-algebra, where $S_0 = A$ is a finitely generated integral domain over k. Then $\Gamma(X,\mathcal{O}_X(n))$ is a finitely generated A-module, for any $n \in \mathbf{Z}$.

Let $x_0, \ldots, x_r \in S_1$ be a set of generators of S_1 as an A-module. Since S is an integral domain, multiplication by x_0 gives an injection $S(n) \to S(n+1)$ for any n. Hence there is an injection $\Gamma(X,\mathcal{O}_X(n)) \to \Gamma(X,\mathcal{O}_X(n+1))$ for any n. Thus it is sufficient to prove $\Gamma(X,\mathcal{O}_X(n))$ finitely generated for all sufficiently large n, say $n \geq 0$.

Let $S' = \bigoplus_{n \geq 0} \Gamma(X,\mathcal{O}_X(n))$. Then S' is a ring, containing S, and contained in the intersection $\bigcap S_{x_i}$ of the localizations of S at the elements $x_0, \ldots, x_r$.

(Use the same argument as in the proof of (5.13).) We will show that S' is integral over S.

Let $s' \in S'$ be homogeneous of degree $d \geqslant 0$. Since $s' \in S_{x_i}$ for each i, we can find an integer n such that $x_i^n s' \in S$. Choose one n that works for all i. Since the x_i generate S_1, the monomials in the x_i of degree m generate S_m for any m. So by taking a larger n, we may assume that $ys' \in S$ for all $y \in S_n$. In fact, since s' has positive degree, we can say that for any $y \in S_{\geqslant n} = \bigoplus_{e \geqslant n} S_e$, $ys' \in S_{\geqslant n}$. Now it follows inductively, for any $q \geqslant 1$ that $y \cdot (s')^q \in S_{\geqslant n}$ for any $y \in S_{\geqslant n}$. Take for example $y = x_0^n$. Then for every $q \geqslant 1$ we have $(s')^q \in (1/x_0^n)S$. This is a finitely generated sub-S-module of the quotient field of S'. It follows by a well-known criterion for integral dependence (Atiyah–Macdonald [1, p. 59]), that s' is integral over S. Thus S' is contained in the integral closure of S in its quotient field.

To complete the proof, we apply the theorem of finiteness of integral closure (I, 3.9A). Since S is a finitely generated k-algebra, S' will be a finitely generated S-module. It follows that for every n, S_n' is a finitely generated S_0-module, which is what we wanted to prove. In fact, our proof shows that $S_n' = S_n$ for all sufficiently large n (Ex. 5.9) and (Ex. 5.14).

Remark 5.19.1. This proof is a generalization of the proof of (I, 3.4a). We will give another proof of this theorem later, using cohomology (III, 5.2.1).

Remark 5.19.2. The hypothesis "A is a finitely generated k-algebra" is used only to be able to apply (I, 3.9A). Thus it would be sufficient to assume only that A is a "Nagata ring" in the sense of Matsumura [2, p. 231]—see also [loc. cit., Th. 72, p. 240].

Corollary 5.20. *Let $f : X \to Y$ be a projective morphism of schemes of finite type over a field k. Let $\mathscr{F}$ be a coherent sheaf on X. Then $f_* \mathscr{F}$ is coherent on Y.*

PROOF. The question is local on Y, so we may assume $Y = \operatorname{Spec} A$, where A is a finitely generated k-algebra. Then in any case, $f_* \mathscr{F}$ is quasi-coherent (5.8c), so $f_* \mathscr{F} = \Gamma(Y, f_* \mathscr{F})^{\sim} = \Gamma(X, \mathscr{F})^{\sim}$. But $\Gamma(X, \mathscr{F})$ is a finitely generated A-module by the theorem, so $f_* \mathscr{F}$ is coherent. See (III, 8.8) for another proof and generalization.

EXERCISES

5.1. Let $(X, \mathcal{O}_X)$ be a ringed space, and let $\mathscr{E}$ be a locally free $\mathcal{O}_X$-module of finite rank. We define the *dual* of $\mathscr{E}$, denoted $\check{\mathscr{E}}$, to be the sheaf $\mathscr{H}om_{\mathcal{O}_X}(\mathscr{E}, \mathcal{O}_X)$.
   (a) Show that $(\check{\mathscr{E}})^{\sim} \cong \mathscr{E}$.
   (b) For any $\mathcal{O}_X$-module $\mathscr{F}$, $\mathscr{H}om_{\mathcal{O}_X}(\mathscr{E}, \mathscr{F}) \cong \check{\mathscr{E}} \otimes_{\mathcal{O}_X} \mathscr{F}$.
   (c) For any $\mathcal{O}_X$-modules $\mathscr{F}, \mathscr{G}$, $\operatorname{Hom}_{\mathcal{O}_X}(\mathscr{E} \otimes \mathscr{F}, \mathscr{G}) \cong \operatorname{Hom}_{\mathcal{O}_X}(\mathscr{F}, \mathscr{H}om_{\mathcal{O}_X}(\mathscr{E}, \mathscr{G}))$.

(d) *(Projection Formula)*. If $f:(X,\mathcal{O}_X) \to (Y,\mathcal{O}_Y)$ is a morphism of ringed spaces, if $\mathcal{F}$ is an $\mathcal{O}_X$-module, and if $\mathcal{E}$ is a locally free $\mathcal{O}_Y$-module of finite rank, then there is a natural isomorphism $f_*(\mathcal{F} \otimes_{\mathcal{O}_X} f^*\mathcal{E}) \cong f_*(\mathcal{F}) \otimes_{\mathcal{O}_Y} \mathcal{E}$.

5.2. Let R be a discrete valuation ring with quotient field K, and let $X = \operatorname{Spec} R$.
  (a) To give an $\mathcal{O}_X$-module is equivalent to giving an R-module M, a K-vector space L, and a homomorphism $\rho : M \otimes_R K \to L$.
  (b) That $\mathcal{O}_X$-module is quasi-coherent if and only if ρ is an isomorphism.

5.3. Let $X = \operatorname{Spec} A$ be an affine scheme. Show that the functors $\sim$ and Γ are adjoint, in the following sense: for any A-module M, and for any sheaf of $\mathcal{O}_X$-modules $\mathcal{F}$, there is a natural isomorphism

$$\operatorname{Hom}_A(M, \Gamma(X,\mathcal{F})) \cong \operatorname{Hom}_{\mathcal{O}_X}(\tilde{M},\mathcal{F}).$$

5.4. Show that a sheaf of $\mathcal{O}_X$-modules $\mathcal{F}$ on a scheme X is quasi-coherent if and only if every point of X has a neighborhood U, such that $\mathcal{F}|_U$ is isomorphic to a cokernel of a morphism of free sheaves on U. If X is noetherian, then $\mathcal{F}$ is coherent if and only if it is locally a cokernel of a morphism of free sheaves of finite rank. (These properties were originally the definition of quasi-coherent and coherent sheaves.)

5.5. Let $f:X \to Y$ be a morphism of schemes.
  (a) Show by example that if $\mathcal{F}$ is coherent on X, then $f_*\mathcal{F}$ need not be coherent on Y, even if X and Y are varieties over a field k.
  (b) Show that a closed immersion is a finite morphism (§3).
  (c) If f is a finite morphism of noetherian schemes, and if $\mathcal{F}$ is coherent on X, then $f_*\mathcal{F}$ is coherent on Y.

5.6. *Support.* Recall the notions of support of a section of a sheaf, support of a sheaf, and subsheaf with supports from (Ex. 1.14) and (Ex. 1.20).
  (a) Let A be a ring, let M be an A-module, let $X = \operatorname{Spec} A$, and let $\mathcal{F} = \tilde{M}$. For any $m \in M = \Gamma(X,\mathcal{F})$, show that $\operatorname{Supp} m = V(\operatorname{Ann} m)$, where $\operatorname{Ann} m$ is the *annihilator* of $m = \{a \in A | am = 0\}$.
  (b) Now suppose that A is noetherian, and M finitely generated. Show that $\operatorname{Supp} \mathcal{F} = V(\operatorname{Ann} M)$.
  (c) The support of a coherent sheaf on a noetherian scheme is closed.
  (d) For any ideal $\mathfrak{a} \subseteq A$, we define a submodule $\Gamma_\mathfrak{a}(M)$ of M by $\Gamma_\mathfrak{a}(M) = \{m \in M | \mathfrak{a}^n m = 0$ for some $n > 0\}$. Assume that A is noetherian, and M any A-module. Show that $\Gamma_\mathfrak{a}(M)^\sim \cong \mathcal{H}^0_Z(\mathcal{F})$, where $Z = V(\mathfrak{a})$ and $\mathcal{F} = \tilde{M}$. [*Hint*: Use (Ex. 1.20) and (5.8) to show a priori that $\mathcal{H}^0_Z(\mathcal{F})$ is quasi-coherent. Then show that $\Gamma_\mathfrak{a}(M) \cong \Gamma_Z(\mathcal{F})$.]
  (e) Let X be a noetherian scheme, and let Z be a closed subset. If $\mathcal{F}$ is a quasi-coherent (respectively, coherent) $\mathcal{O}_X$-module, then $\mathcal{H}^0_Z(\mathcal{F})$ is also quasi-coherent (respectively, coherent).

5.7. Let X be a noetherian scheme, and let $\mathcal{F}$ be a coherent sheaf.
  (a) If the stalk $\mathcal{F}_x$ is a free $\mathcal{O}_x$-module for some point $x \in X$, then there is a neighborhood U of x such that $\mathcal{F}|_U$ is free.
  (b) $\mathcal{F}$ is locally free if and only if its stalks $\mathcal{F}_x$ are free $\mathcal{O}_x$-modules for all $x \in X$.
  (c) $\mathcal{F}$ is invertible (i.e., locally free of rank 1) if and only if there is a coherent sheaf $\mathcal{G}$ such that $\mathcal{F} \otimes \mathcal{G} \cong \mathcal{O}_X$. (This justifies the terminology invertible: it means

that $\mathscr{F}$ is an invertible element of the monoid of coherent sheaves under the operation $\otimes$.)

5.8. Again let X be a noetherian scheme, and $\mathscr{F}$ a coherent sheaf on X. We will consider the function

$$\varphi(x) = \dim_{k(x)} \mathscr{F}_x \otimes_{\mathcal{O}_x} k(x),$$

where $k(x) = \mathcal{O}_x/\mathfrak{m}_x$ is the residue field at the point x. Use Nakayama's lemma to prove the following results.

(a) The function φ is *upper semi-continuous*, i.e., for any $n \in \mathbf{Z}$, the set $\{x \in X | \varphi(x) \geq n\}$ is closed.

(b) If $\mathscr{F}$ is locally free, and X is connected, then φ is a constant function.

(c) Conversely, if X is reduced, and φ is constant, then $\mathscr{F}$ is locally free.

5.9. Let S be a graded ring, generated by S_1 as an S_0-algebra, let M be a graded S-module, and let $X = \operatorname{Proj} S$.

(a) Show that there is a natural homomorphism $\alpha: M \to \Gamma_*(\tilde{M})$.

(b) Assume now that $S_0 = A$ is a finitely generated k-algebra for some field k, that S_1 is a finitely generated A-module, and that M is a finitely generated S-module. Show that the map α is an isomorphism in all large enough degrees, i.e., there is a $d_0 \in \mathbf{Z}$ such that for all $d \geq d_0$, $\alpha_d: M_d \to \Gamma(X, \tilde{M}(d))$ is an isomorphism. [*Hint*: Use the methods of the proof of (5.19).]

(c) With the same hypotheses, we define an equivalence relation $\approx$ on graded S-modules by saying $M \approx M'$ if there is an integer d such that $M_{\geq d} \cong M'_{\geq d}$. Here $M_{\geq d} = \bigoplus_{n \geq d} M_n$. We will say that a graded S-module M is *quasi-finitely generated* if it is equivalent to a finitely generated module. Now show that the functors $\tilde{\ }$ and Γ_* induce an equivalence of categories between the category of quasi-finitely generated graded S-modules modulo the equivalence relation $\approx$, and the category of coherent $\mathcal{O}_X$-modules.

5.10. Let A be a ring, let $S = A[x_0, \dots, x_r]$ and let $X = \operatorname{Proj} S$. We have seen that a homogeneous ideal I in S defines a closed subscheme of X (Ex. 3.12), and that conversely every closed subscheme of X arises in this way (5.16).

(a) For any homogeneous ideal $I \subseteq S$, we define the *saturation* $\bar{I}$ of I to be $\{s \in S | \text{for each } i = 0, \dots, r, \text{ there is an } n \text{ such that } x_i^n s \in I\}$. We say that I is *saturated* if $I = \bar{I}$. Show that $\bar{I}$ is a homogeneous ideal of S.

(b) Two homogeneous ideals I_1 and I_2 of S define the same closed subscheme of X if and only if they have the same saturation.

(c) If Y is any closed subscheme of X, then the ideal $\Gamma_*(\mathscr{I}_Y)$ is saturated. Hence it is the largest homogeneous ideal defining the subscheme Y.

(d) There is a 1-1 correspondence between saturated ideals of S and closed subschemes of X.

5.11. Let S and T be two graded rings with $S_0 = T_0 = A$. We define the *Cartesian product* $S \times_A T$ to be the graded ring $\bigoplus_{d \geq 0} S_d \otimes_A T_d$. If $X = \operatorname{Proj} S$ and $Y = \operatorname{Proj} T$, show that $\operatorname{Proj}(S \times_A T) \cong X \times_A Y$, and show that the sheaf $\mathcal{O}(1)$ on $\operatorname{Proj}(S \times_A T)$ is isomorphic to the sheaf $p_1^*(\mathcal{O}_X(1)) \otimes p_2^*(\mathcal{O}_Y(1))$ on $X \times Y$.

The Cartesian product of rings is related to the *Segre embedding* of projective spaces (I, Ex. 2.14) in the following way. If $x_0, \dots, x_r$ is a set of generators for S_1 over A, corresponding to a projective embedding $X \hookrightarrow \mathbf{P}_A^r$, and if $y_0, \dots, y_s$ is a set of generators for T_1, corresponding to a projective embedding $Y \hookrightarrow \mathbf{P}_A^s$, then $\{x_i \otimes y_j\}$ is a set of generators for $(S \times_A T)_1$, and hence defines a projective embedding $\text{Proj}(S \times_A T) \hookrightarrow \mathbf{P}_A^N$, with $N = rs + r + s$. This is just the image of $X \times Y \subseteq \mathbf{P}^r \times \mathbf{P}^s$ in its Segre embedding.

5.12. (a) Let X be a scheme over a scheme Y, and let $\mathscr{L}, \mathscr{M}$ be two very ample invertible sheaves on X. Show that $\mathscr{L} \otimes \mathscr{M}$ is also very ample. [*Hint:* Use a Segre embedding.]

(b) Let $f: X \to Y$ and $g: Y \to Z$ be two morphisms of schemes. Let $\mathscr{L}$ be a very ample invertible sheaf on X relative to Y, and let $\mathscr{M}$ be a very ample invertible sheaf on Y relative to Z. Show that $\mathscr{L} \otimes f^* \mathscr{M}$ is a very ample invertible sheaf on X relative to Z.

5.13. Let S be a graded ring, generated by S_1 as an S_0-algebra. For any integer $d > 0$, let $S^{(d)}$ be the graded ring $\bigoplus_{n \geqslant 0} S_n^{(d)}$ where $S_n^{(d)} = S_{nd}$. Let $X = \text{Proj } S$. Show that $\text{Proj } S^{(d)} \cong X$, and that the sheaf $\mathcal{O}(1)$ on $\text{Proj } S^{(d)}$ corresponds via this isomorphism to $\mathcal{O}_X(d)$.

This construction is related to the *d-uple embedding* (I, Ex. 2.12) in the following way. If $x_0, \ldots, x_r$ is a set of generators for S_1, corresponding to an embedding $X \hookrightarrow \mathbf{P}_A^r$, then the set of monomials of degree d in the x_i is a set of generators for $S_1^{(d)} = S_d$. These define a projective embedding of $\text{Proj } S^{(d)}$ which is none other than the image of X under the d-uple embedding of $\mathbf{P}_A^r$.

5.14. Let A be a ring, and let X be a closed subscheme of $\mathbf{P}_A^r$. We define the *homogeneous coordinate ring* $S(X)$ of X for the given embedding to be $A[x_0, \ldots, x_r]/I$, where I is the ideal $\Gamma_*(\mathscr{I}_X)$ constructed in the proof of (5.16). (Of course if A is a field and X a variety, this coincides with the definition given in (I, §2)!) Recall that a scheme X is *normal* if its local rings are integrally closed domains. A closed subscheme $X \subseteq \mathbf{P}_A^r$ is *projectively normal* for the given embedding, if its homogeneous coordinate ring $S(X)$ is an integrally closed domain (cf. (I, Ex. 3.18)). Now assume that k is an algebraically closed field, and that X is a connected, normal closed subscheme of $\mathbf{P}_k^r$. Show that for some $d > 0$, the d-uple embedding of X is projectively normal, as follows.

(a) Let S be the homogeneous coordinate ring of X, and let $S' = \bigoplus_{n \geqslant 0} \Gamma(X, \mathcal{O}_X(n))$. Show that S is a domain, and that S' is its integral closure. [*Hint:* First show that X is integral. Then regard S' as the global sections of the sheaf of rings $\mathscr{S} = \bigoplus_{n \geqslant 0} \mathcal{O}_X(n)$ on X, and show that $\mathscr{S}$ is a sheaf of integrally closed domains.]

(b) Use (Ex. 5.9) to show that $S_d = S_d'$ for all sufficiently large d.

(c) Show that $S^{(d)}$ is integrally closed for sufficiently large d, and hence conclude that the d-uple embedding of X is projectively normal.

(d) As a corollary of (a), show that a closed subscheme $X \subseteq \mathbf{P}_A^r$ is projectively normal if and only if it is normal, and for every $n \geqslant 0$ the natural map $\Gamma(\mathbf{P}^r, \mathcal{O}_{\mathbf{P}^r}(n)) \to \Gamma(X, \mathcal{O}_X(n))$ is surjective.

5.15. *Extension of Coherent Sheaves.* We will prove the following theorem in several steps: Let X be a noetherian scheme, let U be an open subset, and let $\mathscr{F}$ be a coherent sheaf on U. Then there is a coherent sheaf $\mathscr{F}'$ on X such that $\mathscr{F}'|_U \cong \mathscr{F}$.

(a) On a noetherian affine scheme, every quasi-coherent sheaf is the union of its coherent subsheaves. We say a sheaf $\mathscr{F}$ is the *union* of its subsheaves $\mathscr{F}_\alpha$ if for every open set U, the group $\mathscr{F}(U)$ is the union of the subgroups $\mathscr{F}_\alpha(U)$.

(b) Let X be an affine noetherian scheme, U an open subset, and $\mathscr{F}$ coherent on U. Then there exists a coherent sheaf $\mathscr{F}'$ on X with $\mathscr{F}'|_U \cong \mathscr{F}$. [*Hint:* Let $i: U \to X$ be the inclusion map. Show that $i_* \mathscr{F}$ is quasi-coherent, then use (a).]

(c) With $X, U, \mathscr{F}$ as in (b), suppose furthermore we are given a quasi-coherent sheaf $\mathscr{G}$ on X such that $\mathscr{F} \subseteq \mathscr{G}|_U$. Show that we can find $\mathscr{F}'$ a coherent subsheaf of $\mathscr{G}$, with $\mathscr{F}'|_U \cong \mathscr{F}$. [*Hint*: Use the same method, but replace $i_* \mathscr{F}$ by $\rho^{-1}(i_* \mathscr{F})$, where ρ is the natural map $\mathscr{G} \to i_*(\mathscr{G}|_U)$.]

(d) Now let X be any noetherian scheme, U an open subset, $\mathscr{F}$ a coherent sheaf on U, and $\mathscr{G}$ a quasi-coherent sheaf on X such that $\mathscr{F} \subseteq \mathscr{G}|_U$. Show that there is a coherent subsheaf $\mathscr{F}' \subseteq \mathscr{G}$ on X with $\mathscr{F}'|_U \cong \mathscr{F}$. Taking $\mathscr{G} = i_* \mathscr{F}$ proves the result announced at the beginning. [*Hint*: Cover X with open affines, and extend over one of them at a time.]

(e) As an extra corollary, show that on a noetherian scheme, any quasi-coherent sheaf $\mathscr{F}$ is the union of its coherent subsheaves. [*Hint*: If s is a section of $\mathscr{F}$ over an open set U, apply (d) to the subsheaf of $\mathscr{F}|_U$ generated by s.]

5.16. *Tensor Operations on Sheaves.* First we recall the definitions of various tensor operations on a module. Let A be a ring, and let M be an A-module. Let $T^n(M)$ be the tensor product $M \otimes \ldots \otimes M$ of M with itself n times, for $n \geqslant 1$. For $n = 0$ we put $T^0(M) = A$. Then $T(M) = \bigoplus_{n \geqslant 0} T^n(M)$ is a (noncommutative) A-algebra, which we call the *tensor algebra* of M. We define the *symmetric algebra* $S(M) = \bigoplus_{n \geqslant 0} S^n(M)$ of M to be the quotient of $T(M)$ by the two-sided ideal generated by all expressions $x \otimes y - y \otimes x$, for all $x, y \in M$. Then $S(M)$ is a commutative A-algebra. Its component $S^n(M)$ in degree n is called the nth *symmetric product* of M. We denote the image of $x \otimes y$ in $S(M)$ by xy, for any $x, y \in M$. As an example, note that if M is a free A-module of rank r, then $S(M) \cong A[x_1, \ldots, x_r]$.

We define the *exterior algebra* $\bigwedge(M) = \bigoplus_{n \geqslant 0} \bigwedge^n(M)$ of M to be the quotient of $T(M)$ by the two-sided ideal generated by all expressions $x \otimes x$ for $x \in M$. Note that this ideal contains all expressions of the form $x \otimes y + y \otimes x$, so that $\bigwedge(M)$ is a *skew commutative* graded A-algebra. This means that if $u \in \bigwedge^r(M)$ and $v \in \bigwedge^s(M)$, then $u \wedge v = (-1)^{rs} v \wedge u$ (here we denote by $\wedge$ the multiplication in this algebra; so the image of $x \otimes y$ in $\bigwedge^2(M)$ is denoted by $x \wedge y$). The nth component $\bigwedge^n(M)$ is called the nth *exterior power* of M.

Now let $(X, \mathcal{O}_X)$ be a ringed space, and let $\mathscr{F}$ be a sheaf of $\mathcal{O}_X$-modules. We define the *tensor algebra, symmetric algebra,* and *exterior algebra* of $\mathscr{F}$ by taking the sheaves associated to the presheaf, which to each open set U assigns the corresponding tensor operation applied to $\mathscr{F}(U)$ as an $\mathcal{O}_X(U)$-module. The results are $\mathcal{O}_X$-algebras, and their components in each degree are $\mathcal{O}_X$-modules.

(a) Suppose that $\mathscr{F}$ is locally free of rank n. Then $T^r(\mathscr{F})$, $S^r(\mathscr{F})$, and $\bigwedge^r(\mathscr{F})$ are also locally free, of ranks n^r, $\binom{n+r-1}{n-1}$, and $\binom{n}{r}$ respectively.

(b) Again let $\mathscr{F}$ be locally free of rank n. Then the multiplication map $\bigwedge^r \mathscr{F} \otimes \bigwedge^{n-r} \mathscr{F} \to \bigwedge^n \mathscr{F}$ is a perfect pairing for any r, i.e., it induces an isomorphism of $\bigwedge^r \mathscr{F}$ with $(\bigwedge^{n-r} \mathscr{F})^{\vee} \otimes \bigwedge^n \mathscr{F}$. As a special case, note if $\mathscr{F}$ has rank 2, then $\mathscr{F} \cong \mathscr{F}^{\vee} \otimes \bigwedge^2 \mathscr{F}$.

(c) Let $0 \to \mathscr{F}' \to \mathscr{F} \to \mathscr{F}'' \to 0$ be an exact sequence of locally free sheaves. Then for any r there is a finite filtration of $S^r(\mathscr{F})$,

$$S^r(\mathscr{F}) = F^0 \supseteq F^1 \supseteq \ldots \supseteq F^r \supseteq F^{r+1} = 0$$

with quotients

$$F^p/F^{p+1} \cong S^p(\mathscr{F}') \otimes S^{r-p}(\mathscr{F}'')$$

for each p.

(d) Same statement as (c), with exterior powers instead of symmetric powers. In particular, if $\mathscr{F}', \mathscr{F}, \mathscr{F}''$ have ranks n', n, n'' respectively, there is an isomorphism $\bigwedge^n \mathscr{F} \cong \bigwedge^{n'} \mathscr{F}' \otimes \bigwedge^{n''} \mathscr{F}''$.

(e) Let $f: X \to Y$ be a morphism of ringed spaces, and let $\mathscr{F}$ be an $\mathcal{O}_Y$-module. Then f^* commutes with all the tensor operations on $\mathscr{F}$, i.e., $f^*(S^n(\mathscr{F})) = S^n(f^*\mathscr{F})$ etc.

5.17. *Affine Morphisms.* A morphism $f: X \to Y$ of schemes is *affine* if there is an open affine cover $\{V_i\}$ of Y such that $f^{-1}(V_i)$ is affine for each i.

(a) Show that $f: X \to Y$ is an affine morphism if and only if for *every* open affine $V \subseteq Y, f^{-1}(V)$ is affine. [*Hint*: Reduce to the case Y affine, and use (Ex. 2.17).]

(b) An affine morphism is quasi-compact and separated. Any finite morphism is affine.

(c) Let Y be a scheme, and let $\mathscr{A}$ be a quasi-coherent sheaf of $\mathcal{O}_Y$-algebras (i.e., a sheaf of rings which is at the same time a quasi-coherent sheaf of $\mathcal{O}_Y$-modules). Show that there is a unique scheme X, and a morphism $f: X \to Y$, such that for every open affine $V \subseteq Y, f^{-1}(V) \cong \operatorname{Spec} \mathscr{A}(V)$, and for every inclusion $U \hookrightarrow V$ of open affines of Y, the morphism $f^{-1}(U) \hookrightarrow f^{-1}(V)$ corresponds to the restriction homomorphism $\mathscr{A}(V) \to \mathscr{A}(U)$. The scheme X is called **Spec** $\mathscr{A}$. [*Hint*: Construct X by glueing together the schemes $\operatorname{Spec} \mathscr{A}(V)$, for V open affine in Y.]

(d) If $\mathscr{A}$ is a quasi-coherent $\mathcal{O}_Y$-algebra, then $f: X = \mathbf{Spec}\, \mathscr{A} \to Y$ is an affine morphism, and $\mathscr{A} \cong f_* \mathcal{O}_X$. Conversely, if $f: X \to Y$ is an affine morphism, then $\mathscr{A} = f_* \mathcal{O}_X$ is a quasi-coherent sheaf of $\mathcal{O}_Y$-algebras, and $X \cong \mathbf{Spec}\, \mathscr{A}$.

(e) Let $f: X \to Y$ be an affine morphism, and let $\mathscr{A} = f_* \mathcal{O}_X$. Show that f_* induces an equivalence of categories from the category of quasi-coherent $\mathcal{O}_X$-modules to the category of quasi-coherent $\mathscr{A}$-modules (i.e., quasi-coherent $\mathcal{O}_Y$-modules having a structure of $\mathscr{A}$-module). [*Hint*: For any quasi-coherent $\mathscr{A}$-module $\mathscr{M}$, construct a quasi-coherent $\mathcal{O}_X$-module $\tilde{\mathscr{M}}$, and show that the functors f_* and $\tilde{}$ are inverse to each other.

5.18. *Vector Bundles.* Let Y be a scheme. A (geometric) *vector bundle* of rank n over Y is a scheme X and a morphism $f: X \to Y$, together with additional data consisting of an open covering $\{U_i\}$ of Y, and isomorphisms $\psi_i: f^{-1}(U_i) \to \mathbf{A}^n_{U_i}$, such that for any i, j, and for any open affine subset $V = \operatorname{Spec} A \subseteq U_i \cap U_j$, the automorphism $\psi = \psi_j \circ \psi_i^{-1}$ of $\mathbf{A}^n_V = \operatorname{Spec} A[x_1, \ldots, x_n]$ is given by a *linear* automorphism θ of $A[x_1, \ldots, x_n]$, i.e., $\theta(a) = a$ for any $a \in A$, and $\theta(x_i) = \sum a_{ij} x_j$ for suitable $a_{ij} \in A$.

An *isomorphism* $g: (X, f, \{U_i\}, \{\psi_i\}) \to (X', f', \{U_i'\}, \{\psi_i'\})$ of one vector bundle of rank n to another one is an isomorphism $g: X \to X'$ of the underlying schemes, such that $f = f' \circ g$, and such that X, f, together with the covering of Y consisting of all the U_i and U_i', and the isomorphisms ψ_i and $\psi_i' \circ g$, is also a vector bundle structure on X.

(a) Let $\mathscr{E}$ be a locally free sheaf of rank n on a scheme Y. Let $S(\mathscr{E})$ be the symmetric algebra on $\mathscr{E}$, and let $X = \mathbf{Spec}\, S(\mathscr{E})$, with projection morphism $f: X \to Y$. For each open affine subset $U \subseteq Y$ for which $\mathscr{E}|_U$ is free, choose a basis of $\mathscr{E}$, and let $\psi: f^{-1}(U) \to \mathbf{A}^n_U$ be the isomorphism resulting from the identification of $S(\mathscr{E}(U))$ with $\mathcal{O}(U)[x_1, \ldots, x_n]$. Then $(X, f, \{U\}, \{\psi\})$ is a vector bundle of rank n over Y, which (up to isomorphism) does not depend on the bases of $\mathscr{E}_U$ chosen. We call it the *geometric vector bundle associated* to $\mathscr{E}$, and denote it by $\mathbf{V}(\mathscr{E})$.

(b) For any morphism $f:X \to Y$, a *section* of f over an open set $U \subseteq Y$ is a morphism $s:U \to X$ such that $f \circ s = \mathrm{id}_U$. It is clear how to restrict sections to smaller open sets, or how to glue them together, so we see that the presheaf $U \mapsto \{\text{set of sections of } f \text{ over } U\}$ is a sheaf of sets on Y, which we denote by $\mathscr{S}(X/Y)$. Show that if $f:X \to Y$ is a vector bundle of rank n, then the sheaf of sections $\mathscr{S}(X/Y)$ has a natural structure of $\mathscr{O}_Y$-module, which makes it a locally free $\mathscr{O}_Y$-module of rank n. [*Hint*: It is enough to define the module structure locally, so we can assume $Y = \mathrm{Spec}\, A$ is affine, and $X = \mathbf{A}_Y^n$. Then a section $s:Y \to X$ comes from an A-algebra homomorphism $\theta:A[x_1,\dots,x_n] \to A$, which in turn determines an ordered n-tuple $\langle \theta(x_1),\dots,\theta(x_n)\rangle$ of elements of A. Use this correspondence between sections s and ordered n-tuples of elements of A to define the module structure.]

(c) Again let $\mathscr{E}$ be a locally free sheaf of rank n on Y, let $X = \mathbf{V}(\mathscr{E})$, and let $\mathscr{S} = \mathscr{S}(X/Y)$ be the sheaf of sections of X over Y. Show that $\mathscr{S} \cong \mathscr{E}^\vee$, as follows. Given a section $s \in \Gamma(V,\mathscr{E}^\vee)$ over any open set V, we think of s as an element of $\mathrm{Hom}(\mathscr{E}|_V,\mathscr{O}_V)$. So s determines an $\mathscr{O}_V$-algebra homomorphism $S(\mathscr{E}|_V) \to \mathscr{O}_V$. This determines a morphism of spectra $V = \mathrm{Spec}\,\mathscr{O}_V \to \mathrm{Spec}\, S(\mathscr{E}|_V) = f^{-1}(V)$, which is a section of X/Y. Show that this construction gives an isomorphism of $\mathscr{E}^\vee$ to $\mathscr{S}$.

(d) Summing up, show that we have established a one-to-one correspondence between isomorphism classes of locally free sheaves of rank n on Y, and isomorphism classes of vector bundles of rank n over Y. Because of this, we sometimes use the words "locally free sheaf" and "vector bundle" interchangeably, if no confusion seems likely to result.

6 Divisors

The notion of divisor forms an important tool for studying the intrinsic geometry on a variety or scheme. In this section we will introduce divisors, linear equivalence and the divisor class group. The divisor class group is an abelian group which is an interesting and subtle invariant of a variety. In §7 we will see that divisors are also important for studying maps from a given variety to a projective space.

There are several different ways of defining divisors, depending on the context. We will begin with Weil divisors, which are easiest to understand geometrically, but which are only defined on certain noetherian integral schemes. For more general schemes there is the notion of Cartier divisor which we treat next. Then we will explain the connection between Weil divisors, Cartier divisors, and invertible sheaves.

We start with an informal example. Let C be a nonsingular projective curve in $\mathbf{P}_k^2$, the projective plane over an algebraically closed field k. For each line L in $\mathbf{P}^2$, we consider $L \cap C$, which is a finite set of points on C. If C is a curve of degree d, and if we count the points with proper multiplicity, then $L \cap C$ will consist of exactly d points (I, Ex. 5.4). We write $L \cap C = \sum n_i P_i$, where $P_i \in C$ are the points, and n_i the multiplicities, and we call this formal sum a divisor on C. As L varies, we obtain a family of divisors on C, parametrized by the set of all lines in $\mathbf{P}^2$, which is the dual projective space $(\mathbf{P}^2)^*$. We call this set of divisors a linear system of divisors on C. Note that the embedding of C in $\mathbf{P}^2$ can be recovered just from knowing this linear system: if P is a point of C, we consider the set of divisors in the linear system which contain P. They correspond to the lines $L \in (\mathbf{P}^2)^*$ passing through P, and this set of lines determines P uniquely as a point of $\mathbf{P}^2$. This connection between linear systems and embeddings in projective space will be studied in detail in §7.

This example should already serve to illustrate the importance of divisors. To see the relation among the different divisors in the linear system, let L and L' be two lines in $\mathbf{P}^2$, and let $D = L \cap C$ and $D' = L' \cap C$ be the corresponding divisors. If L and L' are defined by linear homogeneous equations $f = 0$ and $f' = 0$ in $\mathbf{P}^2$, then f/f' gives a rational function on $\mathbf{P}^2$, which restricts to a rational function g on C. Now by construction, g has zeros at the points of D, and poles at the points of D', counted with multiplicities, in a sense which will be made precise below. We say that D and D' are linearly equivalent, and the existence of such a rational function can be taken as an intrinsic definition of the linear equivalence. We will make these concepts more precise in our formal discussion, starting now.

Weil Divisors

Definition. We say a scheme X is *regular in codimension one* (or sometimes *nonsingular in codimension one*) if every local ring $\mathcal{O}_x$ of X of dimension one is regular.

The most important examples of such schemes are nonsingular varieties over a field (I, §5) and noetherian normal schemes. On a nonsingular variety the local ring of every closed point is regular (I, 5.1), hence all the local rings are regular, since they are localizations of the local rings of closed points. On a noetherian normal scheme, any local ring of dimension one is an integrally closed domain, hence is regular (I, 6.2A).

In this section we will consider schemes satisfying the following condition:

(∗) X is a noetherian integral separated scheme which is regular in codimension one.

Definition. Let X satisfy (∗). A *prime divisor* on X is a closed integral subscheme Y of codimension one. A *Weil divisor* is an element of the free abelian group Div X generated by the prime divisors. We write a divisor as $D = \sum n_i Y_i$, where the Y_i are prime divisors, the n_i are integers, and only finitely many n_i are different from zero. If all the $n_i \geqslant 0$, we say that D is *effective*.

If Y is a prime divisor on X, let $\eta \in Y$ be its generic point. Then the local ring $\mathcal{O}_{\eta,X}$ is a discrete valuation ring with quotient field K, the function field of X. We call the corresponding discrete valuation v_Y the *valuation of Y*. Note that since X is separated, Y is uniquely determined by its valuation (Ex. 4.5). Now let $f \in K^*$ be any nonzero rational function on X. Then $v_Y(f)$ is an integer. If it is positive, we say f has a *zero* along Y, of that order; if it is negative, we say f has a *pole* along Y, of order $-v_Y(f)$.

Lemma 6.1. *Let X satisfy (∗), and let $f \in K^*$ be a nonzero function on X. Then $v_Y(f) = 0$ for all except finitely many prime divisors Y.*

PROOF. Let $U = \operatorname{Spec} A$ be an open affine subset of X on which f is regular. Then $Z = X - U$ is a proper closed subset of X. Since X is noetherian, Z can contain at most finitely many prime divisors of X; all the others must meet U. Thus it will be sufficient to show that there are only finitely many prime divisors Y of U for which $v_Y(f) \neq 0$. Since f is regular on U, we have $v_Y(f) \geqslant 0$ in any case. And $v_Y(f) > 0$ if and only if Y is contained in the closed subset of U defined by the ideal Af in A. Since $f \neq 0$, this is a proper closed subset, hence contains only finitely many closed irreducible subsets of codimension one of U.

Definition. Let X satisfy (∗) and let $f \in K^*$. We define the *divisor* of f, denoted (f), by
$$(f) = \sum v_Y(f) \cdot Y,$$
where the sum is taken over all prime divisors of X. By the lemma, this is a finite sum, hence it is a divisor. Any divisor which is equal to the divisor of a function is called a *principal* divisor.

Note that if $f, g \in K^*$, then $(f/g) = (f) - (g)$ because of the properties of valuations. Therefore sending a function f to its divisor (f) gives a homomorphism of the multiplicative group K^* to the additive group $\operatorname{Div} X$, and the image, which consists of the principal divisors, is a subgroup of $\operatorname{Div} X$.

Definition. Let X satisfy (∗). Two divisors D and D' are said to be *linearly equivalent*, written $D \sim D'$, if $D - D'$ is a principal divisor. The group $\operatorname{Div} X$ of all divisors divided by the subgroup of principal divisors is called the *divisor class group* of X, and is denoted by $\operatorname{Cl} X$.

The divisor class group of a scheme is a very interesting invariant. In general it is not easy to calculate. However, in the following propositions and examples we will calculate a number of special cases to give some idea of what it is like.

Proposition 6.2. *Let A be a noetherian domain. Then A is a unique factorization domain if and only if $X = \operatorname{Spec} A$ is normal and $\operatorname{Cl} X = 0$.*

PROOF. (See also Bourbaki [1, Ch. 7, §3]). It is well-known that a UFD is integrally closed, so X will be normal. On the other hand, A is a UFD if and only if every prime ideal of height 1 is principal (I, 1.12A). So what we must show is that if A is an integrally closed domain, then every prime ideal of height 1 is principal if and only if Cl(Spec A) = 0.

One way is easy: if every prime ideal of height 1 is principal, consider a prime divisor $Y \subseteq X = $ Spec A. Y corresponds to a prime ideal p of height 1. If p is generated by an element $f \in A$, then clearly the divisor of f is $1 \cdot Y$. Thus every prime divisor is principal, so Cl $X = 0$.

For the converse, suppose Cl $X = 0$. Let p be a prime ideal of height 1, and let Y be the corresponding prime divisor. Then there is an $f \in K$, the quotient field of A, with $(f) = Y$. We will show that in fact $f \in A$ and f generates p. Since $v_Y(f) = 1$, we have $f \in A_p$, and f generates pA_p. If $p' \subseteq A$ is any other prime ideal of height 1, then p' corresponds to a prime divisor Y' of X, and $v_{Y'}(f) = 0$, so $f \in A_{p'}$. Now the algebraic result (6.3A) below implies that $f \in A$. In fact, $f \in A \cap pA_p = p$. Now to show that f generates p, let g be any other element of p. Then $v_Y(g) \geqslant 1$ and $v_{Y'}(g) \geqslant 0$ for all $Y' \neq Y$. Hence $v_{Y'}(g/f) \geqslant 0$ for *all* prime divisors Y' (including Y). Thus $g/f \in A_{p'}$ for *all* p' of height 1, so by (6.3A) again, $g/f \in A$. In other words, $g \in Af$, which shows that p is a principal ideal, generated by f.

Proposition 6.3A. *Let A be an integrally closed noetherian domain. Then*

$$A = \bigcap_{\text{ht } p = 1} A_p$$

where the intersection is taken over all prime ideals of height 1.

PROOF. Matsumura [2, Th. 38, p. 124].

Example 6.3.1. If X is affine n-space $\mathbb{A}_k^n$ over a field k, then Cl $X = 0$. Indeed, $X = $ Spec $k[x_1, \ldots, x_n]$, and the polynomial ring is a UFD.

Example 6.3.2. If A is a Dedekind domain, then Cl(Spec A) is just the ideal class group of A, as defined in algebraic number theory. Thus (6.2) generalizes the fact that A is a UFD if and only if its ideal class group is 0.

Proposition 6.4. *Let X be the projective space $\mathbb{P}_k^n$ over a field k. For any divisor $D = \sum n_i Y_i$, define the degree of D by $\deg D = \sum n_i \deg Y_i$, where $\deg Y_i$ is the degree of the hypersurface Y_i. Let H be the hyperplane $x_0 = 0$. Then:*
   (a) *if D is any divisor of degree d, then $D \sim dH$;*
   (b) *for any $f \in K^*$, $\deg(f) = 0$;*
   (c) *the degree function gives an isomorphism $\deg : \text{Cl } X \to \mathbb{Z}$.*

PROOF. Let $S = k[x_0, \ldots, x_n]$ be the homogeneous coordinate ring of X. If g is a homogeneous element of degree d, we can factor it into irreducible polynomials $g = g_1^{n_1} \cdots g_r^{n_r}$. Then g_i defines a hypersurface Y_i of degree $d_i = \deg g_i$, and we can define the divisor of g to be $(g) = \sum n_i Y_i$. Then $\deg(g) = d$. Now a rational function f on X is a quotient g/h of homogeneous polynomials of the same degree. Clearly $(f) = (g) - (h)$, so we see that $\deg(f) = 0$, which proves (b).

If D is any divisor of degree d, we can write it as a difference $D_1 - D_2$ of effective divisors of degrees d_1, d_2 with $d_1 - d_2 = d$. Let $D_1 = (g_1)$ and $D_2 = (g_2)$. This is possible, because an irreducible hypersurface in $\mathbf{P}^n$ corresponds to a homogeneous prime ideal of height 1 in S, which is principal. Taking power products we can get any effective divisor as (g) for some homogeneous g. Now $D - dH = (f)$ where $f = g_1/x_0^d g_2$ is a rational function on X. This proves (a). Statement (c) follows from (a), (b), and the fact that $\deg H = 1$.

Proposition 6.5. *Let X satisfy* (∗), *let Z be a proper closed subset of X, and let $U = X - Z$. Then:*

(a) *there is a surjective homomorphism $\operatorname{Cl} X \to \operatorname{Cl} U$ defined by $D = \sum n_i Y_i \mapsto \sum n_i(Y_i \cap U)$, where we ignore those $Y_i \cap U$ which are empty;*

(b) *if $\operatorname{codim}(Z, X) \geqslant 2$, then $\operatorname{Cl} X \to \operatorname{Cl} U$ is an isomorphism;*

(c) *if Z is an irreducible subset of codimension 1, then there is an exact sequence*

$$\mathbf{Z} \to \operatorname{Cl} X \to \operatorname{Cl} U \to 0,$$

where the first map is defined by $1 \mapsto 1 \cdot Z$.

PROOF.

(a) If Y is a prime divisor on X, then $Y \cap U$ is either empty or a prime divisor on U. If $f \in K^*$, and $(f) = \sum n_i Y_i$, then considering f as a rational function on U, we have $(f)_U = \sum n_i(Y_i \cap U)$, so indeed we have a homomorphism $\operatorname{Cl} X \to \operatorname{Cl} U$. It is surjective because every prime divisor of U is the restriction of its closure in X.

(b) The groups $\operatorname{Div} X$ and $\operatorname{Cl} X$ depend only on subsets of codimension 1, so removing a closed subset Z of codimension $\geqslant 2$ doesn't change anything.

(c) The kernel of $\operatorname{Cl} X \to \operatorname{Cl} U$ consists of divisors whose support is contained in Z. If Z is irreducible, the kernel is just the subgroup of $\operatorname{Cl} X$ generated by $1 \cdot Z$.

Example 6.5.1. Let Y be an irreducible curve of degree d in $\mathbf{P}_k^2$. Then $\operatorname{Cl}(\mathbf{P}^2 - Y) = \mathbf{Z}/d\mathbf{Z}$. This follows immediately from (6.4) and (6.5).

Example 6.5.2. Let k be a field, let $A = k[x, y, z]/(xy - z^2)$, and let $X = \operatorname{Spec} A$. Then X is an affine quadric cone in $\mathbf{A}_k^3$. We will show that $\operatorname{Cl} X = \mathbf{Z}/2\mathbf{Z}$, and that it is generated by a ruling of the cone, say $Y : y = z = 0$ (Fig. 8).

First note that Y is a prime divisor, so by (6.5) we have an exact sequence

$$\mathbf{Z} \to \operatorname{Cl} X \to \operatorname{Cl}(X - Y) \to 0,$$

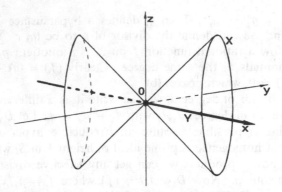

Figure 8.   A ruling on the quadric cone.

where the first map sends $1 \mapsto 1 \cdot Y$. Now Y can be cut out set-theoretically by the function y. In fact, the divisor of y is $2 \cdot Y$, because $y = 0 \Rightarrow z^2 = 0$, and z generates the maximal ideal of the local ring at the generic point of Y. Hence $X - Y = \operatorname{Spec} A_y$. Now $A_y = k[x,y,y^{-1},z]/(xy - z^2)$. In this ring $x = y^{-1}z^2$, so we can eliminate x, and find $A_y \cong k[y,y^{-1},z]$. This is a UFD, so by (6.2), $\operatorname{Cl}(X - Y) = 0$.

Thus we see that $\operatorname{Cl} X$ is generated by Y, and that $2 \cdot Y = 0$. It remains to show that Y itself is not a principal divisor. Since A is integrally closed (Ex. 6.4), it is equivalent to show that the prime ideal of Y, namely $\mathfrak{p} = (y,z)$, is not principal (cf. proof of (6.2)). Let $\mathfrak{m} = (x,y,z)$, and note that $\mathfrak{m}/\mathfrak{m}^2$ is a 3-dimensional vector space over k generated by $\bar{x},\bar{y},\bar{z}$, the images of x,y,z. Now $\mathfrak{p} \subseteq \mathfrak{m}$, and the image of $\mathfrak{p}$ in $\mathfrak{m}/\mathfrak{m}^2$ contains $\bar{y}$ and $\bar{z}$. Hence $\mathfrak{p}$ cannot be a principal ideal.

Proposition 6.6. *Let X satisfy* (*)*. Then $X \times \mathbf{A}^1 (= X \times_{\operatorname{Spec} \mathbf{Z}} \operatorname{Spec} \mathbf{Z}[t])$ also satisfies* (*)*, and $\operatorname{Cl} X \cong \operatorname{Cl}(X \times \mathbf{A}^1)$.*

PROOF. Clearly $X \times \mathbf{A}^1$ is noetherian, integral, and separated. To see that it is regular in codimension one, we note that there are two kinds of points of codimension one on $X \times \mathbf{A}^1$. Type 1 is a point x whose image in X is a point y of codimension one. In this case x is the generic point of $\pi^{-1}(y)$, where $\pi: X \times \mathbf{A}^1 \to X$ is the projection. Its local ring is $\mathcal{O}_x \cong \mathcal{O}_y[t]_{\mathfrak{m}_y}$, which is clearly a discrete valuation ring, since $\mathcal{O}_y$ is. The corresponding prime divisor $\{x\}^-$ is just $\pi^{-1}(\{y\}^-)$.

Type 2 is a point $x \in X \times \mathbf{A}^1$ of codimension one, whose image in X is the generic point of X. In this case $\mathcal{O}_x$ is a localization of $K[t]$ at some maximal ideal, where K is the function field of X. It is a discrete valuation ring because $K[t]$ is a principal ideal domain. Thus $X \times \mathbf{A}^1$ also satisfies (*).

We define a map $\operatorname{Cl} X \to \operatorname{Cl}(X \times \mathbf{A}^1)$ by $D = \sum n_i Y_i \mapsto \pi^* D = \sum n_i \pi^{-1}(Y_i)$. If $f \in K^*$, then $\pi^*((f))$ is the divisor of f considered as an element of $K(t)$, the function field of $X \times \mathbf{A}^1$. Thus we have a homomorphism $\pi^*: \operatorname{Cl} X \to \operatorname{Cl}(X \times \mathbf{A}^1)$.

To show π^* is injective, suppose $D \in \text{Div } X$, and $\pi^*D = (f)$ for some $f \in K(t)$. Since π^*D involves only prime divisors of type 1, f must be in K. For otherwise we could write $f = g/h$, with $g,h \in K[t]$, relatively prime. If g,h are not both in K, then (f) will involve some prime divisor of type 2 on $X \times \mathbf{A}^1$. Now if $f \in K$, it is clear that $D = (f)$, so π^* is injective.

To show that π^* is surjective, it will be sufficient to show that any prime divisor of type 2 on $X \times \mathbf{A}^1$ is linearly equivalent to a linear combination of prime divisors of type 1. So let $Z \subseteq X \times \mathbf{A}^1$ be a prime divisor of type 2. Localizing at the generic point of X, we get a prime divisor in $\text{Spec } K[t]$, which corresponds to a prime ideal $\mathfrak{p} \subseteq K[t]$. This is principal, so let f be a generator. Then $f \in K(t)$, and the divisor of f consists of Z plus perhaps something purely of type 1. It cannot involve any other prime divisors of type 2. Thus Z is linearly equivalent to a divisor purely of type 1. This completes the proof.

Example 6.6.1. Let Q be the nonsingular quadric surface $xy = zw$ in $\mathbf{P}_k^3$. We will show that $\text{Cl } Q \cong \mathbf{Z} \oplus \mathbf{Z}$. We use the fact that Q is isomorphic to $\mathbf{P}_k^1 \times_k \mathbf{P}_k^1$ (I, Ex. 2.15). Let p_1 and p_2 be the projections of Q onto the two factors. Then as in the proof of (6.6) we obtain homomorphisms $p_1^*, p_2^* : \text{Cl } \mathbf{P}^1 \to \text{Cl } Q$. First we show that p_1^* and p_2^* are injective. Let $Y = pt \times \mathbf{P}^1$. Then $Q - Y = \mathbf{A}^1 \times \mathbf{P}^1$, and the composition

$$\text{Cl } \mathbf{P}^1 \xrightarrow{p_2^*} \text{Cl } Q \to \text{Cl}(\mathbf{A}^1 \times \mathbf{P}^1)$$

is the isomorphism of (6.6). Hence p_2^* (and similarly p_1^*) is injective.

Now consider the exact sequence of (6.5) for Y:

$$\mathbf{Z} \to \text{Cl } Q \to \text{Cl}(\mathbf{A}^1 \times \mathbf{P}^1) \to 0.$$

In this sequence the first map sends 1 to Y. But if we identify $\text{Cl } \mathbf{P}^1$ with $\mathbf{Z}$ by letting 1 be the class of a point, then this first map is just p_1^*, hence is injective. Since the image of p_2^* goes isomorphically to $\text{Cl}(\mathbf{A}^1 \times \mathbf{P}^1)$ as we have just seen, we conclude that $\text{Cl } Q \cong \text{Im } p_1^* \oplus \text{Im } p_2^* = \mathbf{Z} \oplus \mathbf{Z}$. If D is any divisor on Q, let (a,b) be the ordered pair of integers in $\mathbf{Z} \oplus \mathbf{Z}$ corresponding to the class of D under this isomorphism. Then we say D is of *type* (a,b) on Q.

Example 6.6.2. Continuing with the quadric surface $Q \subseteq \mathbf{P}^3$, we will show that the embedding induces a homomorphism $\text{Cl } \mathbf{P}^3 \to \text{Cl } Q$, and that the image of a hyperplane H, which generates $\text{Cl } \mathbf{P}^3$, is the element $(1,1)$ in $\text{Cl } Q = \mathbf{Z} \oplus \mathbf{Z}$. Let Y be any irreducible hypersurface of $\mathbf{P}^3$ which does not contain Q. Then we can assign multiplicities to the irreducible components of $Y \cap Q$ so as to obtain a *divisor* $Y \cdot Q$ on Q. Indeed, on each standard open set U_i of $\mathbf{P}^3$, Y is defined by a single function f; we can take the value of this function (restricted to Q) for each valuation of a prime divisor of Q to define the divisor $Y \cdot Q$. By linearity we extend this map to define a divisor $D \cdot Q$ on Q, for each divisor $D = \sum n_i Y_i$ on $\mathbf{P}^3$, such that no Y_i contains Q. Clearly linearly equivalent divisors restrict to linearly equivalent divisors. Since any divisor on $\mathbf{P}^3$ is linearly equivalent to one whose prime divisors don't contain Q by (6.4), we obtain a well-defined homomorphism $\operatorname{Cl} \mathbf{P}^3 \to \operatorname{Cl} Q$. Now if H is the hyperplane $w = 0$, then $H \cap Q$ is the divisor consisting of the two lines $x = w = 0$ and $y = w = 0$. One is in each family (I, Ex. 2.15) so $H \cap Q$ is of type $(1,1)$ in $\operatorname{Cl} Q = \mathbf{Z} \oplus \mathbf{Z}$. Note that the two families of lines correspond to $pt \times \mathbf{P}^1$ and $\mathbf{P}^1 \times pt$, so they are of type $(1,0)$ and $(0,1)$.

Example 6.6.3. Carrying this example one step further, let C be the twisted cubic curve $x = t^3$, $y = u^3$, $z = t^2 u$, $w = t u^2$ which lies on Q. If Y is the quadric cone $yz = w^2$, then $Y \cap Q = C \cup L$ where L is the line $y = w = 0$. Since $Y \sim 2H$ on $\mathbf{P}^3$, $Y \cap Q$ is a divisor of type $(2,2)$. The line L has type $(1,0)$, so C is of type $(1,2)$. It follows that there does not exist any surface $Y \subseteq \mathbf{P}^3$, not containing Q, such that $Y \cap Q = C$, even set-theoretically! For in that case the divisor $Y \cap Q$ would be rC for some integer $r > 0$. This is a divisor of type $(r,2r)$ in $\operatorname{Cl} Q$. But if Y is a surface of degree d, then $Y \cap Q$ is of type (d,d), which can never equal $(r,2r)$. Thus Y does not exist.

Example 6.6.4. We will see later (V, 4.8) that if X is a nonsingular cubic surface in $\mathbf{P}^3$, then $\operatorname{Cl} X \cong \mathbf{Z}^7$.

Divisors on Curves

We will illustrate the notion of the divisor class group further by paying special attention to the case of divisors on curves. We will define the degree of a divisor on a curve, and we will show that on a complete nonsingular curve, the degree is stable under linear equivalence. Further study of divisors on curves will be found in Chapter IV.

To begin with, we need some preliminary information about curves and morphisms of curves. Recall our conventions about terminology from the end of Section 4:

Definition. Let k be an algebraically closed field. A *curve* over k is an integral separated scheme X of finite type over k, of dimension one. If X is proper over k, we say that X is *complete*. If all the local rings of X are regular local rings, we say that X is *nonsingular*.

Proposition 6.7. *Let X be a nonsingular curve over k with function field K. Then the following conditions are equivalent:*

   (i) X *is projective;*

  (ii) X *is complete;*

 (iii) $X \cong t(C_K)$, *where C_K is the abstract nonsingular curve of (I, §6), and t is the functor from varieties to schemes of (2.6).*

PROOF.

(i) ⇒ (ii) follows from (4.9).

(ii) ⇒ (iii). If X is complete, then every discrete valuation ring of K/k has a unique center on X (Ex. 4.5). Since the local rings of X at the closed points are all discrete valuation rings, this implies that the closed points of X are in 1-1 correspondence with the discrete valuation rings of K/k, namely the points of C_K. Thus it is clear that $X \cong t(C_K)$.

(iii) ⇒ (i) follows from (I, 6.9).

Proposition 6.8. *Let X be a complete nonsingular curve over k, let Y be any curve over k, and let $f:X \to Y$ be a morphism. Then either* (1) $f(X) =$ *a point, or* (2) $f(X) = Y$. *In case* (2), $K(X)$ *is a finite extension field of $K(Y)$, f is a finite morphism, and Y is also complete.*

PROOF. Since X is complete, $f(X)$ must be closed in Y, and proper over Spec k (Ex. 4.4). On the other hand, $f(X)$ is irreducible. Thus either (1) $f(X) = pt$, or (2) $f(X) = Y$, and in case (2), Y is also complete.

In case (2), f is dominant, so it induces an inclusion $K(Y) \subseteq K(X)$ of function fields. Since both fields are finitely generated extension fields of transcendence degree 1 of k, $K(X)$ must be a finite algebraic extension of $K(Y)$. To show that f is a finite morphism, let $V = $ Spec B be any open affine subset of Y. Let A be the integral closure of B in $K(X)$. Then A is a finite B-module (I, 3.9A), and Spec A is isomorphic to an open subset U of X (I, 6.7). Clearly $U = f^{-1}V$, so this shows that f is a finite morphism.

Definition. If $f:X \to Y$ is a finite morphism of curves, we define the *degree* of f to be the degree of the field extension $[K(X):K(Y)]$.

Now we come to the study of divisors on curves. If X is a nonsingular curve, then X satisfies the condition (∗) used above, so we can talk about divisors on X. A prime divisor is just a closed point, so an arbitrary divisor can be written $D = \sum n_i P_i$, where the P_i are closed points, and $n_i \in \mathbf{Z}$. We define the *degree* of D to be $\sum n_i$.

Definition. If $f:X \to Y$ is a finite morphism of nonsingular curves, we define a homomorphism $f^*:$ Div $Y \to$ Div X as follows. For any point $Q \in Y$, let $t \in \mathcal{O}_Q$ be a *local parameter* at Q, i.e., t is an element of $K(Y)$ with $v_Q(t) = 1$, where v_Q is the valuation corresponding to the discrete valuation ring $\mathcal{O}_Q$. We define $f^*Q = \sum_{f(P)=Q} v_P(t) \cdot P$. Since f is a finite morphism, this is a finite sum, so we get a divisor on X. Note that f^*Q is independent of the choice of the local parameter t. Indeed, if t' is another local parameter at Q, then $t' = ut$ where u is a unit in $\mathcal{O}_Q$. For any point $P \in X$ with $f(P) = Q$, u will be a unit in $\mathcal{O}_P$, so $v_P(t) = v_P(t')$. We extend the definition by linearity to all divisors on Y. One sees easily that f^* preserves linear equivalence, so it induces a homomorphism $f^*:$ Cl $Y \to$ Cl X.

Proposition 6.9. *Let* $f : X \to Y$ *be a finite morphism of nonsingular curves. Then for any divisor* D *on* Y *we have* $\deg f^* D = \deg f \cdot \deg D$.

PROOF. It will be sufficient to show that for any closed point $Q \in Y$ we have $\deg f^* Q = \deg f$. Let $V = \operatorname{Spec} B$ be an open affine subset of Y containing Q. Let A be the integral closure of B in $K(X)$. Then, as in the proof of (6.8), $U = \operatorname{Spec} A$ is the open subset $f^{-1} V$ of X. Let $\mathfrak{m}_Q$ be the maximal ideal of Q in B. We localize both B and A with respect to the multiplicative system $S = B - \mathfrak{m}_Q$, and we obtain a ring extension $\mathcal{O}_Q \hookrightarrow A'$, where A' is a finitely generated $\mathcal{O}_Q$-module. Now A' is torsion-free, and has rank equal to $r = [K(X):K(Y)]$, so A' is a free $\mathcal{O}_Q$-module of rank $r = \deg f$. If t is a local parameter at Q, it follows that A'/tA' is a k-vector space of dimension r.

On the other hand, the points P_i of X such that $f(P_i) = Q$ are in 1-1 correspondence with the maximal ideals $\mathfrak{m}_i$ of A', and for each i, $A'_{\mathfrak{m}_i} = \mathcal{O}_{P_i}$. Clearly $tA' = \bigcap_i (tA'_{\mathfrak{m}_i} \cap A')$, so by the Chinese remainder theorem,

$$\dim_k A'/tA' = \sum_i \dim_k A'/(tA'_{\mathfrak{m}_i} \cap A').$$

But

$$A'/(tA'_{\mathfrak{m}_i} \cap A') \cong A'_{\mathfrak{m}_i}/tA'_{\mathfrak{m}_i} = \mathcal{O}_{P_i}/t\mathcal{O}_{P_i},$$

so the dimensions in the sum above are just equal to $v_{P_i}(t)$. But $f^* Q = \sum v_{P_i}(t) \cdot P_i$, so we have shown that $\deg f^* Q = \deg f$ as required.

Corollary 6.10. *A principal divisor on a complete nonsingular curve* X *has degree zero. Consequently the degree function induces a surjective homomorphism* $\deg : \operatorname{Cl} X \to \mathbf{Z}$.

PROOF. Let $f \in K(X)^*$. If $f \in k$, then $(f) = 0$, so there is nothing to prove. If $f \notin k$, then the inclusion of fields $k(f) \subseteq K(X)$ induces a finite morphism $\varphi : X \to \mathbf{P}^1$. It is a morphism by (I, 6.12), and it is finite by (6.8). Now $(f) = \varphi^*(\{0\} - \{\infty\})$. Since $\{0\} - \{\infty\}$ is a divisor of degree 0 on $\mathbf{P}^1$, we conclude that (f) has degree 0 on X.

Thus the degree of a divisor on X depends only on its linear equivalence class, and we obtain a homomorphism $\operatorname{Cl} X \to \mathbf{Z}$ as stated. It is surjective, because the degree of a single point is 1.

Example 6.10.1. A complete nonsingular curve X is rational if and only if there exist two distinct points $P, Q \in X$ with $P \sim Q$. Recall that *rational* means birational to $\mathbf{P}^1$. If X is rational, then in fact it is isomorphic to $\mathbf{P}^1$ by (6.7). And on $\mathbf{P}^1$ we have already seen that any two points are linearly equivalent (6.4). Conversely, suppose X has two points $P \neq Q$ with $P \sim Q$. Then there is a rational function $f \in K(X)$ with $(f) = P - Q$. Consider the morphism $\varphi : X \to \mathbf{P}^1$ determined by f as in the proof of (6.10). We have $\varphi^*(\{0\}) = P$, so φ must be a morphism of degree 1. In other words, φ is birational, so X is rational.

Example 6.10.2. Let X be the nonsingular cubic curve $y^2z = x^3 - xz^2$ in $\mathbf{P}_k^2$, with char $k \neq 2$. We have already seen that X is not rational (I, Ex. 6.2). Let $\mathrm{Cl}^\circ X$ be the kernel of the degree map $\mathrm{Cl}\, X \to \mathbf{Z}$. Then from the previous example we know that $\mathrm{Cl}^\circ X \neq 0$. We will show in fact that there is a natural 1-1 correspondence between the set of closed points of X and the elements of the group $\mathrm{Cl}^\circ X$. On the one hand this elucidates the structure of the group $\mathrm{Cl}^\circ X$. On the other hand it gives us a group structure on the set of closed points of X, which makes X into a group variety (Fig. 9).

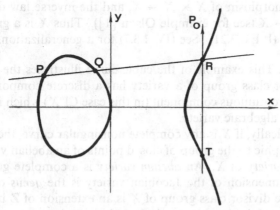

Figure 9. The group law on a cubic curve.

Let P_0 be the point $(0,1,0)$ on X. It is an inflection point, so the tangent line $z = 0$ at that point meets the curve in the divisor $3P_0$. If L is any other line in $\mathbf{P}^2$, meeting X in three points P,Q,R (which may coincide), then since L is linearly equivalent to the line $z = 0$ in $\mathbf{P}^2$, we have $P + Q + R \sim 3P_0$ on X, as in (6.6.2) above.

Now to any closed point $P \in X$, we associate the divisor $P - P_0 \in \mathrm{Cl}^\circ X$. This map is injective, because if $P - P_0 \sim Q - P_0$, then $P \sim Q$, and X would be rational by the previous example, which is impossible.

To show that the map from the closed points of X to $\mathrm{Cl}^\circ X$ is surjective, we proceed in several steps. Let $D \in \mathrm{Cl}^\circ X$. Then $D = \sum n_i P_i$, with $\sum n_i = 0$. Hence we can also write $D = \sum n_i(P_i - P_0)$. Now for any point R, let the line $P_0 R$ meet X further in the point T (always counting intersections with multiplicities—for example, if $R = P_0$, we take the line $P_0 R$ to be the tangent line at P_0, and then the third intersection T is also P_0). Then $P_0 + R + T \sim 3P_0$, so $R - P_0 \sim -(T - P_0)$. If i is an index such that $n_i < 0$ in D, we take $P_i = R$. Then replacing P_i by T, we get a linearly equivalent divisor with the ith coefficient $-n_i > 0$. Repeating this process, we may assume that $D = \sum n_i(P_i - P_0)$ with all $n_i > 0$. We now show by induction on $\sum n_i$ that $D \sim P - P_0$ for some point P. If $\sum n_i = 1$, there is nothing to prove. So suppose $\sum n_i \geq 2$, and let P,Q be two of the points P_i (maybe the same) which occur in D. Let the line PQ meet X in R, and let the line $P_0 R$ meet X in T.

Then we have

$$P + Q + R \sim 3P_0 \qquad \text{and} \qquad P_0 + R + T \sim 3P_0$$

so

$$(P - P_0) + (Q - P_0) \sim (T - P_0).$$

Replacing P and Q by T, we get D linearly equivalent to another divisor of the same form whose $\sum n_i$ is one less, so by induction $D \sim P - P_0$ for some P.

Thus we have shown that the group $\text{Cl}^\circ X$ is in 1-1 correspondence with the set of closed points of X. One can show directly that the addition law determines a morphism of $X \times X \to X$, and the inverse law determines a morphism $X \to X$ (see for example Olson [1]). Thus X is a group variety in the sense of (I, Ex. 3.21). See (IV, 1.3.7) for a generalization.

Remark 6.10.3. This example of the cubic curve illustrates the general fact that the divisor class group of a variety has a discrete component (in this case $\mathbf{Z}$) and a continuous component (in this case $\text{Cl}^\circ X$) which itself has the structure of an algebraic variety.

More specifically, if X is any complete nonsingular curve, then the group $\text{Cl}^\circ X$ is isomorphic to the group of closed points of an abelian variety called the *Jacobian variety* of X. An *abelian variety* is a complete group variety over k. The dimension of the Jacobian variety is the *genus* of the curve. Thus the whole divisor class group of X is an extension of $\mathbf{Z}$ by the group of closed points of the Jacobian variety of X.

If X is a nonsingular projective variety of dimension $\geqslant 2$, then one can define a subgroup $\text{Cl}^\circ X$ of $\text{Cl}\ X$, namely the subgroup of divisor classes *algebraically equivalent* to zero, such that $\text{Cl}\ X/\text{Cl}^\circ X$ is a finitely generated abelian group, called the *Néron-Severi group* of X, and $\text{Cl}^\circ X$ is isomorphic to the group of closed points of an abelian variety called the *Picard variety* of X.

Unfortunately we do not have space in this book to develop the theory of abelian varieties and to study the Jacobian and Picard varieties of a given variety. For more information and further references on this beautiful subject, see Lang [1], Mumford [2], Mumford [5], and Hartshorne [6]. See also (IV, §4), (V, Ex. 1.7), and Appendix B.

Cartier Divisors

Now we want to extend the notion of divisor to an arbitrary scheme. It turns out that using the irreducible subvarieties of codimension one doesn't work very well. So instead, we take as our point of departure the idea that a divisor should be something which locally looks like the divisor of a rational function. This is not exactly a generalization of the Weil divisors (as we will see), but it gives a good notion to use on arbitrary schemes.

Definition. Let X be a scheme. For each open affine subset $U = \text{Spec}\ A$, let S be the set of elements of A which are not zero divisors, and let $K(U)$ be the localization of A by the multiplicative system S. We call $K(U)$ the *total quotient ring* of A. For each open set U, let $S(U)$ denote the set of elements of $\Gamma(U, \mathcal{O}_X)$ which are not zero divisors in each local ring $\mathcal{O}_x$ for $x \in U$. Then the rings $S(U)^{-1} \Gamma(U, \mathcal{O}_X)$ form a presheaf, whose associated sheaf of rings $\mathcal{K}$ we call the *sheaf of total quotient rings* of $\mathcal{O}$. On an arbitrary scheme, the sheaf $\mathcal{K}$ replaces the concept of function field of an integral scheme. We denote by $\mathcal{K}^*$ the sheaf (of multiplicative groups) of invertible elements in the sheaf of rings $\mathcal{K}$. Similarly $\mathcal{O}^*$ is the sheaf of invertible elements in $\mathcal{O}$.

Definition. A *Cartier divisor* on a scheme X is a global section of the sheaf $\mathcal{K}^*/\mathcal{O}^*$. Thinking of the properties of quotient sheaves, we see that a Cartier divisor on X can be described by giving an open cover $\{U_i\}$ of X, and for each i an element $f_i \in \Gamma(U_i, \mathcal{K}^*)$, such that for each i, j, $f_i/f_j \in \Gamma(U_i \cap U_j, \mathcal{O}^*)$. A Cartier divisor is *principal* if it is in the image of the natural map $\Gamma(X, \mathcal{K}^*) \to \Gamma(X, \mathcal{K}^*/\mathcal{O}^*)$. Two Cartier divisors are *linearly equivalent* if their difference is principal. (Although the group operation on $\mathcal{K}^*/\mathcal{O}^*$ is multiplication, we will use the language of additive groups when speaking of Cartier divisors, so as to preserve the analogy with Weil divisors.)

Proposition 6.11. *Let X be an integral, separated noetherian scheme, all of whose local rings are unique factorization domains (in which case we say X is* locally factorial). *Then the group* Div X *of Weil divisors on X is isomorphic to the group of Cartier divisors $\Gamma(X, \mathcal{K}^*/\mathcal{O}^*)$, and furthermore, the principal Weil divisors correspond to the principal Cartier divisors under this isomorphism.*

PROOF. First note that X is normal, hence satisfies (*), since a UFD is integrally closed. Thus it makes sense to talk about Weil divisors. Since X is integral, the sheaf $\mathcal{K}$ is just the constant sheaf corresponding to the function field K of X. Now let a Cartier divisor be given by $\{(U_i, f_i)\}$ where $\{U_i\}$ is an open cover of X, and $f_i \in \Gamma(U_i, \mathcal{K}^*) = K^*$. We define the associated Weil divisor as follows. For each prime divisor Y, take the coefficient of Y to be $v_Y(f_i)$, where i is any index for which $Y \cap U_i \neq \varnothing$. If j is another such index, then f_i/f_j is invertible on $U_i \cap U_j$, so $v_Y(f_i/f_j) = 0$ and $v_Y(f_i) = v_Y(f_j)$. Thus we obtain a well-defined Weil divisor $D = \sum v_Y(f_i)Y$ on X. (The sum is finite because X is noetherian!)

Conversely, if D is a Weil divisor on X, let $x \in X$ be any point. Then D induces a Weil divisor D_x on the local scheme Spec $\mathcal{O}_x$. Since $\mathcal{O}_x$ is a UFD, D_x is a principal divisor, by (6.2), so let $D_x = (f_x)$ for some $f_x \in K$. Now the principal divisor (f_x) on X has the same restriction to Spec $\mathcal{O}_x$ as D, hence they differ only at prime divisors which do not pass through x. There are only finitely many of these which have a non-zero coefficient in D or (f_x), so there is an open neighborhood U_x of x such that D and (f_x) have the same restriction to U_x. Covering X with such open sets U_x, the functions f_x give a Cartier divisor on X. Note that if f, f' give the same Weil divisor on an open set U, then $f/f' \in \Gamma(U, \mathcal{O}^*)$, since X is normal (cf. proof of (6.2)). Thus we have a well-defined Cartier divisor.

These two constructions are inverse to each other, so we see that the groups of Weil divisors and Cartier divisors are isomorphic. Furthermore it is clear that the principal divisors correspond to each other.

Remark 6.11.1A. Since a regular local ring is UFD (Matsumura [2, Th. 48, p. 142]), this proposition applies in particular to any *regular* integral separated noetherian scheme. A scheme is *regular* if all of its local rings are regular local rings.

Remark 6.11.2. If X is a normal scheme, which is not necessarily locally factorial, we can define a subgroup of Div X consisting of the locally principal Weil divisors: D is *locally principal* if X can be covered by open sets U such that $D|_U$ is principal for each U. Then the above proof shows that the Cartier divisors are the same as the locally principal Weil divisors.

Example 6.11.3. Let X be the affine quadric cone Spec $k[x,y,z]/(xy - z^2)$ treated above (6.5.2). The ruling Y is a Weil divisor which is not locally principal in the neighborhood of the vertex of the cone. Indeed, our earlier proof shows that its prime ideal pA_m is not a principal ideal even in the local ring A_m. Thus Y does not correspond to a Cartier divisor. On the other hand $2Y$ is locally principal, and in fact principal. So in this case the group of Cartier divisors modulo principal divisors is 0, whereas Cl $X \cong$ $\mathbf{Z}/2\mathbf{Z}$.

Example 6.11.4. Let X be the cuspidal cubic curve $y^2z = x^3$ in $\mathbf{P}_k^2$, with char $k \neq 2$. In this case X does not satisfy (*), so we cannot talk about Weil divisors on X. However, we can talk about the group CaCl X of Cartier divisor classes modulo principal divisors. Imitating the case of the nonsingular cubic curve (6.10.2) we will show:

(a) there is a surjective degree homomorphism deg: CaCl $X \to \mathbf{Z}$;

(b) there is a 1-1 correspondence between the set of nonsingular closed points of X and the kernel CaCl$^{\circ}X$ of the degree map, which makes it into a group variety; and in fact

(c) there is a natural isomorphism of group varieties between CaCl$^{\circ}X$ and the additive group $\mathbf{G}_a$ of the field k (I, Ex. 3.21a).

To define the degree of a Cartier divisor on X, note that any Cartier divisor is linearly equivalent to one whose local function is invertible in some neighborhood of the singular point $Z = (0,0,1)$. Then this Cartier divisor corresponds to a Weil divisor $D = \sum n_i P_i$ on $X - Z$, and we define the degree of the original divisor to be deg $D = \sum n_i$. The proof of (6.10) shows that if $f \in K$ is invertible at Z, then the principal divisor (f) on $X - Z$ has degree 0. Thus the degree of a Cartier divisor on X is well-defined, and it passes to linear equivalence classes to give a surjective homomorphism deg: CaCl $X \to \mathbf{Z}$.

Now let P_0 be the point $(0,1,0)$ as in the case of the nonsingular cubic curve. To each closed point $P \in X - Z$, we associate the Cartier divisor D_P which is 1 in a neighborhood of Z, and which corresponds to the Weil divisor $P - P_0$ on $X - Z$. First note this map is injective: if $P \neq Q$ are two points in $X - Z$, and if $D_P \sim D_Q$, then there is an $f \in K^*$, which is invertible at Z, and such that $(f) = P - Q$ on $X - Z$. Then f gives a morphism of X to $\mathbf{P}^1$, which must be birational. But then the local ring of Z on X would dominate some discrete valuation ring of $\mathbf{P}^1$, and this is impossible, because Z is a singular point.

To show that every divisor in $\mathrm{CaCl}^{\circ} X$ is linearly equivalent to D_P for some closed point $P \in X - Z$, we proceed exactly as in the case of the nonsingular cubic curve above. The only difference is to note that the geometric constructions $R \mapsto T$ and $P,Q \mapsto R,T$ described above remain inside of $X - Z$. Thus the group $\mathrm{CaCl}^{\circ} X$ is in 1-1 correspondence with the set of closed points of $X - Z$, making it into a group variety.

In this case we are able to identify the group variety as $\mathbf{G}_a$. Of course, we know that X is a rational curve, and so $X - Z \cong \mathbf{A}_k^1$ (I, Ex. 3.2). But in fact, if we use the right parametrization, the group law corresponds. So define a morphism of $\mathbf{G}_a = \operatorname{Spec} k[t]$ to $X - Z$ by $t \mapsto (t,1,t^3)$. This is clearly an isomorphism of varieties. Using a little elementary analytic geometry (left to reader!) one shows that if $P = (t,1,t^3)$ and if $Q = (u,1,u^3)$, then the point T constructed above is just $(t + u,1,(t + u)^3)$. So we have an isomorphism of group varieties of $\mathbf{G}_a$ to $X - Z$ with the group structure of $\mathrm{CaCl}^{\circ} X$.

Invertible Sheaves

Recall that an *invertible sheaf* on a ringed space X is defined to be a locally free $\mathcal{O}_X$-module of rank 1. We will see now that invertible sheaves on a scheme are closely related to divisor classes modulo linear equivalence.

Proposition 6.12. *If $\mathscr{L}$ and $\mathscr{M}$ are invertible sheaves on a ringed space X, so is $\mathscr{L} \otimes \mathscr{M}$. If $\mathscr{L}$ is any invertible sheaf on X, then there exists an invertible sheaf $\mathscr{L}^{-1}$ on X such that $\mathscr{L} \otimes \mathscr{L}^{-1} \cong \mathcal{O}_X$.*

PROOF. The first statement is clear, since $\mathscr{L}$ and $\mathscr{M}$ are both locally free of rank 1, and $\mathcal{O}_X \otimes \mathcal{O}_X \cong \mathcal{O}_X$. For the second statement, let $\mathscr{L}$ be any invertible sheaf, and take $\mathscr{L}^{-1}$ to be the dual sheaf $\mathscr{L}^{\vee} = \mathcal{H}om(\mathscr{L}, \mathcal{O}_X)$. Then $\mathscr{L}^{\vee} \otimes \mathscr{L} \cong \mathcal{H}om(\mathscr{L}, \mathscr{L}) = \mathcal{O}_X$ by (Ex. 5.1).

Definition. For any ringed space X, we define the *Picard group* of X, Pic X, to be the group of isomorphism classes of invertible sheaves on X, under the operation $\otimes$. The proposition shows that in fact it is a group.

Remark 6.12.1. We will see later (III, Ex. 4.5) that Pic X can be expressed as the cohomology group $H^1(X, \mathcal{O}_X^*)$.

Definition. Let D be a Cartier divisor on a scheme X, represented by $\{(U_i, f_i)\}$ as above. We define a subsheaf $\mathscr{L}(D)$ of the sheaf of total quotient rings $\mathscr{K}$ by taking $\mathscr{L}(D)$ to be the sub-$\mathcal{O}_X$-module of $\mathscr{K}$ generated by f_i^{-1} on U_i. This is well-defined, since f_i/f_j is invertible on $U_i \cap U_j$, so f_i^{-1} and f_j^{-1} generate the same $\mathcal{O}_X$-module. We call $\mathscr{L}(D)$ the *sheaf associated* to D.

Proposition 6.13. *Let X be a scheme. Then:*

(a) *for any Cartier divisor D, $\mathscr{L}(D)$ is an invertible sheaf on X. The map $D \mapsto \mathscr{L}(D)$ gives a 1-1 correspondence between Cartier divisors on X and invertible subsheaves of $\mathscr{K}$;*

(b) $\mathscr{L}(D_1 - D_2) \cong \mathscr{L}(D_1) \otimes \mathscr{L}(D_2)^{-1}$;

(c) $D_1 \sim D_2$ *if and only if* $\mathscr{L}(D_1) \cong \mathscr{L}(D_2)$ *as abstract invertible sheaves* (i.e., *disregarding the embedding in $\mathscr{K}$*).

PROOF.

(a) Since each $f_i \in \Gamma(U_i, \mathscr{K}^*)$, the map $\mathcal{O}_{U_i} \to \mathscr{L}(D)|_{U_i}$ defined by $1 \mapsto f_i^{-1}$ is an isomorphism. Thus $\mathscr{L}(D)$ is an invertible sheaf. The Cartier divisor D can be recovered from $\mathscr{L}(D)$ together with its embedding in $\mathscr{K}$, by taking f_i on U_i to be the inverse of a local generator of $\mathscr{L}(D)$. For any invertible subsheaf of $\mathscr{K}$, this construction gives a Cartier divisor, so we have a 1-1 correspondence as claimed.

(b) If D_1 is locally defined by f_i and D_2 is locally defined by g_i, then $\mathscr{L}(D_1 - D_2)$ is locally generated by $f_i^{-1} g_i$, so $\mathscr{L}(D_1 - D_2) = \mathscr{L}(D_1) \cdot \mathscr{L}(D_2)^{-1}$ as subsheaves of $\mathscr{K}$. This product is clearly isomorphic to the abstract tensor product $\mathscr{L}(D_1) \otimes \mathscr{L}(D_2)^{-1}$.

(c) Using (b), it will be sufficient to show that $D = D_1 - D_2$ is principal if and only if $\mathscr{L}(D) \cong \mathcal{O}_X$. If D is principal, defined by $f \in \Gamma(X, \mathscr{K}^*)$, then $\mathscr{L}(D)$ is globally generated by f^{-1}, so sending $1 \mapsto f^{-1}$ gives an isomorphism $\mathcal{O}_X \cong \mathscr{L}(D)$. Conversely, given such an isomorphism, the image of 1 gives an element of $\Gamma(X, \mathscr{K}^*)$ whose inverse will define D as a principal divisor.

Corollary 6.14. *On any scheme X, the map $D \mapsto \mathscr{L}(D)$ gives an injective homomorphism of the group* CaCl X *of Cartier divisors modulo linear equivalence to* Pic X.

Remark 6.14.1. The map CaCl $X \to$ Pic X may not be surjective, because there may be invertible sheaves on X which are not isomorphic to any invertible subsheaf of $\mathscr{K}$. For an example of Kleiman, see Hartshorne [5, I.1.3, p. 9]. On the other hand, this map is an isomorphism in most common situations. Nakai [2, p. 301] has shown that it is an isomorphism whenever X is projective over a field. We will show now that it is an isomorphism if X is integral.

Proposition 6.15. *If X is an integral scheme, the homomorphism* CaCl $X \rightarrow$ Pic X *of (6.14) is an isomorphism.*

PROOF. We have only to show that every invertible sheaf is isomorphic to a subsheaf of $\mathcal{K}$, which in this case is the constant sheaf K, where K is the function field of X. So let $\mathcal{L}$ be any invertible sheaf, and consider the sheaf $\mathcal{L} \otimes_{\mathcal{O}_X} \mathcal{K}$. On any open set U where $\mathcal{L} \cong \mathcal{O}_X$, we have $\mathcal{L} \otimes \mathcal{K} \cong \mathcal{K}$, so it is a constant sheaf on U. Now because X is irreducible, it follows that any sheaf whose restriction to each open set of a covering of X is constant, is in fact a constant sheaf. Thus $\mathcal{L} \otimes \mathcal{K}$ is isomorphic to the constant sheaf $\mathcal{K}$, and the natural map $\mathcal{L} \rightarrow \mathcal{L} \otimes \mathcal{K} \cong \mathcal{K}$ expresses $\mathcal{L}$ as a subsheaf of $\mathcal{K}$.

Corollary 6.16. *If X is a noetherian, integral, separated locally factorial scheme, then there is a natural isomorphism* Cl $X \cong$ Pic X.

PROOF. This follows from (6.11) and (6.15).

Corollary 6.17. *If $X = \mathbf{P}_k^n$ for some field k, then every invertible sheaf on X is isomorphic to $\mathcal{O}(l)$ for some $l \in \mathbf{Z}$.*

PROOF. By (6.4), Cl $X \cong \mathbf{Z}$, so by (6.16), Pic $X \cong \mathbf{Z}$. Furthermore the generator of Cl X is a hyperplane, which corresponds to the invertible sheaf $\mathcal{O}(1)$. Hence Pic X is the free group generated by $\mathcal{O}(1)$, and any invertible sheaf $\mathcal{L}$ is isomorphic to $\mathcal{O}(l)$ for some $l \in \mathbf{Z}$.

We conclude this section with some remarks about closed subschemes of codimension one of a scheme X.

Definition. A Cartier divisor on a scheme X is *effective* if it can be represented by $\{(U_i, f_i)\}$ where all the $f_i \in \Gamma(U_i, \mathcal{O}_{U_i})$. In that case we define the *associated subscheme of codimension 1*, Y, to be the closed subscheme defined by the sheaf of ideals $\mathcal{I}$ which is locally generated by f_i.

Remark 6.17.1. Clearly this gives a 1-1 correspondence between effective Cartier divisors on X and *locally principal* closed subschemes Y, i.e., subschemes whose sheaf of ideals is locally generated by a single element. Note also that if X is an integral separated noetherian locally factorial scheme, so that the Cartier divisors correspond to Weil divisors by (6.11), then the effective Cartier divisors correspond exactly to the effective Weil divisors.

Proposition 6.18. *Let D be an effective Cartier divisor on a scheme X, and let Y be the associated locally principal closed subscheme. Then $\mathcal{I}_Y \cong \mathcal{L}(-D)$.*

145

PROOF. $\mathscr{L}(-D)$ is the subsheaf of $\mathscr{K}$ generated locally by f_i. Since D is effective, this is actually a subsheaf of $\mathscr{O}_X$, which is none other than the ideal sheaf $\mathscr{I}_Y$ of Y.

EXERCISES

6.1. Let X be a scheme satisfying $(*)$. Then $X \times \mathbf{P}^n$ also satisfies $(*)$, and $\mathrm{Cl}(X \times \mathbf{P}^n) \cong (\mathrm{Cl}\, X) \times \mathbf{Z}$.

*6.2. *Varieties in Projective Space.* Let k be an algebraically closed field, and let X be a closed subvariety of $\mathbf{P}^n_k$ which is nonsingular in codimension one (hence satisfies $(*)$). For any divisor $D = \sum n_i Y_i$ on X, we define the *degree* of D to be $\sum n_i \deg Y_i$, where $\deg Y_i$ is the degree of Y_i, considered as a projective variety itself (I, §7).

(a) Let V be an irreducible hypersurface in $\mathbf{P}^n$ which does not contain X, and let Y_i be the irreducible components of $V \cap X$. They all have codimension 1 by (I, Ex. 1.8). For each i, let f_i be a local equation for V on some open set U_i of $\mathbf{P}^n$ for which $Y_i \cap U_i \neq \varnothing$, and let $n_i = v_{Y_i}(\bar{f_i})$, where $\bar{f_i}$ is the restriction of f_i to $U_i \cap X$. Then we define the *divisor V.X* to be $\sum n_i Y_i$. Extend by linearity, and show that this gives a well-defined homomorphism from the subgroup of Div $\mathbf{P}^n$ consisting of divisors, none of whose components contain X, to Div X.

(b) If D is a principal divisor on $\mathbf{P}^n$, for which $D.X$ is defined as in (a), show that $D.X$ is principal on X. Thus we get a homomorphism $\mathrm{Cl}\, \mathbf{P}^n \to \mathrm{Cl}\, X$.

(c) Show that the integer n_i defined in (a) is the same as the intersection multiplicity $i(X, V; Y_i)$ defined in (I, §7). Then use the generalized Bézout theorem (I, 7.7) to show that for any divisor D on $\mathbf{P}^n$, none of whose components contain X,

$$\deg(D.X) = (\deg D) \cdot (\deg X).$$

(d) If D is a principal divisor on X, show that there is a rational function f on $\mathbf{P}^n$ such that $D = (f).X$. Conclude that $\deg D = 0$. Thus the degree function defines a homomorphism deg: $\mathrm{Cl}\, X \to \mathbf{Z}$. (This gives another proof of (6.10), since any complete nonsingular curve is projective.) Finally, there is a commutative diagram

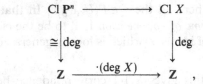

and in particular, we see that the map $\mathrm{Cl}\, \mathbf{P}^n \to \mathrm{Cl}\, X$ is injective.

*6.3. *Cones.* In this exercise we compare the class group of a projective variety V to the class group of its cone (I, Ex. 2.10). So let V be a projective variety in $\mathbf{P}^n$, which is of dimension $\geqslant 1$ and nonsingular in codimension 1. Let $X = C(V)$ be the affine cone over V in $\mathbf{A}^{n+1}$, and let $\bar{X}$ be its projective closure in $\mathbf{P}^{n+1}$. Let $P \in X$ be the vertex of the cone.

(a) Let $\pi: \bar{X} - P \to V$ be the projection map. Show that V can be covered by open subsets U_i such that $\pi^{-1}(U_i) \cong U_i \times \mathbf{A}^1$ for each i, and then show as in (6.6) that $\pi^*: \mathrm{Cl}\, V \to \mathrm{Cl}(\bar{X} - P)$ is an isomorphism. Since $\mathrm{Cl}\, \bar{X} \cong \mathrm{Cl}(\bar{X} - P)$, we have also $\mathrm{Cl}\, V \cong \mathrm{Cl}\, \bar{X}$.

(b) We have $V \subseteq \bar{X}$ as the hyperplane section at infinity. Show that the class of the divisor V in Cl $\bar{X}$ is equal to π^* (class of $V.H$) where H is any hyperplane of $\mathbf{P}^n$ not containing V. Thus conclude using (6.5) that there is an exact sequence

$$0 \to \mathbf{Z} \to \text{Cl } V \to \text{Cl } X \to 0,$$

where the first arrow sends $1 \mapsto V.H$, and the second is π^* followed by the restriction to $X - P$ and inclusion in X. (The injectivity of the first arrow follows from the previous exercise.)

(c) Let $S(V)$ be the homogeneous coordinate ring of V (which is also the affine coordinate ring of X). Show that $S(V)$ is a unique factorization domain if and only if (1) V is projectively normal (Ex. 5.14), and (2) Cl $V \cong \mathbf{Z}$ and is generated by the class of $V.H$.

(d) Let $\mathcal{O}_P$ be the local ring of P on X. Show that the natural restriction map induces an isomorphism Cl $X \to$ Cl(Spec $\mathcal{O}_P$).

6.4. Let k be a field of characteristic $\neq 2$. Let $f \in k[x_1, \ldots, x_n]$ be a *square-free* nonconstant polynomial, i.e., in the unique factorization of f into irreducible polynomials, there are no repeated factors. Let $A = k[x_1, \ldots, x_n, z]/(z^2 - f)$. Show that A is an integrally closed ring. [*Hint*: The quotient field K of A is just $k(x_1, \ldots, x_n)[z]/(z^2 - f)$. It is a Galois extension of $k(x_1, \ldots, x_n)$ with Galois group $\mathbf{Z}/2\mathbf{Z}$ generated by $z \mapsto -z$. If $\alpha = g + hz \in K$, where $g, h \in k(x_1, \ldots, x_n)$, then the minimal polynomial of α is $X^2 - 2gX + (g^2 - h^2 f)$. Now show that α is integral over $k[x_1, \ldots, x_n]$ if and only if $g, h \in k[x_1, \ldots, x_n]$. Conclude that A is the integral closure of $k[x_1, \ldots, x_n]$ in K.]

*6.5. *Quadric Hypersurfaces.* Let char $k \neq 2$, and let X be the affine quadric hypersurface Spec $k[x_0, \ldots, x_n]/(x_0^2 + x_1^2 + \ldots + x_r^2)$—cf. (I, Ex. 5.12).

(a) Show that X is normal if $r \geqslant 2$ (use (Ex. 6.4)).

(b) Show by a suitable linear change of coordinates that the equation of X could be written as $x_0 x_1 = x_2^2 + \ldots + x_r^2$. Now imitate the method of (6.5.2) to show that:

  (1) If $r = 2$, then Cl $X \cong \mathbf{Z}/2\mathbf{Z}$;
  (2) If $r = 3$, then Cl $X \cong \mathbf{Z}$ (use (6.6.1) and (Ex. 6.3) above);
  (3) If $r \geqslant 4$ then Cl $X = 0$.

(c) Now let Q be the projective quadric hypersurface in $\mathbf{P}^n$ defined by the same equation. Show that:

  (1) If $r = 2$, Cl $Q \cong \mathbf{Z}$, and the class of a hyperplane section $Q.H$ is twice the generator;
  (2) If $r = 3$, Cl $Q \cong \mathbf{Z} \oplus \mathbf{Z}$;
  (3) If $r \geqslant 4$, Cl $Q \cong \mathbf{Z}$, generated by $Q.H$.

(d) Prove Klein's theorem, which says that if $r \geqslant 4$, and if Y is an irreducible subvariety of codimension 1 on Q, then there is an irreducible hypersurface $V \subseteq \mathbf{P}^n$ such that $V \cap Q = Y$, with multiplicity one. In other words, Y is a complete intersection. (First show that for $r \geqslant 4$, the homogeneous coordinate ring $S(Q) = k[x_0, \ldots, x_n]/(x_0^2 + \ldots + x_r^2)$ is a UFD.)

6.6. Let X be the nonsingular plane cubic curve $y^2 z = x^3 - xz^2$ of (6.10.2).

(a) Show that three points P, Q, R of X are collinear if and only if $P + Q + R = 0$ in the group law on X. (Note that the point $P_0 = (0,1,0)$ is the zero element in the group structure on X.)

(b) A point $P \in X$ has order 2 in the group law on X if and only if the tangent line at P passes through P_0.

(c) A point $P \in X$ has order 3 in the group law on X if and only if P is an inflection point. (An *inflection point* of a plane curve is a nonsingular point P of the curve, whose tangent line (I, Ex. 7.3) has intersection multiplicity $\geqslant 3$ with the curve at P.)

(d) Let $k = \mathbf{C}$. Show that the points of X with coordinates in $\mathbf{Q}$ form a subgroup of the group X. Can you determine the structure of this subgroup explicitly?

*6.7. Let X be the nodal cubic curve $y^2z = x^3 + x^2z$ in $\mathbf{P}^2$. Imitate (6.11.4) and show that the group of Cartier divisors of degree 0, $\operatorname{CaCl}^\circ X$, is naturally isomorphic to the multiplicative group $\mathbf{G}_m$.

6.8. (a) Let $f : X \to Y$ be a morphism of schemes. Show that $\mathscr{L} \mapsto f^*\mathscr{L}$ induces a homomorphism of Picard groups, $f^* : \operatorname{Pic} Y \to \operatorname{Pic} X$.

(b) If f is a finite morphism of nonsingular curves, show that this homomorphism corresponds to the homomorphism $f^* : \operatorname{Cl} Y \to \operatorname{Cl} X$ defined in the text, via the isomorphisms of (6.16).

(c) If X is a locally factorial integral closed subscheme of $\mathbf{P}^n_k$, and if $f : X \to \mathbf{P}^n$ is the inclusion map, then f^* on Pic agrees with the homomorphism on divisor class groups defined in (Ex. 6.2) via the isomorphisms of (6.16).

*6.9. *Singular Curves.* Here we give another method of calculating the Picard group of a singular curve. Let X be a projective curve over k, let $\tilde{X}$ be its normalization, and let $\pi : \tilde{X} \to X$ be the projection map (Ex. 3.8). For each point $P \in X$, let $\mathcal{O}_P$ be its local ring, and let $\tilde{\mathcal{O}}_P$ be the integral closure of $\mathcal{O}_P$. We use a $*$ to denote the group of units in a ring.

(a) Show there is an exact sequence
$$0 \to \bigoplus_{P \in X} \tilde{\mathcal{O}}_P^* / \mathcal{O}_P^* \to \operatorname{Pic} X \xrightarrow{\pi^*} \operatorname{Pic} \tilde{X} \to 0.$$

[*Hint*: Represent $\operatorname{Pic} X$ and $\operatorname{Pic} \tilde{X}$ as the groups of Cartier divisors modulo principal divisors, and use the exact sequence of sheaves on X
$$0 \to \pi_* \mathcal{O}_{\tilde{X}}^* / \mathcal{O}_X^* \to \mathscr{K}^* / \mathcal{O}_X^* \to \mathscr{K}^* / \pi_* \mathcal{O}_{\tilde{X}}^* \to 0.]$$

(b) Use (a) to give another proof of the fact that if X is a plane cuspidal cubic curve, then there is an exact sequence
$$0 \to \mathbf{G}_a \to \operatorname{Pic} X \to \mathbf{Z} \to 0,$$
and if X is a plane nodal cubic curve, there is an exact sequence
$$0 \to \mathbf{G}_m \to \operatorname{Pic} X \to \mathbf{Z} \to 0.$$

6.10. *The Grothendieck Group $K(X)$.* Let X be a noetherian scheme. We define $K(X)$ to be the quotient of the free abelian group generated by all the coherent sheaves on X, by the subgroup generated by all expressions $\mathscr{F} - \mathscr{F}' - \mathscr{F}''$, whenever there is an exact sequence $0 \to \mathscr{F}' \to \mathscr{F} \to \mathscr{F}'' \to 0$ of coherent sheaves on X. If $\mathscr{F}$ is a coherent sheaf, we denote by $\gamma(\mathscr{F})$ its image in $K(X)$.

(a) If $X = \mathbf{A}^1_k$, then $K(X) \cong \mathbf{Z}$.

(b) If X is any integral scheme, and $\mathscr{F}$ a coherent sheaf, we define the *rank* of $\mathscr{F}$ to be $\dim_K \mathscr{F}_\xi$, where ξ is the generic point of X, and $K = \mathcal{O}_\xi$ is the function field of X. Show that the rank function defines a surjective homomorphism rank$: K(X) \to \mathbf{Z}$.

(c) If Y is a closed subscheme of X, there is an exact sequence

$$K(Y) \to K(X) \to K(X - Y) \to 0,$$

where the first map is extension by zero, and the second map is restriction. [*Hint*: For exactness in the middle, show that if $\mathscr{F}$ is a coherent sheaf on X, whose support is contained in Y, then there is a finite filtration $\mathscr{F} = \mathscr{F}_0 \supseteq \mathscr{F}_1 \supseteq \ldots \supseteq \mathscr{F}_n = 0$, such that each $\mathscr{F}_i/\mathscr{F}_{i+1}$ is an $\mathcal{O}_Y$-module. To show surjectivity on the right, use (Ex. 5.15).]

For further information about $K(X)$, and its applications to the generalized Riemann–Roch theorem, see Borel–Serre [1], Manin [1], and Appendix A.

*6.11. *The Grothendieck Group of a Nonsingular Curve.* Let X be a nonsingular curve over an algebraically closed field k. We will show that $K(X) \cong \operatorname{Pic} X \oplus \mathbf{Z}$, in several steps.

(a) For any divisor $D = \sum n_i P_i$ on X, let $\psi(D) = \sum n_i \gamma(k(P_i)) \in K(X)$, where $k(P_i)$ is the skyscraper sheaf k at P_i and 0 elsewhere. If D is an effective divisor, let $\mathcal{O}_D$ be the structure sheaf of the associated subscheme of codimension 1, and show that $\psi(D) = \gamma(\mathcal{O}_D)$. Then use (6.18) to show that for any D, $\psi(D)$ depends only on the linear equivalence class of D, so ψ defines a homomorphism $\psi: \operatorname{Cl} X \to K(X)$.

(b) For any coherent sheaf $\mathscr{F}$ on X, show that there exist locally free sheaves $\mathscr{E}_0$ and $\mathscr{E}_1$ and an exact sequence $0 \to \mathscr{E}_1 \to \mathscr{E}_0 \to \mathscr{F} \to 0$. Let $r_0 = \operatorname{rank} \mathscr{E}_0$, $r_1 = \operatorname{rank} \mathscr{E}_1$, and define det $\mathscr{F} = (\bigwedge^{r_0}\mathscr{E}_0) \otimes (\bigwedge^{r_1}\mathscr{E}_1)^{-1} \in \operatorname{Pic} X$. Here $\bigwedge$ denotes the exterior power (Ex. 5.16). Show that det $\mathscr{F}$ is independent of the resolution chosen, and that it gives a homomorphism det$: K(X) \to \operatorname{Pic} X$. Finally show that if D is a divisor, then det$(\psi(D)) = \mathscr{L}(D)$.

(c) If $\mathscr{F}$ is any coherent sheaf of rank r, show that there is a divisor D on X and an exact sequence $0 \to \mathscr{L}(D)^{\oplus r} \to \mathscr{F} \to \mathscr{T} \to 0$, where $\mathscr{T}$ is a torsion sheaf. Conclude that if $\mathscr{F}$ is a sheaf of rank r, then $\gamma(\mathscr{F}) - r\gamma(\mathcal{O}_X) \in \operatorname{Im} \psi$.

(d) Using the maps ψ, det, rank, and $1 \mapsto \gamma(\mathcal{O}_X)$ from $\mathbf{Z} \to K(X)$, show that $K(X) \cong \operatorname{Pic} X \oplus \mathbf{Z}$.

6.12. Let X be a complete nonsingular curve. Show that there is a unique way to define the *degree* of any coherent sheaf on X, deg $\mathscr{F} \in \mathbf{Z}$, such that:

(1) If D is a divisor, deg $\mathscr{L}(D) = \deg D$;
(2) If $\mathscr{F}$ is a *torsion sheaf* (meaning a sheaf whose stalk at the generic point is zero), then deg $\mathscr{F} = \sum_{P \in X} \operatorname{length} (\mathscr{F}_P)$; and
(3) If $0 \to \mathscr{F}' \to \mathscr{F} \to \mathscr{F}'' \to 0$ is an exact sequence, then deg $\mathscr{F} = \deg \mathscr{F}' + \deg \mathscr{F}''$.

7 Projective Morphisms

In this section we gather together several topics concerned with morphisms of a given scheme to projective space. We will show how a morphism of a scheme X to a projective space is determined by giving an invertible sheaf

$\mathscr{L}$ on X and a set of its global sections. We will give some criteria for this morphism to be an immersion. Then we study the closely connected topic of ample invertible sheaves. We also introduce the more classical language of linear systems, which from the point of view of schemes is hardly more than another set of terminology for dealing with invertible sheaves and their global sections. However, the geometric understanding furnished by the concept of linear system is often very valuable. At the end of this section we define the **Proj** of a graded sheaf of algebras over a scheme X, and we give two important examples, namely the projective bundle $\mathbf{P}(\mathscr{E})$ associated with a locally free sheaf $\mathscr{E}$, and the definition of blowing up with respect to a coherent sheaf of ideals.

Morphisms to $\mathbf{P}^n$

Let A be a fixed ring, and consider the projective space $\mathbf{P}_A^n = \operatorname{Proj} A[x_0,\ldots,x_n]$ over A. On $\mathbf{P}_A^n$ we have the invertible sheaf $\mathcal{O}(1)$, and the homogeneous coordinates $x_0,\ldots,x_n$ give rise to global sections $x_0,\ldots,x_n \in \Gamma(\mathbf{P}_A^n,\mathcal{O}(1))$. One sees easily that the sheaf $\mathcal{O}(1)$ is generated by the global sections $x_0,\ldots,x_n$, i.e., the images of these sections generate the stalk $\mathcal{O}(1)_P$ of the sheaf $\mathcal{O}(1)$ as a module over the local ring $\mathcal{O}_P$, for each point $P \in \mathbf{P}_A^n$.

Now let X be any scheme over A, and let $\varphi: X \to \mathbf{P}_A^n$ be an A-morphism of X to $\mathbf{P}_A^n$. Then $\mathscr{L} = \varphi^*(\mathcal{O}(1))$ is an invertible sheaf on X, and the global sections $s_0,\ldots,s_n$, where $s_i = \varphi^*(x_i)$, $s_i \in \Gamma(X,\mathscr{L})$, generate the sheaf $\mathscr{L}$. Conversely, we will see that $\mathscr{L}$ and the sections s_i determine φ.

Theorem 7.1. *Let A be a ring, and let X be a scheme over A.*

(a) *If $\varphi: X \to \mathbf{P}_A^n$ is an A-morphism, then $\varphi^*(\mathcal{O}(1))$ is an invertible sheaf on X, which is generated by the global sections $s_i = \varphi^*(x_i)$, $i = 0,1,\ldots,n$.*

(b) *Conversely, if $\mathscr{L}$ is an invertible sheaf on X, and if $s_0,\ldots,s_n \in \Gamma(X,\mathscr{L})$ are global sections which generate $\mathscr{L}$, then there exists a unique A-morphism $\varphi: X \to \mathbf{P}_A^n$ such that $\mathscr{L} \cong \varphi^*(\mathcal{O}(1))$ and $s_i = \varphi^*(x_i)$ under this isomorphism.*

PROOF. Part (a) is clear from the discussion above. To prove (b), suppose given $\mathscr{L}$ and the global sections $s_0,\ldots,s_n$ which generate it. For each i, let $X_i = \{P \in X | (s_i)_P \notin \mathfrak{m}_P \mathscr{L}_P\}$. Then (as we have seen before) X_i is an open subset of X, and since the s_i generate $\mathscr{L}$, the open sets X_i must cover X. We define a morphism from X_i to the standard open set $U_i = \{x_i \neq 0\}$ of $\mathbf{P}_A^n$ as follows. Recall that $U_i \cong \operatorname{Spec} A[y_0,\ldots,y_n]$ where $y_j = x_j/x_i$, with $y_i = 1$ omitted. We define a ring homomorphism $A[y_0,\ldots,y_n] \to \Gamma(X_i,\mathcal{O}_{X_i})$ by sending $y_j \to s_j/s_i$ and making it A-linear. This makes sense, because for each $P \in X_i$, $(s_i)_P \notin \mathfrak{m}_P \mathscr{L}_P$, and $\mathscr{L}$ is locally free of rank 1, so the quotient s_j/s_i is a well-defined element of $\Gamma(X_i,\mathcal{O}_{X_i})$. Now by (Ex. 2.4) this ring homomorphism determines a morphism of schemes (over A) $X_i \to U_i$. Clearly these morphisms glue (cf. Step 3 of proof of (3.3)), so we obtain a morphism $\varphi: X \to \mathbf{P}_A^n$. It is clear from the construction that φ is an A-

morphism, that $\mathscr{L} \cong \varphi^*(\mathcal{O}(1))$, and that the sections s_i correspond to $\varphi^*(x_i)$ under this isomorphism. It is clear that any morphism with these properties must be the one given by the construction, so φ is unique.

Example 7.1.1 (Automorphisms of $\mathbf{P}_k^n$). If $\|a_{ij}\|$ is an invertible $(n+1) \times (n+1)$ matrix of elements of a field k, then $x_i' = \sum a_{ij}x_j$ determines an automorphism of the polynomial ring $k[x_0, \ldots, x_n]$ and hence also an automorphism of $\mathbf{P}_k^n$. If $\lambda \in k$ is a nonzero element, then $\|\lambda a_{ij}\|$ determines the same automorphism of $\mathbf{P}_k^n$. So we are led to consider the group $\mathrm{PGL}(n,k) = \mathrm{GL}(n+1,k)/k^*$, which acts as a group of automorphisms of $\mathbf{P}_k^n$. By considering the points $(1,0,\ldots,0)$, $(0,1,0,\ldots,0),\ldots,(0,0,\ldots,1)$, and $(1,1,\ldots,1)$, one sees easily that this group acts faithfully, i.e., if $g \in \mathrm{PGL}(n,k)$ induces the trivial automorphism of $\mathbf{P}_k^n$, then g is the identity.

Now we will show conversely that every k-automorphism of $\mathbf{P}_k^n$ is an element of $\mathrm{PGL}(n,k)$. This generalizes an earlier result for $\mathbf{P}_k^1$ (I, Ex. 6.6). So let φ be a k-automorphism of $\mathbf{P}_k^n$. We have seen (6.17) that $\mathrm{Pic}\, \mathbf{P}_k^n \cong \mathbf{Z}$ and is generated by $\mathcal{O}(1)$. The automorphism φ induces an automorphism of $\mathrm{Pic}\, \mathbf{P}^n$, so $\varphi^*(\mathcal{O}(1))$ must be a generator of that group, hence isomorphic to either $\mathcal{O}(1)$ or $\mathcal{O}(-1)$. But $\mathcal{O}(-1)$ has no global sections, so we conclude that $\varphi^*(\mathcal{O}(1)) \cong \mathcal{O}(1)$. Now $\Gamma(\mathbf{P}^n, \mathcal{O}(1))$ is a k-vector space with basis $x_0, \ldots, x_n$, by (5.13). Since φ is an automorphism, the $s_i = \varphi^*(x_i)$ must be another basis of this vector space, so we can write $s_i = \sum a_{ij}x_j$, where $\|a_{ij}\|$ is an invertible matrix of elements of k. Since φ is uniquely determined by the s_i according to the theorem, we see that φ coincides with the automorphism given by $\|a_{ij}\|$ as an element of $\mathrm{PGL}(n,k)$.

Example 7.1.2. If X is a scheme over A, $\mathscr{L}$ an invertible sheaf, and $s_0, \ldots, s_n$ any set of global sections, which do not necessarily generate $\mathscr{L}$, we can always consider the open set $U \subseteq X$ (possibly empty) over which the s_i do generate $\mathscr{L}$. Then $\mathscr{L}|_U$ and the $s_i|_U$ give a morphism $U \to \mathbf{P}_A^n$. Such is the case for example, if we take $X = \mathbf{P}_k^{n+1}$, $\mathscr{L} = \mathcal{O}(1)$, and $s_i = x_i$, $i = 0, \ldots, n$ (omitting x_{n+1}). These sections generate everywhere except at the point $(0,0,\ldots,0,1) = P_0$. Thus $U = \mathbf{P}^{n+1} - P_0$, and the corresponding morphism $U \to \mathbf{P}^n$ is nothing other than the projection from the point P_0 to $\mathbf{P}^n$ (I, Ex. 3.14).

Next we give some criteria for a morphism to a projective space to be a closed immersion.

Proposition 7.2. Let $\varphi : X \to \mathbf{P}_A^n$ be a morphism of schemes over A, corresponding to an invertible sheaf $\mathscr{L}$ on X and sections $s_0, \ldots, s_n \in \Gamma(X, \mathscr{L})$ as above. Then φ is a closed immersion if and only if

(1) each open set $X_i = X_{s_i}$ is affine, and
(2) for each i, the map of rings $A[y_0, \ldots, y_n] \to \Gamma(X_i, \mathcal{O}_{X_i})$ defined by $y_j \mapsto s_j/s_i$ is surjective.

PROOF. First suppose φ is a closed immersion. Then $X_i = X \cap U_i$ is a closed subscheme of U_i. Therefore X_i is affine and the corresponding map of rings is surjective by (5.10). Conversely, suppose (1) and (2) satisfied. Then each X_i is a closed subscheme of U_i. Since in any case $X_i = \varphi^{-1}(U_i)$, and the X_i cover X, it is clear that X is a closed subscheme of $\mathbf{P}_A^n$.

With more hypotheses, we can give a more local criterion.

Proposition 7.3. *Let k be an algebraically closed field, let X be a projective scheme over k, and let $\varphi: X \to \mathbf{P}_k^n$ be a morphism (over k) corresponding to $\mathscr{L}$ and $s_0, \ldots, s_n \in \Gamma(X,\mathscr{L})$ as above. Let $V \subseteq \Gamma(X,\mathscr{L})$ be the subspace spanned by the s_i. Then φ is a closed immersion if and only if*

(1) *elements of V separate points, i.e., for any two distinct closed points $P,Q \in X$, there is an $s \in V$ such that $s \in \mathfrak{m}_P \mathscr{L}_P$ but $s \notin \mathfrak{m}_Q \mathscr{L}_Q$, or vice versa, and*

(2) *elements of V separate tangent vectors, i.e., for each closed point $P \in X$, the set $\{s \in V \mid s_P \in \mathfrak{m}_P \mathscr{L}_P\}$ spans the k-vector space $\mathfrak{m}_P \mathscr{L}_P / \mathfrak{m}_P^2 \mathscr{L}_P$.*

PROOF. If φ is a closed immersion, we think of X as a closed subscheme of $\mathbf{P}_k^n$. In this case $\mathscr{L} = \mathcal{O}_X(1)$, and the vector space $V \subseteq \Gamma(X,\mathcal{O}_X(1))$ is just spanned by the images of $x_0, \ldots, x_n \in \Gamma(\mathbf{P}^n,\mathcal{O}(1))$. Given closed points $P \neq Q$ in X, there is a hyperplane containing P but not Q. If its equation is $\sum a_i x_i = 0$, $a_i \in k$, then $s = \sum a_i x_i$ restricted to X has the right property for (1). For (2), the hyperplanes passing through P give rise to sections which generate $\mathfrak{m}_P \mathscr{L}_P / \mathfrak{m}_P^2 \mathscr{L}_P$. For simplicity suppose that P is the point $(1,0,0, \ldots ,0)$. Then on the open affine $U_0 \cong \operatorname{Spec} k[y_1, \ldots, y_n]$, $\mathscr{L}$ is trivial, P is the point $(0, \ldots ,0)$, and $\mathfrak{m}_P / \mathfrak{m}_P^2$ is exactly the vector space spanned by $y_1, \ldots, y_n$. We use the hypothesis k algebraically closed to ensure that every closed point of $\mathbf{P}_k^n$ is of the form $(a_0, \ldots ,a_n)$ for suitable $a_i \in k$, hence points can be separated by hyperplanes with coefficients in k.

For the converse, let $\varphi: X \to \mathbf{P}^n$ satisfy (1) and (2). Since the elements of V are pull-backs of sections of $\mathcal{O}(1)$ on $\mathbf{P}^n$, it is clear from (1) that the map φ is injective as a map of sets. Since X is projective over k, it is proper over k (4.9), so the image $\varphi(X)$ in $\mathbf{P}^n$ is closed (Ex. 4.4), and φ is a proper morphism (4.8e). In particular, φ is a closed map. But, being a morphism, it is also continuous, so we see that φ is a homeomorphism of X onto its image $\varphi(X)$ which is a closed subset of $\mathbf{P}^n$. To show that φ is a closed immersion, it remains only to show that the morphism of sheaves $\mathcal{O}_{\mathbf{P}^n} \to \varphi_* \mathcal{O}_X$ is surjective. This can be checked on the stalks. So it is sufficient to show, for each closed point P, that $\mathcal{O}_{\mathbf{P}^n,P} \to \mathcal{O}_{X,P}$ is surjective. Both local rings have the same residue field k, and our hypothesis (2) implies that the image of the maximal ideal $\mathfrak{m}_{\mathbf{P}^n,P}$ generates $\mathfrak{m}_{X,P}/\mathfrak{m}_{X,P}^2$. We also need to use (5.20), which implies that $\varphi_* \mathcal{O}_X$ is a coherent sheaf on $\mathbf{P}^n$, and hence that $\mathcal{O}_{X,P}$ is a finitely generated $\mathcal{O}_{\mathbf{P}^n,P}$-module. Now our result is a consequence of the following lemma.

Lemma 7.4. *Let* $f : A \to B$ *be a local homomorphism of local noetherian rings, such that*

(1) $A/\mathfrak{m}_A \to B/\mathfrak{m}_B$ *is an isomorphism,*
(2) $\mathfrak{m}_A \to \mathfrak{m}_B/\mathfrak{m}_B^2$ *is surjective, and*
(3) B *is a finitely generated A-module.*

Then f is surjective.

PROOF. Consider the ideal $\mathfrak{a} = \mathfrak{m}_A B$ of B. We have $\mathfrak{a} \subseteq \mathfrak{m}_B$, and by (2), $\mathfrak{a}$ contains a set of generators for $\mathfrak{m}_B/\mathfrak{m}_B^2$. Hence by Nakayama's lemma for the local ring B and the B-module $\mathfrak{m}_B$, we conclude that $\mathfrak{a} = \mathfrak{m}_B$. Now apply Nakayama's lemma to the A-module B. By (3), B is a finitely generated A-module. The element $1 \in B$ gives a generator for $B/\mathfrak{m}_A B = B/\mathfrak{m}_B = A/\mathfrak{m}_A$ by (1), so we conclude that 1 also generates B as an A-module, i.e., f is surjective.

Ample Invertible Sheaves

Now that we have seen that a morphism of a scheme X to a projective space can be characterized by giving an invertible sheaf on X and a suitable set of its global sections, we can reduce the study of varieties in projective space to the study of schemes with certain invertible sheaves and given global sections. Recall that in §5 we defined a sheaf $\mathscr{L}$ on X to be *very ample relative to Y* (where X is a scheme over Y) if there is an immersion $i : X \to \mathbf{P}_Y^n$ for some n such that $\mathscr{L} \cong i^* \mathcal{O}(1)$. In case $Y = \operatorname{Spec} A$, this is the same thing as saying that $\mathscr{L}$ admits a set of global sections $s_0, \ldots, s_n$ such that the corresponding morphism $X \to \mathbf{P}_A^n$ is an immersion. We have also seen (5.17) that if $\mathscr{L}$ is a very ample invertible sheaf on a projective scheme X over a noetherian ring A, then for any coherent sheaf $\mathscr{F}$ on X, there is an integer $n_0 > 0$ such that for all $n \geq n_0$, $\mathscr{F} \otimes \mathscr{L}^n$ is generated by global sections. We will use this last property of being generated by global sections to define the notion of an ample invertible sheaf, which is more general, and in many ways is more convenient to work with than the notion of very ample sheaf.

Definition. An invertible sheaf $\mathscr{L}$ on a noetherian scheme X is said to be *ample* if for every coherent sheaf $\mathscr{F}$ on X, there is an integer $n_0 > 0$ (depending on $\mathscr{F}$) such that for every $n \geq n_0$, the sheaf $\mathscr{F} \otimes \mathscr{L}^n$ is generated by its global sections. (Here $\mathscr{L}^n = \mathscr{L}^{\otimes n}$ denotes the n-fold tensor power of $\mathscr{L}$ with itself.)

Remark 7.4.1. Note that "ample" is an absolute notion, i.e., it depends only on the scheme X, whereas "very ample" is a relative notion, depending on a morphism $X \to Y$.

153

Example 7.4.2. If X is affine, then any invertible sheaf is ample, because every coherent sheaf on an affine scheme is generated by its global sections (5.16.2).

Remark 7.4.3. Serre's theorem (5.17) asserts that a very ample sheaf $\mathscr{L}$ on a projective scheme X over a noetherian ring A is ample. The converse is false, but we will see below (7.6) that if $\mathscr{L}$ is ample, then some tensor power $\mathscr{L}^m$ of $\mathscr{L}$ is very ample. Thus "ample" can be viewed as a stable version of "very ample."

Remark 7.4.4. In Chapter III we will give a characterization of ample invertible sheaves in terms of the vanishing of certain cohomology groups (III, 5.3).

Proposition 7.5. *Let $\mathscr{L}$ be an invertible sheaf on a noetherian scheme X. Then the following conditions are equivalent:*

  (i) $\mathscr{L}$ *is ample;*
  (ii) $\mathscr{L}^m$ *is ample for all $m > 0$;*
  (iii) $\mathscr{L}^m$ *is ample for some $m > 0$.*

PROOF. (i) $\Rightarrow$ (ii) is immediate from the definition of ample; (ii) $\Rightarrow$ (iii) is trivial. To prove (iii) $\Rightarrow$ (i), assume that $\mathscr{L}^m$ is ample. Given a coherent sheaf $\mathscr{F}$ on X, there exists an $n_0 > 0$ such that $\mathscr{F} \otimes (\mathscr{L}^m)^n$ is generated by global sections for all $n \geqslant n_0$. Considering the coherent sheaf $\mathscr{F} \otimes \mathscr{L}$, there exists an $n_1 > 0$ such that $\mathscr{F} \otimes \mathscr{L} \otimes (\mathscr{L}^m)^n$ is generated by global sections for all $n \geqslant n_1$. Similarly, for each $k = 1, 2, \ldots, m - 1$, there is an $n_k > 0$ such that $\mathscr{F} \otimes \mathscr{L}^k \otimes (\mathscr{L}^m)^n$ is generated by global sections for all $n \geqslant n_k$. Now if we take $N = m \cdot \max\{n_i | i = 0, 1, \ldots, m - 1\}$, then $\mathscr{F} \otimes \mathscr{L}^n$ is generated by global sections for all $n \geqslant N$. Hence $\mathscr{L}$ is ample.

Theorem 7.6. *Let X be a scheme of finite type over a noetherian ring A, and let $\mathscr{L}$ be an invertible sheaf on X. Then $\mathscr{L}$ is ample if and only if $\mathscr{L}^m$ is very ample over $\operatorname{Spec} A$ for some $m > 0$.*

PROOF. First suppose $\mathscr{L}^m$ is very ample for some $m > 0$. Then there is an immersion $i : X \to \mathbf{P}_A^n$ such that $\mathscr{L}^m \cong i^*(\mathcal{O}(1))$. Let $\bar{X}$ be the closure of X in $\mathbf{P}_A^n$. Then $\bar{X}$ is a projective scheme over A, so by (5.17), $\mathcal{O}_{\bar{X}}(1)$ is ample on $\bar{X}$. Now given any coherent sheaf $\mathscr{F}$ on X, it extends by (Ex. 5.15) to a coherent sheaf $\bar{\mathscr{F}}$ on $\bar{X}$. If $\bar{\mathscr{F}} \otimes \mathcal{O}_{\bar{X}}(l)$ is generated by global sections, then a fortiori $\mathscr{F} \otimes \mathcal{O}_X(l)$ is also generated by global sections. Thus we see that $\mathscr{L}^m$ is ample on X, and so by (7.5), $\mathscr{L}$ is also ample on X.

For the converse, suppose that $\mathscr{L}$ is ample on X. Given any $P \in X$, let U be an open affine neighborhood of P such that $\mathscr{L}|_U$ is free on U. Let Y be the closed set $X - U$, and let $\mathscr{I}_Y$ be its sheaf of ideals with the reduced induced scheme structure. Then $\mathscr{I}_Y$ is a coherent sheaf on X, so for some

$n > 0$, $\mathscr{I}_Y \otimes \mathscr{L}^n$ is generated by global sections. In particular, there is a section $s \in \Gamma(X, \mathscr{I}_Y \otimes \mathscr{L}^n)$ such that $s_P \notin \mathfrak{m}_P(\mathscr{I}_Y \otimes \mathscr{L}^n)_P$. Now $\mathscr{I}_Y \otimes \mathscr{L}^n$ is a subsheaf of $\mathscr{L}^n$, so we can think of s as an element of $\Gamma(X, \mathscr{L}^n)$. If X_s is the open set $\{Q \in X | s_Q \notin \mathfrak{m}_Q \mathscr{L}^n_Q\}$, then it follows from our choice of s that $P \in X_s$ and that $X_s \subseteq U$. Now U is affine, and $\mathscr{L}|_U$ is trivial, so s induces an element $f \in \Gamma(U, \mathcal{O}_U)$, and then $X_s = U_f$ is also affine.

Thus we have shown that for any point $P \in X$, there is an $n > 0$ and a section $s \in \Gamma(X, \mathscr{L}^n)$ such that $P \in X_s$ and X_s is affine. Since X is quasi-compact, we can cover X by a finite number of such open affines, corresponding to sections $s_i \in \Gamma(X, \mathscr{L}^{n_i})$. Replacing each s_i by a suitable power $s_i^k \in \Gamma(X, \mathscr{L}^{kn_i})$, which doesn't change X_{s_i}, we may assume that all n_i are equal to one n. Finally, since $\mathscr{L}^n$ is also ample, and since we are only trying to show that some power of $\mathscr{L}$ is very ample, we may replace $\mathscr{L}$ by $\mathscr{L}^n$. Thus we may assume now that we have global sections $s_1, \ldots, s_k \in \Gamma(X, \mathscr{L})$ such that each $X_i = X_{s_i}$ is affine, and the X_i cover X.

Now for each i, let $B_i = \Gamma(X_i, \mathcal{O}_{X_i})$. Since X is a scheme of finite type over A, each B_i is a finitely generated A-algebra (Ex. 3.3). So let $\{b_{ij} | j = 1, \ldots, k_i\}$ be a set of generators for B_i as an A-algebra. By (5.14), for each i, j, there is an integer n such that $s_i^n b_{ij}$ extends to a global section $c_{ij} \in \Gamma(X, \mathscr{L}^n)$. We can take one n large enough to work for all i, j. Now we take the invertible sheaf $\mathscr{L}^n$ on X, and the sections $\{s_i^n | i = 1, \ldots, k\}$ and $\{c_{ij} | i = 1, \ldots, k;$ $j = 1, \ldots, k_i\}$ and use all these sections to define a morphism (over A) $\varphi: X \to \mathbf{P}_A^N$ as in (7.1) above. Since X is covered by the X_i, the sections s_i^n already generate the sheaf $\mathscr{L}^n$, so this is indeed a morphism.

Let $\{x_i | i = 1, \ldots, k\}$ and $\{x_{ij} | i = 1, \ldots, k; j = 1, \ldots, k_i\}$ be the homogeneous coordinates of $\mathbf{P}_A^N$ corresponding to the sections of $\mathscr{L}^n$ mentioned above. For each $i = 1, \ldots, k$, let $U_i \subseteq \mathbf{P}_A^N$ be the open subset $x_i \neq 0$. Then $\varphi^{-1}(U_i) = X_i$, and the corresponding map of affine rings

$$A[\{y_i\}; \{y_{ij}\}] \to B_i$$

is surjective, because $y_{ij} \mapsto c_{ij}/s_i^n = b_{ij}$, and we chose the b_{ij} so as to generate B_i as an A-algebra. Thus X_i is mapped onto a closed subscheme of U_i. It follows that φ gives an isomorphism of X with a closed subscheme of $\bigcup_{i=1}^k U_i \subseteq \mathbf{P}_A^N$, so φ is an immersion. Hence $\mathscr{L}^n$ is very ample relative to Spec A, as required.

Example 7.6.1. Let $X = \mathbf{P}_k^n$, where k is a field. Then $\mathcal{O}(1)$ is very ample by definition. For any $d > 0$, $\mathcal{O}(d)$ corresponds to the d-uple embedding (Ex. 5.13), so $\mathcal{O}(d)$ is also very ample. Hence $\mathcal{O}(d)$ is ample for all $d > 0$. On the other hand, since the sheaf $\mathcal{O}(l)$ has no global sections for $l < 0$, one sees easily that the sheaves $\mathcal{O}(l)$ for $l \leq 0$ cannot be ample. So on $\mathbf{P}_k^n$, we have $\mathcal{O}(l)$ is ample $\Leftrightarrow$ very ample $\Leftrightarrow l > 0$.

Example 7.6.2. Let Q be the nonsingular quadric surface $xy = zw$ in $\mathbf{P}_k^3$ over a field k. We have seen (6.6.1) that Pic $Q \cong \mathbf{Z} \oplus \mathbf{Z}$, and so we speak of the *type* (a,b), $a,b \in \mathbf{Z}$, of an invertible sheaf. Now $Q \cong \mathbf{P}^1 \times \mathbf{P}^1$. If $a,b > 0$, then we consider an a-uple embedding $\mathbf{P}^1 \to \mathbf{P}^{n_1}$ and a b-uple embedding $\mathbf{P}^1 \to \mathbf{P}^{n_2}$. Taking their product, and following with a Segre embedding, we obtain a closed immersion

$$Q = \mathbf{P}^1 \times \mathbf{P}^1 \to \mathbf{P}^{n_1} \times \mathbf{P}^{n_2} \to \mathbf{P}^n,$$

which corresponds to an invertible sheaf of type (a,b) on Q. Thus for any $a,b > 0$, the corresponding invertible sheaf is very ample, and hence ample. On the other hand, if $\mathscr{L}$ is of type (a,b) with either $a < 0$ or $b < 0$, then by restricting to a fibre of the product $\mathbf{P}^1 \times \mathbf{P}^1$, one sees that $\mathscr{L}$ is not generated by global sections. Hence if $a \leqslant 0$ or $b \leqslant 0$, $\mathscr{L}$ cannot be ample. So on Q, an invertible sheaf $\mathscr{L}$ of type (a,b) is ample $\Leftrightarrow$ very ample $\Leftrightarrow a,b > 0$.

Example 7.6.3. Let X be the nonsingular cubic curve $y^2z = x^3 - xz^2$ in $\mathbf{P}_k^2$, which was studied in (6.10.2). Let $\mathscr{L}$ be the invertible sheaf $\mathscr{L}(P_0)$. Then $\mathscr{L}$ is ample, because $\mathscr{L}(3P_0) \cong \mathcal{O}_X(1)$ is very ample. On the other hand, $\mathscr{L}$ is not very ample, because $\mathscr{L}(P_0)$ is not generated by global sections. If it were, then P_0 would be linearly equivalent to some other point $Q \in X$, which is impossible, since X is not rational (6.10.1). This shows that an ample sheaf need not be very ample.

Example 7.6.4. We will see later (IV, 3.3) that if D is a divisor on a complete nonsingular curve X, then $\mathscr{L}(D)$ is ample if and only if $\deg D > 0$. This is a consequence of the Riemann–Roch theorem.

Linear Systems

We will see in a minute how global sections of an invertible sheaf correspond to effective divisors on a variety. Thus giving an invertible sheaf and a set of its global sections is the same as giving a certain set of effective divisors, all linearly equivalent to each other. This leads to the notion of linear system, which is the historically older notion. For simplicity, we will employ this terminology only when dealing with nonsingular projective varieties over an algebraically closed field. Over more general schemes the geometrical intuition associated with the concept of linear system may lead one astray, so it is safer to deal with invertible sheaves and their global sections in that case.

So let X be a nonsingular projective variety over an algebraically closed field k. In this case the notions of Weil divisor and Cartier divisor are equivalent (6.11). Furthermore, we have a one-to-one correspondence between linear equivalence classes of divisors and isomorphism classes of invertible sheaves (6.15). Another useful fact in this situation is that for any invertible sheaf $\mathscr{L}$ on X, the global sections $\Gamma(X,\mathscr{L})$ form a finite-dimensional k-vector space (5.19).

Let $\mathscr{L}$ be an invertible sheaf on X, and let $s \in \Gamma(X,\mathscr{L})$ be a nonzero section of $\mathscr{L}$. We define an effective divisor $D = (s)_0$, the *divisor of zeros* of s, as follows. Over any open set $U \subseteq X$ where $\mathscr{L}$ is trivial, let $\varphi: \mathscr{L}|_U \xrightarrow{\sim} \mathcal{O}_U$ be an isomorphism. Then $\varphi(s) \in \Gamma(U,\mathcal{O}_U)$. As U ranges over a covering of X, the collection $\{U,\varphi(s)\}$ determines an effective Cartier divisor D on X. Indeed, φ is determined up to multiplication by an element of $\Gamma(U,\mathcal{O}_U^*)$, so we get a well-defined Cartier divisor.

Proposition 7.7. *Let X be a nonsingular projective variety over the algebraically closed field k. Let D_0 be a divisor on X and let $\mathscr{L} \cong \mathscr{L}(D_0)$ be the corresponding invertible sheaf. Then:*

(a) *for each nonzero $s \in \Gamma(X,\mathscr{L})$, the divisor of zeros $(s)_0$ is an effective divisor linearly equivalent to D_0;*

(b) *every effective divisor linearly equivalent to D_0 is $(s)_0$ for some $s \in \Gamma(X,\mathscr{L})$; and*

(c) *two sections $s,s' \in \Gamma(X,\mathscr{L})$ have the same divisor of zeros if and only if there is a $\lambda \in k^*$ such that $s' = \lambda s$.*

PROOF.

(a) We may identify $\mathscr{L}$ with the subsheaf $\mathscr{L}(D_0)$ of $\mathscr{K}$. Then s corresponds to a rational function $f \in K$. If D_0 is locally defined as a Cartier divisor by $\{U_i, f_i\}$ with $f_i \in K^*$, then $\mathscr{L}(D_0)$ is locally generated by f_i^{-1}, so we get a local isomorphism $\varphi: \mathscr{L}(D_0) \to \mathcal{O}$ by multiplying by f_i. So $D = (s)_0$ is locally defined by $f_i f$. Thus $D = D_0 + (f)$, showing that $D \sim D_0$.

(b) If $D > 0$ and $D = D_0 + (f)$, then $(f) \geq -D_0$. Thus f gives a global section of $\mathscr{L}(D_0)$ whose divisor of zeros is D.

(c) Still using the same construction, if $(s)_0 = (s')_0$, then s and s' correspond to rational functions $f, f' \in K$ such that $(f/f') = 0$. Therefore $f/f' \in \Gamma(X,\mathcal{O}_X^*)$. But since X is a projective variety over k algebraically closed, $\Gamma(X,\mathcal{O}_X) = k$, and so $f/f' \in k^*$ (I, 3.4).

Definition. A *complete linear system* on a nonsingular projective variety is defined as the set (maybe empty) of all effective divisors linearly equivalent to some given divisor D_0. It is denoted by $|D_0|$.

We see from the proposition that the set $|D_0|$ is in one-to-one correspondence with the set $(\Gamma(X,\mathscr{L}) - \{0\})/k^*$. This gives $|D_0|$ a structure of the set of closed points of a projective space over k.

Definition. A *linear system* $\mathfrak{d}$ on X is a subset of a complete linear system $|D_0|$ which is a linear subspace for the projective space structure of $|D_0|$. Thus $\mathfrak{d}$ corresponds to a sub-vector space $V \subseteq \Gamma(X,\mathscr{L})$, where $V = \{s \in \Gamma(X,\mathscr{L}) | (s)_\sigma \in \mathfrak{d}\} \cup \{0\}$. The *dimension* of the linear system $\mathfrak{d}$ is its dimension as a linear projective variety. Hence $\dim \mathfrak{d} = \dim V - 1$. (Note these dimensions are finite because $\Gamma(X,\mathscr{L})$ is a finite-dimensional vector space.)

Definition. A point $P \in X$ is a *base point* of a linear system $\mathfrak{d}$ if $P \in \operatorname{Supp} D$ for all $D \in \mathfrak{d}$. Here $\operatorname{Supp} D$ means the union of the prime divisors of D.

Lemma 7.8. *Let $\mathfrak{d}$ be a linear system on X corresponding to the subspace $V \subseteq \Gamma(X,\mathcal{L})$. Then a point $P \in X$ is a base point of $\mathfrak{d}$ if and only if $s_P \in \mathfrak{m}_P \mathcal{L}_P$ for all $s \in V$. In particular, $\mathfrak{d}$ is base-point-free if and only if $\mathcal{L}$ is generated by the global sections in V.*

PROOF. This follows immediately from the fact that for any $s \in \Gamma(X,\mathcal{L})$, the support of the divisor of zeros $(s)_0$ is the complement of the open set X_s.

Remark 7.8.1. We can rephrase (7.1) in terms of linear systems as follows: to give a morphism from X to $\mathbf{P}_k^n$ it is equivalent to give a linear system $\mathfrak{d}$ without base points on X, and a set of elements $s_0, \ldots, s_n \in V$, which span the vector space V. Often we will simply talk about the morphism to projective space determined by a linear system without base points $\mathfrak{d}$. In this case we understand that $s_0, \ldots, s_n$ should be chosen as a basis of V. If we chose a different basis, the corresponding morphism of $X \to \mathbf{P}^n$ would only differ by an automorphism of $\mathbf{P}^n$.

Remark 7.8.2. We can rephrase (7.3) in terms of linear systems as follows: Let $\varphi: X \to \mathbf{P}^n$ be a morphism corresponding to the linear system (without base points) $\mathfrak{d}$. Then φ is a closed immersion if and only if

(1) $\mathfrak{d}$ *separates points*, i.e., for any two distinct closed points $P, Q \in X$, there is a $D \in \mathfrak{d}$ such that $P \in \operatorname{Supp} D$ and $Q \notin \operatorname{Supp} D$, and
(2) $\mathfrak{d}$ *separates tangent vectors*, i.e., given a closed point $P \in X$ and a tangent vector $t \in T_P(X) = (\mathfrak{m}_P/\mathfrak{m}_P^2)'$, there is a $D \in \mathfrak{d}$ such that $P \in \operatorname{Supp} D$, but $t \notin T_P(D)$. Here we think of D as a locally principal closed subscheme, in which case the Zariski tangent space $T_P(D) = (\mathfrak{m}_{P,D}/\mathfrak{m}_{P,D}^2)'$ is naturally a subspace of $T_P(X)$.

The terminology of "separating points" and "separating tangent vectors" is perhaps somewhat explained by this geometrical interpretation.

Definition. Let $i: Y \hookrightarrow X$ be a closed immersion of nonsingular projective varieties over k. If $\mathfrak{d}$ is a linear system on X, we define the *trace* of $\mathfrak{d}$ on Y, denoted $\mathfrak{d}|_Y$, as follows. The linear system $\mathfrak{d}$ corresponds to an invertible sheaf $\mathcal{L}$ on X, and a sub-vector space $V \subseteq \Gamma(X,\mathcal{L})$. We take the invertible sheaf $i^*\mathcal{L} = \mathcal{L} \otimes \mathcal{O}_Y$ on Y, and we let $W \subseteq \Gamma(Y,i^*\mathcal{L})$ be the image of V under the natural map $\Gamma(X,\mathcal{L}) \to \Gamma(Y,i^*\mathcal{L})$. Then $i^*\mathcal{L}$ and W define the linear system $\mathfrak{d}|_Y$.

One can also describe $\mathfrak{d}|_Y$ geometrically as follows: it consists of all divisors $D.Y$ (defined as in (6.6.2)), where $D \in \mathfrak{d}$ is a divisor whose support does not contain Y.

Note that even if $\mathfrak{d}$ is a complete linear system, $\mathfrak{d}|_Y$ may not be complete.

Example 7.8.3. If $X = \mathbf{P}^n$, then the set of all effective divisors of degree $d > 0$ is a complete linear system of dimension $\binom{n+d}{n} - 1$. Indeed, it corresponds to the invertible sheaf $\mathcal{O}(d)$, whose global sections consist exactly of the space of all homogeneous polynomials in $x_0, \ldots, x_n$ of degree d. This is a vector space of dimension $\binom{n+d}{n}$, so the dimension of the complete linear system is one less.

Example 7.8.4. We can rephrase (Ex. 5.14d) in terms of linear systems as follows: a nonsingular projective variety $X \hookrightarrow \mathbf{P}^n_k$ is projectively normal if and only if for every $d > 0$, the trace on X of the linear system of all divisors of degree d on $\mathbf{P}^n$, is a complete linear system. By slight abuse of language, we say that "the linear system on X, cut out by the hypersurfaces of degree d in $\mathbf{P}^n$, is complete."

Example 7.8.5. Recall that the *twisted cubic curve* in $\mathbf{P}^3$ was defined by the parametric equations $x_0 = t^3$, $x_1 = t^2 u$, $x_2 = tu^2$, $x_3 = u^3$. In other words, it is just the 3-uple embedding of $\mathbf{P}^1$ in $\mathbf{P}^3$ (I, Ex. 2.9, Ex. 2.12). We will now show that any nonsingular curve X in $\mathbf{P}^3$, of degree 3, which is not contained in any $\mathbf{P}^2$, and which is abstractly isomorphic to $\mathbf{P}^1$, can be obtained from the given twisted cubic curve by an automorphism of $\mathbf{P}^3$. So we will refer to any such curve as a twisted cubic curve.

Let X be such a curve. The embedding of X in $\mathbf{P}^3$ is determined by the linear system $\mathfrak{d}$ of hyperplane sections of X (7.1). This is a linear system on X of dimension 3, because the planes in $\mathbf{P}^3$ form a linear system of dimension 3, and by hypothesis X is not contained in any plane, so the map $\Gamma(\mathbf{P}^3, \mathcal{O}(1)) \to \Gamma(X, i^*\mathcal{O}(1))$ is injective. On the other hand, $\mathfrak{d}$ is a linear system of degree 3, since X is a curve of degree 3. By the *degree* of a linear system on a complete nonsingular curve, we mean the degree of any of its divisors, which is independent of the divisor chosen (6.10). Now thinking of X as $\mathbf{P}^1$, the linear system $\mathfrak{d}$ must correspond to a 4-dimensional subspace $V \subseteq \Gamma(\mathbf{P}^1, \mathcal{O}(3))$. But $\Gamma(\mathbf{P}^1, \mathcal{O}(3))$ itself has dimension 4, so $V = \Gamma(\mathbf{P}^1, \mathcal{O}(3))$ and $\mathfrak{d}$ is a complete linear system. Since the embedding is determined by the linear system and the choice of basis of V by (7.1), we conclude that X is the same as the 3-uple embedding of $\mathbf{P}^1$, except for the choice of basis of V. This shows that there is an automorphism of $\mathbf{P}^3$ sending the given twisted cubic curve to X. (See (IV, Ex. 3.4) for generalization.)

Example 7.8.6. We define a *nonsingular rational quartic curve* in $\mathbf{P}^3$ to be a nonsingular curve X in $\mathbf{P}^3$, of degree 4, not contained in any $\mathbf{P}^2$, and which is abstractly isomorphic to $\mathbf{P}^1$. In this case we will see that two such curves need not be obtainable one from the other by an automorphism of $\mathbf{P}^3$. To give a morphism of $\mathbf{P}^1$ to $\mathbf{P}^3$ whose image has degree 4 and is not contained in any $\mathbf{P}^2$, we need a 4-dimensional subspace $V \subseteq \Gamma(\mathbf{P}^1, \mathcal{O}(4))$. This latter vector space has dimension 5. So if we choose two different subspaces V, V', the corresponding curves in $\mathbf{P}^3$ may not be related by an automorphism of $\mathbf{P}^3$. To be sure the image is nonsingular, we use the criterion of (7.3). Thus for example, one sees easily that the subspaces $V = (t^4, t^3 u, tu^3, u^4)$ and $V' = (t^4, t^3 u + at^2 u^2, tu^3, u^4)$ for $a \in k^*$ give nonsingular rational quartic curves in $\mathbf{P}^3$ which are not equivalent by an automorphism of $\mathbf{P}^3$.

Proj, $\mathbf{P}(\mathscr{E})$, and Blowing Up

Earlier we have defined the Proj of a graded ring. Now we introduce a relative version of this construction, which is the **Proj** of a sheaf of graded algebras $\mathscr{S}$ over a scheme X. This construction is useful in particular because it allows us to construct the projective space bundle associated to a locally free sheaf $\mathscr{E}$, and it allows us to give a definition of blowing up with respect to an arbitrary sheaf of ideals. This generalizes the notion of blowing up a point introduced in (I, §4).

For simplicity, we will always impose the following conditions on a scheme X and a sheaf of graded algebras $\mathscr{S}$ before we define a **Proj**:

(†) X is a noetherian scheme, $\mathscr{S}$ is a quasi-coherent sheaf of $\mathcal{O}_X$-modules, which has a structure of a sheaf of graded $\mathcal{O}_X$-algebras. Thus $\mathscr{S} \cong \bigoplus_{d \geq 0} \mathscr{S}_d$, where $\mathscr{S}_d$ is the homogeneous part of degree d. We assume furthermore that $\mathscr{S}_0 = \mathcal{O}_X$, that $\mathscr{S}_1$ is a coherent $\mathcal{O}_X$-module, and that $\mathscr{S}$ is locally generated by $\mathscr{S}_1$ as an $\mathcal{O}_X$-algebra. (It follows that $\mathscr{S}_d$ is coherent for all $d \geq 0$.)

Construction. Let X be a scheme and $\mathscr{S}$ a sheaf of graded $\mathcal{O}_X$-algebras satisfying (†). For each open affine subset $U = \operatorname{Spec} A$ of X, let $\mathscr{S}(U)$ be the graded A-algebra $\Gamma(U, \mathscr{S}|_U)$. Then we consider Proj $\mathscr{S}(U)$ and its natural morphism $\pi_U : \operatorname{Proj} \mathscr{S}(U) \to U$. If $f \in A$, and $U_f = \operatorname{Spec} A_f$, then since $\mathscr{S}$ is quasi-coherent, we see that Proj $\mathscr{S}(U_f) \cong \pi_U^{-1}(U_f)$. It follows that if U, V are two open affine subsets of X, then $\pi_U^{-1}(U \cap V)$ is naturally isomorphic to $\pi_V^{-1}(U \cap V)$—here we leave some technical details to the reader. These isomorphisms allow us to glue the schemes Proj $\mathscr{S}(U)$ together (Ex. 2.12). Thus we obtain a scheme **Proj** $\mathscr{S}$ together with a morphism $\pi : \mathbf{Proj}\, \mathscr{S} \to X$ such that for each open affine $U \subseteq X$, $\pi^{-1}(U) \cong$ Proj $\mathscr{S}(U)$. Furthermore the invertible sheaves $\mathcal{O}(1)$ on each Proj $\mathscr{S}(U)$ are compatible under this construction (5.12c), so they glue together to give an invertible sheaf $\mathcal{O}(1)$ on **Proj** $\mathscr{S}$, canonically determined by this construction.

Thus to any $X, \mathscr{S}$ satisfying (†), we have constructed the scheme **Proj** $\mathscr{S}$, the morphism $\pi : \mathbf{Proj}\, \mathscr{S} \to X$, and the invertible sheaf $\mathcal{O}(1)$ on **Proj** $\mathscr{S}$. Everything we have said about the Proj of a graded ring S can be extended to this relative situation. We will not attempt to do this exhaustively, but will only mention certain aspects of the new situation.

Example 7.8.7. If $\mathscr{S}$ is the polynomial algebra $\mathscr{S} = \mathcal{O}_X[T_0, \ldots, T_n]$, then **Proj** $\mathscr{S}$ is just the relative projective space $\mathbf{P}_X^n$ with its twisting sheaf $\mathcal{O}(1)$ defined earlier (§5).

Caution 7.8.8. In general, $\mathcal{O}(1)$ may not be very ample on **Proj** $\mathscr{S}$ relative to X. See (7.10) and (Ex. 7.14).

Lemma 7.9. *Let $\mathscr{S}$ be a sheaf of graded algebras on a scheme X satisfying* (†). *Let $\mathscr{L}$ be an invertible sheaf on X, and define a new sheaf of graded algebras $\mathscr{S}' = \mathscr{S} * \mathscr{L}$ by $\mathscr{S}'_d = \mathscr{S}_d \otimes \mathscr{L}^d$ for each $d \geqslant 0$. Then $\mathscr{S}'$ also satisfies* (†), *and there is a natural isomorphism $\varphi : P' = $ **Proj** $\mathscr{S}' \overset{\sim}{\to} P = $ **Proj** $\mathscr{S}$, commuting with the projections π and π' to X, and having the property that*

$$\mathcal{O}_{P'}(1) \cong \varphi^* \mathcal{O}_P(1) \otimes \pi'^* \mathscr{L}.$$

PROOF. Let $\theta : \mathcal{O}_U \overset{\sim}{\to} \mathscr{L}|_U$ be a local isomorphism of $\mathcal{O}_U$ with $\mathscr{L}|_U$ over a small open affine subset U of X. Then θ induces an isomorphism of graded rings $\mathscr{S}(U) \cong \mathscr{S}'(U)$ and hence an isomorphism $\theta^* : \text{Proj } \mathscr{S}'(U) \cong \text{Proj } \mathscr{S}(U)$. If $\theta_1 : \mathcal{O}_U \cong \mathscr{L}|_U$ is a different local isomorphism, then θ and θ_1 differ by an element $f \in \Gamma(U, \mathcal{O}_U^*)$, and the corresponding isomorphism $\mathscr{S}(U) \cong \mathscr{S}'(U)$ differs by an automorphism ψ of $\mathscr{S}(U)$ which consists of multiplying by f^d in degree d. This does not affect the set of homogeneous prime ideals in $\mathscr{S}(U)$, and furthermore, since the structure sheaf of Proj $\mathscr{S}(U)$ is formed by elements of degree *zero* in various localizations of $\mathscr{S}(U)$, the automorphism ψ of $\mathscr{S}(U)$ induces the *identity* automorphism of Proj $\mathscr{S}(U)$. In other words, the isomorphism θ^* is independent of the choice of θ. So these local isomorphisms θ^* glue together to give a natural isomorphism $\varphi : \text{\textbf{Proj }} \mathscr{S}' \overset{\sim}{\to}$ **Proj** $\mathscr{S}$, commuting with π and π'. When we form the sheaf $\mathcal{O}(1)$, however, the automorphism ψ of $\mathscr{S}(U)$ induces multiplication by f in $\mathcal{O}(1)$. Thus $\mathcal{O}_{P'}(1)$ looks like $\mathcal{O}_P(1)$ modified by the transition functions of $\mathscr{L}$. Stated precisely, this says $\mathcal{O}_{P'}(1) \cong \varphi^* \mathcal{O}_P(1) \otimes \pi'^* \mathscr{L}$.

Proposition 7.10. *Let $X, \mathscr{S}$ satisfy* (†), *let $P = $ **Proj** $\mathscr{S}$, with projection $\pi : P \to X$ and invertible sheaf $\mathcal{O}_P(1)$ constructed above. Then:*

(a) *π is a proper morphism. In particular, it is separated and of finite type;*

(b) *if X admits an ample invertible sheaf $\mathscr{L}$, then π is a projective morphism, and we can take $\mathcal{O}_P(1) \otimes \pi^* \mathscr{L}^n$ to be a very ample invertible sheaf on P over X, for suitable $n > 0$.*

PROOF.
(a) For each open affine $U \subseteq X$, the morphism $\pi_U : \text{Proj } \mathscr{S}(U) \to U$ is a projective morphism (4.8.1), hence proper (4.9). But the condition for a morphism to be proper is local on the base (4.8f), so π is proper.

(b) Let $\mathscr{L}$ be an ample invertible sheaf on X. Then for some $n > 0$, $\mathscr{S}_1 \otimes \mathscr{L}^n$ is generated by global sections. Since X is noetherian and $\mathscr{S}_1 \otimes \mathscr{L}^n$ is coherent, we can find a finite number of global sections which generate it, in other words we can find a surjective morphism of sheaves $\mathscr{O}_X^{N+1} \to \mathscr{S}_1 \otimes \mathscr{L}^n$ for some N. This allows us to define a surjective map of sheaves of graded $\mathscr{O}_X$-algebras $\mathscr{O}_X[T_0, \ldots, T_N] \to \mathscr{S} * \mathscr{L}^n$, which gives rise to a closed immersion $\mathbf{Proj}\ \mathscr{S} * \mathscr{L}^n \hookrightarrow \mathbf{Proj}\ \mathscr{O}_X[T_0, \ldots, T_N] = \mathbf{P}_X^N$ (Ex. 3.12). But $\mathbf{Proj}\ \mathscr{S} * \mathscr{L}^n \cong \mathbf{Proj}\ \mathscr{S}$ by (7.9), and the very ample invertible sheaf induced by this embedding is just $\mathscr{O}_P(1) \otimes \pi^* \mathscr{L}^n$.

Definition. Let X be a noetherian scheme, and let $\mathscr{E}$ be a locally free coherent sheaf on X. We define the associated *projective space bundle* $\mathbf{P}(\mathscr{E})$ as follows. Let $\mathscr{S} = S(\mathscr{E})$ be the symmetric algebra of $\mathscr{E}$, $\mathscr{S} = \bigoplus_{d>0} S^d(\mathscr{E})$ (Ex. 5.16). Then $\mathscr{S}$ is a sheaf of graded $\mathscr{O}_X$-algebras satisfying (†), and we define $\mathbf{P}(\mathscr{E}) = \mathbf{Proj}\ \mathscr{S}$. As such, it comes with a projection morphism $\pi : \mathbf{P}(\mathscr{E}) \to X$, and an invertible sheaf $\mathscr{O}(1)$.

Note that if $\mathscr{E}$ is free of rank $n + 1$ over an open set U, then $\pi^{-1}(U) \cong \mathbf{P}_U^n$, so $\mathbf{P}(\mathscr{E})$ is a "relative projective space" over X.

Proposition 7.11. *Let $X, \mathscr{E}, \mathbf{P}(\mathscr{E})$ be as in the definition. Then:*

(a) if rank $\mathscr{E} \geq 2$, there is a canonical isomorphism of graded $\mathscr{O}_X$-algebras $\mathscr{S} \cong \bigoplus_{l \in \mathbf{Z}} \pi_*(\mathscr{O}(l))$, with the grading on the right hand side given by l. In particular, for $l < 0$, $\pi_*(\mathscr{O}(l)) = 0$; for $l = 0$, $\pi_*(\mathscr{O}_{\mathbf{P}(\mathscr{E})}) = \mathscr{O}_X$, and for $l = 1$, $\pi_*(\mathscr{O}(1)) = \mathscr{E}$;

(b) there is a natural surjective morphism $\pi^* \mathscr{E} \to \mathscr{O}(1)$.

PROOF.

(a) is just a relative version of (5.13), and follows immediately from it.

(b) is a relative version of the fact that $\mathscr{O}(1)$ on $\mathbf{P}^n$ is generated by the global sections $x_0, \ldots, x_n$ (5.16.2).

Proposition 7.12. *Let $X, \mathscr{E}, \mathbf{P}(\mathscr{E})$ be as above. Let $g : Y \to X$ be any morphism. Then to give a morphism of Y to $\mathbf{P}(\mathscr{E})$ over X, it is equivalent to give an invertible sheaf $\mathscr{L}$ on Y and a surjective map of sheaves on Y, $g^* \mathscr{E} \to \mathscr{L}$.*

PROOF. This is a local version of (7.1). First note that if $f : Y \to \mathbf{P}(\mathscr{E})$ is a morphism over X, then the surjective map $\pi^* \mathscr{E} \to \mathscr{O}(1)$ on $\mathbf{P}(\mathscr{E})$ pulls back to give a surjective map $g^* \mathscr{E} = f^* \pi^* \mathscr{E} \to f^* \mathscr{O}(1)$, so we take $\mathscr{L} = f^* \mathscr{O}(1)$.

Conversely, given an invertible sheaf $\mathscr{L}$ on Y, and a surjective morphism $g^* \mathscr{E} \to \mathscr{L}$, I claim there is a unique morphism $f : Y \to \mathbf{P}(\mathscr{E})$ over X, such that $\mathscr{L} \cong f^* \mathscr{O}(1)$, and the map $g^* \mathscr{E} \to \mathscr{L}$ is obtained from $\pi^* \mathscr{E} \to \mathscr{O}(1)$ by applying f^*. In view of the claimed uniqueness of f, it is sufficient to verify this statement locally on X. Taking open affine subsets $U = \operatorname{Spec} A$ of X which are small enough so that $\mathscr{E}|_U$ is free, the statement reduces to (7.1). Indeed, if $\mathscr{E} \cong \mathscr{O}_X^{n+1}$, then to give a surjective morphism $g^* \mathscr{E} \to \mathscr{L}$ is the same as giving $n + 1$ global sections of $\mathscr{L}$ which generate.

Note. We refer to the exercises for further properties of $\mathbf{P}(\mathscr{E})$ and for the general notion of projective space bundle over a scheme X. Cf. (Ex. 5.18) for the notion of a vector bundle associated to a locally free sheaf.

Now we come to the generalized notion of blowing up. In (I, §4) we defined the blowing-up of a variety with respect to a point. Now we will define the blowing-up of a noetherian scheme with respect to any closed subscheme. Since a closed subscheme corresponds to a coherent sheaf of ideals (5.9), we may as well speak of blowing up a coherent sheaf of ideals.

Definition. Let X be a noetherian scheme, and let $\mathscr{I}$ be a coherent sheaf of ideals on X. Consider the sheaf of graded algebras $\mathscr{S} = \bigoplus_{d \geqslant 0} \mathscr{I}^d$, where $\mathscr{I}^d$ is the dth power of the ideal $\mathscr{I}$, and we set $\mathscr{I}^0 = \mathscr{O}_X$. Then $X, \mathscr{S}$ clearly satisfy (†), so we can consider $\tilde{X} = \mathbf{Proj}\ \mathscr{S}$. We define $\tilde{X}$ to be the *blowing-up* of X *with respect to the coherent sheaf of ideals* $\mathscr{I}$. If Y is the closed subscheme of X corresponding to $\mathscr{I}$, then we also call $\tilde{X}$ the *blowing-up* of X *along* Y, or *with center* Y.

Example 7.12.1. If X is $\mathbf{A}_k^n$ and $P \in X$ is the origin, then the blowing-up of P just defined is isomorphic to the one defined in (I, §4). Indeed, in this case $X = \operatorname{Spec} A$, where $A = k[x_1, \dots, x_n]$, and P corresponds to the ideal $I = (x_1, \dots, x_n)$. So $\tilde{X} = \operatorname{Proj} S$, where $S = \bigoplus_{d \geqslant 0} I^d$. We can define a surjective map of graded rings $\varphi : A[y_1, \dots, y_n] \to S$ by sending y_i to the element $x_i \in I$ considered as an element of S in degree 1. Thus $\tilde{X}$ is isomorphic to a closed subscheme of $\operatorname{Proj} A[y_1, \dots, y_n] = \mathbf{P}_A^{n-1}$. It is defined by the homogeneous polynomials in the y_i which generate the kernel of φ, and one sees easily that $\{x_i y_j - x_j y_i | i, j = 1, \dots, n\}$ will do.

Definition. Let $f : X \to Y$ be a morphism of schemes, and let $\mathscr{I} \subseteq \mathscr{O}_Y$ be a sheaf of ideals on Y. We define the *inverse image ideal sheaf* $\mathscr{I}' \subseteq \mathscr{O}_X$ as follows. First consider f as a continuous map of topological spaces $X \to Y$ and let $f^{-1}\mathscr{I}$ be the inverse image of the sheaf $\mathscr{I}$, as defined in §1. Then $f^{-1}\mathscr{I}$ is a sheaf of ideals in the sheaf of rings $f^{-1}\mathscr{O}_Y$ on the topological space X. Now there is a natural homomorphism of sheaves of rings on X, $f^{-1}\mathscr{O}_Y \to \mathscr{O}_X$, so we define $\mathscr{I}'$ to be the ideal sheaf in $\mathscr{O}_X$ generated by the image of $f^{-1}\mathscr{I}$. We will denote $\mathscr{I}'$ by $f^{-1}\mathscr{I} \cdot \mathscr{O}_X$ or simply $\mathscr{I} \cdot \mathscr{O}_X$, if no confusion seems likely to result.

Caution 7.12.2. If we consider $\mathscr{I}$ as a sheaf of $\mathscr{O}_Y$-modules, then in §5 we have defined the inverse image $f^*\mathscr{I}$ as a sheaf of $\mathscr{O}_X$-modules. It may happen that $f^*\mathscr{I} \neq f^{-1}\mathscr{I} \cdot \mathscr{O}_X$. The reason is that $f^*\mathscr{I}$ is defined as

$$f^{-1}\mathscr{I} \otimes_{f^{-1}\mathscr{O}_Y} \mathscr{O}_X.$$

Since the tensor product functor is not in general left exact, $f^*\mathscr{I}$ may not be a subsheaf of $\mathscr{O}_X$. However, there is a natural map $f^*\mathscr{I} \to \mathscr{O}_X$ coming from the inclusion $\mathscr{I} \hookrightarrow \mathscr{O}_Y$, and $f^{-1}\mathscr{I} \cdot \mathscr{O}_X$ is just the image of $f^*\mathscr{I}$ under this map.

Proposition 7.13. *Let X be a noetherian scheme, $\mathscr{I}$ a coherent sheaf of ideals, and let $\pi:\tilde{X} \to X$ be the blowing-up of $\mathscr{I}$. Then:*

(a) *the inverse image ideal sheaf $\tilde{\mathscr{I}} = \pi^{-1}\mathscr{I} \cdot \mathscr{O}_{\tilde{X}}$ is an invertible sheaf on $\tilde{X}$.*

(b) *if Y is the closed subscheme corresponding to $\mathscr{I}$, and if $U = X - Y$, then $\pi:\pi^{-1}(U) \to U$ is an isomorphism.*

PROOF.

(a) Since $\tilde{X}$ is defined as **Proj** $\mathscr{S}$, where $\mathscr{S} = \bigoplus_{d \geq 0} \mathscr{I}^d$, it comes equipped with a natural invertible sheaf $\mathscr{O}(1)$. For any open affine $U \subseteq X$, this sheaf $\mathscr{O}(1)$ on Proj $\mathscr{S}(U)$ is the sheaf associated to the graded $\mathscr{S}(U)$-module $\mathscr{S}(U)(1) = \bigoplus_{d \geq 0} \mathscr{I}^{d+1}(U)$. But this is clearly equal to the ideal $\mathscr{I} \cdot \mathscr{S}(U)$ generated by $\mathscr{I}$ in $\mathscr{S}(U)$, so we see that the inverse image ideal sheaf $\tilde{\mathscr{I}} = \pi^{-1}\mathscr{I} \cdot \mathscr{O}_{\tilde{X}}$ is in fact equal to $\mathscr{O}_{\tilde{X}}(1)$. Hence it is an invertible sheaf.

(b) If $U = X - Y$, then $\mathscr{I}|_U \cong \mathscr{O}_U$, so $\pi^{-1}U = $ **Proj** $\mathscr{O}_U[T] = U$.

Proposition 7.14 (Universal Property of Blowing Up). *Let X be a noetherian scheme, $\mathscr{I}$ a coherent sheaf of ideals, and $\pi:\tilde{X} \to X$ the blowing-up with respect to $\mathscr{I}$. If $f:Z \to X$ is any morphism such that $f^{-1}\mathscr{I} \cdot \mathscr{O}_Z$ is an invertible sheaf of ideals on Z, then there exists a unique morphism $g:Z \to \tilde{X}$ factoring f.*

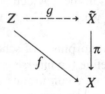

PROOF. In view of the asserted uniqueness of g, the question is local on X. So we may assume that $X = \operatorname{Spec} A$ is affine, A is noetherian, and that $\mathscr{I}$ corresponds to an ideal $I \subseteq A$. Then $\tilde{X} = \operatorname{Proj} S$, where $S = \bigoplus_{d \geq 0} I^d$. Let $a_0, \ldots, a_n \in I$ be a set of generators for the ideal I. Then we can define a surjective map of graded rings $\varphi: A[x_0, \ldots, x_n] \to S$ by sending x_i to $a_i \in I$, considered as an element of degree one in S. This homomorphism gives rise to a closed immersion $\tilde{X} \hookrightarrow \mathbf{P}^n_A$. The kernel of φ is the homogeneous ideal in $A[x_0, \ldots, x_n]$ generated by all homogeneous polynomials $F(x_0, \ldots, x_n)$ such that $F(a_0, \ldots, a_n) = 0$ in A.

Now let $f:Z \to X$ be a morphism such that the inverse image ideal sheaf $f^{-1}\mathscr{I} \cdot \mathscr{O}_Z$ is an invertible sheaf $\mathscr{L}$ on Z. Since I is generated by $a_0, \ldots, a_n$, the inverse images of these elements, considered as global sections of $\mathscr{I}$, give global sections $s_0, \ldots, s_n$ of $\mathscr{L}$ which generate. Then by (7.1) there is a unique morphism $g:Z \to \mathbf{P}^n_A$ with the property that $\mathscr{L} \cong g^*\mathcal{O}(1)$ and that $s_i = g^{-1}x_i$ under this isomorphism. Now I claim that g factors through the closed subscheme $\tilde{X}$ of $\mathbf{P}^n_A$. This follows easily from the fact that if $F(x_0, \ldots, x_n)$ is a homogeneous element of degree d of ker φ, where ker φ is the homogeneous ideal described above which determines $\tilde{X}$, then $F(a_0, \ldots, a_n) = 0$ in A and so $F(s_0, \ldots, s_n) = 0$ in $\Gamma(Z, \mathscr{L}^d)$.

Thus we have constructed a morphism $g:Z \to \tilde{X}$ factoring f. For any such morphism, we must necessarily have $f^{-1}\mathscr{I} \cdot \mathcal{O}_Z = g^{-1}(\pi^{-1}\mathscr{I} \cdot \mathcal{O}_{\tilde{X}}) \cdot \mathcal{O}_Z$ which is just $g^{-1}(\mathcal{O}_{\tilde{X}}(1)) \cdot \mathcal{O}_Z$. Therefore we have a surjective map $g^*\mathcal{O}_{\tilde{X}}(1) \to f^{-1}\mathscr{I} \cdot \mathcal{O}_Z = \mathscr{L}$. Now a surjective map of invertible sheaves on a locally ringed space is necessarily an isomorphism (Ex. 7.1), so we have $g^*\mathcal{O}_{\tilde{X}}(1) \cong \mathscr{L}$. Clearly the sections s_i of $\mathscr{L}$ must be the pull-backs of the sections x_i of $\mathcal{O}(1)$ on $\mathbf{P}^n_A$. Hence the uniqueness of g under our conditions follows from the uniqueness assertion of (7.1).

Corollary 7.15. *Let $f:Y \to X$ be a morphism of noetherian schemes, and let $\mathscr{I}$ be a coherent sheaf of ideals on X. Let $\tilde{X}$ be the blowing-up of $\mathscr{I}$, and let $\tilde{Y}$ be the blowing-up of the inverse image ideal sheaf $\mathscr{J} = f^{-1}\mathscr{I} \cdot \mathcal{O}_Y$ on Y. Then there is a unique morphism $\tilde{f}:\tilde{Y} \to \tilde{X}$*

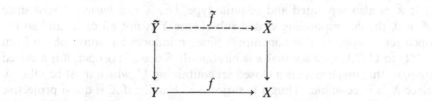

making a commutative diagram as shown. Moreover, if f is a closed immersion, so is $\tilde{f}$.

PROOF. The existence and uniqueness of $\tilde{f}$ follow immediately from the proposition. To show that $\tilde{f}$ is a closed immersion if f is, we go back to the definition of blowing up. $\tilde{X} = \mathbf{Proj}\,\mathscr{S}$ where $\mathscr{S} = \bigoplus_{d \geqslant 0} \mathscr{I}^d$, and $\tilde{Y} = \mathbf{Proj}\,\mathscr{S}'$, where $\mathscr{S}' = \bigoplus_{d \geqslant 0} \mathscr{J}^d$. Since Y is a closed subscheme of X, we can consider $\mathscr{S}'$ as a sheaf of graded algebras on X. Then there is a natural surjective homomorphism of graded rings $\mathscr{S} \to \mathscr{S}'$, which gives rise to the closed immersion $\tilde{f}$.

Definition. In the situation of (7.15), if Y is a closed subscheme of X, we call the closed subscheme $\tilde{Y}$ of $\tilde{X}$ the *strict transform* of Y under the blowing-up $\pi:\tilde{X} \to X$.

Example 7.15.1. If Y is a closed subvariety of $X = \mathbf{A}^n_k$ passing through the origin P, then the strict transform $\tilde{Y}$ of Y in $\tilde{X}$ is a closed subvariety. Hence, provided Y is not just P itself, we can recover $\tilde{Y}$ as the closure of $\pi^{-1}(Y - P)$, where $\pi: \pi^{-1}(X - P) \to X - P$ is the isomorphism of (7.13b). This shows that our new definition of blowing up coincides with the one given in (I, §4) for any closed subvariety of $\mathbf{A}_k^n$. In particular, this shows that blowing up as defined earlier is intrinsic.

Now we will study blowing up in the special case that X is a variety. Recall (§4) that a variety is defined to be an integral separated scheme of finite type over an algebraically closed field k.

Proposition 7.16. *Let X be a variety over k, let $\mathscr{I} \subseteq \mathcal{O}_X$ be a nonzero coherent sheaf of ideals on X, and let $\pi: \tilde{X} \to X$ be the blowing-up with respect to $\mathscr{I}$. Then*:
  (a) $\tilde{X}$ *is also a variety;*
  (b) π *is a birational, proper, surjective morphism;*
  (c) *if X is quasi-projective (respectively, projective) over k, then $\tilde{X}$ is also, and π is a projective morphism.*

PROOF. First of all, since X is integral, the sheaf $\mathscr{S} = \bigoplus_{d \geqslant 0} \mathscr{I}^d$ is a sheaf of integral domains on X, so $\tilde{X}$ is also integral. Next, we have already seen that π is proper (7.10). In particular, π is separated and of finite type, so it follows that $\tilde{X}$ is also separated and of finite type, i.e., $\tilde{X}$ is a variety. Now since $\mathscr{I} \neq 0$, the corresponding closed subscheme Y is not all of X, and so the open set $U = X - Y$ is nonempty. Since π induces an isomorphism from $\pi^{-1}U$ to U (7.13), we see that π is birational. Since π is proper, it is a closed map, so the image $\pi(\tilde{X})$ is a closed set containing U, which must be all of X since X is irreducible. Thus π is surjective. Finally, if X is quasi-projective (respectively, projective), then X admits an ample invertible sheaf (7.6), so by (7.10b) π is a projective morphism. It follows that $\tilde{X}$ is also quasi-projective (respectively, projective) (Ex. 4.9).

Theorem 7.17. *Let X be a quasi-projective variety over k. If Z is another variety and $f: Z \to X$ is any birational projective morphism, then there exists a coherent sheaf of ideals $\mathscr{I}$ on X such that Z is isomorphic to the blowing-up $\tilde{X}$ of X with respect to $\mathscr{I}$, and f corresponds to $\pi: \tilde{X} \to X$ under this isomorphism.*

PROOF. The proof is somewhat difficult, so we divide it into steps.

Step 1. Since f is assumed to be a projective morphism, there exists a closed immersion $i: Z \to \mathbf{P}_X^n$ for some n.

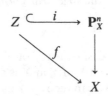

Let $\mathscr{L}$ be the invertible sheaf $i^*\mathcal{O}(1)$ on Z. Now we consider the sheaf of graded $\mathcal{O}_X$-algebras $\mathscr{S} = \mathcal{O}_X \oplus \bigoplus_{d \geqslant 1} f_*\mathscr{L}^d$. Each $f_*\mathscr{L}^d$ is a coherent sheaf on X, by (5.20), so $\mathscr{S}$ is quasi-coherent. However, $\mathscr{S}$ may not be generated by $\mathscr{S}_1$ as an $\mathcal{O}_X$-algebra.

Step 2. For any integer $e > 0$, let $\mathscr{S}^{(e)} = \bigoplus_{d \geqslant 0} \mathscr{S}_d^{(e)}$, where $\mathscr{S}_d^{(e)} = \mathscr{S}_{de}$ (cf. Ex. 5.13). I claim that for e sufficiently large, $\mathscr{S}^{(e)}$ is generated as an $\mathcal{O}_X$-algebra by $\mathscr{S}_1^{(e)}$. Since X is quasi-compact, this question is local on X, so we may assume $X = \operatorname{Spec} A$ is affine, where A is a finitely generated k-algebra. Then Z is a closed subscheme of $\mathbf{P}_A^n$, and $\mathscr{S}$ corresponds to the graded A-algebra $S = A \oplus \bigoplus_{d \geqslant 1} \Gamma(Z, \mathcal{O}_Z(d))$. Let $T = A[x_0, \ldots, x_n]/I_Z$, where I_Z is a homogeneous ideal defining Z. Then, using the technique of (Ex. 5.9, Ex. 5.14), one can show that the A-algebras S, T agree in all large enough degrees (details left to reader). But T is generated as an A-algebra by T_1, so $T^{(e)}$ is generated by $T_1^{(e)}$, and this is the same as $S^{(e)}$ for e sufficiently large.

Step 3. Now let us replace our original embedding $i: Z \to \mathbf{P}_X^n$ by i followed by an e-uple embedding for e sufficiently large. This has the effect of replacing $\mathscr{L}$ by $\mathscr{L}^e$ and $\mathscr{S}$ by $\mathscr{S}^{(e)}$ (Ex. 5.13). Thus we may now assume that $\mathscr{S}$ is generated by $\mathscr{S}_1$ as an $\mathcal{O}_X$-algebra. Note also by construction that $Z \cong \mathbf{Proj}\,\mathscr{S}$ (cf. (5.16)). So at least we have Z isomorphic to $\mathbf{Proj}$ of something. If $\mathscr{S}_1 = f_*\mathscr{L}$ were a sheaf of ideals in $\mathcal{O}_X$ we would be done. So in the next step, we try to make it into one.

Step 4. Now $\mathscr{L}$ is an invertible sheaf on the integral scheme Z, so we can find an embedding $\mathscr{L} \hookrightarrow \mathscr{K}_Z$ where $\mathscr{K}_Z$ is the constant sheaf of the function field of Z (proof of 6.15). Hence $f_*\mathscr{L} \subseteq f_*\mathscr{K}_Z$. But since f is assumed to be birational, we have $f_*\mathscr{K}_Z = \mathscr{K}_X$, and so $f_*\mathscr{L} \subseteq \mathscr{K}_X$. Now let $\mathscr{M}$ be an ample invertible sheaf on X, which exists because X is assumed to be quasi-projective. Then I claim that there is an $n > 0$ and an embedding $\mathscr{M}^{-n} \subseteq \mathscr{K}_X$ such that $\mathscr{M}^{-n} \cdot f_*\mathscr{L} \subseteq \mathcal{O}_X$. Indeed, let $\mathscr{J}$ be the *ideal sheaf of denominators* of $f_*\mathscr{L}$, defined locally as $\{a \in \mathcal{O}_X \mid a \cdot f_*\mathscr{L} \subseteq \mathcal{O}_X\}$. This is a nonzero coherent sheaf of ideals on X, because $f_*\mathscr{L}$ is a coherent subsheaf of $\mathscr{K}_X$, so locally one can just take common denominators for a set of generators of the corresponding finitely generated module. Since $\mathscr{M}$ is ample, $\mathscr{J} \otimes \mathscr{M}^n$ is generated by global sections for n sufficiently large. In particular, for suitable $n > 0$, there is a nonzero map $\mathcal{O}_X \to \mathscr{J} \otimes \mathscr{M}^n$, and hence a nonzero map $\mathscr{M}^{-n} \to \mathscr{J}$. Then by construction $\mathscr{M}^{-n} \cdot f_*\mathscr{L} \subseteq \mathcal{O}_X$.

Step 5. Since $\mathscr{M}^{-n} \cdot f_*\mathscr{L} \subseteq \mathcal{O}_X$, it is a coherent sheaf of ideals on X, which we call $\mathscr{I}$. This is the required ideal sheaf, as we will now show that Z is isomorphic to the blowing up of X with respect to $\mathscr{I}$. We already know that $Z \cong \mathbf{Proj}\,\mathscr{S}$. Therefore by (7.9) Z is also isomorphic to $\mathbf{Proj}\,\mathscr{S} * \mathscr{M}^{-n}$. So to complete the proof, it will be sufficient to identify $(\mathscr{S} * \mathscr{M}^{-n})_d = \mathscr{M}^{-dn} \otimes f_*\mathscr{L}^d$ with $\mathscr{I}^d$ for any $d \geqslant 1$. First note that $f_*\mathscr{L}^d \subseteq \mathscr{K}_X$ for any d (same reason as above for $d = 1$), and since $\mathscr{M}$ is invertible, we can write $\mathscr{M}^{-dn} \cdot f_*\mathscr{L}^d$ instead of $\otimes$. Now since $\mathscr{S}$ is locally generated by $\mathscr{S}_1$ as an $\mathcal{O}_X$-algebra, we have a natural surjective map $\mathscr{I}^d \to \mathscr{M}^{-dn} \cdot f_*\mathscr{L}^d$ for each

167

$d \geqslant 1$. It must also be injective, since both are subsheaves of $\mathscr{K}_X$, so it is an isomorphism. This shows finally that $Z \cong \mathbf{Proj} \bigoplus_{d \geqslant 0} \mathscr{I}^d$, which completes the proof.

Remark 7.17.1. Of course the sheaf of ideals $\mathscr{I}$ in the theorem is not unique. This is clear from the construction, but see also (Ex. 7.11).

Remark 7.17.2. We see from this theorem that blowing up arbitrary coherent sheaves of ideals is a very general process. Accordingly in most applications one learns more by blowing up only along some restricted class of sub-varieties. For example, in his paper on resolution of singularities [4], Hironaka uses only blowing up along a nonsingular subvariety which is "normally flat" in its ambient space. In studying birational geometry of surfaces in Chapter V, we will use only blowing up at a point. In fact one of our main results there will be that any birational transformation of non-singular projective surfaces can be factored into a finite number of blowings up (and blowings down) of points. One important application of the more general blowing-up we have been studying here is Nagata's theorem [6] that any (abstract) variety can be embedded as an open subset of a complete variety.

Example 7.17.3. As an example of the general concept of blowing up a co-herent sheaf of ideals, we show how to eliminate the points of indeterminacy of a rational map determined by an invertible sheaf. So let A be a ring, let X be a noetherian scheme over A, let $\mathscr{L}$ be an invertible sheaf on X, and let $s_0, \ldots, s_n \in \Gamma(X, \mathscr{L})$ be a set of global sections of $\mathscr{L}$. Let U be the open subset of X where the s_i generate the sheaf $\mathscr{L}$. Then according to (7.1) the invertible sheaf $\mathscr{L}|_U$ on U and the global sections $s_0, \ldots, s_n$ determine an A-morphism $\varphi: U \to \mathbf{P}_A^n$. We will now show how to blow up a certain sheaf of ideals $\mathscr{I}$ on X, whose corresponding closed subscheme Y has support equal to $X - U$ (i.e., the underlying topological space of Y is $X - U$), so that the morphism φ extends to a morphism $\tilde{\varphi}$ of $\tilde{X}$ to $\mathbf{P}_A^n$.

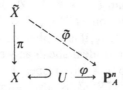

So let $\mathscr{F}$ be the coherent subsheaf of $\mathscr{L}$ generated by $s_0, \ldots, s_n$. We define a coherent sheaf of ideals $\mathscr{I}$ on X as follows: for any open set $V \subseteq X$, such that $\mathscr{L}|_V$ is free, let $\psi: \mathscr{L}|_V \xrightarrow{\sim} \mathcal{O}_V$ be an isomorphism, and take $\mathscr{I}|_V = \psi(\mathscr{F}|_V)$. Clearly the ideal sheaf $\mathscr{I}|_V$ is independent of the choice of ψ, so we get a well-defined coherent sheaf of ideals $\mathscr{I}$ on X. Note also that $\mathscr{I}_x = \mathcal{O}_x$ if and only if $x \in U$, so the corresponding closed subscheme Y has support equal to $X - U$. Let $\pi: \tilde{X} \to X$ be the blowing-up of $\mathscr{I}$. Then by (7.13a), $\pi^{-1}\mathscr{I} \cdot \mathscr{O}_{\tilde{X}}$ is an invertible sheaf of ideals, so we see that the global sections $\pi^* s_i$ of $\pi^* \mathscr{L}$ generate an *invertible* coherent subsheaf $\mathscr{L}'$ of $\pi^* \mathscr{L}$. Now $\mathscr{L}'$ and the sections $\pi^* s_i$ define a morphism $\tilde{\varphi}: \tilde{X} \to \mathbf{P}^n_A$ whose restriction to $\pi^{-1}(U)$ corresponds to φ under the natural isomorphism $\pi: \pi^{-1}(U) \xrightarrow{\sim} U$ (7.13b).

In case X is a nonsingular projective variety over a field, we can rephrase this example in terms of linear systems. The given $\mathscr{L}$ and sections s_i determine a linear system $\mathfrak{d}$ on X. The base points of $\mathfrak{d}$ are just the points of the closed set $X - U$, and $\varphi: U \to \mathbf{P}^n_k$ is the morphism determined by the base-point-free linear system $\mathfrak{d}|_U$ on U. We call Y the *scheme* of base points of $\mathfrak{d}$. So our example shows that if we blow up Y, then $\mathfrak{d}$ extends to a base-point-free linear system $\tilde{\mathfrak{d}}$ on all of $\tilde{X}$.

EXERCISES

7.1. Let $(X, \mathscr{O}_X)$ be a locally ringed space, and let $f: \mathscr{L} \to \mathscr{M}$ be a surjective map of invertible sheaves on X. Show that f is an isomorphism. [*Hint*: Reduce to a question of modules over a local ring by looking at the stalks.]

7.2. Let X be a scheme over a field k. Let $\mathscr{L}$ be an invertible sheaf on X, and let $\{s_0, \ldots, s_n\}$ and $\{t_0, \ldots, t_m\}$ be two sets of sections of $\mathscr{L}$, which generate the same subspace $V \subseteq \Gamma(X, \mathscr{L})$, and which generate the sheaf $\mathscr{L}$ at every point. Suppose $n \leqslant m$. Show that the corresponding morphisms $\varphi: X \to \mathbf{P}^n_k$ and $\psi: X \to \mathbf{P}^m_k$ differ by a suitable linear projection $\mathbf{P}^m - L \to \mathbf{P}^n$ and an automorphism of $\mathbf{P}^n$, where L is a linear subspace of $\mathbf{P}^m$ of dimension $m - n - 1$.

7.3. Let $\varphi: \mathbf{P}^n_k \to \mathbf{P}^m_k$ be a morphism. Then:
 (a) either $\varphi(\mathbf{P}^n) = pt$ or $m \geqslant n$ and dim $\varphi(\mathbf{P}^n) = n$;
 (b) in the second case, φ can be obtained as the composition of (1) a d-uple embedding $\mathbf{P}^n \to \mathbf{P}^N$ for a uniquely determined $d \geqslant 1$, (2) a linear projection $\mathbf{P}^N - L \to \mathbf{P}^m$, and (3) an automorphism of $\mathbf{P}^m$. Also, φ has finite fibres.

7.4. (a) Use (7.6) to show that if X is a scheme of finite type over a noetherian ring A, and if X admits an ample invertible sheaf, then X is separated.
 (b) Let X be the affine line over a field k with the origin doubled (4.0.1). Calculate Pic X, determine which invertible sheaves are generated by global sections, and then show directly (without using (a)) that there is no ample invertible sheaf on X.

7.5. Establish the following properties of ample and very ample invertible sheaves on a noetherian scheme X. $\mathscr{L}, \mathscr{M}$ will denote invertible sheaves, and for (d), (e) we assume furthermore that X is of finite type over a noetherian ring A.
 (a) If $\mathscr{L}$ is ample and $\mathscr{M}$ is generated by global sections, then $\mathscr{L} \otimes \mathscr{M}$ is ample.
 (b) If $\mathscr{L}$ is ample and $\mathscr{M}$ is arbitrary, then $\mathscr{M} \otimes \mathscr{L}^n$ is ample for sufficiently large n.
 (c) If $\mathscr{L}, \mathscr{M}$ are both ample, so is $\mathscr{L} \otimes \mathscr{M}$.
 (d) If $\mathscr{L}$ is very ample and $\mathscr{M}$ is generated by global sections, then $\mathscr{L} \otimes \mathscr{M}$ is very ample.
 (e) If $\mathscr{L}$ is ample, then there is an $n_0 > 0$ such that $\mathscr{L}^n$ is very ample for all $n \geqslant n_0$.

7.6. *The Riemann–Roch Problem.* Let X be a nonsingular projective variety over an algebraically closed field, and let D be a divisor on X. For any $n > 0$ we consider the complete linear system $|nD|$. Then the Riemann–Roch problem is to determine $\dim|nD|$ as a function of n, and, in particular, its behavior for large n. If $\mathscr{L}$ is the corresponding invertible sheaf, then $\dim|nD| = \dim \Gamma(X,\mathscr{L}^n) - 1$, so an equivalent problem is to determine $\dim \Gamma(X,\mathscr{L}^n)$ as a function of n.

   (a) Show that if D is very ample, and if $X \hookrightarrow \mathbf{P}_k^n$ is the corresponding embedding in projective space, then for all n sufficiently large, $\dim|nD| = P_X(n) - 1$, where P_X is the *Hilbert polynomial* of X (I, §7). Thus in this case $\dim|nD|$ is a polynomial function of n, for n large.

   (b) If D corresponds to a torsion element of Pic X, of order r, then $\dim|nD| = 0$ if $r|n$, -1 otherwise. In this case the function is periodic of period r.

   It follows from the general Riemann–Roch theorem that $\dim|nD|$ is a polynomial function for n large, whenever D is an *ample* divisor. See (IV, 1.3.2), (V, 1.6), and Appendix A. In the case of algebraic surfaces, Zariski [7] has shown for any effective divisor D, that there is a finite set of polynomials $P_1, \ldots, P_r$, such that for all n sufficiently large, $\dim|nD| = P_{i(n)}(n)$, where $i(n) \in \{1,2,\ldots,r\}$ is a function of n.

7.7. *Some Rational Surfaces.* Let $X = \mathbf{P}_k^2$, and let $|D|$ be the complete linear system of all divisors of degree 2 on X (conics). D corresponds to the invertible sheaf $\mathcal{O}(2)$, whose space of global sections has a basis $x^2, y^2, z^2, xy, xz, yz$, where x,y,z are the homogeneous coordinates of X.

   (a) The complete linear system $|D|$ gives an embedding of $\mathbf{P}^2$ in $\mathbf{P}^5$, whose image is the Veronese surface (I, Ex. 2.13).

   (b) Show that the subsystem defined by $x^2, y^2, z^2, y(x - z), (x - y)z$ gives a closed immersion of X into $\mathbf{P}^4$. The image is called the *Veronese surface* in $\mathbf{P}^4$. Cf. (IV, Ex. 3.11).

   (c) Let $\mathfrak{d} \subseteq |D|$ be the linear system of all conics passing through a fixed point P. Then $\mathfrak{d}$ gives an immersion of $U = X - P$ into $\mathbf{P}^4$. Furthermore, if we blow up P, to get a surface $\tilde{X}$, then this map extends to give a closed immersion of $\tilde{X}$ in $\mathbf{P}^4$. Show that $\tilde{X}$ is a surface of degree 3 in $\mathbf{P}^4$, and that the lines in X through P are transformed into straight lines in $\tilde{X}$ which do not meet. $\tilde{X}$ is the union of all these lines, so we say $\tilde{X}$ is a *ruled surface* (V, 2.19.1).

7.8. Let X be a noetherian scheme, let $\mathscr{E}$ be a coherent locally free sheaf on X, and let $\pi : \mathbf{P}(\mathscr{E}) \to X$ be the corresponding projective space bundle. Show that there is a natural 1-1 correspondence between *sections* of π (i.e., morphisms $\sigma : X \to \mathbf{P}(\mathscr{E})$ such that $\pi \circ \sigma = \mathrm{id}_X$) and quotient invertible sheaves $\mathscr{E} \to \mathscr{L} \to 0$ of $\mathscr{E}$.

7.9. Let X be a regular noetherian scheme, and $\mathscr{E}$ a locally free coherent sheaf of rank ≥ 2 on X.

   (a) Show that Pic $\mathbf{P}(\mathscr{E}) \cong$ Pic $X \times \mathbf{Z}$.

   (b) If $\mathscr{E}'$ is another locally free coherent sheaf on X, show that $\mathbf{P}(\mathscr{E}) \cong \mathbf{P}(\mathscr{E}')$ (over X) if and only if there is an invertible sheaf $\mathscr{L}$ on X such that $\mathscr{E}' \cong \mathscr{E} \otimes \mathscr{L}$.

7.10. $\mathbf{P}^n$-*Bundles Over a Scheme.* Let X be a noetherian scheme.

   (a) By analogy with the definition of a vector bundle (Ex. 5.18), define the notion of a *projective n-space bundle* over X, as a scheme P with a morphism $\pi : P \to X$ such that P is locally isomorphic to $U \times \mathbf{P}^n$, $U \subseteq X$ open, and the transition automorphisms on Spec $A \times \mathbf{P}^n$ are given by A-linear automorphisms of the homogeneous coordinate ring $A[x_0, \ldots, x_n]$ (e.g., $x_i' = \sum a_{ij}x_j$, $a_{ij} \in A$).

   (b) If $\mathscr{E}$ is a locally free sheaf of rank $n + 1$ on X, then $\mathbf{P}(\mathscr{E})$ is a $\mathbf{P}^n$-bundle over X.

*(c) Assume that X is regular, and show that every $\mathbf{P}^n$-bundle P over X is iso-
   morphic to $\mathbf{P}(\mathscr{E})$ for some locally free sheaf $\mathscr{E}$ on X. [*Hint*: Let $U \subseteq X$ be an
   open set such that $\pi^{-1}(U) \cong U \times \mathbf{P}^n$, and let $\mathscr{L}_0$ be the invertible sheaf $\mathscr{O}(1)$
   on $U \times \mathbf{P}^n$. Show that $\mathscr{L}_0$ extends to an invertible sheaf $\mathscr{L}$ on P. Then show
   that $\pi_*\mathscr{L} = \mathscr{E}$ is a locally free sheaf on X and that $P \cong \mathbf{P}(\mathscr{E})$.] Can you
   weaken the hypothesis "X regular"?

(d) Conclude (in the case X regular) that we have a 1-1 correspondence between
   $\mathbf{P}^n$-bundles over X, and equivalence classes of locally free sheaves $\mathscr{E}$ of rank
   $n + 1$ under the equivalence relation $\mathscr{E}' \sim \mathscr{E}$ if and only if $\mathscr{E}' \cong \mathscr{E} \otimes \mathscr{M}$ for
   some invertible sheaf $\mathscr{M}$ on X.

7.11. On a noetherian scheme X, different sheaves of ideals can give rise to isomorphic
blown up schemes.

(a) If $\mathscr{I}$ is any coherent sheaf of ideals on X, show that blowing up $\mathscr{I}^d$ for any
   $d \geqslant 1$ gives a scheme isomorphic to the blowing up of $\mathscr{I}$ (cf. Ex. 5.13).

(b) If $\mathscr{I}$ is any coherent sheaf of ideals, and if $\mathscr{J}$ is an invertible sheaf of ideals,
   then $\mathscr{I}$ and $\mathscr{I} \cdot \mathscr{J}$ give isomorphic blowings-up.

(c) If X is regular, show that (7.17) can be strengthened as follows. Let $U \subseteq X$
   be the largest open set such that $f: f^{-1}U \to U$ is an isomorphism. Then $\mathscr{I}$
   can be chosen such that the corresponding closed subscheme Y has support
   equal to $X - U$.

7.12. Let X be a noetherian scheme, and let Y, Z be two closed subschemes, neither
one containing the other. Let $\tilde{X}$ be obtained by blowing up $Y \cap Z$ (defined by
the ideal sheaf $\mathscr{I}_Y + \mathscr{I}_Z$). Show that the strict transforms $\tilde{Y}$ and $\tilde{Z}$ of Y and Z
in $\tilde{X}$ do not meet.

*7.13. *A Complete Nonprojective Variety.* Let k be an algebraically closed field of
char $\neq 2$. Let $C \subseteq \mathbf{P}_k^2$ be the nodal cubic curve $y^2z = x^3 + x^2z$. If $P_0 = (0,0,1)$
is the singular point, then $C - P_0$ is isomorphic to the multiplicative group
$\mathbf{G}_m = \operatorname{Spec} k[t,t^{-1}]$ (Ex. 6.7). For each $a \in k$, $a \neq 0$, consider the translation of
$\mathbf{G}_m$ given by $t \mapsto at$. This induces an automorphism of C which we denote by φ_a.

Now consider $C \times (\mathbf{P}^1 - \{0\})$ and $C \times (\mathbf{P}^1 - \{\infty\})$. We glue their open
subsets $C \times (\mathbf{P}^1 - \{0, \infty\})$ by the isomorphism $\varphi: \langle P,u \rangle \mapsto \langle \varphi_u(P),u \rangle$ for
$P \in C, u \in \mathbf{G}_m = \mathbf{P}^1 - \{0,\infty\}$. Thus we obtain a scheme X, which is our example.
The projections to the second factor are compatible with φ, so there is a natural
morphism $\pi: X \to \mathbf{P}^1$.

(a) Show that π is a proper morphism, and hence that X is a complete variety
   over k.

(b) Use the method of (Ex. 6.9) to show that $\operatorname{Pic}(C \times \mathbf{A}^1) \cong \mathbf{G}_m \times \mathbf{Z}$ and
   $\operatorname{Pic}(C \times (\mathbf{A}^1 - \{0\})) \cong \mathbf{G}_m \times \mathbf{Z} \times \mathbf{Z}$. [*Hint*: If A is a domain and if *
   denotes the group of units, then $(A[u])^* \cong A^*$ and $(A[u,u^{-1}])^* \cong A^* \times \mathbf{Z}$.]

(c) Now show that the restriction map $\operatorname{Pic}(C \times \mathbf{A}^1) \to \operatorname{Pic}(C \times (\mathbf{A}^1 - \{0\}))$
   is of the form $\langle t,n \rangle \mapsto \langle t,0,n \rangle$, and that the automorphism φ of $C \times (\mathbf{A}^1 - \{0\})$
   induces a map of the form $\langle t,d,n \rangle \mapsto \langle t,d + n,n \rangle$ on its Picard group.

(d) Conclude that the image of the restriction map $\operatorname{Pic} X \to \operatorname{Pic}(C \times \{0\})$
   consists entirely of divisors of degree 0 on C. Hence X is not projective
   over k and π is not a projective morphism.

7.14. (a) Give an example of a noetherian scheme X and a locally free coherent sheaf $\mathscr{E}$,
   such that the invertible sheaf $\mathscr{O}(1)$ on $\mathbf{P}(\mathscr{E})$ is *not* very ample relative to X.

(b) Let $f: X \to Y$ be a morphism of finite type, let $\mathscr{L}$ be an ample invertible sheaf on X, and let $\mathscr{S}$ be a sheaf of graded $\mathcal{O}_X$-algebras satisfying (†). Let $P = \mathbf{Proj}\, \mathscr{S}$, let $\pi: P \to X$ be the projection, and let $\mathcal{O}_P(1)$ be the associated invertible sheaf. Show that for all $n \gg 0$, the sheaf $\mathcal{O}_P(1) \otimes \pi^* \mathscr{L}^n$ is very ample on P relative to Y. [Hint: Use (7.10) and (Ex. 5.12).]

8 Differentials

In this section we will define the sheaf of relative differential forms of one scheme over another. In the case of a nonsingular variety over $\mathbf{C}$, which is like a complex manifold, the sheaf of differential forms is essentially the same as the dual of the tangent bundle defined in differential geometry. However, in abstract algebraic geometry, we will define the sheaf of differentials first, by a purely algebraic method, and then define the tangent bundle as its dual. Hence we will begin this section with a review of the module of differentials of one ring over another. As applications of the sheaf of differentials, we will give a characterization of nonsingular varieties among schemes of finite type over a field. We will also use the sheaf of differentials on a nonsingular variety to define its tangent sheaf, its canonical sheaf, and its geometric genus. This latter is an important numerical invariant of a variety.

Kähler Differentials

Here we will review the algebraic theory of Kähler differentials. We will use Matsumura [2, Ch. 10] as our main reference, but proofs can also be found in the exposés of Cartier and Godement in Cartan and Chevalley [1, exposés 13, 17], or in Grothendieck [EGA 0_{IV}, §20.5].

Let A be a ring (commutative with identity as always), let B be an A-algebra, and let M be a B-module.

Definition. An *A-derivation* of B into M is a map $d: B \to M$ such that (1) d is additive, (2) $d(bb') = b\,db' + b'\,db$, and (3) $da = 0$ for all $a \in A$.

Definition. We define the *module of relative differential forms* of B over A to be a B-module $\Omega_{B/A}$, together with an A-derivation $d: B \to \Omega_{B/A}$, which satisfies the following universal property: for any B-module M, and for any A-derivation $d': B \to M$, there exists a unique B-module homomorphism $f: \Omega_{B/A} \to M$ such that $d' = f \circ d$.

Clearly one way to construct such a module $\Omega_{B/A}$ is to take the free B-module F generated by the symbols $\{db \mid b \in B\}$, and to divide out by the submodule generated by all expressions of the form (1) $d(b + b') - db - db'$ for $b, b' \in B$, (2) $d(bb') - b\,db' - b'\,db$ for $b, b' \in B$, and (3) da for $a \in A$. The derivation $d: B \to \Omega_{B/A}$ is defined by sending b to db. Thus we see that $\Omega_{B/A}$ exists. It follows from the definition that the pair $\langle \Omega_{B/A}, d \rangle$ is unique up to unique isomorphism. As a corollary of this construction, we see that $\Omega_{B/A}$ is generated as a B-module by $\{db|b \in B\}$.

Proposition 8.1A. *Let B be an A-algebra. Let $f:B \otimes_A B \to B$ be the "diagonal" homomorphism defined by $f(b \otimes b') = bb'$, and let $I = \ker f$. Consider $B \otimes_A B$ as a B-module by multiplication on the left. Then I/I^2 inherits a structure of B-module. Define a map $d:B \to I/I^2$ by $db = 1 \otimes b - b \otimes 1$ (modulo I^2). Then $\langle I/I^2, d \rangle$ is a module of relative differentials for B/A.*

PROOF. Matsumura [2, p. 182].

Proposition 8.2A. *If A' and B are A-algebras, let $B' = B \otimes_A A'$. Then $\Omega_{B'/A'} \cong \Omega_{B/A} \otimes_B B'$. Furthermore, if S is a multiplicative system in B, then $\Omega_{S^{-1}B/A} \cong S^{-1}\Omega_{B/A}$.*

PROOF. Matsumura [2, p. 186].

Example 8.2.1 If $B = A[x_1, \ldots, x_n]$ is a polynomial ring over A, then $\Omega_{B/A}$ is the free B-module of rank n generated by $dx_1, \ldots, dx_n$ (Matsumura [2, p. 184]).

Proposition 8.3A (First Exact Sequence). *Let $A \to B \to C$ be rings and homomorphisms. Then there is a natural exact sequence of C-modules*

$$\Omega_{B/A} \otimes_B C \to \Omega_{C/A} \to \Omega_{C/B} \to 0.$$

PROOF. Matsumura [2, Th. 57 p. 186].

Proposition 8.4A (Second Exact Sequence). *Let B be an A-algebra, let I be an ideal of B, and let $C = B/I$. Then there is a natural exact sequence of C-modules*

$$I/I^2 \xrightarrow{\delta} \Omega_{B/A} \otimes_B C \to \Omega_{C/A} \to 0,$$

where for any $b \in I$, if $\bar{b}$ is its image in I/I^2, then $\delta\bar{b} = db \otimes 1$. Note in particular that I/I^2 has a natural structure of C-module, and that δ is a C-linear map, even though it is defined via the derivation d.

PROOF. Matsumura [2, Th. 58, p. 187].

Corollary 8.5. *If B is a finitely generated A-algebra, or if B is a localization of a finitely generated A-algebra, then $\Omega_{B/A}$ is a finitely generated B-module.*

PROOF. Indeed, B is a quotient of a polynomial ring (or its localization) so the result follows from (8.4A), (8.2A), and the example of the polynomial ring itself.

Now we will consider the module of differentials in the case of field extensions and local rings. Recall (I, §4) that an extension field K of a field k

is *separably generated* if there exists a transcendence base $\{x_\lambda\}$ for K/k such that K is a separable algebraic extension of $k(\{x_\lambda\})$.

Theorem 8.6A. *Let K be a finitely generated extension field of a field k. Then $\dim_K \Omega_{K/k} \geqslant \mathrm{tr.d.}\ K/k$, and equality holds if and only if K is separably generated over k. (Here $\dim_K$ denotes the dimension as a K-vector space.)*

PROOF. Matsumura [2, Th. 59, p. 191]. Note in particular that if K/k is a finite algebraic extension, then $\Omega_{K/k} = 0$ if and only if K/k is separable.

Proposition 8.7. *Let B be a local ring which contains a field k isomorphic to its residue field $B/\mathfrak{m}$. Then the map $\delta: \mathfrak{m}/\mathfrak{m}^2 \to \Omega_{B/k} \otimes_B k$ of (8.4A) is an isomorphism.*

PROOF. According to (8.4A), the cokernel of δ is $\Omega_{k/k} = 0$, so δ is surjective. To show that δ is injective, it will be sufficient to show that the map

$$\delta': \mathrm{Hom}_k(\Omega_{B/k} \otimes k, k) \to \mathrm{Hom}_k(\mathfrak{m}/\mathfrak{m}^2, k)$$

of dual vector spaces is surjective. The term on the left is isomorphic to $\mathrm{Hom}_B(\Omega_{B/k}, k)$, which by definition of the differentials, can be identified with the set $\mathrm{Der}_k(B, k)$ of k-derivations of B to k. If $d: B \to k$ is a derivation, then $\delta'(d)$ is obtained by restricting to $\mathfrak{m}$, and noting that $d(\mathfrak{m}^2) = 0$. Now, to show that δ' is surjective, let $h \in \mathrm{Hom}(\mathfrak{m}/\mathfrak{m}^2, k)$. For any $b \in B$, we can write $b = \lambda + c, \lambda \in k, c \in \mathfrak{m}$, in a unique way. Define $db = h(\bar{c})$, where $\bar{c} \in \mathfrak{m}/\mathfrak{m}^2$ is the image of c. Then one verifies immediately that d is a k-derivation of B to k, and that $\delta'(d) = h$. Thus δ' is surjective, as required.

Theorem 8.8. *Let B be a local ring containing a field k isomorphic to its residue field. Assume furthermore that k is perfect, and that B is a localization of a finitely generated k-algebra. Then $\Omega_{B/k}$ is a free B-module of rank equal to $\dim B$ if and only if B is a regular local ring.*

PROOF. First suppose $\Omega_{B/k}$ is free of rank $= \dim B$. Then by (8.7) we have $\dim_k \mathfrak{m}/\mathfrak{m}^2 = \dim B$, which says by definition that B is a regular local ring (I, §5). Note in particular that this implies that B is an integral domain.

Now conversely, suppose that B is regular local of dimension r. Then $\dim_k \mathfrak{m}/\mathfrak{m}^2 = r$, so by (8.7) we have $\dim_k \Omega_{B/k} \otimes k = r$. On the other hand, let K be the quotient field of B. Then by (8.2A) we have $\Omega_{B/k} \otimes_B K = \Omega_{K/k}$. Now since k is perfect, K is a separably generated extension field of k (I, 4.8A), and so $\dim_K \Omega_{K/k} = \mathrm{tr.d.}\ K/k$ by (8.6A). But we also have $\dim B = \mathrm{tr.d.}\ K/k$ by (I, 1.8A). Finally, note that by (8.5), $\Omega_{B/k}$ is a finitely generated B-module. We conclude that $\Omega_{B/k}$ is a free B module of rank r by using the following well-known lemma.

Lemma 8.9. *Let A be a noetherian local integral domain, with residue field k and quotient field K. If M is a finitely generated A-module and if $\dim_k M \otimes_A k = \dim_K M \otimes_A K = r$, then M is free of rank r.*

PROOF. Since $\dim_k M \otimes k = r$, Nakayama's lemma tells us that M can be generated by r elements. So there is a surjective map $\varphi : A^r \to M \to 0$. Let R be its kernel. Then we obtain an exact sequence

$$0 \to R \otimes K \to K^r \to M \otimes K \to 0,$$

and since $\dim_K M \otimes K = r$, we have $R \otimes K = 0$. But R is torsion-free, so $R = 0$, and M is isomorphic to A^r.

Sheaves of Differentials

We now carry the definition of the module of differentials over to schemes. Let $f : X \to Y$ be a morphism of schemes. We consider the diagonal morphism $\varDelta : X \to X \times_Y X$. It follows from the proof of (4.2) that $\varDelta$ gives an isomorphism of X onto its image $\varDelta(X)$, which is a *locally closed* subscheme of $X \times_Y X$, i.e., a closed subscheme of an open subset W of $X \times_Y X$.

Definition. Let $\mathscr{I}$ be the sheaf of ideals of $\varDelta(X)$ in W. Then we define the *sheaf of relative differentials* of X over Y to be the sheaf $\Omega_{X/Y} = \varDelta^*(\mathscr{I}/\mathscr{I}^2)$ on X.

Remark 8.9.1. First note that $\mathscr{I}/\mathscr{I}^2$ has a natural structure of $\mathcal{O}_{\varDelta(X)}$-module. Then since $\varDelta$ induces an isomorphism of X to $\varDelta(X)$, $\Omega_{X/Y}$ has a natural structure of $\mathcal{O}_X$-module. Furthermore, it follows from (5.9) that $\Omega_{X/Y}$ is quasi-coherent; if Y is noetherian and f is a morphism of finite type, then $X \times_Y X$ is also noetherian, and so $\Omega_{X/Y}$ is coherent.

Remark 8.9.2. Now if $U = \operatorname{Spec} A$ is an open affine subset of Y and $V = \operatorname{Spec} B$ is an open affine subset of X such that $f(V) \subseteq U$, then $V \times_U V$ is an open affine subset of $X \times_Y X$ isomorphic to $\operatorname{Spec}(B \otimes_A B)$, and $\varDelta(X) \cap (V \times_U V)$ is the closed subscheme defined by the kernel of the diagonal homomorphism $B \otimes_A B \to B$. Thus $\mathscr{I}/\mathscr{I}^2$ is the sheaf associated to the module I/I^2 of (8.1A). It follows that $\Omega_{V/U} \cong (\Omega_{B/A})^{\sim}$. Thus our definition of the sheaf of differentials of X/Y is compatible, in the affine case, with the module of differentials defined above, via the functor $\sim$. This also shows that we could have defined $\Omega_{X/Y}$ by covering X and Y with open affine subsets V and U as above, and glueing the corresponding sheaves $(\Omega_{B/A})^{\sim}$. The derivations $d : B \to \Omega_{B/A}$ glue together to give a map $d : \mathcal{O}_X \to \Omega_{X/Y}$ of sheaves of abelian groups on X, which is a derivation of the local rings at each point.

Therefore, we can carry over our algebraic results to sheaves, and we obtain the following results.

Proposition 8.10. *Let $f : X \to Y$ be a morphism, let $g : Y' \to Y$ be another morphism, and let $f' : X' = X \times_Y Y' \to Y'$ be obtained by base extension. Then $\Omega_{X'/Y'} \cong g'^*(\Omega_{X/Y})$ where $g' : X' \to X$ is the first projection.*

PROOF. Follows from (8.2A).

175

Proposition 8.11. *Let $f: X \to Y$ and $g: Y \to Z$ be morphisms of schemes. Then there is an exact sequence of sheaves on X,*

$$f^* \Omega_{Y/Z} \to \Omega_{X/Z} \to \Omega_{X/Y} \to 0.$$

PROOF. Follows from (8.3A).

Proposition 8.12. *Let $f: X \to Y$ be a morphism, and let Z be a closed subscheme of X, with ideal sheaf $\mathscr{I}$. Then there is an exact sequence of sheaves on Z,*

$$\mathscr{I}/\mathscr{I}^2 \xrightarrow{\delta} \Omega_{X/Y} \otimes \mathcal{O}_Z \to \Omega_{Z/Y} \to 0.$$

PROOF. Follows from (8.4A).

Example 8.12.1. If $X = \mathbf{A}_Y^n$, then $\Omega_{X/Y}$ is a free $\mathcal{O}_X$-module of rank n, generated by the global sections $dx_1, \ldots, dx_n$, where $x_1, \ldots, x_n$ are affine coordinates for $\mathbf{A}^n$.

Next we will give an exact sequence relating the sheaf of differentials on a projective space to sheaves we already know. This is a fundamental result, upon which we will base all future calculations involving differentials on projective varieties.

Theorem 8.13. *Let A be a ring, let $Y = \operatorname{Spec} A$, and let $X = \mathbf{P}_A^n$. Then there is an exact sequence of sheaves on X,*

$$0 \to \Omega_{X/Y} \to \mathcal{O}_X(-1)^{n+1} \to \mathcal{O}_X \to 0.$$

(The exponent $n+1$ in the middle means a direct sum of $n+1$ copies of $\mathcal{O}_X(-1)$.)

PROOF. Let $S = A[x_0, \ldots, x_n]$ be the homogeneous coordinate ring of X. Let E be the graded S-module $S(-1)^{n+1}$, with basis $e_0, \ldots, e_n$ in degree 1. Define a (degree 0) homomorphism of graded S-modules $E \to S$ by sending $e_i \mapsto x_i$, and let M be the kernel. Then the exact sequence

$$0 \to M \to E \to S$$

of graded S-modules gives rise to an exact sequence of sheaves on X,

$$0 \to \tilde{M} \to \mathcal{O}_X(-1)^{n+1} \to \mathcal{O}_X \to 0.$$

Note that $E \to S$ is not surjective, but it is surjective in all degrees $\geqslant 1$, so the corresponding map of sheaves is surjective.

We will now proceed to show that $\tilde{M} \cong \Omega_{X/Y}$. First note that if we localize at x_i, then $E_{x_i} \to S_{x_i}$ is a surjective homomorphism of free S_{x_i}-modules, so M_{x_i} is free of rank n, generated by $\{e_j - (x_j/x_i)e_i \mid j \neq i\}$. It follows that if U_i is the standard open set of X defined by x_i, then $\tilde{M}|_{U_i}$ is a free $\mathcal{O}_{U_i}$-module generated by the sections $(1/x_i)e_j - (x_j/x_i^2)e_i$ for $j \neq i$. (Here we need the additional factor $1/x_i$ to get elements of degree 0 in the module M_{x_i}.)

We define a map $\varphi_i: \Omega_{X/Y}|_{U_i} \to \tilde{M}|_{U_i}$ as follows. Recall that $U_i \cong$ Spec $A[x_0/x_i, \ldots, x_n/x_i]$, so $\Omega_{X/Y}|_{U_i}$ is a free $\mathcal{O}_{U_i}$-module generated by $d(x_0/x_i), \ldots, d(x_n/x_i)$. So we define φ_i by

$$\varphi_i(d(x_j/x_i)) = (1/x_i^2)(x_i e_j - x_j e_i).$$

Thus φ_i is an isomorphism. I claim now that the isomorphisms φ_i glue together to give an isomorphism $\varphi: \Omega_{X/Y} \to \tilde{M}$ on all of X. This is a simple calculation. On $U_i \cap U_j$, we have, for any k, $(x_k/x_i) = (x_k/x_j) \cdot (x_j/x_i)$. Hence in $\Omega|_{U_i \cap U_j}$ we have

$$d\left(\frac{x_k}{x_i}\right) - \frac{x_k}{x_j} d\left(\frac{x_j}{x_i}\right) = \frac{x_j}{x_i} d\left(\frac{x_k}{x_j}\right).$$

Now applying φ_i to the left-hand side and φ_j to the right-hand side, we get the same thing both ways, namely $(1/x_i x_j)(x_j e_k - x_k e_j)$. Thus the isomorphisms φ_i glue, which completes our proof.

Nonsingular Varieties

Our principal application of the sheaf of differentials is to nonsingular varieties. In (I, §5) we defined a nonsingular quasi-projective variety to be one whose local rings were all regular local rings. Here we extend that definition to abstract varieties.

Definition. An (abstract) variety X over an algebraically closed field k is *nonsingular* if all its local rings are regular local rings.

Note that we are apparently requiring more here, because in Chapter I we had only closed points, but now our varieties also have nonclosed points. However, the two definitions are equivalent, because every local ring at a nonclosed point is the localization of a local ring at a closed point, and we have the following algebraic result.

Theorem 8.14A. *Any localization of a regular local ring at a prime ideal is again a regular local ring.*

PROOF. Matsumura [2, p. 139].

The connection between nonsingularity and differentials is given by the following result.

Theorem 8.15. *Let X be an irreducible separated scheme of finite type over an algebraically closed field k. Then $\Omega_{X/k}$ is a locally free sheaf of rank $n = \dim X$ if and only if X is a nonsingular variety over k.*

PROOF. If $x \in X$ is a closed point, then the local ring $B = \mathcal{O}_{x,x}$ has dimension n, residue field k, and is a localization of a k-algebra of finite type. Furthermore the module $\Omega_{B/k}$ of differentials of B over k is equal to the stalk $(\Omega_{X/k})_x$ of the sheaf $\Omega_{X/k}$. Thus we can apply (8.8) and we see that $(\Omega_{X/k})_x$ is free of rank n if and only if B is a regular local ring. Now the theorem follows in view of (8.14A) and (Ex. 5.7).

Corollary 8.16. *If X is a variety over k, then there is an open dense subset U of X which is nonsingular.*

PROOF. (This gives a new proof of (I, 5.3).) If $n = \dim X$, then the function field K of X has transcendence degree n over k, and it is a finitely generated extension field, which is separably generated by (I, 4.8A). Therefore by (8.6A), $\Omega_{K/k}$ is a K-vector space of dimension n. Now $\Omega_{K/k}$ is just the stalk of the sheaf $\Omega_{X/k}$ at the generic point of X. Thus by (Ex. 5.7), $\Omega_{X/k}$ is locally free of rank n in some neighborhood of the generic point, i.e., on a nonempty open set U. Then U is nonsingular by the theorem.

Theorem 8.17. *Let X be a nonsingular variety over k. Let $Y \subseteq X$ be an irreducible closed subscheme defined by a sheaf of ideals $\mathscr{I}$. Then Y is nonsingular if and only if*

(1) *$\Omega_{Y/k}$ is locally free, and*
(2) *the sequence of (8.12) is exact on the left also:*

$$0 \to \mathscr{I}/\mathscr{I}^2 \to \Omega_{X/k} \otimes \mathcal{O}_Y \to \Omega_{Y/k} \to 0.$$

Furthermore, in this case, $\mathscr{I}$ is locally generated by $r = \operatorname{codim}(Y,X)$ elements, and $\mathscr{I}/\mathscr{I}^2$ is a locally free sheaf of rank r on Y.

PROOF. First suppose (1) and (2) hold. Then $\Omega_{Y/k}$ is locally free, so by (8.15) we have only to show that rank $\Omega_{Y/k} = \dim Y$. Let rank $\Omega_{Y/k} = q$. We know that $\Omega_{X/k}$ is locally free of rank n, so it follows from (2) that $\mathscr{I}/\mathscr{I}^2$ is locally free on Y of rank $n - q$. Hence by Nakayama's lemma, $\mathscr{I}$ can be locally generated by $n - q$ elements, and it follows that $\dim Y \geqslant n - (n - q) = q$ (I, Ex. 1.9). On the other hand, considering any closed point $y \in Y$, we have $q = \dim_k(\mathfrak{m}_y/\mathfrak{m}_y^2)$ by (8.7), and so $q \geqslant \dim Y$ by (I, 5.2A). Thus $q = \dim Y$. This shows that Y is nonsingular, and at the same time establishes the statements at the end of the theorem, since we now have $n - q = \operatorname{codim}(Y,X)$.

Conversely, assume that Y is nonsingular. Then $\Omega_{Y/k}$ is locally free of rank $q = \dim Y$, so (1) is immediate. From (8.12) we have the exact sequence

$$\mathscr{I}/\mathscr{I}^2 \xrightarrow{\delta} \Omega_{X/k} \otimes \mathcal{O}_Y \xrightarrow{\varphi} \Omega_{Y/k} \to 0.$$

We consider a closed point $y \in Y$. Then ker φ is locally free of rank $r = n - q$ at y, so it is possible to choose sections $x_1, \dots, x_r \in \mathscr{I}$ in a suitable neighborhood of y, such that $dx_1, \dots, dx_r$ generate ker φ. Let $\mathscr{I}'$ be the ideal sheaf generated by $x_1, \dots, x_r$, and let Y' be the corresponding closed subscheme. Then by construction, the $dx_1, \dots, dx_r$ generate a free subsheaf of rank r of $\Omega_{X/k} \otimes \mathcal{O}_{Y'}$ in a neighborhood of y. It follows that in the exact sequence of (8.12) for Y',

$$\mathscr{I}'/\mathscr{I}'^2 \xrightarrow{\delta} \Omega_{X/k} \otimes \mathcal{O}_{Y'} \to \Omega_{Y'/k} \to 0,$$

we have δ injective (since its image is free of rank r), and $\Omega_{Y'/k}$ is locally free of rank $n - r$. The previous part of the proof now shows that Y' is irreducible and nonsingular of dimension $n - r$ (in a neighborhood of y). But $Y \subseteq Y'$, both are integral schemes of the same dimension, so we must have $Y = Y'$, $\mathscr{I} = \mathscr{I}'$, and this shows that $\mathscr{I}/\mathscr{I}^2 \xrightarrow{\delta} \Omega_{X/k} \otimes \mathcal{O}_Y$ is injective, as required.

Next we include a result which tells us that under suitable conditions, a hyperplane section of a nonsingular variety in projective space is again nonsingular. There is actually a large class of such results, which say that if a projective variety has a certain property, then a sufficiently general hyperplane section has the same property. The result we give here is not the strongest, but it is sufficient for many applications. See also (III, 10.9) for another version in characteristic 0.

Theorem 8.18 (Bertini's Theorem). *Let X be a nonsingular closed subvariety of $\mathbf{P}_k^n$, where k is an algebraically closed field. Then there exists a hyperplane $H \subseteq \mathbf{P}_k^n$, not containing X, and such that the scheme $H \cap X$ is regular at every point. (In fact, we will see later (III, 7.9.1) that if $\dim X \geq 2$, then $H \cap X$ is connected, hence irreducible, and so $H \cap X$ is a nonsingular variety.) Furthermore, the set of hyperplanes with this property forms an open dense subset of the complete linear system $|H|$, considered as a projective space.*

PROOF. For a closed point $x \in X$, let us consider the set $B_x = \{$hyperplanes $H|H \supseteq X$ or $H \not\supseteq X$ but $x \in H \cap X$, and x is not a regular point of $H \cap X.\}$ (Fig. 10). These are the bad hyperplanes with respect to the point x. Now a

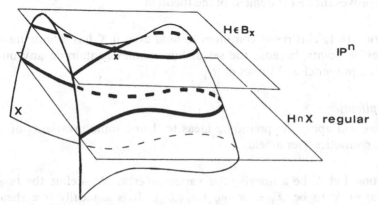

Figure 10. Hyperplane sections of a nonsingular variety.

hyperplane H is determined by a nonzero global section $f \in V = \Gamma(\mathbf{P}^n, \mathcal{O}_{\mathbf{P}^n}(1))$. Let us fix an $f_0 \in V$ such that $x \notin H_0$, the hyperplane defined by f_0. Then we can define a map of k-vector spaces

$$\varphi_x : V \to \mathcal{O}_{x,x}/\mathfrak{m}_x^2$$

as follows. Given $f \in V$, then f/f_0 is a regular function on $\mathbf{P}^n - H_0$, which induces a regular function on $X - X \cap H_0$. We take $\varphi_x(f)$ to be the image of f/f_0 in the local ring $\mathcal{O}_{x,x}$ modulo $\mathfrak{m}_x^2$. Now the scheme $H \cap X$ is defined at x by the ideal generated by f/f_0 in $\mathcal{O}_x$. So $x \in H \cap X$ if and only if $\varphi_x(f) \in \mathfrak{m}_x$, and x is nonregular on $H \cap X$ if and only if $\varphi_x(f) \in \mathfrak{m}_x^2$, because in.that case, the local ring $\mathcal{O}_x/(\varphi(f))$ will not be regular. Thus we see that the hyperplanes $H \in B_x$ correspond exactly to those $f \in \ker \varphi_x$ (note also that $\varphi_x(f) = 0 \Leftrightarrow H \supseteq X$.)

Since x is a closed point and k is algebraically closed, $\mathfrak{m}_x$ is generated by linear forms in the coordinates, so we see that φ_x is surjective. If $\dim X = r$, then $\dim_k \mathcal{O}_x/\mathfrak{m}_x^2 = r + 1$. We have $\dim V = n + 1$, so $\dim \ker \varphi_x = n - r$. This shows that B_x is a linear system of hyperplanes (in the sense of §7) of dimension $n - r - 1$.

Now, considering the complete linear system $|H|$ as a projective space, consider the subset $B \subseteq X \times |H|$ consisting of all pairs $\langle x, H \rangle$ such that $x \in X$ is a closed point and $H \in B_x$. Clearly B is the set of closed points of a closed subset of $X \times |H|$, which we denote also by B, and which we give a reduced induced scheme structure. We have just seen that the first projection $p_1 : B \to X$ is surjective, with fibre a projective space of dimension $n - r - 1$. Hence B is irreducible, and has dimension $(n - r - 1) + r = n - 1$. Therefore, considering the second projection $p_2 : B \to |H|$, we have $\dim p_2(B) \leqslant n - 1$. Since $\dim |H| = n$, we conclude that $p_2(B) < |H|$. If $H \in |H| - p_2(B)$, then $H \not\supseteq X$ and every point of $H \cap X$ is regular, so that H satisfies the requirements of the theorem. Finally note that since X is projective, $p_2 : X \times |H| \to |H|$ is a proper morphism; B is closed in $X \times |H|$, so $p_2(B)$ is closed in $|H|$. Thus $|H| - p_2(B)$ is an open dense subset of $|H|$, which proves the last statement of the theorem.

Remark 8.18.1. This result continues to hold even if X has a finite number of singular points, because the set of hyperplanes containing any one of them is a proper closed subset of $|H|$.

Applications

Now we will apply the preceding ideas to define some invariants of non-singular varieties over a field.

Definition. Let X be a nonsingular variety over k. We define the *tangent sheaf* of X to be $\mathcal{T}_X = \mathcal{H}om_{\mathcal{O}_X}(\Omega_{X/k}, \mathcal{O}_X)$. It is a locally free sheaf of rank $n = \dim X$. We define the *canonical sheaf* of X to be $\omega_X = \bigwedge^n \Omega_{X/k}$, the nth exterior power of the sheaf of differentials, where $n = \dim X$. It is an invertible sheaf on X. If X is projective and nonsingular, we define the *geometric genus* of X to be $p_g = \dim_k \Gamma(X, \omega_X)$. It is a nonnegative integer.

Remark 8.18.2. Earlier (I, Ex. 7.2) we defined the arithmetic genus p_a of a variety in projective space. In the case of a projective nonsingular curve, the arithmetic genus and the geometric genus coincide. This is a consequence of the Serre duality theorem which we will prove later (III, 7.12.2). For varieties of dimension $\geqslant 2$, however, p_a and p_g need not be equal (Ex. 8.3). See also (III, 7.12.3).

Remark 8.18.3. Since the sheaf of differentials, the tangent sheaf, and the canonical sheaf are all defined intrinsically, any numbers which we can define from them, such as the geometric genus, are invariants of X up to isomorphism. In fact, we will now show that the geometric genus is a *birational invariant* of a nonsingular projective variety. This makes it extremely important for the classification problem.

Theorem 8.19. *Let X and X' be two birationally equivalent nonsingular projective varieties over k. Then $p_g(X) = p_g(X')$.*

PROOF. Recall from (I, §4) that for X and X' to be birationally equivalent means that there are rational maps from X to X' and from X' to X which are inverses to each other. Considering the rational map from X to X', let $V \subseteq X$ be the largest open set for which there is a morphism $f : V \to X'$ representing this rational map. Then from (8.11) we have a map $f^* \Omega_{X'/k} \to \Omega_{V/k}$. These are locally free sheaves of the same rank $n = \dim X$, so we get an induced map on the exterior powers: $f^* \omega_{X'} \to \omega_V$. This map in turn induces a map on the space of global sections $f^* : \Gamma(X', \omega_{X'}) \to \Gamma(V, \omega_V)$. Now since f is birational, by (I, 4.5), there is an open set $U \subseteq V$ such that $f(U)$ is open in X', and f induces an isomorphism from U to $f(U)$. Thus $\omega_V|_U \cong \omega_{X'}|_{f(U)}$ via f. Since a nonzero global section of an invertible sheaf cannot vanish on a dense open set, we conclude that the map of vector spaces $f^* : \Gamma(X', \omega_{X'}) \to \Gamma(V, \omega_V)$ must be injective.

Next we will compare $\Gamma(V, \omega_V)$ with $\Gamma(X, \omega_X)$. First I claim that $X - V$ has codimension $\geqslant 2$ in X. Indeed, this follows from the valuative criterion of properness (4.7). If $P \in X$ is a point of codimension 1, then $\mathcal{O}_{P,X}$ is a discrete valuation ring (because X is nonsingular). We already have a map of the generic point of X to X'; and X' is projective, hence proper over k, so there exists a unique morphism $\operatorname{Spec} \mathcal{O}_{P,X} \to X'$ compatible with the given birational map. This extends to a morphism of some neighborhood of P to X', so we must have $P \in V$ by definition of V.

Now we can show that the natural restriction map $\Gamma(X, \omega_X) \to \Gamma(V, \omega_V)$ is bijective. It is enough to show, for any open affine subset $U \subseteq X$ such that

181

$\omega_X|_U \cong \mathcal{O}_U$, that $\Gamma(U,\mathcal{O}_U) \to \Gamma(U \cap V,\mathcal{O}_{U \cap V})$ is bijective. Since X is nonsingular, hence normal, and since $U - U \cap V$ has codimension ≥ 2 in U, this is an immediate consequence of (6.3A).

Combining our results, we see that $p_g(X') \leq p_g(X)$. We obtain the reverse inequality by symmetry, and thus conclude that $p_g(X) = p_g(X')$.

Next we study the behavior of the tangent sheaf and the canonical sheaf for a nonsingular subvariety of a variety X.

Definition. Let Y be a nonsingular subvariety of a nonsingular variety X over k. The locally free sheaf $\mathscr{I}/\mathscr{I}^2$ of (8.17) we call the *conormal* sheaf of Y in X. Its dual $\mathscr{N}_{Y/X} = \mathscr{H}om_{\mathcal{O}_Y}(\mathscr{I}/\mathscr{I}^2,\mathcal{O}_Y)$ is called the *normal sheaf* of Y in X. It is locally free of rank $r = \mathrm{codim}(Y,X)$.

Note that if we take the dual on Y of the exact sequence of locally free sheaves on Y given in (8.17), then we obtain an exact sequence

$$0 \to \mathscr{T}_Y \to \mathscr{T}_X \otimes \mathcal{O}_Y \to \mathscr{N}_{Y/X} \to 0.$$

This shows that the normal sheaf we have just defined corresponds to the usual geometric notion of normal vectors being tangent vectors of the ambient space modulo tangent vectors of the subspace.

Proposition 8.20. *Let Y be a nonsingular subvariety of codimension r in a nonsingular variety X over k. Then $\omega_Y \cong \omega_X \otimes \wedge^r \mathscr{N}_{Y/X}$. In case $r = 1$, consider Y as a divisor, and let $\mathscr{L}$ be the associated invertible sheaf on X. Then $\omega_Y \cong \omega_X \otimes \mathscr{L} \otimes \mathcal{O}_Y$.*

PROOF. We take the highest exterior powers of the locally free sheaves in the exact sequence

$$0 \to \mathscr{I}/\mathscr{I}^2 \to \Omega_X \otimes \mathcal{O}_Y \to \Omega_Y \to 0$$

(Ex. 5.16d). Thus we find that $\omega_X \otimes \mathcal{O}_Y \cong \omega_Y \otimes \wedge^r(\mathscr{I}/\mathscr{I}^2)$. Since formation of the highest exterior power commutes with taking the dual sheaf, we find $\omega_Y \cong \omega_X \otimes \wedge^r \mathscr{N}_{Y/X}$. In the special case $r = 1$, we have $\mathscr{I}_Y \cong \mathscr{L}^{-1}$ by (6.18). Thus $\mathscr{I}/\mathscr{I}^2 \cong \mathscr{L}^{-1} \otimes \mathcal{O}_Y$, and $\mathscr{N}_{Y/X} \cong \mathscr{L} \otimes \mathcal{O}_Y$. So applying the previous result with $r = 1$, we obtain $\omega_Y \cong \omega_X \otimes \mathscr{L} \otimes \mathcal{O}_Y$.

Example 8.20.1. Let $X = \mathbf{P}_k^n$. Taking the dual of the exact sequence of (8.13) gives us this exact sequence involving the tangent sheaf of $\mathbf{P}^n$:

$$0 \to \mathcal{O}_X \to \mathcal{O}_X(1)^{n+1} \to \mathscr{T}_X \to 0.$$

To obtain the canonical sheaf of $\mathbf{P}^n$, we take the highest exterior powers of the exact sequence of (8.13) and we find $\omega_X \cong \mathcal{O}_X(-n-1)$. Since $\mathcal{O}(l)$ has no global sections for $l < 0$, we find that $p_g(\mathbf{P}^n) = 0$ for any $n \geq 1$. Recall that a *rational variety* is defined as a variety birational to $\mathbf{P}^n$ for some n

(I, Ex. 4.4). We conclude from (8.19) that if X is any nonsingular projective rational variety, then $p_g(X) = 0$. This fact will enable us to demonstrate the existence of nonrational varieties in all dimensions.

Example 8.20.2. Let $X = \mathbf{P}_k^n$, with $n \geqslant 2$. For any integer $d \geqslant 1$, the divisor dH, where H is a hyperplane, is a very ample divisor (7.6.1). Thus dH becomes a hyperplane section of X in a suitable projective embedding (the d-uple embedding), and we can apply Bertini's theorem (8.18). We find that there is a subscheme $Y \in |dH|$ which is regular at every one of its points. If Y had at least two irreducible components, say Y_1 and Y_2, then since $n \geqslant 2$, their intersection $Y_1 \cap Y_2$ would be nonempty (I, 7.2). But this cannot happen because Y would be singular at any point of $Y_1 \cap Y_2$, so we conclude in fact that Y is irreducible, hence a nonsingular variety. Thus we see for any $d \geqslant 1$ that there are nonsingular hypersurfaces of degree d in $\mathbf{P}^n$. In fact, they form a dense open subset of the complete linear system $|dH|$. (This generalizes (I, Ex. 5.5).)

Example 8.20.3. Let Y be a nonsingular hypersurface of degree d in $\mathbf{P}^n$, $n \geqslant 2$. Then from (8.20) and the first example above, we conclude that $\omega_Y \cong \mathcal{O}_Y(d - n - 1)$. Let's look at some particular cases.

$n = 2, d = 1$. Y is a line in $\mathbf{P}^2$, so $Y \cong \mathbf{P}^1$, and we have $\omega_Y \cong \mathcal{O}_Y(-2)$ which we already knew.

$n = 2, d = 2$. Y is a conic in $\mathbf{P}^2$, and $\omega_Y \cong \mathcal{O}_Y(-1)$. In this case Y is the 2-uple embedding of $\mathbf{P}^1$, so pulling ω_Y back to $\mathbf{P}^1$ gives $\omega_{\mathbf{P}^1} \cong \mathcal{O}_{\mathbf{P}^1}(-2)$, which is again what we already knew.

$n = 2, d = 3$. Y is a nonsingular plane cubic curve, and $\omega_Y \cong \mathcal{O}_Y$. Therefore $p_g(Y) = \dim \Gamma(Y, \mathcal{O}_Y) = 1$, and we see that Y is not rational! This generalizes (I, Ex. 6.2), where we gave just one example of a nonsingular cubic curve, and showed by a different method that it was not rational.

$n = 2, d \geqslant 4$. Y is a nonsingular plane curve of degree d, $\omega_Y \cong \mathcal{O}_Y(d - 3)$, and $d - 3 > 0$. Hence $p_g > 0$, and Y is not rational. In fact, $p_g = \frac{1}{2}(d - 1)(d - 2)$ (Ex. 8.4f), so we see that plane curves of different degrees $d, d' \geqslant 3$ are not birational to each other. Another way of seeing this is as follows. For any nonsingular projective curve, we can consider the *degree* of the canonical sheaf. Since a nonsingular projective curve is unique in its birational equivalence class (I, §6), this number is in fact a birational invariant. In the present case its value is $d(d - 3)$, since $\mathcal{O}(1)$ has degree d on Y. These numbers are also distinct for different $d, d' \geqslant 3$. This shows the existence of infinitely many mutually nonbirational curves.

$n = 3, d = 1$. This gives $Y \cong \mathbf{P}^2$, $\omega_Y \cong \mathcal{O}_Y(-3)$ which we knew.

$n = 3, d = 2$. Here Y is a nonsingular quadric surface, and $\omega_Y \cong \mathcal{O}_Y(-2)$. We have $p_g(Y) = 0$, which is consistent with the fact that Y is rational (I, Ex. 4.5). In terms of the isomorphism $Y \cong \mathbf{P}^1 \times \mathbf{P}^1$, ω_Y corresponds to a divisor class of type $(-2, -2)$—see (6.6.1). This illustrates the general fact

(Ex. 8.3) that the canonical sheaf on a direct product of nonsingular varieties is the tensor product of the pull-backs of the canonical sheaves on the two factors.

$n = 3, d = 3$. Y is a nonsingular cubic surface in $\mathbf{P}^3$, $\omega_Y \cong \mathcal{O}_Y(-1)$ and so $p_g(Y) = 0$. In this case also, Y is a rational surface, as we will see later (Chapter V).

$n = 3, d = 4$. In this case $\omega_Y \cong \mathcal{O}_Y$. The canonical sheaf is trivial so $p_g = 1$. This is a nonrational surface which belongs to the class of "K3 surfaces."

$n = 3, d \geqslant 5$. Here $\omega_Y \cong \mathcal{O}_Y(d - 4)$ with $d - 4 > 0$. Hence $p_g > 0$, and Y is not rational. Surfaces such as these on which the canonical sheaf is very ample belong to the class of "surfaces of general type."

$n = 4, d = 3,4$. The cubic and the quartic threefold in $\mathbf{P}^4$ both have $p_g = 0$, but it has recently been shown (by different methods) that they are not in general rational varieties. For the cubic threefold, see Clemens and Griffiths [1]. For the quartic threefold, see Iskovskih and Manin [1].

$n\ arbitrary, d \geqslant n + 1$. In this case we obtain a nonsingular hypersurface Y in $\mathbf{P}^n$, with $\omega_Y \cong \mathcal{O}_Y(d - n - 1)$ and $d - n - 1 \geqslant 0$. Hence $p_g(Y) \geqslant 1$, and so Y is not rational. This shows the existence of nonrational varieties in all dimensions.

Some Local Algebra

Here we will gather some results from local algebra, mainly concerning depth and Cohen–Macaulay rings, which are useful in algebraic geometry. Then we relate them to the geometric notion of local complete intersection, and give an application to blowing up. We refer to Matsumura [2, Ch. 6] for proofs.

If A is a ring, and M is an A-module, recall that a sequence $x_1, \ldots, x_r$ of elements of A is called a *regular sequence* for M if x_1 is not a zero divisor in M, and for all $i = 2, \ldots, r$, x_i is not a zero divisor in $M/(x_1, \ldots, x_{i-1})M$. If A is a local ring with maximal ideal $\mathfrak{m}$, then the *depth* of M is the maximum length of a regular sequence $x_1, \ldots, x_r$ for M with all $x_i \in \mathfrak{m}$. These definitions apply to the ring A itself, and we say that a local noetherian ring A is *Cohen–Macaulay* if depth $A = \dim A$. Now we list some properties of Cohen–Macaulay rings.

Theorem 8.21A. *Let A be a local noetherian ring with maximal ideal $\mathfrak{m}$.*

(a) *If A is regular, then it is Cohen–Macaulay.*

(b) *If A is Cohen–Macaulay, then any localization of A at a prime ideal is also Cohen–Macaulay.*

(c) *If A is Cohen–Macaulay, then a set of elements $x_1, \ldots, x_r \in \mathfrak{m}$ forms a regular sequence for A if and only if $\dim A/(x_1, \ldots, x_r) = \dim A - r$.*

(d) *If A is Cohen–Macaulay, and $x_1, \ldots, x_r \in \mathfrak{m}$ is a regular sequence for A, then $A/(x_1, \ldots, x_r)$ is also Cohen–Macaulay.*

(e) *If A is Cohen–Macaulay, and $x_1,\ldots,x_r \in \mathfrak{m}$ is a regular sequence, let I be the ideal $(x_1,\ldots,x_r)$. Then the natural map $(A/I)[t_1,\ldots,t_r] \to \mathrm{gr}_I A = \bigoplus_{n\geq 0} I^n/I^{n+1}$, defined by sending $t_i \mapsto x_i$, is an isomorphism. In other words, I/I^2 is a free A/I-module of rank r, and for each $n \geq 1$, the natural map $S^n(I/I^2) \to I^n/I^{n+1}$ is an isomorphism, where S^n denotes the nth symmetric power.*

PROOFS. Matsumura [2: (a) p. 121; (b) p. 104; (c) p. 105; (d) p. 104; (e) p. 110].

In keeping with the terminology for schemes (Ex. 3.8), we will say that a noetherian ring A is *normal* if for every prime ideal $\mathfrak{p}$, the localization $A_\mathfrak{p}$ is an integrally closed domain. A normal ring is a finite direct product of integrally closed domains.

Theorem 8.22A (Serre). *A noetherian ring A is normal if and only if it satisfies the following two conditions:*

(1) *for every prime ideal $\mathfrak{p} \subseteq A$ of height ≤ 1, $A_\mathfrak{p}$ is regular (hence a field or a discrete valuation ring); and*

(2) *for every prime ideal $\mathfrak{p} \subseteq A$ of height ≥ 2, we have depth $A_\mathfrak{p} \geq 2$.*

PROOF. Matsumura [2, Th. 39, p. 125]. Condition (1) is sometimes called "R_1", or "regular in codimension 1". Condition (2), supplemented by the requirement that for ht $\mathfrak{p} = 1$, depth $A_\mathfrak{p} = 1$, which is a consequence of (1) in our case, is called the "condition S_2 of Serre".

Now we apply these results to algebraic geometry. We will say that a scheme is *Cohen–Macaulay* if all of its local rings are Cohen–Macaulay.

Definition. Let Y be a closed subscheme of a nonsingular variety X over k. We say that Y is a *local complete intersection* in X if the ideal sheaf $\mathscr{I}_Y$ of Y in X can be locally generated by $r = \mathrm{codim}(Y,X)$ elements at every point.

Example 8.22.1. If Y itself is nonsingular, then by (8.17) it is a local complete intersection inside any nonsingular X which contains it.

Remark 8.22.2. In fact, the notion of being a local complete intersection is an intrinsic property of the scheme Y, i.e., independent of the nonsingular variety containing it. This is proved using the cotangent complex of a morphism, which extends the concept of relative differentials introduced above—see Lichtenbaum and Schlessinger [1]. We will not use this fact in the sequel.

Proposition 8.23. *Let Y be a locally complete intersection subscheme of a nonsingular variety X over k. Then:*
  (a) *Y is Cohen–Macaulay;*
  (b) *Y is normal if and only if it is regular in codimension 1.*

PROOF.
  (a) Since X is nonsingular, it is Cohen–Macaulay by (8.21Aa). Since $\mathscr{I}_Y$ is locally generated by $r = \mathrm{codim}(Y,X)$ elements, those elements locally form a regular sequence in $\mathcal{O}_X$, by (8.21Ac), and so Y is Cohen–Macaulay by (8.21Ad).
  (b) We already know that normal implies regular in codimension 1 (I, 6.2A). For the converse, we use (8.22A) applied to the local rings of Y. Condition (1) is our hypothesis, and condition (2) holds automatically because Y is Cohen–Macaulay.

As our last application, we consider the blowing-up of a nonsingular variety along a nonsingular subvariety (cf. §7 for definition of blowing-up). The following theorem will be useful in comparing invariants of X and $\tilde{X}$ (Ex. 8.5).

Theorem 8.24. *Let X be a nonsingular variety over k, and let $Y \subseteq X$ be a nonsingular closed subvariety, with ideal sheaf $\mathscr{I}$. Let $\pi \colon \tilde{X} \to X$ be the blowing-up of $\mathscr{I}$, and let $Y' \subseteq \tilde{X}$ be the subscheme defined by the inverse image ideal sheaf $\mathscr{I}' = \pi^{-1}\mathscr{I} \cdot \mathcal{O}_{\tilde{X}}$. Then:*
  (a) *$\tilde{X}$ is also nonsingular;*
  (b) *Y', together with the induced projection map $\pi \colon Y' \to Y$, is isomorphic to $\mathbf{P}(\mathscr{I}/\mathscr{I}^2)$, the projective space bundle associated to the (locally free) sheaf $\mathscr{I}/\mathscr{I}^2$ on Y;*
  (c) *under this isomorphism, the normal sheaf $\mathscr{N}_{Y'/\tilde{X}}$ corresponds to $\mathcal{O}_{\mathbf{P}(\mathscr{I}/\mathscr{I}^2)}(-1)$.*

PROOF. We prove (b) first. Since $\tilde{X} = \mathbf{Proj} \bigoplus \mathscr{I}^d$, we have

$$Y' \cong \mathbf{Proj} \bigoplus (\mathscr{I}^d \otimes \mathcal{O}_X/\mathscr{I}) = \mathbf{Proj} \bigoplus \mathscr{I}^d/\mathscr{I}^{d+1}.$$

But Y is nonsingular, so $\mathscr{I}$ is locally generated by a regular sequence in $\mathcal{O}_X$, and we can apply (8.21Ae). This implies that $\mathscr{I}/\mathscr{I}^2$ is locally free and that for each $n \geqslant 1$, $\mathscr{I}^n/\mathscr{I}^{n+1} \cong S^n(\mathscr{I}/\mathscr{I}^2)$. Thus $Y' \cong \mathbf{Proj} \bigoplus S^d(\mathscr{I}/\mathscr{I}^2)$, which by definition is $\mathbf{P}(\mathscr{I}/\mathscr{I}^2)$.
  In particular, Y' is locally isomorphic to $Y \times \mathbf{P}^{r-1}$, where $r = \mathrm{codim}(Y,X)$, so Y' is also nonsingular. Since Y' is locally principal in $\tilde{X}$ (7.13a), it follows that $\tilde{X}$ is also nonsingular: if a quotient of a noetherian local ring by an element which is not a zero divisor is regular, then the local ring itself is regular.
  To prove (c), we recall from the proof of (7.13) that $\mathscr{I}' = \pi^{-1}(\mathscr{I}) \cdot \mathcal{O}_{\tilde{X}}$ is isomorphic to $\mathcal{O}_{\tilde{X}}(1)$. It follows that $\mathscr{I}'/\mathscr{I}'^2 \cong \mathcal{O}_{Y'}(1)$, and hence $\mathscr{N}_{Y'/\tilde{X}} \cong \mathcal{O}_{Y'}(-1)$.

We will use the following algebraic result in the exercises.

Theorem 8.25A (I. S. Cohen). *Let A be a complete local ring containing a field k. Assume that the residue field $k(A) = A/\mathfrak{m}$ is a separably generated extension of k. Then there is a subfield $K \subseteq A$, containing k, such that $K \to A/\mathfrak{m}$ is an isomorphism. (The subfield K is called a field of representatives for A.)*

PROOF. Matsumura [2, p. 205].

EXERCISES

8.1 Here we will strengthen the results of the text to include information about the sheaf of differentials at a not necessarily closed point of a scheme X.

(a) Generalize (8.7) as follows. Let B be a local ring containing a field k, and assume that the residue field $k(B) = B/\mathfrak{m}$ of B is a separably generated extension of k. Then the exact sequence of (8.4A),

$$0 \to \mathfrak{m}/\mathfrak{m}^2 \xrightarrow{\delta} \Omega_{B/k} \otimes k(B) \to \Omega_{k(B)/k} \to 0$$

is exact on the left also. [*Hint*: In copying the proof of (8.7), first pass to $B/\mathfrak{m}^2$, which is a complete local ring, and then use (8.25A) to choose a field of representatives for $B/\mathfrak{m}^2$.]

(b) Generalize (8.8) as follows. With B, k as above, assume furthermore that k is perfect, and that B is a localization of an algebra of finite type over k. Then show that B is a regular local ring if and only if $\Omega_{B/k}$ is free of rank $= \dim B + $ tr.d. $k(B)/k$.

(c) Strengthen (8.15) as follows. Let X be an irreducible scheme of finite type over a perfect field k, and let $\dim X = n$. For any point $x \in X$, not necessarily closed, show that the local ring $\mathcal{O}_{x,X}$ is a regular local ring if and only if the stalk $(\Omega_{X/k})_x$ of the sheaf of differentials at x is free of rank n.

(d) Strengthen (8.16) as follows. If X is a variety over an algebraically closed field k, then $U = \{x \in X | \mathcal{O}_x \text{ is a regular local ring}\}$ is an open dense subset of X.

8.2. Let X be a variety of dimension n over k. Let $\mathscr{E}$ be a locally free sheaf of rank $> n$ on X, and let $V \subseteq \Gamma(X,\mathscr{E})$ be a vector space of global sections which generate $\mathscr{E}$. Then show that there is an element $s \in V$, such that for each $x \in X$, we have $s_x \notin \mathfrak{m}_x\mathscr{E}_x$. Conclude that there is a morphism $\mathcal{O}_X \to \mathscr{E}$ giving rise to an exact sequence

$$0 \to \mathcal{O}_X \to \mathscr{E} \to \mathscr{E}' \to 0$$

where $\mathscr{E}'$ is also locally free. [*Hint*: Use a method similar to the proof of Bertini's theorem (8.18).]

8.3. *Product Schemes.*

(a) Let X and Y be schemes over another scheme S. Use (8.10) and (8.11) to show that $\Omega_{X\times_S Y/S} \cong p_1^*\Omega_{X/S} \oplus p_2^*\Omega_{Y/S}$.

(b) If X and Y are nonsingular varieties over a field k, show that $\omega_{X\times Y} \cong p_1^*\omega_X \otimes p_2^*\omega_Y$.

(c) Let Y be a nonsingular plane cubic curve, and let X be the surface $Y \times Y$. Show that $p_g(X) = 1$ but $p_a(X) = -1$ (I, Ex. 7.2). This shows that the arithmetic genus and the geometric genus of a nonsingular projective variety may be different.

8.4. *Complete Intersections in* $\mathbf{P}^n$. A closed subscheme Y of $\mathbf{P}^n_k$ is called a (*strict, global*) *complete intersection* if the homogeneous ideal I of Y in $S = k[x_0, \ldots, x_n]$ can be generated by $r = \text{codim}(Y, \mathbf{P}^n)$ elements (I, Ex. 2.17).

(a) Let Y be a closed subscheme of codimension r in $\mathbf{P}^n$. Then Y is a complete intersection if and only if there are hypersurfaces (i.e., locally principal subschemes of codimension 1) $H_1, \ldots, H_r$, such that $Y = H_1 \cap \ldots \cap H_r$ *as schemes*, i.e., $\mathscr{I}_Y = \mathscr{I}_{H_1} + \ldots + \mathscr{I}_{H_r}$. [*Hint*: Use the fact that the unmixedness theorem holds in S (Matsumura [2, p. 107]).]

(b) If Y is a complete intersection of dimension $\geqslant 1$ in $\mathbf{P}^n$, and if Y is normal, then Y is projectively normal (Ex. 5.14). [*Hint*: Apply (8.23) to the affine cone over Y.]

(c) With the same hypotheses as (b), conclude that for all $l \geqslant 0$, the natural map $\Gamma(\mathbf{P}^n, \mathcal{O}_{\mathbf{P}^n}(l)) \to \Gamma(Y, \mathcal{O}_Y(l))$ is surjective. In particular, taking $l = 0$, show that Y is connected.

(d) Now suppose given integers $d_1, \ldots, d_r \geqslant 1$, with $r < n$. Use Bertini's theorem (8.18) to show that there exist nonsingular hypersurfaces $H_1, \ldots, H_r$ in $\mathbf{P}^n$, with $\deg H_i = d_i$, such that the scheme $Y = H_1 \cap \ldots \cap H_r$ is irreducible and nonsingular of codimension r in $\mathbf{P}^n$.

(e) If Y is a nonsingular complete intersection as in (d), show that $\omega_Y \cong \mathcal{O}_Y(\sum d_i - n - 1)$.

(f) If Y is a nonsingular hypersurface of degree d in $\mathbf{P}^n$, use (c) and (e) above to show that $p_g(Y) = \binom{d-1}{n}$. Thus $p_g(Y) = p_a(Y)$ (I, Ex. 7.2). In particular, if Y is a nonsingular plane curve of degree d, then $p_g(Y) = \frac{1}{2}(d-1)(d-2)$.

(g) If Y is a nonsingular curve in $\mathbf{P}^3$, which is a complete intersection of nonsingular surfaces of degrees d, e, then $p_g(Y) = \frac{1}{2}de(d + e - 4) + 1$. Again the geometric genus is the same as the arithmetic genus (I, Ex. 7.2).

8.5. *Blowing up a Nonsingular Subvariety*. As in (8.24), let X be a nonsingular variety, let Y be a nonsingular subvariety of codimension $r \geqslant 2$, let $\pi : \tilde{X} \to X$ be the blowing-up of X along Y, and let $Y' = \pi^{-1}(Y)$.

(a) Show that the maps $\pi^* : \text{Pic } X \to \text{Pic } \tilde{X}$, and $\mathbf{Z} \to \text{Pic } \tilde{X}$ defined by $n \mapsto$ class of nY', give rise to an isomorphism $\text{Pic } \tilde{X} \cong \text{Pic } X \oplus \mathbf{Z}$.

(b) Show that $\omega_{\tilde{X}} \cong f^* \omega_X \otimes \mathscr{L}((r-1)Y')$. [*Hint*: By (a) we can write in any case $\omega_{\tilde{X}} \cong f^* \mathscr{M} \otimes \mathscr{L}(qY')$ for some invertible sheaf $\mathscr{M}$ on X, and some integer q. By restricting to $\tilde{X} - Y' \cong X - Y$, show that $\mathscr{M} \cong \omega_X$. To determine q, proceed as follows. First show that $\omega_{Y'} \cong f^* \omega_X \otimes \mathcal{O}_{Y'}(-q-1)$. Then take a closed point $y \in Y$ and let Z be the fibre of Y' over y. Then show that $\omega_Z \cong \mathcal{O}_Z(-q-1)$. But since $Z \cong \mathbf{P}^{r-1}$, we have $\omega_Z \cong \mathcal{O}_Z(-r)$, so $q = r - 1$.]

8.6. *The Infinitesimal Lifting Property*. The following result is very important in studying deformations of nonsingular varieties. Let k be an algebraically closed field, let A be a finitely generated k-algebra such that Spec A is a nonsingular variety over k. Let $0 \to I \to B' \to B \to 0$ be an exact sequence, where B' is a k-algebra, and I is an ideal with $I^2 = 0$. Finally suppose given a k-algebra homomorphism $f : A \to B$. Then there exists a k-algebra homomorphism $g : A \to B'$ making a commutative diagram

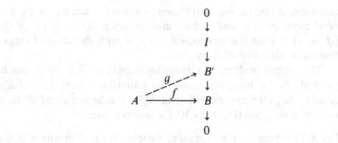

We call this result the *infinitesimal lifting property* for A. We prove this result in several steps.

(a) First suppose that $g: A \to B'$ is a given homomorphism lifting f. If $g': A \to B'$ is another such homomorphism, show that $\theta = g - g'$ is a k-derivation of A into I, which we can consider as an element of $\operatorname{Hom}_A(\Omega_{A/k}, I)$. Note that since $I^2 = 0$, I has a natural structure of B-module and hence also of A-module. Conversely, for any $\theta \in \operatorname{Hom}_A(\Omega_{A/k}, I)$, $g' = g + \theta$ is another homomorphism lifting f. (For this step, you do not need the hypothesis about Spec A being nonsingular.)

(b) Now let $P = k[x_1, \ldots, x_n]$ be a polynomial ring over k of which A is a quotient, and let J be the kernel. Show that there does exist a homomorphism $h: P \to B'$ making a commutative diagram,

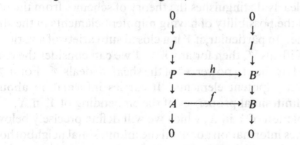

and show that h induces an A-linear map $\bar{h}: J/J^2 \to I$.

(c) Now use the hypothesis Spec A nonsingular and (8.17) to obtain an exact sequence

$$0 \to J/J^2 \to \Omega_{P/k} \otimes A \to \Omega_{A/k} \to 0.$$

Show furthermore that applying the functor $\operatorname{Hom}_A(\cdot, I)$ gives an exact sequence

$$0 \to \operatorname{Hom}_A(\Omega_{A/k}, I) \to \operatorname{Hom}_P(\Omega_{P/k}, I) \to \operatorname{Hom}_A(J/J^2, I) \to 0.$$

Let $\theta \in \operatorname{Hom}_P(\Omega_{P/k}, I)$ be an element whose image gives $\bar{h} \in \operatorname{Hom}_A(J/J^2, I)$. Consider θ as a derivation of P to B'. Then let $h' = h - \theta$, and show that h' is a homomorphism of $P \to B'$ such that $h'(J) = 0$. Thus h' induces the desired homomorphism $g: A \to B'$.

8.7. As an application of the infinitesimal lifting property, we consider the following general problem. Let X be a scheme of finite type over k, and let $\mathscr{F}$ be a coherent sheaf on X. We seek to classify schemes X' over k, which have a sheaf of ideals $\mathscr{I}$ such that $\mathscr{I}^2 = 0$ and $(X', \mathcal{O}_{X'}/\mathscr{I}) \cong (X, \mathcal{O}_X)$, and such that $\mathscr{I}$ with its resulting structure of $\mathcal{O}_X$-module is isomorphic to the given sheaf $\mathscr{F}$. Such a pair $X', \mathscr{I}$ we call an *infinitesimal extension* of the scheme X by the sheaf $\mathscr{F}$. One such extension, the *trivial* one, is obtained as follows. Take $\mathcal{O}_{X'} = \mathcal{O}_X \oplus \mathcal{F}$ as sheaves of abelian groups, and define multiplication by $(a \oplus f) \cdot (a' \oplus f') = aa' \oplus (af' + a'f)$. Then the topological space X with the sheaf of rings $\mathcal{O}_{X'}$ is an infinitesimal extension of X by $\mathcal{F}$.

The general problem of classifying extensions of X by $\mathcal{F}$ can be quite complicated. So for now, just prove the following special case: if X is affine and nonsingular, then any extension of X by a coherent sheaf $\mathcal{F}$ is isomorphic to the trivial one. See (III, Ex. 4.10) for another case.

8.8. Let X be a projective nonsingular variety over k. For any $n > 0$ we define the nth *plurigenus* of X to be $P_n = \dim_k \Gamma(X, \omega_X^{\otimes n})$. Thus in particular $P_1 = p_g$. Also, for any q, $0 \leqslant q \leqslant \dim X$ we define an integer $h^{q,0} = \dim_k \Gamma(X, \Omega_{X/k}^q)$ where $\Omega_{X/k}^q = \bigwedge^q \Omega_{X/k}$ is the sheaf of regular q-forms on X. In particular, for $q = \dim X$, we recover the geometric genus again. The integers $h^{q,0}$ are called *Hodge numbers.*

Using the method of (8.19), show that P_n and $h^{q,0}$ are *birational* invariants of X, i.e., if X and X' are birationally equivalent nonsingular projective varieties, then $P_n(X) = P_n(X')$ and $h^{q,0}(X) = h^{q,0}(X')$.

9 Formal Schemes

One feature which clearly distinguishes the theory of schemes from the older theory of varieties is the possibility of having nilpotent elements in the structure sheaf of a scheme. In particular, if Y is a closed subvariety of a variety X, defined by a sheaf of ideals $\mathcal{I}$, then for any $n \geqslant 1$ we can consider the closed subscheme Y_n defined by the nth power $\mathcal{I}^n$ of the sheaf of ideals $\mathcal{I}$. For $n \geqslant 2$, this is a scheme with nilpotent elements. It carries information about Y together with the infinitesimal properties of the embedding of Y in X.

The formal completion of Y in X, which we will define precisely below, is an object which carries information about all the infinitesimal neighborhoods Y_n of Y at once. Thus it is thicker than any Y_n, but it is contained inside any actual open neighborhood of Y in X. We might call it the formal neighborhood of Y in X.

The idea of considering these formal completions is already implicit in the memoir of Zariski [3], where he uses the "holomorphic functions along a subvariety" for his proof of the connectedness principle. We will give different proofs of some of Zariski's results, using cohomology, in (III, §11). A striking application of formal schemes as something in between a subvariety and an ambient variety is in Grothendieck's proof of the Lefschetz theorems on Pic and π_1 [SGA 2]. This material is also explained in Hartshorne [5, Ch. IV].

We will define an arbitrary formal scheme as something which looks locally like the completion of a usual scheme along a closed subscheme.

Inverse Limits of Abelian Groups

First we recall the notion of inverse limit. An *inverse system* of abelian groups is a collection of abelian groups A_n, for each $n \geqslant 1$, together with homomorphisms $\varphi_{n'n}: A_{n'} \to A_n$ for each $n' \geqslant n$, such that for each $n'' \geqslant n' \geqslant n$ we have $\varphi_{n''n} = \varphi_{n'n} \circ \varphi_{n''n'}$. We will denote the inverse system by $(A_n, \varphi_{n'n})$, or simply (A_n), with the φ being understood. If (A_n) is an inverse system of abelian groups, we define the *inverse limit* $A = \varprojlim A_n$ to be the set of sequences $\{a_n\} \in \prod A_n$ such that $\varphi_{n'n}(a_{n'}) = a_n$ for all $n' \geqslant n$. Clearly A is a group. The inverse limit A can be characterized by the following universal property: given a group B, and homomorphisms $\psi_n: B \to A_n$ for each n, such that for any $n' \geqslant n$, $\psi_n = \varphi_{n'n} \circ \psi_{n'}$, then there exists a unique homomorphism $\psi: B \to A$ such that $\psi_n = p_n \circ \psi$ for each n, where $p_n: A \to A_n$ is the restriction of the nth projection map $\prod A_n \to A_n$.

If the groups A_n have the additional structure of vector spaces over a field k, or modules over a ring R, then the above discussion makes sense in the category of k-vector spaces or R-modules.

Next we study exactness properties of the inverse limit (cf. Atiyah–Macdonald [1, Ch.10]). A *homomorphism* $(A_n) \to (B_n)$ of inverse systems of abelian groups is a collection of homomorphisms $f_n: A_n \to B_n$ for each n, which are compatible with the maps of the inverse system, i.e., for each $n' \geqslant n$, we have a commutative diagram

$$
\begin{array}{ccc}
A_{n'} & \xrightarrow{\;f_{n'}\;} & B_{n'} \\
\downarrow{\scriptstyle \varphi_{n'n}} & & \downarrow{\scriptstyle \psi_{n'n}} \\
A_n & \xrightarrow{\;f_n\;} & B_n.
\end{array}
$$

A sequence

$$0 \to (A_n) \to (B_n) \to (C_n) \to 0$$

of homomorphisms of inverse systems is *exact* if the corresponding sequence of groups is exact for each n. Given such a short exact sequence of inverse systems, one sees easily that the sequence of inverse limits

$$0 \to \varprojlim A_n \to \varprojlim B_n \to \varprojlim C_n$$

is also exact. However, the last map need not be surjective. So we say that $\varprojlim$ is a left exact functor.

To give a criterion for exactness of $\varprojlim$ on the right, we make the following definition: an inverse system $(A_n, \varphi_{n'n})$ satisfies the *Mittag–Leffler condition* (ML) if for each n, the decreasing family $\{\varphi_{n'n}(A_{n'}) \subseteq A_n | n' \geqslant n\}$ of subgroups of A_n is stationary. In other words, for each n, there is an $n_0 \geqslant n$, such that for all n', $n'' \geqslant n_0$, $\varphi_{n'n}(A_{n'}) = \varphi_{n''n}(A_{n''})$ as subgroups of A_n.

Suppose an inverse system (A_n) satisfies (ML). Then for each n, we let $A_n' \subseteq A_n$ be the *stable image* $\varphi_{n'n}(A_{n'})$ for any $n' \geqslant n_0$, which exists by the definition. Then one sees easily that (A_n') is also an inverse system, with the induced maps, and that the maps of the new system (A_n') are all surjective. Furthermore, it is clear that $\varprojlim A_n' = \varprojlim A_n$. So we see that $A = \varprojlim A_n$ maps surjectively to each A_n'.

Proposition 9.1. *Let*

$$0 \to (A_n) \xrightarrow{f} (B_n) \xrightarrow{g} (C_n) \to 0$$

be a short exact sequence of inverse systems of abelian groups. Then:
  (a) *if* (B_n) *satisfies* (ML), *so does* (C_n).
  (b) *if* (A_n) *satisfies* (ML), *then the sequence of inverse limits*

$$0 \to \varprojlim A_n \to \varprojlim B_n \to \varprojlim C_n \to 0$$

is exact.

PROOF. (See also Grothendieck [EGA 0_{III}, 13.2].)
  (a) For each $n' \geqslant n$, the image of $B_{n'}$ in B_n maps surjectively to the image of $C_{n'}$ in C_n, so (ML) for (B_n) implies (ML) for (C_n) immediately.
  (b) The only nonobvious part is to show that the last map is surjective. So let $\{c_n\} \in \varprojlim C_n$. For each n, let $E_n = g^{-1}(c_n)$. Then E_n is a subset of B_n, and (E_n) is an inverse system of sets. Furthermore, each E_n is bijective, in a noncanonical way, with A_n, because of the exactness of the sequence $0 \to A_n \to B_n \to C_n \to 0$. Thus since (A_n) satisfies (ML), one sees easily that (E_n) satisfies the Mittag–Leffler condition as an inverse system of sets (same definition). Since each E_n is nonempty, it follows from considering the inverse system of stable images as above, that $\varprojlim E_n$ is also nonempty. Taking any element of this set gives an element of $\varprojlim B_n$ which maps to $\{c_n\}$.

Example 9.1.1. If all the maps $\varphi_{n'n} \colon A_{n'} \to A_n$ are surjective, then (A_n) satisfies (ML), so (9.1b) applies.

Example 9.1.2. If (A_n) is an inverse system of finite-dimensional vector spaces over a field, or more generally, an inverse system of modules with descending chain condition over a ring, then (A_n) satisfies (ML).

Inverse Limits of Sheaves

In any category $\mathfrak{C}$, we define the notion of inverse limit by analogy with the universal property of the inverse limit of abelian groups above. Thus if $(A_n, \varphi_{n'n})$ is an inverse system of objects of $\mathfrak{C}$ (same definition as above), then an *inverse limit* $A = \varprojlim A_n$ is an object A of $\mathfrak{C}$, together with morphisms $p_n \colon A \to A_n$ for each n, such that for each $n' \geqslant n$, $p_n = \varphi_{n'n} \circ p_{n'}$, satisfying the following universal property: given any object B of $\mathfrak{C}$, together with morphisms $\psi_n \colon B \to A_n$ for each n, such that for each $n' \geqslant n$, $\psi_n = \varphi_{n'n} \circ \psi_{n'}$, there exists a unique morphism $\psi \colon B \to A$ such that for each n, $\psi_n = p_n \circ \psi$. Clearly the inverse limit is unique if it exists. But the question of existence depends on the particular category considered.

Proposition 9.2. *Let X be a topological space, and let $\mathfrak{C}$ be the category of sheaves of abelian groups on X. Then inverse limits exist in $\mathfrak{C}$. Furthermore, if $(\mathscr{F}_n)$ is an inverse system of sheaves on X, and $\mathscr{F} = \varprojlim \mathscr{F}_n$ is its inverse limit, then for any open set U, we have $\Gamma(U, \mathscr{F}) = \varprojlim \Gamma(U, \mathscr{F}_n)$ in the category of abelian groups.*

PROOF. Given an inverse system of sheaves $(\mathscr{F}_n)$ on X, we consider the presheaf $U \to \varprojlim \Gamma(U, \mathscr{F}_n)$, where this inverse limit is taken in the category of abelian groups. Now using the sheaf property for each $\mathscr{F}_n$, one verifies immediately that this presheaf is a sheaf. Call it $\mathscr{F}$. Now given any other sheaf $\mathscr{G}$, and a system of compatible maps $\psi_n : \mathscr{G} \to \mathscr{F}_n$ for each n, it follows from the universal property of an inverse limit of abelian groups that we obtain unique maps, for each U, $\Gamma(U, \mathscr{G}) \to \Gamma(U, \mathscr{F})$. These give a sheaf map $\mathscr{G} \to \mathscr{F}$, thus verifying that $\mathscr{F}$ is the inverse limit of the $\mathscr{F}_n$ in $\mathfrak{C}$.

Caution 9.2.1. Even though inverse limits exist in the category $\mathfrak{C}$ of abelian sheaves on a topological space, one must beware of using intuition derived from the category of abelian groups. In particular, the statement of (9.1b) is *false* in $\mathfrak{C}$, even if all maps in the inverse system (A_n) are surjective. So in studying exactness questions, we will always pass to sections over an open set, and thus reduce to questions about abelian groups. For more details about exactness of $\varprojlim$ in $\mathfrak{C}$, see Hartshorne [7, I, §4].

Completion of a Ring

One important application of inverse limits is to define the completion of a ring with respect to an ideal. This generalizes the notion of completion of a local ring which was discussed in (I, §5). It also forms the algebraic model for the completion of a scheme along a closed subscheme which will come next.

So let A be a commutative ring with identity (as always), and let I be an ideal of A. We denote by I^n the nth power of the ideal I. Then we have natural homomorphisms

$$\ldots \to A/I^3 \to A/I^2 \to A/I,$$

which make (A/I^n) into an inverse system of rings. The inverse limit ring $\varprojlim A/I^n$ is denoted by $\hat{A}$ and is called the *completion of A with respect to I* or the *I-adic completion of A*. For each n we have a natural map $A \to A/I^n$, so by the universal property we obtain a homomorphism $A \to \hat{A}$.

Similarly, if M is any A-module, we define $\hat{M} = \varprojlim M/I^n M$, and call it the *$I$-adic completion of M*. It has a natural structure of $\hat{A}$-module.

Theorem 9.3A. *Let A be a noetherian ring, and I an ideal of A. We denote by $\hat{}$ the I-adic completion as above. Then:*

(a) $\hat{I} = \varprojlim I/I^n$ *is an ideal of $\hat{A}$. For any n, $\hat{I}^n = I^n \hat{A}$, and $\hat{A}/\hat{I}^n \cong A/I^n$;*

(b) *if M is a finitely generated A-module, then $\hat{M} \cong M \otimes_A \hat{A}$;*

(c) *the functor $M \mapsto \hat{M}$ is an exact functor on the category of finitely generated A-modules;*

(d) $\hat{A}$ *is a noetherian ring;*

(e) *if (M_n) is an inverse system, where each M_n is a finitely generated A/I^n-module, each $\varphi_{n'n} : M_{n'} \mapsto M_n$ is surjective, and $\ker \varphi_{n'n} = I^n M_{n'}$, then $M = \varprojlim M_n$ is a finitely generated $\hat{A}$-module, and for each n, $M_n \cong M/I^n M$.*

193

PROOFS.
- (a) Atiyah–Macdonald [1, p. 109].
- (b) [Ibid., p. 108].
- (c) [Ibid., p. 108].
- (d) [Ibid., p. 113].
- (e) Bourbaki [1, Ch. III, §2, no. 11, Prop. & Cor. 14].

Formal Schemes

We begin by defining the completion of a scheme along a closed subscheme. For technical reasons we will limit our discussion to noetherian schemes.

Definition. Let X be a noetherian scheme, and let Y be a closed subscheme, defined by a sheaf of ideals $\mathscr{I}$. Then we define the *formal completion of X along Y*, denoted $(\hat{X}, \mathcal{O}_{\hat{X}})$, to be the following ringed space. We take the topological space Y, and on it the sheaf of rings $\mathcal{O}_{\hat{X}} = \varprojlim \mathcal{O}_X/\mathscr{I}^n$. Here we consider each $\mathcal{O}_X/\mathscr{I}^n$ as a sheaf of rings on Y, and make them into an inverse system in the natural way.

Remark 9.3.1. The structure sheaf $\mathcal{O}_{\hat{X}}$ of $\hat{X}$ actually depends only on the closed subset Y, and not on the particular scheme structure on Y. For if $\mathscr{J}$ is another sheaf of ideals defining a closed subscheme structure on Y, then since X is a noetherian scheme, there are integers m, n such that $\mathscr{I} \supseteq \mathscr{J}^m$ and $\mathscr{J} \supseteq \mathscr{I}^n$. Thus the inverse systems $(\mathcal{O}_X/\mathscr{J}^n)$ and $(\mathcal{O}_X/\mathscr{J}^m)$ are cofinal with each other, and hence have the same inverse limit.

One sees easily that the stalks of the sheaf $\mathcal{O}_{\hat{X}}$ are local rings, so in fact $(\hat{X}, \mathcal{O}_{\hat{X}})$ is a locally ringed space. If $U = \operatorname{Spec} A$ is an open affine subset of X, and if $I \subseteq A$ is the ideal $\Gamma(U, \mathscr{I})$, then from (9.2) we see that $\Gamma(\hat{X} \cap U, \mathcal{O}_{\hat{X}}) = \hat{A}$, the I-adic completion of A. Thus the process of completing X along Y is analogous to the I-adic completion of a ring discussed above. However, one should note that the local rings of $\hat{X}$ are in general *not* complete, and their dimension ($= \dim X$) is *not* equal to the dimension of the underlying topological space Y.

Definition. With $X, Y, \mathscr{I}$ as in the previous definition, let $\mathscr{F}$ be a coherent sheaf on X. We define the *completion of $\mathscr{F}$ along Y*, denoted $\hat{\mathscr{F}}$, to be the sheaf $\varprojlim \mathscr{F}/\mathscr{I}^n\mathscr{F}$ on Y. It has a natural structure of $\mathcal{O}_{\hat{X}}$-module.

Definition. A *noetherian formal scheme* is a locally ringed space $(\mathfrak{X}, \mathcal{O}_{\mathfrak{X}})$ which has a finite open cover $\{\mathfrak{U}_i\}$ such that for each i, the pair $(\mathfrak{U}_i, \mathcal{O}_{\mathfrak{X}}|_{\mathfrak{U}_i})$ is isomorphic, as a locally ringed space, to the completion of some noetherian scheme X_i along a closed subscheme Y_i. A *morphism* of noetherian formal schemes is a morphism as locally ringed spaces. A sheaf $\mathfrak{F}$ of $\mathcal{O}_{\mathfrak{X}}$-modules is said to be *coherent* if there is a finite open cover $\mathfrak{U}_i$ as above, with $\mathfrak{U}_i \cong \hat{X}_i$, and for each i there is a coherent sheaf $\mathscr{F}_i$ on X_i such that $\mathfrak{F}|_{\mathfrak{U}_i} \cong \hat{\mathscr{F}}_i$ as $\mathcal{O}_{\hat{X}_i}$-modules via the given isomorphism $\mathfrak{U}_i \cong \hat{X}_i$.

Examples 9.3.2. If X is any noetherian scheme, and Y a closed subscheme, then its completion $\hat{X}$ is a formal scheme. Such a formal scheme, which can be obtained by completing a *single* noetherian scheme along a closed subscheme, is called *algebraizable*. It is not so easy to give examples, but there are nonalgebraizable noetherian formal schemes—see Hironaka and Matsumura [1, §5] or Hartshorne [5, V, 3.3, p. 205].

Example 9.3.3. If X is a noetherian scheme, and we take $Y = X$, then $\hat{X} = X$. Thus the category of noetherian formal schemes includes all noetherian schemes.

Example 9.3.4. If X is a noetherian scheme, and Y is a closed point P, then $\hat{X}$ is a one point space $\{P\}$ with the completion $\hat{\mathcal{O}}_P$ of the local ring at P as its structure sheaf. An $\hat{\mathcal{O}}_P$-module M, considered as a sheaf on $\hat{X}$, is coherent if and only if M is a finitely generated module. Indeed, clearly coherent implies finitely generated. But the converse is also true since we can obtain $\hat{X}$ by completing the scheme $\operatorname{Spec} \hat{\mathcal{O}}_P$ at its closed point, and any finitely generated $\hat{\mathcal{O}}_P$-module M corresponds to a coherent sheaf on $\operatorname{Spec} \hat{\mathcal{O}}_P$.

Next we will study the structure of coherent sheaves on a formal scheme. As in the study of coherent sheaves on usual schemes in §5, we first analyze what happens in the affine case.

Definition. An *affine (noetherian) formal scheme* is a formal scheme obtained by completing a single affine noetherian scheme along a closed subscheme. If $X = \operatorname{Spec} A$, $Y = V(I)$, and $\mathfrak{X} = \hat{X}$, then for any finitely generated A-module M, we define the sheaf M^Δ on $\mathfrak{X}$ to be the completion of the coherent sheaf $\tilde{M}$ on X. Thus by definition, M^Δ is a coherent sheaf on $\mathfrak{X}$.

Proposition 9.4. *Let A be a noetherian ring, I an ideal of A, let $X = \operatorname{Spec} A$, $Y = V(I)$, and let $\mathfrak{X} = \hat{X}$. Then:*

(a) $\mathfrak{I} = I^\Delta$ *is a sheaf of ideals in $\mathcal{O}_{\mathfrak{X}}$, and for any n, $\mathcal{O}_{\mathfrak{X}}/\mathfrak{I}^n \cong (A/I^n)^\sim$ as sheaves on Y;*

(b) *if M is a finitely generated A-module, then $M^\Delta = \tilde{M} \otimes_{\mathcal{O}_X} \mathcal{O}_{\mathfrak{X}}$.*

(c) *The functor $M \mapsto M^\Delta$ is an exact functor from the category of finitely generated A-modules to the category of coherent $\mathcal{O}_{\mathfrak{X}}$-modules.*

PROOF. In each case we have a statement about sheaves on $\mathfrak{X}$. Since the open affine subsets of X form a base for the topology of X, and their intersections with Y a base for the topology of Y, it will be sufficient to establish the corresponding property of the sections over any such open set. So let $U = \operatorname{Spec} B$ be an open affine subset of X, let $J = \Gamma(U, \tilde{I})$, and for any finitely generated A-module M, let $N = \Gamma(U, \tilde{M})$. Then B is a noetherian ring (3.2), N is a finitely generated B-module (5.4), and the functor $M \mapsto N$ is an exact functor from A-modules to B-modules (5.5).

We prove (c) first. So let M be a finitely generated A-module. Then $M^\Delta = \varprojlim \tilde{M}/\tilde{I}^n\tilde{M}$ by definition, so by (9.2), $\Gamma(U,M^\Delta) = \varprojlim \Gamma(U,\tilde{M}/\tilde{I}^n\tilde{M})$. But this is equal to $\varprojlim N/J^nN = \hat{N}$, where $\hat{\ }$ now denotes the J-adic completion of a B-module. Now $M \mapsto N$ is exact as we saw above, and $N \mapsto \hat{N}$ is exact by (9.3A). Thus $M \mapsto \Gamma(U,M^\Delta)$ is exact for each U, and so $M \mapsto M^\Delta$ is exact.

(a) For any U as above, $\Gamma(U,I^\Delta) = \varprojlim \Gamma(U,\tilde{I}/\tilde{I}^n) = \hat{J}$. Furthermore $\Gamma(U,\mathcal{O}_{\mathfrak{x}}) = \hat{B}$ similarly. But by (9.3A), $\hat{J}$ is an ideal of $\hat{B}$, so this shows that $\mathfrak{J} = I^\Delta$ is a sheaf of ideals in $\mathcal{O}_{\mathfrak{x}}$.

Now we consider the exact sequence of A-modules

$$0 \to I^n \to A \to A/I^n \to 0.$$

According to (c) which we have already proved, this gives an exact sequence of $\mathcal{O}_{\mathfrak{x}}$-modules

$$0 \to \mathfrak{J}^n \to \mathcal{O}_{\mathfrak{x}} \to (A/I^n)^\Delta \to 0.$$

Observe that the inverse system which defines $(A/I^n)^\Delta$ as the completion of $(A/I^n)^\sim$ is eventually stationary, since this sheaf is annihilated by $\tilde{I}^n$. Hence $(A/I^n)^\Delta = (A/I^n)^\sim$, and we conclude that $\mathcal{O}_{\mathfrak{x}}/\mathfrak{J}^n \cong (A/I^n)^\sim$ as required.

(b) We have a slight abuse of notation in our statement: since $\tilde{M}$ and $\mathcal{O}_X$ are sheaves on X, we should actually write $M^\Delta \cong \tilde{M}|_Y \otimes_{\mathcal{O}_{X|Y}} \mathcal{O}_{\mathfrak{x}}$. But we will simply regard M^Δ and $\mathcal{O}_{\mathfrak{x}}$ as sheaves on X, by extending by zero outside of Y (Ex. 1.19). For any finitely generated A module M, and for U an open set as above, we have $\Gamma(U,M^\Delta) = \hat{N}$ as before. On the other hand, $\tilde{M} \otimes_{\mathcal{O}_X} \mathcal{O}_{\mathfrak{x}}$ is the sheaf associated to the presheaf

$$U \mapsto \Gamma(U,\tilde{M}) \otimes_{\Gamma(U,\mathcal{O}_X)} \Gamma(U,\mathcal{O}_{\mathfrak{x}}) = N \otimes_B \hat{B}.$$

Since $\hat{N} \cong N \otimes_B \hat{B}$ by (9.3A), we conclude that the corresponding sheaves are isomorphic too: $M^\Delta \cong \tilde{M} \otimes_{\mathcal{O}_X} \mathcal{O}_{\mathfrak{x}}$.

Definition. Let $(\mathfrak{X},\mathcal{O}_{\mathfrak{x}})$ be a noetherian formal scheme. A sheaf of ideals $\mathfrak{J} \subseteq \mathcal{O}_{\mathfrak{x}}$ is called an *ideal of definition* for $\mathfrak{X}$ if Supp $\mathcal{O}_{\mathfrak{x}}/\mathfrak{J} = \mathfrak{X}$ and the locally ringed space $(\mathfrak{X},\mathcal{O}_{\mathfrak{x}}/\mathfrak{J})$ is a noetherian scheme.

Proposition 9.5. *Let $(\mathfrak{X},\mathcal{O}_{\mathfrak{x}})$ be a noetherian formal scheme.*

(a) *If $\mathfrak{J}_1$ and $\mathfrak{J}_2$ are two ideals of definition, then there are integers $m,n > 0$ such that $\mathfrak{J}_1 \supseteq \mathfrak{J}_2^m$ and $\mathfrak{J}_2 \supseteq \mathfrak{J}_1^n$.*

(b) *There is a unique largest ideal of definition $\mathfrak{J}$, characterized by the fact that $(\mathfrak{X},\mathcal{O}_{\mathfrak{x}}/\mathfrak{J})$ is a reduced scheme. In particular, ideals of definition exist.*

(c) *If $\mathfrak{J}$ is an ideal of definition, so is $\mathfrak{J}^n$, for any $n > 0$.*

PROOF.

(a) Let $\mathfrak{J}_1$ and $\mathfrak{J}_2$ be two ideals of definition. Then on the topological space $\mathfrak{X}$, we have surjective maps of sheaves of rings $f_1:\mathcal{O}_{\mathfrak{x}} \to \mathcal{O}_{\mathfrak{x}}/\mathfrak{J}_1$ and $f_2:\mathcal{O}_{\mathfrak{x}} \to \mathcal{O}_{\mathfrak{x}}/\mathfrak{J}_2$. For any point $P \in \mathfrak{X}$, the stalk $(\mathfrak{J}_2)_P$ of $\mathfrak{J}_2$ at P is contained in m_P, the maximal ideal of the local ring $\mathcal{O}_{\mathfrak{X},P}$. Indeed, $\mathcal{O}_{\mathfrak{X},P}/(\mathfrak{I}_2)_P$ is the local ring of P on the scheme $(\mathfrak{X},\mathcal{O}_{\mathfrak{X}}/\mathfrak{I}_2)$. In particular, it is nonzero, so $(\mathfrak{I}_2)_P \subseteq m_P$. Now we consider the sheaf of ideals $f_1(\mathfrak{I}_2)$ on the scheme $(\mathfrak{X},\mathcal{O}_{\mathfrak{X}}/\mathfrak{I}_1)$. For each point P, its stalk is contained inside the maximal ideal of the local ring. Hence every local section of $f_1(\mathfrak{I}_2)$ is nilpotent (Ex. 2.18), and since $(\mathfrak{X},\mathcal{O}_{\mathfrak{X}}/\mathfrak{I}_1)$ is a noetherian scheme, $f_1(\mathfrak{I}_2)$ itself is nilpotent. This shows that for some $m > 0$, $\mathfrak{I}_1 \supseteq \mathfrak{I}_2^m$. The other way follows by symmetry.

(b) Suppose $(\mathfrak{X},\mathcal{O}_{\mathfrak{X}}/\mathfrak{I}_1)$ is a reduced scheme. Then in the proof of (a), we find $f_1(\mathfrak{I}_2) = 0$, so $\mathfrak{I}_1 \supseteq \mathfrak{I}_2$. Thus such an $\mathfrak{I}_1$ is largest, if it exists. Since it is unique, the existence becomes a local question. Thus we may assume that $\mathfrak{X}$ is the completion of an affine noetherian scheme X along a closed subscheme Y. By (9.3.1) we may assume that Y has the reduced induced structure. Let $X = \operatorname{Spec} A$, $Y = V(I)$. Then by (9.4), $\mathfrak{I} = I^{\triangle}$ is an ideal in $\mathcal{O}_{\mathfrak{X}}$, and $\mathcal{O}_{\mathfrak{X}}/\mathfrak{I} \cong (A/I)^{\sim} = \mathcal{O}_Y$. Thus $\mathfrak{I}$ is an ideal of definition for which $(\mathfrak{X},\mathcal{O}_{\mathfrak{X}}/\mathfrak{I})$ is reduced. This shows the existence of the largest ideal of definition.

(c) Let $\mathfrak{I}$ be any ideal of definition, and suppose given $n > 0$. Let $\mathfrak{I}_0$ be the unique largest ideal of definition. Then by (a), there is an integer r such that $\mathfrak{I} \supseteq \mathfrak{I}_0^r$, and hence $\mathfrak{I}^n \supseteq \mathfrak{I}_0^{nr}$. First note that $\mathfrak{I}_0^{nr}$ is an ideal of definition. Indeed, this can be checked locally. If $\mathfrak{I}_0 = I^{\triangle}$ on an affine, using the notation of (b), then $\mathcal{O}_{\mathfrak{X}}/\mathfrak{I}_0^{nr} \cong (A/I^{nr})^{\sim}$ by (9.4), so $(\mathfrak{X},\mathcal{O}_{\mathfrak{X}}/\mathfrak{I}_0^{nr})$ is a scheme with support Y. Let's call this scheme Y', and let $f:\mathcal{O}_{\mathfrak{X}} \to \mathcal{O}_{Y'}$ be the corresponding map of sheaves. Then $(Y',\mathcal{O}_{Y'}/f(\mathfrak{I})) = (\mathfrak{X},\mathcal{O}_{\mathfrak{X}}/\mathfrak{I})$ is a noetherian scheme, by hypothesis, so $f(\mathfrak{I})$ is a coherent sheaf. Therefore $f(\mathfrak{I}^n) = f(\mathfrak{I})^n$ is also coherent, and we conclude that $(Y',\mathcal{O}_{Y'}/f(\mathfrak{I}^n)) = (\mathfrak{X},\mathcal{O}_{\mathfrak{X}}/\mathfrak{I}^n)$ is also a noetherian scheme.

Proposition 9.6. *Let $\mathfrak{X}$ be a noetherian formal scheme and let $\mathfrak{I}$ be an ideal of definition. For each $n > 0$ we denote by Y_n the scheme $(\mathfrak{X},\mathcal{O}_{\mathfrak{X}}/\mathfrak{I}^n)$.*

(a) If $\mathfrak{F}$ is a coherent sheaf of $\mathcal{O}_{\mathfrak{X}}$-modules, then for each n, $\mathfrak{F}_n = \mathfrak{F}/\mathfrak{I}^n\mathfrak{F}$ is a coherent sheaf of $\mathcal{O}_{Y_n}$-modules, and $\mathfrak{F} \cong \varprojlim \mathfrak{F}_n$.

(b) Conversely, suppose given for each n a coherent $\mathcal{O}_{Y_n}$-module $\mathfrak{F}_n$, together with surjective maps $\varphi_{n'n}:\mathfrak{F}_{n'} \to \mathfrak{F}_n$ for each $n' \geqslant n$, making $\{\mathfrak{F}_n\}$ into an inverse system of sheaves. Assume furthermore that for each $n' \geqslant n$, $\ker \varphi_{n'n} = \mathfrak{I}^n\mathfrak{F}_{n'}$. Then $\mathfrak{F} = \varprojlim \mathfrak{F}_n$ is a coherent $\mathcal{O}_{\mathfrak{X}}$-module, and for each n, $\mathfrak{F}_n \cong \mathfrak{F}/\mathfrak{I}^n\mathfrak{F}$.

PROOF.

(a) The question is local, so we may assume that $\mathfrak{X}$ is affine, equal to the completion of $X = \operatorname{Spec} A$ along $Y = V(I)$, and that $\mathfrak{F} = M^{\triangle}$ for some finitely generated A-module M. Then as in the proof of (9.4a) we see that $\mathfrak{F}/\mathfrak{I}^n\mathfrak{F} \cong (M/I^nM)^{\sim}$ for each n. Thus $\mathfrak{F}_n$ is coherent on $Y_n = \operatorname{Spec}(A/I^n)$, and $\mathfrak{F} \cong \varprojlim \mathfrak{F}_n$.

(b) Again the question is local, so we may assume that $\mathfrak{X}$ is affine as above. Furthermore, we may assume that A is I-adically complete, because replacing A by $\hat{A}$ does not change $\hat{X}$. For each n, let $M_n = \Gamma(Y_n, \mathfrak{F}_n)$. Then

197

(M_n) is an inverse system of modules satisfying the hypotheses of (9.3Ae). Therefore we conclude that $M = \varprojlim M_n$ is a finitely generated A-module (since A is complete), and that for each n, $M_n \cong M/I^n M$. But then $\mathfrak{F} = \varprojlim \mathscr{F}_n$ is just $M^\triangle$, hence it is a coherent $\mathcal{O}_{\mathfrak{x}}$-module. Furthermore $\mathfrak{F}/\mathfrak{I}^n \mathfrak{F} \cong (M/I^n M)^\sim$ as in (a), so $\mathfrak{F}/\mathfrak{I}^n \mathfrak{F} \cong \mathscr{F}_n$.

Theorem 9.7. *Let A be a noetherian ring, I an ideal, and assume that A is I-adically complete. Let $X = \operatorname{Spec} A$, $Y = V(I)$, and $\mathfrak{X} = \hat{X}$. Then the functors $M \mapsto M^\triangle$ and $\mathfrak{F} \mapsto \Gamma(\mathfrak{X}, \mathfrak{F})$ are exact, and inverse to each other, on the categories of finitely generated A-modules and coherent $\mathcal{O}_{\mathfrak{x}}$-modules respectively. Thus they establish an equivalence of categories. In particular, every coherent $\mathcal{O}_{\mathfrak{x}}$-module $\mathfrak{F}$ is of the form $M^\triangle$ for some M.*

PROOF. We have already seen that $M \mapsto M^\triangle$ is exact (9.4). If M is an A-module of finite type, then $\Gamma(\mathfrak{X}, M^\triangle) = \varprojlim M/I^n M = \hat{M}$, and $\hat{M} = M$ because A is complete (9.3Ab). Thus one composition of our two functors is the identity.

Conversely, let $\mathfrak{F}$ be a coherent $\mathcal{O}_{\mathfrak{x}}$-module, and let $\mathfrak{I} = I^\triangle$. Then by (9.6a), $\mathfrak{F} \cong \varprojlim \mathscr{F}_n$, where for each $n > 0$, $\mathscr{F}_n = \mathfrak{F}/\mathfrak{I}^n \mathfrak{F}$. Now the inverse system of sheaves $(\mathscr{F}_n)$ satisfies the hypotheses of (9.6b), and the proof of (9.6b) shows in fact that $\mathfrak{F} \cong M^\triangle$, for some finitely generated A-module M. Furthermore, by (9.2), $\Gamma(\mathfrak{X}, \mathfrak{F}) = \varprojlim \Gamma(Y, (M/I^n M)^\sim) = \varprojlim M/I^n M = \hat{M}$, which is equal to M since A is complete. This shows that $\Gamma(\mathfrak{X}, \mathfrak{F})$ is a finitely generated A-module, and $\mathfrak{F} \cong \Gamma(\mathfrak{X}, \mathfrak{F})^\triangle$. Thus the other composition of our two functors is the identity.

It remains to show that the functor $\Gamma(\mathfrak{X}, \cdot)$ is exact on the category of coherent $\mathcal{O}_{\mathfrak{x}}$-modules. So let

$$0 \to \mathfrak{F}_1 \to \mathfrak{F}_2 \to \mathfrak{F}_3 \to 0$$

be an exact sequence of coherent $\mathcal{O}_{\mathfrak{x}}$-modules. For each i, let $M_i = \Gamma(\mathfrak{X}, \mathfrak{F}_i)$. Then the M_i are finitely generated A-modules, and we have at least a left-exact sequence

$$0 \to M_1 \to M_2 \to M_3.$$

Let R be the cokernel on the right. Then applying the functor $\triangle$ we obtain an exact sequence

$$0 \to M_1^\triangle \to M_2^\triangle \to M_3^\triangle \to R^\triangle \to 0$$

on $\mathfrak{X}$. But for each i, $M_i^\triangle \cong \mathfrak{F}_i$ as we saw above, so we conclude that $R^\triangle = 0$. But also by the above, $R = \Gamma(\mathfrak{X}, R^\triangle)$, so $R = 0$. This shows that $\Gamma(\mathfrak{X}, \cdot)$ is exact, which concludes the proof.

Corollary 9.8. *If X is any noetherian scheme, Y a closed subscheme, and $\mathfrak{X} = \hat{X}$ the completion along Y, then the functor $\mathscr{F} \mapsto \hat{\mathscr{F}}$ is an exact functor from coherent $\mathcal{O}_X$-modules to coherent $\mathcal{O}_{\mathfrak{x}}$-modules. Furthermore, if $\mathscr{I}$ is the sheaf of ideals of Y, and $\hat{\mathscr{I}}$ its completion, then we have $\hat{\mathscr{F}}/\hat{\mathscr{I}}^n \hat{\mathscr{F}} \cong \mathscr{F}/\mathscr{I}^n \mathscr{F}$ for each n, and $\hat{\mathscr{F}} \cong \mathscr{F} \otimes_{\mathcal{O}_X} \mathcal{O}_{\mathfrak{x}}$.*

PROOF. These questions are all local, in which case they reduce to (9.4).

Corollary 9.9. *Any kernel, cokernel, or image of a morphism of coherent sheaves on a noetherian formal scheme is again coherent.*

PROOF. These questions are also local, in which case they follow from (9.7).

Remark 9.9.1. It is also true that an extension of coherent sheaves on a noetherian formal scheme is coherent (Ex. 9.4). On the other hand, some properties of coherent sheaves on usual schemes do not carry over to formal schemes. For example, if $\mathfrak{X}$ is the completion of a projective variety $X \subseteq \mathbf{P}_k^n$ along a closed subvariety Y, and if $\mathcal{O}_{\mathfrak{X}}(1) = \mathcal{O}_X(1)^{\wedge}$, then there may be nonzero coherent sheaves $\mathfrak{F}$ on $\mathfrak{X}$ such that $\Gamma(\mathfrak{X},\mathfrak{F}(v)) = 0$ for all $v \in \mathbf{Z}$. In particular, no twist of $\mathfrak{F}$ is generated by global sections (III, Ex. 11.7).

EXERCISES

9.1. Let X be a noetherian scheme, Y a closed subscheme, and $\hat{X}$ the completion of X along Y. We call the ring $\Gamma(\hat{X},\mathcal{O}_{\hat{X}})$ the ring of *formal-regular functions* on X along Y. In this exercise we show that if Y is a connected, nonsingular, positive-dimensional subvariety of $X = \mathbf{P}_k^n$ over an algebraically closed field k, then $\Gamma(\hat{X},\mathcal{O}_{\hat{X}}) = k$.
   (a) Let $\mathscr{I}$ be the ideal sheaf of Y. Use (8.13) and (8.17) to show that there is an inclusion of sheaves on Y, $\mathscr{I}/\mathscr{I}^2 \hookrightarrow \mathcal{O}_Y(-1)^{n+1}$.
   (b) Show that for any $r \geqslant 1$, $\Gamma(Y,\mathscr{I}^r/\mathscr{I}^{r+1}) = 0$.
   (c) Use the exact sequences

$$0 \to \mathscr{I}^r/\mathscr{I}^{r+1} \to \mathcal{O}_X/\mathscr{I}^{r+1} \to \mathcal{O}_X/\mathscr{I}^r \to 0$$

   and induction on r to show that $\Gamma(Y,\mathcal{O}_X/\mathscr{I}^r) = k$ for all $r \geqslant 1$. (Use (8.21Ae).)
   (d) Conclude that $\Gamma(\hat{X},\mathcal{O}_{\hat{X}}) = k$. (Actually, the same result holds without the hypothesis Y nonsingular, but the proof is more difficult—see Hartshorne [3, (7.3)].)

9.2. Use the result of (Ex. 9.1) to prove the following geometric result. Let $Y \subseteq X = \mathbf{P}_k^n$ be as above, and let $f:X \to Z$ be a morphism of k-varieties. Suppose that $f(Y)$ is a single closed point $P \in Z$. Then $f(X) = P$ also.

9.3. Prove the analogue of (5.6) for formal schemes, which says, if $\mathfrak{X}$ is an affine formal scheme, and if

$$0 \to \mathfrak{F}' \to \mathfrak{F} \to \mathfrak{F}'' \to 0$$

is an exact sequence of $\mathcal{O}_{\mathfrak{X}}$-modules, and if $\mathfrak{F}'$ is coherent, then the sequence of global sections

$$0 \to \Gamma(\mathfrak{X},\mathfrak{F}') \to \Gamma(\mathfrak{X},\mathfrak{F}) \to \Gamma(\mathfrak{X},\mathfrak{F}'') \to 0$$

is exact. For the proof, proceed in the following steps.
   (a) Let $\mathfrak{I}$ be an ideal of definition for $\mathfrak{X}$, and for each $n > 0$ consider the exact sequence

$$0 \to \mathfrak{F}'/\mathfrak{I}^n\mathfrak{F}' \to \mathfrak{F}/\mathfrak{I}^n\mathfrak{F}' \to \mathfrak{F}'' \to 0.$$

199

Use (5.6), slightly modified, to show that for every open affine subset $\mathfrak{U} \subseteq \mathfrak{X}$, the sequence

$$0 \to \Gamma(\mathfrak{U}, \mathfrak{F}'/\mathfrak{I}^n\mathfrak{F}') \to \Gamma(\mathfrak{U}, \mathfrak{F}/\mathfrak{I}^n\mathfrak{F}') \to \Gamma(\mathfrak{U}, \mathfrak{F}'') \to 0$$

is exact.

(b) Now pass to the limit, using (9.1), (9.2), and (9.6). Conclude that $\mathfrak{F} \cong \varprojlim \mathfrak{F}/\mathfrak{I}^n\mathfrak{F}'$ and that the sequence of global sections above is exact.

9.4. Use (Ex. 9.3) to prove that if

$$0 \to \mathfrak{F}' \to \mathfrak{F} \to \mathfrak{F}'' \to 0$$

is an exact sequence of $\mathcal{O}_{\mathfrak{X}}$-modules on a noetherian formal scheme $\mathfrak{X}$, and if $\mathfrak{F}', \mathfrak{F}''$ are coherent, then $\mathfrak{F}$ is coherent also.

9.5. If $\mathfrak{F}$ is a coherent sheaf on a noetherian formal scheme $\mathfrak{X}$, which can be generated by global sections, show in fact that it can be generated by a finite number of its global sections.

9.6. Let $\mathfrak{X}$ be a noetherian formal scheme, let $\mathfrak{I}$ be an ideal of definition, and for each n, let Y_n be the scheme $(\mathfrak{X}, \mathcal{O}_{\mathfrak{X}}/\mathfrak{I}^n)$. Assume that the inverse system of groups $(\Gamma(Y_n, \mathcal{O}_{Y_n}))$ satisfies the Mittag–Leffler condition. Then prove that $\operatorname{Pic} \mathfrak{X} = \varprojlim \operatorname{Pic} Y_n$. As in the case of a scheme, we define $\operatorname{Pic} \mathfrak{X}$ to be the group of locally free $\mathcal{O}_{\mathfrak{X}}$-modules of rank 1 under the operation $\otimes$. Proceed in the following steps.

(a) Use the fact that $\ker(\Gamma(Y_{n+1}, \mathcal{O}_{Y_{n+1}}) \to \Gamma(Y_n, \mathcal{O}_{Y_n}))$ is a nilpotent ideal to show that the inverse system $(\Gamma(Y_n, \mathcal{O}_{Y_n}^*))$ of units in the respective rings also satisfies (ML).

(b) Let $\mathfrak{F}$ be a coherent sheaf of $\mathcal{O}_{\mathfrak{X}}$-modules, and assume that for each n, there is some isomorphism $\varphi_n: \mathfrak{F}/\mathfrak{I}^n\mathfrak{F} \cong \mathcal{O}_{Y_n}$. Then show that there is an isomorphism $\mathfrak{F} \cong \mathcal{O}_{\mathfrak{X}}$. Be careful, because the φ_n may not be compatible with the maps in the two inverse systems $(\mathfrak{F}/\mathfrak{I}^n\mathfrak{F})$ and $(\mathcal{O}_{Y_n})$! Conclude that the natural map $\operatorname{Pic} \mathfrak{X} \to \varprojlim \operatorname{Pic} Y_n$ is injective.

(c) Given an invertible sheaf $\mathscr{L}_n$ on Y_n for each n, and given isomorphisms $\mathscr{L}_{n+1} \otimes \mathcal{O}_{Y_n} \cong \mathscr{L}_n$, construct maps $\mathscr{L}_{n'} \to \mathscr{L}_n$ for each $n' \geq n$ so as to make an inverse system, and show that $\mathfrak{L} = \varprojlim \mathscr{L}_n$ is a coherent sheaf on $\mathfrak{X}$. Then show that $\mathfrak{L}$ is locally free of rank 1, and thus conclude that the map $\operatorname{Pic} \mathfrak{X} \to \varprojlim \operatorname{Pic} Y_n$ is surjective. Again be careful, because even though each $\mathscr{L}_n$ is locally free of rank 1, the open sets needed to make them free might get smaller and smaller with n.

(d) Show that the hypothesis "$(\Gamma(Y_n, \mathcal{O}_{Y_n}))$ satisfies (ML)" is satisfied if either $\mathfrak{X}$ is affine, or each Y_n is projective over a field k.

Note: See (III, Ex. 11.5–11.7) for further examples and applications.

CHAPTER III

Cohomology

In this chapter we define the general notion of cohomology of a sheaf of abelian groups on a topological space, and then study in detail the cohomology of coherent and quasi-coherent sheaves on a noetherian scheme.

Although the end result is usually the same, there are many different ways of introducing cohomology. There are the fine resolutions often used in several complex variables—see Gunning and Rossi [1]; the Čech cohomology used by Serre [3], who first introduced cohomology into abstract algebraic geometry; the canonical flasque resolutions of Godement [1]; and the derived functor approach of Grothendieck [1]. Each is important in its own way.

We will take as our basic definition the derived functors of the global section functor (§1, 2). This definition is the most general, and also best suited for theoretical questions, such as the proof of Serre duality in §7. However, it is practically impossible to calculate, so we introduce Čech cohomology in §4, and use it in §5 to compute explicitly the cohomology of the sheaves $\mathcal{O}(n)$ on a projective space $\mathbf{P}^r$. This calculation is the basis of many later results on projective varieties.

In order to prove that the Čech cohomology agrees with the derived functor cohomology, we need to know that the higher cohomology of a quasi-coherent sheaf on an affine scheme is zero. We prove this in §3 in the noetherian case only, because it is technically much simpler than the case of an arbitrary affine scheme ([EGA III, §1]). Hence we are bound to include noetherian hypotheses in all theorems involving cohomology.

As applications, we show for example that the arithmetic genus of a projective variety X, whose definition in (I, §7) depended on a projective embedding of X, can be computed in terms of the cohomology groups $H^i(X, \mathcal{O}_X)$, and hence is intrinsic (Ex. 5.3). We also show that the arithmetic genus is constant in a family of normal projective varieties (9.13).

Another application is Zariski's main theorem (11.4) which is important in the birational study of varieties.

The latter part of the chapter (§8–12) is devoted to families of schemes, i.e., the study of the fibres of a morphism. In particular, we include a section on flat morphisms and a section on smooth morphisms. While these can be treated without cohomology, it seems to be an appropriate place to include them, because flatness can be understood better using cohomology (9.9).

1   Derived Functors

In this chapter we will assume familiarity with the basic techniques of homological algebra. Since notation and terminology vary from one source to another, we will assemble in this section (without proofs) the basic definitions and results we will need. More details can be found in the following sources: Godement [1, esp. Ch. I, §1.1–1.8, 2.1–2.4, 5.1–5.3], Hilton and Stammbach [1, Ch. II,IV,IX], Grothendieck [1, Ch. II, §1,2,3], Cartan and Eilenberg [1, Ch. III,V], Rotman [1, §6].

Definition. An *abelian category* is a category $\mathfrak{A}$, such that: for each $A,B \in$ Ob $\mathfrak{A}$, Hom(A,B) has a structure of an abelian group, and the composition law is linear; finite direct sums exist; every morphism has a kernel and a cokernel; every monomorphism is the kernel of its cokernel, every epimorphism is the cokernel of its kernel; and finally, every morphism can be factored into an epimorphism followed by a monomorphism. (Hilton and Stammbach [1, p. 78].)

The following are all abelian categories.

Example 1.0.1. $\mathfrak{Ab}$, the category of abelian groups.

Example 1.0.2. $\mathfrak{Mod}(A)$, the category of modules over a ring A (commutative with identity as always).

Example 1.0.3. $\mathfrak{Ab}(X)$, the category of sheaves of abelian groups on a topological space X.

Example 1.0.4. $\mathfrak{Mod}(X)$, the category of sheaves of $\mathcal{O}_X$-modules on a ringed space $(X,\mathcal{O}_X)$.

Example 1.0.5. $\mathfrak{Qco}(X)$, the category of quasi-coherent sheaves of $\mathcal{O}_X$-modules on a scheme X (II, 5.7).

Example 1.0.6. $\mathfrak{Coh}(X)$, the category of coherent sheaves of $\mathcal{O}_X$-modules on a noetherian scheme X (II, 5.7).

Example 1.0.7. $\mathfrak{Coh}(\mathfrak{X})$, the category of coherent sheaves of $\mathcal{O}_{\mathfrak{x}}$-modules on a noetherian formal scheme $(\mathfrak{X}, \mathcal{O}_{\mathfrak{x}})$ (II, 9.9).

In the rest of this section, we will be stating some basic results of homological algebra in the context of an arbitrary abelian category. However, in most books, these results are proved only for the category of modules over a ring, and proofs are often done by "diagram-chasing": you pick an element and chase its images and pre-images through a diagram. Since diagram-chasing doesn't make sense in an arbitrary abelian category, the conscientious reader may be disturbed. There are at least three ways to handle this difficulty. (1) Provide intrinsic proofs for all the results, starting from the axioms of an abelian category, and without even mentioning an element. This is cumbersome, but can be done—see, e.g., Freyd [1]. Or (2), note that in each of the categories we use (most of which are in the above list of examples), one can in fact carry out proofs by diagram-chasing. Or (3), accept the "full embedding theorem" (Freyd [1, Ch. 7]), which states roughly that any abelian category is equivalent to a subcategory of $\mathfrak{Ab}$. This implies that any category-theoretic statement (e.g., the 5-lemma) which can be proved in $\mathfrak{Ab}$ (e.g., by diagram-chasing) also holds in any abelian category.

Now we begin our review of homological algebra. A *complex* $A^{\cdot}$ in an abelian category $\mathfrak{A}$ is a collection of objects A^i, $i \in \mathbf{Z}$, and morphisms $d^i : A^i \to A^{i+1}$, such that $d^{i+1} \circ d^i = 0$ for all i. If the objects A^i are specified only in a certain range, e.g., $i \geqslant 0$, then we set $A^i = 0$ for all other i. A *morphism* of complexes, $f : A^{\cdot} \to B^{\cdot}$ is a set of morphisms $f^i : A^i \to B^i$ for each i, which commute with the coboundary maps d^i.

The ith *cohomology object* $h^i(A^{\cdot})$ of the complex $A^{\cdot}$ is defined to be $\ker d^i / \operatorname{im} d^{i-1}$. If $f : A^{\cdot} \to B^{\cdot}$ is a morphism of complexes, then f induces a natural map $h^i(f) : h^i(A^{\cdot}) \to h^i(B^{\cdot})$. If $0 \to A^{\cdot} \to B^{\cdot} \to C^{\cdot} \to 0$ is a short exact sequence of complexes, then there are natural maps $\delta^i : h^i(C^{\cdot}) \to h^{i+1}(A^{\cdot})$ giving rise to a long exact sequence

$$\ldots \to h^i(A^{\cdot}) \to h^i(B^{\cdot}) \to h^i(C^{\cdot}) \xrightarrow{\delta^i} h^{i+1}(A^{\cdot}) \to \ldots$$

Two morphisms of complexes $f, g : A^{\cdot} \to B^{\cdot}$ are *homotopic* (written $f \sim g$) if there is a collection of morphisms $k^i : A^i \to B^{i-1}$ for each i (which need not commute with the d^i) such that $f - g = dk + kd$. The collection of morphisms, $k = (k^i)$ is called a *homotopy operator*. If $f \sim g$, then f and g induce the *same* morphism $h^i(A^{\cdot}) \to h^i(B^{\cdot})$ on the cohomology objects, for each i.

A covariant functor $F : \mathfrak{A} \to \mathfrak{B}$ from one abelian category to another is *additive* if for any two objects A, A' in $\mathfrak{A}$, the induced map $\operatorname{Hom}(A, A') \to \operatorname{Hom}(FA, FA')$ is a homomorphism of abelian groups. F is *left exact* if it is additive and for every short exact sequence

$$0 \to A' \to A \to A'' \to 0$$

in $\mathfrak{A}$, the sequence

$$0 \to FA' \to FA \to FA''$$

is exact in $\mathfrak{B}$. If we can write a 0 on the right instead of the left, we say F is *right exact*. If it is both left and right exact, we say it is *exact*. If only the middle part $FA' \to FA \to FA''$ is exact, we say F is *exact in the middle*.

For a contravariant functor we make analogous definitions. For example, $F:\mathfrak{A} \to \mathfrak{B}$ is *left exact* if it is additive, and for every short exact sequence as above, the sequence

$$0 \to FA'' \to FA \to FA'$$

is exact in $\mathfrak{B}$.

Example 1.0.8. If $\mathfrak{A}$ is an abelian category, and A is a fixed object, then the functor $B \to \mathrm{Hom}(A,B)$, usually denoted $\mathrm{Hom}(A,\cdot)$, is a covariant left exact functor from $\mathfrak{A}$ to $\mathfrak{A}b$. The functor $\mathrm{Hom}(\cdot,A)$ is a contravariant left exact functor from $\mathfrak{A}$ to $\mathfrak{A}b$.

Next we come to resolutions and derived functors. An object I of $\mathfrak{A}$ is *injective* if the functor $\mathrm{Hom}(\cdot,I)$ is exact. An *injective resolution* of an object A of $\mathfrak{A}$ is a complex $I^{\cdot}$, defined in degrees $i \geq 0$, together with a morphism $\varepsilon:A \to I^0$, such that I^i is an injective object of $\mathfrak{A}$ for each $i \geq 0$, and such that the sequence

$$0 \to A \xrightarrow{\varepsilon} I^0 \to I^1 \to \ldots$$

is exact.

If every object of $\mathfrak{A}$ is isomorphic to a subobject of an injective object of $\mathfrak{A}$, then we say $\mathfrak{A}$ *has enough injectives*. If $\mathfrak{A}$ has enough injectives, then every object has an injective resolution. Furthermore, a well-known lemma states that any two injective resolutions are homotopy equivalent.

Now let $\mathfrak{A}$ be an abelian category with enough injectives, and let $F:\mathfrak{A} \to \mathfrak{B}$ be a covariant left exact functor. Then we construct the *right derived functors* R^iF, $i \geq 0$, of F as follows. For each object A of $\mathfrak{A}$, choose once and for all an injective resolution $I^{\cdot}$ of A. Then we define $R^iF(A) = h^i(F(I^{\cdot}))$.

Theorem 1.1A. *Let $\mathfrak{A}$ be an abelian category with enough injectives, and let $F:\mathfrak{A} \to \mathfrak{B}$ be a covariant left exact functor to another abelian category $\mathfrak{B}$. Then*

(a) For each $i \geq 0$, R^iF as defined above is an additive functor from $\mathfrak{A}$ to $\mathfrak{B}$. Furthermore, it is independent (up to natural isomorphism of functors) of the choices of injective resolutions made.

(b) There is a natural isomorphism $F \cong R^0F$.

(c) For each short exact sequence $0 \to A' \to A \to A'' \to 0$ and for each $i \geq 0$ there is a natural morphism $\delta^i:R^iF(A'') \to R^{i+1}F(A')$, such that we obtain a long exact sequence

$$\ldots \to R^iF(A') \to R^iF(A) \to R^iF(A'') \xrightarrow{\delta^i} R^{i+1}F(A') \to R^{i+1}F(A) \to \ldots .$$

(d) *Given a morphism of the exact sequence of* (c) *to another* $0 \to B' \to$ $B \to B'' \to 0$, *the δ's give a commutative diagram*

$$R^i F(A'') \overset{\delta^i}{\to} R^{i+1} F(A')$$
$$\downarrow \qquad\qquad \downarrow$$
$$R^i F(B'') \overset{\delta^i}{\to} R^{i+1} F(B').$$

(e) *For each injective object* I *of* $\mathfrak{A}$, *and for each* $i > 0$, *we have* $R^i F(I) = 0$.

Definition. With $F: \mathfrak{A} \to \mathfrak{B}$ as in the theorem, an object J of $\mathfrak{A}$ is *acyclic* for F if $R^i F(J) = 0$ for all $i > 0$.

Proposition 1.2A. *With* $F: \mathfrak{A} \to \mathfrak{B}$ *as in* (1.1A), *suppose there is an exact sequence*

$$0 \to A \to J^0 \to J^1 \to \cdots$$

where each J^i *is acyclic for* F, $i \geqslant 0$. (*We say* $J^{\cdot}$ *is an F-acyclic resolution of* A.) *Then for each* $i \geqslant 0$ *there is a natural isomorphism* $R^i F(A) \cong h^i(F(J^{\cdot}))$.

We leave to the reader the analogous definitions of projective objects, projective resolutions, an abelian category having enough projectives, and the left derived functors of a covariant right exact functor. Also, the right derived functors of a left exact contravariant functor (use projective resolutions) and the left derived functors of a right exact contravariant functor (use injective resolutions).

Next we will give a universal property of derived functors. For this purpose, we generalize slightly with the following definition.

Definition. Let $\mathfrak{A}$ and $\mathfrak{B}$ be abelian categories. A *(covariant) δ-functor* from $\mathfrak{A}$ to $\mathfrak{B}$ is a collection of functors $T = (T^i)_{i \geqslant 0}$, together with a morphism $\delta^i: T^i(A'') \to T^{i+1}(A')$ for each short exact sequence $0 \to A' \to A \to A'' \to 0$, and each $i \geqslant 0$, such that:

(1) For each short exact sequence as above, there is a long exact sequence

$$0 \to T^0(A') \to T^0(A) \to T^0(A'') \overset{\delta^0}{\to} T^1(A') \to \cdots$$
$$\cdots \to T^i(A) \to T^i(A'') \overset{\delta^i}{\to} T^{i+1}(A') \to T^{i+1}(A) \to \cdots;$$

(2) for each morphism of one short exact sequence (as above) into another $0 \to B' \to B \to B'' \to 0$, the δ's give a commutative diagram

$$T^i(A'') \overset{\delta^i}{\to} T^{i+1}(A')$$
$$\downarrow \qquad\qquad \downarrow$$
$$T^i(B'') \overset{\delta^i}{\to} T^{i+1}(B').$$

Definition. The δ-functor $T = (T^i): \mathfrak{A} \to \mathfrak{B}$ is said to be *universal* if, given any other δ-functor $T' = (T'^i): \mathfrak{A} \to \mathfrak{B}$, and given any morphism of functors $f^0: T^0 \to T'^0$, there exists a unique sequence of morphisms $f^i: T^i \to T'^i$ for each $i \geqslant 0$, starting with the given f^0, which commute with the δ^i for each short exact sequence.

Remark 1.2.1. If $F: \mathfrak{A} \to \mathfrak{B}$ is a covariant additive functor, then by definition there can exist at most one (up to unique isomorphism) universal δ-functor T with $T^0 = F$. If T exists, the T^i are sometimes called the *right satellite functors* of F.

Definition. An additive functor $F: \mathfrak{A} \to \mathfrak{B}$ is *effaceable* if for each object A of $\mathfrak{A}$, there is a monomorphism $u: A \to M$, for some M, such that $F(u) = 0$. It is *coeffaceable* if for each A there exists an epimorphism $u: P \to A$ such that $F(u) = 0$.

Theorem 1.3A. *Let $T = (T^i)_{i \geqslant 0}$ be a covariant δ-functor from $\mathfrak{A}$ to $\mathfrak{B}$. If T^i is effaceable for each $i > 0$, then T is universal.*

PROOF. Grothendieck [1, II, 2.2.1]

Corollary 1.4. *Assume that $\mathfrak{A}$ has enough injectives. Then for any left exact functor $F: \mathfrak{A} \to \mathfrak{B}$, the derived functors $(R^i F)_{i \geqslant 0}$ form a universal δ-functor with $F \cong R^0 F$. Conversely, if $T = (T^i)_{i \geqslant 0}$ is any universal δ-functor, then T^0 is left exact, and the T^i are isomorphic to $R^i T^0$ for each $i \geqslant 0$.*

PROOF. If F is a left exact functor, then the $(R^i F)_{i \geqslant 0}$ form a δ-functor by (1.1A). Furthermore, for any object A, let $u: A \to I$ be a monomorphism of A into an injective. Then $R^i F(I) = 0$ for $i > 0$ by (1.1A), so $R^i F(u) = 0$. Thus $R^i F$ is effaceable for each $i > 0$. It follows from the theorem that $(R^i F)$ is universal.

On the other hand, given a universal δ-functor T, we have T^0 left exact by the definition of δ-functor. Since $\mathfrak{A}$ has enough injectives, the derived functors $R^i T^0$ exist. We have just seen that $(R^i T^0)$ is another universal δ-functor. Since $R^0 T^0 = T^0$, we find $R^i T^0 \cong T^i$ for each i, by (1.2.1).

2 Cohomology of Sheaves

In this section we define cohomology of sheaves by taking the derived functors of the global section functor. Then as an application of general techniques of cohomology we prove Grothendieck's theorem about the vanishing of cohomology on a noetherian topological space. To begin with, we must verify that the categories we use have enough injectives.

Proposition 2.1A. *If A is a ring, then every A-module is isomorphic to a submodule of an injective A-module.*

PROOF. Godement [1, I, 1.2.2] or Hilton and Stammbach [1, I, 8.3].

Proposition 2.2. *Let $(X,\mathcal{O}_X)$ be a ringed space. Then the category $\mathfrak{Mod}(X)$ of sheaves of $\mathcal{O}_X$-modules has enough injectives.*

PROOF. Let $\mathscr{F}$ be a sheaf of $\mathcal{O}_X$-modules. For each point $x \in X$, the stalk $\mathscr{F}_x$ is an $\mathcal{O}_{x,x}$-module. Therefore there is an injection $\mathscr{F}_x \to I_x$, where I_x is an injective $\mathcal{O}_{x,x}$-module (2.1A). For each point x, let j denote the inclusion of the one-point space $\{x\}$ into X, and consider the sheaf $\mathscr{I} = \prod_{x \in X} j_*(I_x)$. Here we consider I_x as a sheaf on the one-point space $\{x\}$, and j_* is the direct image functor (II, §1).

Now for any sheaf $\mathscr{G}$ of $\mathcal{O}_X$-modules, we have $\text{Hom}_{\mathcal{O}_X}(\mathscr{G},\mathscr{I}) = \prod \text{Hom}_{\mathcal{O}_X}(\mathscr{G},j_*(I_x))$ by definition of the direct product. On the other hand, for each point $x \in X$, we have $\text{Hom}_{\mathcal{O}_X}(\mathscr{G},j_*(I_x)) \cong \text{Hom}_{\mathcal{O}_{x,x}}(\mathscr{G}_x,I_x)$ as one sees easily. Thus we conclude first that there is a natural morphism of sheaves of $\mathcal{O}_X$-modules $\mathscr{F} \to \mathscr{I}$ obtained from the local maps $\mathscr{F}_x \to I_x$. It is clearly injective. Second, the functor $\text{Hom}_{\mathcal{O}_X}(\cdot,\mathscr{I})$ is the direct product over all $x \in X$ of the stalk functor $\mathscr{G} \mapsto \mathscr{G}_x$, which is exact, followed by $\text{Hom}_{\mathcal{O}_{x,x}}(\cdot,I_x)$, which is exact, since I_x is an injective $\mathcal{O}_{x,x}$-module. Hence $\text{Hom}(\cdot,\mathscr{I})$ is an exact functor, and therefore $\mathscr{I}$ is an injective $\mathcal{O}_X$-module.

Corollary 2.3. *If X is any topological space, then the category $\mathfrak{Ab}(X)$ of sheaves of abelian groups on X has enough injectives.*

PROOF. Indeed, if we let $\mathcal{O}_X$ be the constant sheaf of rings $\mathbf{Z}$, then $(X,\mathcal{O}_X)$ is a ringed space, and $\mathfrak{Mod}(X) = \mathfrak{Ab}(X)$.

Definition. Let X be a topological space. Let $\Gamma(X,\cdot)$ be the global section functor from $\mathfrak{Ab}(X)$ to $\mathfrak{Ab}$. We define the *cohomology functors* $H^i(X,\cdot)$ to be the right derived functors of $\Gamma(X,\cdot)$. For any sheaf $\mathscr{F}$, the groups $H^i(X,\mathscr{F})$ are the *cohomology groups* of $\mathscr{F}$. Note that even if X and $\mathscr{F}$ have some additional structure, e.g., X a scheme and $\mathscr{F}$ a quasi-coherent sheaf, we always take cohomology in this sense, regarding $\mathscr{F}$ simply as a sheaf of abelian groups on the underlying topological space X.

We let the reader write out the long exact sequences which follow from the general properties of derived functors (1.1A).

Recall (II, Ex. 1.16) that a sheaf $\mathscr{F}$ on a topological space X is *flasque* if for every inclusion of open sets $V \subseteq U$, the restriction map $\mathscr{F}(U) \to \mathscr{F}(V)$ is surjective.

Lemma 2.4. *If $(X,\mathcal{O}_X)$ is a ringed space, any injective $\mathcal{O}_X$-module is flasque.*

PROOF. For any open subset $U \subseteq X$, let $\mathcal{O}_U$ denote the sheaf $j_!(\mathcal{O}_X|_U)$, which is the restriction of $\mathcal{O}_X$ to U, extended by zero outside U (II, Ex. 1.19). Now let $\mathscr{I}$ be an injective $\mathcal{O}_X$-module, and let $V \subseteq U$ be open sets. Then we have an inclusion $0 \to \mathcal{O}_V \to \mathcal{O}_U$ of sheaves of $\mathcal{O}_X$-modules. Since $\mathscr{I}$ is injective, we get a surjection $\text{Hom}(\mathcal{O}_U,\mathscr{I}) \to \text{Hom}(\mathcal{O}_V,\mathscr{I}) \to 0$. But $\text{Hom}(\mathcal{O}_U,\mathscr{I}) = \mathscr{I}(U)$ and $\text{Hom}(\mathcal{O}_V,\mathscr{I}) = \mathscr{I}(V)$, so $\mathscr{I}$ is flasque.

Proposition 2.5. *If $\mathscr{F}$ is a flasque sheaf on a topological space X, then $H^i(X,\mathscr{F}) = 0$ for all $i > 0$.*

PROOF. Embed $\mathscr{F}$ in an injective object $\mathscr{I}$ of $\mathfrak{Ab}(X)$ and let $\mathscr{G}$ be the quotient:

$$0 \to \mathscr{F} \to \mathscr{I} \to \mathscr{G} \to 0.$$

Then $\mathscr{F}$ is flasque by hypothesis, $\mathscr{I}$ is flasque by (2.4), and so $\mathscr{G}$ is flasque by (II, Ex. 1.16c). Now since $\mathscr{F}$ is flasque, we have an exact sequence (II, Ex. 1.16b)

$$0 \to \Gamma(X,\mathscr{F}) \to \Gamma(X,\mathscr{I}) \to \Gamma(X,\mathscr{G}) \to 0.$$

On the other hand, since $\mathscr{I}$ is injective, we have $H^i(X,\mathscr{I}) = 0$ for $i > 0$ (1.1Ae). Thus from the long exact sequence of cohomology, we get $H^1(X,\mathscr{F}) = 0$ and $H^i(X,\mathscr{F}) \cong H^{i-1}(X,\mathscr{G})$ for each $i \geqslant 2$. But $\mathscr{G}$ is also flasque, so by induction on i we get the result.

Remark 2.5.1. This result tells us that flasque sheaves are acyclic for the functor $\Gamma(X,\cdot)$. Hence we can calculate cohomology using flasque resolutions (1.2A). In particular, we have the following result.

Proposition 2.6. *Let $(X,\mathcal{O}_X)$ be a ringed space. Then the derived functors of the functor $\Gamma(X,\cdot)$ from $\mathfrak{Mod}(X)$ to $\mathfrak{Ab}$ coincide with the cohomology functors $H^i(X,\cdot)$.*

PROOF. Considering $\Gamma(X,\cdot)$ as a functor from $\mathfrak{Mod}(X)$ to $\mathfrak{Ab}$, we calculate its derived functors by taking injective resolutions in the category $\mathfrak{Mod}(X)$. But any injective is flasque (2.4), and flasques are acyclic (2.5) so this resolution gives the usual cohomology functors (1.2A).

Remark 2.6.1. Let $(X,\mathcal{O}_X)$ be a ringed space, and let $A = \Gamma(X,\mathcal{O}_X)$. Then for any sheaf of $\mathcal{O}_X$-modules $\mathscr{F}$, $\Gamma(X,\mathscr{F})$ has a natural structure of A-module. In particular, since we can calculate cohomology using resolutions in the category $\mathfrak{Mod}(X)$, all the cohomology groups of $\mathscr{F}$ have a natural structure of A-module; the associated exact sequences are sequences of A-modules, and so forth. Thus for example, if X is a scheme over Spec B for some ring B, the cohomology groups of any $\mathcal{O}_X$-module $\mathscr{F}$ have a natural structure of B-module.

A Vanishing Theorem of Grothendieck

Theorem 2.7 (Grothendieck [1]). *Let X be a noetherian topological space of dimension n. Then for all $i > n$ and all sheaves of abelian groups $\mathscr{F}$ on X, we have $H^i(X,\mathscr{F}) = 0$.*

Before proving the theorem, we need some preliminary results, mainly concerning direct limits. If $(\mathscr{F}_\alpha)$ is a direct system of sheaves on X, indexed by a directed set A, then we have defined the direct limit $\varinjlim \mathscr{F}_\alpha$ (II, Ex. 1.10).

Lemma 2.8. *On a noetherian topological space, a direct limit of flasque sheaves is flasque.*

PROOF. Let $(\mathscr{F}_\alpha)$ be a directed system of flasque sheaves. Then for any inclusion of open sets $V \subseteq U$, and for each α, we have $\mathscr{F}_\alpha(U) \to \mathscr{F}_\alpha(V)$ is surjective. Since $\varinjlim$ is an exact functor, we get

$$\varinjlim \mathscr{F}_\alpha(U) \to \varinjlim \mathscr{F}_\alpha(V)$$

is also surjective. But on a noetherian topological space, $\varinjlim \mathscr{F}_\alpha(U) = (\varinjlim \mathscr{F}_\alpha)(U)$ for any open set (II, Ex. 1.11). So we have

$$(\varinjlim \mathscr{F}_\alpha)(U) \to (\varinjlim \mathscr{F}_\alpha)(V)$$

is surjective, and so $\varinjlim \mathscr{F}_\alpha$ is flasque.

Proposition 2.9. *Let X be a noetherian topological space, and let $(\mathscr{F}_\alpha)$ be a direct system of abelian sheaves. Then there are natural isomorphisms, for each $i \geqslant 0$*

$$\varinjlim H^i(X, \mathscr{F}_\alpha) \to H^i(X, \varinjlim \mathscr{F}_\alpha).$$

PROOF. For each α we have a natural map $\mathscr{F}_\alpha \to \varinjlim \mathscr{F}_\alpha$. This induces a map on cohomology, and then we take the direct limit of these maps. For $i = 0$, the result is already known (II, Ex. 1.11). For the general case, we consider the category $\mathrm{in}\mathfrak{d}_A(\mathfrak{Ab}(X))$ consisting of all directed systems of objects of $\mathfrak{Ab}(X)$, indexed by A. This is an abelian category. Furthermore, since $\varinjlim$ is an exact functor, we have a natural transformation of δ-functors

$$\varinjlim H^i(X, \cdot) \to H^i(X, \varinjlim \cdot)$$

from $\mathrm{in}\mathfrak{d}_A(\mathfrak{Ab}(X))$ to $\mathfrak{Ab}$. They agree for $i = 0$, so to prove they are the same, it will be sufficient to show they are both effaceable for $i > 0$. For in that case, they are both universal by (1.3A), and so must be isomorphic.

So let $(\mathscr{F}_\alpha) \in \mathrm{in}\mathfrak{d}_A(\mathfrak{Ab}(X))$. For each α, let $\mathscr{G}_\alpha$ be the sheaf of discontinuous sections of $\mathscr{F}_\alpha$ (II, Ex. 1.16e). Then $\mathscr{G}_\alpha$ is flasque, and there is a natural inclusion $\mathscr{F}_\alpha \to \mathscr{G}_\alpha$. Furthermore, the construction of $\mathscr{G}_\alpha$ is functorial, so the $\mathscr{G}_\alpha$ also form a direct system, and we obtain a monomorphism $u : (\mathscr{F}_\alpha) \to (\mathscr{G}_\alpha)$ in the category $\mathrm{in}\mathfrak{d}_A(\mathfrak{Ab}(X))$. Now the $\mathscr{G}_\alpha$ are all flasque, so $H^i(X, \mathscr{G}_\alpha) = 0$ for $i > 0$ (2.5). Thus $\varinjlim H^i(X, \mathscr{G}_\alpha) = 0$, and the functor on the left-hand side is effaceable for $i > 0$. On the other hand, $\varinjlim \mathscr{G}_\alpha$ is also flasque by (2.8). So $H^i(X, \varinjlim \mathscr{G}_\alpha) = 0$ for $i > 0$, and we see that the functor on the right-hand side is also effaceable. This completes the proof.

Remark 2.9.1. As a special case we see that cohomology commutes with infinite direct sums.

Lemma 2.10. *Let Y be a closed subset of X, let $\mathscr{F}$ be a sheaf of abelian groups on Y, and let $j : Y \to X$ be the inclusion. Then $H^i(Y, \mathscr{F}) = H^i(X, j_* \mathscr{F})$, where $j_* \mathscr{F}$ is the extension of $\mathscr{F}$ by zero outside Y (II, Ex. 1.19).*

PROOF. If $\mathscr{J}^{\cdot}$ is a flasque resolution of $\mathscr{F}$ on Y, then $j_*\mathscr{J}^{\cdot}$ is a flasque resolution of $j_*\mathscr{F}$ on X, and for each i, $\Gamma(Y,\mathscr{J}^i) = \Gamma(X,j_*\mathscr{J}^i)$. So we get the same cohomology groups.

Remark 2.10.1. Continuing our earlier abuse of notation (II, Ex. 1.19), we often write $\mathscr{F}$ instead of $j_*\mathscr{F}$. This lemma shows there will be no ambiguity about the cohomology groups.

PROOF OF (2.7). First we fix some notation. If Y is a closed subset of X, then for any sheaf $\mathscr{F}$ on X we let $\mathscr{F}_Y = j_*(\mathscr{F}|_Y)$, where $j: Y \to X$ is the inclusion. If U is an open subset of X, we let $\mathscr{F}_U = i_!(\mathscr{F}|_U)$, where $i: U \to X$ is the inclusion. In particular, if $U = X - Y$, we have an exact sequence (II, Ex. 1.19)

$$0 \to \mathscr{F}_U \to \mathscr{F} \to \mathscr{F}_Y \to 0.$$

We will prove the theorem by induction on $n = \dim X$, in several steps.

Step 1. Reduction to the case X irreducible. If X is reducible, let Y be one of its irreducible components, and let $U = X - Y$. Then for any $\mathscr{F}$ we have an exact sequence

$$0 \to \mathscr{F}_U \to \mathscr{F} \to \mathscr{F}_Y \to 0.$$

From the long exact sequence of cohomology, it will be sufficient to prove that $H^i(X,\mathscr{F}_Y) = 0$ and $H^i(X,\mathscr{F}_U) = 0$ for $i > n$. But Y is closed and irreducible, and $\mathscr{F}_U$ can be regarded as a sheaf on the closed subset $\bar{U}$, which has one fewer irreducible components than X. Thus using (2.10) and induction on the number of irreducible components, we reduce to the case X irreducible.

Step . Suppose X is irreducible of dimension 0. Then the only open subsets of X are X and the empty set. For otherwise, X would have a proper irreducible closed subset, and $\dim X$ would be $\geqslant 1$. Thus $\Gamma(X,\cdot)$ induces an equivalence of categories $\mathfrak{Ab}(X) \to \mathfrak{Ab}$. In particular, $\Gamma(X,\cdot)$ is an exact functor, so $H^i(X,\mathscr{F}) = 0$ for $i > 0$, and for all $\mathscr{F}$.

Step 3. Now let X be irreducible of dimension n, and let $\mathscr{F} \in \mathfrak{Ab}(X)$. Let $B = \bigcup_{U \subseteq X} \mathscr{F}(U)$, and let A be the set of all finite subsets of B. For each $\alpha \in A$, let $\mathscr{F}_\alpha$ be the subsheaf of $\mathscr{F}$ generated by the sections in α (over various open sets). Then A is a directed set, and $\mathscr{F} = \varinjlim \mathscr{F}_\alpha$. So by (2.9), it will be sufficient to prove vanishing of cohomology for each $\mathscr{F}_\alpha$. If α' is a subset of α, then we have an exact sequence

$$0 \to \mathscr{F}_{\alpha'} \to \mathscr{F}_\alpha \to \mathscr{G} \to 0,$$

where $\mathscr{G}$ is a sheaf generated by $\#(\alpha - \alpha')$ sections over suitable open sets. Thus, using the long exact sequence of cohomology, and induction on $\#(\alpha)$, we reduce to the case that $\mathscr{F}$ is generated by a single section over some open set U. In that case $\mathscr{F}$ is a quotient of the sheaf $\mathbf{Z}_U$ (where $\mathbf{Z}$ denotes the constant sheaf $\mathbf{Z}$ on X). Letting $\mathscr{R}$ be the kernel, we have an exact sequence

$$0 \to \mathscr{R} \to \mathbf{Z}_U \to \mathscr{F} \to 0.$$

Again using the long exact sequence of cohomology, it will be sufficient to prove vanishing for $\mathscr{R}$ and for $\mathbf{Z}_U$.

Step 4. Let U be an open subset of X and let $\mathscr{R}$ be a subsheaf of $\mathbf{Z}_U$. For each $x \in U$, the stalk $\mathscr{R}_x$ is a subgroup of $\mathbf{Z}$. If $\mathscr{R} = 0$, skip to Step 5. If not, let d be the least positive integer which occurs in any of the groups $\mathscr{R}_x$. Then there is a nonempty open subset $V \subseteq U$ such that $\mathscr{R}|_V \cong d \cdot \mathbf{Z}|_V$ as a subsheaf of $\mathbf{Z}|_V$. Thus $\mathscr{R}_V \cong \mathbf{Z}_V$ and we have an exact sequence

$$0 \to \mathbf{Z}_V \to \mathscr{R} \to \mathscr{R}/\mathbf{Z}_V \to 0.$$

Now the sheaf $\mathscr{R}/\mathbf{Z}_V$ is supported on the closed subset $(U - V)^-$ of X, which has dimension $<n$, since X is irreducible. So using (2.10) and the induction hypothesis, we know $H^i(X, \mathscr{R}/\mathbf{Z}_V) = 0$ for $i \geqslant n$. So by the long exact sequence of cohomology we need only show vanishing for $\mathbf{Z}_V$.

Step 5. To complete the proof, we need only show that for any open subset $U \subseteq X$, we have $H^i(X, \mathbf{Z}_U) = 0$ for $i > n$. Let $Y = X - U$. Then we have an exact sequence

$$0 \to \mathbf{Z}_U \to \mathbf{Z} \to \mathbf{Z}_Y \to 0.$$

Now $\dim Y < \dim X$ since X is irreducible, so using (2.10) and the induction hypothesis, we have $H^i(X, \mathbf{Z}_Y) = 0$ for $i \geqslant n$. On the other hand, $\mathbf{Z}$ is flasque, since it is a constant sheaf on an irreducible space (II, Ex. 1.16a). Hence $H^i(X, \mathbf{Z}) = 0$ for $i > 0$ by (2.5). So from the long exact sequence of cohomology we have $H^i(X, \mathbf{Z}_U) = 0$ for $i > n$.　　　　　q.e.d.

Historical Note: The derived functor cohomology which we defined in this section was introduced by Grothendieck [1]. It is the theory which is used in [EGA]. The use of sheaf cohomology in algebraic geometry started with Serre [3]. In that paper, and in the later paper [4], Serre used Čech cohomology for coherent sheaves on an algebraic variety with its Zariski topology. The equivalence of this theory with the derived functor theory follows from the "theorem of Leray" (Ex. 4.11). The same argument, using Cartan's "Theorem B" shows that the Čech cohomology of a coherent analytic sheaf on a complex analytic space is equal to the derived functor cohomology. Gunning and Rossi [1] use a cohomology theory computed by fine resolutions of a sheaf on a paracompact Hausdorff space. The equivalence of this theory with ours is shown by Godement [1, Thm. 4.7.1, p. 181 and Ex. 7.2.1, p. 263], who shows at the same time that both theories coincide with his theory which is defined by a canonical flasque resolution. Godement also shows [1, Thm. 5.10.1, p. 228] that on a paracompact Hausdorff space, his theory coincides with Čech cohomology. This provides a bridge to the standard topological theories with constant coefficients, as developed in the book of Spanier [1]. He shows that on a paracompact Hausdorff space, Čech cohomology and Alexander cohomology and singular cohomology all agree (see Spanier [1, pp. 314, 327, 334]).

The vanishing theorem (2.7) was proved by Serre [3] for coherent sheaves on algebraic curves and projective algebraic varieties, and later [5] for abstract algebraic varieties. It is analogous to the theorem that singular cohomology on a (real) manifold of dimension n vanishes in degrees $i > n$.

EXERCISES

2.1. (a) Let $X = \mathbf{A}_k^1$ be the affine line over an infinite field k. Let P, Q be distinct closed points of X, and let $U = X - \{P, Q\}$. Show that $H^1(X, \mathbf{Z}_U) \neq 0$.

   *(b) More generally, let $Y \subseteq X = \mathbf{A}_k^n$ be the union of $n + 1$ hyperplanes in suitably general position, and let $U = X - Y$. Show that $H^n(X, \mathbf{Z}_U) \neq 0$. Thus the result of (2.7) is the best possible.

2.2. Let $X = \mathbf{P}_k^1$ be the projective line over an algebraically closed field k. Show that the exact sequence $0 \to \mathcal{O} \to \mathcal{K} \to \mathcal{K}/\mathcal{O} \to 0$ of (II, Ex. 1.21d) is a flasque resolution of $\mathcal{O}$. Conclude from (II, Ex. 1.21e) that $H^i(X, \mathcal{O}) = 0$ for all $i > 0$.

2.3. *Cohomology with Supports* (Grothendieck [7]). Let X be a topological space, let Y be a closed subset, and let $\mathcal{F}$ be a sheaf of abelian groups. Let $\Gamma_Y(X, \mathcal{F})$ denote the group of sections of $\mathcal{F}$ with support in Y (II, Ex. 1.20).

   (a) Show that $\Gamma_Y(X, \cdot)$ is a left exact functor from $\mathfrak{Ab}(X)$ to $\mathfrak{Ab}$.

   We denote the right derived functors of $\Gamma_Y(X, \cdot)$ by $H_Y^i(X, \cdot)$. They are the *cohomology groups* of X with *supports* in Y, and coefficients in a given sheaf.

   (b) If $0 \to \mathcal{F}' \to \mathcal{F} \to \mathcal{F}'' \to 0$ is an exact sequence of sheaves, with $\mathcal{F}'$ flasque, show that
$$0 \to \Gamma_Y(X, \mathcal{F}') \to \Gamma_Y(X, \mathcal{F}) \to \Gamma_Y(X, \mathcal{F}'') \to 0$$
   is exact.

   (c) Show that if $\mathcal{F}$ is flasque, then $H_Y^i(X, \mathcal{F}) = 0$ for all $i > 0$.

   (d) If $\mathcal{F}$ is flasque, show that the sequence
$$0 \to \Gamma_Y(X, \mathcal{F}) \to \Gamma(X, \mathcal{F}) \to \Gamma(X - Y, \mathcal{F}) \to 0$$
   is exact.

   (e) Let $U = X - Y$. Show that for any $\mathcal{F}$, there is a long exact sequence of cohomology groups
$$0 \to H_Y^0(X, \mathcal{F}) \to H^0(X, \mathcal{F}) \to H^0(U, \mathcal{F}|_U) \to$$
$$\to H_Y^1(X, \mathcal{F}) \to H^1(X, \mathcal{F}) \to H^1(U, \mathcal{F}|_U) \to$$
$$\to H_Y^2(X, \mathcal{F}) \to \dots.$$

   (f) *Excision.* Let V be an open subset of X containing Y. Then there are natural functorial isomorphisms, for all i and $\mathcal{F}$,
$$H_Y^i(X, \mathcal{F}) \cong H_Y^i(V, \mathcal{F}|_V).$$

2.4. *Mayer–Vietoris Sequence.* Let Y_1, Y_2 be two closed subsets of X. Then there is a long exact sequence of cohomology with supports
$$\dots \to H_{Y_1 \cap Y_2}^i(X, \mathcal{F}) \to H_{Y_1}^i(X, \mathcal{F}) \oplus H_{Y_2}^i(X, \mathcal{F}) \to H_{Y_1 \cup Y_2}^i(X, \mathcal{F}) \to$$
$$\to H_{Y_1 \cap Y_2}^{i+1}(X, \mathcal{F}) \to \dots.$$

2.5. Let X be a Zariski space (II, Ex. 3.17). Let $P \in X$ be a closed point, and let X_P be the subset of X consisting of all points $Q \in X$ such that $P \in \{Q\}^-$. We call X_P the *local space* of X at P, and give it the induced topology. Let $j: X_P \to X$ be the inclusion, and for any sheaf $\mathscr{F}$ on X, let $\mathscr{F}_P = j^*\mathscr{F}$. Show that for all i, $\mathscr{F}$, we have

$$H^i_P(X, \mathscr{F}) = H^i_P(X_P, \mathscr{F}_P).$$

2.6. Let X be a noetherian topological space, and let $\{\mathscr{I}_\alpha\}_{\alpha \in A}$ be a direct system of injective sheaves of abelian groups on X. Then $\varinjlim \mathscr{I}_\alpha$ is also injective. [*Hints:* First show that a sheaf $\mathscr{I}$ is injective if and only if for every open set $U \subseteq X$, and for every subsheaf $\mathscr{R} \subseteq \mathbf{Z}_U$, and for every map $f: \mathscr{R} \to \mathscr{I}$, there exists an extension of f to a map of $\mathbf{Z}_U \to \mathscr{I}$. Secondly, show that any such sheaf $\mathscr{R}$ is finitely generated, so any map $\mathscr{R} \to \varinjlim \mathscr{I}_\alpha$ factors through one of the $\mathscr{I}_\alpha$.]

2.7. Let S^1 be the circle (with its usual topology), and let $\mathbf{Z}$ be the constant sheaf $\mathbf{Z}$.
 (a) Show that $H^1(S^1, \mathbf{Z}) \cong \mathbf{Z}$, using our definition of cohomology.
 (b) Now let $\mathscr{R}$ be the sheaf of germs of continuous real-valued functions on S^1. Show that $H^1(S^1, \mathscr{R}) = 0$.

3 Cohomology of a Noetherian Affine Scheme

In this section we will prove that if $X = \operatorname{Spec} A$ is a noetherian affine scheme, then $H^i(X, \mathscr{F}) = 0$ for all $i > 0$ and all quasi-coherent sheaves $\mathscr{F}$ of $\mathcal{O}_X$-modules. The key point is to show that if I is an injective A-module, then the sheaf $\tilde{I}$ on $\operatorname{Spec} A$ is flasque. We begin with some algebraic preliminaries.

Proposition 3.1A (Krull's Theorem). *Let A be a noetherian ring, let $M \subseteq N$ be finitely generated A-modules, and let $\mathfrak{a}$ be an ideal of A. Then the $\mathfrak{a}$-adic topology on M is induced by the $\mathfrak{a}$-adic topology on N. In particular, for any $n > 0$, there exists an $n' \geqslant n$ such that $\mathfrak{a}^n M \supseteq M \cap \mathfrak{a}^{n'} N$.*
PROOF. Atiyah–Macdonald [1, 10.11] or Zariski–Samuel [1, vol. II, Ch. VIII, Th. 4].

Recall (II, Ex. 5.6) that for any ring A, and any ideal $\mathfrak{a} \subseteq A$, and any A-module M, we have defined the submodule $\Gamma_\mathfrak{a}(M)$ to be $\{m \in M \,|\, \mathfrak{a}^n m = 0$ for some $n > 0\}$.

Lemma 3.2. *Let A be a noetherian ring, let $\mathfrak{a}$ be an ideal of A, and let I be an injective A-module. Then the submodule $J = \Gamma_\mathfrak{a}(I)$ is also an injective A-module.*

PROOF. To show that J is injective, it will be sufficient to show that for any ideal $\mathfrak{b} \subseteq A$, and for any homomorphism $\varphi: \mathfrak{b} \to J$, there exists a homomorphism $\psi: A \to J$ extending φ. (This is a well-known criterion for an injective module—Godement [1, I, 1.4.1]). Since A is noetherian, $\mathfrak{b}$ is finitely generated. On the other hand, every element of J is annihilated by some

213

power of $\mathfrak{a}$, so there exists an $n > 0$ such that $\mathfrak{a}^n\varphi(b) = 0$, or equivalently, $\varphi(\mathfrak{a}^n b) = 0$. Now applying (3.1A) to the inclusion $\mathfrak{b} \subseteq A$, we find that there is an $n' \geqslant n$ such that $\mathfrak{a}^{n'}\mathfrak{b} \supseteq \mathfrak{b} \cap \mathfrak{a}^{n'}$. Hence $\varphi(\mathfrak{b} \cap \mathfrak{a}^{n'}) = 0$, and so the map $\varphi:\mathfrak{b} \to J$ factors through $\mathfrak{b}/(\mathfrak{b} \cap \mathfrak{a}^{n'})$. Now we consider the following diagram:

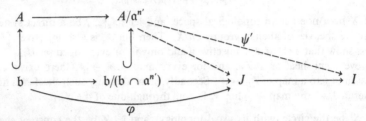

Since I is injective, the composed map of $\mathfrak{b}/(\mathfrak{b} \cap \mathfrak{a}^{n'})$ to I extends to a map $\psi':A/\mathfrak{a}^{n'} \to I$. But the image of ψ' is annihilated by $\mathfrak{a}^{n'}$, so it is contained in J. Composing with the natural map $A \to A/\mathfrak{a}^{n'}$, we obtain the required map $\psi:A \to J$ extending φ.

Lemma 3.3. *Let I be an injective module over a noetherian ring A. Then for any $f \in A$, the natural map of I to its localization I_f is surjective.*

PROOF. For each $i > 0$, let $\mathfrak{b}_i$ be the annihilator of f^i in A. Then $\mathfrak{b}_1 \subseteq \mathfrak{b}_2 \subseteq \ldots$, and since A is noetherian, there is an r such that $\mathfrak{b}_r = \mathfrak{b}_{r+1} = \ldots$. Now let $\theta:I \to I_f$ be the natural map, and let $x \in I_f$ be any element. Then by definition of localization, there is a $y \in I$ and an $n \geqslant 0$ such that $x = \theta(y)/f^n$. We define a map φ from the ideal (f^{n+r}) of A to I by sending f^{n+r} to $f^r y$. This is possible, because the annihilator of f^{n+r} is $\mathfrak{b}_{n+r} = \mathfrak{b}_r$, and $\mathfrak{b}_r$ annihilates $f^r y$. Since I is injective, φ extends to a map $\psi:A \to I$. Let $\psi(1) = z$. Then $f^{n+r}z = f^r y$. But this implies that $\theta(z) = \theta(y)/f^n = x$. Hence θ is surjective.

Proposition 3.4. *Let I be an injective module over a noetherian ring A. Then the sheaf $\tilde{I}$ on $X = \operatorname{Spec} A$ is flasque.*

PROOF. We will use noetherian induction on $Y = (\operatorname{Supp} \tilde{I})^-$. See (II, Ex. 1.14) for the notion of support. If Y consists of a single closed point of X, then $\tilde{I}$ is a skyscraper sheaf (II, Ex. 1.17) which is obviously flasque.

In the general case, to show that $\tilde{I}$ is flasque, it will be sufficient to show, for any open set $U \subseteq X$, that $\Gamma(X,\tilde{I}) \to \Gamma(U,\tilde{I})$ is surjective. If $Y \cap U = \varnothing$, there is nothing to prove. If $Y \cap U \neq \varnothing$, we can find an $f \in A$ such that the open set $X_f = D(f)$ (II, §2) is contained in U and $X_f \cap Y \neq \varnothing$. Let $Z = X - X_f$, and consider the following diagram:

$$\Gamma(X,\tilde{I}) \to \Gamma(U,\tilde{I}) \to \Gamma(X_f,\tilde{I})$$
$$\uparrow \qquad\qquad \uparrow$$
$$\Gamma_Z(X,\tilde{I}) \to \Gamma_Z(U,\tilde{I}),$$

where Γ_Z denotes sections with support in Z (II, Ex. 1.20). Now given a section $s \in \Gamma(U, \tilde{I})$, we consider its image s' in $\Gamma(X_f, \tilde{I})$. But $\Gamma(X_f, \tilde{I}) = I_f$ (II, 5.1), so by (3.3), there is a $t \in I = \Gamma(X, \tilde{I})$ restricting to s'. Let t' be the restriction of t to $\Gamma(U, \tilde{I})$. Then $s - t'$ goes to 0 in $\Gamma(X_f, \tilde{I})$, so it has support in Z. Thus to complete the proof, it will be sufficient to show that $\Gamma_Z(X, \tilde{I}) \to \Gamma_Z(U, \tilde{I})$ is surjective.

Let $J = \Gamma_Z(X, \tilde{I})$. If $\mathfrak{a}$ is the ideal generated by f, then $J = \Gamma_{\mathfrak{a}}(I)$ (II, Ex. 5.6), so by (3.2), J is also an injective A-module. Furthermore, the support of $\tilde{J}$ is contained in $Y \cap Z$, which is strictly smaller than Y. Hence by our induction hypothesis, $\tilde{J}$ is flasque. Since $\Gamma(U, \tilde{J}) = \Gamma_Z(U, \tilde{I})$ (II, Ex. 5.6), we conclude that $\Gamma_Z(X, \tilde{I}) \to \Gamma_Z(U, \tilde{I})$ is surjective, as required.

Theorem 3.5. *Let $X = \operatorname{Spec} A$ be the spectrum of a noetherian ring A. Then for all quasi-coherent sheaves $\mathscr{F}$ on X, and for all $i > 0$, we have $H^i(X, \mathscr{F}) = 0$.*

PROOF. Given $\mathscr{F}$, let $M = \Gamma(X, \mathscr{F})$, and take an injective resolution $0 \to M \to I^\cdot$ of M in the category of A-modules. Then we obtain an exact sequence of sheaves $0 \to \tilde{M} \to \tilde{I}^\cdot$ on X. Now $\mathscr{F} = \tilde{M}$ (II, 5.5) and each $\tilde{I}^i$ is flasque by (3.4), so we can use this resolution of $\mathscr{F}$ to calculate cohomology (2.5.1). Applying the functor Γ, we recover the exact sequence of A-modules $0 \to M \to I^\cdot$. Hence $H^0(X, \mathscr{F}) = M$, and $H^i(X, \mathscr{F}) = 0$ for $i > 0$.

Remark 3.5.1. This result is also true without the noetherian hypothesis, but the proof is more difficult [EGA III, 1.3.1].

Corollary 3.6. *Let X be a noetherian scheme, and let $\mathscr{F}$ be a quasi-coherent sheaf on X. Then $\mathscr{F}$ can be embedded in a flasque, quasi-coherent sheaf $\mathscr{G}$.*

PROOF. Cover X with a finite number of open affines $U_i = \operatorname{Spec} A_i$, and let $\mathscr{F}|_{U_i} = \tilde{M}_i$ for each i. Embed M_i in an injective A_i-module I_i. For each i, let $f: U_i \to X$ be the inclusion, and let $\mathscr{G} = \bigoplus f_*(\tilde{I}_i)$. For each i we have an injective map of sheaves $\mathscr{F}|_{U_i} \to \tilde{I}_i$. Hence we obtain a map $\mathscr{F} \to f_*(\tilde{I}_i)$. Taking the direct sum over i gives a map $\mathscr{F} \to \mathscr{G}$ which is clearly injective. On the other hand, for each i, $\tilde{I}_i$ is flasque (3.4) and quasi-coherent on U_i. Hence $f_*(\tilde{I}_i)$ is also flasque (II, Ex. 1.16d) and quasi-coherent (II, 5.8). Taking the direct sum of these, we see that $\mathscr{G}$ is flasque and quasi-coherent.

Theorem 3.7 (Serre [5]). *Let X be a noetherian scheme. Then the following conditions are equivalent:*

  (i) *X is affine;*
  (ii) *$H^i(X, \mathscr{F}) = 0$ for all $\mathscr{F}$ quasi-coherent and all $i > 0$;*
  (iii) *$H^1(X, \mathscr{I}) = 0$ for all coherent sheaves of ideals $\mathscr{I}$.*

PROOF. (i) $\Rightarrow$ (ii) is (3.5). (ii) $\Rightarrow$ (iii) is trivial, so we have only to prove (iii) $\Rightarrow$ (i). We use the criterion of (II, Ex. 2.17). First we show that X can be covered by open affine subsets of the form X_f, with $f \in A = \Gamma(X, \mathcal{O}_X)$. Let P be a closed point of X, let U be an open affine neighborhood of P, and let $Y = X - U$. Then we have an exact sequence

$$0 \to \mathcal{I}_{Y \cup \{P\}} \to \mathcal{I}_Y \to k(P) \to 0,$$

where $\mathcal{I}_Y$ and $\mathcal{I}_{Y \cup \{P\}}$ are the ideal sheaves of the closed sets Y and $Y \cup \{P\}$, respectively. The quotient is the skyscraper sheaf $k(P) = \mathcal{O}_P / \mathfrak{m}_P$ at P. Now from the exact sequence of cohomology, and hypothesis (iii), we get an exact sequence

$$\Gamma(X, \mathcal{I}_Y) \to \Gamma(X, k(P)) \to H^1(X, \mathcal{I}_{Y \cup \{P\}}) = 0.$$

So there is an element $f \in \Gamma(X, \mathcal{I}_Y)$ which goes to 1 in $k(P)$, i.e., $f_P \equiv 1$ (mod $\mathfrak{m}_P$). Since $\mathcal{I}_Y \subseteq \mathcal{O}_X$, we can consider f as an element of A. Then by construction, we have $P \in X_f \subseteq U$. Furthermore, $X_f = U_{\bar{f}}$, where $\bar{f}$ is the image of f in $\Gamma(U, \mathcal{O}_U)$, so X_f is affine.

Thus every closed point of X has an open affine neighborhood of the form X_f. By quasi-compactness, we can cover X with a finite number of these, corresponding to $f_1, \ldots, f_r \in A$.

Now by (II, Ex. 2.17), to show that X is affine, we need only verify that $f_1, \ldots, f_r$ generate the unit ideal in A. We use $f_1, \ldots, f_r$ to define a map $\alpha : \mathcal{O}_X^r \to \mathcal{O}_X$ by sending $\langle a_1, \ldots, a_r \rangle$ to $\sum f_i a_i$. Since the X_{f_i} cover X, this is a surjective map of sheaves. Let $\mathcal{F}$ be the kernel:

$$0 \to \mathcal{F} \to \mathcal{O}_X^r \xrightarrow{\alpha} \mathcal{O}_X \to 0.$$

We filter $\mathcal{F}$ as follows:

$$\mathcal{F} = \mathcal{F} \cap \mathcal{O}_X^r \supseteq \mathcal{F} \cap \mathcal{O}_X^{r-1} \supseteq \ldots \supseteq \mathcal{F} \cap \mathcal{O}_X$$

for a suitable ordering of the factors of $\mathcal{O}_X^r$. Each of the quotients of this filtration is a coherent sheaf of ideals in $\mathcal{O}_X$. Thus using our hypothesis (iii) and the long exact sequence of cohomology, we climb up the filtration and deduce that $H^1(X, \mathcal{F}) = 0$. But then $\Gamma(X, \mathcal{O}_X^r) \xrightarrow{\alpha} \Gamma(X, \mathcal{O}_X)$ is surjective, which tells us that $f_1, \ldots, f_r$ generate the unit ideal in A. q.e.d.

Remark 3.7.1. This result is analogous to another theorem of Serre in complex analytic geometry, which characterizes Stein spaces by the vanishing of coherent analytic sheaf cohomology.

EXERCISES

3.1. Let X be a noetherian scheme. Show that X is affine if and only if X_{red} (II, Ex. 2.3) is affine. [*Hint:* Use (3.7), and for any coherent sheaf $\mathcal{F}$ on X, consider the filtration $\mathcal{F} \supseteq \mathcal{N} \cdot \mathcal{F} \supseteq \mathcal{N}^2 \cdot \mathcal{F} \supseteq \ldots$, where $\mathcal{N}$ is the sheaf of nilpotent elements on X.]

3.2. Let X be a reduced noetherian scheme. Show that X is affine if and only if each irreducible component is affine.

3.3. Let A be a noetherian ring, and let $\mathfrak{a}$ be an ideal of A.

(a) Show that $\Gamma_\mathfrak{a}(\cdot)$ (II, Ex. 5.6) is a left-exact functor from the category of A-modules to itself. We denote its right derived functors, calculated in $\mathfrak{Mod}(A)$, by $H^i_\mathfrak{a}(\cdot)$.

(b) Now let $X = \operatorname{Spec} A$, $Y = V(\mathfrak{a})$. Show that for any A-module M,

$$H^i_\mathfrak{a}(M) = H^i_Y(X, \tilde{M}),$$

where $H^i_Y(X, \cdot)$ denotes cohomology with supports in Y (Ex. 2.3).

(c) For any i, show that $\Gamma_\mathfrak{a}(H^i_\mathfrak{a}(M)) = H^i_\mathfrak{a}(M)$.

3.4. *Cohomological Interpretation of Depth.* If A is a ring, $\mathfrak{a}$ an ideal, and M an A-module, then $depth_\mathfrak{a} M$ is the maximum length of an M-regular sequence $x_1, \ldots, x_r$, with all $x_i \in \mathfrak{a}$. This generalizes the notion of depth introduced in (II, §8).

(a) Assume that A is noetherian. Show that if $depth_\mathfrak{a} M \geqslant 1$, then $\Gamma_\mathfrak{a}(M) = 0$, and the converse is true if M is finitely generated. [*Hint*: When M is finitely generated, both conditions are equivalent to saying that $\mathfrak{a}$ is not contained in any associated prime of M.]

(b) Show inductively, for M finitely generated, that for any $n \geqslant 0$, the following conditions are equivalent:

(i) $depth_\mathfrak{a} M \geqslant n$;

(ii) $H^i_\mathfrak{a}(M) = 0$ for all $i < n$.

For more details, and related results, see Grothendieck [7].

3.5. Let X be a noetherian scheme, and let P be a closed point of X. Show that the following conditions are equivalent:

(i) $depth\ \mathcal{O}_P \geqslant 2$;

(ii) if U is any open neighborhood of P, then every section of $\mathcal{O}_X$ over $U - P$ extends uniquely to a section of $\mathcal{O}_X$ over U.

This generalizes (I, Ex. 3.20), in view of (II, 8.22A).

3.6. Let X be a noetherian scheme.

(a) Show that the sheaf $\mathcal{G}$ constructed in the proof of (3.6) is an injective object in the category $\mathfrak{Qco}(X)$ of quasi-coherent sheaves on X. Thus $\mathfrak{Qco}(X)$ has enough injectives.

*(b) Show that any injective object of $\mathfrak{Qco}(X)$ is flasque. [*Hints*: The method of proof of (2.4) will *not* work, because $\mathcal{O}_U$ is not quasi-coherent on X in general. Instead, use (II, Ex. 5.15) to show that if $\mathscr{I} \in \mathfrak{Qco}(X)$ is injective, and if $U \subseteq X$ is an open subset, then $\mathscr{I}|_U$ is an injective object of $\mathfrak{Qco}(U)$. Then cover X with open affines ...]

(c) Conclude that one can compute cohomology as the derived functors of $\Gamma(X, \cdot)$, considered as a functor from $\mathfrak{Qco}(X)$ to $\mathfrak{Ab}$.

3.7. Let A be a noetherian ring, let $X = \operatorname{Spec} A$, let $\mathfrak{a} \subseteq A$ be an ideal, and let $U \subseteq X$ be the open set $X - V(\mathfrak{a})$.

(a) For any A-module M, establish the following formula of Deligne:

$$\Gamma(U, \tilde{M}) \cong \varinjlim_n \operatorname{Hom}_A(\mathfrak{a}^n, M).$$

(b) Apply this in the case of an injective A-module I, to give another proof of (3.4).

3.8. Without the noetherian hypothesis, (3.3) and (3.4) are false. Let $A = k[x_0, x_1, x_2, \ldots]$ with the relations $x_0^n x_n = 0$ for $n = 1, 2, \ldots$. Let I be an injective A-module containing A. Show that $I \to I_{x_0}$ is not surjective.

4 Čech Cohomology

In this section we construct the Čech cohomology groups for a sheaf of abelian groups on a topological space X, with respect to a given open covering of X. We will prove that if X is a noetherian separated scheme, the sheaf is quasi-coherent, and the covering is an open affine covering, then these Čech cohomology groups coincide with the cohomology groups defined in §2. The value of this result is that it gives a practical method for computing cohomology of quasi-coherent sheaves on a scheme.

Let X be a topological space, and let $\mathfrak{U} = (U_i)_{i \in I}$ be an open covering of X. Fix, once and for all, a well-ordering of the index set I. For any finite set of indices $i_0, \ldots, i_p \in I$ we denote the intersection $U_{i_0} \cap \ldots \cap U_{i_p}$ by $U_{i_0, \ldots, i_p}$.

Now let $\mathscr{F}$ be a sheaf of abelian groups on X. We define a complex $C^{\cdot}(\mathfrak{U}, \mathscr{F})$ of abelian groups as follows. For each $p \geqslant 0$, let

$$C^p(\mathfrak{U}, \mathscr{F}) = \prod_{i_0 < \ldots < i_p} \mathscr{F}(U_{i_0, \ldots, i_p}).$$

Thus an element $\alpha \in C^p(\mathfrak{U}, \mathscr{F})$ is determined by giving an element

$$\alpha_{i_0, \ldots, i_p} \in \mathscr{F}(U_{i_0, \ldots, i_p}),$$

for each $(p + 1)$-tuple $i_0 < \ldots < i_p$ of elements of I. We define the coboundary map $d : C^p \to C^{p+1}$ by setting

$$(d\alpha)_{i_0, \ldots, i_{p+1}} = \sum_{k=0}^{p+1} (-1)^k \alpha_{i_0, \ldots, \hat{i}_k, \ldots, i_{p+1}}\big|_{U_{i_0, \ldots, i_{p+1}}}.$$

Here the notation $\hat{i}_k$ means omit i_k. Then since $\alpha_{i_0, \ldots, \hat{i}_k, \ldots, i_{p+1}}$ is an element of $\mathscr{F}(U_{i_0, \ldots, \hat{i}_k, \ldots, i_{p+1}})$, we restrict to $U_{i_0, \ldots, i_{p+1}}$ to get an element of $\mathscr{F}(U_{i_0, \ldots, i_{p+1}})$. One checks easily that $d^2 = 0$, so we have indeed defined a complex of abelian groups.

Remark 4.0.1. If $\alpha \in C^p(\mathfrak{U}, \mathscr{F})$, it is sometimes convenient to have the symbol $\alpha_{i_0, \ldots, i_p}$ defined for *all* $(p + 1)$-tuples of elements of I. If there is a repeated index in the set $\{i_0, \ldots, i_p\}$, we define $\alpha_{i_0, \ldots, i_p} = 0$. If the indices are all distinct, we define $\alpha_{i_0, \ldots, i_p} = (-1)^\sigma \alpha_{\sigma i_0, \ldots, \sigma i_p}$, where σ is the permutation for which $\sigma i_0 < \ldots < \sigma i_p$. With these conventions, one can check that the formula given above for $d\alpha$ remains correct for any $(p + 2)$-tuple $i_0, \ldots, i_{p+1}$ of elements of I.

Definition. Let X be a topological space and let $\mathfrak{U}$ be an open covering of X. For any sheaf of abelian groups $\mathscr{F}$ on X, we define the pth Čech cohomology group of $\mathscr{F}$, with respect to the covering $\mathfrak{U}$, to be

$$\check{H}^p(\mathfrak{U},\mathscr{F}) = h^p(C^{\cdot}(\mathfrak{U},\mathscr{F})).$$

Caution 4.0.2. Keeping X and $\mathfrak{U}$ fixed, if $0 \to \mathscr{F}' \to \mathscr{F} \to \mathscr{F}'' \to 0$ is a short exact sequence of sheaves of abelian groups on X, we do *not* in general get a long exact sequence of Čech cohomology groups. In other words, the functors $\check{H}^p(\mathfrak{U},\cdot)$ do not form a δ-functor (§1). For example, if $\mathfrak{U}$ consists of the single open set X, then this results from the fact that the global section functor $\Gamma(X,\cdot)$ is not exact.

Example 4.0.3. To illustrate how well suited Čech cohomology is for computations, we will compute some examples. Let $X = \mathbf{P}_k^1$, let $\mathscr{F}$ be the sheaf of differentials Ω (II, §8), and let $\mathfrak{U}$ be the open covering by the two open sets $U = \mathbf{A}^1$ with affine coordinate x, and $V = \mathbf{A}^1$ with affine coordinate $y = 1/x$. Then the Čech complex has only two terms:

$$C^0 = \Gamma(U,\Omega) \times \Gamma(V,\Omega)$$
$$C^1 = \Gamma(U \cap V,\Omega).$$

Now

$$\Gamma(U,\Omega) = k[x]\,dx$$
$$\Gamma(V,\Omega) = k[y]\,dy$$
$$\Gamma(U \cap V,\Omega) = k\left[x,\frac{1}{x}\right]dx,$$

and the map $d:C^0 \to C^1$ is given by

$$x \mapsto x$$
$$y \mapsto \frac{1}{x}$$
$$dy \mapsto -\frac{1}{x^2}\,dx.$$

So ker d is the set of pairs $\langle f(x)\,dx, g(y)\,dy \rangle$ such that

$$f(x) = -\frac{1}{x^2}\,g\left(\frac{1}{x}\right).$$

This can happen only if $f = g = 0$, since one side is a polynomial in x and the other side is a polynomial in $1/x$ with no constant term. So $\check{H}^0(\mathfrak{U},\Omega) = 0$. To compute H^1, note that the image of d is the set of all expressions

$$\left(f(x) + \frac{1}{x^2}\,g\left(\frac{1}{x}\right)\right)dx,$$

where f and g are polynomials. This gives the subvector space of $k[x, 1/x]\,dx$ generated by all $x^n\,dx$, $n \in \mathbf{Z}$, $n \neq -1$. Therefore $\check{H}^1(\mathfrak{U}, \Omega) \cong k$, generated by the image of $x^{-1}\,dx$.

Example 4.0.4. Let S^1 be the circle (in its usual topology), let $\mathbf{Z}$ be the constant sheaf $\mathbf{Z}$, and let $\mathfrak{U}$ be the open covering by two connected open semi-circles U, V, which overlap at each end, so that $U \cap V$ consists of two small intervals. Then

$$C^0 = \Gamma(U, \mathbf{Z}) \times \Gamma(V, \mathbf{Z}) = \mathbf{Z} \times \mathbf{Z}$$
$$C^1 = \Gamma(U \cap V, \mathbf{Z}) = \mathbf{Z} \times \mathbf{Z}$$

and the map $d: C^0 \to C^1$ takes $\langle a,b \rangle$ to $\langle b - a, b - a \rangle$. Thus $\check{H}^0(\mathfrak{U}, \mathbf{Z}) = \mathbf{Z}$ and $\check{H}^1(\mathfrak{U}, \mathbf{Z}) = \mathbf{Z}$. Since we know this is the right answer (Ex. 2.7), this illustrates the general principle that Čech cohomology agrees with the usual cohomology provided the open covering is taken fine enough so that there is no cohomology on any of the open sets (Ex. 4.11).

Now we will study some properties of the Čech cohomology groups.

Lemma 4.1. *For any* $X, \mathfrak{U}, \mathscr{F}$ *as above, we have* $\check{H}^0(\mathfrak{U}, \mathscr{F}) \cong \Gamma(X, \mathscr{F})$.

PROOF. $\check{H}^0(\mathfrak{U}, \mathscr{F}) = \ker(d: C^0(\mathfrak{U}, \mathscr{F}) \to C^1(\mathfrak{U}, \mathscr{F}))$. If $\alpha \in C^0$ is given by $\{\alpha_i \in \mathscr{F}(U_i)\}$, then for each $i < j$, $(d\alpha)_{ij} = \alpha_j - \alpha_i$. So $d\alpha = 0$ says the sections α_i and α_j agree on $U_i \cap U_j$. Thus it follows from the sheaf axioms that $\ker d = \Gamma(X, \mathscr{F})$.

Next we define a "sheafified" version of the Čech complex. For any open set $V \subseteq X$, let $f: V \to X$ denote the inclusion map. Now given $X, \mathfrak{U}, \mathscr{F}$ as above, we construct a complex $\mathscr{C}^{\cdot}(\mathfrak{U}, \mathscr{F})$ of sheaves on X as follows. For each $p \geqslant 0$, let

$$\mathscr{C}^p(\mathfrak{U}, \mathscr{F}) = \prod_{i_0 < \ldots < i_p} f_*(\mathscr{F}|_{U_{i_0, \ldots, i_p}}),$$

and define

$$d: \mathscr{C}^p \to \mathscr{C}^{p+1}$$

by the same formula as above. Note by construction that for each p we have $\Gamma(X, \mathscr{C}^p(\mathfrak{U}, \mathscr{F})) = C^p(\mathfrak{U}, \mathscr{F})$.

Lemma 4.2. *For any sheaf of abelian groups* $\mathscr{F}$ *on* X, *the complex* $\mathscr{C}^{\cdot}(\mathfrak{U}, \mathscr{F})$ *is a resolution of* $\mathscr{F}$, *i.e., there is a natural map* $\varepsilon: \mathscr{F} \to \mathscr{C}^0$ *such that the sequence of sheaves*

$$0 \to \mathscr{F} \xrightarrow{\varepsilon} \mathscr{C}^0(\mathfrak{U}, \mathscr{F}) \to \mathscr{C}^1(\mathfrak{U}, \mathscr{F}) \to \ldots$$

is exact.

PROOF. We define $\varepsilon: \mathscr{F} \to \mathscr{C}^0$ by taking the product of the natural maps $\mathscr{F} \to f_*(\mathscr{F}|_{U_i})$ for $i \in I$. Then the exactness at the first step follows from the sheaf axioms for $\mathscr{F}$.

To show the exactness of the complex $\mathscr{C}^{\boldsymbol{\cdot}}$ for $p \geqslant 1$, it is enough to check exactness on the stalks. So let $x \in X$, and suppose $x \in U_j$. For each $p \geqslant 1$, we define a map

$$k: \mathscr{C}^p(\mathfrak{U}, \mathscr{F})_x \to \mathscr{C}^{p-1}(\mathfrak{U}, \mathscr{F})_x$$

as follows. Given $\alpha_x \in \mathscr{C}^p(\mathfrak{U}, \mathscr{F})_x$, it is represented by a section $\alpha \in \Gamma(V, \mathscr{C}^p(\mathfrak{U}, \mathscr{F}))$ over a neighborhood V of x, which we may choose so small that $V \subseteq U_j$. Now for any p-tuple $i_0 < \ldots < i_{p-1}$, we set

$$(k\alpha)_{i_0, \ldots, i_{p-1}} = \alpha_{j, i_0, \ldots, i_{p-1}},$$

using the notational convention of (4.0.1). This makes sense because $V \cap U_{i_0, \ldots, i_{p-1}} = V \cap U_{j, i_0, \ldots, i_{p-1}}$. Then take the stalk of $k\alpha$ at x to get the required map k. Now one checks that for any $p \geqslant 1$, $\alpha \in \mathscr{C}^p_x$,

$$(dk + kd)(\alpha) = \alpha.$$

Thus k is a homotopy operator for the complex $\mathscr{C}^{\boldsymbol{\cdot}}_x$, showing that the identity map is homotopic to the zero map. It follows (§1) that the cohomology groups $h^p(\mathscr{C}^{\boldsymbol{\cdot}}_x)$ of this complex are 0 for $p \geqslant 1$.

Proposition 4.3. *Let X be a topological space, let $\mathfrak{U}$ be an open covering, and let $\mathscr{F}$ be a flasque sheaf of abelian groups on X. Then for all $p > 0$ we have $\check{H}^p(\mathfrak{U}, \mathscr{F}) = 0$.*

PROOF. Consider the resolution $0 \to \mathscr{F} \to \mathscr{C}^{\boldsymbol{\cdot}}(\mathfrak{U}, \mathscr{F})$ given by (4.2). Since $\mathscr{F}$ is flasque, the sheaves $\mathscr{C}^p(\mathfrak{U}, \mathscr{F})$ are flasque for each $p \geqslant 0$. Indeed, for any $i_0, \ldots, i_p$, $\mathscr{F}|_{U_{i_0, \ldots, i_p}}$ is a flasque sheaf on $U_{i_0, \ldots, i_p}$; f_* preserves flasque sheaves (II, Ex. 1.16d), and a product of flasque sheaves is flasque. So by (2.5.1) we can use this resolution to compute the usual cohomology groups of $\mathscr{F}$. But $\mathscr{F}$ is flasque, so $H^p(X, \mathscr{F}) = 0$ for $p > 0$ by (2.5). On the other hand, the answer given by this resolution is

$$h^p(\Gamma(X, \mathscr{C}^{\boldsymbol{\cdot}}(\mathfrak{U}, \mathscr{F}))) = \check{H}^p(\mathfrak{U}, \mathscr{F}).$$

So we conclude that $\check{H}^p(\mathfrak{U}, \mathscr{F}) = 0$ for $p > 0$.

Lemma 4.4. *Let X be a topological space, and $\mathfrak{U}$ an open covering. Then for each $p \geqslant 0$ there is a natural map, functorial in $\mathscr{F}$,*

$$\check{H}^p(\mathfrak{U}, \mathscr{F}) \to H^p(X, \mathscr{F}).$$

PROOF. Let $0 \to \mathscr{F} \to \mathscr{I}^{\boldsymbol{\cdot}}$ be an injective resolution of $\mathscr{F}$ in $\mathfrak{Ab}(X)$. Comparing with the resolution $0 \to \mathscr{F} \to \mathscr{C}^{\boldsymbol{\cdot}}(\mathfrak{U}, \mathscr{F})$ of (4.2), it follows from a general result on complexes (Hilton and Stammbach [1, IV, 4.4]) that there is a morphism of complexes $\mathscr{C}^{\boldsymbol{\cdot}}(\mathfrak{U}, \mathscr{F}) \to \mathscr{I}^{\boldsymbol{\cdot}}$, inducing the identity map on $\mathscr{F}$, and unique up to homotopy. Applying the functors $\Gamma(X, \cdot)$ and h^p, we get the required map.

Theorem 4.5. *Let X be a noetherian separated scheme, let $\mathfrak{U}$ be an open affine cover of X, and let $\mathscr{F}$ be a quasi-coherent sheaf on X. Then for all $p \geqslant 0$, the natural maps of (4.4) give isomorphisms*

$$\check{H}^p(\mathfrak{U},\mathscr{F}) \xrightarrow{\sim} H^p(X,\mathscr{F}).$$

PROOF. For $p = 0$ we have an isomorphism by (4.1). For the general case, embed $\mathscr{F}$ in a flasque, quasi-coherent sheaf $\mathscr{G}$ (3.6), and let $\mathscr{R}$ be the quotient:

$$0 \to \mathscr{F} \to \mathscr{G} \to \mathscr{R} \to 0.$$

For each $i_0 < \ldots < i_p$, the open set $U_{i_0,\ldots,i_p}$ is affine, since it is an intersection of affine open subsets of a separated scheme (II, Ex. 4.3). Since $\mathscr{F}$ is quasi-coherent, we therefore have an exact sequence

$$0 \to \mathscr{F}(U_{i_0,\ldots,i_p}) \to \mathscr{G}(U_{i_0,\ldots,i_p}) \to \mathscr{R}(U_{i_0,\ldots,i_p}) \to 0$$

of abelian groups, by (3.5) or (II, 5.6). Taking products, we find that the corresponding sequence of Čech complexes

$$0 \to C^{\cdot}(\mathfrak{U},\mathscr{F}) \to C^{\cdot}(\mathfrak{U},\mathscr{G}) \to C^{\cdot}(\mathfrak{U},\mathscr{R}) \to 0$$

is exact. Therefore we get a long exact sequence of Čech cohomology groups. Since $\mathscr{G}$ is flasque, its Čech cohomology vanishes for $p > 0$ by (4.3), so we have an exact sequence

$$0 \to \check{H}^0(\mathfrak{U},\mathscr{F}) \to \check{H}^0(\mathfrak{U},\mathscr{G}) \to \check{H}^0(\mathfrak{U},\mathscr{R}) \to \check{H}^1(\mathfrak{U},\mathscr{F}) \to 0$$

and isomorphisms

$$\check{H}^p(\mathfrak{U},\mathscr{R}) \xrightarrow{\sim} \check{H}^{p+1}(\mathfrak{U},\mathscr{F})$$

for each $p \geqslant 1$. Now comparing with the long exact sequence of usual cohomology for the above short exact sequence, using the case $p = 0$, and (2.5), we conclude that the natural map

$$\check{H}^1(\mathfrak{U},\mathscr{F}) \to H^1(X,\mathscr{F})$$

is an isomorphism. But $\mathscr{R}$ is also quasi-coherent (II, 5.7), so we obtain the result for all p by induction.

EXERCISES

4.1. Let $f:X \to Y$ be an affine morphism of noetherian separated schemes (II, Ex. 5.17). Show that for any quasi-coherent sheaf $\mathscr{F}$ on X, there are natural isomorphisms for all $i \geqslant 0$,

$$H^i(X,\mathscr{F}) \cong H^i(Y,f_*\mathscr{F}).$$

[*Hint*: Use (II, 5.8).]

4.2. Prove Chevalley's theorem: Let $f:X \to Y$ be a finite surjective morphism of noetherian separated schemes, with X affine. Then Y is affine.

(a) Let $f:X \to Y$ be a finite surjective morphism of integral noetherian schemes. Show that there is a coherent sheaf $\mathscr{M}$ on X, and a morphism of sheaves $\alpha:\mathscr{O}_Y^r \to f_*\mathscr{M}$ for some $r > 0$, such that α is an isomorphism at the generic point of Y.

(b) For any coherent sheaf $\mathscr{F}$ on Y, show that there is a coherent sheaf $\mathscr{G}$ on X, and a morphism $\beta: f_* \mathscr{G} \to \mathscr{F}'$ which is an isomorphism at the generic point of Y. [*Hint*: Apply $\mathscr{H}om(\cdot, \mathscr{F})$ to α and use (II, Ex. 5.17e).]

(c) Now prove Chevalley's theorem. First use (Ex. 3.1) and (Ex. 3.2) to reduce to the case X and Y integral. Then use (3.7), (Ex. 4.1), consider ker β and coker β, and use noetherian induction on Y.

4.3. Let $X = \mathbf{A}_k^2 = \operatorname{Spec} k[x, y]$, and let $U = X - \{(0,0)\}$. Using a suitable cover of U by open affine subsets, show that $H^1(U, \mathcal{O}_U)$ is isomorphic to the k-vector space spanned by $\{x^i y^j | i, j < 0\}$. In particular, it is infinite-dimensional. (Using (3.5), this provides another proof that U is not affine—cf. (I, Ex. 3.6).)

4.4. On an arbitrary topological space X with an arbitrary abelian sheaf $\mathscr{F}$, Čech cohomology may not give the same result as the derived functor cohomology. But here we show that for H^1, there is an isomorphism if one takes the limit over all coverings.

(a) Let $\mathfrak{U} = (U_i)_{i \in I}$ be an open covering of the topological space X. A *refinement* of $\mathfrak{U}$ is a covering $\mathfrak{V} = (V_j)_{j \in J}$, together with a map $\lambda: J \to I$ of the index sets, such that for each $j \in J$, $V_j \subseteq U_{\lambda(j)}$. If $\mathfrak{V}$ is a refinement of $\mathfrak{U}$, show that there is a natural induced map on Čech cohomology, for any abelian sheaf $\mathscr{F}$, and for each i,

$$\lambda^i: \check{H}^i(\mathfrak{U}, \mathscr{F}) \to \check{H}^i(\mathfrak{V}, \mathscr{F}).$$

The coverings of X form a partially ordered set under refinement, so we can consider the Čech cohomology in the limit

$$\varinjlim_{\mathfrak{U}} \check{H}^i(\mathfrak{U}, \mathscr{F}).$$

(b) For any abelian sheaf $\mathscr{F}$ on X, show that the natural maps (4.4) for each covering

$$\check{H}^i(\mathfrak{U}, \mathscr{F}) \to H^i(X, \mathscr{F})$$

are compatible with the refinement maps above.

(c) Now prove the following theorem. Let X be a topological space, $\mathscr{F}$ a sheaf of abelian groups. Then the natural map

$$\varinjlim_{\mathfrak{U}} \check{H}^1(\mathfrak{U}, \mathscr{F}) \to H^1(X, \mathscr{F})$$

is an isomorphism. [*Hint*: Embed $\mathscr{F}$ in a flasque sheaf $\mathscr{G}$, and let $\mathscr{R} = \mathscr{G}/\mathscr{F}$, so that we have an exact sequence

$$0 \to \mathscr{F} \to \mathscr{G} \to \mathscr{R} \to 0.$$

Define a complex $D^{\cdot}(\mathfrak{U})$ by

$$0 \to C^{\cdot}(\mathfrak{U}, \mathscr{F}) \to C^{\cdot}(\mathfrak{U}, \mathscr{G}) \to D^{\cdot}(\mathfrak{U}) \to 0.$$

Then use the exact cohomology sequence of this sequence of complexes, and the natural map of complexes

$$D^{\cdot}(\mathfrak{U}) \to C^{\cdot}(\mathfrak{U}, \mathscr{R}),$$

and see what happens under refinement.]

4.5. For any ringed space $(X, \mathcal{O}_X)$, let Pic X be the group of isomorphism classes of invertible sheaves (II, §6). Show that Pic $X \cong H^1(X, \mathcal{O}_X^*)$, where $\mathcal{O}_X^*$ denotes the sheaf whose sections over an open set U are the units in the ring $\Gamma(U, \mathcal{O}_X)$, with multiplication as the group operation. [*Hint*: For any invertible sheaf $\mathscr{L}$ on X, cover X by open sets U_i on which $\mathscr{L}$ is free, and fix isomorphisms $\varphi_i : \mathcal{O}_{U_i} \xrightarrow{\sim} \mathscr{L}|_{U_i}$. Then on $U_i \cap U_j$, we get an isomorphism $\varphi_i^{-1} \circ \varphi_j$ of $\mathcal{O}_{U_i \cap U_j}$ with itself. These isomorphisms give an element of $\check{H}^1(\mathfrak{U}, \mathcal{O}_X^*)$. Now use (Ex. 4.4).]

4.6. Let $(X, \mathcal{O}_X)$ be a ringed space, let $\mathscr{I}$ be a sheaf of ideals with $\mathscr{I}^2 = 0$, and let X_0 be the ringed space $(X, \mathcal{O}_X/\mathscr{I})$. Show that there is an exact sequence of sheaves of abelian groups on X,

$$0 \to \mathscr{I} \to \mathcal{O}_X^* \to \mathcal{O}_{X_0}^* \to 0,$$

where $\mathcal{O}_X^*$ (respectively, $\mathcal{O}_{X_0}^*$) denotes the sheaf of (multiplicative) groups of units in the sheaf of rings $\mathcal{O}_X$ (respectively, $\mathcal{O}_{X_0}$); the map $\mathscr{I} \to \mathcal{O}_X^*$ is defined by $a \mapsto 1 + a$, and $\mathscr{I}$ has its usual (additive) group structure. Conclude there is an exact sequence of abelian groups

$$\ldots \to H^1(X, \mathscr{I}) \to \text{Pic } X \to \text{Pic } X_0 \to H^2(X, \mathscr{I}) \to \ldots .$$

4.7. Let X be a subscheme of $\mathbf{P}_k^2$ defined by a single homogeneous equation $f(x_0, x_1, x_2) = 0$ of degree d. (Do not assume f is irreducible.) Assume that $(1,0,0)$ is not on X. Then show that X can be covered by the two open affine subsets $U = X \cap \{x_1 \neq 0\}$ and $V = X \cap \{x_2 \neq 0\}$. Now calculate the Čech complex

$$\Gamma(U, \mathcal{O}_X) \oplus \Gamma(V, \mathcal{O}_X) \to \Gamma(U \cap V, \mathcal{O}_X)$$

explicitly, and thus show that

$$\dim H^0(X, \mathcal{O}_X) = 1,$$
$$\dim H^1(X, \mathcal{O}_X) = \frac{1}{2}(d-1)(d-2).$$

4.8. *Cohomological Dimension* (Hartshorne [3]). Let X be a noetherian separated scheme. We define the *cohomological dimension* of X, denoted cd(X), to be the least integer n such that $H^i(X, \mathscr{F}) = 0$ for all quasi-coherent sheaves $\mathscr{F}$ and all $i > n$. Thus for example, Serre's theorem (3.7) says that cd$(X) = 0$ if and only if X is affine. Grothendieck's theorem (2.7) implies that cd$(X) \leqslant \dim X$.

(a) In the definition of cd(X), show that it is sufficient to consider only coherent sheaves on X. Use (II, Ex. 5.15) and (2.9).

(b) If X is quasi-projective over a field k, then it is even sufficient to consider only locally free coherent sheaves on X. Use (II, 5.18).

(c) Suppose X has a covering by $r + 1$ open affine subsets. Use Čech cohomology to show that cd$(X) \leqslant r$.

*(d) If X is a quasi-projective scheme of dimension r over a field k, then X can be covered by $r + 1$ open affine subsets. Conclude (independently of (2.7)) that cd$(X) \leqslant \dim X$.

(e) Let Y be a set-theoretic complete intersection (I, Ex. 2.17) of codimension r in $X = \mathbf{P}_k^n$. Show that cd$(X - Y) \leqslant r - 1$.

4.9. Let $X = \text{Spec } k[x_1, x_2, x_3, x_4]$ be affine four-space over a field k. Let Y_1 be the plane $x_1 = x_2 = 0$ and let Y_2 be the plane $x_3 = x_4 = 0$. Show that $Y = Y_1 \cup Y_2$ is not a set-theoretic complete intersection in X. Therefore the projective closure

$\bar{Y}$ in $\mathbf{P}_k^4$ is also not a set-theoretic complete intersection. [*Hints:* Use an affine analogue of (Ex. 4.8e). Then show that $H^2(X - Y, \mathcal{O}_X) \neq 0$, by using (Ex. 2.3) and (Ex. 2.4). If $P = Y_1 \cap Y_2$, imitate (Ex. 4.3) to show $H^3(X - P, \mathcal{O}_X) \neq 0$.]

*4.10. Let X be a nonsingular variety over an algebraically closed field k, and let $\mathcal{F}$ be a coherent sheaf on X. Show that there is a one-to-one correspondence between the set of infinitesimal extensions of X by $\mathcal{F}$ (II, Ex. 8.7) up to isomorphism, and the group $H^1(X, \mathcal{F} \otimes \mathcal{T})$, where $\mathcal{T}$ is the tangent sheaf of X (II,§8). [*Hint:* Use (II, Ex. 8.6) and (4.5).]

4.11. This exercise shows that Čech cohomology will agree with the usual cohomology whenever the sheaf has no cohomology on any of the open sets. More precisely, let X be a topological space, $\mathcal{F}$ a sheaf of abelian groups, and $\mathfrak{U} = (U_i)$ an open cover. Assume for any finite intersection $V = U_{i_0} \cap \ldots \cap U_{i_p}$ of open sets of the covering, and for any $k > 0$, that $H^k(V, \mathcal{F}|_V) = 0$. Then prove that for all $p \geqslant 0$, the natural maps

$$\check{H}^p(\mathfrak{U}, \mathcal{F}) \to H^p(X, \mathcal{F})$$

of (4.4) are isomorphisms. Show also that one can recover (4.5) as a corollary of this more general result.

5   The Cohomology of Projective Space

In this section we make explicit calculations of the cohomology of the sheaves $\mathcal{O}(n)$ on a projective space, by using Čech cohomology for a suitable open affine covering. These explicit calculations form the basis for various general results about cohomology of coherent sheaves on projective varieties.

Let A be a noetherian ring, let $S = A[x_0, \ldots, x_r]$, and let $X = \operatorname{Proj} S$ be the projective space $\mathbf{P}_A^r$ over A. Let $\mathcal{O}_X(1)$ be the twisting sheaf of Serre (II, §5). For any sheaf of $\mathcal{O}_X$-modules $\mathcal{F}$, we denote by $\Gamma_*(\mathcal{F})$ the graded S-module $\bigoplus_{n \in \mathbf{Z}} \Gamma(X, \mathcal{F}(n))$ (see II, §5).

Theorem 5.1. *Let A be a noetherian ring, and let $X = \mathbf{P}_A^r$, with $r \geqslant 1$. Then:*
(a) *the natural map $S \to \Gamma_*(\mathcal{O}_X) = \bigoplus_{n \in \mathbf{Z}} H^0(X, \mathcal{O}_X(n))$ is an isomorphism of graded S-modules;*
(b) *$H^i(X, \mathcal{O}_X(n)) = 0$ for $0 < i < r$ and all $n \in \mathbf{Z}$;*
(c) *$H^r(X, \mathcal{O}_X(-r-1)) \cong A$;*
(d) *The natural map*

$$H^0(X, \mathcal{O}_X(n)) \times H^r(X, \mathcal{O}_X(-n-r-1)) \to H^r(X, \mathcal{O}_X(-r-1)) \cong A$$

is a perfect pairing of finitely generated free A-modules, for each $n \in \mathbf{Z}$.

PROOF. Let $\mathcal{F}$ be the quasi-coherent sheaf $\bigoplus_{n \in \mathbf{Z}} \mathcal{O}_X(n)$. Since cohomology commutes with arbitrary direct sums on a noetherian topological space (2.9.1), the cohomology of $\mathcal{F}$ will be the direct sum of the cohomology of the sheaves $\mathcal{O}(n)$. So we will compute the cohomology of $\mathcal{F}$, and keep track of the grading by n, so that we can sort out the pieces at the end. Note that all the cohomology groups in question have a natural structure of A-module (2.6.1).

For each $i = 0, \ldots, r$, let U_i be the open set $D_+(x_i)$. Then each U_i is an open affine subset of X, and the U_i cover X, so we can compute the cohomology of $\mathscr{F}$ by using Čech cohomology for the covering $\mathfrak{U} = (U_i)$, by (4.5). For any set of indices $i_0, \ldots, i_p$, the open set $U_{i_0, \ldots, i_p}$ is just $D_+(x_{i_0} \cdots x_{i_p})$, so by (II, 5.11) we have

$$\mathscr{F}(U_{i_0, \ldots, i_p}) \cong S_{x_{i_0} \cdots x_{i_p}},$$

the localization of S with respect to the element $x_{i_0} \cdots x_{i_p}$. Furthermore, the grading on $\mathscr{F}$ corresponds to the natural grading of $S_{x_{i_0} \cdots x_{i_p}}$ under this isomorphism. Thus the Čech complex of $\mathscr{F}$ is given by

$$C^{\cdot}(\mathfrak{U}, \mathscr{F}): \prod S_{x_{i_0}} \to \prod S_{x_{i_0} x_{i_1}} \to \ldots \to S_{x_0 \cdots x_r},$$

and the modules all have a natural grading compatible with the grading on $\mathscr{F}$.

Now $H^0(X, \mathscr{F})$ is the kernel of the first map, which is just S, as we have seen earlier (II, 5.13). This proves (a).

Next we consider $H^r(X, \mathscr{F})$. It is the cokernel of the last map in the Čech complex, which is

$$d^{r-1}: \prod_k S_{x_0 \cdots \hat{x}_k \cdots x_r} \to S_{x_0 \cdots x_r}.$$

We think of $S_{x_0 \cdots x_r}$ as a free A-module with basis $x_0^{l_0} \cdots x_r^{l_r}$, with $l_i \in \mathbf{Z}$. The image of d^{r-1} is the free submodule generated by those basis elements for which at least one $l_i \geq 0$. Thus $H^r(X, \mathscr{F})$ is a free A-module with basis consisting of the "negative" monomials

$$\{x_0^{l_0} \cdots x_r^{l_r} | l_i < 0 \text{ for each } i\}.$$

Furthermore the grading is given by $\sum l_i$. There is only one such monomial of degree $-r - 1$, namely $x_0^{-1} \cdots x_r^{-1}$, so we see that $H^r(X, \mathcal{O}_X(-r - 1))$ is a free A-module of rank 1. This proves (c).

To prove (d), first note that if $n < 0$, then $H^0(X, \mathcal{O}_X(n)) = 0$ by (a), and $H^r(X, \mathcal{O}_X(-n - r - 1)) = 0$ by what we have just seen, since in that case $-n - r - 1 > -r - 1$, and there are no negative monomials of that degree. So the statement is trivial for $n < 0$. For $n \geq 0$, $H^0(X, \mathcal{O}_X(n))$ has a basis consisting of the usual monomials of degree n, i.e., $\{x_0^{m_0} \cdots x_r^{m_r} | m_i \geq 0$ and $\sum m_i = n\}$. The natural pairing with $H^r(X, \mathcal{O}_X(-n - r - 1))$ into $H^r(X, \mathcal{O}_X(-r - 1))$ is determined by

$$(x_0^{m_0} \cdots x_r^{m_r}) \cdot (x_0^{l_0} \cdots x_r^{l_r}) = x_0^{m_0 + l_0} \cdots x_r^{m_r + l_r},$$

where $\sum l_i = -n - r - 1$, and the object on the right becomes 0 if any $m_i + l_i \geq 0$. So it is clear that we have a perfect pairing, under which $x_0^{-m_0 - 1} \cdots x_r^{-m_r - 1}$ is the dual basis element of $x_0^{m_0} \cdots x_r^{m_r}$.

It remains to prove statement (b), which we will do by induction on r. If $r = 1$ there is nothing to prove, so let $r > 1$. If we localize the complex $C^{\cdot}(\mathfrak{U},\mathscr{F})$ with respect to x_r, as graded S-modules, we get the Čech complex for the sheaf $\mathscr{F}|_{U_r}$ on the space U_r, with respect to the open affine covering $\{U_i \cap U_r | i = 0, \dots, r\}$. By (4.5), this complex gives the cohomology of $\mathscr{F}|_{U_r}$ on U_r, which is 0 for $i > 0$ by (3.5). Since localization is an exact functor, we conclude that $H^i(X,\mathscr{F})_{x_r} = 0$ for $i > 0$. In other words, every element of $H^i(X,\mathscr{F})$, for $i > 0$, is annihilated by some power of x_r.

To complete the proof of (b), we will show that for $0 < i < r$, multiplication by x_r induces a bijective map of $H^i(X,\mathscr{F})$ into itself. Then it will follow that this module is 0.

Consider the exact sequence of graded S-modules

$$0 \to S(-1) \xrightarrow{x_r} S \to S/(x_r) \to 0.$$

This gives the exact sequence of sheaves

$$0 \to \mathcal{O}_X(-1) \to \mathcal{O}_X \to \mathcal{O}_H \to 0$$

on X, where H is the hyperplane $x_r = 0$. Twisting by all $n \in \mathbf{Z}$ and taking the direct sum, we have

$$0 \to \mathscr{F}(-1) \to \mathscr{F} \to \mathscr{F}_H \to 0,$$

where $\mathscr{F}_H = \bigoplus_{n \in \mathbf{Z}} \mathcal{O}_H(n)$. Taking cohomology, we get a long exact sequence

$$\dots \to H^i(X,\mathscr{F}(-1)) \to H^i(X,\mathscr{F}) \to H^i(X,\mathscr{F}_H) \to \dots .$$

Considered as graded S-modules, $H^i(X,\mathscr{F}(-1))$ is just $H^i(X,\mathscr{F})$ shifted one place, and the map $H^i(X,\mathscr{F}(-1)) \to H^i(X,\mathscr{F})$ of the exact sequence is multiplication by x_r.

Now H is isomorphic to $\mathbf{P}_A^{r-1}$, and $H^i(X,\mathscr{F}_H) = H^i(H,\bigoplus\mathcal{O}_H(n))$ by (2.10). So we can apply our induction hypothesis to $\mathscr{F}_H$, and find that $H^i(X,\mathscr{F}_H) = 0$ for $0 < i < r - 1$. Furthermore, for $i = 0$ we have an exact sequence

$$0 \to H^0(X,\mathscr{F}(-1)) \to H^0(X,\mathscr{F}) \to H^0(X,\mathscr{F}_H) \to 0$$

by (a), since $H^0(X,\mathscr{F}_H)$ is just $S/(x_r)$. At the other end of the exact sequence we have

$$0 \to H^{r-1}(X,\mathscr{F}_H) \xrightarrow{\delta} H^r(X,\mathscr{F}(-1)) \xrightarrow{x_r} H^r(X,\mathscr{F}) \to 0.$$

Indeed, we have described $H^r(X,\mathscr{F})$ above as the free A-module with basis formed by the negative monomials in $x_0, \dots, x_r$. So it is clear that x_r is surjective. On the other hand, the kernel of x_r is the free submodule generated by those negative monomials $x_0^{l_0} \cdots x_r^{l_r}$ with $l_r = -1$. Since $H^{r-1}(X,\mathscr{F}_H)$ is the free A-module with basis consisting of the negative monomials in $x_0, \dots, x_{r-1}$, and δ is division by x_r, the sequence is exact. In particular, δ is injective.

Putting these results all together, the long exact sequence of cohomology shows that the map multiplication by $x_r: H^i(X, \mathcal{F}(-1)) \to H^i(X, \mathcal{F})$ is bijective for $0 < i < r$, as required.   q.e.d.

Theorem 5.2 (Serre [3]). *Let X be a projective scheme over a noetherian ring A, and let $\mathcal{O}_X(1)$ be a very ample invertible sheaf on X over $\operatorname{Spec} A$. Let $\mathcal{F}$ be a coherent sheaf on X. Then:*
   (a) *for each $i \geqslant 0$, $H^i(X, \mathcal{F})$ is a finitely generated A-module;*
   (b) *there is an integer n_0, depending on $\mathcal{F}$, such that for each $i > 0$ and each $n \geqslant n_0$, $H^i(X, \mathcal{F}(n)) = 0$.*

PROOF. Since $\mathcal{O}_X(1)$ is a very ample sheaf on X over $\operatorname{Spec} A$, there is a closed immersion $i: X \to \mathbf{P}_A^r$ of schemes over A, for some r, such that $\mathcal{O}_X(1) = i^* \mathcal{O}_{\mathbf{P}^r}(1)$—cf. (II, 5.16.1). If $\mathcal{F}$ is coherent on X, then $i_* \mathcal{F}$ is coherent on $\mathbf{P}_A^r$ (II, Ex. 5.5), and the cohomology is the same (2.10). Thus we reduce to the case $X = \mathbf{P}_A^r$.

For $X = \mathbf{P}_A^r$, we observe that (a) and (b) are true for any sheaf of the form $\mathcal{O}_X(q)$, $q \in \mathbf{Z}$. This follows immediately from the explicit calculations (5.1). Hence the same is true for any finite direct sum of such sheaves.

To prove the theorem for arbitrary coherent sheaves, we use descending induction on i. For $i > r$, we have $H^i(X, \mathcal{F}) = 0$, since X can be covered by $r + 1$ open affines (Ex. 4.8), so the result is trivial in this case.

In general, given a coherent sheaf $\mathcal{F}$ on X, we can write $\mathcal{F}$ as a quotient of a sheaf $\mathcal{E}$, which is a finite direct sum of sheaves $\mathcal{O}(q_i)$, for various integers q_i (II, 5.18). Let $\mathcal{R}$ be the kernel,

$$0 \to \mathcal{R} \to \mathcal{E} \to \mathcal{F} \to 0.$$

Then $\mathcal{R}$ is also coherent. We get an exact sequence of A-modules

$$\ldots \to H^i(X, \mathcal{E}) \to H^i(X, \mathcal{F}) \to H^{i+1}(X, \mathcal{R}) \to \ldots.$$

Now the module on the left is finitely generated because $\mathcal{E}$ is a sum of $\mathcal{O}(q_i)$, as remarked above. The module on the right is finitely generated by the induction hypothesis. Since A is a noetherian ring, we conclude that the one in the middle is also finitely generated. This proves (a).

To prove (b), we twist and again write down a piece of the long exact sequence

$$\ldots \to H^i(X, \mathcal{E}(n)) \to H^i(X, \mathcal{F}(n)) \to H^{i+1}(X, \mathcal{R}(n)) \to \ldots.$$

Now for $n \gg 0$, the module on the left vanishes because $\mathcal{E}$ is a sum of $\mathcal{O}(q_i)$. The module on the right also vanishes for $n \gg 0$ because of the induction hypothesis. Hence $H^i(X, \mathcal{F}(n)) = 0$ for $n \gg 0$. Note since there are only finitely many i involved in statement (b), namely $0 < i \leqslant r$, it is sufficient to determine n_0 separately for each i. This proves (b).

Remark 5.2.1. As a special case of (a), we see that for any coherent sheaf $\mathcal{F}$ on X, $\Gamma(X, \mathcal{F})$ is a finitely generated A-module. This generalizes, and gives another proof of (II, 5.19).

As an application of these results, we give a cohomological criterion for an invertible sheaf to be ample (II, §7).

Proposition 5.3. *Let A be a noetherian ring, and let X be a proper scheme over* Spec A. *Let $\mathscr{L}$ be an invertible sheaf on X. Then the following conditions are equivalent:*

(i) *$\mathscr{L}$ is ample;*

(ii) *For each coherent sheaf $\mathscr{F}$ on X, there is an integer n_0, depending on $\mathscr{F}$, such that for each $i > 0$ and each $n \geq n_0$, $H^i(X, \mathscr{F} \otimes \mathscr{L}^n) = 0$.*

PROOF. (i) $\Rightarrow$ (ii). If $\mathscr{L}$ is ample on X, then for some $m > 0$, $\mathscr{L}^m$ is very ample on X over Spec A, by (II, 7.6). Since X is proper over Spec A, it is necessarily projective (II, 5.16.1). Now applying (5.2) to each of the sheaves $\mathscr{F}, \mathscr{F} \otimes \mathscr{L}, \mathscr{F} \otimes \mathscr{L}^2, \ldots, \mathscr{F} \otimes \mathscr{L}^{m-1}$ gives (ii). Cf. (II, 7.5) for a similar technique of proof.

(ii) $\Rightarrow$ (i). To show that $\mathscr{L}$ is ample, we will show that for any coherent sheaf $\mathscr{F}$ on X, there is an integer n_0 such that $\mathscr{F} \otimes \mathscr{L}^n$ is generated by global sections for all $n \geq n_0$. This is the definition of ampleness (II, §7).

Given $\mathscr{F}$, let P be a closed point of X, and let $\mathscr{I}_P$ be the ideal sheaf of the closed subset $\{P\}$. Then there is an exact sequence

$$0 \to \mathscr{I}_P \mathscr{F} \to \mathscr{F} \to \mathscr{F} \otimes k(P) \to 0,$$

where $k(P)$ is the skyscraper sheaf $\mathcal{O}_X/\mathscr{I}_P$. Tensoring with $\mathscr{L}^n$, we get

$$0 \to \mathscr{I}_P \mathscr{F} \otimes \mathscr{L}^n \to \mathscr{F} \otimes \mathscr{L}^n \to \mathscr{F} \otimes \mathscr{L}^n \otimes k(P) \to 0.$$

Now by our hypothesis (ii), there is an n_0 such that $H^1(X, \mathscr{I}_P \mathscr{F} \otimes \mathscr{L}^n) = 0$ for all $n \geq n_0$. Therefore

$$\Gamma(X, \mathscr{F} \otimes \mathscr{L}^n) \to \Gamma(X, \mathscr{F} \otimes \mathscr{L}^n \otimes k(P))$$

is surjective for all $n \geq n_0$. It follows from Nakayama's lemma over the local ring $\mathcal{O}_P$, that the stalk of $\mathscr{F} \otimes \mathscr{L}^n$ at P is generated by global sections. Since it is a coherent sheaf, we conclude that for each $n \geq n_0$, there is an open neighborhood U of P, depending on n, such that the global sections of $\mathscr{F} \otimes \mathscr{L}^n$ generate the sheaf at every point of U.

In particular, taking $\mathscr{F} = \mathcal{O}_X$, we find there is an integer $n_1 > 0$ and an open neighborhood V of P such that $\mathscr{L}^{n_1}$ is generated by global sections over V. On the other hand, for each $r = 0, 1, \ldots, n_1 - 1$, the above argument gives a neighborhood U_r of P such that $\mathscr{F} \otimes \mathscr{L}^{n_0 + r}$ is generated by global sections over U_r. Now let

$$U_P = V \cap U_0 \cap \ldots \cap U_{n_1 - 1}.$$

Then over U_P, all of the sheaves $\mathscr{F} \otimes \mathscr{L}^n$, for $n \geq n_0$, are generated by global sections. Indeed, any such sheaf can be written as a tensor product

$$(\mathscr{F} \otimes \mathscr{L}^{n_0 + r}) \otimes (\mathscr{L}^{n_1})^m$$

for suitable $0 \leq r < n_1$ and $m \geq 0$.

Now cover X by a finite number of the open sets U_P, for various closed points P, and let the new n_0 be the maximum of the n_0 corresponding to those points P. Then $\mathscr{F} \otimes \mathscr{L}^n$ is generated by global sections over all of X, for all $n \geqslant n_0$.   q.e.d.

EXERCISES

5.1. Let X be a projective scheme over a field k, and let $\mathscr{F}$ be a coherent sheaf on X. We define the *Euler characteristic* of $\mathscr{F}$ by

$$\chi(\mathscr{F}) = \sum(-1)^i \dim_k H^i(X,\mathscr{F}).$$

If

$$0 \to \mathscr{F}' \to \mathscr{F} \to \mathscr{F}'' \to 0$$

is a short exact sequence of coherent sheaves on X, show that $\chi(\mathscr{F}) = \chi(\mathscr{F}') + \chi(\mathscr{F}'')$.

5.2. (a) Let X be a projective scheme over a field k, let $\mathcal{O}_X(1)$ be a very ample invertible sheaf on X over k, and let $\mathscr{F}$ be a coherent sheaf on X. Show that there is a polynomial $P(z) \in \mathbf{Q}[z]$, such that $\chi(\mathscr{F}(n)) = P(n)$ for all $n \in \mathbf{Z}$. We call P the *Hilbert polynomial* of $\mathscr{F}$ with respect to the sheaf $\mathcal{O}_X(1)$. [*Hints*: Use induction on dim Supp $\mathscr{F}$, general properties of numerical polynomials (I, 7.3), and suitable exact sequences

$$0 \to \mathscr{R} \to \mathscr{F}(-1) \to \mathscr{F} \to \mathscr{Q} \to 0.]$$

(b) Now let $X = \mathbf{P}_k^r$, and let $M = \Gamma_*(\mathscr{F})$, considered as a graded $S = k[x_0,\dots,x_r]$-module. Use (5.2) to show that the Hilbert polynomial of $\mathscr{F}$ just defined is the same as the Hilbert polynomial of M defined in (I, §7).

5.3. *Arithmetic Genus.* Let X be a projective scheme of dimension r over a field k. We define the *arithmetic genus* p_a of X by

$$p_a(X) = (-1)^r(\chi(\mathcal{O}_X)-1).$$

Note that it depends only on X, not on any projective embedding.

(a) If X is integral, and k algebraically closed, show that $H^0(X,\mathcal{O}_X) \cong k$, so that

$$p_a(X) = \sum_{i=0}^{r-1} (-1)^i \dim_k H^{r-i}(X,\mathcal{O}_X).$$

In particular, if X is a curve, we have

$$p_a(X) = \dim_k H^1(X,\mathcal{O}_X).$$

[*Hint*: Use (I, 3.4).]

(b) If X is a closed subvariety of $\mathbf{P}_k^r$, show that this $p_a(X)$ coincides with the one defined in (I, Ex. 7.2), which apparently depended on the projective embedding.

(c) If X is a nonsingular projective curve over an algebraically closed field k, show that $p_a(X)$ is in fact a *birational* invariant. Conclude that a nonsingular plane curve of degree $d \geqslant 3$ is not rational. (This gives another proof of (II, 8.20.3) where we used the geometric genus.)

5.4. Recall from (II, Ex. 6.10) the definition of the Grothendieck group $K(X)$ of a noetherian scheme X.

(a) Let X be a projective scheme over a field k, and let $\mathcal{O}_X(1)$ be a very ample invertible sheaf on X. Show that there is a (unique) additive homomorphism

$$P:K(X) \to \mathbf{Q}[z]$$

such that for each coherent sheaf $\mathscr{F}$ on X, $P(\gamma(\mathscr{F}))$ is the Hilbert polynomial of $\mathscr{F}$ (Ex. 5.2).

(b) Now let $X = \mathbf{P}_k^r$. For each $i = 0, 1, \ldots, r$, let L_i be a linear space of dimension i in X. Then show that

(1) $K(X)$ is the free abelian group generated by $\{\gamma(\mathcal{O}_{L_i})|i = 0, \ldots, r\}$, and
(2) the map $P:K(X) \to \mathbf{Q}[z]$ is injective.

[*Hint*: Show that $(1) \Rightarrow (2)$. Then prove (1) and (2) simultaneously, by induction on r, using (II, Ex. 6.10c).]

5.5. Let k be a field, let $X = \mathbf{P}_k^r$, and let Y be a closed subscheme of dimension $q \geqslant 1$, which is a complete intersection (II, Ex. 8.4). Then:

(a) for all $n \in \mathbf{Z}$, the natural map

$$H^0(X, \mathcal{O}_X(n)) \to H^0(Y, \mathcal{O}_Y(n))$$

is surjective. (This gives a generalization and another proof of (II, Ex. 8.4c), where we assumed Y was normal.)

(b) Y is connected;
(c) $H^i(Y, \mathcal{O}_Y(n)) = 0$ for $0 < i < q$ and all $n \in \mathbf{Z}$;
(d) $p_a(Y) = \dim_k H^q(Y, \mathcal{O}_Y)$.

[*Hint*: Use exact sequences and induction on the codimension, starting from the case $Y = X$ which is (5.1).]

5.6. *Curves on a Nonsingular Quadric Surface.* Let Q be the nonsingular quadric surface $xy = zw$ in $X = \mathbf{P}_k^3$ over a field k. We will consider locally principal closed subschemes Y of Q. These correspond to Cartier divisors on Q by (II, 6.17.1). On the other hand, we know that $\operatorname{Pic} Q \cong \mathbf{Z} \oplus \mathbf{Z}$, so we can talk about the *type* (a,b) of Y (II, 6.16) and (II, 6.6.1). Let us denote the invertible sheaf $\mathscr{L}(Y)$ by $\mathcal{O}_Q(a,b)$. Thus for any $n \in \mathbf{Z}$, $\mathcal{O}_Q(n) = \mathcal{O}_Q(n,n)$.

(a) Use the special cases $(q,0)$ and $(0,q)$, with $q > 0$, when Y is a disjoint union of q lines $\mathbf{P}^1$ in Q, to show:

(1) if $|a - b| \leqslant 1$, then $H^1(Q, \mathcal{O}_Q(a,b)) = 0$;
(2) if $a,b < 0$, then $H^1(Q, \mathcal{O}_Q(a,b)) = 0$;
(3) If $a \leqslant -2$, then $H^1(Q, \mathcal{O}_Q(a,0)) \neq 0$.

(b) Now use these results to show:

(1) if Y is a locally principal closed subscheme of type (a,b), with $a,b > 0$, then Y is connected;
(2) now assume k is algebraically closed. Then for any $a,b > 0$, there exists an irreducible nonsingular curve Y of type (a,b). Use (II, 7.6.2) and (II, 8.18).
(3) an irreducible nonsingular curve Y of type (a,b), $a,b > 0$ on Q is projectively normal (II, Ex. 5.14) if and only if $|a - b| \leqslant 1$. In particular, this gives lots of examples of nonsingular, but not projectively normal curves in $\mathbf{P}^3$. The simplest is the one of type $(1,3)$, which is just the rational quartic curve (I, Ex. 3.18).

III Cohomology

(c) If Y is a locally principal subscheme of type (a,b) in Q, show that $p_a(Y) = ab - a - b + 1$. [*Hint*: Calculate Hilbert polynomials of suitable sheaves, and again use the special case $(q,0)$ which is a disjoint union of q copies of $\mathbf{P}^1$. See (V, 1.5.2) for another method.]

5.7. Let X (respectively, Y) be proper schemes over a noetherian ring A. We denote by $\mathscr{L}$ an invertible sheaf.
(a) If $\mathscr{L}$ is ample on X, and Y is any closed subscheme of X, then $i^*\mathscr{L}$ is ample on Y, where $i: Y \to X$ is the inclusion.
(b) $\mathscr{L}$ is ample on X if and only if $\mathscr{L}_{red} = \mathscr{L} \otimes \mathcal{O}_{X_{red}}$ is ample on X_{red}.
(c) Suppose X is reduced. Then $\mathscr{L}$ is ample on X if and only if $\mathscr{L} \otimes \mathcal{O}_{X_i}$ is ample on X_i, for each irreducible component X_i of X.
(d) Let $f: X \to Y$ be a finite surjective morphism, and let $\mathscr{L}$ be an invertible sheaf on Y. Then $\mathscr{L}$ is ample on Y if and only if $f^*\mathscr{L}$ is ample on X. [*Hints*: Use (5.3) and compare (Ex. 3.1, Ex. 3.2, Ex. 4.1, Ex. 4.2). See also Hartshorne [5, Ch. I §4] for more details.]

5.8. Prove that every one-dimensional proper scheme X over an algebraically closed field k is projective.
(a) If X is irreducible and nonsingular, then X is projective by (II, 6.7).
(b) If X is integral, let $\tilde{X}$ be its normalization (II, Ex. 3.8). Show that $\tilde{X}$ is complete and nonsingular, hence projective by (a). Let $f: \tilde{X} \to X$ be the projection. Let $\mathscr{L}$ be a very ample invertible sheaf on $\tilde{X}$. Show there is an effective divisor $D = \sum P_i$ on $\tilde{X}$ with $\mathscr{L}(D) \cong \mathscr{L}$, and such that $f(P_i)$ is a nonsingular point of X, for each i. Conclude that there is an invertible sheaf $\mathscr{L}_0$ on X with $f^*\mathscr{L}_0 \cong \mathscr{L}$. Then use (Ex. 5.7d), (II, 7.6) and (II, 5.16.1) to show that X is projective.
(c) If X is reduced, but not necessarily irreducible, let $X_1, \ldots, X_r$ be the irreducible components of X. Use (Ex. 4.5) to show Pic $X \to \bigoplus$ Pic X_i is surjective. Then use (Ex. 5.7c) to show X is projective.
(d) Finally, if X is any one-dimensional proper scheme over k, use (2.7) and (Ex. 4.6) to show that Pic $X \to$ Pic X_{red} is surjective. Then use (Ex. 5.7b) to show X is projective.

5.9. *A Nonprojective Scheme.* We show the result of (Ex. 5.8) is false in dimension 2. Let k be an algebraically closed field of characteristic 0, and let $X = \mathbf{P}_k^2$. Let ω be the sheaf of differential 2-forms (II, §8). Define an infinitesimal extension X' of X by ω by giving the element $\xi \in H^1(X, \omega \otimes \mathscr{T})$ defined as follows (Ex. 4.10). Let x_0, x_1, x_2 be the homogeneous coordinates of X, let U_0, U_1, U_2 be the standard open covering, and let $\xi_{ij} = (x_j/x_i)d(x_i/x_j)$. This gives a Čech 1-cocycle with values in Ω_X^1, and since dim $X = 2$, we have $\omega \otimes \mathscr{T} \cong \Omega^1$ (II, Ex. 5.16b). Now use the exact sequence

$$\ldots \to H^1(X,\omega) \to \text{Pic } X' \to \text{Pic } X \xrightarrow{\delta} H^2(X,\omega) \to \ldots$$

of (Ex. 4.6) and show δ is injective. We have $\omega \cong \mathcal{O}_X(-3)$ by (II, 8.20.1), so $H^2(X,\omega) \cong k$. Since char $k = 0$, you need only show that $\delta(\mathcal{O}(1)) \neq 0$, which can be done by calculating in Čech cohomology. Since $H^1(X,\omega) = 0$, we see that Pic $X' = 0$. In particular, X' has no ample invertible sheaves, so it is not projective.

Note. In fact, this result can be generalized to show that for any nonsingular projective surface X over an algebraically closed field k of characteristic 0, there is an infinitesimal extension X' of X by ω, such that X' is not projective over k.

Indeed, let D be an ample divisor on X. Then D determines an element $c_1(D) \in H^1(X, \Omega^1)$ which we use to define X', as above. Then for any divisor E on X one can show that $\delta(\mathscr{L}(E)) = (D.E)$, where $(D.E)$ is the intersection number (Chapter V), considered as an element of k. Hence if E is ample, $\delta(\mathscr{L}(E)) \neq 0$. Therefore X' has no ample divisors.

On the other hand, over a field of characteristic $p > 0$, a proper scheme X is projective if and only if X_{red} is!

5.10. Let X be a projective scheme over a noetherian ring A, and let $\mathscr{F}^1 \to \mathscr{F}^2 \to \ldots \to \mathscr{F}^r$ be an exact sequence of coherent sheaves on X. Show that there is an integer n_0, such that for all $n \geqslant n_0$, the sequence of global sections

$$\Gamma(X, \mathscr{F}^1(n)) \to \Gamma(X, \mathscr{F}^2(n)) \to \ldots \to \Gamma(X, \mathscr{F}^r(n))$$

is exact.

6 Ext Groups and Sheaves

In this section we develop the properties of Ext groups and sheaves, which we will need for the duality theorem. We work on a ringed space $(X, \mathcal{O}_X)$, and all sheaves will be sheaves of $\mathcal{O}_X$-modules.

If $\mathscr{F}$ and $\mathscr{G}$ are $\mathcal{O}_X$-modules, we denote by $\mathrm{Hom}(\mathscr{F}, \mathscr{G})$ the group of $\mathcal{O}_X$-module homomorphisms, and by $\mathscr{H}om(\mathscr{F}, \mathscr{G})$ the sheaf Hom (II, §5). If necessary, we put a subscript X to indicate which space we are on: $\mathrm{Hom}_X(\mathscr{F}, \mathscr{G})$. For fixed $\mathscr{F}$, $\mathrm{Hom}(\mathscr{F}, \cdot)$ is a left exact covariant functor from $\mathfrak{Mod}(X)$ to $\mathfrak{Ab}$, and $\mathscr{H}om(\mathscr{F}, \cdot)$ is a left exact covariant functor from $\mathfrak{Mod}(X)$ to $\mathfrak{Mod}(X)$. Since $\mathfrak{Mod}(X)$ has enough injectives (2.2) we can make the following definition.

Definition. Let $(X, \mathcal{O}_X)$ be a ringed space, and let $\mathscr{F}$ be an $\mathcal{O}_X$-module. We define the functors $\mathrm{Ext}^i(\mathscr{F}, \cdot)$ as the right derived functors of $\mathrm{Hom}(\mathscr{F}, \cdot)$, and $\mathscr{E}xt^i(\mathscr{F}, \cdot)$ as the right derived functors of $\mathscr{H}om(\mathscr{F}, \cdot)$.

Consequently, according to the general properties of derived functors (1.1A) we have $\mathrm{Ext}^0 = \mathrm{Hom}$, a long exact sequence for a short exact sequence in the second variable, $\mathrm{Ext}^i(\mathscr{F}, \mathscr{G}) = 0$ for $i > 0$, $\mathscr{G}$ injective in $\mathfrak{Mod}(X)$, and ditto for the $\mathscr{E}xt$ sheaves.

Lemma 6.1. *If $\mathscr{I}$ is an injective object of $\mathfrak{Mod}(X)$, then for any open subset $U \subseteq X$, $\mathscr{I}|_U$ is an injective object of $\mathfrak{Mod}(U)$.*

PROOF. Let $j: U \to X$ be the inclusion map. Then given an inclusion $\mathscr{F} \subseteq \mathscr{G}$ in $\mathfrak{Mod}(U)$, and given a map $\mathscr{F} \to \mathscr{I}|_U$, we get an inclusion $j_! \mathscr{F} \subseteq j_! \mathscr{G}$, and a map $j_! \mathscr{F} \to j_!(\mathscr{I}|_U)$, where $j_!$ is extension by zero (II, Ex. 1.19). But $j_!(\mathscr{I}|_U)$ is a subsheaf of $\mathscr{I}$, so we have a map $j_! \mathscr{F} \to \mathscr{I}$. Since $\mathscr{I}$ is injective in $\mathfrak{Mod}(X)$, this extends to a map of $j_! \mathscr{G}$ to $\mathscr{I}$. Restricting to U gives the required map of $\mathscr{G}$ to $\mathscr{I}|_U$.

Proposition 6.2. *For any open subset* $U \subseteq X$ *we have*

$$\mathcal{E}xt_X^i(\mathcal{F},\mathcal{G})|_U \cong \mathcal{E}xt_U^i(\mathcal{F}|_U,\mathcal{G}|_U).$$

PROOF. We use (1.3A). Both sides give δ-functors in $\mathcal{G}$ from $\mathfrak{Mod}(X)$ to $\mathfrak{Mod}(U)$. They agree for $i = 0$, both sides vanish for $i > 0$ and $\mathcal{G}$ injective, by (6.1), so they are equal.

Proposition 6.3. *For any* $\mathcal{G} \in \mathfrak{Mod}(X)$, *we have*:
  (a) $\mathcal{E}xt^0(\mathcal{O}_X,\mathcal{G}) = \mathcal{G}$;
  (b) $\mathcal{E}xt^i(\mathcal{O}_X,\mathcal{G}) = 0$ *for* $i > 0$;
  (c) $\operatorname{Ext}^i(\mathcal{O}_X,\mathcal{G}) \cong H^i(X,\mathcal{G})$ *for all* $i \geqslant 0$.

PROOF. The functor $\mathcal{H}om(\mathcal{O}_X,\cdot)$ is the identity functor, so its derived functors are 0 for $i > 0$. This proves (a) and (b). The functors $\operatorname{Hom}(\mathcal{O}_X,\cdot)$ and $\Gamma(X,\cdot)$ are equal, so their derived functors (as functors from $\mathfrak{Mod}(X)$ to $\mathfrak{Ab}$) are the same. Then use (2.6).

Proposition 6.4. *If* $0 \to \mathcal{F}' \to \mathcal{F} \to \mathcal{F}'' \to 0$ *is a short exact sequence in* $\mathfrak{Mod}(X)$, *then for any* $\mathcal{G}$ *we have a long exact sequence*

$$0 \to \operatorname{Hom}(\mathcal{F}'',\mathcal{G}) \to \operatorname{Hom}(\mathcal{F},\mathcal{G}) \to \operatorname{Hom}(\mathcal{F}',\mathcal{G}) \to$$
$$\to \operatorname{Ext}^1(\mathcal{F}'',\mathcal{G}) \to \operatorname{Ext}^1(\mathcal{F},\mathcal{G}) \to \ldots,$$

and similarly for the $\mathcal{E}xt$ *sheaves.*

PROOF. Let $0 \to \mathcal{G} \to \mathcal{I}^{\cdot}$ be an injective resolution of $\mathcal{G}$. For any injective sheaf $\mathcal{I}$, the functor $\operatorname{Hom}(\cdot,\mathcal{I})$ is exact, so we get a short exact sequence of complexes

$$0 \to \operatorname{Hom}(\mathcal{F}'',\mathcal{I}^{\cdot}) \to \operatorname{Hom}(\mathcal{F},\mathcal{I}^{\cdot}) \to \operatorname{Hom}(\mathcal{F}',\mathcal{I}^{\cdot}) \to 0.$$

Taking the associated long exact sequence of cohomology groups h^i gives the sequence of Ext^i.

Similarly, using (6.1) we see that $\mathcal{H}om(\cdot,\mathcal{I})$ is an exact functor from $\mathfrak{Mod}(X)$ to $\mathfrak{Mod}(X)$. Thus the same argument gives the exact sequence of $\mathcal{E}xt^i$.

Proposition 6.5. *Suppose there is an exact sequence*

$$\ldots \to \mathcal{L}_1 \to \mathcal{L}_0 \to \mathcal{F} \to 0$$

in $\mathfrak{Mod}(X)$, *where the* $\mathcal{L}_i$ *are locally free sheaves of finite rank* (*in this case we say* $\mathcal{L}.$ *is a locally free resolution of* $\mathcal{F}$). *Then for any* $\mathcal{G} \in \mathfrak{Mod}(X)$ *we have*

$$\mathcal{E}xt^i(\mathcal{F},\mathcal{G}) \cong h^i(\mathcal{H}om(\mathcal{L}.,\mathcal{G})).$$

PROOF. Both sides are δ-functors in $\mathcal{G}$ from $\mathfrak{Mod}(X)$ to $\mathfrak{Mod}(X)$. For $i = 0$ they are equal, because then $\mathcal{H}om(\cdot,\mathcal{G})$ is contravariant and left exact. Both sides vanish for $i > 0$ and $\mathcal{G}$ injective, because then $\mathcal{H}om(\cdot,\mathcal{G})$ is exact. So by (1.3A) they are equal.

Example 6.5.1. If X is a scheme, which is quasi-projective over Spec A, where A is a noetherian ring, then by (II, 5.18), any coherent sheaf $\mathscr{F}$ on X is a quotient of a locally free sheaf. Thus any coherent sheaf on X has a locally free resolution $\mathscr{L}. \to \mathscr{F} \to 0$. So (6.5) tells us that we can calculate $\mathscr{E}xt$ by taking locally free resolutions in the first variable.

Caution 6.5.2. The results (6.4) and (6.5) do *not* imply that $\mathscr{E}xt$ can be construed as a derived functor in its first variable. In fact, we cannot even define the right derived functors of Hom or $\mathscr{H}om$ in the first variable because the category $\mathfrak{Mod}(X)$ does not have enough projectives (Ex. 6.2). However, see (Ex. 6.4) for a universal property.

Lemma 6.6. *If $\mathscr{L} \in \mathfrak{Mod}(X)$ is locally free of finite rank, and $\mathscr{I} \in \mathfrak{Mod}(X)$ is injective, then $\mathscr{L} \otimes \mathscr{I}$ is also injective.*

PROOF. We must show that the functor $\mathrm{Hom}(\cdot, \mathscr{L} \otimes \mathscr{I})$ is exact. But it is the same as the functor $\mathrm{Hom}(\cdot \otimes \mathscr{L}^\vee, \mathscr{I})$ (II, Ex. 5.1), which is exact because $\cdot \otimes \mathscr{L}^\vee$ is exact, and $\mathscr{I}$ is injective.

Proposition 6.7. *Let $\mathscr{L}$ be a locally free sheaf of finite rank, and let $\mathscr{L}^\vee = \mathscr{H}om(\mathscr{L}, \mathcal{O}_X)$ be its dual. Then for any $\mathscr{F}, \mathscr{G} \in \mathfrak{Mod}(X)$ we have*

$$\mathrm{Ext}^i(\mathscr{F} \otimes \mathscr{L}, \mathscr{G}) \cong \mathrm{Ext}^i(\mathscr{F}, \mathscr{L}^\vee \otimes \mathscr{G}),$$

and for the sheaf $\mathscr{E}xt$ we have

$$\mathscr{E}xt^i(\mathscr{F} \otimes \mathscr{L}, \mathscr{G}) \cong \mathscr{E}xt^i(\mathscr{F}, \mathscr{L}^\vee \otimes \mathscr{G}) \cong \mathscr{E}xt^i(\mathscr{F}, \mathscr{G}) \otimes \mathscr{L}^\vee.$$

PROOF. The case $i = 0$ follows from (II, Ex. 5.1). For the general case, note that all of them are δ-functors in $\mathscr{G}$ from $\mathfrak{Mod}(X)$ to $\mathfrak{Ab}$ (respectively, $\mathfrak{Mod}(X)$), since tensoring with $\mathscr{L}^\vee$ is an exact functor. For $i > 0$ and $\mathscr{G}$ injective they all vanish, by (6.6), so by (1.3A) they are equal.

Next we will give some properties which are more particular to the case of schemes.

Proposition 6.8. *Let X be a noetherian scheme, let $\mathscr{F}$ be a coherent sheaf on X, let $\mathscr{G}$ be any $\mathcal{O}_X$-module, and let $x \in X$ be a point. Then we have*

$$\mathscr{E}xt^i(\mathscr{F}, \mathscr{G})_x \cong \mathrm{Ext}^i_{\mathcal{O}_x}(\mathscr{F}_x, \mathscr{G}_x)$$

for any $i \geqslant 0$, where the right-hand side is Ext over the local ring $\mathcal{O}_x$.

PROOF. Of course, Ext over a ring A is defined as the right derived functor of $\mathrm{Hom}_A(M, \cdot)$ for any A-module M, considered as a functor from $\mathfrak{Mod}(A)$ to $\mathfrak{Mod}(A)$. Or, by considering a one-point space with the ring A attached, it becomes a special case of the Ext of a ringed space defined above.

Our question is local, by (6.2), so we may assume that X is affine. Then $\mathscr{F}$ has a locally free (or even a free) resolution $\mathscr{L}. \to \mathscr{F} \to 0$, which on the stalks at x gives a free resolution $(\mathscr{L}.)_x \to \mathscr{F}_x \to 0$. So by (6.5) we can calculate both sides by these resolutions. Since $\mathscr{H}om(\mathscr{L},\mathscr{G})_x = \text{Hom}_{\mathcal{O}_x}(\mathscr{L}_x,\mathscr{G}_x)$ for a locally free sheaf $\mathscr{L}$, and since the stalk functor is exact, we get the equality of Ext's.

Note that even the case $i = 0$ is not true without some special hypothesis on $\mathscr{F}$, such as $\mathscr{F}$ coherent.

Proposition 6.9. *Let X be a projective scheme over a noetherian ring A, let $\mathcal{O}_X(1)$ be a very ample invertible sheaf, and let $\mathscr{F},\mathscr{G}$ be coherent sheaves on X. Then there is an integer $n_0 > 0$, depending on $\mathscr{F}$, $\mathscr{G}$, and i, such that for every $n \geqslant n_0$ we have*

$$\text{Ext}^i(\mathscr{F},\mathscr{G}(n)) \cong \Gamma(X,\mathscr{E}xt^i(\mathscr{F},\mathscr{G}(n))).$$

PROOF. If $i = 0$, this is true for any $\mathscr{F},\mathscr{G},n$. If $\mathscr{F} = \mathcal{O}_X$, then the left-hand side is $H^i(X,\mathscr{G}(n))$ by (6.3). So for $n \gg 0$ and $i > 0$ it is 0 by (5.2). On the other hand, the right-hand side is always 0 for $i > 0$ by (6.3), so we have the result for $\mathscr{F} = \mathcal{O}_X$.

If $\mathscr{F}$ is a locally free sheaf, we reduce to the case $\mathscr{F} = \mathcal{O}_X$ by (6.7).

Finally, if $\mathscr{F}$ is an arbitrary coherent sheaf, write it as a quotient of a locally free sheaf $\mathscr{E}$ (II, 5.18), and let $\mathscr{R}$ be the kernel:

$$0 \to \mathscr{R} \to \mathscr{E} \to \mathscr{F} \to 0.$$

Since $\mathscr{E}$ is locally free, by the earlier results, for $n \gg 0$, we have an exact sequence

$$0 \to \text{Hom}(\mathscr{F},\mathscr{G}(n)) \to \text{Hom}(\mathscr{E},\mathscr{G}(n)) \to \text{Hom}(\mathscr{R},\mathscr{G}(n)) \to \text{Ext}^1(\mathscr{F},\mathscr{G}(n)) \to 0$$

and isomorphisms, for all $i > 0$

$$\text{Ext}^i(\mathscr{R},\mathscr{G}(n)) \xrightarrow{\sim} \text{Ext}^{i+1}(\mathscr{F},\mathscr{G}(n)),$$

and similarly for the sheaf $\mathscr{H}om$ and $\mathscr{E}xt$. Now by (Ex. 5.10), the sequence of global sections of the sheaf sequence is exact after twisting a little more, so from the case $i = 0$, using (6.7), we get the case $i = 1$ for $\mathscr{F}$. But $\mathscr{R}$ is also coherent, so by induction we get the general result.

Remark 6.9.1. More generally, on any ringed space X, the relation between the global Ext and the sheaf $\mathscr{E}xt$ can be expressed by a spectral sequence (see Grothendieck [1] or Godement [1, II, 7.3.3]).

Now, for future reference, we recall the notion of projective dimension of a module over a ring. Let A be a ring, and let M be an A-module. A *projective resolution* of M is a complex $L.$ of projective A-modules, such that

$$\dots \to L_2 \to L_1 \to L_0 \to M \to 0$$

is exact. If $L_i = 0$ for $i > n$, and $L_n \neq 0$, we say it has *length n*. Then we define the *projective dimension* of M, denoted pd(M), to be the least length of a projective resolution of M (or $+\infty$ if there is no finite projective resolution).

Proposition 6.10A. *Let A be a ring, and M an A-module. Then:*
    (a) *M is projective if and only if* $\text{Ext}^1(M,N) = 0$ *for all A-modules N*;
    (b) $\text{pd}(M) \leqslant n$ *if and only if* $\text{Ext}^i(M,N) = 0$ *for all* $i > n$ *and all A-modules N.*

PROOF. Matsumura [2, pp. 127–128].

Proposition 6.11A. *If A is a regular local ring, then:*
    (a) *for every M*, $\text{pd}(M) \leqslant \dim A$;
    (b) *If* $k = A/\mathfrak{m}$, *then* $\text{pd}(k) = \dim A$.

PROOF. Matsumura [2, Th. 42, p. 131].

Proposition 6.12A. *Let A be a regular local ring of dimension n, and let M be a finitely generated A-module. Then we have*

$$\text{pd } M + \text{depth } M = n.$$

PROOF. Matsumura [2, p. 113, Ex. 4] or Serre [11, IVD, Prop. 21].

EXERCISES

6.1. Let $(X,\mathcal{O}_X)$ be a ringed space, and let $\mathscr{F}',\mathscr{F}'' \in \mathfrak{Mod}(X)$. An *extension* of $\mathscr{F}''$ by $\mathscr{F}'$ is a short exact sequence

$$0 \to \mathscr{F}' \to \mathscr{F} \to \mathscr{F}'' \to 0$$

in $\mathfrak{Mod}(X)$. Two extensions are *isomorphic* if there is an isomorphism of the short exact sequences, inducing the identity maps on $\mathscr{F}'$ and $\mathscr{F}''$. Given an extension as above consider the long exact sequence arising from $\text{Hom}(\mathscr{F}'',\cdot)$, in particular the map

$$\delta : \text{Hom}(\mathscr{F}'',\mathscr{F}'') \to \text{Ext}^1(\mathscr{F}'',\mathscr{F}'),$$

and let $\xi \in \text{Ext}^1(\mathscr{F}'',\mathscr{F}')$ be $\delta(1_{\mathscr{F}''})$. Show that this process gives a one-to-one correspondence between isomorphism classes of extensions of $\mathscr{F}''$ by $\mathscr{F}'$, and elements of the group $\text{Ext}^1(\mathscr{F}'',\mathscr{F}')$. For more details, see, e.g., Hilton and Stammbach [1, Ch. III].

6.2. Let $X = \mathbf{P}_k^1$, with k an infinite field.
    (a) Show that there does not exist a projective object $\mathscr{P} \in \mathfrak{Mod}(X)$, together with a surjective map $\mathscr{P} \to \mathcal{O}_X \to 0$. [*Hint*: Consider surjections of the form $\mathcal{O}_V \to k(x) \to 0$, where $x \in X$ is a closed point, V is an open neighborhood of x, and $\mathcal{O}_V = j_!(\mathcal{O}_X|_V)$, where $j : V \to X$ is the inclusion.]
    (b) Show that there does not exist a projective object $\mathscr{P}$ in either $\mathfrak{Qco}(X)$ or $\mathfrak{Coh}(X)$ together with a surjection $\mathscr{P} \to \mathcal{O}_X \to 0$. [*Hint*: Consider surjections of the form $\mathscr{L} \to \mathscr{L} \otimes k(x) \to 0$, where $x \in X$ is a closed point, and $\mathscr{L}$ is an invertible sheaf on X.]

6.3. Let X be a noetherian scheme, and let $\mathscr{F},\mathscr{G} \in \mathfrak{Mod}(X)$.
(a) If $\mathscr{F},\mathscr{G}$ are both coherent, then $\mathscr{E}xt^i(\mathscr{F},\mathscr{G})$ is coherent, for all $i \geqslant 0$.
(b) If $\mathscr{F}$ is coherent and $\mathscr{G}$ is quasi-coherent, then $\mathscr{E}xt^i(\mathscr{F},\mathscr{G})$ is quasi-coherent, for all $i \geqslant 0$.

6.4. Let X be a noetherian scheme, and suppose that every coherent sheaf on X is a quotient of a locally free sheaf. In this case we say $\mathfrak{Coh}(X)$ has *enough locally frees*. Then for any $\mathscr{G} \in \mathfrak{Mod}(X)$, show that the δ-functor $(\mathscr{E}xt^i(\cdot,\mathscr{G}))$, from $\mathfrak{Coh}(X)$ to $\mathfrak{Mod}(X)$, is a contravariant universal δ-functor. [*Hint*: Show $\mathscr{E}xt^i(\cdot,\mathscr{G})$ is coefface-able (§1) for $i > 0$.]

6.5. Let X be a noetherian scheme, and assume that $\mathfrak{Coh}(X)$ has enough locally frees (Ex. 6.4). Then for any coherent sheaf $\mathscr{F}$ we define the *homological dimension* of $\mathscr{F}$, denoted $\mathrm{hd}(\mathscr{F})$, to be the least length of a locally free resolution of $\mathscr{F}$ (or $+\infty$ if there is no finite one). Show:
(a) $\mathscr{F}$ is locally free $\Leftrightarrow \mathscr{E}xt^1(\mathscr{F},\mathscr{G}) = 0$ for all $\mathscr{G} \in \mathfrak{Mod}(X)$;
(b) $\mathrm{hd}(\mathscr{F}) \leqslant n \Leftrightarrow \mathscr{E}xt^i(\mathscr{F},\mathscr{G}) = 0$ for all $i > n$ and all $\mathscr{G} \in \mathfrak{Mod}(X)$;
(c) $\mathrm{hd}(\mathscr{F}) = \sup_x \mathrm{pd}_{\mathcal{O}_x} \mathscr{F}_x$.

6.6. Let A be a regular local ring, and let M be a finitely generated A-module. In this case, strengthen the result (6.10A) as follows.
(a) M is projective if and only if $\mathrm{Ext}^i(M,A) = 0$ for all $i > 0$. [*Hint*: Use (6.11A) and descending induction on i to show that $\mathrm{Ext}^i(M,N) = 0$ for all $i > 0$ and all finitely generated A-modules N. Then show M is a direct summand of a free A-module (Matsumura [2, p. 129]).]
(b) Use (a) to show that for any n, $\mathrm{pd}\, M \leqslant n$ if and only if $\mathrm{Ext}^i(M,A) = 0$ for all $i > n$.

6.7. Let $X = \mathrm{Spec}\, A$ be an affine noetherian scheme. Let M, N be A-modules, with M finitely generated. Then

$$\mathrm{Ext}^i_X(\tilde{M},\tilde{N}) \cong \mathrm{Ext}^i_A(M,N)$$

and

$$\mathscr{E}xt^i_X(\tilde{M},\tilde{N}) \cong \mathrm{Ext}^i_A(M.N)^{\sim}.$$

6.8. Prove the following theorem of Kleiman (see Borelli [1]): if X is a noetherian, integral, separated, locally factorial scheme, then every coherent sheaf on X is a quotient of a locally free sheaf (of finite rank).
(a) First show that open sets of the form X_s, for various $s \in \Gamma(X,\mathscr{L})$, and various invertible sheaves $\mathscr{L}$ on X, form a base for the topology of X. [*Hint*: Given a closed point $x \in X$ and an open neighborhood U of x, to show there is an $\mathscr{L},s$ such that $x \in X_s \subseteq U$, first reduce to the case that $Z = X - U$ is irreducible. Then let ζ be the generic point of Z. Let $f \in K(X)$ be a rational function with $f \in \mathcal{O}_x, f \notin \mathcal{O}_\zeta$. Let $D = (f)_\infty$, and let $\mathscr{L} = \mathscr{L}(D), s \in \Gamma(X,\mathscr{L}(D))$ correspond to D (II, §6).]
(b) Now use (II, 5.14) to show that any coherent sheaf is a quotient of a direct sum $\bigoplus \mathscr{L}_i^{n_i}$ for various invertible sheaves $\mathscr{L}_i$ and various integers n_i.

6.9. Let X be a noetherian, integral, separated, regular scheme. (We say a scheme is *regular* if all of its local rings are regular local rings.) Recall the definition of the *Grothendieck group* $K(X)$ from (II, Ex. 6.10). We define similarly another group $K_1(X)$ using locally free sheaves: it is the quotient of the free abelian group generated by all locally free (coherent) sheaves, by the subgroup generated by all expressions of the form $\mathscr{E} - \mathscr{E}' - \mathscr{E}''$, whenever $0 \to \mathscr{E}' \to \mathscr{E} \to \mathscr{E}'' \to 0$ is a short exact sequence of locally free sheaves. Clearly there is a natural group homomorphism $\varepsilon : K_1(X) \to K(X)$. Show that ε is an isomorphism (Borel and Serre [1, §4]) as follows.

(a) Given a coherent sheaf $\mathscr{F}$, use (Ex. 6.8) to show that it has a locally free resolution $\mathscr{E}. \to \mathscr{F} \to 0$. Then use (6.11A) and (Ex. 6.5) to show that it has a finite locally free resolution

$$0 \to \mathscr{E}_n \to \ldots \to \mathscr{E}_1 \to \mathscr{E}_0 \to \mathscr{F} \to 0.$$

(b) For each $\mathscr{F}$, choose a finite locally free resolution $\mathscr{E}. \to \mathscr{F} \to 0$, and let $\delta(\mathscr{F}) = \sum (-1)^i \gamma(\mathscr{E}_i)$ in $K_1(X)$. Show that $\delta(\mathscr{F})$ is independent of the resolution chosen, that it defines a homomorphism of $K(X)$ to $K_1(X)$, and finally, that it is an inverse to ε.

6.10. *Duality for a Finite Flat Morphism.*

(a) Let $f : X \to Y$ be a finite morphism of noetherian schemes. For any quasi-coherent $\mathcal{O}_Y$-module $\mathscr{G}$, $\mathscr{H}om_Y(f_* \mathcal{O}_X, \mathscr{G})$ is a quasi-coherent $f_* \mathcal{O}_X$-module, hence corresponds to a quasi-coherent $\mathcal{O}_X$-module, which we call $f^! \mathscr{G}$ (II, Ex. 5.17e).

(b) Show that for any coherent $\mathscr{F}$ on X and any quasi-coherent $\mathscr{G}$ on Y, there is a natural isomorphism

$$f_* \mathscr{H}om_X(\mathscr{F}, f^! \mathscr{G}) \xrightarrow{\sim} \mathscr{H}om_Y(f_* \mathscr{F}, \mathscr{G}).$$

(c) For each $i \geqslant 0$, there is a natural map

$$\varphi_i : \operatorname{Ext}^i_X(\mathscr{F}, f^! \mathscr{G}) \to \operatorname{Ext}^i_Y(f_* \mathscr{F}, \mathscr{G}).$$

[*Hint:* First construct a map

$$\operatorname{Ext}^i_X(\mathscr{F}, f^! \mathscr{G}) \to \operatorname{Ext}^i_Y(f_* \mathscr{F}, f_* f^! \mathscr{G}).$$

Then compose with a suitable map from $f_* f^! \mathscr{G}$ to $\mathscr{G}$.]

(d) Now assume that X and Y are separated, $\mathfrak{Coh}(X)$ has enough locally frees, and assume that $f_* \mathcal{O}_X$ is locally free on Y (this is equivalent to saying f flat—see §9). Show that φ_i is an isomorphism for all i, all $\mathscr{F}$ coherent on X, and all $\mathscr{G}$ quasi-coherent on Y. [*Hints:* First do $i = 0$. Then do $\mathscr{F} = \mathcal{O}_X$, using (Ex. 4.1). Then do $\mathscr{F}$ locally free. Do the general case by induction on i, writing $\mathscr{F}$ as a quotient of a locally free sheaf.]

7 The Serre Duality Theorem

In this section we prove the Serre duality theorem for the cohomology of coherent sheaves on a projective scheme. First we do the case of projective space itself, which follows easily from the explicit calculations of §5. Then on an arbitrary projective scheme X, we show that there is a coherent sheaf ω_X°, which plays a role in duality theory similar to the canonical sheaf of a nonsingular variety. In particular, if X is Cohen–Macaulay, it gives a duality theorem just like the one on projective space. Finally, if X is a nonsingular variety over an algebraically closed field, we show that the dualizing sheaf ω_X° coincides with the canonical sheaf ω_X. At the end of the section, we mention the connection between duality and residues of differential forms.

Let k be a field, let $X = \mathbf{P}^n_k$ be the n-dimensional projective space over k, and let $\omega_X = \wedge^n \Omega_{X/k}$ be the canonical sheaf on X (II, §8).

Theorem 7.1 (Duality for $\mathbf{P}^n_k$). *Let $X = \mathbf{P}^n_k$ over a field k. Then:*

(a) $H^n(X,\omega_X) \cong k$. *Fix one such isomorphism;*

(b) *for any coherent sheaf $\mathscr{F}$ on X, the natural pairing*

$$\mathrm{Hom}(\mathscr{F},\omega) \times H^n(X,\mathscr{F}) \to H^n(X,\omega) \cong k$$

is a perfect pairing of finite-dimensional vector spaces over k;

(c) *for every $i \geqslant 0$ there is a natural functorial isomorphism*

$$\mathrm{Ext}^i(\mathscr{F},\omega) \xrightarrow{\sim} H^{n-i}(X,\mathscr{F})',$$

where $'$ denotes the dual vector space, which for $i = 0$ is the one induced by the pairing of (b).

PROOF.

(a) It follows from (II, 8.13) that $\omega_X \cong \mathcal{O}_X(-n-1)$ (see II, 8.20.1). Thus (a) follows from (5.1c).

(b) A homomorphism of $\mathscr{F}$ to ω induces a map of cohomology groups $H^n(X,\mathscr{F}) \to H^n(X,\omega)$. This gives the natural pairing. If $\mathscr{F} \cong \mathcal{O}(q)$ for some $q \in \mathbf{Z}$, then $\mathrm{Hom}(\mathscr{F},\omega) \cong H^0(X,\omega(-q))$, so the result follows from (5.1d). Hence (b) holds also for a finite direct sum of sheaves of the form $\mathcal{O}(q_i)$. If $\mathscr{F}$ is an arbitrary coherent sheaf, we can write it as a cokernel $\mathscr{E}_1 \to \mathscr{E}_0 \to \mathscr{F} \to 0$ of a map of sheaves $\mathscr{E}_i$, each $\mathscr{E}_i$ being a direct sum of sheaves $\mathcal{O}(q_i)$. Now $\mathrm{Hom}(\cdot,\omega)$ and $H^n(X,\cdot)'$ are both left-exact contravariant functors, so by the 5-lemma we get an isomorphism $\mathrm{Hom}(\mathscr{F},\omega) \xrightarrow{\sim} H^n(X,\mathscr{F})'$.

(c) Both sides are contravariant δ-functors, for $\mathscr{F} \in \mathfrak{Coh}(X)$, indexed by $i \geqslant 0$. For $i = 0$ we have an isomorphism by (b). Thus to show they are isomorphic, by (1.3A), it will be sufficient to show both sides are coeffaceable for $i > 0$. Given $\mathscr{F}$ coherent, it follows from (II, 5.18) and its proof that we can write $\mathscr{F}$ as a quotient of a sheaf $\mathscr{E} = \bigoplus_{i=1}^N \mathcal{O}(-q)$, with $q \gg 0$. Then $\mathrm{Ext}^i(\mathscr{E},\omega) = \bigoplus H^i(X,\omega(q)) = 0$ for $i > 0$ by (5.1). On the other hand, $H^{n-i}(X,\mathscr{E})' = \bigoplus H^{n-i}(X,\mathcal{O}(-q))'$, which is 0 for $i > 0$, $q > 0$, as we see again from (5.1) by inspection. Thus both sides are coeffaceable for $i > 0$, so the δ-functors are universal, hence isomorphic.

Remark 7.1.1. One may ask, why bother phrasing (7.1) with the sheaf ω_X, rather than simply writing $\mathcal{O}_X(-n-1)$, which is what we use in the proof? One reason is that this is the form of the theorem which generalizes well. But a more intrinsic reason is that when written this way, the isomorphism of (a) can be made independent of the choice of basis of $\mathbf{P}^n$, hence stable under automorphisms of $\mathbf{P}^n$. Thus it is truly a *natural* isomorphism. To do this, consider the Čech cocycle

$$\alpha = \frac{x_0^n}{x_1 \cdots x_n} d\left(\frac{x_1}{x_0}\right) \wedge \cdots \wedge d\left(\frac{x_n}{x_0}\right)$$

in $C^n(\mathfrak{U},\omega)$, where $\mathfrak{U}$ is the standard open covering. Then one can show that α determines a generator of $H^n(X,\omega)$, which is stable under change of variables.

To generalize (7.1) to other schemes, we take properties (a) and (b) as our guide, and make the following definition.

Definition. Let X be a proper scheme of dimension n over a field k. A *dualizing sheaf* for X is a coherent sheaf ω_X° on X, together with a *trace* morphism $t: H^n(X,\omega_X^\circ) \to k$, such that for all coherent sheaves $\mathscr{F}$ on X, the natural pairing

$$\mathrm{Hom}(\mathscr{F},\omega_X^\circ) \times H^n(X,\mathscr{F}) \to H^n(X,\omega_X^\circ)$$

followed by t gives an *isomorphism*

$$\mathrm{Hom}(\mathscr{F},\omega_X^\circ) \xrightarrow{\sim} H^n(X,\mathscr{F})'.$$

Proposition 7.2. *Let X be a proper scheme over k. Then a dualizing sheaf for X, if it exists, is unique. More precisely, if ω° is one, with its trace map t, and if ω',t' is another, then there is a unique isomorphism $\varphi: \omega^\circ \xrightarrow{\sim} \omega'$ such that $t = t' \circ H^n(\varphi)$.*

PROOF. Since ω' is dualizing, we get an isomorphism $\mathrm{Hom}(\omega^\circ,\omega') \cong H^n(\omega^\circ)'$. So there is a unique morphism $\varphi: \omega^\circ \to \omega'$ corresponding to the element $t \in H^n(\omega^\circ)'$, i.e., such that $t' \circ H^n(\varphi) = t$. Similarly, using the fact that ω° is dualizing, there is a unique morphism $\psi: \omega' \to \omega^\circ$ such that $t \circ H^n(\psi) = t'$. It follows that $t \circ H^n(\psi \circ \varphi) = t$. But again since ω° is dualizing, this implies that $\psi \circ \varphi$ is the identity map of ω°. Similarly $\varphi \circ \psi$ is the identity map of ω', so φ is an isomorphism. (This proof is a special case of the uniqueness of an object representing a functor (see Grothendieck [EGA I, new ed., Ch. 0, §1]). For by definition (ω°,t) represents the functor $\mathscr{F} \mapsto H^n(X,\mathscr{F})'$ from $\mathfrak{Coh}(X)$ to $\mathfrak{Mod}(k)$.)

The question of existence of dualizing sheaves is more difficult. In fact they exist for any X proper over k, but we will prove the existence here only for projective schemes. First we need some preliminary results.

Lemma 7.3. *Let X be a closed subscheme of codimension r of $P = \mathbf{P}_k^N$. Then $\mathscr{E}xt_P^i(\mathscr{O}_X,\omega_P) = 0$ for all $i < r$.*

PROOF. For any i, the sheaf $\mathscr{F}^i = \mathscr{E}xt_P^i(\mathscr{O}_X,\omega_P)$ is a coherent sheaf on P (Ex. 6.3), so after twisting by a suitably large integer q, it will be generated by global sections (II, 5.17). Thus to show $\mathscr{F}^i$ is zero, it will be sufficient to show that $\Gamma(P,\mathscr{F}^i(q)) = 0$ for all $q \gg 0$. But by (6.7) and (6.9) we have

$$\Gamma(P,\mathscr{F}^i(q)) \cong \mathrm{Ext}_P^i(\mathscr{O}_X,\omega_P(q))$$

for $q \gg 0$. On the other hand, by (7.1) this last Ext group is dual to $H^{N-i}(P, \mathcal{O}_X(-q))$. For $i < r$, $N - i > \dim X$, so this group is 0 by (2.7) or (Ex. 4.8d).

Lemma 7.4. *With the same hypotheses as (7.3), let* $\omega_X^\circ = \mathcal{E}xt_P^r(\mathcal{O}_X, \omega_P)$. *Then for any* $\mathcal{O}_X$-*module* $\mathcal{F}$, *there is a functorial isomorphism*

$$\operatorname{Hom}_X(\mathcal{F}, \omega_X^\circ) \cong \operatorname{Ext}_P^r(\mathcal{F}, \omega_P).$$

PROOF. Let $0 \to \omega_P \to \mathcal{I}^\cdot$ be an injective resolution of ω_P in $\mathfrak{Mod}(P)$. Then we calculate $\operatorname{Ext}_P^i(\mathcal{F}, \omega_P)$ as the cohomology groups h^i of the complex $\operatorname{Hom}_P(\mathcal{F}, \mathcal{I}^\cdot)$. But since $\mathcal{F}$ is an $\mathcal{O}_X$-module, any morphism $\mathcal{F} \to \mathcal{I}^i$ factors through $\mathcal{J}^i = \mathcal{H}om_P(\mathcal{O}_X, \mathcal{I}^i)$. Thus we have

$$\operatorname{Ext}_P^i(\mathcal{F}, \omega_P) = h^i(\operatorname{Hom}_X(\mathcal{F}, \mathcal{J}^\cdot)).$$

Now each $\mathcal{J}^i$ is an *injective* $\mathcal{O}_X$-module. Indeed, for $\mathcal{F} \in \mathfrak{Mod}(X)$, $\operatorname{Hom}_X(\mathcal{F}, \mathcal{J}^i) = \operatorname{Hom}_P(\mathcal{F}, \mathcal{I}^i)$, so $\operatorname{Hom}_X(\cdot, \mathcal{J}^i)$ is an exact functor. Furthermore, by (7.3) we have $h^i(\mathcal{J}^\cdot) = 0$ for $i < r$, so the complex $\mathcal{J}^\cdot$ is exact up to the rth step. Since the $\mathcal{J}^i$ are injective, it is actually split exact up to the rth step. This implies that we can write the complex as a direct sum of two injective complexes, $\mathcal{J}^\cdot = \mathcal{J}_1^\cdot \oplus \mathcal{J}_2^\cdot$, where $\mathcal{J}_1^\cdot$ is in degrees $0 \leqslant i \leqslant r$ and is exact, and $\mathcal{J}_2^\cdot$ is in degrees $i \geqslant r$. It follows that $\omega_X^\circ = \ker(d^r : \mathcal{J}_2^r \to \mathcal{J}_2^{r+1})$, and that for any $\mathcal{O}_X$-module $\mathcal{F}$,

$$\operatorname{Hom}_X(\mathcal{F}, \omega_X^\circ) \cong \operatorname{Ext}_P^r(\mathcal{F}, \omega_P).$$

(It also follows that $\operatorname{Ext}_P^i(\mathcal{F}, \omega_P) = 0$ for $i < r$, which we won't need.)

Proposition 7.5. *Let X be a projective scheme over a field k. Then X has a dualizing sheaf.*

PROOF. Embed X as a closed subscheme of $P = \mathbf{P}_k^N$ for some N, let r be its codimension, and let $\omega_X^\circ = \mathcal{E}xt_P^r(\mathcal{O}_X, \omega_P)$. Then by (7.4) we have an isomorphism for any $\mathcal{O}_X$-module $\mathcal{F}$,

$$\operatorname{Hom}_X(\mathcal{F}, \omega_X^\circ) \cong \operatorname{Ext}_P^r(\mathcal{F}, \omega_P).$$

On the other hand, when $\mathcal{F}$ is coherent, the duality theorem for P (7.1) gives an isomorphism

$$\operatorname{Ext}_P^r(\mathcal{F}, \omega_P) \cong H^{N-r}(P, \mathcal{F})'.$$

But $N - r = n$, the dimension of X, and $\mathcal{F}$ is a sheaf on X, so we obtain a functorial isomorphism, for $\mathcal{F} \in \mathfrak{Coh}(X)$,

$$\operatorname{Hom}_X(\mathcal{F}, \omega_X^\circ) \cong H^n(X, \mathcal{F})'.$$

In particular, taking $\mathcal{F} = \omega_X^\circ$, the element $1 \in \operatorname{Hom}(\omega_X^\circ, \omega_X^\circ)$ gives us a homomorphism $t : H^n(X, \omega_X^\circ) \to k$, which we take as our trace map. Then it is clear by functoriality that (ω_X°, t) is a dualizing sheaf for X.

Now we can prove the duality theorem for a projective scheme X. Recall that a scheme is *Cohen–Macaulay* if all of its local rings are Cohen–Macaulay rings (II, §8).

Theorem 7.6 (Duality for a Projective Scheme). *Let X be a projective scheme of dimension n over an algebraically closed field k. Let ω_X° be a dualizing sheaf on X, and let $\mathcal{O}(1)$ be a very ample sheaf on X. Then:*

(a) *for all $i \geqslant 0$ and $\mathcal{F}$ coherent on X, there are natural functorial maps*

$$\theta^i : \operatorname{Ext}^i(\mathcal{F}, \omega_X^\circ) \to H^{n-i}(X, \mathcal{F})',$$

such that θ^0 is the map given in the definition of dualizing sheaf above;

(b) *the following conditions are equivalent:*

(i) *X is Cohen–Macaulay and equidimensional (i.e., all irreducible components have the same dimension);*

(ii) *for any $\mathcal{F}$ locally free on X, we have $H^i(X, \mathcal{F}(-q)) = 0$ for $i < n$ and $q \gg 0$;*

(iii) *the maps θ^i of (a) are isomorphisms for all $i \geqslant 0$ and all $\mathcal{F}$ coherent on X.*

PROOF.

(a) As in the proof of (7.1c), we can write any coherent sheaf $\mathcal{F}$ as a quotient of a sheaf $\mathcal{E} = \bigoplus_{i=1}^{N} \mathcal{O}_X(-q)$, with $q \gg 0$. Then $\operatorname{Ext}^i(\mathcal{E}, \omega_X^\circ) \cong \bigoplus H^i(X, \omega_X^\circ(q))$, which is 0 for $i > 0$ and $q \gg 0$ by (5.2). Thus the functor $\operatorname{Ext}^i(\cdot, \omega_X^\circ)$ is coeffaceable for $i > 0$, so we have a universal contravariant δ-functor by (1.3A). On the right-hand side we have a contravariant δ-functor, indexed by $i \geqslant 0$, so there is a unique morphism of δ-functors (θ^i) reducing to the given θ^0 for $i = 0$.

(b) (i) $\Rightarrow$ (ii). Embed X as a closed subscheme of $P = \mathbf{P}_k^N$. Then for any $\mathcal{F}$ locally free on X, and any closed point $x \in X$, we have depth $\mathcal{F}_x = n$, since X is Cohen–Macaulay and equidimensional of dimension n. Let $A = \mathcal{O}_{P,x}$ be the local ring of x on P. Then A is a regular local ring of dimension N. (Since k is algebraically closed, x is rational over k, so the fact that A is regular can be seen directly. Or it follows from the fact that P is a nonsingular variety over k (II, §8).) Now depth $\mathcal{F}_x$ is the same, whether calculated over $\mathcal{O}_{X,x}$ or over A. Thus we conclude from (6.12A) that $\operatorname{pd}_A \mathcal{F}_x = N - n$. Therefore by (6.8) and (6.10A) we have

$$\mathcal{E}xt_P^i(\mathcal{F}, \cdot) = 0$$

for $i > N - n$.

On the other hand, using (7.1), we find that $H^i(X, \mathcal{F}(-q))$ is dual to $\operatorname{Ext}_P^{N-i}(\mathcal{F}, \omega_P(q))$. For $q \gg 0$, this Ext is isomorphic to $\Gamma(P, \mathcal{E}xt_P^{N-i}(\mathcal{F}, \omega_P(q)))$ by (6.9). But this is 0 for $N - i > N - n$, as we have just seen. In other words, $H^i(X, \mathcal{F}(-q)) = 0$ for $i < n$ and $q \gg 0$.

(ii) $\Rightarrow$ (i). Running the above argument backwards, using condition (ii) with $\mathcal{F} = \mathcal{O}_X$, we find that

$$\mathcal{E}xt_P^i(\mathcal{O}_X, \omega_P) = 0$$

243

for $i > N - n$. This implies that over a local ring $A = \mathcal{O}_{P,x}$ as above, we have $\operatorname{Ext}_A^i(\mathcal{O}_{X,x}, A) = 0$ for all $i > N - n$. Therefore by (Ex. 6.6) we have $\operatorname{pd}_A \mathcal{O}_{X,x} \leqslant N - n$, and so by (6.12A), depth $\mathcal{O}_{X,x} \geqslant n$. But since $\dim X = n$, we must have equality for every closed point of X. This shows, using (II, 8.21Ab), that X is Cohen–Macaulay and equidimensional.

(ii) $\Rightarrow$ (iii). Since we have already seen that $\operatorname{Ext}^i(\cdot, \omega_X^\circ))$ is a universal contravariant δ-functor, to show that the θ^i are isomorphisms, it will be sufficient to show that the δ-functor $(H^{n-i}(X, \cdot)')$ is universal also. For this it suffices by (1.3A) to show that $H^{n-i}(X, \cdot)'$ is coeffaceable for $i > 0$. So given a coherent sheaf $\mathscr{F}$, write $\mathscr{F}$ as a quotient of $\mathscr{E} = \bigoplus \mathcal{O}(-q)$ with $q \gg 0$. Then $H^{n-i}(X, \mathscr{E})' = 0$ for $i > 0$ by (ii), so the functor is coeffaceable.

(iii) $\Rightarrow$ (ii). If θ^i is an isomorphism, then for any $\mathscr{F}$ locally free, we have

$$H^i(X, \mathscr{F}(-q)) \cong \operatorname{Ext}^{n-i}(\mathscr{F}(-q), \omega_X^\circ)'.$$

But this Ext is isomorphic to $H^{n-i}(X, \mathscr{F}^\vee \otimes \omega_X^\circ(q))$ by (6.3) and (6.7), so it is 0 for $n - i > 0$ and $q \gg 0$ by (5.2). q.e.d.

Remark 7.6.1. In particular, if X is nonsingular over k, or more generally a local complete intersection, then X is Cohen–Macaulay (II, 8.21A) and (II, 8.23), so the θ^i are isomorphisms. In these two cases, one can show directly (cf. proof of (7.11) below) that $\operatorname{pd}_P \mathcal{O}_X = N - n$, and thus avoid use of the algebraic results (6.12A) and (Ex. 6.6).

Corollary 7.7. Let X be a projective Cohen–Macaulay scheme of equidimension n over k. Then for any locally free sheaf $\mathscr{F}$ on X there are natural isomorphisms

$$H^i(X, \mathscr{F}) \cong H^{n-i}(X, \mathscr{F}^\vee \otimes \omega_X^\circ)'.$$

PROOF. Use (6.3) and (6.7).

Corollary 7.8 (Lemma of Enriques–Severi–Zariski (Zariski [4])). Let X be a normal projective scheme of dimension $\geqslant 2$. Then for any locally free sheaf $\mathscr{F}$ on X,

$$H^1(X, \mathscr{F}(-q)) = 0$$

for $q \gg 0$.

PROOF. Since X is normal of dimension $\geqslant 2$, we have depth $\mathscr{F}_x \geqslant 2$ for every closed point $x \in X$ by (II, 8.22A). So the result follows by the same method as the proof of (i) $\Rightarrow$ (ii) in (7.6b).

Corollary 7.9. Let X be an integral, normal projective variety of dimension $\geqslant 2$ over an algebraically closed field k. Let Y be a closed subset of codimension 1 which is the support of an effective ample divisor. Then Y is connected.

PROOF. By (II, 7.6) we may assume that Y is the support of a very ample divisor D. Let $\mathcal{O}(1)$ be the corresponding very ample invertible sheaf. For each $q > 0$, let Y_q be the closed subscheme supported on Y corresponding to the divisor qD (II, 6.17.1). Then we have an exact sequence (II, 6.18)

$$0 \to \mathcal{O}_X(-q) \to \mathcal{O}_X \to \mathcal{O}_{Y_q} \to 0.$$

Taking cohomology and applying (7.8), we find that for $q \gg 0$,

$$H^0(X,\mathcal{O}_X) \to H^0(Y,\mathcal{O}_{Y_q}) \to 0$$

is surjective. But $H^0(X,\mathcal{O}_X) = k$ (I, 3.4a), and $H^0(Y,\mathcal{O}_{Y_q})$ contains k, so we conclude that $H^0(Y,\mathcal{O}_{Y_q}) = k$. Hence Y is connected. (If not, there would be at least one copy of k for each connected component.)

Remark 7.9.1. This implies that the schemes $H \cap X$ mentioned in Bertini's theorem (II, 8.18) are in fact *irreducible* and nonsingular when $\dim X \geqslant 2$. Indeed, they are connected by (7.9). On the other hand, they are regular by (II, 8.18). Hence the local rings are all integral domains, so we could not have two irreducible components meeting at a point.

Now that we have proved the duality theorem (7.6), our next task is to give more information about the dualizing sheaf ω_X° in some special cases. Again we need some algebraic preliminaries.

Let A be a ring, and let $f_1,\ldots,f_r \in A$. We define the *Koszul complex* $K.(f_1,\ldots,f_r)$ as follows: K_1 is a free A-module of rank r with basis $e_1,\ldots,e_r$. For each $p = 0,\ldots,r$, $K_p = \wedge^p K_1$. We define the boundary map $d: K_p \to K_{p-1}$ by its action on the basis vectors:

$$d(e_{i_1} \wedge \ldots \wedge e_{i_p}) = \sum(-1)^{j-1} f_{i_j} e_{i_1} \wedge \ldots \wedge \hat{e}_{i_j} \wedge \ldots \wedge e_{i_p}.$$

Thus $K.(f_1,\ldots,f_r)$ is a (homological) complex of A-modules. If M is any A-module, we set $K.(f_1,\ldots,f_r; M) = K.(f_1,\ldots,f_r) \otimes_A M$.

Proposition 7.10A. *Let A be a ring, $f_1,\ldots,f_r \in A$, and let M be an A-module. If the f_i form a regular sequence for M, then*

$$h_i(K.(f_1,\ldots,f_r; M)) = 0 \quad \text{for } i > 0$$

and

$$h_0(K.(f_1,\ldots,f_r; M)) \cong M/(f_1,\ldots,f_r)M.$$

PROOF. Matsumura [2, Th. 43, p. 135] or Serre [11, IV.A].

Theorem 7.11. *Let X be a closed subscheme of $P = \mathbf{P}_k^N$ which is a local complete intersection of codimension r. Let $\mathscr{I}$ be the ideal sheaf of X. Then $\omega_X^\circ \cong \omega_P \otimes \wedge^r(\mathscr{I}/\mathscr{I}^2)\check{\ }$. In particular, ω_X° is an invertible sheaf on X.*

PROOF. We have to calculate $\omega_X^\circ = \mathscr{E}xt_P^r(\mathcal{O}_X, \omega_P)$. Let U be an open affine subset over which $\mathscr{I}$ can be generated by r elements $f_1, \ldots, f_r \in A = \Gamma(U, \mathcal{O}_U)$ and let $x \in X \cap U$ be a point corresponding to an ideal $\mathfrak{m} \subseteq A$. Because X has codimension r and $A_{\mathfrak{m}}$ is Cohen-Macaulay, $f_1, \ldots, f_r$ form a regular sequence for $A_{\mathfrak{m}}$ (II, 8.21A). Therefore the localized Koszul complex $K.(f_1, \ldots, f_r; A_{\mathfrak{m}})$ gives a free resolution of $A_{\mathfrak{m}}/(f_1, \ldots, f_r)A_{\mathfrak{m}}$ over $A_{\mathfrak{m}}$, so replacing U by a smaller neighborhood of x if necessary, $K.(f_1, \ldots, f_r)$ gives a free resolution of $A/(f_1, \ldots, f_r)$ over A. Sheafifying gives a free resolution $K.(f_1, \ldots, f_r; \mathcal{O}_P)$ of $\mathcal{O}_X$ over U with which we can calculate $\mathscr{E}xt_P^r(\mathcal{O}_X, \omega_P)$ (6.5). We get

$$h^r(\mathscr{H}om(K.(f_1, \ldots, f_r; \mathcal{O}_P), \omega_P)) \cong \omega_P/(f_1, \ldots, f_r)\omega_P.$$

In other words,

$$\mathscr{E}xt_P^r(\mathcal{O}_X, \omega_P) \cong \omega_P \otimes \mathcal{O}_X$$

over U. However, this isomorphism depends on the choice of basis $f_1, \ldots, f_r$ for $\mathscr{I}$. If $g_i = \sum c_{ij} f_j, i = 1, \ldots, r$, is another basis, then the exterior powers of the matrix $\|c_{ij}\|$ give an isomorphism of Koszul complexes. In particular, we have a factor of $\det|c_{ij}|$ on K_r, so our isomorphism of $\mathscr{E}xt^r$ changes by $\det|c_{ij}|$.

To remedy this situation, we consider the sheaf $\mathscr{I}/\mathscr{I}^2$ on X, which is locally free of rank r (II, 8.21A). In particular, it is free over U, with basis $f_1, \ldots, f_r$. Therefore $\wedge^r(\mathscr{I}/\mathscr{I}^2)$ is free of rank 1, with basis $f_1 \wedge \cdots \wedge f_r$. If we change to the basis $g_1, \ldots, g_r$, this element changes by $\det|c_{ij}|$. Therefore, we can obtain an intrinsic isomorphism above by tensoring with this free sheaf of rank 1 (check variance!)

$$\mathscr{E}xt_P^r(\mathcal{O}_X, \omega_P) \cong \omega_P \otimes \mathcal{O}_X \otimes \wedge^r(\mathscr{I}/\mathscr{I}^2)^\vee.$$

This isomorphism, defined over U, is independent of the choice of basis. Therefore when we cover P with such open sets, these isomorphisms glue together, and we obtain the required isomorphism $\omega_X^\circ \cong \omega_P \otimes \wedge^r(\mathscr{I}/\mathscr{I}^2)^\vee$.

Corollary 7.12. *If X is a projective nonsingular variety over an algebraically closed field k, then the dualizing sheaf ω_X° is isomorphic to the canonical sheaf ω_X.*

PROOF. Embed X in $P = \mathbf{P}_k^N$. Then X is a local complete intersection in P (II, 8.17), and $\omega_X \cong \omega_P \otimes \wedge^r(\mathscr{I}/\mathscr{I}^2)^\vee$ by (II, 8.20).

Remark 7.12.1. Thus for a projective nonsingular variety X, the duality theorem (7.6) and its corollary (7.7) hold with ω_X in place of ω_X°. In particular, we obtain an isomorphism $H^n(X, \omega_X) \cong k$, whose existence is by no means obvious a priori.

Remark 7.12.2. If X is a projective nonsingular curve, we find that $H^1(X, \mathcal{O}_X)$ and $H^0(X, \omega_X)$ are dual vector spaces. Hence the arithmetic genus $p_a = \dim H^1(X, \mathcal{O}_X)$ and the geometric genus $p_g = \dim \Gamma(X, \omega_X)$ are equal— cf. (Ex. 5.3a) and (II. 8.18.2).

Remark 7.12.3. If X is a projective nonsingular surface, then $H^0(X,\omega)$ is dual to $H^2(X,\mathcal{O}_X)$, so $p_g = \dim H^2(X,\mathcal{O}_X)$. On the other hand $p_a = \dim H^2(X,\mathcal{O}_X) - \dim H^1(X,\mathcal{O}_X)$ by (Ex. 5.3a). Thus $p_g \geqslant p_a$. The difference, $p_g - p_a = \dim H^1(X,\mathcal{O}_X)$ is usually denoted by q, and is called the *irregularity* of X. For example, the surface of (II, Ex. 8.3c) has irregularity 2.

Corollary 7.13. *Let X be a nonsingular projective variety of dimension n. For any $p = 0,1,\ldots,n$, let $\Omega^p = \wedge^p\Omega_{X/k}$ be the sheaf of differential p-forms. Then for each $p,q = 0,1,\ldots,n$, we have a natural isomorphism*

$$H^q(X,\Omega^p) \cong H^{n-q}(X,\Omega^{n-p})'.$$

PROOF. Indeed, for any p, $\Omega^{n-p} \cong (\Omega^p)^{\check{}} \otimes \omega$ (II, Ex. 5.16b). Then use (7.7).

Remark 7.13.1. The numbers $h^{p,q} = \dim H^q(X,\Omega^p)$ are important biregular invariants of the variety X.

Remark 7.14 (Residues of Differentials on Curves). A weakness of the duality theorem as we have proved it is that even for a nonsingular projective variety X, we don't have much information about the trace map $t:H^n(X,\omega) \to k$. We know only that it exists. In the case of curves, there is another way of proving the duality theorem, using residues, which improves this situation.

Let X be a complete nonsingular curve over an algebraically closed field k, and let K be the function field of X. Let Ω_X be the sheaf of differentials of X over k, and for a closed point $P \in X$, let Ω_P be its stalk at P. Let Ω_K be the module of differentials of K over k. Then one first proves:

Theorem 7.14.1 (Existence of Residues). *For each closed point $P \in X$, there is a unique k-linear map $\operatorname{res}_P:\Omega_K \to k$ with the following properties:*
   (a) $\operatorname{res}_P(\tau) = 0$ *for all* $\tau \in \Omega_P$;
   (b) $\operatorname{res}_P(f^n df) = 0$ *for all* $f \in K^*$, *all* $n \neq -1$;
   (c) $\operatorname{res}_P(f^{-1} df) = v_P(f) \cdot 1$, *where v_P is the valuation associated to P.*

From these properties we see immediately how to calculate the residue of any differential. Indeed, let $t \in \mathcal{O}_P$ be a uniformizing parameter. Then dt is a generator for Ω_K as a K-vector space, so we can write any $\tau \in \Omega_K$ as $g\,dt$ for some $g \in K$. Furthermore, since $\mathcal{O}_P$ is a valuation ring, we can write $g = \sum_{i<0} a_i t^i + h$ with $a_i \in k$, $h \in \mathcal{O}_P$, and the sum finite. Thus $\tau = \sum a_i t^i dt + h\,dt$. Now from linearity and (a), (b), (c) we find
   (d) $\operatorname{res}_P \tau = a_{-1}$.
Thus the uniqueness of res_P is clear.

The existence is more difficult. One approach by Serre [7, Ch. II] is to take (d) as the definition of the residue. Then one has an awkward time proving that it is independent of the choice of the uniformizing parameter t, especially in the case of characteristic $p > 0$. Another approach by Tate [2] gives an intrinsic construction of the residue map by a clever use of certain k-linear transformations of K.

The basic result about residues is:

Theorem 7.14.2 (Residue Theorem). *For any* $\tau \in \Omega_K$, *we have* $\sum_{P \in X} \operatorname{res}_P \tau = 0$.

In Serre's approach this theorem is first proved on $\mathbf{P}^1$, by explicit calculation. Then the general case is obtained by using a finite morphism $X \to \mathbf{P}^1$ and studying the relationship between the residues in both places. In Tate's approach the residue theorem follows directly from the construction of the residue map.

Once one has the theory of residues, the duality theorem for X can be proved by a method of Weil using repartitions. We refer to the lucid expositions of Serre and Tate mentioned above for the details of this classic story.

The connection with our approach can be explained as follows. The exact sequence

$$0 \to \mathcal{O}_X \to \mathcal{K}_X \to \mathcal{K}_X/\mathcal{O}_X \to 0,$$

where $\mathcal{K}_X$ is the constant sheaf K_X, is a flasque resolution of $\mathcal{O}_X$ (cf. Ex. 2.2). Furthermore,

$$\mathcal{K}_X/\mathcal{O}_X \cong \bigoplus_{P \in X} i_*(K_X/\mathcal{O}_P)$$

where we consider $K_X/\mathcal{O}_P$ as an $\mathcal{O}_P$-module, and $i : \{P\} \to X$ is the inclusion map. Tensoring with Ω_X, we get a flasque resolution of Ω_X:

$$0 \to \Omega_X \to \Omega_X \otimes \mathcal{K}_X \to \bigoplus_{P \in X} i_*(\Omega_K/\Omega_P) \to 0.$$

Taking cohomology, we get an exact sequence

$$\Omega_K \to \bigoplus_{P \in X} \Omega_K/\Omega_P \to H^1(X, \Omega_X) \to 0.$$

We define a map

$$\bigoplus_{P \in X} \Omega_K/\Omega_P \to k$$

by taking the sum of all the maps $\operatorname{res}_P : \Omega_K/\Omega_P \to k$. Then by (7.14.2) this map vanishes on the image of Ω_K, hence it passes to the quotient and gives a map $t : H^1(X, \Omega_X) \to k$. This is the trace map of our duality theorem, which appears now in a much more explicit form.

Remark 7.15 (The Kodaira Vanishing Theorem). Our discussion of the cohomology of projective varieties would not be complete without mentioning the Kodaira vanishing theorem. It says if X is a projective nonsingular variety of dimension n over $\mathbf{C}$, and if $\mathcal{L}$ is an ample invertible sheaf on X, then:

(a) $H^i(X, \mathcal{L} \otimes \omega) = 0$ for $i > 0$;
(b) $H^i(X, \mathcal{L}^{-1}) = 0$ for $i < n$.

Of course (a) and (b) are equivalent to each other by Serre duality. The theorem is proved using methods of complex analytic differential geometry. At present there is no purely algebraic proof. On the other hand, Raynaud has recently shown that this result does not hold over fields of characteristic $p > 0$.

The first proof was given by Kodaira [1]. For other proofs, including the generalization by Nakano, see Wells [1, Ch. VI, §2], Mumford [3], and Ramanujam [1]. For a relative version of the theorem, see Grauert and Riemenschneider [1].

References for the Duality Theorem. The duality theorem was first proved by Serre [2] (in the form of (7.7)) for locally free sheaves on a compact complex manifold, and in the case of abstract algebraic geometry by Serre [1]. Our proof follows Grothendieck [5] and Grothendieck [SGA 2, exp. XII], with some improvements suggested by Lipman. The duality theorem and the theory of residues have been generalized to the case of an arbitrary proper morphism by Grothendieck—see Grothendieck [4] and Hartshorne [2]. Deligne has given another proof of the existence of a dualizing sheaf, and Verdier [1] has shown that this one agrees with the sheaf ω for a non-singular variety. Kunz [1] gives another construction, using differentials, of the dualizing sheaf ω_X° for an integral projective scheme X over k.

The duality theorem has also been generalized to the case of a proper morphism of complex analytic spaces—see Ramis and Ruget [1] and Ramis, Ruget, and Verdier [1]. For a generalization to noncompact complex manifolds, see Suominen [1].

In the case of curves, the duality theorem is the most important ingredient in the proof of the Riemann–Roch theorem (IV, §1). See Serre [7, Ch. II] for the history of this approach, and also Gunning [1] for a proof in the language of compact Riemann surfaces.

EXERCISES

7.1. Let X be an integral projective scheme of dimension $\geqslant 1$ over a field k, and let $\mathscr{L}$ be an ample invertible sheaf on X. Then $H^0(X, \mathscr{L}^{-1}) = 0$. (This is an easy special case of Kodaira's vanishing theorem.)

7.2. Let $f: X \to Y$ be a finite morphism of projective schemes of the same dimension over a field k, and let ω_Y° be a dualizing sheaf for Y.
  (a) Show that $f^! \omega_Y^\circ$ is a dualizing sheaf for X, where $f^!$ is defined as in (Ex. 6.10).
  (b) If X and Y are both nonsingular, and k algebraically closed, conclude that there is a natural trace map $t: f_* \omega_X \to \omega_Y$.

7.3. Let $X = \mathbf{P}_k^n$. Show that $H^q(X, \Omega_X^p) = 0$ for $p \neq q$, k for $p = q$, $0 \leqslant p, q \leqslant n$.

*7.4. *The Cohomology Class of a Subvariety.* Let X be a nonsingular projective variety of dimension n over an algebraically closed field k. Let Y be a nonsingular subvariety of codimension p (hence dimension $n - p$). From the natural map $\Omega_X \otimes \mathcal{O}_Y \to \Omega_Y$ of (II, 8.12) we deduce a map $\Omega_X^{n-p} \to \Omega_Y^{n-p}$. This induces a map on cohomology $H^{n-p}(X, \Omega_X^{n-p}) \to H^{n-p}(Y, \Omega_Y^{n-p})$. Now $\Omega_Y^{n-p} = \omega_Y$ is a dualizing sheaf

249

for Y, so we have the trace map $t_Y: H^{n-p}(Y, \Omega_Y^{n-p}) \to k$. Composing, we obtain a linear map $H^{n-p}(X, \Omega_X^{n-p}) \to k$. By (7.13) this corresponds to an element $\eta(Y) \in H^p(X, \Omega_X^p)$, which we call the *cohomology class* of Y.

(a) If $P \in X$ is a closed point, show that $t_X(\eta(P)) = 1$, where $\eta(P) \in H^n(X, \Omega^n)$ and t_X is the trace map.

(b) If $X = \mathbf{P}^n$, identify $H^p(X, \Omega^p)$ with k by (Ex. 7.3), and show that $\eta(Y) = (\deg Y) \cdot 1$, where $\deg Y$ is its *degree* as a projective variety (I, §7). [*Hint*: Cut with a hyperplane $H \subseteq X$, and use Bertini's theorem (II, 8.18) to reduce to the case Y is a finite set of points.]

(c) For any scheme X of finite type over k, we define a homomorphism of sheaves of abelian groups $d\log: \mathcal{O}_X^* \to \Omega_X$ by $d\log(f) = f^{-1}df$. Here $\mathcal{O}^*$ is a group under multiplication, and Ω_X is a group under addition. This induces a map on cohomology $\operatorname{Pic} X = H^1(X, \mathcal{O}_X^*) \to H^1(X, \Omega_X)$ which we denote by c—see (Ex. 4.5).

(d) Returning to the hypotheses above, suppose $p = 1$. Show that $\eta(Y) = c(\mathscr{L}(Y))$, where $\mathscr{L}(Y)$ is the invertible sheaf corresponding to the divisor Y.

See Matsumura [1] for further discussion.

8 Higher Direct Images of Sheaves

For the remainder of this chapter we will be studying families of schemes. Recall (II, §3) that a family of schemes is simply a morphism $f: X \to Y$, and the members of the family are the fibres $X_y = X \times_Y \operatorname{Spec} k(y)$ for various points $y \in Y$. To study a family, we need some form of "relative cohomology of X over Y," or "cohomology along the fibres of X over Y." This notion is provided by the higher direct image functors $R^i f_*$ which we define below. The precise relationship between these functors and the cohomology of the fibres X_y will be studied in §11, 12.

Definition. Let $f: X \to Y$ be a continuous map of topological spaces. Then we define the *higher direct image* functors $R^i f_*: \mathfrak{Ab}(X) \to \mathfrak{Ab}(Y)$ to be the right derived functors of the direct image functor f_* (II, §1).

This makes sense because f_* is obviously left exact, and $\mathfrak{Ab}(X)$ has enough injectives (2.3).

Proposition 8.1. *For each $i \geqslant 0$ and each $\mathscr{F} \in \mathfrak{Ab}(X)$, $R^i f_*(\mathscr{F})$ is the sheaf associated to the presheaf*

$$V \mapsto H^i(f^{-1}(V), \mathscr{F}|_{f^{-1}(V)})$$

on Y.

PROOF. Let us denote the sheaf associated to the above presheaf by $\mathscr{H}^i(X, \mathscr{F})$. Then, since the operation of taking the sheaf associated to a presheaf is exact, the functors $\mathscr{H}^i(X, \cdot)$ form a δ-functor from $\mathfrak{Ab}(X)$ to $\mathfrak{Ab}(Y)$. For $i = 0$ we have $f_* \mathscr{F} = \mathscr{H}^0(X, \mathscr{F})$ by definition of f_*. For an injective object $\mathscr{I} \in \mathfrak{Ab}(X)$ we have $R^i f_*(\mathscr{I}) = 0$ for $i > 0$ because $R^i f_*$ is a derived functor.

On the other hand, for each V, $\mathscr{I}|_{f^{-1}(V)}$ is injective in $\mathfrak{Ab}(f^{-1}(V))$ by (6.1) (think of X as a ringed space with the constant sheaf $\mathbf{Z}$), so $\mathscr{H}^i(X,\mathscr{I}) = 0$ for $i > 0$ also. Hence there is a unique isomorphism of δ-functors $R^i f_*(\cdot) \cong \mathscr{H}^i(X,\cdot)$ by (1.3A).

Corollary 8.2. *If $V \subseteq Y$ is any open subset, then*

$$R^i f_*(\mathscr{F})|_V = R^i f'_*(\mathscr{F}|_{f^{-1}(V)})$$

where $f':f^{-1}(V) \to V$ is the restricted map.

PROOF. Obvious.

Corollary 8.3. *If $\mathscr{F}$ is a flasque sheaf on X, then $R^i f_*(\mathscr{F}) = 0$ for all $i > 0$.*

PROOF. Since the restriction of a flasque sheaf to an open subset is flasque, this follows from (2.5).

Proposition 8.4. *Let $f:X \to Y$ be a morphism of ringed spaces. Then the functors $R^i f_*$ can be calculated on $\mathfrak{Mod}(X)$ as the derived functors of $f_*:\mathfrak{Mod}(X) \to \mathfrak{Mod}(Y)$.*

PROOF. To calculate the derived functors of f_* on $\mathfrak{Mod}(X)$, we use resolutions by injective objects of $\mathfrak{Mod}(X)$. Any injective of $\mathfrak{Mod}(X)$ is flasque by (2.4), hence acyclic for f_* on $\mathfrak{Ab}(X)$ by (8.3), so they can be used to calculate $R^i f_*$ by (1.2A).

Proposition 8.5. *Let X be a noetherian scheme, and let $f:X \to Y$ be a morphism of X to an affine scheme $Y = \mathrm{Spec}\,A$. Then for any quasi-coherent sheaf $\mathscr{F}$ on X, we have*

$$R^i f_*(\mathscr{F}) \cong H^i(X,\mathscr{F})^\sim .$$

PROOF. By (II, 5.8), $f_*\mathscr{F}$ is a quasi-coherent sheaf on Y. Hence $f_*\mathscr{F} \cong \Gamma(Y,f_*\mathscr{F})^\sim$. But $\Gamma(Y,f_*\mathscr{F}) = \Gamma(X,\mathscr{F})$. So we have an isomorphism for $i = 0$.

Since $\sim$ is an exact functor from $\mathfrak{Mod}(A)$ to $\mathfrak{Mod}(Y)$, both sides are δ-functors from $\mathfrak{Qco}(X)$ to $\mathfrak{Mod}(Y)$. Furthermore, by (3.6), any quasi-coherent sheaf $\mathscr{F}$ on X can be embedded in a flasque, quasi-coherent sheaf. Hence both sides are effaceable for $i > 0$. We conclude from (1.3A) that there is a unique isomorphism of δ-functors as above, reducing to the given one for $i = 0$.

Note that we must work in the category $\mathfrak{Qco}(X)$, because already the case $i = 0$ fails if $\mathscr{F}$ is not quasi-coherent.

Corollary 8.6. *Let $f:X \to Y$ be a morphism of schemes, with X noetherian. Then for any quasi-coherent sheaf $\mathscr{F}$ on X, the sheaves $R^i f_*(\mathscr{F})$ are quasi-coherent on Y.*

PROOF. The question is local on Y, so we may use (8.5).

Proposition 8.7. *Let $f: X \to Y$ be a morphism of separated noetherian schemes. Let $\mathscr{F}$ be a quasi-coherent sheaf on X, let $\mathfrak{U} = (U_i)$ be an open affine cover of X, and let $\mathscr{C}^{\cdot}(\mathfrak{U}, \mathscr{F})$ be the Čech resolution of $\mathscr{F}$ given by (4.2). Then for each $p \geqslant 0$,*

$$R^p f_*(\mathscr{F}) \cong h^p(f_* \mathscr{C}^{\cdot}(\mathfrak{U}, \mathscr{F})).$$

PROOF. For any open affine subset $V \subseteq Y$, the open subsets $U_i \cap f^{-1}(V)$ of X are all affine (check!—cf. (II, Ex. 4.3)). Hence we may reduce to the case Y affine. The sheaves $\mathscr{C}^p(\mathfrak{U}, \mathscr{F})$ are all quasi-coherent, so we have

$$f_* \mathscr{C}^{\cdot}(\mathfrak{U}, \mathscr{F}) \cong C^{\cdot}(\mathfrak{U}, \mathscr{F})^{\sim}$$

by (II, 5.8). Now the result follows from (4.5) and (8.5).

Theorem 8.8. *Let $f: X \to Y$ be a projective morphism of noetherian schemes, let $\mathcal{O}_X(1)$ be a very ample invertible sheaf on X over Y, and let $\mathscr{F}$ be a coherent sheaf on X. Then:*
  (a) *for all $n \gg 0$, the natural map $f^* f_*(\mathscr{F}(n)) \to \mathscr{F}(n)$ is surjective;*
  (b) *for all $i \geqslant 0$, $R^i f_*(\mathscr{F})$ is a coherent sheaf on Y;*
  (c) *for $i > 0$ and $n \gg 0$, $R^i f_*(\mathscr{F}(n)) = 0$.*

PROOF. Since Y is quasi-compact, the question is local on Y, so we may assume Y is affine, say $Y = \operatorname{Spec} A$. Then, using (8.5), (a) says that $\mathscr{F}(n)$ is generated by global sections, which is (II, 5.17). (b) says that $H^i(X, \mathscr{F})$ is a finitely generated A-module, which is (5.2a). Finally, (c) says that $H^i(X, \mathscr{F}(n)) = 0$, which is (5.2b).

Remark 8.8.1. Part (b) of this theorem is true more generally for a proper morphism of noetherian schemes—see Grothendieck [EGA III, 3.2.1]. The analogous theorem for a proper morphism of complex analytic spaces was proved by Grauert [1].

EXERCISES

8.1. Let $f: X \to Y$ be a continuous map of topological spaces. Let $\mathscr{F}$ be a sheaf of abelian groups on X, and assume that $R^i f_*(\mathscr{F}) = 0$ for all $i > 0$. Show that there are natural isomorphisms, for each $i \geqslant 0$,

$$H^i(X, \mathscr{F}) \cong H^i(Y, f_* \mathscr{F}).$$

(This is a degenerate case of the Leray spectral sequence—see Godement [1, II, 4.17.1].)

8.2. Let $f: X \to Y$ be an affine morphism of schemes (II, Ex. 5.17) with X noetherian, and let $\mathscr{F}$ be a quasi-coherent sheaf on X. Show that the hypotheses of (Ex. 8.1) are satisfied, and hence that $H^i(X, \mathscr{F}) \cong H^i(Y, f_* \mathscr{F})$ for each $i \geqslant 0$. (This gives another proof of (Ex. 4.1).)

8.3. Let $f : X \to Y$ be a morphism of ringed spaces, let $\mathscr{F}$ be an $\mathcal{O}_X$-module, and let $\mathscr{E}$ be a locally free $\mathcal{O}_Y$-module of finite rank. Prove the *projection formula* (cf. (II, Ex. 5.1))

$$R^i f_*(\mathscr{F} \otimes f^* \mathscr{E}) \cong R^i f_*(\mathscr{F}) \otimes \mathscr{E}.$$

8.4. Let Y be a noetherian scheme, and let $\mathscr{E}$ be a locally free $\mathcal{O}_Y$-module of rank $n + 1$, $n \geqslant 1$. Let $X = \mathbf{P}(\mathscr{E})$ (II, §7), with the invertible sheaf $\mathcal{O}_X(1)$ and the projection morphism $\pi : X \to Y$.

(a) Then $\pi_*(\mathcal{O}(l)) \cong S^l(\mathscr{E})$ for $l \geqslant 0$, $\pi_*(\mathcal{O}(l)) = 0$ for $l < 0$ (II, 7.11); $R^i \pi_*(\mathcal{O}(l)) = 0$ for $0 < i < n$ and all $l \in \mathbf{Z}$; and $R^n \pi_*(\mathcal{O}(l)) = 0$ for $l > -n - 1$.

(b) Show there is a natural exact sequence

$$0 \to \Omega_{X/Y} \to (\pi^* \mathscr{E})(-1) \to \mathcal{O} \to 0,$$

cf. (II, 8.13), and conclude that the *relative canonical sheaf* $\omega_{X/Y} = \wedge^n \Omega_{X/Y}$ is isomorphic to $(\pi^* \wedge^{n+1} \mathscr{E})(-n - 1)$. Show furthermore that there is a natural isomorphism $R^n \pi_*(\omega_{X/Y}) \cong \mathcal{O}_Y$ (cf. (7.1.1)).

(c) Now show, for any $l \in \mathbf{Z}$, that

$$R^n \pi_*(\mathcal{O}(l)) \cong \pi_*(\mathcal{O}(-l - n - 1))^{\vee} \otimes (\wedge^{n+1} \mathscr{E})^{\vee}.$$

(d) Show that $p_a(X) = (-1)^n p_a(Y)$ (use (Ex. 8.1)) and $p_g(X) = 0$ (use (II, 8.11)).

(e) In particular, if Y is a nonsingular projective curve of genus g, and $\mathscr{E}$ a locally free sheaf of rank 2, then X is a projective surface with $p_a = -g$, $p_g = 0$, and irregularity g (7.12.3). This kind of surface is called a *geometrically ruled surface* (V, §2).

9 Flat Morphisms

In this section we introduce the notion of a flat morphism of schemes. By taking the fibres of a flat morphism, we get the notion of a flat family of schemes. This provides a concise formulation of the intuitive idea of a "continuous family of schemes." We will show, through various results and examples, why flatness is a natural as well as a convenient condition to put on a family of schemes.

First we recall the algebraic notion of a flat module. Let A be a ring, and let M be an A-module. We say that M is *flat* over A if the functor $N \mapsto M \otimes_A N$ is an exact functor for $N \in \mathfrak{Mod}(A)$. If $A \to B$ is a ring homomorphism, we say that B is *flat* over A if it is flat as a module.

Proposition 9.1A.

(a) *An A-module M is flat if and only if for every finitely generated ideal $\mathfrak{a} \subseteq A$, the map $\mathfrak{a} \otimes M \to M$ is injective.*

(b) *Base extension: If M is a flat A-module, and $A \to B$ is a homomorphism, then $M \otimes_A B$ is a flat B-module.*

(c) *Transitivity: If B is a flat A-algebra, and N is a flat B-module, then N is also flat as an A-module.*

(d) *Localization: M is flat over A if and only if $M_{\mathfrak{p}}$ is flat over $A_{\mathfrak{p}}$ for all $\mathfrak{p} \in \operatorname{Spec} A$.*

(e) *Let $0 \to M' \to M \to M'' \to 0$ be an exact sequence of A-modules. If M' and M'' are both flat then M is flat; if M and M'' are both flat, then M' is flat.*

(f) *A finitely generated module M over a local noetherian ring A is flat if and only if it is free.*

PROOFS. Matsumura [2, Ch. 2, §3] or Bourbaki [1, Ch. I.].

Example 9.1.1. If A is a ring and $S \subseteq A$ is a multiplicative system, then the localization $S^{-1}A$ is a flat A-algebra. If $A \to B$ is a ring homomorphism, if M is a B-module which is flat over A, and if S is a multiplicative system in B, then $S^{-1}M$ is flat over A.

Example 9.1.2. If A is a noetherian ring and $\mathfrak{a} \subseteq A$ an ideal, then the $\mathfrak{a}$-adic completion $\hat{A}$ is a flat A-algebra (II, 9.3A).

Example 9.1.3. Let A be a principal ideal domain. Then an A-module M is flat if and only if it is torsion-free. Indeed, by (9.1Aa) we must check that for every ideal $\mathfrak{a} \subseteq A$, $\mathfrak{a} \otimes M \to M$ is injective. But $\mathfrak{a}$ is principal, say generated by t, so this just says that t is not a zero divisor in M, i.e., M is torsion-free.

Definition. Let $f : X \to Y$ be a morphism of schemes, and let $\mathscr{F}$ be an $\mathcal{O}_X$-module. We say that $\mathscr{F}$ is *flat over Y at a point* $x \in X$, if the stalk $\mathscr{F}_x$ is a flat $\mathcal{O}_{y,Y}$-module, where $y = f(x)$ and we consider $\mathscr{F}_x$ as an $\mathcal{O}_{y,Y}$-module via the natural map $f^\# : \mathcal{O}_{y,Y} \to \mathcal{O}_{x,X}$. We say simply $\mathscr{F}$ is *flat over Y* if it is flat at every point of X. We say X is *flat over Y* if $\mathcal{O}_X$ is.

Proposition 9.2.

(a) *An open immersion is flat.*

(b) *Base change: let $f : X \to Y$ be a morphism, let $\mathscr{F}$ be an $\mathcal{O}_X$-module which is flat over Y, and let $g : Y' \to Y$ be any morphism. Let $X' = X \times_Y Y'$, let $f' : X' \to Y'$ be the second projection, and let $\mathscr{F}' = p_1^*(\mathscr{F})$. Then $\mathscr{F}'$ is flat over Y'.*

(c) *Transitivity: let $f : X \to Y$ and $g : Y \to Z$ be morphisms. Let $\mathscr{F}$ be an $\mathcal{O}_X$-module which is flat over Y, and assume also that Y is flat over Z. Then $\mathscr{F}$ is flat over Z.*

(d) *Let $A \to B$ be a ring homomorphism, and let M be a B-module. Let $f : X = \mathrm{Spec}\, B \to Y = \mathrm{Spec}\, A$ be the corresponding morphism of affine schemes, and let $\mathscr{F} = \tilde{M}$. Then $\mathscr{F}$ is flat over Y if and only if M is flat over A.*

(e) *Let X be a noetherian scheme, and $\mathscr{F}$ a coherent $\mathcal{O}_X$-module. Then $\mathscr{F}$ is flat over X if and only if it is locally free.*

PROOF. These properties all follow from the corresponding properties of modules, taking into account that the functor $\sim$ is compatible with $\otimes$ (II, 5.2).

Next, as an illustration of the convenience of flat morphisms, we show that "cohomology commutes with flat base extension":

Proposition 9.3. *Let $f: X \to Y$ be a separated morphism of finite type of noetherian schemes, and let $\mathscr{F}$ be a quasi-coherent sheaf on X. Let $u: Y' \to Y$ be a flat morphism of noetherian schemes.*

$$
\begin{array}{ccc}
X' & \xrightarrow{\ v\ } & X \\
\downarrow{g} & & \downarrow{f} \\
Y' & \xrightarrow{\ u\ } & Y
\end{array}
$$

Then for all $i \geq 0$ there are natural isomorphisms

$$u^* R^i f_*(\mathscr{F}) \cong R^i g_*(v^* \mathscr{F}).$$

PROOF. The question is local on Y and on Y', so we may assume they are both affine, say $Y = \operatorname{Spec} A$ and $Y' = \operatorname{Spec} A'$. Then by (8.5) what we have to show is that

$$H^i(X, \mathscr{F}) \otimes_A A' \cong H^i(X', \mathscr{F}').$$

Since X is separated and noetherian, and $\mathscr{F}$ is quasi-coherent, we can calculate $H^i(X, \mathscr{F})$ by Čech cohomology with respect to an open affine cover $\mathfrak{U}$ of X (4.5). On the other hand, $\{v^{-1}(U) | U \in \mathfrak{U}\}$ forms an open affine cover $\mathfrak{U}'$ of X', and clearly the Čech complex $C^{\cdot}(\mathfrak{U}', v^* \mathscr{F})$ is just $C^{\cdot}(\mathfrak{U}, \mathscr{F}) \otimes_A A'$. Since A' is flat over A, the functor $\cdot \otimes_A A'$ commutes with taking cohomology groups of the Čech complex, so we get our result. Note that g is also separated and of finite type by base extension, so X' is also noetherian and separated, allowing us to apply (4.5) on X'.

Remark 9.3.1. Even if u is not flat, this proof shows that there is a natural map $u^* R^i f_*(\mathscr{F}) \to R^i g_*(v^* \mathscr{F})$.

Corollary 9.4. *Let $f: X \to Y$ and $\mathscr{F}$ be as in (9.3), and assume Y affine. For any point $y \in Y$, let X_y be the fibre over y, and let $\mathscr{F}_y$ be the induced sheaf. On the other hand, let $k(y)$ denote the constant sheaf $k(y)$ on the closed subset $\{y\}^-$ of Y. Then for all $i \geq 0$ there are natural isomorphisms*

$$H^i(X_y, \mathscr{F}_y) \cong H^i(X, \mathscr{F} \otimes k(y)).$$

PROOF. First let $Y' \subseteq Y$ be the reduced induced subscheme structure on $\{y\}^-$, and let $X' = X \times_Y Y'$, which is a closed subscheme of X. Then both sides of our desired isomorphism depend only on the sheaf $\mathscr{F}' = \mathscr{F} \otimes k(y)$ on X'. Thus we can replace $X, Y, \mathscr{F}$ by $X', Y', \mathscr{F}'$, i.e., we can assume that Y is an integral affine scheme and that $y \in Y$ is its generic point. In that case, $\operatorname{Spec} k(y) \to Y$ is a flat morphism, so we can apply (9.3) and conclude that

$$H^i(X_y, \mathscr{F}_y) \cong H^i(X, \mathscr{F}) \otimes k(y).$$

But after our reduction, $H^i(X,\mathscr{F})$ is already a $k(y)$-module, so tensoring with $k(y)$ has no effect, and we obtain the desired result. (This result is used in §12.)

Flat Families

For many reasons it is important to have a good notion of an algebraic family of varieties or schemes. The most naive definition would be just to take the fibres of a morphism. To get a good notion, however, we should require that certain numerical invariants remain constant in a family, such as the dimension of the fibres. It turns out that if we are dealing with non-singular (or even normal) varieties over a field, then the naive definition is already a good one. Evidence for this is the theorem (9.13) that in such a family, the arithmetic genus is constant.

On the other hand, if we deal with nonnormal varieties, or more general schemes, the naive definition will not do. So we consider a flat family of schemes, which means the fibres of a flat morphism, and this is a very good notion. Why the algebraic condition of flatness on the structure sheaves should give a good definition of a family is something of a mystery. But at least we will justify this choice by showing that flat families have many good properties, and by giving necessary and sufficient conditions for flatness in some special cases. In particular, we will show that a family of closed subschemes of projective space (over an integral scheme) is flat if and only if the Hilbert polynomials of the fibres are the same.

Proposition 9.5. *Let $f : X \to Y$ be a flat morphism of schemes of finite type over a field k. For any point $x \in X$, let $y = f(x)$. Then*

$$\dim_x(X_y) = \dim_x X - \dim_y Y.$$

Here for any scheme X and any point $x \in X$, by $\dim_x X$ we mean the dimension of the local ring $\mathcal{O}_{x,X}$.

PROOF. First we make a base change $Y' \to Y$ where $Y' = \operatorname{Spec} \mathcal{O}_{y,Y}$, and consider the new morphism $f' : X' \to Y'$ where $X' = X \times_Y Y'$. Then f' is also flat by (9.2), x lifts to X', and the three numbers in question are the same. Thus we may assume that y is a closed point of Y, and $\dim_y Y = \dim Y$.

Now we use induction on $\dim Y$. If $\dim Y = 0$, then X_y is defined by a nilpotent ideal in X, so we have $\dim_x(X_y) = \dim_x X$, and $\dim_y Y = 0$.

If $\dim Y > 0$, we make a base extension to Y_{red}. Nothing changes, so we may assume that Y is reduced. Then we can find an element $t \in \mathfrak{m}_y \subseteq \mathcal{O}_{y,Y}$ such that t is not a zero divisor. Let $Y' = \operatorname{Spec} \mathcal{O}_{y,Y}/(t)$, and make the base extension $Y' \to Y$. Then $\dim Y' = \dim Y - 1$ by (I, 1.8A) and (I, 1.11A). Since f is flat, $f^\# t \in \mathfrak{m}_x$ is also not a zero divisor. So for the same reason, $\dim_x X' = \dim_x X - 1$. Of course the fibre X_y does not change under base extension, so we have only to prove our formula for $f' : X' \to Y'$. But this follows from the induction hypothesis, so we are done.

Corollary 9.6. *Let* $f:X \to Y$ *be a flat morphism of schemes of finite type over a field* k, *and assume that* Y *is irreducible. Then the following conditions are equivalent:*

(i) *every irreducible component of* X *has dimension equal to* $\dim Y + n$;

(ii) *for any point* $y \in Y$ *(closed or not), every irreducible component of the fibre* X_y *has dimension* n.

PROOF.

(i) $\Rightarrow$ (ii). Given $y \in Y$, let $Z \subseteq X_y$ be an irreducible component, and let $x \in Z$ be a closed point, which is not in any other irreducible component of X_y. Applying (9.5) we have

$$\dim_x Z = \dim_x X - \dim_y Y.$$

Now $\dim_x Z = \dim Z$ since x is a closed point (II, Ex. 3.20). On the other hand, since Y is irreducible and X is equidimensional, and both are of finite type over k, we have (II, Ex. 3.20)

$$\dim_x X = \dim X - \dim\{x\}^-$$
$$\dim_y Y = \dim Y - \dim\{y\}^-.$$

Finally, since x is a closed point of the fibre X_y, $k(x)$ is a finite algebraic extension of $k(y)$ and so

$$\dim\{x\}^- = \dim\{y\}^-.$$

Combining all these, and using (i) we find $\dim Z = n$.

(ii) $\Rightarrow$ (i). This time let Z be an irreducible component of X, and let $x \in Z$ be a closed point which is not contained in any other irreducible component of X. Then applying (9.5), we have

$$\dim_x(X_y) = \dim_x X - \dim_y Y.$$

But $\dim_x(X_y) = n$ by (ii), $\dim_x X = \dim Z$, and $\dim_y Y = \dim Y$, since $y = f(x)$ must be a closed point of Y. Thus

$$\dim Z = \dim Y + n$$

as required.

Definition. A point x of a scheme X is an *associated point* of X if the maximal ideal $\mathfrak{m}_x$ is an associated prime of 0 in the local ring $\mathcal{O}_{x,x}$, or in other words, if every element of $\mathfrak{m}_x$ is a zero divisor.

Proposition 9.7. *Let* $f:X \to Y$ *be a morphism of schemes, with* Y *integral and regular of dimension* 1. *Then* f *is flat if and only if every associated point* $x \in X$ *maps to the generic point of* Y. *In particular, if* X *is reduced, this says that every irreducible component of* X *dominates* Y.

257

PROOF. First suppose that f is flat, and let $x \in X$ be a point whose image $y = f(x)$ is a closed point of Y. Then $\mathcal{O}_{y,Y}$ is a discrete valuation ring. Let $t \in \mathfrak{m}_y - \mathfrak{m}_y^2$ be a uniformizing parameter. Then t is not a zero divisor in $\mathcal{O}_{y,Y}$. Since f is flat, $f^\# t \in \mathfrak{m}_x$ is not a zero divisor, so x is not an associated point of X.

Conversely, suppose that every associated point of X maps to the generic point of Y. To show f is flat, we must show that for any $x \in X$, letting $y = f(x)$, the local ring $\mathcal{O}_{x,X}$ is flat over $\mathcal{O}_{y,Y}$. If y is the generic point, $\mathcal{O}_{y,Y}$ is a field, so there is nothing to prove. If y is a closed point, $\mathcal{O}_{y,Y}$ is a discrete valuation ring, so by (9.1.3) we must show that $\mathcal{O}_{x,X}$ is a torsion-free module. If it is not, then $f^\# t$ must be a zero divisor in $\mathfrak{m}_x$, where t is a uniformizing parameter of $\mathcal{O}_{y,Y}$. Therefore $f^\# t$ is contained in some associated prime ideal $\mathfrak{p}$ of (0) in $\mathcal{O}_x$ (Matsumura [2, Cor. 2, p. 50]). Then $\mathfrak{p}$ determines a point $x' \in X$, which is an associated point of X, and whose image by f is y, which is a contradiction.

Finally, note that if X is reduced, its associated points are just the generic points of its irreducible components, so our condition says that each irreducible component of X dominates Y.

Example 9.7.1. Let Y be a curve with a node, and let $f: X \to Y$ be the map of its normalization to it. Then f is not flat. For if it were, then $f_* \mathcal{O}_X$ would be a flat sheaf of $\mathcal{O}_Y$-modules. Since it is coherent, it would be locally free by (9.2e). And finally, since its rank is 1, it would be an invertible sheaf on Y. But there are two points P_1, P_2 of X going to the node Q of Y, so $(f_* \mathcal{O}_X)_Q$ needs two generators as an $\mathcal{O}_Y$-module, hence it cannot be locally free.

Example 9.7.2. The result of (9.7) also fails if Y is regular of dimension > 1. For example, let $Y = \mathbf{A}^2$, and let X be obtained by blowing up a point. Then X and Y are both nonsingular, and X dominates Y, but f is not flat, because the dimension of the fibre over the blown-up point is too big (9.5).

Proposition 9.8. *Let Y be a regular, integral scheme of dimension 1, let $P \in Y$ be a closed point, and let $X \subseteq \mathbf{P}_{Y-P}^n$ be a closed subscheme which is flat over $Y - P$. Then there exists a unique closed subscheme $\bar{X} \subseteq \mathbf{P}_Y^n$, flat over Y, whose restriction to $\mathbf{P}_{Y-P}^n$ is X.*

PROOF. Take $\bar{X}$ to be the scheme-theoretic closure of X in $\mathbf{P}_Y^n$ (II, Ex. 3.11d). Then the associated points of $\bar{X}$ are just those of X, so by (9.7), $\bar{X}$ is flat over Y. Furthermore, $\bar{X}$ is unique, because any other extension of X to $\mathbf{P}_Y^n$ would have some associated points mapping to P.

Remark 9.8.1. This proposition says that we can "pass to the limit," when we have a flat family of closed subschemes of $\mathbf{P}^n$ over a punctured curve. Hence it implies that "the Hilbert scheme is proper." The Hilbert scheme is a scheme H which parametrizes all closed subschemes of $\mathbf{P}_k^n$. It has the property that to give a closed subscheme $X \subseteq \mathbf{P}_T^n$, flat over T, for any scheme T, is equivalent to giving a morphism $\varphi: T \to H$. Here, naturally, for any $t \in T$, $\varphi(t)$ is the point of H corresponding to the fibre $X_t \subseteq \mathbf{P}_{k(t)}^n$.

Now once one knows that the Hilbert scheme exists (see Grothendieck [5, exp. 221]) then the question of its properness can be decided using the valuative criterion of properness (II, 4.7). And the result just proved is the essential point needed to show that each connected component of H is proper over k.

Example 9.8.2. Even though the dimension of the fibres is constant in a flat family, we cannot expect properties such as "irreducible" or "reduced" to be preserved in a flat family. Take for example, the families given in (II, 3.3.1) and (II, 3.3.2). In each case the total space X is integral, the base Y is a nonsingular curve and the morphism $f: X \to Y$ is surjective, so the family is flat. Also most fibres are integral in both families. However, the special fibre in one is a doubled line (not reduced), and the special fibre in the other is two lines (not irreducible).

Example 9.8.3 (Projection from a Point). We get some new insight into the geometric process of projection from a point (I, Ex. 3.14) using (9.8). Let $P = (0,0,\ldots,0,1) \in \mathbf{P}^{n+1}$, and consider the projection $\varphi: \mathbf{P}^{n+1} - \{P\} \to \mathbf{P}^n$, which is defined by $(x_0,\ldots,x_{n+1}) \mapsto (x_0,\ldots,x_n)$. For each $a \in k$, $a \neq 0$, consider the automorphism σ_a of $\mathbf{P}^{n+1}$ defined by $(x_0,\ldots,x_{n+1}) \mapsto (x_0,\ldots,x_n,ax_{n+1})$. Now let X_1 be a closed subscheme of $\mathbf{P}^{n+1}$, not containing P. For each $a \neq 0$, let $X_a = \sigma_a(X_1)$. Then the X_a form a flat family parametrized by $\mathbf{A}^1 - \{0\}$. It is flat, because the X_a are all isomorphic as abstract schemes, and in fact, the whole family is isomorphic to $X_1 \times (\mathbf{A}^1 - \{0\})$ if we forget the embedding in $\mathbf{P}^{n+1}$.

Now according to (9.8) this family extends uniquely to a flat family defined over all of $\mathbf{A}^1$, and clearly the fibre X_0 over 0 agrees, at least set-theoretically, with the projection $\varphi(X_1)$ of X_1. Thus we see that there is a flat family over $\mathbf{A}^1$, whose fibres for all $a \neq 0$ are isomorphic to X_1, and whose fibre at 0 is some scheme with the same underlying space as $\varphi(X_1)$.

Example 9.8.4. We will now calculate the flat family just described in the special case where X_1 is a twisted cubic curve in $\mathbf{P}^3$, φ is a projection to $\mathbf{P}^2$, and $\varphi(X_1)$ is a nodal cubic curve in $\mathbf{P}^2$. The remarkable result of this calculation is that the special fibre X_0 of our flat family consists of the curve $\varphi(X_1)$ together with some nilpotent elements at the double point! We say that X_0 is a scheme with an embedded point. It seems as if the scheme X_0 is retaining the information that it is the limit of a family of space curves, by having these nilpotent elements which point out of the plane. In particular, X_0 is not a closed subscheme of $\mathbf{P}^2$ (Fig. 11).

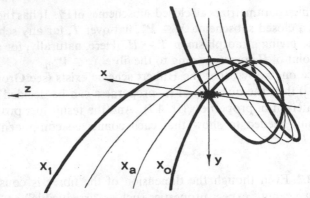

Figure 11.   A flat family of subschemes of $\mathbf{P}^3$.

Now for the calculation. We are just interested in what happens near the double point, so we will use affine coordinates x, y in $\mathbf{A}^2$ and x, y, z in $\mathbf{A}^3$. Let X_1 be given by the parametric equations

$$
\begin{cases}
x = t^2 - 1 \\
y = t^3 - t \\
z = t.
\end{cases}
$$

Then since $t = z$, $t^2 = x + 1$, $t^3 = y + z$, we recognize this as a twisted cubic curve in $\mathbf{A}^3$ (I, Ex. 1.2).

Now for any $a \neq 0$, the scheme X_a is given by

$$
\begin{cases}
x = t^2 - 1 \\
y = t^3 - t \\
z = at.
\end{cases}
$$

To get the ideal $I \subseteq k[a, x, y, z]$ of the total family $\bar{X}$ extended over all of $\mathbf{A}^1$, we eliminate t from the parametric equations, and make sure a is not a zero divisor in $k[a, x, y, z]/I$, so that $\bar{X}$ will be flat. We find

$$
I = (a^2(x + 1) - z^2, \, ax(x + 1) - yz, \, xz - ay, \, y^2 - x^2(x + 1)).
$$

From this, setting $a = 0$, we obtain the ideal $I_0 \subseteq k[x, y, z]$ of X_0, which is

$$
I_0 = (z^2, yz, xz, y^2 - x^2(x + 1)).
$$

So we see that X_0 is a scheme with support equal to the nodal cubic curve $y^2 = x^2(x + 1)$. At any point where $x \neq 0$, we get z in the ideal, so X_0 is reduced there. But in the local ring at the node $(0,0,0)$, we have the element z with $z^2 = 0$, a nonzero nilpotent element.

So here we have an example of a flat family of curves, whose general member is nonsingular, but whose special member is singular, with an embedded point. See also (9.10.1) and (IV, Ex. 3.5).

Example 9.8.5 (Algebraic Families of Divisors). Let X be a scheme of finite type over an algebraically closed field k, let T be a nonsingular curve over k, and let D be an effective Cartier divisor on $X \times T$ (II, §6). Then we can think of D as a closed subscheme of $X \times T$, which is locally described on a small open set U as the zeros of a single element $f \in \Gamma(U, \mathcal{O}_U)$ such that f is not a zero divisor. For any closed point $t \in T$, let $X_t (\cong X)$ be the fibre of $X \times T$ over t. We say that the intersection divisor $D_t = D.X_t$ is *defined* if at every point of X_t, the image $\bar{f} \in \Gamma(U \cap X_t, \mathcal{O}_{X_t})$ of a local equation f of D is not a zero divisor. In that case the covering $\{U \cap X_t\}$ and the elements $\bar{f}$ define a Cartier divisor D_t on X_t. If D_t is defined for all t, we say that the divisors $\{D_t | t \in T\}$ form an *algebraic family of divisors on X parametrized by T.*

This definition, which is natural in the context of Cartier divisors, is connected with flatness in the following way: the original Cartier divisor D, considered as a scheme over T, is flat over T if and only if $D_t = D.X_t$ is defined for each $t \in T$. Indeed, let $x \in D$ be any point, let $A = \mathcal{O}_{x, X \times T}$ be the local ring of x on $X \times T$, let $f \in A$ be a local equation for D, let $p_2(x) = t$, and let $u \in \mathcal{O}_{t,T}$ be a uniformizing parameter. Then $D.X_t$ is defined at x if and only if $\bar{f} \in A/uA$ is not a zero divisor. Since u is automatically not a zero divisor in A, this is equivalent to saying that (u, f) is a regular sequence (II, §8) in A. On the other hand, D is flat over T at x if and only if $\mathcal{O}_{x,D}$ is flat over $\mathcal{O}_{t,T}$. By (9.1.3) this is equivalent to $\mathcal{O}_{x,D}$ being torsion-free, i.e., u not being a zero divisor in $\mathcal{O}_{x,D}$. But $\mathcal{O}_{x,D} \cong A/fA$, so this says that (f, u) is a regular sequence in A. Since the property of being a regular sequence is independent of the order of the sequence (Matsumura [2, Th. 28, p. 102]), the two conditions are equivalent.

Theorem 9.9. *Let T be an integral noetherian scheme. Let $X \subseteq \mathbf{P}_T^n$ be a closed subscheme. For each point $t \in T$, we consider the Hilbert polynomial $P_t \in \mathbf{Q}[z]$ of the fibre X_t considered as a closed subscheme of $\mathbf{P}_{k(t)}^n$. Then X is flat over T if and only if the Hilbert polynomial P_t is independent of t.*

PROOF. Recall that the Hilbert polynomial was defined in (I, §7), and computed another way in (Ex. 5.2). We will use the defining property that

$$P_t(m) = \dim_{k(t)} H^0(X_t, \mathcal{O}_{X_t}(m))$$

for all $m \gg 0$.

First we generalize, replacing $\mathcal{O}_X$ by any coherent sheaf $\mathcal{F}$ on $\mathbf{P}_T^n$, and using the Hilbert polynomial of $\mathcal{F}_t$. Thus we may assume $X = \mathbf{P}_T^n$. Second, the question is local on T. In fact, by comparing any point to the generic point, we see that it is sufficient to consider the case $T = \operatorname{Spec} A$, with A a *local* noetherian ring.

So now let $T = \operatorname{Spec} A$ with A a local noetherian domain, let $X = \mathbf{P}_T^n$, and let $\mathcal{F}$ be a coherent sheaf on X. We will show that the following conditions are equivalent:

(i) $\mathcal{F}$ is flat over T;

(ii) $H^0(X, \mathcal{F}(m))$ is a free A-module of finite rank, for all $m \gg 0$;

261

(iii) the Hilbert polynomial P_t of $\mathscr{F}_t$ on $X_t = \mathbf{P}^n_{k(t)}$ is independent of t, for any $t \in T$.

(i) $\Rightarrow$ (ii). We compute $H^i(X,\mathscr{F}(m))$ by Čech cohomology using the standard open affine cover $\mathfrak{U}$ of X. Then

$$H^i(X,\mathscr{F}(m)) = h^i(C^{\cdot}(\mathfrak{U},\mathscr{F}(m))).$$

Since $\mathscr{F}$ is flat, each term $C^i(\mathfrak{U},\mathscr{F}(m))$ of the Čech complex is a flat A-module. On the other hand, if $m \gg 0$, then $H^i(X,\mathscr{F}(m)) = 0$ for $i > 0$, by (5.2). Thus the complex $C^{\cdot}(\mathfrak{U},\mathscr{F}(m))$ is a resolution of the A-module $H^0(X,\mathscr{F}(m))$: we have an exact sequence

$$0 \to H^0(X,\mathscr{F}(m)) \to C^0(\mathfrak{U},\mathscr{F}(m)) \to C^1(\mathfrak{U},\mathscr{F}(m)) \to \ldots \to C^n(\mathfrak{U},\mathscr{F}(m)) \to 0.$$

Splitting this into short exact sequences, using (9.1Ae) and the fact that the C^i are all flat, we conclude that $H^0(X,\mathscr{F}(m))$ is a flat A-module. But it is also finitely generated (5.2), and hence free of finite rank by (9.1Af).

(ii) $\Rightarrow$ (i). Let $S = A[x_0,\ldots,x_n]$, and let M be the graded S-module

$$M = \bigoplus_{m \geq m_0} H^0(X,\mathscr{F}(m)),$$

where m_0 is chosen large enough so that the $H^0(X,\mathscr{F}(m))$ are all free for $m \geq m_0$. Then $\mathscr{F} = \tilde{M}$ by (II, 5.15). Note that M is the same as $\Gamma_*(\mathscr{F})$ in degrees $m \geq m_0$, so $\tilde{M} = \Gamma_*(\mathscr{F})^{\sim}$. Since M is a free (and hence flat) A-module, we see that $\mathscr{F}$ is flat over A (9.1.1).

(ii) $\Rightarrow$ (iii). It will be enough to show that

$$P_t(m) = \mathrm{rank}_A\, H^0(X,\mathscr{F}(m))$$

for $m \gg 0$. To prove this we will show, for any $t \in T$, that

$$H^0(X_t,\mathscr{F}_t(m)) \cong H^0(X,\mathscr{F}(m)) \otimes_A k(t)$$

for all $m \gg 0$.

First we let $T' = \mathrm{Spec}\, A_\mathfrak{p}$, where $\mathfrak{p}$ is the prime ideal corresponding to t, and we make the flat base extension $T' \to T$. Thus by (9.3) we reduce to the case where t is the closed point of T. Denote the closed fibre X_t by X_0, $\mathscr{F}_t$ by $\mathscr{F}_0$, and $k(t)$ by k. Take a presentation of k over A,

$$A^q \to A \to k \to 0.$$

Then we get an exact sequence of sheaves on X,

$$\mathscr{F}^q \to \mathscr{F} \to \mathscr{F}_0 \to 0.$$

Now by (Ex. 5.10) for $m \gg 0$ we get an exact sequence

$$H^0(X,\mathscr{F}(m)^q) \to H^0(X,\mathscr{F}(m)) \to H^0(X_0,\mathscr{F}_0(m)) \to 0.$$

On the other hand, we can tensor the sequence $A^q \to A \to k \to 0$ with $H^0(X,\mathscr{F}(m))$. Comparing, we deduce that

$$H^0(X_0,\mathscr{F}_0(m)) \cong H^0(X,\mathscr{F}(m)) \otimes_A k$$

for all $m \gg 0$, as required.

(iii) $\Rightarrow$ (ii). According to (II, 8.9) we can check the freeness of $H^0(X,\mathscr{F}(m))$ by comparing its rank at the generic point and the closed point of T. Hence the argument of (ii) $\Rightarrow$ (iii) above is reversible.

Corollary 9.10. *Let T be a connected noetherian scheme, and let $X \subseteq \mathbf{P}_T^n$ be a closed subscheme which is flat over T. For any $t \in T$, let X_t be the fibre, considered as a closed subscheme of $\mathbf{P}_{k(t)}^n$. Then the dimension of X_t, the degree of X_t, and the arithmetic genus of X_t are all independent of t.*

PROOF. By base extension to the irreducible components of T with their reduced induced structure, we reduce to the case T integral. Now the result follows from the theorem and the facts (I, §7) and (Ex. 5.3) that

$$\dim X_t = \deg P_t,$$
$$\deg X_t = (r!) \cdot (\text{leading coefficient of } P_t),$$

where $r = \dim X$, and

$$p_a(X_t) = (-1)^r(P_t(0) - 1).$$

Definition. Let k be an algebraically closed field, let $f: X \to T$ be a surjective map of varieties over k, and assume that for each closed point $t \in T$, we have

(1) $f^{-1}(t)$ is irreducible of dimension equal to $\dim X - \dim T$, and
(2) if $\mathfrak{m}_t \subseteq \mathcal{O}_{t,T}$ is the maximal ideal, and if $\zeta \in f^{-1}(t)$ is the generic point, then $f^\# \mathfrak{m}_t$ generates the maximal ideal $\mathfrak{m}_\zeta \subseteq \mathcal{O}_{\zeta,X}$.

Under these circumstances, we let $X_{(t)}$ be the variety $f^{-1}(t)$ (with the reduced induced structure) and we say that the $X_{(t)}$ form an *algebraic family of varieties*, parametrized by T. The second condition is necessary to be sure that $X_{(t)}$ occurs with "multiplicity one" in the family. It is equivalent to saying that the scheme-theoretic fibre X_t is reduced at its generic point.

Example 9.10.1. In the flat family of (9.8.4), if we take the fibres with their reduced induced structures, we get an algebraic family of varieties $X_{(t)}$ parametrized by $\mathbf{A}^1$. For $t \neq 0$ it is a nonsingular rational curve, and for $t = 0$ it is the plane nodal cubic curve. Note that the arithmetic genus is not constant in this family: $p_a(X_{(t)}) = 0$ for $t \neq 0$ and $p_a(X_{(0)}) = 1$. This accounts for the appearance of nilpotent elements in the scheme-theoretic fibre X_0, since in a flat family of schemes p_a is constant by (9.10). The embedded point at 0 alters the constant term of the Hilbert polynomial so that we get $p_a(X_0) = 0$.

Theorem 9.11. *Let $X_{(t)}$ be an algebraic family of normal varieties parametrized by a nonsingular curve T over an algebraically closed field k. Then $X_{(t)}$ is a flat family of schemes.*

PROOF. Let $f: X \to T$ be the defining morphism of the family. Then f is a flat morphism by (9.7). So we have only to show that for each closed point $t \in T$, the scheme-theoretic fibre X_t coincides with the variety $X_{(t)}$. In other words, we must show that X_t is reduced. For any point $x \in X$, let $A = \mathcal{O}_{x,X}$ be its local ring; let $f(x) = t$, and denote also by t a uniformizing parameter in the local ring $\mathcal{O}_{t,T}$. Then A/tA is the local ring of x on X_t. By hypothesis X_t is irreducible, so t has a unique minimal prime ideal $\mathfrak{p}$ in A. Furthermore, t generates the maximal ideal of the local ring of the generic point of X_t on X, which says that t generates the maximal ideal of $A_\mathfrak{p}$. Finally, the local ring of x on $X_{(t)}$ is $A/\mathfrak{p}$, so our hypothesis says that $A/\mathfrak{p}$ is normal. Now our result is a consequence of the following lemma, which tells us that $\mathfrak{p} = tA$, so $X_{(t)} = X_t$.

Lemma 9.12 (Lemma of Hironaka [1]). *Let A be a local noetherian domain, which is a localization of an algebra of finite type over a field k. Let $t \in A$, and assume*

(1) *tA has only one minimal associated prime ideal $\mathfrak{p}$,*

(2) *t generates the maximal ideal of $A_\mathfrak{p}$,*

(3) *$A/\mathfrak{p}$ is normal.*

Then $\mathfrak{p} = tA$ and A is normal.

PROOF. Let $\tilde{A}$ be the normalization of A. Then $\tilde{A}$ is a finitely generated A-module by (I, 3.9A). We will show that the maps

$$\varphi: A/tA \to \tilde{A}/t\tilde{A}$$

and

$$\psi: A/tA \to A/\mathfrak{p}$$

are both isomorphisms.

First we localize at $\mathfrak{p}$. Then ψ is an isomorphism by hypothesis. Therefore $A_\mathfrak{p}$ is a discrete valuation ring, hence normal. So $\tilde{A}_\mathfrak{p} = A_\mathfrak{p}$ and φ is also an isomorphism.

Now suppose that at least one of φ, ψ is not an isomorphism. Then, after localizing A at a suitable prime ideal, we may assume that φ and ψ are isomorphisms at every localization $A_\mathfrak{q}$ with $\mathfrak{q} \neq \mathfrak{m}$, but that at least one of φ, ψ is not an isomorphism at $\mathfrak{m}$. By the previous step, we have $\mathfrak{p} < \mathfrak{m}$, so $\dim A \geqslant 2$. Now $\tilde{A}$ is normal of dimension $\geqslant 2$, so it has depth $\geqslant 2$ (II, 8.22A), so $\tilde{A}/t\tilde{A}$ has depth $\geqslant 1$. Therefore it does not have $\mathfrak{m}$ as an associated prime. On the other hand, $\tilde{A}/t\tilde{A}$ agrees with $A/\mathfrak{p}$ outside of $\mathfrak{m}$, so we conclude that $\tilde{A}/t\tilde{A}$ is an integral domain. Thus we have a natural map $(A/tA)_{\mathrm{red}} \to \tilde{A}/t\tilde{A}$. But $(A/tA)_{\mathrm{red}} \cong A/\mathfrak{p}$ since ψ is an isomorphism outside $\mathfrak{m}$. Thus $\tilde{A}/t\tilde{A}$ is a finitely generated $(A/\mathfrak{p})$-module with the same quotient field. Since $A/\mathfrak{p}$ is normal by hypothesis, we conclude that $\tilde{A}/t\tilde{A} \cong A/\mathfrak{p}$. Therefore φ is surjective, so we can write $\tilde{A} = A + t\tilde{A}$. By Nakayama's lemma, this implies that $A = \tilde{A}$. Thus $A/tA \cong A/\mathfrak{p}$ and both φ and ψ are isomorphisms. But this is a contradiction, so we conclude that φ and ψ were already isomorphisms on the original ring A before localization.

To conclude, we find that $p = tA$ because ψ is an isomorphism. Since φ is an isomorphism, we have $\tilde{A} = A + t\tilde{A}$, so by Nakayama's lemma as before, we find that $A = \tilde{A}$, so A is normal.

Corollary 9.13 (Igusa [1]). *Let $X_{(t)}$ be an algebraic family of normal varieties in $\mathbf{P}_k^n$, parametrized by a variety T. Then the Hilbert polynomial of $X_{(t)}$, and hence also the arithmetic genus $p_a(X_{(t)})$, are independent of t.*

PROOF. Any two closed points of T lie in the image of a morphism $g: T' \to T$, where T' is a nonsingular curve, or can be connected by a finite number of such curves, so by base extension, we reduce to the case where T is a nonsingular curve. Then the result follows from (9.10) and (9.11).

Example 9.13.1 (Infinitesimal Deformations). Now that we have seen that flatness is a natural condition for algebraic families of varieties, we come to an important nonclassical example of flatness in the category of schemes. Let X_0 be a scheme of finite type over a field k. Let $D = k[t]/t^2$ be the *ring of dual numbers* over k. An *infinitesimal deformation* of X_0 is a scheme X', flat over D, and such that $X' \otimes_D k \cong X_0$.

These arise geometrically in the following way. If $f: X \to T$ is any flat family, having a point $t \in T$ with $X_t \cong X_0$, we say that X is a (global) *deformation* of X_0. Now given an element of the Zariski tangent space of T at t, we obtain a morphism $\operatorname{Spec} D \to T$ (II, Ex. 2.8). Then by base extension we obtain an X' flat over $\operatorname{Spec} D$ with closed fibre X_0. Thus the study of the infinitesimal deformations of X_0 ultimately will help in the study of global deformations.

Example 9.13.2. Continuing the same ideas, it is often possible to classify the infinitesimal deformations of a scheme X. In particular, if X is nonsingular over an algebraically closed field k, we will show that the set of infinitesimal deformations of X, up to isomorphism, is in one-to-one correspondence with the elements of the cohomology group $H^1(X, \mathscr{T}_X)$, where $\mathscr{T}_X$ is the tangent sheaf.

Indeed, given X' flat over D, we consider the exact sequence

$$0 \to k \xrightarrow{t} D \to k \to 0$$

of D-modules. By flatness, we obtain an exact sequence

$$0 \to \mathcal{O}_X \xrightarrow{t} \mathcal{O}_{X'} \to \mathcal{O}_X \to 0$$

of $\mathcal{O}_{X'}$-modules. Thus X' is an infinitesimal extension of the scheme X by the sheaf $\mathcal{O}_X$, in the sense of (II, Ex. 8.7). Conversely, such an extension gives X' flat over D. Now these extensions are classified by $H^1(X, \mathscr{T}_X)$ by (Ex. 4.10).

Remark 9.13.3. There is a whole subject called *deformation theory* devoted to the study of deformations of a given scheme (or variety) X_0 over a field k.

It is closely related to the moduli problem. There one attempts to classify all varieties, and put them into algebraic families. Here we study only those that are close to a given one X_0.

Deformation theory is one area of algebraic geometry where the influence of schemes has been enormous. Because even if one's primary interest is in a variety X_0 over k, by working in the category of schemes, one can consider flat families over arbitrary Artin rings with residue field k, whose closed fibre is X_0. Taking the limit of Artin rings, one can study flat families over a complete local ring. Both of these types of families are intermediate between X_0 itself and a global deformation $f: X \to T$ where T is another variety. Thus they form a powerful tool for studying all deformations of X_0. For some references on deformation theory see Schlessinger [1], or Morrow and Kodaira [1, Ch. 4].

EXERCISES

9.1. A flat morphism $f: X \to Y$ of finite type of noetherian schemes is open, i.e, for every open subset $U \subseteq X$, $f(U)$ is open in Y. [*Hint*: Show that $f(U)$ is constructible and stable under generization (II, Ex. 3.18) and (II, Ex. 3.19).]

9.2. Do the calculation of (9.8.4) for the curve of (I, Ex. 3.14). Show that you get an embedded point at the cusp of the plane cubic curve.

9.3. Some examples of flatness and nonflatness.
    (a) If $f: X \to Y$ is a finite surjective morphism of nonsingular varieties over an algebraically closed field k, then f is flat.
    (b) Let X be a union of two planes meeting at a point, each of which maps isomorphically to a plane Y. Show that f is not flat. For example, let $Y = \operatorname{Spec} k[x, y]$ and $X = \operatorname{Spec} k[x, y, z, w]/(z, w) \cap (x + z, y + w)$.
    (c) Again let $Y = \operatorname{Spec} k[x, y]$, but take $X = \operatorname{Spec} k[x, y, z, w]/(z^2, zw, w^2, xz - yw)$. Show that $X_{\text{red}} \cong Y$, X has no embedded points, but that f is not flat.

9.4. *Open Nature of Flatness.* Let $f: X \to Y$ be a morphism of finite type of noetherian schemes. Then $\{x \in X | f \text{ is flat at } x\}$ is an open subset of X (possibly empty)—see Grothendieck [EGA IV$_3$, 11.1.1].

9.5. *Very Flat Families.* For any closed subscheme $X \subseteq \mathbf{P}^n$, we denote by $C(X) \subseteq \mathbf{P}^{n+1}$ the projective cone over X (I, Ex. 2.10). If $I \subseteq k[x_0, \ldots, x_n]$ is the (largest) homogeneous ideal of X, then $C(X)$ is defined by the ideal generated by I in $k[x_0, \ldots, x_{n+1}]$.
    (a) Give an example to show that if $\{X_t\}$ is a flat family of closed subschemes of $\mathbf{P}^n$, then $\{C(X_t)\}$ need not be a flat family in $\mathbf{P}^{n+1}$.
    (b) To remedy this situation, we make the following definition. Let $X \subseteq \mathbf{P}^n_T$ be a closed subscheme, where T is a noetherian integral scheme. For each $t \in T$, let $I_t \subseteq S_t = k(t)[x_0, \ldots, x_n]$ be the homogeneous ideal of X_t in $\mathbf{P}^n_{k(t)}$. We say that the family $\{X_t\}$ is *very flat* if for all $d \geqslant 0$,

$$\dim_{k(t)}(S_t/I_t)_d$$

is independent of t. Here $(\ )_d$ means the homogeneous part of degree d.

(c) If $\{X_t\}$ is a very flat family in $\mathbf{P}^n$, show that it is flat. Show also that $\{C(X_t)\}$ is a very flat family in $\mathbf{P}^{n+1}$, and hence flat.

(d) If $\{X_{(t)}\}$ is an algebraic family of projectively normal varieties in $\mathbf{P}^n_k$, parametrized by a nonsingular curve T over an algebraically closed field k, then $\{X_{(t)}\}$ is a very flat family of schemes.

9.6. Let $Y \subseteq \mathbf{P}^n$ be a nonsingular variety of dimension ≥ 2 over an algebraically closed field k. Suppose $\mathbf{P}^{n-1}$ is a hyperplane in $\mathbf{P}^n$ which does not contain Y, and such that the scheme $Y' = Y \cap \mathbf{P}^{n-1}$ is also nonsingular. Prove that Y is a complete intersection in $\mathbf{P}^n$ if and only if Y' is a complete intersection in $\mathbf{P}^{n-1}$. [*Hint*: See (II, Ex. 8.4) and use (9.12) applied to the affine cones over Y and Y'.]

9.7. Let $Y \subseteq X$ be a closed subscheme, where X is a scheme of finite type over a field k. Let $D = k[t]/t^2$ be the ring of dual numbers, and define an *infinitesimal deformation* of Y *as a closed subscheme of* X, to be a closed subscheme $Y' \subseteq X \times_k D$, which is flat over D, and whose closed fibre is Y. Show that these Y' are classified by $H^0(Y, \mathcal{N}_{Y/X})$, where

$$\mathcal{N}_{Y/X} = \mathcal{H}om_{\mathcal{O}_Y}(\mathcal{I}_Y/\mathcal{I}_Y^2, \mathcal{O}_Y).$$

*9.8. Let A be a finitely generated k-algebra. Write A as a quotient of a polynomial ring P over k, and let J be the kernel:

$$0 \to J \to P \to A \to 0.$$

Consider the exact sequence of (II, 8.4A)

$$J/J^2 \to \Omega_{P/k} \otimes_P A \to \Omega_{A/k} \to 0.$$

Apply the functor $\mathrm{Hom}_A(\cdot, A)$, and let $T^1(A)$ be the cokernel:

$$\mathrm{Hom}_A(\Omega_{P/k} \otimes A, A) \to \mathrm{Hom}_A(J/J^2, A) \to T^1(A) \to 0.$$

Now use the construction of (II, Ex. 8.6) to show that $T^1(A)$ classifies infinitesimal deformations of A, i.e., algebras A' flat over $D = k[t]/t^2$, with $A' \otimes_D k \cong A$. It follows that $T^1(A)$ is independent of the given representation of A as a quotient of a polynomial ring P. (For more details, see Lichtenbaum and Schlessinger [1].)

9.9. A k-algebra A is said to be *rigid* if it has no infinitesimal deformations, or equivalently, by (Ex. 9.8) if $T^1(A) = 0$. Let $A = k[x,y,z,w]/(x,y) \cap (z,w)$, and show that A is rigid. This corresponds to two planes in $\mathbf{A}^4$ which meet at a point.

9.10. A scheme X_0 over a field k is *rigid* if it has no infinitesimal deformations.

(a) Show that $\mathbf{P}^1_k$ is rigid, using (9.13.2).

(b) One might think that if X_0 is rigid over k, then every global deformation of X_0 is locally trivial. Show that this is not so, by constructing a proper, flat morphism $f : X \to \mathbf{A}^2$ over k algebraically closed, such that $X_0 \cong \mathbf{P}^1_k$, but there is no open neighborhood U of 0 in $\mathbf{A}^2$ for which $f^{-1}(U) \cong U \times \mathbf{P}^1$.

*(c) Show, however, that one can trivialize a global deformation of $\mathbf{P}^1$ after a flat base extension, in the following sense: let $f : X \to T$ be a flat projective morphism, where T is a nonsingular curve over k algebraically closed. Assume there is a closed point $t \in T$ such that $X_t \cong \mathbf{P}^1_k$. Then there exists a nonsingular curve T', and a flat morphism $g : T' \to T$, whose image contains t, such that if $X' = X \times_T T'$ is the base extension, then the new family $f' : X' \to T'$ is isomorphic to $\mathbf{P}^1_{T'} \to T'$.

9.11. Let Y be a nonsingular curve of degree d in $\mathbf{P}_k^n$, over an algebraically closed field k. Show that

$$0 \leqslant p_a(Y) \leqslant \tfrac{1}{2}(d-1)(d-2).$$

[*Hint*: Compare Y to a suitable projection of Y into $\mathbf{P}^2$, as in (9.8.3) and (9.8.4).]

10 Smooth Morphisms

The notion of smooth morphism is a relative version of the notion of non-singular variety over a field. In this section we will give some basic results about smooth morphisms. As an application, we give Kleiman's elegant proof of the characteristic 0 Bertini theorem. For further information about smooth and étale morphisms, see Altman and Kleiman [1, Ch. VI, VII], Matsumura [2, Ch. 11], and Grothendieck [SGA 1, exp. I, II, III].

For simplicity, we assume that all schemes in this section are of finite type over a field k.

Definition. A morphism $f : X \to Y$ of schemes of finite type over k is *smooth of relative dimension n* if:

(1) f is flat;
(2) if $X' \subseteq X$ and $Y' \subseteq Y$ are irreducible components such that $f(X') \subseteq Y'$, then $\dim X' = \dim Y' + n$;
(3) for each point $x \in X$ (closed or not),

$$\dim_{k(x)}(\Omega_{X/Y} \otimes k(x)) = n.$$

Example 10.0.1. For any Y, $\mathbf{A}_Y^n$ and $\mathbf{P}_Y^n$ are smooth of relative dimension n over Y.

Example 10.0.2. If X is integral, then condition (3) is equivalent to saying $\Omega_{X/Y}$ is locally free on X of rank n (II, 8.9).

Example 10.0.3. If $Y = \operatorname{Spec} k$ and k is algebraically closed, then X is smooth over k if and only if X is regular of dimension n. In particular, if X is irreducible and separated over k, then it is smooth if and only if it is a nonsingular variety. Cf. (II, 8.8) and (II, 8.15).

Proposition 10.1.

(a) *An open immersion is smooth of relative dimension 0.*

(b) Base change. *If $f : X \to Y$ is smooth of relative dimension n, and $g : Y' \to Y$ is any morphism, then the morphism $f' : X' \to Y'$ obtained by base extension is also smooth of relative dimension n.*

(c) Composition. *If $f : X \to Y$ is smooth of relative dimension n, and $g : Y \to Z$ is smooth of relative dimension m, then $g \circ f : X \to Z$ is smooth of relative dimension $n + m$.*

(d) Product. *If X and Y are smooth over Z, of relative dimensions n and m, respectively, then $X \times_Z Y$ is smooth over Z of relative dimension $n + m$.*

PROOFS.

(a) is trivial.

(b) f' is flat by (9.2). According to (9.6), the condition (2) in the definition of smoothness is equivalent to saying that every irreducible component of every fibre X_y of f has dimension n. This condition is preserved under base extension (II, Ex. 3.20). Finally, $\Omega_{X/Y}$ is stable under base extension (II, 8.10), so the number $\dim_{k(x)}(\Omega_{X/Y} \otimes k(x))$ is also. Hence f' is smooth.

(c) $g \circ f$ is flat by (9.2). If $X' \subseteq X$, $Y' \subseteq Y$, and $Z' \subseteq Z$ are irreducible components such that $f(X') \subseteq Y'$ and $g(Y') \subseteq Z'$, then clearly $\dim X' = \dim Z' + n + m$ by hypothesis. For the last condition, we use the exact sequence of (II, 8.11)

$$f^*\Omega_{Y/Z} \to \Omega_{X/Z} \to \Omega_{X/Y} \to 0.$$

Tensoring with $k(x)$ we have

$$f^*\Omega_{Y/Z} \otimes k(x) \to \Omega_{X/Z} \otimes k(x) \to \Omega_{X/Y} \otimes k(x) \to 0.$$

Now the first has dimension m, and the last has dimension n, by hypothesis. So the middle one has dimension $\leqslant n + m$.

On the other hand, let $z = g(f(x))$. Then

$$\Omega_{X/Z} \otimes k(x) = \Omega_{X_z/k(z)} \otimes k(x),$$

since relative differentials commute with base extension. Let X' be an irreducible component of X_z containing x, with its reduced induced structure. Then we have a surjective map·

$$\Omega_{X_z/k(z)} \otimes k(x) \to \Omega_{X'/k(z)} \otimes k(x) \to 0$$

by (II, 8.12). But X' is an integral scheme of finite type over $k(z)$, of dimension $n + m$, by (9.6), so $\Omega_{X'/k(z)}$ is a coherent sheaf of rank $\geqslant n + m$ by (II, 8.6A). Hence it requires at least $n + m$ generators at every point, so

$$\dim_{k(x)}(\Omega_{X'/k(z)} \otimes k(x)) \geqslant n + m.$$

Combining our inequalities, we find that

$$\dim_{k(x)}(\Omega_{X/Z} \otimes k(x)) = n + m$$

as required.

(d) This statement is a consequence of (b) and (c) since we can factor into $X \times_Z Y \overset{p_2}{\to} Y \to Z$.

Theorem 10.2. *Let $f : X \to Y$ be a morphism of schemes of finite type over k. Then f is smooth of relative dimension n if and only if:*

(1) *f is flat; and*

(2) *for each point $y \in Y$, let $X_{\bar{y}} = X_y \otimes_{k(y)} k(y)^-$, where $k(y)^-$ is the algebraic closure of $k(y)$. Then $X_{\bar{y}}$ is equidimensional of dimension n and*

regular. (We say "the fibres of f are geometrically regular of equi-dimension n.")

PROOF. If f is smooth of relative dimension n, so is any base extension. In particular, $X_{\bar{y}}$ is smooth of relative dimension n over $k(y)^-$, so is regular (10.0.3).

Conversely, suppose (1) and (2) satisfied. Then f is flat by (1). From (2) we conclude that every irreducible component of X_y has dimension n, which gives condition (2) of the definition of smoothness by (9.6). Finally, since $k(y)^-$ is algebraically closed, regularity of $X_{\bar{y}}$ implies that $\Omega_{X_{\bar{y}}/k(y)^-}$ is locally free of rank n (10.0.3). This in turn implies that $\Omega_{X_y/k(y)}$ is locally free of rank n (see e.g. Matsumura [2, (4.E), p. 29]), and so for any $x \in X$,

$$\dim_{k(x)}(\Omega_{X/Y} \otimes k(x)) = \dim_{k(x)}(\Omega_{X_y/k(y)} \otimes k(x)) = n$$

as required.

Next we will study when a morphism of nonsingular varieties is smooth. Recall (II, Ex. 2.8) that for a point x in a scheme X we define the *Zariski tangent space* T_x to be the dual of the $k(x)$-vector space m_x/m_x^2. If $f : X \to Y$ is a morphism, and $y = f(x)$, then there is a natural induced mapping on the tangent spaces

$$T_f : T_x \to T_y \otimes_{k(y)} k(x).$$

Before stating our criterion, we recall an algebraic fact.

Lemma 10.3.A. *Let $A \to B$ be a local homomorphism of local noetherian rings. Let M be a finitely generated B-module, and let $t \in A$ be a nonunit that is not a zero divisor. Then M is flat over A if and only if:*

(1) *t is not a zero divisor in M; and*
(2) *M/tM is flat over A/tA.*

PROOF. This is a special case of the "Local criterion of flatness." See Bourbaki [1, III, §5] or Altman and Kleiman [1, V, §3].

Proposition 10.4. *Let $f : X \to Y$ be a morphism of nonsingular varieties over an algebraically closed field k. Let $n = \dim X - \dim Y$. Then the following conditions are equivalent:*

(i) *f is smooth of relative dimension n;*
(ii) *$\Omega_{X/Y}$ is locally free of rank n on X;*
(iii) *for every closed point $x \in X$, the induced map on the Zariski tangent spaces $T_f : T_x \to T_y$ is surjective.*

PROOF.
(i) $\Rightarrow$ (ii) follows from the definition of smoothness, since X is integral (10.0.2).

(ii) $\Rightarrow$ (iii). From the exact sequence of (II, 8.11), tensoring with $k(x)$, we have

$$f^*\Omega_{Y/k} \otimes k(x) \to \Omega_{X/k} \otimes k(x) \to \Omega_{X/Y} \otimes k(x) \to 0.$$

Now X and Y are both smooth over k, so the dimensions of these vector spaces are equal to dim Y, dim X, and n respectively. Therefore the map on the left is injective. But for a closed point x, $k(x) \cong k$, so using (II, 8.7) we see that this map is just the natural map

$$\mathfrak{m}_y/\mathfrak{m}_y^2 \to \mathfrak{m}_x/\mathfrak{m}_x^2$$

induced by f. Taking dual vector spaces over k, we find that T_f is surjective.

(iii) $\Rightarrow$ (i). First we show f is flat. For this, it is enough to show that $\mathcal{O}_x$ is flat over $\mathcal{O}_y$ for every *closed* point $x \in X$, where $y = f(x)$, by localization of flatness. Since X and Y are nonsingular, these are both regular local rings. Furthermore, since T_f is surjective, we have $\mathfrak{m}_y/\mathfrak{m}_y^2 \to \mathfrak{m}_x/\mathfrak{m}_x^2$ injective as above. So let $t_1, \ldots, t_r$ be a regular system of parameters for $\mathcal{O}_y$. Then their images in $\mathcal{O}_x$ form part of a regular system of parameters of $\mathcal{O}_x$. Since $\mathcal{O}_x/(t_1, \ldots, t_r)$ is automatically flat over $\mathcal{O}_y/(t_1, \ldots, t_r) = k$, we can use (10.3A) to show by descending induction on i that $\mathcal{O}_x/(t_1, \ldots, t_i)$ is flat over $\mathcal{O}_y/(t_1, \ldots, t_i)$ for each i. In particular, for $i = 0$, $\mathcal{O}_x$ is flat over $\mathcal{O}_y$. Thus f is flat.

Now we can read the argument of (ii) $\Rightarrow$ (iii) backwards to conclude that

$$\dim_{k(x)}(\Omega_{X/Y} \otimes k(x)) = n$$

for each closed point $x \in X$. On the other hand since f is flat, it is dominant, so for the generic point $\zeta \in X$, we have

$$\dim_{k(\zeta)}(\Omega_{X/Y} \otimes k(\zeta)) \geqslant n$$

by (II, 8.6A). We conclude that $\Omega_{X/Y}$ is coherent of rank $\geqslant n$, so it must be locally free of rank $= n$ by (II, 8.9). Therefore $\Omega_{X/Y} \otimes k(x)$ has dimension n at every point of X, so f is smooth of relative dimension n.

Next we will give some special results about smoothness which hold only in characteristic zero.

Lemma 10.5. *Let $f: X \to Y$ be a dominant morphism of integral schemes of finite type over an algebraically closed field k of characteristic 0. Then there is a nonempty open set $U \subseteq X$ such that $f: U \to Y$ is smooth.*

PROOF. Replacing X and Y by suitable open subsets, we may assume that they are both nonsingular varieties over k (II, 8.16). Next, since we are in characteristic 0, $K(X)$ is a separably generated field extension of $K(Y)$ (I, 4.8A). So by (II, 8.6A), $\Omega_{X/Y}$ is free of rank $n = \dim X - \dim Y$ at the generic point of X. Therefore it is locally free of rank n on some nonempty open set $U \subseteq X$. We conclude that $f: U \to Y$ is smooth by (10.4).

Example 10.5.1. Let k be an algebraically closed field of characteristic p, let $X = Y = \mathbf{P}_k^1$, and let $f: X \to Y$ be the Frobenius morphism (I, Ex. 3.2). Then f is not smooth on any open set. Indeed, since $d(t^p) = 0$, the natural map $f^*\Omega_{Y/k} \to \Omega_{X/k}$ is the zero map, and so $\Omega_{X/Y} \cong \Omega_{X/k}$ is locally free of rank 1. But f has relative dimension 0, so it is nowhere smooth.

Proposition 10.6. *Let $f: X \to Y$ be a morphism of schemes of finite type over an algebraically closed field k of characteristic 0. For any r, let*

$$X_r = \{\text{closed points } x \in X \,|\, \text{rank } T_{f,x} \leqslant r\}.$$

Then

$$\dim \overline{f(X_r)} \leqslant r.$$

PROOF. Let Y' be any irreducible component of $\overline{f(X_r)}$, and let X' be an irreducible component of $\overline{X}_r$ which dominates Y'. We give X' and Y' their reduced induced structures, and consider the induced dominant morphism $f': X' \to Y'$. Then by (10.5) there is a nonempty open subset $U' \subseteq X'$ such that $f': U' \to Y'$ is smooth. Now let $x \in U' \cap X_r$, and consider the commutative diagram of maps of Zariski tangent spaces

$$
\begin{array}{ccc}
T_{x,U'} & \longrightarrow & T_{x,X} \\
\Big\downarrow{\scriptstyle T_{f',x}} & & \Big\downarrow{\scriptstyle T_{f,x}} \\
T_{y,Y'} & \longrightarrow & T_{y,Y}
\end{array}
$$

The horizontal arrows are injective, because U' and Y' are locally closed subschemes of X and Y, respectively. On the other hand, rank $T_{f,x} \leqslant r$ since $x \in X_r$, and $T_{f',x}$ is surjective because f' is smooth (10.4). We conclude that $\dim T_{y,Y'} \leqslant r$, and therefore $\dim Y' \leqslant r$.

Corollary 10.7 (Generic Smoothness). *Let $f: X \to Y$ be a morphism of varieties over an algebraically closed field k of characteristic 0, and assume that X is nonsingular. Then there is a nonempty open subset $V \subseteq Y$ such that $f: f^{-1}V \to V$ is smooth.*

PROOF. We may assume Y is nonsingular by (II, 8.16). Let $r = \dim Y$. Let $X_{r-1} \subseteq X$ be the subset defined in (10.6). Then $\dim \overline{f(X_{r-1})} \leqslant r - 1$ by (10.6), so removing it from Y, we may assume that rank $T_f \geqslant r$ for every closed point of X. But since Y is nonsingular of dimension r, this implies that T_f is surjective for every closed point of X. Hence f is smooth by (10.4).

Note that if the original f was not dominant, then $V \subseteq Y - \overline{f(X)}$, and $f^{-1}V$ will be empty.

For the next results, we recall the notion of a group variety (I, Ex. 3.21). A *group variety* G over an algebraically closed field k is a variety G, together with morphisms $\mu: G \times G \to G$ and $\rho: G \to G$, such that the set $G(k)$ of k-rational points of G (which is just the set of all closed points of G, since k is algebraically closed) becomes a group under the operation induced by μ, with ρ giving the inverses.

We say that a group variety G *acts* on a variety X if we have a morphism $\theta: G \times X \to X$ which induces a homomorphism $G(k) \to \operatorname{Aut} X$ of groups.

A *homogeneous space* is a variety X, together with a group variety G acting on it, such that the group $G(k)$ acts transitively on the set $X(k)$ of k-rational points of X.

Remark 10.7.1. Any group variety is a homogeneous space if we let it act on itself by left multiplication.

Example 10.7.2. The projective space $\mathbf{P}_k^n$ is a homogeneous space for the action of $G = \operatorname{PGL}(n)$—cf. (II, 7.1.1).

Example 10.7.3. A homogeneous space is necessarily a nonsingular variety. Indeed, it has an open subset which is nonsingular by (II, 8.16). But we have a transitive group of automorphisms acting, so it is nonsingular everywhere.

Theorem 10.8 (Kleiman [3]). *Let X be a homogeneous space with group variety G over an algebraically closed field k of characteristic 0. Let $f: Y \to X$ and $g: Z \to X$ be morphisms of nonsingular varieties Y, Z to X. For any $\sigma \in G(k)$, let Y^σ be Y with the morphism $\sigma \circ f$ to X. Then there is a nonempty open subset $V \subseteq G$ such that for every $\sigma \in V(k)$, $Y^\sigma \times_X Z$ is nonsingular and either empty or of dimension exactly*

$$\dim Y + \dim Z - \dim X.$$

PROOF. First we consider the morphism

$$h: G \times Y \to X$$

defined by composing f with the group action $\theta: G \times X \to X$. Now G is nonsingular since it is a group variety (10.7.3), and Y is nonsingular by hypothesis, so $G \times Y$ is nonsingular by (10.1). Since char $k = 0$, we can apply generic smoothness (10.7) to h, and conclude that there is a nonempty open subset $U \subseteq X$ such that $h: h^{-1}(U) \to U$ is smooth. Now G acts on $G \times Y$ by left multiplication on G; G acts on X by θ, and these two actions are compatible with the morphism h, by construction. Therefore, for any $\sigma \in G(k)$, $h: h^{-1}(U^\sigma) \to U^\sigma$ is also smooth. Since the U^σ cover X, we conclude that h is smooth everywhere.

Next, we consider the fibred product

$$W = (G \times Y) \times_X Z,$$

with maps g' and h' to $G \times Y$ and Z as shown.

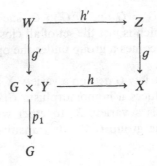

Since h is smooth, h' is also smooth by base extension (10.1). Since Z is nonsingular, it is smooth over k, so by composition (10.1), W is also smooth over k, so W is nonsingular.

Now we consider the morphism

$$q = p_1 \circ g': W \to G.$$

Applying generic smoothness (10.7) again, we find there is a nonempty open subset $V \subseteq G$ such that $q: q^{-1}(V) \to V$ is smooth. Therefore, if $\sigma \in V(k)$ is any closed point, the fibre W_σ will be nonsingular. But W_σ is just $Y^\sigma \times_X Z$, so this is what we wanted to show. Note that W_σ may not be irreducible, but our result shows that each connected component is a nonsingular variety.

To find the dimension of W_σ, we first note that h is smooth of relative dimension

$$\dim G + \dim Y - \dim X.$$

Hence h' has the same relative dimension, and we see that

$$\dim W = \dim G + \dim Y - \dim X + \dim Z.$$

If W is nonempty, then q on $q^{-1}(V)$ has relative dimension equal to $\dim W - \dim G$, so for each σ,

$$\dim W_\sigma = \dim Y + \dim Z - \dim X.$$

Corollary 10.9 (Bertini). *Let X be a nonsingular projective variety over an algebraically closed field k of characteristic 0. Let $\mathfrak{d}$ be a linear system without base points. Then almost every element of $\mathfrak{d}$, considered as a closed subscheme of X, is nonsingular (but maybe reducible).*

PROOF. Let $f: X \to \mathbf{P}^n$ be the morphism to $\mathbf{P}^n$ determined by $\mathfrak{d}$ (II, 7.8.1). We consider $\mathbf{P}^n$ as a homogeneous space under the action of $G = \mathrm{PGL}(n)$ (10.7.2). We apply the theorem taking $g: H \to \mathbf{P}^n$ to be the inclusion map of a hyperplane $H \cong \mathbf{P}^{n-1}$. We conclude that for almost all $\sigma \in G(k)$, $X \times_{\mathbf{P}^n} H^\sigma = f^{-1}(H^\sigma)$ is nonsingular. But the divisors $f^{-1}(H^\sigma)$ are just the elements of the linear system $\mathfrak{d}$, by construction of f. Thus almost all elements of $\mathfrak{d}$ are nonsingular.

Remark 10.9.1. We will see later (Ex. 11.3) that if $\dim f(X) \geqslant 2$, then all the divisors in $\mathfrak{d}$ are connected. Hence almost all of them are irreducible and nonsingular.

Remark 10.9.2. The hypothesis "X projective" is not necessary if we talk about a finite-dimensional linear system $\mathfrak{d}$. In particular, if X was projective, and $\mathfrak{d}$ was a linear system with base points Σ, then by considering the base-point-free linear system $\mathfrak{d}$ on $X - \Sigma$ we obtain the more general statement that "a general member of $\mathfrak{d}$ can have singularities only at the base points."

Remark 10.9.3. This result fails in characteristic $p > 0$. For example, in (10.5.1) the morphism f corresponds to the one-dimensional linear system $\{pP | P \in \mathbf{P}^1\}$. Thus every divisor in $\mathfrak{d}$ is a point with multiplicity p.

Remark 10.9.4. Compare this result to the earlier Bertini theorem (II, 8.18).

EXERCISES

10.1. Over a nonperfect field, smooth and regular are not equivalent. For example, let k_0 be a field of characteristic $p > 0$, let $k = k_0(t)$, and let $X \subseteq \mathbf{A}_k^2$ be the curve defined by $y^2 = x^p - t$. Show that every local ring of X is a regular local ring, but X is not smooth over k.

10.2. Let $f : X \to Y$ be a proper, flat morphism of varieties over k. Suppose for some point $y \in Y$ that the fibre X_y is smooth over $k(y)$. Then show that there is an open neighborhood U of y in Y such that $f : f^{-1}(U) \to U$ is smooth.

10.3. A morphism $f : X \to Y$ of schemes of finite type over k is *étale* if it is smooth of relative dimension 0. It is *unramified* if for every $x \in X$, letting $y = f(x)$, we have $\mathfrak{m}_y \cdot \mathcal{O}_x = \mathfrak{m}_x$, and $k(x)$ is a separable algebraic extension of $k(y)$. Show that the following conditions are equivalent:

  (i) f is étale;
  (ii) f is flat, and $\Omega_{X/Y} = 0$;
  (iii) f is flat and unramified.

10.4. Show that a morphism $f : X \to Y$ of schemes of finite type over k is étale if and only if the following condition is satisfied: for each $x \in X$, let $y = f(x)$. Let $\hat{\mathcal{O}}_x$ and $\hat{\mathcal{O}}_y$ be the completions of the local rings at x and y. Choose fields of representatives (II, 8.25A) $k(x) \subseteq \hat{\mathcal{O}}_x$ and $k(y) \subseteq \hat{\mathcal{O}}_y$ so that $k(y) \subseteq k(x)$ via the natural map $\hat{\mathcal{O}}_y \to \hat{\mathcal{O}}_x$. Then our condition is that for every $x \in X$, $k(x)$ is a separable algebraic extension of $k(y)$, and the natural map

$$\hat{\mathcal{O}}_y \otimes_{k(y)} k(x) \to \hat{\mathcal{O}}_x$$

is an isomorphism.

10.5. If x is a point of a scheme X, we define an *étale neighborhood* of x to be an étale morphism $f : U \to X$, together with a point $x' \in U$ such that $f(x') = x$. As an example of the use of étale neighborhoods, prove the following: if $\mathscr{F}$ is a coherent sheaf on X, and if every point of X has an étale neighborhood $f : U \to X$ for which $f^* \mathscr{F}$ is a free $\mathcal{O}_U$-module, then $\mathscr{F}$ is locally free on X.

10.6. Let Y be the plane nodal cubic curve $y^2 = x^2(x + 1)$. Show that Y has a finite étale covering X of degree 2, where X is a union of two irreducible components, each one isomorphic to the normalization of Y (Fig. 12).

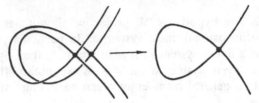

Figure 12.   A finite étale covering.

10.7. (*Serre*). *A linear system with moving singularities.* Let k be an algebraically closed field of characteristic 2. Let $P_1, \ldots, P_7 \in \mathbf{P}_k^2$ be the seven points of the projective plane over the prime field $\mathbf{F}_2 \subseteq k$. Let $\mathfrak{d}$ be the linear system of all cubic curves in X passing through $P_1, \ldots, P_7$.

   (a) $\mathfrak{d}$ is a linear system of dimension 2 with base points $P_1, \ldots, P_7$, which determines an inseparable morphism of degree 2 from $X - \{P_i\}$ to $\mathbf{P}^2$.

   (b) Every curve $C \in \mathfrak{d}$ is singular. More precisely, either C consists of 3 lines all passing through one of the P_i, or C is an irreducible cuspidal cubic with cusp $P \neq$ any P_i. Furthermore, the correspondence $C \mapsto$ the singular point of C is a 1-1 correspondence between $\mathfrak{d}$ and $\mathbf{P}^2$. Thus the singular points of elements of $\mathfrak{d}$ move all over.

10.8. *A linear system with moving singularities contained in the base locus (any characteristic).* In affine 3-space with coordinates x, y, z, let C be the conic $(x - 1)^2 + y^2 = 1$ in the xy-plane, and let P be the point $(0,0,t)$ on the z-axis. Let Y_t be the closure in $\mathbf{P}^3$ of the cone over C with vertex P. Show that as t varies, the surfaces $\{Y_t\}$ form a linear system of dimension 1, with a moving singularity at P. The base locus of this linear system is the conic C plus the z-axis.

10.9. Let $f : X \to Y$ be a morphism of varieties over k. Assume that Y is regular, X is Cohen–Macaulay, and that every fibre of f has dimension equal to $\dim X - \dim Y$. Then f is flat. [*Hint*: Imitate the proof of (10.4), using (II, 8.21A).]

11   The Theorem on Formal Functions

In this section we prove the so-called theorem on formal functions, and its important corollaries, Zariski's Main Theorem, and the Stein factorization theorem. The theorem itself compares the cohomology of the infinitesimal neighborhoods of a fibre of a projective morphism to the stalk of the higher direct image sheaves. While the corollaries use only the case $i = 0$ of the theorem (which could be stated without cohomology), the proof is by descending induction on i, and thus makes essential use of the cohomological machinery. Zariski's first proof [2] of his "Main Theorem" was by an entirely different method which did not use cohomology.

Let $f : X \to Y$ be a projective morphism of noetherian schemes, let $\mathscr{F}$ be a coherent sheaf on X, and let $y \in Y$ be a point. For each $n \geqslant 1$ we

define
$$X_n = X \times_Y \operatorname{Spec} \mathcal{O}_y/\mathfrak{m}_y^n.$$
Then for $n = 1$, we get the fibre X_y, and for $n > 1$, we get a scheme with nilpotent elements having the same underlying space as X_y. It is a kind of "thickened fibre" of X over the point y.

$$
\begin{array}{ccc}
X_n & \xrightarrow{\;v\;} & X \\
{\scriptstyle f'}\downarrow & & \downarrow{\scriptstyle f} \\
\operatorname{Spec} \mathcal{O}_y/\mathfrak{m}_y^n & \longrightarrow & Y
\end{array}
$$

Let $\mathcal{F}_n = v^*\mathcal{F}$, where $v: X_n \to X$ is the natural map. Then by (9.3.1) we have natural maps, for each n,
$$R^i f_*(\mathcal{F}) \otimes \mathcal{O}_y/\mathfrak{m}_y^n \to R^i f'_*(\mathcal{F}_n).$$
Since $\operatorname{Spec} \mathcal{O}_y/\mathfrak{m}_y^n$ is affine, concentrated at one point, the right-hand side is just the group $H^i(X_n, \mathcal{F}_n)$ by (8.5). As n varies, both sides form inverse systems (see (II, §9) for generalities on inverse systems and inverse limits). Thus we can take inverse limits and get a natural map
$$R^i f_*(\mathcal{F})_y^\wedge \to \varprojlim H^i(X_n, \mathcal{F}_n).$$

Theorem 11.1 (Theorem on Formal Functions). *Let $f: X \to Y$ be a projective morphism of noetherian schemes, let $\mathcal{F}$ be a coherent sheaf on X, and let $y \in Y$. Then the natural map*
$$R^i f_*(\mathcal{F})_y^\wedge \to \varprojlim H^i(X_n, \mathcal{F}_n)$$
is an isomorphism, for all $i \geq 0$.

PROOF. As a first step, we embed X in some projective space $\mathbf{P}_Y^N$, and consider $\mathcal{F}$ as a coherent sheaf on $\mathbf{P}_Y^N$. Thus we reduce to the case $X = \mathbf{P}_Y^N$.

Next we let $A = \mathcal{O}_y$, and make the flat base extension $\operatorname{Spec} A \to Y$. Thus, using (9.3), we reduce to the case where Y is affine, equal to the spectrum of a local noetherian ring A, and y is the closed point of Y. Then using (8.5) again, we can restate our result as an isomorphism of A-modules,
$$H^i(X, \mathcal{F})^\wedge \xrightarrow{\sim} \varprojlim H^i(X_n, \mathcal{F}_n).$$

Now suppose $\mathcal{F}$ is a sheaf of the form $\mathcal{O}(q)$ on $X = \mathbf{P}_A^N$, for some $q \in \mathbf{Z}$. Then $\mathcal{F}_n$ is just $\mathcal{O}(q)$ on $X_n = \mathbf{P}_{A/\mathfrak{m}^n}^N$. So by the explicit calculations of (5.1) we see that
$$H^i(X_n, \mathcal{F}_n) \cong H^i(X, \mathcal{F}) \otimes_A A/\mathfrak{m}^n$$
for each n. Therefore by definition of completion, we have
$$H^i(X, \mathcal{F})^\wedge \cong \varprojlim H^i(X_n, \mathcal{F}_n)$$
in this case. Clearly the same calculation holds for any finite direct sum of sheaves of the form $\mathcal{O}(q_i)$.

We will now prove the theorem for an arbitrary coherent sheaf $\mathscr{F}$ on X, by descending induction on i. For $i > N$, both sides are 0, because X can be covered by $N + 1$ open affine subsets (Ex. 4.8). So we assume the theorem has been proved for $i + 1$, and for all coherent sheaves.

Given $\mathscr{F}$ coherent on X, it follows from (II, 5.18) that we can write $\mathscr{F}$ as a quotient of a sheaf $\mathscr{E}$ which is a finite direct sum of sheaves $\mathcal{O}(q_i)$ for suitable $q_i \in \mathbf{Z}$. Let $\mathscr{R}$ be the kernel:

$$0 \to \mathscr{R} \to \mathscr{E} \to \mathscr{F} \to 0. \tag{1}$$

Now unfortunately, tensoring with $\mathcal{O}_{X_n}$ is not an exact functor—it is only right exact, so we have an exact sequence

$$\mathscr{R}_n \to \mathscr{E}_n \to \mathscr{F}_n \to 0$$

of sheaves on X_n for each n. We introduce the image $\mathscr{T}_n$ and the kernel $\mathscr{S}_n$ of the map $\mathscr{R}_n \to \mathscr{E}_n$, so that we have exact sequences

$$0 \to \mathscr{S}_n \to \mathscr{R}_n \to \mathscr{T}_n \to 0 \tag{2}$$

and

$$0 \to \mathscr{T}_n \to \mathscr{E}_n \to \mathscr{F}_n \to 0. \tag{3}$$

We now consider the following diagram:

$$
\begin{array}{ccccccccc}
H^i(X,\mathscr{R})^{\widehat{}} & \longrightarrow & H^i(X,\mathscr{E})^{\widehat{}} & \longrightarrow & H^i(X,\mathscr{F})^{\widehat{}} & \longrightarrow & H^{i+1}(X,\mathscr{R})^{\widehat{}} & \longrightarrow & H^{i+1}(X \\
\downarrow{\alpha_1} & & \downarrow{\alpha_2} & & \downarrow{\alpha_3} & & \downarrow{\alpha_4} & & \downarrow{\alpha} \\
\varprojlim H^i(X_n,\mathscr{R}_n) & & & & & & \varprojlim H^{i+1}(X_n,\mathscr{R}_n) & & \\
\downarrow{\beta_1} & & & & & & \downarrow{\beta_2} & & \\
\varprojlim H^i(X_n,\mathscr{T}_n) & \to & \varprojlim H^i(X_n,\mathscr{E}_n) & \to & \varprojlim H^i(X_n,\mathscr{F}_n) & \to & \varprojlim H^{i+1}(X_n,\mathscr{T}_n) & \to & \varprojlim H^{i+1}(X_n,
\end{array}
$$

The top row comes from the cohomology sequence of (1) by completion. Since they are all finitely generated A-modules (5.2), completion is an exact functor (II, 9.3A). The bottom row comes from the cohomology sequence of (3) by taking inverse limits. These groups are all finitely generated $A/\mathfrak{m}^n$-modules, and so satisfy d.c.c. for submodules. Therefore the inverse systems all satisfy the Mittag–Leffler condition (II, 9.1.2), and so the bottom row is exact (II, 9.1). The vertical arrows $\alpha_1, \ldots, \alpha_5$ are the maps of the theorem. We have α_2 an isomorphism because $\mathscr{E}$ is a sum of sheaves $\mathcal{O}(q_i)$, and α_4 and α_5 are isomorphisms by the induction hypothesis. Finally, β_1 and β_2 are maps induced from the sequence (2). We will show below that β_1 and β_2 are isomorphisms.

Admitting this for the moment, it follows from the subtle 5-lemma that α_3 is surjective. But this is then true for any coherent sheaf on X, so α_1 must also be surjective. This in turn implies α_3 is an isomorphism, which is what we want.

It remains to prove that β_1 and β_2 are isomorphisms. Taking the cohomology sequence of (2), and passing to the inverse limit, using (II, 9.1) again, it will be sufficient to show that

$$\varprojlim H^i(X_n, \mathscr{S}_n) = 0$$

for all $i \geqslant 0$. To accomplish this, we will show that for any n, there is an $n' > n$ such that the map of sheaves $\mathscr{S}_{n'} \to \mathscr{S}_n$ is the zero map. By quasi-compactness, the question is local on X, so we may assume that X is affine, $X = \operatorname{Spec} B$. We denote by R, E, S_n the B-modules corresponding to the sheaves $\mathscr{R}, \mathscr{E}, \mathscr{S}_n$, and we denote by $\mathfrak{a}$ the ideal $\mathfrak{m}B$.

Recall that R is a submodule of E, and that

$$S_n = \ker(R/\mathfrak{a}^n R \to E/\mathfrak{a}^n E).$$

Thus

$$S_n = (R \cap \mathfrak{a}^n E)/\mathfrak{a}^n R.$$

But by Krull's theorem (3.1A), the $\mathfrak{a}$-adic topology on R is induced by the $\mathfrak{a}$-adic topology on E. In other words, for any n, there is an $n' > n$ such that

$$R \cap \mathfrak{a}^{n'} E \subseteq \mathfrak{a}^n R.$$

In that case the map $S_{n'} \to S_n$ is zero.   q.e.d.

Remark 11.1.1. This theorem is proved more generally for a proper morphism in Grothendieck [EGA III, §4].

Remark 11.1.2. Many applications of this theorem use only the case $i = 0$. In that case the right-hand side is equal to $\Gamma(\hat{X}, \hat{\mathscr{F}})$, where $\hat{X}$ is the formal completion of X along X_y, and $\hat{\mathscr{F}} = \mathscr{F} \otimes \mathcal{O}_{\hat{X}}$ (II, 9.2). In particular, if $\mathscr{F} = \mathcal{O}_X$, we have $\Gamma(\hat{X}, \mathcal{O}_{\hat{X}})$, which is the ring of *formal-regular functions* (also called *holomorphic functions*) on X along X_y. Hence the name of the theorem.

Remark 11.1.3. One can also introduce the cohomology $H^i(\hat{X}, \hat{\mathscr{F}})$ of $\hat{\mathscr{F}}$ on the formal scheme $\hat{X}$, and prove that it is isomorphic to the two other quantities in the theorem [EGA III, §4].

Corollary 11.2. *Let* $f: X \to Y$ *be a projective morphism of noetherian schemes, and let* $r = \max\{\dim X_y | y \in Y\}$. *Then* $R^i f_*(\mathscr{F}) = 0$ *for all* $i > r$, *and for all coherent sheaves* $\mathscr{F}$ *on* X.

PROOF. For any $y \in Y$, X_n is a scheme whose underlying topological space is the same as X_y. Hence

$$H^i(X_n, \mathscr{F}_n) = 0$$

for $i > r$ by (2.7). If follows that $R^i f_*(\mathscr{F})_y^{\hat{}} = 0$ for all $y \in Y, i > r$, and therefore since $R^i f_*(\mathscr{F})$ is coherent (8.8) it must be 0.

Corollary 11.3. *Let* $f: X \to Y$ *be a projective morphism of noetherian schemes, and assume that* $f_* \mathcal{O}_X = \mathcal{O}_Y$. *Then* $f^{-1}(y)$ *is connected, for every* $y \in Y$.

PROOF. Suppose to the contrary that $f^{-1}(y) = X' \cup X''$, where X' and X'' are disjoint closed subsets. Then for each n, we would have

$$H^0(X_n, \mathcal{O}_{X_n}) = H^0(X'_n, \mathcal{O}_{X_n}) \oplus H^0(X''_n, \mathcal{O}_{X_n}).$$

By the theorem, we have

$$\hat{\mathcal{O}}_y = (f_* \mathcal{O}_X)\hat{\;}_y = \varprojlim H^0(X_n, \mathcal{O}_{X_n}).$$

Therefore $\hat{\mathcal{O}}_y = A' \oplus A''$, where

$$A' = \varprojlim H^0(X'_n, \mathcal{O}_{X_n})$$

and

$$A'' = \varprojlim H^0(X''_n, \mathcal{O}_{X_n}).$$

But this is impossible, because a local ring cannot be a direct sum of two other rings. Indeed, let e', e'' be the unit elements of A' and A''. Then $e' + e'' = 1$ in $\hat{\mathcal{O}}_y$. On the other hand, $e'e'' = 0$, so e', e'' are nonunits, hence contained in the maximal ideal of $\hat{\mathcal{O}}_y$, so their sum cannot be 1 (cf. (II, Ex. 2.19)).

Corollary 11.4 (Zariski's Main Theorem). *Let $f: X \to Y$ be a birational projective morphism of noetherian integral schemes, and assume that Y is normal. Then for every $y \in Y, f^{-1}(y)$ is connected. (See also (V, 5.2).)*

PROOF. By the previous result, we have only to verify that $f_* \mathcal{O}_X = \mathcal{O}_Y$. The question is local on Y, so we may assume Y is affine, equal to Spec A. Then $f_* \mathcal{O}_X$ is a coherent sheaf of $\mathcal{O}_Y$-algebras, so $B = \Gamma(Y, f_* \mathcal{O}_X)$ is a finitely generated A-module. But A and B are integral domains with the same quotient field, and A is integrally closed, so we must have $A = B$. Thus $f_* \mathcal{O}_X = \mathcal{O}_Y$.

Corollary 11.5 (Stein Factorization). *Let $f: X \to Y$ be a projective morphism of noetherian schemes. Then one can factor f into $g \circ f'$, where $f': X \to Y'$ is a projective morphism with connected fibres, and $g: Y' \to Y$ is a finite morphism.*

PROOF. Let $Y' = \mathbf{Spec}\, f_* \mathcal{O}_X$ (II, Ex. 5.17). Then since $f_* \mathcal{O}_X$ is a coherent sheaf of $\mathcal{O}_Y$-algebras, the natural map $g: Y' \to Y$ is finite. On the other hand f clearly factors through g, so we get a morphism $f': X \to Y'$. Since g is separated, we conclude that f' is projective by (II, Ex. 4.9). By construction $f'_* \mathcal{O}_X = \mathcal{O}_{Y'}$ so f' has connected fibres by (11.3).

EXERCISES

11.1. Show that the result of (11.2) is false without the projective hypothesis. For example, let $X = \mathbf{A}^n_k$, let $P = (0, \dots, 0)$, let $U = X - P$, and let $f: U \to X$ be the inclusion. Then the fibres of f all have dimension 0, but $R^{n-1} f_* \mathcal{O}_U \neq 0$.

11.2. Show that a projective morphism with finite fibres ($=$ quasi-finite (II, Ex. 3.5)) is a finite morphism.

11.3. Let X be a normal, projective variety over an algebraically closed field k. Let $\mathfrak{d}$ be a linear system (of effective Cartier divisors) without base points, and assume that $\mathfrak{d}$ is *not composite with a pencil*, which means that if $f: X \to \mathbf{P}^n_k$ is the morphism determined by $\mathfrak{d}$, then dim $f(X) \geqslant 2$. Then show that every divisor in $\mathfrak{d}$ is connected. This improves Bertini's theorem (10.9.1). [*Hints*: Use (11.5), (Ex. 5.7) and (7.9).]

11.4. *Principle of Connectedness.* Let $\{X_t\}$ be a flat family of closed subschemes of $\mathbf{P}_k^n$ parametrized by an irreducible curve T of finite type over k. Suppose there is a nonempty open set $U \subseteq T$, such that for all closed points $t \in U$, X_t is connected. Then prove that X_t is connected for all $t \in T$.

*11.5. Let Y be a hypersurface in $X = \mathbf{P}_k^N$ with $N \geqslant 4$. Let $\hat{X}$ be the formal completion of X along Y (II, §9). Prove that the natural map Pic $\hat{X} \to$ Pic Y is an isomorphism. [*Hint*: Use (II, Ex. 9.6), and then study the maps Pic $X_{n+1} \to$ Pic X_n for each n using (Ex. 4.6) and (Ex. 5.5).]

11.6. Again let Y be a hypersurface in $X = \mathbf{P}_k^N$, this time with $N \geqslant 2$.
  (a) If $\mathscr{F}$ is a locally free sheaf on X, show that the natural map
  $$H^0(X,\mathscr{F}) \to H^0(\hat{X},\mathscr{F})$$
  is an isomorphism.
  (b) Show that the following conditions are equivalent:

   (i) for each locally free sheaf $\mathfrak{F}$ on $\hat{X}$, there exists a coherent sheaf $\mathscr{F}$ on X such that $\mathfrak{F} \cong \mathscr{F}$ (i.e., $\mathfrak{F}$ is *algebraizable*);
   (ii) for each locally free sheaf $\mathfrak{F}$ on $\hat{X}$, there is an integer n_0 such that $\mathfrak{F}(n)$ is generated by global sections for all $n \geqslant n_0$.

   [*Hint*: For (ii) $\Rightarrow$ (i), show that one can find sheaves $\mathscr{E}_0, \mathscr{E}_1$ on X, which are direct sums of sheaves of the form $\mathcal{O}(-q_i)$, and an exact sequence $\hat{\mathscr{E}}_1 \to \hat{\mathscr{E}}_0 \to \mathfrak{F} \to 0$ on $\hat{X}$. Then apply (a) to the sheaf $\mathscr{H}om(\mathscr{E}_1,\mathscr{E}_0)$.]
  (c) Show that the conditions (i) and (ii) of (b) imply that the natural map Pic $X \to$ Pic $\hat{X}$ is an isomorphism.
    Note. In fact, (i) and (ii) always hold if $N \geqslant 3$. This fact, coupled with (Ex. 11.5) leads to Grothendieck's proof [SGA 2] of the Lefschetz theorem which says that if Y is a hypersurface in $\mathbf{P}_k^N$ with $N \geqslant 4$, then Pic $Y \cong \mathbf{Z}$, and it is generated by $\mathcal{O}_Y(1)$. See Hartshorne [5, Ch. IV] for more details.

11.7. Now let Y be a curve in $X = \mathbf{P}_k^2$.
  (a) Use the method of (Ex. 11.5) to show that Pic $\hat{X} \to$ Pic Y is surjective, and its kernel is an infinite-dimensional vector space over k.
  (b) Conclude that there is an invertible sheaf $\mathfrak{L}$ on $\hat{X}$ which is not algebraizable.
  (c) Conclude also that there is a locally free sheaf $\mathfrak{F}$ on $\hat{X}$ so that no twist $\mathfrak{F}(n)$ is generated by global sections. Cf. (II, 9.9.1).

11.8. Let $f: X \to Y$ be a projective morphism, let $\mathscr{F}$ be a coherent sheaf on X which is flat over Y, and assume that $H^i(X_y, \mathscr{F}_y) = 0$ for some i and some $y \in Y$. Then show that $R^i f_*(\mathscr{F})$ is 0 in a neighborhood of y.

12 The Semicontinuity Theorem

In this section we consider a projective morphism $f: X \to Y$ and a coherent sheaf $\mathscr{F}$ on X, flat over Y. We ask, how does the cohomology along the fibre $H^i(X_y, \mathscr{F}_y)$ vary as a function of $y \in Y$? Our technique is to find some relation between these groups and the sheaves $R^i f_*(\mathscr{F})$. The main results are the semicontinuity theorem (12.8), and the theorem on cohomology and base change (12.11).

Since the question is local on Y, we will usually restrict our attention to the case $Y = \operatorname{Spec} A$ is affine. Then we compare the A-modules $H^i(X, \mathscr{F})$ and $H^i(X_y, \mathscr{F}_y)$. Using (9.4), the cohomology of the fibre is equal to $H^i(X, \mathscr{F} \otimes k(y))$. Grothendieck's idea is to study more generally $H^i(X, \mathscr{F} \otimes_A M)$ for any A-module M, and consider it as a functor on A-modules.

Definition. Let A be a noetherian ring, let $Y = \operatorname{Spec} A$, let $f : X \to Y$ be a projective morphism, and let $\mathscr{F}$ be a coherent sheaf on X, flat over Y. (This data will remain fixed throughout this section.) Then for each A-module M, define

$$T^i(M) = H^i(X, \mathscr{F} \otimes_A M)$$

for all $i \geqslant 0$.

Proposition 12.1. *Each T^i is an additive, covariant functor from A-modules to A-modules which is exact in the middle. The collection $(T^i)_{i \geqslant 0}$ forms a δ-functor (§1).*

PROOF. Clearly each T^i is an additive, covariant functor. Since $\mathscr{F}$ is flat over Y, for any exact sequence

$$0 \to M' \to M \to M'' \to 0$$

of A-modules, we get an exact sequence

$$0 \to \mathscr{F} \otimes M' \to \mathscr{F} \otimes M \to \mathscr{F} \otimes M'' \to 0$$

of sheaves on X. Now the long exact sequence of cohomology shows that each T^i is exact in the middle, and that together they form a δ-functor.

We reduce the calculation of the functors T^i to a process involving only A-modules, by the following result.

Proposition 12.2. *With the hypotheses above, there exists a complex $L^{\cdot}$ of finitely generated free A-modules, bounded above (i.e., $L^n = 0$ for $n \gg 0$), such that*

$$T^i(M) \cong h^i(L^{\cdot} \otimes_A M)$$

for any A-module M, any $i \geqslant 0$, and this gives an isomorphism of δ-functors.

PROOF. For any A-module M, the sheaf $\mathscr{F} \otimes_A M$ is quasi-coherent on X, so we can use Čech cohomology to compute $H^i(X, \mathscr{F} \otimes_A M)$. Let $\mathfrak{U} = (U_i)$ be an open cover of X. Let $C^{\cdot} = C^{\cdot}(\mathfrak{U}, \mathscr{F})$ be the Čech complex of $\mathscr{F}$ (§4). Then for any $i_0, \ldots, i_p$, we have

$$\Gamma(U_{i_0, \ldots, i_p}, \mathscr{F} \otimes_A M) = \Gamma(U_{i_0, \ldots, i_p}, \mathscr{F}) \otimes_A M,$$

so

$$C^{\cdot}(\mathfrak{U}, \mathscr{F} \otimes_A M) = C^{\cdot} \otimes_A M.$$

Hence we have

$$T^i(M) = h^i(C^{\cdot} \otimes_A M)$$

for each M.

This is a step in the right direction, since $C^{\cdot}$ is a bounded complex of A-modules. However, the C^i will almost never be finitely generated A-modules. But the complex $C^{\cdot}$ does have the good properties that for each i, C^i is a *flat* A-module (since $\mathscr{F}$ is flat over Y), and for each i, $h^i(C^{\cdot}) = H^i(X, \mathscr{F})$ is a finitely generated A-module, since $\mathscr{F}$ is coherent and f is projective. Now the result of the proposition is a consequence of the following algebraic lemma.

Lemma 12.3. *Let A be a noetherian ring, and let $C^{\cdot}$ be a complex of A-modules, bounded above, such that for each i, $h^i(C^{\cdot})$ is a finitely generated A-module. Then there is a complex $L^{\cdot}$ of finitely generated free A-modules, also bounded above, and a morphism of complexes $g: L^{\cdot} \to C^{\cdot}$, such that the induced map $h^i(L^{\cdot}) \to h^i(C^{\cdot})$ is an isomorphism for all i. Furthermore, if each C^i is a flat A-module, then the map*

$$h^i(L^{\cdot} \otimes M) \to h^i(C^{\cdot} \otimes M)$$

is an isomorphism for any A-module M.

PROOF. First we fix our notation. For any complex $N^{\cdot}$, we let

$$Z^n(N^{\cdot}) = \ker(d^n: N^n \to N^{n+1})$$

and

$$B^n(N^{\cdot}) = \operatorname{im}(d^{n-1}: N^{n-1} \to N^n).$$

Thus we have

$$h^n(N^{\cdot}) = Z^n(N^{\cdot})/B^n(N^{\cdot}).$$

Now for large n, we have $C^n = 0$, so we define $L^n = 0$ there also. Suppose inductively that the complex $L^{\cdot}$ and the morphism of complexes $g: L^{\cdot} \to C^{\cdot}$ has been defined in degrees $i > n$ in such a way that

$$h^i(L^{\cdot}) \xrightarrow{\sim} h^i(C^{\cdot}) \text{ for all } i > n+1, \text{ and} \tag{1}$$

$$Z^{n+1}(L^{\cdot}) \to h^{n+1}(C^{\cdot}) \text{ is surjective.} \tag{2}$$

Then we will construct L^n, $d: L^n \to L^{n+1}$, and $g: L^n \to C^n$ to propagate these properties one step further.

Choose a set of generators $\bar{x}_1, \ldots, \bar{x}_r$ of $h^n(C^{\cdot})$, which is possible since $h^n(C^{\cdot})$ is finitely generated. Lift them to a set of elements $x_1, \ldots, x_r \in Z^n(C^{\cdot})$. On the other hand, let $y_{r+1}, \ldots, y_s$ be a set of generators of $g^{-1}(B^{n+1}(C^{\cdot}))$, which is a submodule of L^{n+1}, hence finitely generated. Let $g(y_i) = \bar{y}_i \in B^{n+1}(C^{\cdot})$, and lift the $\bar{y}_i$ to a set of elements $x_{r+1}, \ldots, x_s$ of C^n.

Now take L^n to be a free A-module on s generators $e_1, \ldots, e_s$. Define $d: L^n \to L^{n+1}$ by $de_i = 0$ for $i = 1, \ldots, r$, and $de_i = y_i$ for $i = r+1, \ldots, s$. Define $g: L^n \to C^n$ by $ge_i = x_i$ for all i.

Then one checks easily that g commutes with d, that $h^{n+1}(L^{\cdot}) \to h^{n+1}(C^{\cdot})$ is an isomorphism, and that $Z^n(L^{\cdot}) \to h^n(C^{\cdot})$ is surjective. So inductively, we construct the complex $L^{\cdot}$ required.

Now suppose that each C^i is a flat A-module. Then we will prove, by descending induction on i, that

$$h^i(L^{\cdot} \otimes M) \to h^i(C^{\cdot} \otimes M)$$

is an isomorphism for all A-modules M. For $i \gg 0$, both L^i and C^i are 0, so both sides are 0. So suppose this is true for $i + 1$. It is sufficient to prove the result for finitely generated A-modules, because any A-module is a direct limit of finitely generated ones, and both $\otimes$ and h^i commute with direct limits. So given M finitely generated, write it as a quotient of a free finitely generated A-module E, and let R be the kernel:

$$0 \to R \to E \to M \to 0.$$

Since each C^i is flat by hypothesis, and each L^i is flat, because it is free, we get an exact, commutative diagram of complexes

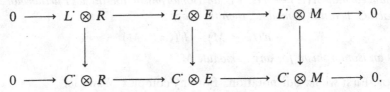

Applying h^i, we get a commutative diagram of long exact sequences. Since the result holds for $i + 1$ by induction, and for E, since E is free, and $h^i(L^{\cdot}) \to h^i(C^{\cdot})$ is an isomorphism, the result for any M follows from the subtle 5-lemma.

Now we will study conditions under which one of the functors T^i is left exact, right exact, or exact. For any complex $N^{\cdot}$, we define

$$W^i(N^{\cdot}) = \mathrm{coker}(d^{i-1} : N^{i-1} \to N^i)$$

so that we have an exact sequence

$$0 \to h^i(N^{\cdot}) \to W^i(N^{\cdot}) \to N^{i+1}.$$

Proposition 12.4. *The following conditions are equivalent:*

   (i) *T^i is left exact;*
  (ii) *$W^i = W^i(L^{\cdot})$ is a projective A-module;*
 (iii) *there is a finitely generated A-module Q, such that*

$$T^i(M) = \mathrm{Hom}_A(Q, M)$$

    for all M.

    Furthermore the Q in (iii) *is unique.*

PROOF. Since tensor product is right exact, we have

$$W^i(L' \otimes M) = W^i(L') \otimes M,$$

for any A-module M. We will write simply W^i for $W^i(L')$. Hence

$$T^i(M) = \ker(W^i \otimes M \to L^{i+1} \otimes M).$$

Let $0 \to M' \to M$ be an inclusion. Then we obtain an exact, commutative diagram

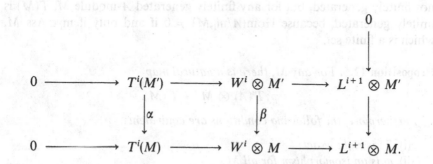

The third vertical arrow is injective, since L^{i+1} is free. A simple diagram chase shows that α is injective if and only if β is. Since this is true for any choice of $0 \to M' \to M$, we see that T^i is left exact if and only if W^i is flat. (Recall that in any case T^i is exact in the middle (12.1).) But since W^i is finitely generated, this is equivalent to W^i being projective (9.1A). This shows (i) $\Leftrightarrow$ (ii).

(iii) $\Rightarrow$ (i) is obvious.

To prove (ii) $\Rightarrow$ (iii), let $\check{L}^{i+1}$ and $\check{W}^i$ be the dual projective modules. Define

$$Q = \mathrm{coker}(\check{L}^{i+1} \to \check{W}^i).$$

Then for every A-module M, we have

$$0 \to \mathrm{Hom}(Q,M) \to \mathrm{Hom}(\check{W}^i,M) \to \mathrm{Hom}(\check{L}^{i+1},M).$$

But the last two groups are $W^i \otimes M$, and $L^{i+1} \otimes M$, respectively, so $\mathrm{Hom}(Q,M) = T^i(M)$.

To see the uniqueness of Q, let Q' be another module such that $T^i(M) = \mathrm{Hom}(Q',M)$ for all M. Then

$$\mathrm{Hom}(Q,M) = \mathrm{Hom}(Q',M)$$

for all M. In particular, the elements

$$1 \in \mathrm{Hom}(Q,Q) = \mathrm{Hom}(Q',Q)$$

and

$$1' \in \mathrm{Hom}(Q',Q') = \mathrm{Hom}(Q,Q')$$

give isomorphisms of Q and Q', inverse to each other, and canonically defined.

Remark 12.4.1. There is a general theorem to the effect that any left-exact functor T on A-modules, which commutes with direct sums, is of the form $\mathrm{Hom}(Q,\cdot)$ for some A-module Q. But even if T takes finitely generated modules into finitely generated modules, Q need not be finitely generated. Thus the fact that our Q in (iii) above is finitely generated is a strong fact about the functor T^i.

For example, let A be a noetherian ring with infinitely many maximal ideals $\mathfrak{m}_i$. Let $Q = \sum A/\mathfrak{m}_i$, and let T be the functor $\mathrm{Hom}(Q,\cdot)$. Then Q is not finitely generated, but for any finitely generated A-module M, $T(M)$ is finitely generated, because $\mathrm{Hom}(A/\mathfrak{m}_i,M) \neq 0$ if and only if $\mathfrak{m}_i \in \mathrm{Ass}\ M$, which is a finite set.

Proposition 12.5. *For any M, there is a natural map*

$$\varphi : T^i(A) \otimes M \to T^i(M).$$

Furthermore, the following conditions are equivalent:

(i) *T^i is right exact;*
(ii) *φ is an isomorphism for all M;*
(iii) *φ is surjective for all M.*

PROOF. Since T^i is a functor, we have a natural map, for any M,

$$M = \mathrm{Hom}(A,M) \xrightarrow{\psi} \mathrm{Hom}(T^i(A),T^i(M)).$$

This gives φ, by setting

$$\varphi(\textstyle\sum a_i \otimes m_i) = \sum \psi(m_i)a_i.$$

Since T^i and $\otimes$ commute with direct limits, it will be sufficient to consider finitely generated A-modules M. Write

$$A^r \to A^s \to M \to 0.$$

Then we have a diagram

$$
\begin{array}{ccccccc}
T^i(A) \otimes A^r & \longrightarrow & T^i(A) \otimes A^s & \longrightarrow & T^i(A) \otimes M & \longrightarrow & 0 \\
\downarrow & & \downarrow & & \downarrow{\scriptstyle\varphi} & & \\
T^i(A^r) & \longrightarrow & T^i(A^s) & \longrightarrow & T^i(M) & &
\end{array}
$$

where the bottom row is not necessarily exact. The first two vertical arrows are isomorphisms. Thus, if T^i is right exact, φ is an isomorphism. This proves (i) $\Rightarrow$ (ii). The implication (ii) $\Rightarrow$ (iii) is obvious, so we have only to prove (iii) $\Rightarrow$ (i). We must show if

$$0 \to M' \to M \to M'' \to 0$$

is an exact sequence of A-modules, then

$$T^i(M') \to T^i(M) \to T^i(M'') \to 0$$

is exact. By (12.1) it is exact in the middle, so we have only to show that $T^i(M) \to T^i(M'')$ is surjective. This follows from the diagram

$$
\begin{array}{ccc}
T^i(A) \otimes M & \longrightarrow & T^i(A) \otimes M'' \longrightarrow 0 \\
\downarrow{\scriptstyle \varphi(M)} & & \downarrow{\scriptstyle \varphi(M'')} \\
T^i(M) & \longrightarrow & T^i(M'')
\end{array}
$$

and the fact that $\varphi(M'')$ is surjective.

Corollary 12.6. *The following conditions are equivalent:*

   (i) T^i *is exact;*

   (ii) T^i *is right exact, and* $T^i(A)$ *is a projective A-module.*

PROOF. In any case, T^i is right exact, so by (12.5) we have $T^i(M) \cong T^i(A) \otimes M$ for all A-modules M. Therefore T^i is exact if and only if $T^i(A)$ is flat. But $T^i(A)$ is a finitely generated A-module, so this is equivalent to being locally free (9.1A), hence projective.

Now we wish to localize the above discussion. For any point $y \in Y = \operatorname{Spec} A$, we denote by T^i_y the restriction of the functor T^i to the category of $A_{\mathfrak{p}}$-modules, where $\mathfrak{p} \subseteq A$ is the prime ideal corresponding to y. Then we say "T^i is left exact at y" to mean T^i_y is left exact, and similarly for right exact, or exact. Note that for any $A_{\mathfrak{p}}$-module N, $T^i_y(N) = h^i(L^{\cdot}_{\mathfrak{p}} \otimes N)$. Also note that T^i is left exact if and only if it is left exact at all points $y \in Y$; similarly for right exact, or exact. Finally, since cohomology commutes with flat base extension (9.3), we see that T^i_y is the functor T^i associated with the morphism $f': X' \to Y'$ obtained from f by the flat base extension $Y' = \operatorname{Spec} \mathcal{O}_y \to Y$. So we can apply the results (12.4), (12.5), (12.6) locally to each T^i_y.

Proposition 12.7. *If T^i is left exact (respectively, right exact, exact) at some point $y_0 \in Y$, then the same is true for all points y in a suitable open neighborhood U of y_0.*

PROOF. T^i is left exact at y_0 if and only if $\tilde{W}^i_{y_0}$ is free, by (12.4). But since $\tilde{W}^i$ is a coherent sheaf on Y, this implies that $\tilde{W}^i$ is locally free in some neighborhood U of y_0, and so T^i is left exact at all points of U.

T^i is right exact at a point y if and only if T^{i+1} is left exact there, by the long exact sequence (12.1). So the second statement follows from the first, applied to T^{i+1}.

T^i is exact at a point if and only if it is both left exact and right exact, so the third statement is the conjunction of the first two.

Definition. Let Y be a topological space. A function $\varphi: Y \to \mathbf{Z}$ is *upper semicontinuous* if for each $y \in Y$, there is an open neighborhood U of y, such that for all $y' \in U$, $\varphi(y') \leqslant \varphi(y)$. Intuitively, this means that φ may get bigger at special points.

Remark 12.7.1. A function $\varphi : Y \to \mathbf{Z}$ is upper semicontinuous if and only if for each $n \in \mathbf{Z}$, the set $\{ y \in Y \mid \varphi(y) \geqslant n \}$ is a closed subset of Y.

Example 12.7.2. Let Y be a noetherian scheme, and let $\mathscr{F}$ be a coherent sheaf on Y. Then the function

$$\varphi(y) = \dim_{k(y)}(\mathscr{F} \otimes k(y))$$

is upper semicontinuous. Indeed, by Nakayama's lemma, $\varphi(y)$ is equal to the minimal number of generators of the $\mathscr{O}_y$-module $\mathscr{F}_y$. But if $s_1, \ldots, s_r \in \mathscr{F}_y$ form a minimal set of generators, they extend to sections of $\mathscr{F}$ in some neighborhood of y, and they generate $\mathscr{F}$ in some neighborhood, because $\mathscr{F}$ is coherent. So if y' is in that neighborhood, then $\varphi(y')$, which is the minimal number of generators of $\mathscr{F}_{y'}$, is $\leqslant r = \varphi(y)$.

Theorem 12.8 (Semicontinuity). *Let $f : X \to Y$ be a projective morphism of noetherian schemes, and let $\mathscr{F}$ be a coherent sheaf on X, flat over Y. Then for each $i \geqslant 0$, the function*

$$h^i(y, \mathscr{F}) = \dim_{k(y)} H^i(X_y, \mathscr{F}_y)$$

is an upper semicontinuous function on Y.

PROOF. The question is local on Y, so we may assume $Y = \operatorname{Spec} A$ is affine, with A noetherian. Thus we can apply the earlier results of this section. By (9.4) we have

$$h^i(y, \mathscr{F}) = \dim_{k(y)} T^i(k(y)).$$

As in the proof of (12.4), we have

$$T^i(k(y)) = \ker(W^i \otimes k(y) \to L^{i+1} \otimes k(y)).$$

On the other hand, there is an exact sequence

$$W^i \to L^{i+1} \to W^{i+1} \to 0,$$

so tensoring with $k(y)$, we obtain a four-term exact sequence

$$0 \to T^i(k(y)) \to W^i \otimes k(y) \to L^{i+1} \otimes k(y) \to W^{i+1} \otimes k(y) \to 0.$$

Therefore, counting dimensions, we have

$$h^i(y, \mathscr{F}) = \dim_{k(y)} W^i \otimes k(y) + \dim_{k(y)} W^{i+1} \otimes k(y) - \dim_{k(y)} L^{i+1} \otimes k(y).$$

Now W^i and W^{i+1} are finitely generated A-modules, so by (12.7.2) the first two terms of this sum are upper semicontinuous functions of y. But L^{i+1} is a free A-module, so the last term is a constant function of y. Combining, we see that $h^i(y, \mathscr{F})$ is upper semicontinuous.

Corollary 12.9 (Grauert). *With the same hypotheses as the theorem, suppose furthermore that Y is integral, and that for some i, the function $h^i(y, \mathscr{F})$ is constant on Y. Then $R^i f_*(\mathscr{F})$ is locally free on Y, and for every y the natural* *map*

$$R^i f_*(\mathscr{F}) \otimes k(y) \rightarrow H^i(X_y, \mathscr{F}_y)$$

is an isomorphism.

PROOF. As above, we may assume that Y is affine. Using the expression for $h^i(y, \mathscr{F})$ in the proof of the theorem, we conclude that the functions dim $W^i \otimes k(y)$ and dim $W^{i+1} \otimes k(y)$ must both be constant. But this implies (II, 8.9) that $\tilde{W}^i$ and $\tilde{W}^{i+1}$ are both locally free sheaves on Y. So by (12.4), T^i and T^{i+1} are both left exact, so T^i is exact, so by (12.6), $T^i(A)$ is a projective A-module. But $R^i f_*(\mathscr{F})$ is just $T^i(A)^{\sim}$, so it is a locally free sheaf. Finally by (12.5) we see that

$$R^i f_*(\mathscr{F}) \otimes k(y) \rightarrow H^i(X_y, \mathscr{F}_y)$$

is an isomorphism for all $y \in Y$.

Example 12.9.1. Let $\{X_t\}$ be a flat family of integral curves in $\mathbf{P}_k^n$, with k algebraically closed. Then for every closed point $t \in T$, $H^0(X_t, \mathscr{O}_{X_t}) = k$. On the other hand, the arithmetic genus $p_a = 1 - \chi(\mathscr{O}_{X_t})$ is constant, by (9.10). So we see that in this case the functions $h^0(t, \mathscr{O}_X)$ and $h^1(t, \mathscr{O}_X)$ are both constant on T.

Example 12.9.2. In the flat family of (9.8.4), we have $h^0(X_t, \mathscr{O}_{X_t}) = 1$ if $t \neq 0$, and 2 if $t = 0$, because of the nilpotent elements. On the other hand, $h^1(X_t, \mathscr{O}_{X_t}) = 0$ for $t \neq 0$, since X_t is rational, and $h^1(X_0, \mathscr{O}_{X_0}) = h^1(X_0, (\mathscr{O}_{X_0})_{red}) = 1$, since $(X_0)_{red}$ is a plane cubic curve. So in this case the functions h^0, h^1 both jump up at $t = 0$.

Example 12.9.3. If $\{X_t\}$ is an algebraic family of nonsingular projective varieties over $\mathbf{C}$, parametrized by a variety T, then the functions $h^i(X_t, \mathscr{O}_{X_t})$ are actually constant for all i. The proof of this result requires transcendental methods, namely the degeneration of the Hodge spectral sequence—cf. Deligne[4].

Now we wish to give some more precise information about when the map

$$T^i(A) \otimes k(y) \rightarrow T^i(k(y))$$

is an isomorphism. And here we will use a new ingredient in our proof, namely the theorem on formal functions (11.1).

Proposition 12.10. *Assume that for some i, y, the map*

$$\varphi : T^i(A) \otimes k(y) \rightarrow T^i(k(y))$$

is surjective. Then T^i is right exact at y (and conversely, by (12.5)).

PROOF. By making a flat base extension Spec $\mathscr{O}_y \rightarrow Y$ if necessary (9.3), we may assume that y is a closed point of Y; A is a local ring, with maximal ideal

$\mathfrak{m}$, and $k(y) = k = A/\mathfrak{m}$. By (12.5), it is sufficient to show that

$$\varphi(M): T^i(A) \otimes M \to T^i(M)$$

is surjective for all A-modules M. Since T^i and tensor product commute with direct limits, it is sufficient to consider finitely generated M.

First, we consider A-modules M of finite length, and we show that $\varphi(M)$ is surjective, by induction on the length of M. If the length is 1, then $M = k$, and $\varphi(k)$ is surjective by hypothesis. In general, write

$$0 \to M' \to M \to M'' \to 0,$$

where M' and M'' have length less than length M. Then using (12.1) we have a commutative diagram with exact rows

$$
\begin{array}{ccccccc}
T^i(A) \otimes M' & \longrightarrow & T^i(A) \otimes M & \longrightarrow & T^i(A) \otimes M'' & \longrightarrow & 0 \\
\downarrow & & \downarrow & & \downarrow & & \\
T^i(M') & \longrightarrow & T^i(M) & \longrightarrow & T^i(M'') & &
\end{array}
$$

The two outside vertical arrows are surjective, by the induction hypothesis, so the middle one is surjective also.

Now let M be any finitely generated A-module. For each n, $M/\mathfrak{m}^n M$ is a module of finite length, so that by the previous case,

$$\varphi_n: T^i(A) \otimes M/\mathfrak{m}^n M \to T^i(M/\mathfrak{m}^n M)$$

is surjective. Note that $\ker \varphi_n$ is an A-module of finite length, so the inverse system $(\ker \varphi_n)$ satisfies the Mittag–Leffler condition (II, 9.1.2). Hence by (II, 9.1) the map

$$\varprojlim \varphi_n : (T^i(A) \otimes M)^{\hat{}} \to \varprojlim T^i(M/\mathfrak{m}^n M)$$

is also surjective. But by the theorem on formal functions (11.1), applied to the sheaf $\mathscr{F} \otimes_A M$ on X, the right hand side is just $T^i(M)^{\hat{}}$. So we have a surjection

$$(T^i(A) \otimes M)^{\hat{}} \to T^i(M)^{\hat{}}.$$

Since completion is a faithful exact functor for finitely generated A-modules, it follows that

$$\varphi(M): T^i(A) \otimes M \to T^i(M)$$

is surjective, so we are done.

Combining this with our earlier results, we obtain the following theorem.

Theorem 12.11 (Cohomology and Base Change). *Let $f: X \to Y$ be a projective morphism of noetherian schemes, and let $\mathscr{F}$ be a coherent sheaf on X, flat over Y. Let y be a point of Y. Then:*
   (a) *if the natural map*

$$\varphi^i(y): R^i f_*(\mathscr{F}) \otimes k(y) \to H^i(X_y, \mathscr{F}_y)$$

is surjective, then it is an isomorphism, and the same is true for all y' *in a suitable neighborhood of* y;

   (b) *Assume that* $\varphi^i(y)$ *is surjective. Then the following conditions are equivalent*:

   (i) $\varphi^{i-1}(y)$ *is also surjective*;
   (ii) $R^i f_*(\mathcal{F})$ *is locally free in a neighborhood of* y.

PROOF. (a) follows from (12.10), (12.7), and (12.5). (b) follows from (12.10), (12.6), and (12.5), using the fact that T^i is exact if and only if T^{i-1} and T^i are both right exact.

References for §12. The semicontinuity theorem was first proved by Grauert [1] in the complex-analytic case. These theorems in the algebraic case are due to Grothendieck [EGA III, 7.7]. Our proof follows the main ideas of Grothendieck's proof, with simplifications due to Mumford [5, II, §5].

EXERCISES

12.1. Let Y be a scheme of finite type over an algebraically closed field k. Show that the function
$$\varphi(y) = \dim_k(m_y/m_y^2)$$
is upper semicontinuous on the set of closed points of Y.

12.2. Let $\{X_t\}$ be a family of hypersurfaces of the same degree in $\mathbf{P}_k^n$. Show that for each i, the function $h^i(X_t, \mathcal{O}_{X_t})$ is a constant function of t.

12.3. Let $X_1 \subseteq \mathbf{P}_k^4$ be the *rational normal quartic curve* (which is the 4-uple embedding of $\mathbf{P}^1$ in $\mathbf{P}^4$). Let $X_0 \subseteq \mathbf{P}_k^3$ be a nonsingular rational quartic curve, such as the one in (I, Ex. 3.18b). Use (9.8.3) to construct a flat family $\{X_t\}$ of curves in $\mathbf{P}^4$, parametrized by $T = \mathbf{A}^1$, with the given fibres X_1 and X_0 for $t = 1$ and $t = 0$.
   Let $\mathcal{I} \subseteq \mathcal{O}_{\mathbf{P}^4 \times T}$ be the ideal sheaf of the total family $X \subseteq \mathbf{P}^4 \times T$. Show that $\mathcal{I}$ is flat over T. Then show that
$$h^0(t, \mathcal{I}) = \begin{cases} 0 & \text{for } t \neq 0 \\ 1 & \text{for } t = 0 \end{cases}$$
and also
$$h^1(t, \mathcal{I}) = \begin{cases} 0 & \text{for } t \neq 0 \\ 1 & \text{for } t = 0. \end{cases}$$
This gives another example of cohomology groups jumping at a special point.

12.4. Let Y be an integral scheme of finite type over an algebraically closed field k. Let $f : X \to Y$ be a flat projective morphism whose fibres are all integral schemes. Let $\mathcal{L}, \mathcal{M}$ be invertible sheaves on X, and assume for each $y \in Y$ that $\mathcal{L}_y \cong \mathcal{M}_y$ on the fibre X_y. Then show that there is an invertible sheaf $\mathcal{N}$ on Y such that $\mathcal{L} \cong \mathcal{M} \otimes f^*\mathcal{N}$. [*Hint*: Use the results of this section to show that $f_*(\mathcal{L} \otimes \mathcal{M}^{-1})$ is locally free of rank 1 on Y.]

12.5. Let Y be an integral scheme of finite type over an algebraically closed field k. Let $\mathcal{E}$ be a locally free sheaf on Y, and let $X = \mathbf{P}(\mathcal{E})$—see (II, §7). Then show that Pic $X \cong (\text{Pic } Y) \times \mathbf{Z}$. This strengthens (II, Ex. 7.9).

*12.6. Let X be an integral projective scheme over an algebraically closed field k, and assume that $H^1(X, \mathcal{O}_X) = 0$. Let T be a connected scheme of finite type over k.

(a) If $\mathscr{L}$ is an invertible sheaf on $X \times T$, show that the invertible sheaves $\mathscr{L}_t$ on $X = X \times \{t\}$ are isomorphic, for all closed points $t \in T$.

(b) Show that $\text{Pic}(X \times T) = \text{Pic } X \times \text{Pic } T$. (Do *not* assume that T is reduced!) Cf. (IV, Ex. 4.10) and (V, Ex. 1.6) for examples where $\text{Pic}(X \times T) \neq \text{Pic } X \times \text{Pic } T$. [*Hint*: Apply (12.11) with $i = 0,1$ for suitable invertible sheaves on $X \times T$.]

CHAPTER IV

Curves

In this chapter we apply the techniques we have learned earlier to study curves. But in fact, except for the proof of the Riemann–Roch theorem (1.3), which uses Serre duality, we use very little of the fancy methods of schemes and cohomology. So if a reader is willing to accept the statement of the Riemann–Roch theorem, he can read this chapter at a much earlier stage of his study of algebraic geometry. That may not be a bad idea, pedagogically, because in that way he will see some applications of the general theory, and in particular will gain some respect for the significance of the Riemann–Roch theorem. In contrast, the proof of the Riemann–Roch theorem is not very enlightening.

After reviewing what is needed from the earlier part of the book in §1, we study in §2, 3 various ways of representing a curve explicitly. One way is to represent the curve as a branched covering of $\mathbf{P}^1$. So in §2 we make a general study of one curve as a branched covering of another. The central result here is Hurwitz's theorem (2.4) which compares the canonical divisors on the two curves.

In §3 we give two other ways of representing a curve. We show that any nonsingular projective curve can be embedded in $\mathbf{P}^3$, and it can be mapped birationally into $\mathbf{P}^2$ in such a way that the image has only nodes for singularities. The proof of the latter theorem has an interesting extra twist in characteristic $p > 0$.

In §4 we discuss the special case of curves of genus 1, called elliptic curves. This is a whole subject in itself, quite independent of the rest of the chapter. We have space for only a brief glimpse of some aspects of this fascinating theory.

In §5, 6 we discuss the canonical embedding, and some classification questions, both for abstract curves and for curves in $\mathbf{P}^3$.

1 Riemann–Roch Theorem

In this chapter we will use the word *curve* to mean a complete, nonsingular curve over an algebraically closed field k. In other words (II, §6), a curve is an integral scheme of dimension 1, proper over k, all of whose local rings are regular. Such a curve is necessarily projective (II, 6.7). If we want to consider a more general kind of curve, we will use the word "scheme," appropriately qualified, e.g., "an integral scheme of dimension 1 of finite type over k." We will use the word *point* to mean a closed point, unless we specify the *generic point*.

We begin by reviewing some of the concepts introduced earlier in the book, which we will use in our study of curves.

The most important single invariant of a curve is its genus. There are several ways of defining it, all equivalent. For a curve X in projective space, we have the *arithmetic genus* $p_a(X)$, defined as $1 - P_X(0)$, where P_X is the Hilbert polynomial of X (I, Ex. 7.2). On the other hand, we have the *geometric genus* $p_g(X)$, defined as $\dim_k \Gamma(X, \omega_X)$, where ω_X is the canonical sheaf (II, 8.18.2).

Proposition 1.1. *If X is a curve, then*

$$p_a(X) = p_g(X) = \dim_k H^1(X, \mathcal{O}_X),$$

so we call this number simply the genus *of X, and denote it by g.*

PROOF. The equality $p_a(X) = \dim H^1(X, \mathcal{O}_X)$ has been shown in (III, Ex. 5.3). The equality $p_g = \dim H^1(X, \mathcal{O}_X)$ is a consequence of Serre duality (III, 7.12.2).

Remark 1.1.1. From $g = p_g$, we see that the genus of a curve is always non-negative. Conversely, for any $g \geqslant 0$, there exist curves of genus g. For example, take a divisor of type $(g + 1, 2)$ on a nonsingular quadric surface. There exist such divisors which are irreducible and nonsingular, and they have $p_a = g$ (III, Ex. 5.6).

A (*Weil*) *divisor* on the curve X is an element of the free abelian group generated by the set of points of X (II, §6). We write a divisor as $D = \sum n_i P_i$ with $n_i \in \mathbf{Z}$. Its *degree* is $\sum n_i$. Two divisors are *linearly equivalent* if their difference is the divisor of a rational function. We have seen that the degree of a divisor depends only on its linear equivalence class (II, 6.10). Since X is nonsingular, for every divisor D we have an associated invertible sheaf $\mathscr{L}(D)$, and the correspondence $D \to \mathscr{L}(D)$ gives an isomorphism of the group $\mathrm{Cl}(X)$ of divisors modulo linear equivalence with the group $\mathrm{Pic}\, X$ of invertible sheaves modulo isomorphism (II, 6.16).

A divisor $D = \sum n_i P_i$ on X is *effective* if all $n_i \geqslant 0$. The set of all effective divisors linearly equivalent to a given divisor D is called a *complete linear system* (II, §7) and is denoted by $|D|$. The elements of $|D|$ are in one-to-one correspondence with the space

$$(H^0(X,\mathscr{L}(D)) - \{0\})/k^*,$$

so $|D|$ carries the structure of the set of closed points of a projective space (II, 7.7). We denote $\dim_k H^0(X,\mathscr{L}(D))$ by $l(D)$, so that the *dimension of* $|D|$ is $l(D) - 1$. The number $l(D)$ is finite by (II, 5.19) or (III, 5.2).

As a consequence of this correspondence we have the following elementary, but useful, result.

Lemma 1.2. *Let D be a divisor on a curve X. Then if $l(D) \neq 0$, we must have* $\deg D \geqslant 0$. *Furthermore, if $l(D) \neq 0$ and $\deg D = 0$, we must have $D \sim 0$,* i.e., $\mathscr{L}(D) \cong \mathcal{O}_X$.

PROOF. If $l(D) \neq 0$, then the complete linear system $|D|$ is nonempty. Hence D is linearly equivalent to some effective divisor. Since the degree depends only on the linear equivalence class, and the degree of an effective divisor is nonnegative, we find $\deg D \geqslant 0$. If $\deg D = 0$, then D is linearly equivalent to an effective divisor of degree 0. But there is only one such, namely the zero divisor.

We denote by $\Omega_{X/k}$, or simply Ω_X, the sheaf of relative differentials of X over k (II, §8). Since X has dimension 1, it is an invertible sheaf on X, and so is equal to the canonical sheaf ω_X on X. We call any divisor in the corresponding linear equivalence class a *canonical divisor*, and denote it by K. (We also occasionally use the letter K to denote the function field of X, but it should be clear from the context which meaning is intended.)

Theorem 1.3 (Riemann–Roch). *Let D be a divisor on a curve X of genus g. Then*

$$l(D) - l(K - D) = \deg D + 1 - g.$$

PROOF. The divisor $K - D$ corresponds to the invertible sheaf $\omega_X \otimes \mathscr{L}(D)^{\check{}}$. Since X is projective (II, 6.7), we can apply Serre duality (III, 7.12.1) to conclude that the vector space $H^0(X,\omega_X \otimes \mathscr{L}(D)^{\check{}})$ is dual to $H^1(X,\mathscr{L}(D))$. Thus we have to show that for any D,

$$\chi(\mathscr{L}(D)) = \deg D + 1 - g,$$

where for any coherent sheaf $\mathscr{F}$ on X, $\chi(\mathscr{F})$ is the Euler characteristic

$$\chi(\mathscr{F}) = \dim H^0(X,\mathscr{F}) - \dim H^1(X,\mathscr{F}).$$

First we consider the case $D = 0$. Then our formula says

$$\dim H^0(X,\mathcal{O}_X) - \dim H^1(X,\mathcal{O}_X) = 0 + 1 - g.$$

This is true, because $H^0(X,\mathcal{O}_X) = k$ for any projective variety (I, 3.4), and $\dim H^1(X,\mathcal{O}_X) = g$ by (1.1).

Next, let D be any divisor, and let P be any point. We will show that the formula is true for D if and only if it is true for $D + P$. Since any divisor can be reached from 0 in a finite number of steps by adding or subtracting a point each time, this will show the result holds for all D.

We consider P as a closed subscheme of X. Its structure sheaf is a skyscraper sheaf k sitting at the point P, which we denote by $k(P)$, and its ideal sheaf is $\mathscr{L}(-P)$ by (II, 6.18). Therefore we have an exact sequence

$$0 \to \mathscr{L}(-P) \to \mathcal{O}_X \to k(P) \to 0.$$

Tensoring with $\mathscr{L}(D + P)$ we get

$$0 \to \mathscr{L}(D) \to \mathscr{L}(D + P) \to k(P) \to 0.$$

(Since $\mathscr{L}(D + P)$ is locally free of rank 1, tensoring by it does not affect the sheaf $k(P)$.) Now the Euler characteristic is additive on short exact sequences (III, Ex. 5.1), and $\chi(k(P)) = 1$, so we have

$$\chi(\mathscr{L}(D + P)) = \chi(\mathscr{L}(D)) + 1.$$

On the other hand, $\deg(D + P) = \deg D + 1$, so our formula is true for D if and only if it is true for $D + P$, as required.

Remark 1.3.1. If X is a curve of degree d in $\mathbf{P}^n$, and D is a hyperplane section $X \cap H$, so that $\mathscr{L}(D) = \mathcal{O}_X(1)$, then the Hilbert polynomial (III, Ex. 5.2) tells us that

$$\chi(\mathscr{L}(D)) = d + 1 - p_a.$$

This is a special case of the Riemann–Roch theorem.

Remark 1.3.2. The Riemann–Roch theorem enables us to solve the "Riemann–Roch problem" (II, Ex. 7.6) for a divisor D on a curve X. If $\deg D < 0$, then $\dim|nD| = -1$ for all $n > 0$. If $\deg D = 0$, then $\dim|nD|$ is 0 or -1 depending on whether $nD \sim 0$ or not. If $\deg D > 0$, then $l(K - nD) = 0$ as soon as $n \cdot \deg D > \deg K$, by (1.2), so for $n \gg 0$ we have

$$\dim|nD| = n \cdot \deg D - g.$$

Example 1.3.3. On a curve X of genus g, the canonical divisor K has degree $2g - 2$. Indeed, we apply (1.3) with $D = K$. Since $l(K) = p_g = g$ and $l(0) = 1$, we have

$$g - 1 = \deg K + 1 - g,$$

hence $\deg K = 2g - 2$.

Example 1.3.4. We say a divisor D is *special* if $l(K - D) > 0$, and that $l(K - D)$ is its *index of speciality*. Otherwise D is *nonspecial*. If $\deg D > 2g - 2$, then by (1.3.3), $\deg(K - D) < 0$, so $l(K - D) = 0$ (1.2). Thus D is nonspecial.

Example 1.3.5. Recall that a curve is *rational* if it is birational to $\mathbf{P}^1$ (I, Ex. 6.1). Since curves in this chapter are complete and nonsingular by definition, a curve X is rational if and only if $X \cong \mathbf{P}^1$ (I, 6.12). Now using (1.3) we can show that X is rational if and only if $g = 0$. We already know that $p_a(\mathbf{P}^1) = 0$ (I, Ex. 7.2), so conversely suppose given a curve X of genus 0. Let P,Q be two distinct points of X and apply Riemann–Roch to the divisor $D = P - Q$. Since $\deg(K - D) = -2$, using (1.3.3) above, we have $l(K - D) = 0$, and so we find $l(D) = 1$. But D is a divisor of degree 0, so by (1.2) we have $D \sim 0$, in other words $P \sim Q$. But this implies X is rational (II, 6.10.1).

Example 1.3.6. We say a curve X is *elliptic* if $g = 1$. On an elliptic curve, the canonical divisor K has degree 0, by (1.3.3). On the other hand, $l(K) = p_g = 1$, so from (1.2) we conclude that $K \sim 0$.

Example 1.3.7. Let X be an elliptic curve, let P_0 be a point of X, and let $\mathrm{Pic}^\circ X$ denote the subgroup of $\mathrm{Pic}\, X$ corresponding to divisors of degree 0. Then the map $P \to \mathscr{L}(P - P_0)$ gives a one-to-one correspondence between the set of points of X and the elements of the group $\mathrm{Pic}^\circ X$. Thus we get a group structure (with P_0 as identity) on the set of points of X, generalizing (II, 6.10.2).

To see this, it will be enough to show that if D is any divisor of degree 0, then there exists a unique point $P \in X$ such that $D \sim P - P_0$. We apply Riemann–Roch to $D + P_0$, and get
$$l(D + P_0) - l(K - D - P_0) = 1 + 1 - 1.$$
Now $\deg K = 0$, so $\deg(K - D - P_0) = -1$, and hence $l(K - D - P_0) = 0$. Therefore, $l(D + P_0) = 1$. In other words, $\dim|D + P_0| = 0$. This means there is a unique effective divisor linearly equivalent to $D + P_0$. Since the degree is 1, it must be a single point P. Thus we have shown that there is a unique point $P \sim D + P_0$, i.e., $D \sim P - P_0$.

Remark 1.3.8. For other proofs of the Riemann–Roch theorem, see Serre [7, Ch. II] and Fulton [1].

EXERCISES

1.1. Let X be a curve, and let $P \in X$ be a point. Then there exists a nonconstant rational function $f \in K(X)$, which is regular everywhere except at P.

1.2. Again let X be a curve, and let $P_1, \dots, P_r \in X$ be points. Then there is a rational function $f \in K(X)$ having poles (of some order) at each of the P_i, and regular elsewhere.

1.3. Let X be an integral, separated, regular, one-dimensional scheme of finite type over k, which is *not* proper over k. Then X is affine. [*Hint*: Embed X in a (proper) curve $\bar{X}$ over k, and use (Ex. 1.2) to construct a morphism $f : \bar{X} \to \mathbf{P}^1$ such that $f^{-1}(\mathbf{A}^1) = X$.]

1.4. Show that a separated, one-dimensional scheme of finite type over k, none of whose irreducible components is proper over k, is affine. [*Hint*: Combine (Ex. 1.3) with (III, Ex. 3.1, Ex. 3.2, Ex. 4.2).]

1.5. For an effective divisor D on a curve X of genus g, show that $\dim|D| \leqslant \deg D$. Furthermore, equality holds if and only if $D = 0$ or $g = 0$.

1.6. Let X be a curve of genus g. Show that there is a finite morphism $f: X \to \mathbf{P}^1$ of degree $\leqslant g + 1$. (Recall that the *degree* of a finite morphism of curves $f: X \to Y$ is defined as the degree of the field extension $[K(X):K(Y)]$ (II, §6).)

1.7. A curve X is called *hyperelliptic* if $g \geqslant 2$ and there exists a finite morphism $f: X \to \mathbf{P}^1$ of degree 2.
   (a) If X is a curve of genus $g = 2$, show that the canonical divisor defines a complete linear system $|K|$ of degree 2 and dimension 1, without base points. Use (II, 7.8.1) to conclude that X is hyperelliptic.
   (b) Show that the curves constructed in (1.1.1) all admit a morphism of degree 2 to $\mathbf{P}^1$. Thus there exist hyperelliptic curves of any genus $g \geqslant 2$.
   Note. We will see later (Ex. 3.2) that there exist nonhyperelliptic curves. See also (V, Ex. 2.10).

1.8. p_a *of a Singular Curve*. Let X be an integral projective scheme of dimension 1 over k, and let $\tilde{X}$ be its normalization (II, Ex. 3.8). Then there is an exact sequence of sheaves on X,

$$0 \to \mathcal{O}_X \to f_* \mathcal{O}_{\tilde{X}} \to \sum_{P \in X} \tilde{\mathcal{O}}_P / \mathcal{O}_P \to 0,$$

where $\tilde{\mathcal{O}}_P$ is the integral closure of $\mathcal{O}_P$. For each $P \in X$, let $\delta_P = \text{length}(\tilde{\mathcal{O}}_P / \mathcal{O}_P)$.
   (a) Show that $p_a(X) = p_a(\tilde{X}) + \sum_{P \in X} \delta_P$. [*Hint*: Use (III, Ex. 4.1) and (III, Ex. 5.3).]
   (b) If $p_a(X) = 0$, show that X is already nonsingular and in fact isomorphic to $\mathbf{P}^1$. This strengthens (1.3.5).
   *(c) If P is a node or an ordinary cusp (I, Ex. 5.6, Ex. 5.14), show that $\delta_P = 1$. [*Hint*: Show first that δ_P depends only on the analytic isomorphism class of the singularity at P. Then compute δ_P for the node and cusp of suitable plane cubic curves. See (V, 3.9.3) for another method.]

*1.9. *Riemann–Roch for Singular Curves*. Let X be an integral projective scheme of dimension 1 over k. Let X_{reg} be the set of regular points of X.
   (a) Let $D = \sum n_i P_i$ be a divisor with support in X_{reg}, i.e., all $P_i \in X_{\text{reg}}$. Then define $\deg D = \sum n_i$. Let $\mathcal{L}(D)$ be the associated invertible sheaf on X, and show that

$$\chi(\mathcal{L}(D)) = \deg D + 1 - p_a.$$

   (b) Show that any Cartier divisor on X is the difference of two very ample Cartier divisors. (Use (II, Ex. 7.5).)
   (c) Conclude that every invertible sheaf $\mathcal{L}$ on X is isomorphic to $\mathcal{L}(D)$ for some divisor D with support in X_{reg}.
   (d) Assume furthermore that X is a locally complete intersection in some projective space. Then by (III, 7.11) the dualizing sheaf ω_X° is an invertible sheaf on X, so we can define the *canonical divisor* K to be a divisor with support in X_{reg} corresponding to ω_X°. Then the formula of (a) becomes

$$l(D) - l(K - D) = \deg D + 1 - p_a.$$

1.10. Let X be an integral projective scheme of dimension 1 over k, which is locally complete intersection, and has $p_a = 1$. Fix a point $P_0 \in X_{reg}$. Imitate (1.3.7) to show that the map $P \to \mathscr{L}(P - P_0)$ gives a one-to-one correspondence between the points of X_{reg} and the elements of the group $\mathrm{Pic}^\circ X$. This generalizes (II, 6.11.4) and (II, Ex. 6.7).

2 Hurwitz's Theorem

In this section we consider a finite morphism of curves $f: X \to Y$, and study the relation between their canonical divisors. The resulting formula involving the genus of X, the genus of Y, and the number of ramification points is called Hurwitz's theorem.

Recall that the *degree* of a finite morphism $f: X \to Y$ of curves is defined to be the degree $[K(X):K(Y)]$ of the extension of function fields (II, §6).

For any point $P \in X$ we define the *ramification index* e_P as follows. Let $Q = f(P)$, let $t \in \mathcal{O}_Q$ be a local parameter at Q, consider t as an element of $\mathcal{O}_P$ via the natural map $f^\#: \mathcal{O}_Q \to \mathcal{O}_P$, and define

$$e_P = v_P(t),$$

where v_P is the valuation associated to the valuation ring $\mathcal{O}_P$. If $e_P > 1$ we say f is *ramified* at P, and that Q is a *branch point* of f (Fig. 13). If $e_P = 1$, we say f is *unramified* at P. This definition is consistent with the earlier definition of unramified (III, Ex. 10.3) since our groundfield k is algebraically closed, and so $k(P) = k(Q)$ for any point P of X. In particular, if f is unramified everywhere, it is étale, because in any case it is flat by (III, 9.7).

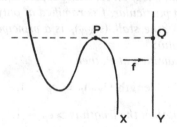

Figure 13.   A finite morphism of curves.

If char $k = 0$, or if char $k = p$, and p does not divide e_P, we say that the ramification is *tame*. If p does divide e_P, it is *wild*.

Recall that we have defined a homomorphism $f^*: \mathrm{Div}\, Y \to \mathrm{Div}\, X$ of the groups of divisors, by setting

$$f^*(Q) = \sum_{P \to Q} e_P \cdot P$$

for any point Q of Y, and extending by linearity (II, §6). If D is a divisor on Y, then $f^*(\mathscr{L}(D)) \cong \mathscr{L}(f^*D)$ (II, Ex. 6.8), so that this f^* on divisors is compatible with the homomorphism $f^*: \mathrm{Pic}\, Y \to \mathrm{Pic}\, X$ on invertible sheaves.

We say the morphism $f : X \to Y$ is *separable* if $K(X)$ is a separable field extension of $K(Y)$.

Proposition 2.1. *Let $f : X \to Y$ be a finite separable morphism of curves. Then there is an exact sequence of sheaves on X,*

$$0 \to f^*\Omega_Y \to \Omega_X \to \Omega_{X/Y} \to 0.$$

PROOF. From (II, 8.11) we have this exact sequence, but without the 0 on the left. So we have only to show that $f^*\Omega_Y \to \Omega_X$ is injective. Since both are invertible sheaves on X, it will be sufficient to show that the map is nonzero at the generic point. But since $K(X)$ is separable over $K(Y)$, the sheaf $\Omega_{X/Y}$ is zero at the generic point of X, by (II, 8.6A). Hence $f^*\Omega_Y \to \Omega_X$ is surjective at the generic point.

Since Ω_Y and Ω_X correspond to the canonical divisors on Y and X, respectively, we see that the sheaf of relative differentials $\Omega_{X/Y}$ measures their difference. So we will study this sheaf. For any point $P \in X$, let $Q = f(P)$, let t be a local parameter at Q, and let u be a local parameter at P. Then dt is a generator of the free $\mathcal{O}_Q$-module $\Omega_{Y,Q}$, and du is a generator of the free $\mathcal{O}_P$-module $\Omega_{X,P}$, by (II, 8.7) and (II, 8.8). In particular, there is a unique element $g \in \mathcal{O}_P$ such that $f^*dt = g \cdot du$. We denote this element by dt/du.

Proposition 2.2. *Let $f : X \to Y$ be a finite, separable morphism of curves. Then:*

  (a) *$\Omega_{X/Y}$ is a torsion sheaf on X, with support equal to the set of ramification points of f. In particular, f is ramified at only finitely many points;*

  (b) *for each $P \in X$, the stalk $(\Omega_{X/Y})_P$ is a principal $\mathcal{O}_P$-module of finite length equal to $v_P(dt/du)$;*

  (c) *if f is tamely ramified at P, then*

$$\text{length}(\Omega_{X/Y})_P = e_P - 1.$$

If f is wildly ramified, then the length is $> e_P - 1$.

PROOF.
  (a) The fact that $\Omega_{X/Y}$ is a torsion sheaf follows from (2.1) since $f^*\Omega_Y$ and Ω_X are both invertible sheaves on X. Now $(\Omega_{X/Y})_P = 0$ if and only if f^*dt is a generator for $\Omega_{X,P}$, using the above notation. But this happens if and only if t is a local parameter for $\mathcal{O}_P$, i.e., f is unramified at P.

  (b) Indeed, from the exact sequence of (2.1), we see that $(\Omega_{X/Y})_P \cong \Omega_{X,P}/f^*\Omega_{Y,Q}$, which is isomorphic as an $\mathcal{O}_P$-module to $\mathcal{O}_P/(dt/du)$.

  (c) If f has ramification index $e = e_P$, then we can write $t = au^e$ for some unit $a \in \mathcal{O}_P$. Then

$$dt = aeu^{e-1}du + u^e da.$$

If the ramification is tame, then e is a nonzero element of k, so we have $v_P(dt/du) = e - 1$. Otherwise $v_P(dt/du) \geqslant e$.

Definition. Let $f: X \to Y$ be a finite, separable morphism of curves. Then we define the *ramification divisor* of f to be

$$R = \sum_{P \in X} \text{length}(\Omega_{X/Y})_P \cdot P.$$

Proposition 2.3. *Let* $f: X \to Y$ *be a finite, separable morphism of curves. Let* K_X *and* K_Y *be the canonical divisors of* X *and* Y, *respectively. Then*

$$K_X \sim f^* K_Y + R.$$

PROOF. Considering the divisor R as a closed subscheme of X, we see from (2.2) that its structure sheaf $\mathcal{O}_R$ is isomorphic to $\Omega_{X/Y}$. Tensoring the exact sequence of (2.1) with Ω_X^{-1}, we can therefore write an exact sequence

$$0 \to f^* \Omega_Y \otimes \Omega_X^{-1} \to \mathcal{O}_X \to \mathcal{O}_R \to 0.$$

But by (II, 6.18), the ideal sheaf of R is isomorphic to $\mathscr{L}(-R)$, so we have

$$f^* \Omega_Y \otimes \Omega_X^{-1} \cong \mathscr{L}(-R).$$

Now the result follows from taking associated divisors. (One can also prove this proposition by applying the operation det of (II, Ex. 6.11) to the exact sequence of (2.1).)

Corollary 2.4 (Hurwitz). *Let* $f: X \to Y$ *be a finite separable morphism of curves. Let* $n = \deg f$. *Then*

$$2g(X) - 2 = n \cdot (2g(Y) - 2) + \deg R.$$

Furthermore, if f *has only tame ramification, then*

$$\deg R = \sum_{P \in X} (e_P - 1).$$

PROOF. We take the degrees of the divisors in (2.3). The canonical divisor has degree $2g - 2$ by (1.3.3); f^* multiplies degrees by n (II, 6.9); and if the ramification is tame, R has degree $\sum(e_P - 1)$ by (2.2).

We complete our discussion of finite morphisms by describing what happens in the purely inseparable case. First we define the Frobenius morphism.

Definition. Let X be a scheme, all of whose local rings are of characteristic p (i.e., contain $\mathbf{Z}/p$). We define the *Frobenius morphism* $F: X \to X$ as follows: F is the identity map on the topological space of X, and $F^\#: \mathcal{O}_X \to \mathcal{O}_X$ is the pth power map. Since the local rings are of characteristic p, $F^\#$ induces a local homomorphism on each local ring, so F is indeed a morphism.

Remark 2.4.1. If $\pi:X \to \operatorname{Spec} k$ is a scheme over a field k of characteristic p, then $F:X \to X$ is not k-linear. On the contrary, we have a commutative diagram

$$
\begin{array}{ccc}
X & \xrightarrow{\ F\ } & X \\
\Big\downarrow{\scriptstyle \pi} & & \Big\downarrow{\scriptstyle \pi} \\
\operatorname{Spec} k & \xrightarrow{\ F\ } & \operatorname{Spec} k
\end{array}
$$

with the Frobenius morphism F of $\operatorname{Spec} k$ (which corresponds to the pth power map of k to itself).

We define a new scheme over k, X_p, to be the same scheme X, but with structural morphism $F \circ \pi$. Thus k acts on $\mathcal{O}_{X_p}$ via pth powers. Then F becomes a k-linear morphism $F':X_p \to X$. We call this the *k-linear Frobenius morphism*.

Example 2.4.2. If X is a scheme over k, then X_p may or may not be isomorphic to X as a scheme over k. For example, if $X = \operatorname{Spec} k[t]$, where k is a perfect field, then X_p is isomorphic to X, because the pth power map $k \to k$ is bijective. Under this identification, the k-linear Frobenius morphism $F':X \to X$ corresponds to the homomorphism $k[t] \to k[t]$ defined by $t \to t^p$. This is the morphism given in (I, Ex. 3.2).

Example 2.4.3. If X is a curve over k, an algebraically closed field of characteristic p, then $F':X_p \to X$ is a finite morphism of degree p. It corresponds to the field inclusion $K \hookrightarrow K^{1/p}$, where K is the function field of X, and $K^{1/p}$ is the field of pth roots of elements of K in some fixed algebraic closure of K.

Proposition 2.5. *Let $f:X \to Y$ be a finite morphism of curves, and suppose that $K(X)$ is a purely inseparable field extension of $K(Y)$. Then X and Y are isomorphic as abstract schemes, and f is a composition of k-linear Frobenius morphisms. In particular, $g(X) = g(Y)$.*

PROOF. Let the degree of f be p^r. Then $K(X)^{p^r} \subseteq K(Y)$, or in other words, $K(X) \subseteq K(Y)^{1/p^r}$. On the other hand, consider the k-linear Frobenius morphisms

$$
Y_{p^r} \xrightarrow{F'} Y_{p^{r-1}} \to \ldots \to Y_p \xrightarrow{F'} Y,
$$

where for each i, $Y_{p^i} = (Y_{p^{i-1}})_p$. The composition of these is a morphism $f':Y_{p^r} \to Y$, also of degree p^r. Since $K(X) \subseteq K(Y)^{1/p^r}$, and both have the same degree over $K(Y)$, we conclude that $K(X) = K(Y)^{1/p^r}$. Since a curve is uniquely determined by its function field (I, 6.12), we have $X \cong Y_{p^r}$, and $f = f'$. Therefore X and Y are isomorphic as abstract schemes, and their genus (which does not depend on the k-structure) is the same.

Example 2.5.1. If $X = Y_p$, and $f:X \to Y$ is the k-linear Frobenius morphism, then f is ramified everywhere, with ramification index p. Indeed, f is the identity on point sets, but pth power on structure sheaves. So if $t \in \mathcal{O}_P$ is a local parameter, $f^\# t = t^p$. Since $d(t^p) = 0$, the map $f^*\Omega_Y \to \Omega_X$ is the zero map, so $\Omega_{X/Y} \cong \Omega_X$.

Example 2.5.2. If $f:X \to Y$ is separable, then the degree of the ramification divisor R is always an even number. This follows from the formula of (2.4).

Example 2.5.3. An *étale covering* of a scheme Y is a scheme X, together with a finite étale morphism $f:X \to Y$. It is called *trivial* if X is isomorphic to a finite disjoint union of copies of Y. Y is called *simply connected* if it has no nontrivial étale coverings.

Now we show that $\mathbf{P}^1$ is simply connected. Indeed, let $f:X \to \mathbf{P}^1$ be an étale covering. We may assume X is connected. Then X is smooth over k since f is étale (III, 10.1), and X is proper over k since f is finite, so X is a curve (note connected and regular imply irreducible). Again since f is étale, f is separable, so we can apply Hurwitz's theorem. Since f is unramified, $R = 0$ so we have

$$2g(X) - 2 = n(-2).$$

Since $g(X) \geqslant 0$, the only way this can happen is for $g(X) = 0$ and $n = 1$. Thus $X = \mathbf{P}^1$.

Example 2.5.4. If $f:X \to Y$ is any finite morphism of curves, then $g(X) \geqslant g(Y)$. We can factor the field extension $K(Y) \subseteq K(X)$ into a separable extension followed by a purely inseparable extension. Since the genus doesn't change for a purely inseparable extension (2.5), we reduce to the case f separable. If $g(Y) = 0$, there is nothing to prove, so we may assume $g(Y) \geqslant 1$. Then we rewrite the formula of (2.4) as

$$g(X) = g(Y) + (n - 1)(g(Y) - 1) + \tfrac{1}{2} \deg R.$$

Since $n - 1 \geqslant 0$, $g(Y) - 1 \geqslant 0$, and $\deg R \geqslant 0$, we are done. By the way, this shows also that equality occurs (for f separable) only if $n = 1$, or $g(Y) = 1$ and f is unramified.

Example 2.5.5 (Lüroth's Theorem). This says that if L is a subfield of a pure transcendental extension $k(t)$ of k, containing k, then L is also pure transcendental. We may assume that $L \neq k$, so that L has transcendence degree 1 over k. Then L is a function field of a curve Y, and the inclusion $L \subseteq k(t)$ corresponds to a finite morphism $f:\mathbf{P}^1 \to Y$. By (2.5.4) we conclude that $g(Y) = 0$, so by (1.3.5), $Y \cong \mathbf{P}^1$. Hence $L \cong k(u)$ for some u.

Note: This proof is only for k algebraically closed, but the theorem is true for any field k. An analogous result over k algebraically closed is true also in dimension 2 (V, 6.2.1). In dimension 3 the corresponding statement is false, because of the existence of nonrational unirational 3-folds—see Clemens and Griffiths [1] and Iskovskih and Manin [1].

EXERCISES

2.1. Use (2.5.3) to show that $\mathbf{P}^n$ is simply connected.

2.2. *Classification of Curves of Genus 2.* Fix an algebraically closed field k of characteristic $\neq 2$.

(a) If X is a curve of genus 2 over k, the canonical linear system $|K|$ determines a finite morphism $f: X \to \mathbf{P}^1$ of degree 2 (Ex. 1.7). Show that it is ramified at exactly 6 points, with ramification index 2 at each one. Note that f is uniquely determined, up to an automorphism of $\mathbf{P}^1$, so X determines an (unordered) set of 6 points of $\mathbf{P}^1$, up to an automorphism of $\mathbf{P}^1$.

(b) Conversely, given six distinct elements $\alpha_1, \ldots, \alpha_6 \in k$, let K be the extension of $k(x)$ determined by the equation $z^2 = (x - \alpha_1) \cdots (x - \alpha_6)$. Let $f: X \to \mathbf{P}^1$ be the corresponding morphism of curves. Show that $g(X) = 2$, the map f is the same as the one determined by the canonical linear system, and f is ramified over the six points $x = \alpha_i$ of $\mathbf{P}^1$, and nowhere else. (Cf. (II, Ex. 6.4).)

(c) Using (I, Ex. 6.6), show that if P_1, P_2, P_3 are three distinct points of $\mathbf{P}^1$, then there exists a unique $\varphi \in \operatorname{Aut} \mathbf{P}^1$ such that $\varphi(P_1) = 0$, $\varphi(P_2) = 1$, $\varphi(P_3) = \infty$. Thus in (a), if we order the six points of $\mathbf{P}^1$, and then normalize by sending the first three to $0, 1, \infty$, respectively, we may assume that X is ramified over $0, 1, \infty, \beta_1, \beta_2, \beta_3$, where $\beta_1, \beta_2, \beta_3$ are three distinct elements of k, $\neq 0, 1$.

(d) Let Σ_6 be the symmetric group on 6 letters. Define an action of Σ_6 on sets of three distinct elements $\beta_1, \beta_2, \beta_3$ of k, $\neq 0, 1$, as follows: reorder the set $0, 1, \infty, \beta_1, \beta_2, \beta_3$ according to a given element $\sigma \in \Sigma_6$, then renormalize as in (c) so that the first three become $0, 1, \infty$ again. Then the last three are the new $\beta_1', \beta_2', \beta_3'$.

(e) Summing up, conclude that there is a one-to-one correspondence between the set of isomorphism classes of curves of genus 2 over k, and triples of distinct elements $\beta_1, \beta_2, \beta_3$ of k, $\neq 0, 1$, modulo the action of Σ_6 described in (d). In particular, there are many non-isomorphic curves of genus 2. We say that curves of genus 2 depend on three parameters, since they correspond to the points of an open subset of $\mathbf{A}_k^3$ modulo a finite group.

2.3. *Plane Curves.* Let X be a curve of degree d in $\mathbf{P}^2$. For each point $P \in X$, let $T_P(X)$ be the tangent line to X at P (I, Ex. 7.3). Considering $T_P(X)$ as a point of the dual projective plane $(\mathbf{P}^2)^*$, the map $P \to T_P(X)$ gives a morphism of X to its *dual curve* X^* in $(\mathbf{P}^2)^*$ (I, Ex. 7.3). Note that even though X is nonsingular, X^* in general will have singularities. We assume char $k = 0$ below.

(a) Fix a line $L \subseteq \mathbf{P}^2$ which is not tangent to X. Define a morphism $\varphi: X \to L$ by $\varphi(P) = T_P(X) \cap L$, for each point $P \in X$. Show that φ is ramified at P if and only if either (1) $P \in L$, or (2) P is an *inflection point* of X, which means that the intersection multiplicity (I, Ex. 5.4) of $T_P(X)$ with X at P is $\geqslant 3$. Conclude that X has only finitely many inflection points.

(b) A line of $\mathbf{P}^2$ is a *multiple tangent* of X if it is tangent to X at more than one point. It is a *bitangent* if it is tangent to X at exactly two points. If L is a multiple tangent of X, tangent to X at the points $P_1, \ldots, P_r$, and if none of the P_i is an inflection point, show that the corresponding point of the dual curve X^* is an *ordinary r-fold point*, which means a point of multiplicity r with distinct tangent directions (I, Ex. 5.3). Conclude that X has only finitely many multiple tangents.

(c) Let $O \in \mathbf{P}^2$ be a point which is not on X, nor on any inflectional or multiple tangent of X. Let L be a line not containing O. Let $\psi: X \to L$ be the morphism defined by projection from O. Show that ψ is ramified at a point $P \in X$ if and only if the line OP is tangent to X at P, and in that case the ramification index is 2. Use Hurwitz's theorem and (I, Ex. 7.2) to conclude that there are exactly $d(d-1)$ tangents of X passing through O. Hence the degree of the dual curve (sometimes called the *class* of X) is $d(d-1)$.

(d) Show that for all but a finite number of points of X, a point O of X lies on exactly $(d+1)(d-2)$ tangents of X, not counting the tangent at O.

(e) Show that the degree of the morphism φ of (a) is $d(d-1)$. Conclude that if $d \geqslant 2$, then X has $3d(d-2)$ inflection points, properly counted. (If $T_P(X)$ has intersection multiplicity r with X at P, then P should be counted $r-2$ times as an inflection point. If $r=3$ we call it an *ordinary inflection point*.) Show that an ordinary inflection point of X corresponds to an ordinary cusp of the dual curve X^*.

(f) Now let X be a plane curve of degree $d \geqslant 2$, and assume that the dual curve X^* has only nodes and ordinary cusps as singularities (which should be true for sufficiently general X). Then show that X has exactly $\frac{1}{2}d(d-2)(d-3)(d+3)$ bitangents. [*Hint*: Show that X is the normalization of X^*. Then calculate $p_a(X^*)$ two ways: once as a plane curve of degree $d(d-1)$, and once using (Ex. 1.8).]

(g) For example, a plane cubic curve has exactly 9 inflection points, all ordinary. The line joining any two of them intersects the curve in a third one.

(h) A plane quartic curve has exactly 28 bitangents. (This holds even if the curve has a tangent with four-fold contact, in which case the dual curve X^* has a tacnode.)

2.4. *A Funny Curve in Characteristic p.* Let X be the plane quartic curve $x^3y + y^3z + z^3x = 0$ over a field of characteristic 3. Show that X is nonsingular, every point of X is an inflection point, the dual curve X^* is isomorphic to X, but the natural map $X \to X^*$ is purely inseparable.

2.5. *Automorphisms of a Curve of Genus $\geqslant 2$.* Prove the theorem of Hurwitz [1] that a curve X of genus $g \geqslant 2$ over a field of characteristic 0 has at most $84(g-1)$ automorphisms. We will see later (Ex. 5.2) or (V, Ex. 1.11) that the group $G = \operatorname{Aut} X$ is finite. So let G have order n. Then G acts on the function field $K(X)$. Let L be the fixed field. Then the field extension $L \subseteq K(X)$ corresponds to a finite morphism of curves $f: X \to Y$ of degree n.

(a) If $P \in X$ is a ramification point, and $e_P = r$, show that $f^{-1}f(P)$ consists of exactly n/r points, each having ramification index r. Let $P_1, \ldots, P_s$ be a maximal set of ramification points of X lying over distinct points of Y, and let $e_{P_i} = r_i$. Then show that Hurwitz's theorem implies that

$$(2g-2)/n = 2g(Y) - 2 + \sum_{i=1}^{s} (1 - 1/r_i).$$

(b) Since $g \geq 2$, the left hand side of the equation is > 0. Show that if $g(Y) \geq 0$, $s \geq 0, r_i \geq 2, i = 1, \ldots, s$ are integers such that

$$2g(Y) - 2 + \sum_{i=1}^{s} (1 - 1/r_i) > 0,$$

then the minimum value of this expression is $1/42$. Conclude that $n \leq 84(g - 1)$. See (Ex. 5.7) for an example where this maximum is achieved.

Note: It is known that this maximum is achieved for infinitely many values of g (Macbeath [1]). Over a field of characteristic $p > 0$, the same bound holds, provided $p > g + 1$, with one exception, namely the hyperelliptic curve $y^2 = x^p - x$, which has $p = 2g + 1$ and $2p(p^2 - 1)$ automorphisms (Roquette [1]). For other bounds on the order of the group of automorphisms in characteristic p, see Singh [1] and Stichtenoth [1].

2.6. f_* *for Divisors.* Let $f : X \to Y$ be a finite morphism of curves of degree n. We define a homomorphism $f_* : \mathrm{Div}\, X \to \mathrm{Div}\, Y$ by $f_*(\sum n_i P_i) = \sum n_i f(P_i)$ for any divisor $D = \sum n_i P_i$ on X.

(a) For any locally free sheaf $\mathscr{E}$ on Y, of rank r, we define $\det \mathscr{E} = \wedge^r \mathscr{E} \in \mathrm{Pic}\, Y$ (II, Ex. 6.11). In particular, for any invertible sheaf $\mathscr{M}$ on X, $f_*\mathscr{M}$ is locally free of rank n on Y, so we can consider $\det f_*\mathscr{M} \in \mathrm{Pic}\, Y$. Show that for any divisor D on X,

$$\det(f_*\mathscr{L}(D)) \cong (\det f_*\mathcal{O}_X) \otimes \mathscr{L}(f_*D).$$

Note in particular that $\det(f_*\mathscr{L}(D)) \neq \mathscr{L}(f_*D)$ in general! [*Hint*: First consider an effective divisor D, apply f_* to the exact sequence $0 \to \mathscr{L}(-D) \to \mathcal{O}_X \to \mathcal{O}_D \to 0$, and use (II, Ex. 6.11).]

(b) Conclude that f_*D depends only on the linear equivalence class of D, so there is an induced homomorphism $f_* : \mathrm{Pic}\, X \to \mathrm{Pic}\, Y$. Show that $f_* \circ f^* : \mathrm{Pic}\, Y \to \mathrm{Pic}\, Y$ is just multiplication by n.

(c) Use duality for a finite flat morphism (III, Ex. 6.10) and (III, Ex. 7.2) to show that

$$\det f_*\Omega_X \cong (\det f_*\mathcal{O}_X)^{-1} \otimes \Omega_Y^{\otimes n}.$$

(d) Now assume that f is separable, so we have the ramification divisor R. We define the *branch divisor* B to be the divisor f_*R on Y. Show that

$$(\det f_*\mathcal{O}_X)^2 \cong \mathscr{L}(-B).$$

2.7. *Étale Covers of Degree* 2. Let Y be a curve over a field k of characteristic $\neq 2$. We show there is a one-to-one correspondence between finite étale morphisms $f : X \to Y$ of degree 2, and 2-torsion elements of $\mathrm{Pic}\, Y$, i.e., invertible sheaves $\mathscr{L}$ on Y with $\mathscr{L}^2 \cong \mathcal{O}_Y$.

(a) Given an étale morphism $f : X \to Y$ of degree 2, there is a natural map $\mathcal{O}_Y \to f_*\mathcal{O}_X$. Let $\mathscr{L}$ be the cokernel. Then $\mathscr{L}$ is an invertible sheaf on Y, $\mathscr{L} \cong \det f_*\mathcal{O}_X$, and so $\mathscr{L}^2 \cong \mathcal{O}_Y$ by (Ex. 2.6). Thus an étale cover of degree 2 determines a 2-torsion element in $\mathrm{Pic}\, Y$.

(b) Conversely, given a 2-torsion element $\mathscr{L}$ in $\mathrm{Pic}\, Y$, define an $\mathcal{O}_Y$-algebra structure on $\mathcal{O}_Y \oplus \mathscr{L}$ by $\langle a, b \rangle \cdot \langle a', b' \rangle = \langle aa' + \varphi(b \otimes b'), ab' + a'b \rangle$, where φ is an isomorphism of $\mathscr{L} \otimes \mathscr{L} \to \mathcal{O}_Y$. Then take $X = \mathbf{Spec}(\mathcal{O}_Y \oplus \mathscr{L})$ (II, Ex. 5.17). Show that X is an étale cover of Y.

(c) Show that these two processes are inverse to each other. [*Hint*: Let $\tau : X \to X$ be the involution which interchanges the points of each fibre of f. Use the trace map $a \mapsto a + \tau(a)$ from $f_* \mathcal{O}_X \to \mathcal{O}_Y$ to show that the sequence of $\mathcal{O}_Y$-modules in (a)

$$0 \to \mathcal{O}_Y \to f_* \mathcal{O}_X \to \mathcal{L} \to 0$$

is split exact.

Note. This is a special case of the more general fact that for $(n, \operatorname{char} k) = 1$, the étale Galois covers of Y with group $\mathbf{Z}/n\mathbf{Z}$ are classified by the étale cohomology group $H^1_{\mathrm{et}}(Y, \mathbf{Z}/n\mathbf{Z})$, which is equal to the group of n-torsion points of Pic Y. See Serre [6].

3 Embeddings in Projective Space

In this section we study embeddings of a curve in projective space. We will show that any curve can be embedded in $\mathbf{P}^3$. Furthermore, any curve can be mapped birationally into $\mathbf{P}^2$ in such a way that the image has at most nodes as singularities.

Recall that an invertible sheaf $\mathcal{L}$ on a curve X is *very ample* (II, §5) if it is isomorphic to $\mathcal{O}_X(1)$ for some immersion of X in a projective space. It is *ample* (II, §7) if for any coherent sheaf $\mathcal{F}$ on X, the sheaf $\mathcal{F} \otimes \mathcal{L}^n$ is generated by global sections for $n \gg 0$. We have seen that $\mathcal{L}$ is ample if and only if $\mathcal{L}^n$ is very ample for some $n > 0$ (II, 7.6). If D is a divisor on X, we will say D is *ample* or *very ample* if $\mathcal{L}(D)$ is.

Recall that a *linear system* is a set $\mathfrak{d}$ of effective divisors, which forms a linear subspace of a complete linear system $|D|$. A point P is a *base point* of the linear system $\mathfrak{d}$ if $P \in \operatorname{Supp} D$ for all $D \in \mathfrak{d}$. We have seen that a complete linear system $|D|$ is base-point free if and only if $\mathcal{L}(D)$ is generated by global sections (II, 7.8).

Our first result is a reinterpretation in the case of curves of the criterion of (II, §7) for when a linear system gives rise to a closed immersion into projective space.

Proposition 3.1. *Let D be a divisor on a curve X. Then:*

(a) *the complete linear system $|D|$ has no base points if and only if for every point $P \in X$,*

$$\dim|D - P| = \dim|D| - 1;$$

(b) *D is very ample if and only if for every two points $P, Q \in X$ (including the case $P = Q$),*

$$\dim|D - P - Q| = \dim|D| - 2.$$

PROOF. First we consider the exact sequence of sheaves

$$0 \to \mathcal{L}(D - P) \to \mathcal{L}(D) \to k(P) \to 0.$$

Taking global sections, we have

$$0 \to \Gamma(X, \mathcal{L}(D - P)) \to \Gamma(X, \mathcal{L}(D)) \to k,$$

so in any case, we see that $\dim|D - P|$ is equal to either $\dim|D|$ or $\dim|D| - 1$. Furthermore, sending a divisor E to $E + P$ defines a linear map

$$\varphi:|D - P| \to |D|$$

which is clearly injective. Therefore, the dimensions of these two linear systems are equal if and only if φ is surjective. On the other hand, φ is surjective if and only if P is a base point of $|D|$, so this proves (a).

To prove (b), we may assume that $|D|$ has no base points. Indeed, this is true if D is very ample. On the other hand, if D satisfies the condition of (b), then we must a fortiori have

$$\dim|D - P| = \dim|D| - 1$$

for every $P \in X$, so $|D|$ has no base points.

This being the case, $|D|$ determines a morphism of X to $\mathbf{P}^n$ (II, 7.1) and (II, 7.8.1), so the question is whether that morphism is a closed immersion. We use the criterion of (II, 7.3) and (II, 7.8.2), so we have to see whether $|D|$ separates points and separates tangent vectors. The first condition says that for any two distinct points $P,Q \in X$, Q is not a base point of $|D - P|$. By (a) this is equivalent to saying

$$\dim|D - P - Q| = \dim|D| - 2.$$

The second condition says that for any point $P \in X$, there is a divisor $D' \in |D|$ such that P occurs with multiplicity 1 in D', because $\dim T_P(X) = 1$, and $\dim T_P(D') = 0$ if P has multiplicity 1 in D', 1 if P has higher multiplicity. But this just says P is not a base point of $|D - P|$, or, using (a) again,

$$\dim|D - 2P| = \dim|D| - 2.$$

Thus our result follows from (II, 7.3).

Corollary 3.2. *Let D be a divisor on a curve X of genus g.*
  (a) *If $\deg D \geqslant 2g$, then $|D|$ has no base points.*
  (b) *If $\deg D \geqslant 2g + 1$, then D is very ample.*

PROOF. In case (a), both D and $D - P$ are nonspecial (1.3.4), so by Riemann–Roch, $\dim|D - P| = \dim|D| - 1$. In case (b), D and $D - P - Q$ are both nonspecial, so $\dim|D - P - Q| = \dim|D| - 2$ again by Riemann–Roch.

Corollary 3.3. *A divisor D on a curve X is ample if and only if $\deg D > 0$.*

PROOF. If D is ample, some multiple is very ample (II, 7.6), so $nD \sim H$ where H is a hyperplane section for a projective embedding, so $\deg H > 0$, hence $\deg D > 0$. Conversely, if $\deg D > 0$, then for $n \gg 0$, $\deg nD \geqslant 2g(X) + 1$, so by (3.2), nD is very ample, and so D is ample (II, 7.6).

Example 3.3.1. If $g = 0$, then D is ample $\Leftrightarrow$ very ample $\Leftrightarrow \deg D > 0$. Since $X \cong \mathbf{P}^1$ (1.3.5), this is just (II, 7.6.1).

Example 3.3.2. Let X be a curve, and let D be a very ample divisor on X, corresponding to a closed immersion $\varphi : X \to \mathbf{P}^n$. Then the degree of $\varphi(X)$, as defined in (I, §7) for a projective variety, is just equal to deg D (II, Ex. 6.2).

Example 3.3.3. Let X be an elliptic curve, i.e., $g = 1$ (1.3.6). Then any divisor D of degree 3 is very ample. Such a divisor is nonspecial, so by Riemann–Roch, $\dim |D| = 2$. Thus we see that any elliptic curve can be embedded in $\mathbf{P}^2$ as a cubic curve. (Conversely, of course, any nonsingular plane cubic is elliptic, by the genus formula (I, Ex. 7.2).)

In the case $g = 1$ we can actually say D very ample $\Leftrightarrow$ deg $D \geqslant 3$. Because if deg $D = 2$, then by Riemann–Roch, $\dim |D| = 1$, so $|D|$ defines a morphism of X to $\mathbf{P}^1$, which cannot be a closed immersion.

Example 3.3.4. If $g = 2$, then any divisor D of degree 5 is very ample. By Riemann–Roch, $\dim |D| = 3$, so any curve of genus 2 can be embedded in $\mathbf{P}^3$ as a curve of degree 5.

Example 3.3.5. The result of (3.2) is not the best possible in general. For example, if X is a plane curve of degree 4, then $D = X.H$ is a very ample divisor of degree 4, but $g = 3$ so $2g + 1 = 7$.

Our next objective is to show that any curve can be embedded in $\mathbf{P}^3$. For this purpose we consider a curve $X \subseteq \mathbf{P}^n$, take a point $O \notin X$, and project X from O into $\mathbf{P}^{n-1}$ (I, Ex. 3.14). This gives a morphism of X into $\mathbf{P}^{n-1}$, and we investigate when it is a closed immersion.

If P,Q are two distinct points of X, we define the *secant line* determined by P and Q to be the line in $\mathbf{P}^n$ joining P and Q. If P is a point of X, we define the *tangent line* to X at P to be the unique line $L \subseteq \mathbf{P}^n$ passing through P, whose tangent space $T_P(L)$ is equal to $T_P(X)$ as a subspace of $T_P(\mathbf{P}^n)$.

Proposition 3.4. *Let X be a curve in $\mathbf{P}^n$, let O be a point not on X, and let $\varphi : X \to \mathbf{P}^{n-1}$ be the morphism determined by projection from O. Then φ is a closed immersion if and only if*

(1) *O is not on any secant line of X, and*
(2) *O is not on any tangent line of X.*

PROOF. The morphism φ corresponds (II, 7.8.1) to the linear system cut out on X by the hyperplanes H of $\mathbf{P}^n$ passing through O. So φ is a closed immersion if and only if this linear system separates points and separates tangent vectors on X (II, 7.8.2). If P,Q are two distinct points on X, then φ separates them if and only if there is an H containing O and P, but not Q. This is possible if and only if O is not on the line PQ. If $P \in X$, then φ separates tangent vectors at P if and only if there is an H containing O and P, and meeting X at P with multiplicity 1. This is possible if and only if O is not on the tangent line at P.

Proposition 3.5. *If X is a curve in* $\mathbf{P}^n$, *with* $n \geqslant 4$, *then there is a point* $O \notin X$ *such that the projection from* O *gives a closed immersion of* X *into* $\mathbf{P}^{n-1}$.

PROOF. Let Sec X be the union of all secant lines of X. We call this the *secant variety* of X. It is a locally closed subset of $\mathbf{P}^n$, of dimension $\leqslant 3$, since (at least locally) it is the image of a morphism from $(X \times X - \varDelta) \times \mathbf{P}^1$ to $\mathbf{P}^n$ which sends $\langle P,Q,t \rangle$ to the point t on the secant line through P and Q, suitably parametrized.

Let Tan X, the *tangent variety* of X, be the union of all tangent lines of X. It is a closed subset of $\mathbf{P}^n$, of dimension $\leqslant 2$, because it is locally an image of $X \times \mathbf{P}^1$.

Since $n \geqslant 4$, Sec $X \cup$ Tan $X \neq \mathbf{P}^n$, so we can find plenty of points O which do not lie on any secant or tangent of X. Then the projection from O gives the required closed immersion, by (3.4).

Corollary 3.6. *Any curve can be embedded in* $\mathbf{P}^3$.

PROOF. First embed X in any projective space $\mathbf{P}^n$. For example, take a divisor D of degree $d \geqslant 2g + 1$ and use (3.2). Since D is very ample, the complete linear system $|D|$ determines an embedding of X in $\mathbf{P}^n$ with $n = \dim|D|$. If $n \leqslant 3$, we can consider $\mathbf{P}^n$ as a subspace of $\mathbf{P}^3$, so there is nothing to prove. If $n \geqslant 4$, we use (3.5) repeatedly to project from points until we have X embedded in $\mathbf{P}^3$.

Next we study the projection of a curve X in $\mathbf{P}^3$ to $\mathbf{P}^2$. In general the secant variety will fill up all of $\mathbf{P}^3$, so we cannot avoid all the secants, and the projected curve will be singular. However, we will see that it is possible to choose the center of projection O so that the resulting morphism φ from X to $\mathbf{P}^2$ is birational onto its image, and the image $\varphi(X)$ has at most nodes as singularities.

Recall (I, Ex. 5.6) that a *node* is a singular point of a plane curve of multiplicity 2, with distinct tangent directions. We define a *multisecant* of X to be a line in $\mathbf{P}^3$ which meets X in three or more distinct points. A *secant with coplanar tangent lines* is a secant joining two points P,Q of X, whose tangent lines L_P, L_Q lie in the same plane, or equivalently, such that L_P meets L_Q.

Proposition 3.7. *Let* X *be a curve in* $\mathbf{P}^3$, *let* O *be a point not on* X, *and let* $\varphi: X \to \mathbf{P}^2$ *be the morphism determined by projection from* O. *Then* φ *is birational onto its image and* $\varphi(X)$ *has at most nodes as singularities, if and only if*

(1) *O lies on only finitely many secants of X,*

(2) *O is not on any tangent line of X,*

(3) *O is not on any multisecant of X, and*

(4) *O is not on any secant with coplanar tangent lines.*

PROOF. Going back to the proof of (II, 7.3), condition (1) says that φ is one-to-one almost everywhere, hence birational. When O does lie on a secant line, conditions (2), (3), (4) tell us that line meets X in exactly two points P,Q, it is not tangent to X at either one, and the tangent lines at P,Q are mapped to distinct lines in $\mathbf{P}^2$. Hence the image $\varphi(X)$ has a node at that point.

To show that a point O exists satisfying (1)–(4) of (3.7), we will count the dimensions of the bad points, as in the proof of (3.5). The hard part is to show that not every secant is a multisecant, and not every secant has coplanar tangent lines. Over $\mathbf{C}$, one could see this from differential geometry. However, we give a different proof, valid in all characteristics, which is achieved by an interesting application of Hurwitz's theorem.

Proposition 3.8. *Let X be a curve in $\mathbf{P}^3$, which is not contained in any plane. Suppose either*
   (a) *every secant of X is a multisecant, or*
   (b) *for any two points $P,Q \in X$, the tangent lines L_P,L_Q are coplanar.*
   Then there is a point $A \in \mathbf{P}^3$, which lies on every tangent line of X.

PROOF. First we show that (a) implies (b). Fix a point R in X, and consider the morphism $\psi: X - R \to \mathbf{P}^2$ induced by projection from R. Since every secant is a multisecant, ψ is a many-to-one map. If ψ is inseparable, then for any $P \in X$, the tangent line L_P at X passes through R. This gives (b) and our conclusion immediately, so we may assume that each such ψ is separable. In that case, let T be a nonsingular point of $\psi(X)$ over which ψ is not ramified. If $P,Q \in \psi^{-1}(T)$, then the tangent lines L_P,L_Q to X are projected into the tangent line L_T to $\psi(X)$ at T. So L_P and L_Q are both in the plane spanned by R and L_T, hence coplanar.

Thus we have shown that for any R, and for almost all P,Q such that P,Q,R are collinear, L_P and L_Q are coplanar. Therefore, there is an open set of $\langle P,Q \rangle$ in $X \times X$ for which L_P and L_Q are coplanar. But the property of L_P and L_Q being coplanar is a closed condition, so we conclude that for all $P,Q \in X$, L_P and L_Q are coplanar. This is (b).

Now assume (b). Take any two points $P,Q \in X$ with distinct tangents, and let $A = L_P \cap L_Q$. By hypothesis, X is not contained in any plane, so in particular, if π is the plane spanned by L_P and L_Q, then $X \cap \pi$ is a finite set of points. For any point $R \in X - X \cap \pi$, the tangent line L_R must meet both L_P and L_Q. But since $L_R \nsubseteq \pi$, it must pass through A. So there is an open set of X consisting of points R such that $A \in L_R$. Since this is a closed condition, we conclude that $A \in L_R$ for all $R \in X$.

Definition. A curve X in $\mathbf{P}^n$ is *strange* if there is a point A which lies on all the tangent lines of X.

Example 3.8.1. $\mathbf{P}^1$ is strange. Indeed, the tangent line at any point is the same $\mathbf{P}^1$, so any point $A \in \mathbf{P}^1$ will do.

Example 3.8.2. A conic in $\mathbf{P}^2$ over a field of characteristic 2 is strange. For example, consider the conic $y = x^2$. Then $dy/dx \equiv 0$, so all the tangent lines are horizontal, so they all pass through the point at infinity on the x-axis.

Theorem 3.9 (Samuel [2]). *The only strange curves in any $\mathbf{P}^n$ are the line* (3.8.1) *and the conic in characteristic 2* (3.8.2).

PROOF. By projecting down if necessary (3.5) we may assume that X lies in $\mathbf{P}^3$. Choose an $\mathbf{A}^3$ in $\mathbf{P}^3$ with affine coordinates x, y, z in such a way that

(1) A is the point at infinity on the x-axis,
(2) if $A \in X$, then its tangent line L_A is not in the xz-plane,
(3) the z-axis does not meet X,
(4) X does not meet the line at infinity of the xz-plane, except possibly at A (Fig. 14).

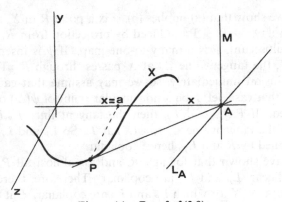

Figure 14. Proof of (3.9).

First we project from A to the yz-plane. Since A lies on every tangent line to X, the corresponding morphism from X to $\mathbf{P}^2$ is ramified everywhere. So either the image is a point (in which case X is a line), or it is inseparable (2.2). We conclude that the functions y and z restricted to X lie in $K(X)^p$, where char $k = p > 0$.

Next, we project from the z-axis to the line M at infinity in the xy-plane. In other words, for each point $P \in X$, we define $\varphi(P)$ to be the intersection of the plane spanned by P and the z-axis with the line M. This gives a morphism $\varphi: X \to M$ of degree $d = \deg X$. Note that φ is ramified exactly at the points of X which lie in the finite part of the xz-plane, but not at A.

We will apply Hurwitz's theorem (2.4) to the morphism φ. For any point $P \in X \cap xz$-plane, we take $u = x - a$ as a local coordinate, where $a \in k$, $a \neq 0$. We take $t = y/x$ as a local coordinate at A on M. Then by (2.2) we have to calculate $v_P(dt/du)$. Write $x = u + a$, so $t = y(u + a)^{-1}$. Since $y \in K(X)^p$, we have $dy/du = 0$, so

$$dt/du = -y(u + a)^{-2}.$$

But $u + a$ is a unit in the local ring $\mathcal{O}_P$, so

$$v_P(dt/du) = v_P(y).$$

If we let $P_1, \ldots, P_r$ be all the finite points of $X \cap xz$-plane, then Hurwitz's theorem tells us that

$$2g - 2 = -2d + \sum_{i=1}^{r} v_{P_i}(y).$$

Now we consider two cases.

Case 1. If $A \notin X$, the xz-plane meets X only at the points P_i. Since this plane is defined by the equation $y = 0$, we can compute the degree of X as the number of intersections of X with this plane, namely

$$d = \sum_{i=1}^{r} v_{P_i}(y).$$

Substituting in the above, we have

$$2g - 2 = -d$$

which is possible only if $g = 0$ and $d = 2$. Thus $X \cong \mathbf{P}^1$ as an abstract curve (1.3.5), and its embedding is by a divisor D of degree 2. We have $\dim|D| = 2$ by Riemann–Roch, so X is a conic in a plane $\mathbf{P}^2$. For the conic to be strange, we must have char $k = 2$.

Case 2. If $A \in X$, then by condition (2) the xz-plane meets X transversally at A, so we see similarly

$$d = \sum_{i=1}^{r} v_{P_i}(y) + 1.$$

So

$$2g - 2 = -d - 1$$

which implies $g = 0$, $d = 1$. This is the line.

Theorem 3.10. *Let X be a curve in $\mathbf{P}^3$. Then there is a point $O \notin X$ such that the projection from O determines a birational morphism φ from X to its image in $\mathbf{P}^2$, and that image has at most nodes for singularities.*

PROOF. If X is contained in a plane already, any O not in that plane will do. So we assume X is not contained in any plane. Then in particular, X is

313

neither a line nor a conic, so by (3.9), X is not strange. Therefore, by (3.8), X has a secant which is not a multisecant, and it has a secant without co-planar tangents. Since the same must be true for nearby secants, we see that there is an open subset of $X \times X$ consisting of pairs $\langle P,Q \rangle$ such that the secant line through P,Q is not a multisecant and does not have coplanar tangents. Hence the subset of $X \times X$ consisting of pairs $\langle P,Q \rangle$ where the secant is a multisecant or has coplanar tangents is a proper subset, has dimension $\leqslant 1$, and so the union in $\mathbf{P}^3$ of the corresponding secant lines has dimension $\leqslant 2$. Combining with the fact that the tangent variety to X has dimension $\leqslant 2$ (see (3.5)), we see that there is an open subset of $\mathbf{P}^3$ consisting of points O which satisfy (2), (3), and (4) of (3.7).

To complete the proof, by (3.7), we must show that O can be chosen to lie on only finitely many secants of X. For this we consider the morphism $(X \times X - \Delta) \times \mathbf{P}^1 \to \mathbf{P}^3$ (defined at least locally) which sends $\langle P,Q,t \rangle$ to the point t on the secant line through P and Q. If the image has dimension < 3, then we can choose O lying on no secant. If the image has dimension $= 3$, then since it is a morphism between two varieties of the same dimension, we can apply (II, Ex. 3.7), and find there is an open set of points in $\mathbf{P}^3$ over which the fibre is finite. These points lie on only finitely many secants, so we are done.

Corollary 3.11. *Any curve is birationally equivalent to a plane curve with at most nodes as singularities.*

PROOF. Combine (3.6) with (3.10).

Remark 3.11.1. In view of (3.11), one way to approach the classification problem for all curves is to study the family of plane curves of degree d with r nodes, for any given d and r. The family of all plane curves of degree d is a linear system of dimension $\frac{1}{2}d(d + 3)$, so it is parametrized by a projective space of that dimension. Inside that projective space, the (irreducible) curves with r nodes form a locally closed subset $V_{d,r}$. If X is such a curve, then the genus g of its normalization $\tilde{X}$ is given by

$$g = \tfrac{1}{2}(d - 1)(d - 2) - r$$

because of (Ex. 1.8). So in order for $V_{d,r}$ to be nonempty, we must have

$$0 \leqslant r \leqslant \tfrac{1}{2}(d - 1)(d - 2).$$

Furthermore, both extremes are possible. We have seen by Bertini's theorem (II, 8.20.2) that for any d, there are irreducible nonsingular curves of degree d in $\mathbf{P}^2$, so this gives the case $r = 0$. On the other hand, for any d, we can embed $\mathbf{P}^1$ in $\mathbf{P}^d$ as a curve of degree d (Ex. 3.4), and then project it into $\mathbf{P}^2$ by (3.5) and (3.10), to get a curve X of degree d in $\mathbf{P}^2$ having only nodes, and with $g(\tilde{X}) = 0$. This gives $r = \frac{1}{2}(d - 1)(d - 2)$.

But the general problem of the structure of the $V_{d,r}$ is very difficult. Severi [2, Anhang F] states that for every d,r, satisfying $0 \leqslant r \leqslant \frac{1}{2}(d - 1)(d - 2)$, the algebraic set $V_{d,r}$ is irreducible and nonempty of dimension $\frac{1}{2}d(d+3)-r$, but a complete proof was given only recently by Joe Harris.

EXERCISES

3.1. If X is a curve of genus 2, show that a divisor D is very ample $\Leftrightarrow$ deg $D \geqslant 5$. This strengthens (3.3.4).

3.2. Let X be a plane curve of degree 4.
  (a) Show that the effective canonical divisors on X are exactly the divisors $X.L$, where L is a line in $\mathbf{P}^2$.
  (b) If D is any effective divisor of degree 2 on X, show that dim$|D|=0$.
  (c) Conclude that X is not hyperelliptic (Ex. 1.7).

3.3. If X is a curve of genus $\geqslant 2$ which is a complete intersection (II, Ex. 8.4) in some $\mathbf{P}^n$, show that the canonical divisor K is very ample. Conclude that a curve of genus 2 can never be a complete intersection in any $\mathbf{P}^n$. Cf. (Ex. 5.1).

3.4. Let X be the d-uple embedding (I, Ex. 2.12) of $\mathbf{P}^1$ in $\mathbf{P}^d$, for any $d \geqslant 1$. We call X the *rational normal curve of degree d* in $\mathbf{P}^d$.
  (a) Show that X is projectively normal, and that its homogeneous ideal can be generated by forms of degree 2.
  (b) If X is any curve of degree d in $\mathbf{P}^n$, with $d \leqslant n$, which is not contained in any $\mathbf{P}^{n-1}$, show that in fact $d = n$, $g(X) = 0$, and X differs from the rational normal curve of degree d only by an automorphism of $\mathbf{P}^d$. Cf. (II. 7.8.5).
  (c) In particular, any curve of degree 2 in any $\mathbf{P}^n$ is a conic in some $\mathbf{P}^2$.
  (d) A curve of degree 3 in any $\mathbf{P}^n$ must be either a plane cubic curve, or the twisted cubic curve in $\mathbf{P}^3$.

3.5. Let X be a curve in $\mathbf{P}^3$, which is not contained in any plane.
  (a) If $O \notin X$ is a point, such that the projection from O induces a birational morphism φ from X to its image in $\mathbf{P}^2$, show that $\varphi(X)$ *must* be singular. [*Hint*: Calculate dim $H^0(X, \mathcal{O}_X(1))$ two ways.]
  (b) If X has degree d and genus g, conclude that $g < \frac{1}{2}(d-1)(d-2)$. (Use (Ex. 1.8).)
  (c) Now let $\{X_t\}$ be the flat family of curves induced by the projection (III, 9.8.3) whose fibre over $t = 1$ is X, and whose fibre X_0 over $t = 0$ is a scheme with support $\varphi(X)$. Show that X_0 always has nilpotent elements. Thus the example (III, 9.8.4) is typical.

3.6. *Curves of Degree* 4.
  (a) If X is a curve of degree 4 in some $\mathbf{P}^n$, show that either

  (1) $g = 0$, in which case X is either the rational normal quartic in $\mathbf{P}^4$ (Ex. 3.4) or the rational quartic curve in $\mathbf{P}^3$ (II, 7.8.6), or
  (2) $X \subseteq \mathbf{P}^2$, in which case $g = 3$, or
  (3) $X \subseteq \mathbf{P}^3$ and $g = 1$.

  (b) In the case $g = 1$, show that X is a complete intersection of two irreducible quadric surfaces in $\mathbf{P}^3$ (I, Ex. 5.11). [*Hint*: Use the exact sequence $0 \to \mathscr{I}_X \to \mathcal{O}_{\mathbf{P}^3} \to \mathcal{O}_X \to 0$ to compute dim $H^0(\mathbf{P}^3, \mathscr{I}_X(2))$, and thus conclude that X is contained in at least two irreducible quadric surfaces.]

3.7. In view of (3.10), one might ask conversely, is every plane curve with nodes a projection of a nonsingular curve in $\mathbf{P}^3$? Show that the curve $xy + x^4 + y^4 = 0$ (assume char $k \neq 2$) gives a counterexample.

3.8. We say a (singular) integral curve in $\mathbf{P}^n$ is *strange* if there is a point which lies on all the tangent lines at nonsingular points of the curve.
  (a) There are many singular strange curves, e.g., the curve given parametrically by $x = t, y = t^p, z = t^{2p}$ over a field of characteristic $p > 0$.
  (b) Show, however, that if char $k = 0$, there aren't even any singular strange curves besides $\mathbf{P}^1$.

3.9. Prove the following lemma of Bertini: if X is a curve of degree d in $\mathbf{P}^3$, not contained in any plane, then for almost all planes $H \subseteq \mathbf{P}^3$ (meaning a Zariski open subset of the dual projective space $(\mathbf{P}^3)^*$), the intersection $X \cap H$ consists of exactly d distinct points, no three of which are collinear.

3.10. Generalize the statement that "not every secant is a multisecant" as follows. If X is a curve in $\mathbf{P}^n$, not contained in any $\mathbf{P}^{n-1}$, and if char $k = 0$, show that for almost all choices of $n - 1$ points $P_1, \ldots, P_{n-1}$ on X, the linear space L^{n-2} spanned by the P_i does not contain any further points of X.

3.11 (a) If X is a nonsingular variety of dimension r in $\mathbf{P}^n$, and if $n > 2r + 1$, show that there is a point $O \notin X$, such that the projection from O induces a closed immersion of X into $\mathbf{P}^{n-1}$.
  (b) If X is the Veronese surface in $\mathbf{P}^5$, which is the 2-uple embedding of $\mathbf{P}^2$ (I, Ex. 2.13), show that each point of every secant line of X lies on infinitely many secant lines. Therefore, the secant variety of X has dimension 4, and so in this case there is a projection which gives a closed immersion of X into $\mathbf{P}^4$ (II, Ex. 7.7). (A theorem of Severi [1] states that the Veronese surface is the only surface in $\mathbf{P}^5$ for which there is a projection giving a closed immersion into $\mathbf{P}^4$. Usually one obtains a finite number of double points with transversal tangent planes.)

3.12. For each value of $d = 2,3,4,5$ and r satisfying $0 \leqslant r \leqslant \frac{1}{2}(d - 1)(d - 2)$, show that there exists an irreducible plane curve of degree d with r nodes and no other singularities.

4 Elliptic Curves

The theory of elliptic curves (curves of genus 1) is varied and rich, and provides a good example of the profound connections between abstract algebraic geometry, complex analysis, and number theory. In this section we will discuss briefly a number of topics concerning elliptic curves, to give some idea of this theory. First we define the *j*-invariant, which classifies elliptic curves up to isomorphism. Then we discuss the group structure on the curve, and show that the elliptic curve is its own Jacobian variety. Next we recall without proof the main results of the theory of elliptic functions of a complex variable, and deduce various results about elliptic curves over **C**. Then we define the Hasse invariant of a curve over a field of characteristic p, and finally we consider the group of rational points of a curve defined over **Q**.

For simplicity, we will omit the case of a ground field k of characteristic 2. Most of the results of this section remain true, but the proofs require special care. See, e.g., Tate [3] or the "formulaire" of Deligne and Tate in Birch and Kuyk [1].

The j-Invariant

Our first topic is to define the j-invariant of an elliptic curve, and to show that it classifies elliptic curves up to isomorphism. Since j can be any element of the ground field k, this will show that the affine line $\mathbf{A}_k^1$ is a variety of moduli for elliptic curves over k.

Let X be an elliptic curve over the algebraically closed field k. Let $P_0 \in X$ be a point, and consider the linear system $|2P_0|$ on X. The divisor $2P_0$ is nonspecial, so by Riemann–Roch, this linear system has dimension 1. It has no base points, because otherwise the curve would be rational. Therefore, it defines a morphism $f : X \to \mathbf{P}^1$ of degree 2, and we can specify that $f(P_0) = \infty$ by a change of coordinates in $\mathbf{P}^1$.

Now if we assume char $k \neq 2$, it follows from Hurwitz's theorem that f is ramified at exactly four points, with P_0 being one of them. If $x = a,b,c$ are the three branch points in $\mathbf{P}^1$ besides ∞, then there is a unique automorphism of $\mathbf{P}^1$ leaving ∞ fixed and sending a to 0 and b to 1, namely $x' = (x - a)/(b - a)$. So after this automorphism, we may assume that f is branched over the points $0,1,\lambda,\infty$ of $\mathbf{P}^1$, where $\lambda \in k$, $\lambda \neq 0,1$. This defines a quantity λ. We define $j = j(\lambda)$ by the formula

$$ j = 2^8 \frac{(\lambda^2 - \lambda + 1)^3}{\lambda^2(\lambda - 1)^2}. $$

This is the j-invariant of the curve X. (The coefficient 2^8 is thrown in to make things work in characteristic 2, despite appearances to the contrary!) Our main result then is the following.

Theorem 4.1. *Let k be an algebraically closed field of characteristic $\neq 2$. Then:*

    (a) *for any elliptic curve X over k, the quantity j defined above depends only on X;*

    (b) *two elliptic curves X and X' over k are isomorphic if and only if $j(X) = j(X')$;*

    (c) *every element of k occurs as the j-invariant of some elliptic curve over k.*

    Thus we have a one-to-one correspondence between the set of elliptic curves over k, up to isomorphism, and the elements of k, given by $X \mapsto j(X)$.

We will prove this theorem after some other preliminary results.

Lemma 4.2. *Given any two points $P,Q \in X$ (including the case $P = Q$), there is an automorphism σ of X such that $\sigma^2 = $ id, $\sigma(P) = Q$, and for any $R \in X$, $R + \sigma(R) \sim P + Q$.*

PROOF. The linear system $|P + Q|$ has dimension 1 and is base-point free, hence defines a morphism $g : X \to \mathbf{P}^1$ of degree 2. It is separable, since $X \not\cong \mathbf{P}^1$ (2.5), so $K(X)$ is a Galois extension of $K(\mathbf{P}^1)$. Let σ be the non-trivial automorphism of order 2 of $K(X)$ over $K(\mathbf{P}^1)$. Then σ interchanges the two points of each fibre of g. Hence $\sigma(P) = Q$, and for any $R \in X$, $R + \sigma(R)$ is a fibre of g, hence $R + \sigma(R) \in |P + Q|$, i.e., $R + \sigma(R) \sim P + Q$.

Corollary 4.3. *The group* Aut X *of automorphisms of X is transitive.*

Lemma 4.4. *If $f_1 : X \to \mathbf{P}^1$ and $f_2 : X \to \mathbf{P}^1$ are any two morphisms of degree 2 from X to $\mathbf{P}^1$, then there are automorphisms $\sigma \in$ Aut X and $\tau \in$ Aut $\mathbf{P}^1$ such that $f_2 \circ \sigma = \tau \circ f_1$.*

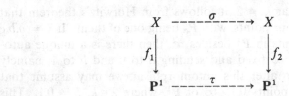

PROOF. Let $P_1 \in X$ be a ramification point of f_1 and let $P_2 \in X$ be a ramification point of f_2. Then by (4.3) there is a $\sigma \in$ Aut X such that $\sigma(P_1) = P_2$. On the other hand, f_1 is determined by the linear system $|2P_1|$ and f_2 is determined by $|2P_2|$. Since σ takes one to the other, f_1 and $f_2 \circ \sigma$ correspond to the same linear system, so they differ only by an automorphism τ of $\mathbf{P}^1$ (II, 7.8.1).

Lemma 4.5. *Let the symmetric group Σ_3 act on $k - \{0,1\}$ as follows: given $\lambda \in k$, $\lambda \neq 0,1$, permute the numbers $0,1,\lambda$ according to $\alpha \in \Sigma_3$, then apply a linear transformation of x to send the first two back to $0,1$, and let $\alpha(\lambda)$ be the image of the third. Then the orbit of λ consists of*

$$\lambda, \frac{1}{\lambda}, 1 - \lambda, \frac{1}{1 - \lambda}, \frac{\lambda}{\lambda - 1}, \frac{\lambda - 1}{\lambda}.$$

PROOF. Since the linear transformation sending a,b to $0,1$ is $x' = (x - a)/(b - a)$, we have only to evaluate $(c - a)/(b - a)$, where $\{a,b,c\} = \{0,1,\lambda\}$ in any order.

Proposition 4.6. *Let X be an elliptic curve over k, with char $k \neq 2$, and let $P_0 \in X$ be a given point. Then there is a closed immersion $X \to \mathbf{P}^2$ such that the image is the curve*

$$y^2 = x(x - 1)(x - \lambda)$$

for some $\lambda \in k$, and the point P_0 goes to the point at infinity $(0,1,0)$ on the y-axis. Furthermore, this λ is the same as the λ defined earlier, up to an element of Σ_3 as in (4.5).

PROOF. We embed X in $\mathbf{P}^2$ by the linear system $|3P_0|$, which gives a closed immersion (3.3.3). We choose our coordinates as follows. Think of the vector spaces $H^0(\mathcal{O}(nP_0))$ as contained in each other,

$$k = H^0(\mathcal{O}) \subseteq H^0(\mathcal{O}(P_0)) \subseteq H^0(\mathcal{O}(2P_0)) \subseteq \ldots .$$

By Riemann–Roch, we have

$$\dim H^0(\mathcal{O}(nP_0)) = n$$

for $n > 0$. Choose $x \in H^0(\mathcal{O}(2P_0))$ so that $1, x$ form a basis of that space, and choose $y \in H^0(\mathcal{O}(3P_0))$ so that $1, x, y$ form a basis for that space. Then the seven quantities

$$1, x, y, x^2, xy, x^3, y^2$$

are in $H^0(\mathcal{O}(6P_0))$, which has dimension 6, so there is a linear relation among them. Furthermore, both x^3 and y^2 occur with coefficient not equal to zero, because they are the only functions with a 6-fold pole at P_0. So replacing x and y by suitable scalar multiples, we may assume they have coefficient 1. Then we have a relation

$$y^2 + a_1 xy + a_3 y = x^3 + a_2 x^2 + a_4 x + a_6$$

for suitable $a_i \in k$.

Now we will make linear changes of coordinates to get the equation in the required form. First we complete the square on the left (here we use char $k \neq 2$), replacing y by

$$y' = y + \frac{1}{2}(a_1 x + a_3).$$

The new equation has y^2 equal to a cubic equation in x, so it can be written

$$y^2 = (x - a)(x - b)(x - c)$$

for suitable $a, b, c \in k$. Now we make a linear change of x to send a, b to $0, 1$, so the equation becomes

$$y^2 = x(x - 1)(x - \lambda)$$

as required.

Since both x and y have a pole at P_0, that point goes to the unique point at infinity on this curve, which is $(0,1,0)$.

If we project from P_0 to the x-axis, we get a finite morphism of degree 2, sending P_0 to ∞, and ramified at $0,1,\lambda,\infty$. So the λ is the same as the one defined earlier.

PROOF OF (4.1).

(a) To show that j depends only on X, suppose we made two choices of base point $P_1, P_2 \in X$. Let $f_1 : X \to \mathbf{P}^1$ and $f_2 : X \to \mathbf{P}^1$ be the corresponding morphisms. Then by (4.4) we can find automorphisms $\sigma \in \operatorname{Aut} X$ and $\tau \in \operatorname{Aut} \mathbf{P}^1$ such that $f_2 \circ \sigma = \tau \circ f_1$. Furthermore, we could choose σ such that $\sigma(P_1) = P_2$, hence $\tau(\infty) = \infty$. So τ sends the branch points $0,1,\lambda_1$ of f_1 to the branch points $0,1,\lambda_2$ of f_2 in some order. Hence by (4.5), λ_1 and λ_2 differ only by an element of Σ_3, via the action of (4.5). So we have only to observe that for any $\alpha \in \Sigma_3$, $j(\lambda) = j(\alpha(\lambda))$. Indeed, since Σ_3 is generated by any two elements of order 2, it is enough to show that

$$j(\lambda) = j\left(\frac{1}{\lambda}\right) \quad \text{and} \quad j(\lambda) = j(1 - \lambda),$$

which is clear by direct computation. Thus j depends only on X.

(b) Now suppose X and X' are two elliptic curves giving rise to λ and λ', such that $j(\lambda) = j(\lambda')$. First we note that j is a rational function of λ of degree 6, i.e., $\lambda \to j$ defines a finite morphism $\mathbf{P}^1 \to \mathbf{P}^1$ of degree 6. Furthermore, this is a Galois covering, with Galois group Σ_3 under the action described above. Therefore, $j(\lambda) = j(\lambda')$ if and only if λ and λ' differ by an element of Σ_3.

Now according to (4.6), X and X' can be embedded in $\mathbf{P}^2$ so as to have the equation $y^2 = x(x-1)(x-\lambda)$, or same with λ'. Since λ and λ' differ by an element of Σ_3 as in (4.5), after a linear change of variable in x, we have $\lambda = \lambda'$. Thus X and X' are both isomorphic to the same curve in $\mathbf{P}^2$.

(c) Given any $j \in k$, we can solve the polynomial equation

$$2^8(\lambda^2 - \lambda + 1)^3 - j\lambda^2(\lambda - 1)^2 = 0$$

for λ, and find a value of λ, necessarily $\neq 0,1$. Then the equation $y^2 = x(x-1)(x-\lambda)$ defines a nonsingular curve of degree 3 in $\mathbf{P}^2$, which is therefore elliptic, and has the given j as its j-invariant.

Example 4.6.1. The curve $y^2 = x^3 - x$ of (I, Ex. 6.2) is nonsingular over any field k with char $k \neq 2$. It has $\lambda = -1$, hence $j = 2^6 \cdot 3^3 = 1728$.

Example 4.6.2. The "Fermat curve" $x^3 + y^3 = z^3$ is nonsingular over any field k with char $k \neq 3$. Making a change of variables $x = x' + z$, and setting $x' = -1/3$, the equation becomes

$$z^2 - \frac{1}{3}z = y^3 - \frac{1}{27}.$$

From here one can reduce it to a standard form, as in the proof of (4.6), with $\lambda = -\omega$ or $-\omega^2$, where $\omega^3 = 1$. Therefore, $j = 0$.

Corollary 4.7. *Let X be an elliptic curve over k with* char $k \neq 2$. *Let $P_0 \in X$, and let $G = \mathrm{Aut}(X,P_0)$ be the group of automorphisms of X leaving P_0 fixed. Then G is a finite group of order*

$$
\begin{array}{ll}
2 & \textit{if } j \neq 0,\ 1728 \\
4 & \textit{if } j = 1728 \textit{ and char } k \neq 3 \\
6 & \textit{if } j = 0 \textit{ and char } k \neq 3 \\
12 & \textit{if } j = 0\ (=1728)\textit{ and char } k = 3.
\end{array}
$$

PROOF. Let $f : X \to \mathbf{P}^1$ be a morphism of degree 2, with $f(P_0) = \infty$, branched over $0, 1, \lambda, \infty$ as above. If $\sigma \in G$, then by (4.4) there is an automorphism τ of $\mathbf{P}^1$, sending ∞ to ∞, such that $f \circ \sigma = \tau \circ f$. In particular, τ sends $\{0,1,\lambda\}$ to $\{0,1,\lambda\}$ in some order. If $\tau = \mathrm{id}$, then either $\sigma = \mathrm{id}$ or σ is the automorphism interchanging the sheets of f. Thus in any case we have two elements in G.

If $\tau \neq \mathrm{id}$, then τ permutes $\{0,1,\lambda\}$, so λ must be equal to one of the other expressions of (4.5). This can happen only in the following cases:

(1) if $\lambda = -1$ or $\frac{1}{2}$ or 2, and char $k \neq 3$, then λ coincides with one other element of its orbit under Σ_3, so G has order 4. This is the case $j = 1728$;
(2) if $\lambda = -\omega$ or $-\omega^2$, and char $k \neq 3$, then λ coincides with two other elements of its orbit under Σ_3, so G has order 6. In this case $j = 0$;
(3) if char $k = 3$ and $\lambda = -1$, then all six elements of the orbit are the same, so G has order 12. In this case $j = 0 = 1728$.

The Group Structure

Let X be an elliptic curve, and let $P_0 \in X$ be a fixed point. We have seen, as a consequence of the Riemann–Roch theorem (1.3.7) that the map $P \mapsto \mathscr{L}(P - P_0)$ induces a bijection between the set of points of X and the group $\mathrm{Pic}^\circ X$. Thus the set of points of X forms a group, with P_0 as the 0 element, and with addition characterized by $P + Q = R$ if and only if $P + Q \sim R + P_0$ as divisors on X. This is the group structure on (X,P_0).

If we embed X in $\mathbf{P}^2$ by the linear system $|3P_0|$, then three points P,Q,R of the image are collinear if and only if $P + Q + R \sim 3P_0$. This in turn is equivalent to saying $P + Q + R = 0$ in the group structure. This shows that the group law can be recovered from the geometry of the embedding. It also generalizes (II, 6.10.2), where we used the geometry to define the group law.

Now we will show that X is a group variety in the sense of (I, Ex. 3.21).

Proposition 4.8. *Let (X,P_0) be an elliptic curve with its group structure. Then the maps $\rho : X \to X$ given by $P \mapsto -P$, and $\mu : X \times X \to X$ given by $\langle P,Q \rangle \mapsto P + Q$ are morphisms.*

PROOF. First we apply (4.2) with $P = Q = P_0$. Thus there is an automorphism σ of X such that for any R, $R + \sigma(R) \sim 2P_0$. In other words, $\sigma(R) = -R$ in the group structure, so this σ is just ρ.

Next we apply (4.2) to P and P_0. So there is an automorphism σ of X with $R + \sigma(R) \sim P + P_0$, i.e., $\sigma(R) = P - R$ in the group. Preceding this σ with ρ, we see that $R \to P + R$, i.e., translation by P, is a morphism, for any P.

Now take two distinct points $P \neq Q$ in X. Embed X in $\mathbf{P}^2$ by $|3P_0|$. Form the equation of the line L joining P and Q. This depends on the coordinates of P and Q. Now intersect L with X. We get a cubic equation in the parameter along L, but we already know two of the intersections, so we obtain the coordinates of the third point of intersection R as rational functions in the coordinates of P and Q. Since $R = -P - Q$ in the group structure, this shows that the map $(X \times X - \Delta) \to X$ defined by $\langle P,Q \rangle \to -P - Q$ is a morphism. Composing with ρ, we see that μ is a morphism for pairs of distinct points of X.

To show that μ is a morphism also at points of the form $\langle P,P \rangle$, take any $Q \neq 0$. Translate one variable by Q, apply μ to $\langle P,P + Q \rangle$, then translate by $-Q$. Since translation is a morphism, we see that μ is also a morphism at these points.

Example 4.8.1. By iterating μ, we see that for any integer n, multiplication by n gives a morphism $n_X : X \to X$. We will see later that for any $n \neq 0$, n_X is a finite morphism of degree n^2; its kernel is a group isomorphic to $\mathbf{Z}/n \times \mathbf{Z}/n$ if $(n,p) = 1$, where $p = \operatorname{char} k$, and is isomorphic to $\mathbf{Z}/p$ or 0 if $n = p$, depending on the Hasse invariant of X. See (4.10), (4.17), (Ex. 4.6), (Ex. 4.7), (Ex. 4.15).

Example 4.8.2. If P is a point of order 2 on X, then $2P \sim 2P_0$, so P is a ramification point of the morphism $f : X \to \mathbf{P}^1$ defined by $|2P_0|$, and f is separable since $X \not\cong \mathbf{P}^1$ (2.5). So there are only finitely many such points, and if $\operatorname{char} k \neq 2$, there are exactly 4. Thus 2_X is always a finite morphism, and if $\operatorname{char} k \neq 2$, we see that it has degree 4, and its kernel is $\mathbf{Z}/2 \times \mathbf{Z}/2$.

Example 4.8.3. If P is a point of order 3 on X, then $3P \sim 3P_0$, so P is an inflection point of the embedding of X in $\mathbf{P}^2$ by $|3P_0|$. If $\operatorname{char} k \neq 2,3$, we see by (Ex. 2.3) that there are exactly 9 inflection points of X. Thus 3_X has degree 9, and its kernel is isomorphic to $\mathbf{Z}/3 \times \mathbf{Z}/3$. By the way, this has the amusing geometric consequence that if P,Q are inflection points of X, then the line PQ meets X in a third inflection point R of X. Indeed, $R = -P - Q$, so it is also a point of order 3.

Lemma 4.9. *If X,P_0 and X',P_0' are two elliptic curves, and if $f : X \to X'$ is a morphism sending P_0 to P_0', then f is a homomorphism of the group structures.*

PROOF. If $P + Q = R$ on X, then $P + Q \sim R + P_0$ as divisors. It follows that $f(P) + f(Q) \sim f(R) + f(P_0)$ by (Ex. 2.6), and since $f(P_0) = P_0'$, we have $f(P) + f(Q) = f(R)$ in the group law on X'.

Definition. If f, g are two morphisms of an elliptic curve X, P_0 to itself, sending P_0 to P_0, we define a morphism $f + g$ by composing $f \times g : X \to X \times X$ with μ. In other words, $(f + g)(P) = f(P) + g(P)$ for all P. We define the morphism $f \cdot g$ to be $f \circ g$. Then the set of all morphisms of X to itself sending P_0 to P_0 forms a ring $R = \text{End}(X, P_0)$, which we call the *ring of endomorphisms* of X, P_0. Its zero element 0 is the morphism sending X to P_0. The unit element 1 is the identity map. The inverse morphism ρ is -1. The distributive law $f \cdot (g + h) = f \cdot g + f \cdot h$ is a consequence of the fact (4.9) that f is a homomorphism.

Proposition 4.10. *Assume* char $k \neq 2$. *The map* $n \mapsto n_X$ *defines an injective ring homomorphism* $\mathbf{Z} \to \text{End}(X, P_0)$. *In particular, for all* $n \neq 0$, n_X *is a finite morphism.*

PROOF. We will show by induction on n that $n_X \neq 0$ for $n \geqslant 1$. It follows that n_X is a finite morphism (II, 6.8). For $n = 1$ it is clear; for $n = 2$ we have seen it above (4.8.2). So let $n > 2$. If n is odd, say $n = 2r + 1$, and if $n_X = 0$, then $(2r)_X = \rho$. But ρ has degree 1, and $(2r)_X = 2_X \cdot r_X$ is a finite morphism (use induction hypothesis for r) of degree $\geqslant 4$, since 2_X has degree 4 (4.8.2). So this is impossible.

If n is even, say $n = 2r$, then $n_X = 2_X \cdot r_X$ is finite by induction.

Remark 4.10.1. The ring of endomorphisms R is an important invariant of the elliptic curve, but it is not easy to calculate. Let us just note for the moment that its group of units R^* is the group $G = \text{Aut}(X, P_0)$ studied above (4.7). In particular, if $j = 0$ or 1728, it is bigger than $\{\pm 1\}$, so R is definitely bigger than $\mathbf{Z}$.

The Jacobian Variety

Now we will give another, perhaps more natural, proof that the group law on the elliptic curve makes it a group variety. Our earlier proof used geometric properties of the embedding in $\mathbf{P}^2$. Now instead, we will show that the group $\text{Pic}^\circ X$ has a structure of algebraic variety which is so natural that it is automatically a group variety. This approach makes sense for a curve of any genus, and leads to the Jacobian variety of a curve. The idea is to find a universal parameter space for divisor classes of degree 0.

Let X be a curve over k. For any scheme T over k, we define $\text{Pic}^\circ(X \times T)$ to be the subgroup of $\text{Pic}(X \times T)$ consisting of invertible sheaves whose restriction to each fibre X_t for $t \in T$ has degree 0. Let $p : X \times T \to T$ be the second projection. For any invertible sheaf $\mathcal{N}$ on T, $p^* \mathcal{N} \in \text{Pic}^\circ(X \times T)$, because it is in fact trivial on each fibre. We define $\text{Pic}^\circ(X/T) = \text{Pic}^\circ(X \times T)/p^* \text{Pic} \, T$, and we regard its elements as "families of invertible sheaves of degree 0 on X, parametrized by T." Justification for this is the fact that if T is integral and of finite type over k, and if $\mathcal{L}, \mathcal{M} \in \text{Pic}(X \times T)$, then $\mathcal{L}_t \cong \mathcal{M}_t$ on X_t for all $t \in T$ if and only if $\mathcal{L} \otimes \mathcal{M}^{-1} \in p^* \text{Pic} \, T$ (III, Ex. 12.4).

Definition. Let X be a curve (of any genus) over k. The *Jacobian variety* of X is a scheme J of finite type over k, together with an element $\mathscr{L} \in \text{Pic}^{\circ}(X/J)$, having the following universal property: for any scheme T of finite type over k, and for any $\mathscr{M} \in \text{Pic}^{\circ}(X/T)$, there is a unique morphism $f : T \to J$ such that $f^* \mathscr{L} \cong \mathscr{M}$ in $\text{Pic}^{\circ}(X/T)$. (Note that $f : X \times T \to X \times J$ induces a homomorphism $f^* : \text{Pic}^{\circ}(X/J) \to \text{Pic}^{\circ}(X/T)$.)

Remark 4.10.2. In the language of representable functors, this definition says that J represents the functor $T \to \text{Pic}^{\circ}(X/T)$.

Remark 4.10.3. Since J is defined by a universal property, it is unique if it exists. We will prove below that if X is an elliptic curve, then J exists, and in fact we can take $J = X$. For curves of genus $\geqslant 2$ the existence is much more difficult. See, for example, Chow [3] or Mumford [2] or Grothendieck [5].

Remark 4.10.4. Assuming J exists, its closed points are in one-to-one correspondence with elements of the group $\text{Pic}^{\circ} X$. Indeed, to give a closed point of J is the same as giving a morphism $\text{Spec } k \to J$, which by the universal property is the same thing as an element of $\text{Pic}^{\circ}(X/k) = \text{Pic}^{\circ} X$.

Definition. A scheme X with a morphism to another scheme S is a *group scheme over* S if there is a section $e : S \to X$ (the identity) and a morphism $\rho : X \to X$ over S (the inverse) and a morphism $\mu : X \times X \to X$ over S (the group operation) such that

(1) the composition $\mu \circ (\text{id} \times \rho) : X \to X$ is equal to the projection $X \to S$ followed by e, and

(2) the two morphisms $\mu \circ (\mu \times \text{id})$ and $\mu \circ (\text{id} \times \mu)$ from $X \times X \times X \to X$ are the same.

Remark 4.10.5. This notion of group scheme generalizes the earlier notion of group variety (I, Ex. 3.21). Indeed, if $S = \text{Spec } k$ and X is a variety over k, taking e to be the 0 point, the properties (1), (2) can be checked on the closed points of X. Then (1) says that ρ gives the inverse of each point, and (2) says that the group law is associative.

Remark 4.10.6. The Jacobian variety J of a curve X is automatically a group scheme over k. Indeed, using the universal property of J, define $e : \text{Spec } k \to J$ by taking the element $0 \in \text{Pic}^{\circ}(X/k)$. Define $\rho : J \to J$ by taking $\mathscr{L}^{-1} \in \text{Pic}^{\circ}(X/J)$. Define $\mu : J \times J \to J$ by taking $p_1^* \mathscr{L} \otimes p_2^* \mathscr{L} \in \text{Pic}^{\circ}(X/J \times J)$. The properties (1) and (2) are verified immediately by the universal property of J.

Remark 4.10.7. We can determine the Zariski tangent space to J at 0 as follows. To give an element of the Zariski tangent space is equivalent to giving a morphism of $T = \operatorname{Spec} k[\varepsilon]/\varepsilon^2$ to J sending $\operatorname{Spec} k$ to 0 (II, Ex. 2.8). By the definition of J, this is equivalent to giving $\mathcal{M} \in \operatorname{Pic}^\circ(X/T)$ whose restriction to $\operatorname{Pic}^\circ(X/k)$ is 0. But according to (III, Ex. 4.6) there is an exact sequence $0 \to H^1(X,\mathcal{O}_X) \to \operatorname{Pic} X[\varepsilon] \to \operatorname{Pic} X \to 0$. So we see that the Zariski tangent space to J at 0 is just $H^1(X,\mathcal{O}_X)$.

Remark 4.10.8. J is proper over k. We apply the valuative criterion of properness (II, 4.7). It is enough to show (II, Ex. 4.11) that if R is any discrete valuation ring containing k, with quotient field K, then a morphism of $\operatorname{Spec} K$ to J extends uniquely to a morphism of $\operatorname{Spec} R$ to J. In other words, we must show that an invertible sheaf $\mathcal{M}$ on $X \times \operatorname{Spec} K$ extends uniquely to an invertible sheaf on $X \times \operatorname{Spec} R$. Since $X \times \operatorname{Spec} R$ is a regular scheme, this follows from (II, 6.5) (note that the closed fibre of $X \times \operatorname{Spec} R$ over $\operatorname{Spec} R$, as a divisor on $X \times \operatorname{Spec} R$, is linearly equivalent to 0).

Remark 4.10.9. If we fix a base point $P_0 \in X$, then for any $n \geqslant 1$ there is a morphism $\varphi_n \colon X^n \to J$ defined by "$\langle P_1,\ldots,P_n \rangle \to \mathcal{L}(P_1 + \ldots + P_n - nP_0)$" (which means cook up the appropriate sheaf on $X \times X^n$ to define φ_n). If g is the genus of X, then φ_n will be surjective for $n \geqslant g$, because by Riemann–Roch, every divisor class of degree $\geqslant g$ contains an effective divisor. The fibre of φ_n over a point of J consists of all n-tuples $\langle P_1,\ldots,P_n \rangle$ such that the divisors $P_1 + \ldots + P_n$ form a complete linear system.

If $n = g$, then for most choices of $P_1,\ldots,P_g$, we have $l(P_1 + \ldots + P_g) = 1$. Indeed, by Riemann–Roch,

$$l(P_1 + \ldots + P_g) = g + 1 - g + l(K - P_1 - \ldots - P_g).$$

But $l(K) = g$. Taking P_1 not a base point of K, $l(K - P_1) = g - 1$. At each step, taking P_i not a base point of $K - P_1 - \ldots - P_{i-1}$, we get $l(K - P_1 - \ldots - P_g) = 0$. Therefore, most fibres of φ_g are finite sets of points. We conclude that J is irreducible and $\dim J = g$. On the other hand, by (4.10.7), the Zariski tangent space to J at 0 is $H^1(X,\mathcal{O}_X)$, which has dimension g, so J is nonsingular at 0. Since it is a group scheme, it is a homogeneous space, hence nonsingular everywhere. Hence J is a nonsingular variety.

Theorem 4.11. *Let X be an elliptic curve, and fix a point $P_0 \in X$. Take $J = X$, and take $\mathcal{L}$ on $X \times J$ to be $\mathcal{L}(\Delta) \otimes p_1^* \mathcal{L}(-P_0)$, where $\Delta \subseteq X \times X$ is the diagonal. Then $J,\mathcal{L}$ is a Jacobian variety for X. Furthermore, the resulting structure of group variety on J (4.10.6) induces the same group structure on X,P_0 as defined earlier.*

PROOF. The last statement is obvious from the definitions. So we have only to show that if T is any scheme of finite type over k, and if $\mathcal{M} \in \operatorname{Pic}^\circ(X/T)$, then there is a unique morphism $f \colon T \to J$ such that $f^* \mathcal{L} \cong \mathcal{M}$.

Let $p \colon X \times T \to T$ be the projection, and let $q \colon X \times T \to X$ be the other projection. Define $\mathcal{M}' = \mathcal{M} \otimes q^* \mathcal{L}(P_0)$. Then $\mathcal{M}'$ has degree 1 along the fibres. Hence, for any closed point $t \in T$, we can apply Riemann–Roch to

325

$\mathcal{M}'_t$ on $X_t = X$, and we find

$$\dim H^0(X, \mathcal{M}'_t) = 1$$
$$\dim H^1(X, \mathcal{M}'_t) = 0.$$

Since p is a projective morphism, and $\mathcal{M}'$ is flat over T, we can apply the theorem of cohomology and base change (III, 12.11). Looking first at $R^1 p_*(\mathcal{M}')$, since the cohomology along the fibres is 0, the map $\varphi^1(t)$ of (III, 12.11) is automatically surjective, hence an isomorphism, so we conclude that $R^1 p_*(\mathcal{M}')$ is identically 0. In particular, it is locally free, so we deduce from part (b) of the theorem that $\varphi^0(t)$ is also surjective. Therefore, it is an isomorphism, and since $\varphi^{-1}(t)$ is always surjective, we see that $p_*(\mathcal{M}')$ is locally free of rank 1.

Now replacing $\mathcal{M}$ by $\mathcal{M} \otimes p^* p_*(\mathcal{M}')^{-1}$ in $\operatorname{Pic}^\circ(X/T)$, we may then assume that $p_*(\mathcal{M}') \cong \mathcal{O}_T$. The section $1 \in \Gamma(T, \mathcal{O}_T)$ gives a section $s \in \Gamma(X \times T, \mathcal{M}')$, which defines an effective Cartier divisor $Z \subseteq X \times T$. By construction, Z intersects each fibre of p in just one point, and in fact one sees easily that the restricted morphism $p : Z \to T$ is an isomorphism. Thus we get a section $s : T \to Z \subseteq X \times T$. Composing with q gives the required morphism $f : T \to X$.

Indeed, since Z is the graph of f, we see that $Z = f^*\Delta$, where $\Delta \subseteq X \times X$ is the diagonal. Hence the corresponding invertible sheaves correspond: $\mathcal{M}' \cong f^*\mathcal{L}(\Delta)$. Now twisting by $-P_0$ shows that $\mathcal{M} \cong f^*\mathcal{L}$, as required. The uniqueness of f is clear for the same reasons.

Elliptic Functions

It is hard to discuss elliptic curves without bringing in the theory of elliptic functions of a complex variable. This classical topic from complex analysis gives an insight into the theory of elliptic curves over **C** which cannot be matched by purely algebraic techniques. So we will recall some of the definitions and results of that theory without proof (signaling those statements with a B in their number), and give some applications to elliptic curves. We refer to the book Hurwitz–Courant [1] for proofs.

Fix a complex number $\tau, \tau \notin \mathbf{R}$. Let Λ be the lattice in the complex plane **C** consisting of all $n + m\tau$, with $n, m \in \mathbf{Z}$ (Fig. 15).

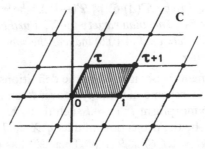

Figure 15. A lattice in **C**, with one period parallelogram.

Definition. An *elliptic function* (with respect to the lattice Λ) is a meromorphic function $f(z)$ of the complex variable z such that $f(z + \omega) = f(z)$ for all $\omega \in \Lambda$. (Sometimes these are called *doubly periodic functions*, since they are periodic with respect to the periods $1, \tau$.)

Because of the periodicity, an elliptic function is determined if one knows its values on a single *period parallelogram*, such as the one bounded by $0, 1, \tau, \tau + 1$ (Fig. 15).

An example of an elliptic function is the *Weierstrass $\wp$-function* defined by

$$\wp(z) = \frac{1}{z^2} + \sum_{\omega \in \Lambda'} \left(\frac{1}{(z - \omega)^2} - \frac{1}{\omega^2} \right),$$

where $\Lambda' = \Lambda - \{0\}$. One shows (Hurwitz–Courant [1, II, 1, §6]) that this series converges at all $z \notin \Lambda$, thus giving a meromorphic function having a double pole at the points of Λ, and which is elliptic. Its derivative

$$\wp'(z) = \sum_{\omega \in \Lambda} \frac{-2}{(z - \omega)^3}$$

is another elliptic function.

If one adds, subtracts, multiplies, or divides two elliptic functions with periods in Λ, one gets another such. Hence the elliptic functions for a given Λ form a field.

Theorem 4.12B. *The field of elliptic functions for given Λ is generated over* $\mathbf{C}$ *by the Weierstrass $\wp$-function and its derivative $\wp'$. They satisfy the algebraic relation*

$$(\wp')^2 = 4\wp^3 - g_2\wp - g_3,$$

where

$$g_2 = 60 \sum_{\omega \in \Lambda'} \frac{1}{\omega^4} \quad and \quad g_3 = 140 \sum_{\omega \in \Lambda'} \frac{1}{\omega^6}.$$

PROOF. Hurwitz–Courant [1, II, 1, §8, 9].

Thus if we define a mapping $\varphi : \mathbf{C} \to \mathbf{P}_{\mathbf{C}}^2$ by sending $z \to (\wp(z), \wp'(z))$ in affine coordinates, we obtain a holomorphic mapping whose image lies inside the curve X with equation

$$y^2 = 4x^3 - g_2 x - g_3.$$

In fact, φ induces a bijective mapping of $\mathbf{C}/\Lambda$ to X (Hurwitz–Courant [1, II, 5, §1]), and X is nonsingular, hence an elliptic curve. Under this mapping the field of elliptic functions is identified with the function field of the curve X. Thus for any elliptic function, we can speak of its *divisor* $\sum n_i(a_i)$, with $a_i \in \mathbf{C}/\Lambda$.

Theorem 4.13B. *Given distinct points $a_1, \ldots, a_q \in \mathbf{C}/\Lambda$, and given integers $n_1, \ldots, n_q$, a necessary and sufficient condition that there exist an elliptic function with divisor $\sum n_i(a_i)$ is that $\sum n_i = 0$ and $\sum n_i a_i = 0$ in the group $\mathbf{C}/\Lambda$.*

PROOF. Hurwitz–Courant [1, II, 1, §5, 14].

In particular, this says that $a_1 + a_2 \equiv b \pmod{\Lambda}$ if and only if there is an elliptic function with zeros at a_1 and a_2, and poles at b and 0. Since this function is a rational function on the curve X, this says that $\varphi(a_1) + \varphi(a_2) \sim \varphi(b) + \varphi(0)$ as divisors on X. If we let $P_0 = \varphi(0)$, which is the point at infinity on the y-axis, and give X the group structure with origin P_0, this says that $\varphi(a_1) + \varphi(a_2) = \varphi(b)$ in the group structure on X. In other words, φ gives a group isomorphism between $\mathbf{C}/\Lambda$ under addition, and X with its group law.

Theorem 4.14B. *Given $c_2, c_3 \in \mathbf{C}$, with $\Delta \neq 0$, where $\Delta = c_2^3 - 27c_3^2$, there exists a $\tau \in \mathbf{C}$, $\tau \notin \mathbf{R}$, and an $\alpha \in \mathbf{C}$, $\alpha \neq 0$, such that the lattice $\Lambda = (1, \tau)$ gives $g_2 = \alpha^4 c_2$ and $g_3 = \alpha^6 c_3$ by the formulas above.*

PROOF. Hurwitz–Courant [1, II, 4, §4].

This shows that every elliptic curve over $\mathbf{C}$ arises in this way. Indeed, if X is any elliptic curve, we can embed X in $\mathbf{P}^2$ to have an equation of the form $y^2 = x(x - 1)(x - \lambda)$, with $\lambda \neq 0,1$ (4.6). By a linear change of variable in x, one can bring this into the form $y^2 = 4x^3 - c_2 x - c_3$, with $c_2 = (\sqrt[3]{4/3})(\lambda^2 - \lambda + 1)$ and $c_3 = (1/27)(\lambda + 1)(2\lambda^2 - 5\lambda + 2)$. Then $\Delta = \lambda^2(\lambda - 1)^2$, which is $\neq 0$ since $\lambda \neq 0,1$. Now the curve determined by the lattice Λ is equivalent to this one by a change of variables $y' = \alpha^3 y$, $x' = \alpha^2 x$.

Next we define $J(\tau) = g_2^3/\Delta$. Then the j-invariant of X which we defined earlier is just $j = 1728 \cdot J(\tau)$. Thus $J(\tau)$ classifies the curve X up to isomorphism.

Theorem 4.15B. *Let τ, τ' be two complex numbers. Then $J(\tau) = J(\tau')$ if and only if there are integers $a,b,c,d \in \mathbf{Z}$ with $ad - bc = \pm 1$ and*

$$\tau' = \frac{a\tau + b}{c\tau + d}.$$

Furthermore, given any τ', there is a unique τ with $J(\tau) = J(\tau')$ such that τ lies in the region G (Fig. 16) defined by

$$-\frac{1}{2} \leqslant \operatorname{Re} \tau < \frac{1}{2}$$

and

$$|\tau| \geqslant 1 \quad \text{if } \operatorname{Re} \tau \leqslant 0$$
$$|\tau| > 1 \quad \text{if } \operatorname{Re} \tau > 0.$$

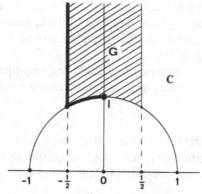

Figure 16.   The region G.

PROOF. Hurwitz–Courant [1, II, 4, §3].

Now we will start drawing consequences from this theory.

Theorem 4.16. *Let X be an elliptic curve over* **C**. *Then as an abstract group, X is isomorphic to* **R**/**Z** $\times$ **R**/**Z**. *In particular, for any n, the subgroup of points of order n is isomorphic to* **Z**/n $\times$ **Z**/n.

PROOF. We have seen that X is isomorphic as a group to C/Λ, which in turn is isomorphic to **R**/**Z** $\times$ **R**/**Z**. The points of order n are represented by $(a/n) + (b/n)\tau$, with $a,b = 0,1,\ldots,n-1$. The points whose coordinates are not rational combinations of $1,\tau$ are of infinite order.

Corollary 4.17. *The morphism multiplication by n, $n_X : X \to X$ is a finite morphism of degree n^2.*

PROOF. Since it is separable, and a group homomorphism, its degree is the order of the kernel, which is n^2.

Next we will investigate the ring of endomorphisms $R = \text{End}(X, P_0)$ of the elliptic curve X determined by the elliptic functions with periods $1,\tau$.

Proposition 4.18. *There is a one-to-one correspondence between endomorphisms $f \in R$ and complex numbers $\alpha \in$ **C** such that $\alpha \cdot \Lambda \subseteq \Lambda$. This correspondence gives an injective ring homomorphism of R to* **C**.

PROOF. Given $f \in R$, we have seen (4.9) that f is a group homomorphism of X to X. Hence under the identification of X with C/Λ it gives a group homomorphism $\bar{f}$ of **C** to **C**, such that $\bar{f}(\Lambda) \subseteq \Lambda$. On the other hand, since f is a morphism, the induced map $\bar{f} : C \to C$ is holomorphic. Now expanding $\bar{f}$ as a power series in a neighborhood of the origin, and expressing the fact that $\bar{f}(z + w) = \bar{f}(z) + \bar{f}(w)$ for any z,w there, we see that $\bar{f}$ must be just multiplication by some complex number α.

Conversely, given $\alpha \in \mathbf{C}$, such that $\alpha \cdot \Lambda \subseteq \Lambda$, clearly multiplication by α induces a group homomorphism f of $\mathbf{C}/\Lambda$ to itself, hence of X to itself. But f is also holomorphic, so in fact it is a morphism of X to itself by GAGA ($=$ Serre [4]): see (App. B, Ex. 6.6).

It is clear under this correspondence that the ring operations of R correspond to addition and multiplication of the corresponding complex numbers α.

Remark 4.18.1. Note in particular that the morphism $n_X \in R$, which is multiplication by n in the group structure (4.8.1) corresponds to multiplication by n in $\mathbf{C}$. This gives another proof of (4.10) for elliptic curves over $\mathbf{C}$.

Definition. If X is an elliptic curve over $\mathbf{C}$, we say it has *complex multiplication* if the ring of endomorphisms R is bigger than $\mathbf{Z}$. This terminology is explained by (4.18).

Theorem 4.19. *If X has complex multiplication, then $\tau \in \mathbf{Q}(\sqrt{-d})$ for some $d \in \mathbf{Z}, d > 0$, and in that case, R is a subring ($\neq \mathbf{Z}$) of the ring of integers of the field $\mathbf{Q}(\sqrt{-d})$. Conversely, if $\tau = r + s\sqrt{-d}$, with $r,s \in \mathbf{Q}$, then X has complex multiplication, and in fact*

$$R = \{a + b\tau \,|\, a,b \in \mathbf{Z}, \text{ and } 2br, b(r^2 + ds^2) \in \mathbf{Z}\}.$$

PROOF. Given τ, we can determine R as the set of all $\alpha \in \mathbf{C}$ such that $\alpha \cdot \Lambda \subseteq \Lambda$. A necessary and sufficient condition for $\alpha \cdot \Lambda \subseteq \Lambda$ is that there exist integers a,b,c,e such that

$$\alpha = a + b\tau$$
$$\alpha\tau = c + e\tau.$$

If $\alpha \in \mathbf{R}$, then $\alpha \in \mathbf{Z}$, so we see that $R \cap \mathbf{R} = \mathbf{Z}$. On the other hand, if X has complex multiplication, then there is an $\alpha \notin \mathbf{R}$, and in this case, $b \neq 0$.

Eliminating α from these equations, we see that

$$b\tau^2 + (a - e)\tau - c = 0,$$

which shows that τ is in a quadratic extension of $\mathbf{Q}$. Since $\tau \notin \mathbf{R}$, it must be an imaginary quadratic extension, so $\tau \in \mathbf{Q}(\sqrt{-d})$ for some $d \in \mathbf{Z}, d > 0$.

Eliminating τ from the same equations, we find that

$$\alpha^2 - (a - e)\alpha + (ae - bc) = 0,$$

which shows that α is integral over $\mathbf{Z}$. Therefore R must be a subring of the ring of integers of the field $\mathbf{Q}(\sqrt{-d})$.

Conversely, suppose $\tau = r + s\sqrt{-d}$, with $r,s \in \mathbf{Q}$. Then we can determine R as the set of all $\alpha = a + b\tau$, with $a,b \in \mathbf{Z}$, such that $\alpha\tau \in \Lambda$. Since

$\alpha\tau = a\tau + b\tau^2$, we must have $b\tau^2 \in \Lambda$. Now

$$\tau^2 = r^2 - ds^2 + 2rs\sqrt{-d},$$

which can be written

$$\tau^2 = -(r^2 + ds^2) + 2r\tau.$$

So in order to have $b\tau^2 \in \Lambda$ we must have $2br \in \mathbf{Z}$ and $b(r^2 + ds^2) \in \mathbf{Z}$. These conditions are necessary and sufficient so we get the required expression for R. In particular, $R > \mathbf{Z}$, so X has complex multiplication.

Corollary 4.20. *There are only countably many values of $j \in \mathbf{C}$ for which the corresponding elliptic curve X has complex multiplication.*

PROOF. Indeed, there are only countably many elements of all quadratic extensions of $\mathbf{Q}$.

Example 4.20.1. If $\tau = i$, then R is the ring of Gaussian integers $\mathbf{Z}[i]$. In this case the group of units R^* of R consists of $\pm 1, \pm i$, so $R^* \cong \mathbf{Z}/4$. This means that the group of automorphisms of X has order 4, so by (4.7) we must have $j = 1728$. So we see in a roundabout way that $\tau = i$ gives $J(\tau) = 1$. Another way to see this is as follows. Since $\Lambda = \mathbf{Z} \oplus \mathbf{Z}i$, the lattice Λ is stable under multiplication by i. Therefore

$$g_3 = 140 \sum_{\omega \in \Lambda'} \omega^{-6} = 140 \sum_{\omega \in \Lambda'} i^{-6}\omega^{-6} = -g_3.$$

So $g_3 = 0$, which implies that $J(\tau) = 1$. The equation of X can be written $y^2 = x^3 - Ax$.

Example 4.20.2. If $\tau = \omega$, where $\omega^3 = 1$, then $R = \mathbf{Z}[\omega]$, which is the ring of integers in the field $\mathbf{Q}(\sqrt{-3})$. In this case $R^* = \{\pm 1, \pm\omega, \pm\omega^2\}$ which is isomorphic to $\mathbf{Z}/6$. So again from (4.7) we conclude that $j = 0$. One can also see this directly as in (4.20.1) by showing that $g_2 = 0$. The equation of X can be written $y^2 = x^3 - B$.

Example 4.20.3. If $\tau = 2i$, then $R = \mathbf{Z}[2i]$. In this case R is a proper subring of the ring of integers in the quadratic field $\mathbf{Q}(i)$, with conductor 2 (Ex. 4.21).

Remark 4.20.4. Even though we have a good criterion for complex multiplication in terms of τ, the connection between τ and j is not easy to compute. Thus if we are given a curve by its equation in $\mathbf{P}^2$, or by its j-invariant, it is not easy to tell whether it has complex multiplication or not. See (Ex. 4.5) and (Ex. 4.12). There is an extensive classical literature relating complex multiplication to class field theory–see, e.g. Deuring [2] or Serre's article in Cassels and Fröhlich [1, Ch. XIII]. Here are some of the principal results: let X be an elliptic curve with complex multiplication, let $R = \mathrm{End}(X,P_0)$, let $K = \mathbf{Q}(\sqrt{-d})$ be the quotient field of R (4.19), and let j be the j-invariant. Then (1) j is an algebraic integer; (2) the field $K(j)$ is an abelian extension of K of degree $h_R = \# \operatorname{Pic} R$; (3) $j \in \mathbf{Z} \Leftrightarrow h_R = 1$, and there are exactly 13 such values of j.

The Hasse Invariant

If X is an elliptic curve over a field k of characteristic $p > 0$, we define an important invariant of X as follows. Let $F: X \to X$ be the Frobenius morphism (2.4.1). Then F induces a map

$$F^*: H^1(X, \mathcal{O}_X) \to H^1(X, \mathcal{O}_X)$$

on cohomology. This map is not linear, but it is *p-linear*, namely $F^*(\lambda a) = \lambda^p F^*(a)$ for all $\lambda \in k$, $a \in H^1(X, \mathcal{O}_X)$. Since X is elliptic, $H^1(X, \mathcal{O}_X)$ is a one-dimensional vector space. Thus, since k is perfect, the map F^* is either 0 or bijective.

Definition. If $F^* = 0$, we say that X has *Hasse invariant* 0 or that X is *super-singular*; otherwise we say that X has *Hasse invariant* 1.

For other interpretations of the Hasse invariant, see (Ex. 4.15), (Ex. 4.16).

Proposition 4.21. *Let the elliptic curve X be embedded as a cubic curve in $\mathbf{P}^2$ with homogeneous equation $f(x, y, z) = 0$. Then the Hasse invariant of X is 0 if and only if the coefficient of $(xyz)^{p-1}$ in f^{p-1} is 0.*

PROOF. The ideal sheaf of X is isomorphic to $\mathcal{O}_\mathbf{P}(-3)$, so we have an exact sequence

$$0 \to \mathcal{O}_\mathbf{P}(-3) \xrightarrow{f} \mathcal{O}_\mathbf{P} \to \mathcal{O}_X \to 0.$$

From this, by taking cohomology, we obtain an isomorphism

$$H^1(X, \mathcal{O}_X) \to H^2(\mathcal{O}_\mathbf{P}(-3)),$$

since $H^i(\mathcal{O}_\mathbf{P}) = 0$ for $i = 1, 2$. Recall also (III, 5.1) that $H^2(\mathcal{O}_\mathbf{P}(-3))$ is a one-dimensional vector space with a natural basis $(xyz)^{-1}$.

Now we can compute the action of Frobenius using this embedding. If F_1 is the Frobenius morphism on $\mathbf{P}^2$, then F_1^* takes $\mathcal{O}_X$ to $\mathcal{O}_{X^p}$, where X^p is the subscheme of $\mathbf{P}^2$ defined by $f^p = 0$. On the other hand, X is a closed subscheme of X^p, so we have a commutative diagram

$$
\begin{array}{ccccccccc}
0 & \longrightarrow & \mathcal{O}_\mathbf{P}(-3p) & \longrightarrow & \mathcal{O}_\mathbf{P} & \longrightarrow & \mathcal{O}_{X^p} & \longrightarrow & 0 \\
 & & \Big\downarrow{\scriptstyle f^{p-1}} & & \Big\downarrow & & \Big\downarrow & & \\
0 & \longrightarrow & \mathcal{O}_\mathbf{P}(-3) & \longrightarrow & \mathcal{O}_\mathbf{P} & \longrightarrow & \mathcal{O}_X & \longrightarrow & 0
\end{array}
$$

Hence we have a commutative diagram

$$
\begin{array}{ccc}
H^1(X,\mathcal{O}_X) & \overset{\sim}{\longrightarrow} & H^2(\mathbf{P}^2,\mathcal{O}_{\mathbf{P}}(-3)) \\
\big\downarrow{\scriptstyle F_1^*} & & \big\downarrow{\scriptstyle F_1^*} \\
H^1(X^p,\mathcal{O}_{X^p}) & \overset{\sim}{\longrightarrow} & H^2(\mathbf{P}^2,\mathcal{O}_{\mathbf{P}}(-3p)) \\
\big\downarrow & & \big\downarrow{\scriptstyle f^{p-1}} \\
H^1(X,\mathcal{O}_X) & \overset{\sim}{\longrightarrow} & H^2(\mathbf{P}^2,\mathcal{O}_{\mathbf{P}}(-3))
\end{array}
$$

F^* (spanning the left bracket)

where F is the Frobenius morphism of X. Now $F_1^*((xyz)^{-1}) = (xyz)^{-p}$, and its image in $H^2(\mathcal{O}_{\mathbf{P}}(-3))$ will be $f^{p-1} \cdot (xyz)^{-p}$. On the other hand, $H^2(\mathcal{O}_{\mathbf{P}}(-3))$ has basis $(xyz)^{-1}$, and any monomial having a nonnegative exponent on x, y, or z is 0. Thus the image is just $(xyz)^{-1}$ times the coefficient of $(xyz)^{p-1}$ in f^{p-1}, and so the Hasse invariant of X is determined by whether or not this coefficient is zero.

Corollary 4.22. *Assume $p \neq 2$, and let X be given by the equation $y^2 = x(x - 1)(x - \lambda)$, with $\lambda \neq 0,1$. Then the Hasse invariant of X is 0 if and only if $h_p(\lambda) = 0$, where*

$$
h_p(\lambda) = \sum_{i=0}^{k} \binom{k}{i}^2 \lambda^i, \qquad k = \frac{1}{2}(p-1).
$$

PROOF. We use the criterion of (4.21). In this case $f = y^2 z - x(x - z)(x - \lambda z)$. To get $(xyz)^{p-1}$ in f^{p-1}, we must have $(y^2 z)^k$ and $(x(x - z)(x - \lambda z))^k$. Then, inside $((x - z)(x - \lambda z))^k$, we need the coefficient of $x^k z^k$. So we take the coefficient of $x^i z^{k-i}$ in $(x - z)^k$, and the coefficient of $x^{k-i} z^i$ in $(x - \lambda z)^k$. Summing up, the coefficient of $(xyz)^{p-1}$ in f^{p-1} is

$$
(-1)^k \binom{p-1}{k} \sum \binom{k}{i}^2 \lambda^i.
$$

Since the outer factor is $\equiv 1 \pmod{p}$, we get $h_p(\lambda)$ as defined above.

Corollary 4.23. *For given p, there are only finitely many elliptic curves (up to isomorphism) over k having Hasse invariant 0. In fact, there are at most $[p/12] + 2$ of them.*

PROOF. The polynomial $h_p(\lambda)$ has degree $k = \frac{1}{2}(p - 1)$ in λ, so it has at most k distinct roots. In particular, there are only finitely many corresponding values of j. Since the correspondence $\lambda \mapsto j$ is 6 to 1 with two exceptions, we can have at most $k/6 + 2$ values of j, hence at most $[p/12] + 2$.

Note. In fact, Igusa [2] has shown that the roots of $h_p(\lambda)$ are always distinct. Using this, one can easily count the exact number of j with Hasse invariant 0: $j = 0$ occurs $\Leftrightarrow p \equiv 2 \pmod 3$ (Ex. 4.14); $j = 1728$ occurs $\Leftrightarrow p \equiv 3 \pmod 4$ (4.23.5); the number of $j \neq 0, 1728$ is exactly $[p/12]$. There are also tables of these j for small values of p—see Deuring [1] or Birch and Kuyk [1, Table 6].

Example 4.23.1. Let $p = 3$. Then $h_p(\lambda) = \lambda + 1$. The only solution is $\lambda = -1$, which corresponds to $j = 0 = 1728$.

Example 4.23.2. If $p = 5$, $h_p(\lambda) = \lambda^2 + 4\lambda + 1 \equiv \lambda^2 - \lambda + 1 \pmod 5$. This has roots $-\omega, -\omega^2$ in a quadratic extension of $\mathbf{F}_p$, with $\omega^3 = 1$. So $j = 0$.

Example 4.23.3. If $p = 7$, then
$$h_p(\lambda) = \lambda^3 + 9\lambda^2 + 9\lambda + 1.$$
This has roots $-1, 2, 4$, which correspond to $j = 1728$.

Remark 4.23.4. A very interesting problem arises if we "fix the curve and vary p." To make sense of this, let $X \subseteq \mathbf{P}_{\mathbf{Z}}^2$ be a cubic curve defined by an equation $f(x, y, z) = 0$ with integer coefficients, and assume that X is nonsingular as a curve over $\mathbf{C}$. Then for almost all primes p, the curve $X_{(p)} \subseteq \mathbf{P}_{\mathbf{F}_p}^2$ obtained by reducing the coefficients of f (mod p) will be nonsingular over $k_{(p)} = \bar{\mathbf{F}}_p$. So it makes sense to consider the set
$$\mathfrak{P} = \{p \text{ prime} | X_{(p)} \text{ is nonsingular over } k_{(p)}, \text{ and } X_{(p)} \text{ has Hasse invariant } 0\}.$$

What can we say about this set? The facts (which we will not prove) are that if X, as a curve over $\mathbf{C}$, has complex multiplication, then $\mathfrak{P}$ has density $\frac{1}{2}$. Here we define the *density* of a set of primes $\mathfrak{P}$ to be
$$\lim_{x \to \infty} \# \{p \in \mathfrak{P} | p \leqslant x\} / \# \{p \text{ prime} | p \leqslant x\}.$$

In fact, assuming $X_{(p)}$ is nonsingular, then $X_{(p)}$ has Hasse invariant 0 if and only if either p is ramified or p remains prime in the imaginary quadratic field containing the ring of complex multiplication of X (Deuring [1]). If X does not have complex multiplication, then $\mathfrak{P}$ has density 0, but Elkies has shown that $\mathfrak{P}$ is infinite (N. Elkies, The existence of infinitely many supersingular primes for every elliptic curve over $\mathbf{Q}$, Invent. Math. *89* (1987) 561–567). There is also ample numerical evidence for the conjecture of Lang and Trotter [1], that more precisely
$$\# \{p \in \mathfrak{P} | p \leqslant x\} \sim c \cdot \sqrt{x}/\log x$$
as $x \to \infty$, for some constant $c > 0$.

Example 4.23.5. Let X be the curve $y^2 = x^3 - x$. Then $j = 1728$, and as we have seen (4.20.1), X has complex multiplication by i. For any $p \neq 2$, $X_{(p)}$ is nonsingular, and we compute its Hasse invariant by the criterion of (4.21). With $k = \frac{1}{2}(p - 1)$, we need the coefficient of x^k in $(x^2 - 1)^k$. If k is odd, it is 0. If k is even, say $k = 2m$, it is $(-1)^m\binom{k}{m}$ which is nonzero. We conclude that

$$\begin{cases} \text{if } p \equiv 1 \pmod{4}, \text{ then Hasse } = 1 \\ \text{if } p \equiv 3 \pmod{4}, \text{ then Hasse } = 0. \end{cases}$$

Thus $\mathfrak{P} = \{p \text{ prime} | p \equiv 3 \pmod{4}\}$. According to Dirichlet's theorem on primes in arithmetic progressions (see, e.g., Serre [14, Ch VI, §4], this is a set of primes of density $\frac{1}{2}$. In particular, there are infinitely many such primes. Note that $p \equiv 3 \pmod{4}$ if and only if p is prime in the ring of Gaussian integers $\mathbf{Z}[i]$.

Example 4.23.6. Let X be the curve $y^2 = x(x - 1)(x + 2)$, so $\lambda = -2$, and $j = 2^6 \cdot 3^{-2} \cdot 7^3$. Then $X_{(p)}$ is nonsingular for $p \neq 2,3$, but one checks by the criterion of (4.22), using a calculator, that the only value of $p \leqslant 73$ giving Hasse $= 0$ is $p = 23$. So we can guess that $\mathfrak{P}$ has density 0. Indeed, j is not an integer, so by (4.20.4), X does not have complex multiplication. See Lang and Trotter [1] for more extensive computations.

Rational Points on an Elliptic Curve

Let X be an elliptic curve over an algebraically closed field k, let P_0 be a fixed point, and let X be embedded in $\mathbf{P}_k^2$ by the linear system $|3P_0|$. Suppose that X can be defined by an equation $f(x,y,z) = 0$ with coefficients in a smaller field $k_0 \subseteq k$, and that the point P_0 has coordinates in k_0. In this case we say (X,P_0) is *defined over* k_0. If this happens, then it is clear from the geometric nature of the group law on X, that the set $X(k_0)$ of points of X with coordinates in k_0 forms a subgroup of the group of all points of X. It is an interesting arithmetic problem to determine the nature of this subgroup.

In particular, if $k = \mathbf{C}$ and $k_0 = \mathbf{Q}$, then because x,y,z are homogeneous coordinates in $\mathbf{P}^2$, we may assume that the equation $f(x,y,z) = 0$ has integer coefficients, and we are looking for integer solutions x,y,z. So we have a cubic Diophantine equation in three variables.

A theorem of Mordell states that the group $X(\mathbf{Q})$ is a finitely generated abelian group. We will not prove this, but just give some examples. See Cassels [1] and Tate [3] for two excellent surveys of the subject.

Example 4.23.7. The Fermat curve $x^3 + y^3 = z^3$ is defined over $\mathbf{Q}$. Because Fermat's theorem is true for exponent 3, the only points of $X(\mathbf{Q})$ are $(1,-1,0)$, $(1,0,1)$, and $(0,1,1)$. These are three inflection points of X. Taking any one as base point, the group $X(\mathbf{Q})$ is isomorphic to $\mathbf{Z}/3$.

Example 4.23.8. The curve $y^2 + y = x^3 - x$ is defined over $\mathbf{Q}$. Take $P_0 = (0,1,0)$ to be the 0 element in the group law, as usual. Then (according to Tate [3]), the group $X(\mathbf{Q})$ is infinite cyclic, generated by the point P with affine coordinates $(0,0)$. Figure 17 shows this curve, with nP labeled as n, for various integers n.

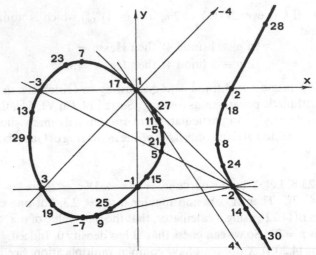

Figure 17. Rational points on the curve $y^2 + y = x^3 - x$.

EXERCISES

4.1. Let X be an elliptic curve over k, with char $k \neq 2$, let $P \in X$ be a point, and let R be the graded ring $R = \bigoplus_{n \geq 0} H^0(X, \mathcal{O}_X(nP))$. Show that for suitable choice of t, x, y,

$$R \cong k[t, x, y]/(y^2 - x(x - t^2)(x - \lambda t^2)),$$

as a graded ring, where $k[t, x, y]$ is graded by setting $\deg t = 1$, $\deg x = 2$, $\deg y = 3$.

4.2. If D is any divisor of degree ≥ 3 on the elliptic curve X, and if we embed X in $\mathbf{P}^n$ by the complete linear system $|D|$, show that the image of X in $\mathbf{P}^n$ is projectively normal.

Note. It is true more generally that if D is a divisor of degree $\geq 2g + 1$ on a curve of genus g, then the embedding of X by $|D|$ is projectively normal (Mumford [4, p. 55]).

4.3. Let the elliptic curve X be embedded in $\mathbf{P}^2$ so as to have the equation $y^2 = x(x - 1)(x - \lambda)$. Show that any automorphism of X leaving $P_0 = (0,1,0)$ fixed is induced by an automorphism of $\mathbf{P}^2$ coming from the automorphism of the affine (x, y)-plane given by

$$\begin{cases} x' = ax + b \\ y' = cy. \end{cases}$$

In each of the four cases of (4.7), describe these automorphisms of $\mathbf{P}^2$ explicitly, and hence determine the structure of the group $G = \mathrm{Aut}(X, P_0)$.

4.4. Let X be an elliptic curve in $\mathbf{P}^2$ given by an equation of the form

$$y^2 + a_1 xy + a_3 y = x^3 + a_2 x^2 + a_4 x + a_6.$$

Show that the j-invariant is a rational function of the a_i, with coefficients in $\mathbf{Q}$. In particular, if the a_i are all in some field $k_0 \subseteq k$, then $j \in k_0$ also. Furthermore, for every $\alpha \in k_0$, there exists an elliptic curve defined over k_0 with j-invariant equal to α.

4.5. Let X, P_0 be an elliptic curve having an endomorphism $f : X \to X$ of degree 2.
  (a) If we represent X as a 2-1 covering of $\mathbf{P}^1$ by a morphism $\pi : X \to \mathbf{P}^1$ ramified at P_0, then as in (4.4), show that there is another morphism $\pi' : X \to \mathbf{P}^1$ and a morphism $g : \mathbf{P}^1 \to \mathbf{P}^1$, also of degree 2, such that $\pi \circ f = g \circ \pi'$.
  (b) For suitable choices of coordinates in the two copies of $\mathbf{P}^1$, show that g can be taken to be the morphism $x \to x^2$.
  (c) Now show that g is branched over two of the branch points of π, and that g^{-1} of the other two branch points of π consists of the four branch points of π'. Deduce a relation involving the invariant λ of X.
  (d) Solving the above, show that there are just three values of j corresponding to elliptic curves with an endomorphism of degree 2, and find the corresponding values of λ and j. [Answers: $j = 2^6 \cdot 3^3$; $j = 2^6 \cdot 5^3$; $j = -3^3 \cdot 5^3$.]

4.6. (a) Let X be a curve of genus g embedded birationally in $\mathbf{P}^2$ as a curve of degree d with r nodes. Generalize the method of (Ex. 2.3) to show that X has $6(g-1) + 3d$ inflection points. A node does not count as an inflection point. Assume char $k = 0$.
  (b) Now let X be a curve of genus g embedded as a curve of degree d in $\mathbf{P}^n$, $n \geqslant 3$, not contained in any $\mathbf{P}^{n-1}$. For each point $P \in X$, there is a hyperplane H containing P, such that P counts at least n times in the intersection $H \cap X$. This is called an *osculating hyperplane* at P. It generalizes the notion of tangent line for curves in $\mathbf{P}^2$. If P counts at least $n+1$ times in $H \cap X$, we say H is a *hyperosculating hyperplane*, and that P is a *hyperosculation point*. Use Hurwitz's theorem as above, and induction on n, to show that X has $n(n+1)(g-1) + (n+1)d$ hyperosculation points.
  (c) If X is an elliptic curve, for any $d \geqslant 3$, embed X as a curve of degree d in $\mathbf{P}^{d-1}$, and conclude that X has exactly d^2 points of order d in its group law.

4.7. *The Dual of a Morphism.* Let X and X' be elliptic curves over k, with base points P_0, P'_0.
  (a) If $f : X \to X'$ is any morphism, use (4.11) to show that $f^* : \operatorname{Pic} X' \to \operatorname{Pic} X$ induces a homomorphism $\hat{f} : (X', P'_0) \to (X, P_0)$. We call this the *dual* of f.
  (b) If $f : X \to X'$ and $g : X' \to X''$ are two morphisms, then $(g \circ f)\hat{\ } = \hat{f} \circ \hat{g}$.
  (c) Assume $f(P_0) = P'_0$, and let $n = \deg f$. Show that if $Q \in X$ is any point, and $f(Q) = Q'$, then $\hat{f}(Q') = n_X(Q)$. (Do the separable and purely inseparable cases separately, then combine.) Conclude that $f \circ \hat{f} = n_{X'}$ and $\hat{f} \circ f = n_X$.
  *(d) If $f, g : X \to X'$ are two morphisms preserving the base points P_0, P'_0, then $(f+g)\hat{\ } = \hat{f} + \hat{g}$. [*Hints:* It is enough to show for any $\mathscr{L} \in \operatorname{Pic} X'$, that $(f+g)^* \mathscr{L} \cong f^* \mathscr{L} \otimes g^* \mathscr{L}$. For any f, let $\Gamma_f : X \to X \times X'$ be the graph morphism. Then it is enough to show (for $\mathscr{L}' = p_2^* \mathscr{L}$) that

$$\Gamma^*_{f+g}(\mathscr{L}') = \Gamma^*_f \mathscr{L}' \otimes \Gamma^*_g \mathscr{L}'.$$

Let $\sigma : X \to X \times X'$ be the section $x \to (x, P'_0)$. Define a subgroup of $\operatorname{Pic}(X \times X')$ as follows:

$$\operatorname{Pic}^o_\sigma = \{ \mathscr{L} \in \operatorname{Pic}(X \times X') | \mathscr{L} \text{ has degree 0 along each fibre of } p_1, \text{ and } \sigma^* \mathscr{L} = 0 \text{ in } \operatorname{Pic} X \}.$$

Note that this subgroup is isomorphic to the group $\text{Pic}^\circ(X'/X)$ used in the definition of the Jacobian variety. Hence there is a 1-1 correspondence between morphisms $f : X \to X'$ and elements $\mathscr{L}_f \in \text{Pic}_\sigma$ (this defines $\mathscr{L}_f$). Now compute explicitly to show that $\Gamma_g^*(\mathscr{L}_f) = \Gamma_f^*(\mathscr{L}_g)$ for any f, g.

Use the fact that $\mathscr{L}_{f+g} = \mathscr{L}_f \otimes \mathscr{L}_g$, and the fact that for any $\mathscr{L}$ on X', $p_2^* \mathscr{L} \in \text{Pic}_\sigma^\circ$ to prove the result.]

(e) Using (d), show that for any $n \in \mathbf{Z}$, $\hat{n}_X = n_X$. Conclude that $\deg n_X = n^2$.

(f) Show for any f that $\deg \hat{f} = \deg f$.

4.8. For any curve X, the *algebraic fundamental group* $\pi_1(X)$ is defined as $\varprojlim \text{Gal}(K'/K)$, where K is the function field of X, and K' runs over all Galois extensions of K such that the corresponding curve X' is étale over X (III, Ex. 10.3). Thus, for example, $\pi_1(\mathbf{P}^1) = 1$ (2.5.3). Show that for an elliptic curve X,

$$\pi_1(X) = \prod_{l \text{ prime}} \mathbf{Z}_l \times \mathbf{Z}_l \qquad \text{if char } k = 0;$$

$$\pi_1(X) = \prod_{l \neq p} \mathbf{Z}_l \times \mathbf{Z}_l \qquad \text{if char } k = p \text{ and Hasse } X = 0;$$

$$\pi_1(X) = \mathbf{Z}_p \times \prod_{l \neq p} \mathbf{Z}_l \times \mathbf{Z}_l \qquad \text{if char } k = p \text{ and Hasse } X \neq 0,$$

where $\mathbf{Z}_l = \varprojlim \mathbf{Z}/l^r$ is the l-adic integers.

[*Hints*: Any Galois étale cover X' of an elliptic curve is again an elliptic curve. If the degree of X' over X is relatively prime to p, then X' can be dominated by the cover $n_X : X \to X'$ for some integer n with $(n, p) = 1$. The Galois group of the covering n_X is $\mathbf{Z}/n \times \mathbf{Z}/n$. Étale covers of degree divisible by p can occur only if the Hasse invariant of X is not zero.]

Note: More generally, Grothendieck has shown [SGA 1, X, 2.6, p. 272] that the algebraic fundamental group of any curve of genus g is isomorphic to a quotient of the completion, with respect to subgroups of finite index, of the ordinary topological fundamental group of a compact Riemann surface of genus g, i.e., a group with $2g$ generators $a_1, \ldots, a_g, b_1, \ldots, b_g$ and the relation $(a_1 b_1 a_1^{-1} b_1^{-1}) \cdots (a_g b_g a_g^{-1} b_g^{-1}) = 1$.

4.9. We say two elliptic curves X, X' are *isogenous* if there is a finite morphism $f : X \to X'$.

(a) Show that isogeny is an equivalence relation.

(b) For any elliptic curve X, show that the set of elliptic curves X' isogenous to X, up to isomorphism, is countable. [*Hint*: X' is uniquely determined by X and $\ker f$.]

4.10. If X is an elliptic curve, show that there is an exact sequence

$$0 \to p_1^* \text{Pic } X \oplus p_2^* \text{Pic } X \to \text{Pic}(X \times X) \to R \to 0,$$

where $R = \text{End}(X, P_0)$. In particular, we see that $\text{Pic}(X \times X)$ is bigger than the sum of the Picard groups of the factors. Cf. (III, Ex. 12.6), (V, Ex. 1.6).

4.11. Let X be an elliptic curve over $\mathbf{C}$, defined by the elliptic functions with periods $1, \tau$. Let R be the ring of endomorphisms of X.

(a) If $f \in R$ is a nonzero endomorphism corresponding to complex multiplication by α, as in (4.18), show that $\deg f = |\alpha|^2$.

(b) If $f \in R$ corresponds to $\alpha \in \mathbf{C}$ again, show that the dual $\hat{f}$ of (Ex. 4.7) corresponds to the complex conjugate $\bar{\alpha}$ of α.

(c) If $\tau \in \mathbf{Q}(\sqrt{-d})$ happens to be integral over $\mathbf{Z}$, show that $R = \mathbf{Z}[\tau]$.

4.12. Again let X be an elliptic curve over $\mathbf{C}$ determined by the elliptic functions with periods $1, \tau$, and assume that τ lies in the region G of (4.15B).
(a) If X has any automorphisms leaving P_0 fixed other than ± 1, show that either $\tau = i$ or $\tau = \omega$, as in (4.20.1) and (4.20.2). This gives another proof of the fact (4.7) that there are only two curves, up to isomorphism, having automorphisms other than ± 1.
(b) Now show that there are exactly three values of τ for which X admits an endomorphism of degree 2. Can you match these with the three values of j determined in (Ex. 4.5)? [Answers: $\tau = i$; $\tau = \sqrt{-2}$; $\tau = \frac{1}{2}(-1 + \sqrt{-7})$.]

4.13. If $p = 13$, there is just one value of j for which the Hasse invariant of the corresponding curve is 0. Find it. [Answer: $j = 5 \pmod{13}$.]

4.14. The Fermat curve $X : x^3 + y^3 = z^3$ gives a nonsingular curve in characteristic p for every $p \neq 3$. Determine the set $\mathfrak{P} = \{p \neq 3 | X_{(p)}$ has Hasse invariant $0\}$, and observe (modulo Dirichlet's theorem) that it is a set of primes of density $\frac{1}{2}$.

4.15. Let X be an elliptic curve over a field k of characteristic p. Let $F' : X_p \to X$ be the k-linear Frobenius morphism (2.4.1). Use (4.10.7) to show that the dual morphism $\hat{F}' : X \to X_p$ is separable if and only if the Hasse invariant of X is 1. Now use (Ex. 4.7) to show that if the Hasse invariant is 1, then the subgroup of points of order p on X is isomorphic to $\mathbf{Z}/p$; if the Hasse invariant is 0, it is 0.

4.16. Again let X be an elliptic curve over k of characteristic p, and suppose X is defined over the field $\mathbf{F}_q$ of $q = p^r$ elements, i.e., $X \subseteq \mathbf{P}^2$ can be defined by an equation with coefficients in $\mathbf{F}_q$. Assume also that X has a rational point over $\mathbf{F}_q$. Let $F' : X_q \to X$ be the k-linear Frobenius with respect to q.
(a) Show that $X_q \cong X$ as schemes over k, and that under this identification, $F' : X \to X$ is the map obtained by the qth-power map on the coordinates of points of X, embedded in $\mathbf{P}^2$.
(b) Show that $1_X - F'$ is a separable morphism and its kernel is just the set $X(\mathbf{F}_q)$ of points of X with coordinates in $\mathbf{F}_q$.
(c) Using (Ex. 4.7), show that $F' + \hat{F}' = a_X$ for some integer a, and that $N = q - a + 1$, where $N = \#X(\mathbf{F}_q)$.
(d) Use the fact that $\deg(m + nF') > 0$ for all $m, n \in \mathbf{Z}$ to show that $|a| \leqslant 2\sqrt{q}$. This is Hasse's proof of the analogue of the Riemann hypothesis for elliptic curves (App. C, Ex. 5.6).
(e) Now assume $q = p$, and show that the Hasse invariant of X is 0 if and only if $a \equiv 0 \pmod{p}$. Conclude for $p \geqslant 5$ that X has Hasse invariant 0 if and only if $N = p + 1$.

4.17. Let X be the curve $y^2 + y = x^3 - x$ of (4.23.8).
(a) If $Q = (a,b)$ is a point on the curve, compute the coordinates of the point $P + Q$, where $P = (0,0)$, as a function of a,b. Use this formula to find the coordinates of nP, $n = 1, 2, \ldots, 10$. [Check: $6P = (6,14)$.]
(b) This equation defines a nonsingular curve over $\mathbf{F}_p$ for all $p \neq 37$.

4.18. Let X be the curve $y^2 = x^3 - 7x + 10$. This curve has at least 26 points with integer coordinates. Find them (use a calculator), and verify that they are all contained in the subgroup (maybe equal to all of $X(\mathbf{Q})$?) generated by $P = (1,2)$ and $Q = (2,2)$.

4.19. Let X, P_0 be an elliptic curve defined over $\mathbf{Q}$, represented as a curve in $\mathbf{P}^2$ defined by an equation with integer coefficients. Then X can be considered as the fibre over the generic point of a scheme $\bar{X}$ over Spec $\mathbf{Z}$. Let $T \subseteq$ Spec $\mathbf{Z}$ be the open subset consisting of all primes $p \neq 2$ such that the fibre $X_{(p)}$ of $\bar{X}$ over p is nonsingular. For any n, show that $n_X : X \to X$ is defined over T, and is a flat morphism. Show that the kernel of n_X is also flat over T. Conclude that for any $p \in T$, the natural map $X(\mathbf{Q}) \to X_{(p)}(\mathbf{F}_p)$ induced on the groups of rational points, maps the n-torsion points of $X(\mathbf{Q})$ *injectively* into the torsion subgroup of $X_{(p)}(\mathbf{F}_p)$, for any $(n,p) = 1$.

By this method one can show easily that the groups $X(\mathbf{Q})$ in (Ex. 4.17) and (Ex. 4.18) are torsion-free.

4.20. Let X be an elliptic curve over a field k of characteristic $p > 0$, and let $R = \text{End}(X,P_0)$ be its ring of endomorphisms.
   (a) Let X_p be the curve over k defined by changing the k-structure of X (2.4.1). Show that $j(X_p) = j(X)^{1/p}$. Thus $X \cong X_p$ over k if and only if $j \in \mathbf{F}_p$.
   (b) Show that p_X in R factors into a product $\pi\hat{\pi}$ of two elements of degree p if and only if $X \cong X_p$. In this case, the Hasse invariant of X is 0 if and only if π and $\hat{\pi}$ are associates in R (i.e., differ by a unit). (Use (2.5).)
   (c) If Hasse $(X) = 0$ show in any case $j \in \mathbf{F}_{p^2}$.
   (d) For any $f \in R$, there is an induced map $f^* : H^1(\mathcal{O}_X) \to H^1(\mathcal{O}_X)$. This must be multiplication by an element $\lambda_f \in k$. So we obtain a ring homomorphism $\varphi : R \to k$ by sending f to λ_f. Show that any $f \in R$ commutes with the (nonlinear) Frobenius morphism $F : X \to X$, and conclude that if Hasse $(X) \neq 0$, then the image of φ is $\mathbf{F}_p$. Therefore, R contains a prime ideal $\mathfrak{p}$ with $R/\mathfrak{p} \cong \mathbf{F}_p$.

4.21. Let O be the ring of integers in a quadratic number field $\mathbf{Q}(\sqrt{-d})$. Show that any subring $R \subseteq O$, $R \neq \mathbf{Z}$, is of the form $R = \mathbf{Z} + f \cdot O$, for a uniquely determined integer $f \geqslant 1$. This integer f is called the *conductor* of the ring R.

*4.22. If $X \to \mathbf{A}^1_{\mathbf{C}}$ is a family of elliptic curves having a section, show that the family is trivial. [*Hints*: Use the section to fix the group structure on the fibres. Show that the points of order 2 on the fibres form an étale cover of $\mathbf{A}^1_{\mathbf{C}}$, which must be trivial, since $\mathbf{A}^1_{\mathbf{C}}$ is simply connected. This implies that λ can be defined on the family, so it gives a map $\mathbf{A}^1_{\mathbf{C}} \to \mathbf{A}^1_{\mathbf{C}} - \{0,1\}$. Any such map is constant, so λ is constant, so the family is trivial.]

5 The Canonical Embedding

We return now to the study of curves of arbitrary genus, and we study the rational map to a projective space determined by the canonical linear system. For nonhyperelliptic curves of genus $g \geqslant 3$, we will see that it is an embedding, which we call the canonical embedding. Closely related to this discussion is Clifford's theorem about the dimension of a special linear system, which we prove below. Using these results, we will say something about the classification of curves.

Throughout this section, X will denote a curve of genus g over the algebraically closed field k. We will consider the canonical linear system $|K|$. If $g = 0$, $|K|$ is empty. If $g = 1$, $|K| = 0$, so it determines the constant map of X to a point. For $g \geq 2$, however, $|K|$ is an effective linear system without base points, as we will see, so it determines a morphism to projective space which we call the *canonical morphism*.

Lemma 5.1. *If $g \geq 2$, then the canonical linear system $|K|$ has no base points.*

PROOF. According to (3.1), we must show that for each $P \in X$, $\dim |K - P| = \dim |K| - 1$. Now $\dim |K| = \dim H^0(X, \omega_X) - 1 = g - 1$. On the other hand, since X is not rational, for any point P, $\dim |P| = 0$, so by Riemann–Roch we find that $\dim |K - P| = g - 2$, as required.

Recall that a curve X of genus $g \geq 2$ is called *hyperelliptic* (Ex. 1.7) if there is a finite morphism $f : X \to \mathbf{P}^1$ of degree 2. Considering the corresponding linear system, we see that X is hyperelliptic if and only if it has a linear system of dimension 1 and degree 2. It is convenient here to introduce a classical notation. The *symbol g_d^r* will stand for "a linear system of dimension r and degree d." Thus we say X is hyperelliptic if it has a g_2^1.

If X is a curve of genus 2, then the canonical linear system $|K|$ is a g_2^1 (Ex. 1.7). So X is necessarily hyperelliptic, and the canonical morphism $f : X \to \mathbf{P}^1$ is the 2-1 map of the definition.

Proposition 5.2. *Let X be a curve of genus $g \geq 2$. Then $|K|$ is very ample if and only if X is not hyperelliptic.*

PROOF. We use the criterion of (3.1). Since $\dim |K| = g - 1$ we see that $|K|$ is very ample if and only if for every $P, Q \in X$, possibly equal, $\dim |K - P - Q| = g - 3$. Applying Riemann–Roch to the divisor $P + Q$, we have
$$\dim |P + Q| - \dim |K - P - Q| = 2 + 1 - g.$$
So the question is whether $\dim |P + Q| = 0$. If X is hyperelliptic, then for any divisor $P + Q$ of the g_2^1 we have $\dim |P + Q| = 1$. Conversely, if $\dim |P + Q| > 0$ for some P, Q, then the linear system $|P + Q|$ contains a g_2^1 (in fact is a g_2^1), so X is hyperelliptic. This completes the proof.

Definition. If X is nonhyperelliptic of genus $g \geq 3$, the embedding $X \to \mathbf{P}^{g-1}$ determined by the canonical linear system is the *canonical embedding* of X (determined up to an automorphism of $\mathbf{P}^{g-1}$), and its image, which is a curve of degree $2g - 2$, is a *canonical curve*.

Example 5.2.1. If X is a nonhyperelliptic curve of genus 3, then its canonical embedding is a quartic curve in $\mathbf{P}^2$. Conversely, any nonsingular quartic curve X in $\mathbf{P}^2$ has $\omega_X \cong \mathcal{O}_X(1)$ (II, 8.20.3), so it is a canonical curve. In particular, there exist nonhyperelliptic curves of genus 3 (see also (Ex. 3.2)).

Example 5.2.2. If X is a nonhyperelliptic curve of genus 4, then its canonical embedding is a curve of degree 6 in $\mathbf{P}^3$. We will show that X is contained in a unique irreducible quadric surface Q, and that X is the complete intersection of Q with an irreducible cubic surface F. Conversely, if X is a nonsingular curve in $\mathbf{P}^3$ which is a complete intersection of a quadric and a cubic surface, then $\deg X = 6$, and $\omega_X = \mathcal{O}_X(1)$ (II, Ex. 8.4), so X is a canonical curve of genus 4. In particular, there exist such nonsingular complete intersections by Bertini's theorem (II, Ex. 8.4), so there exist nonhyperelliptic curves of genus 4.

To prove the above assertions, let X be a canonical curve of genus 4 in $\mathbf{P}^3$, and let $\mathscr{I}$ be its ideal sheaf. Then we have an exact sequence

$$0 \to \mathscr{I} \to \mathcal{O}_{\mathbf{P}} \to \mathcal{O}_X \to 0.$$

Twisting by 2 and taking cohomology, we have

$$0 \to H^0(\mathbf{P},\mathscr{I}(2)) \to H^0(\mathbf{P},\mathcal{O}_{\mathbf{P}}(2)) \to H^0(X,\mathcal{O}_X(2)) \to \cdots$$

Now the middle vector space has dimension 10 by (III, 5.1), and the right hand vector space has dimension 9 by Riemann–Roch on X (note that $\mathcal{O}_X(2)$ corresponds to the divisor $2K$, which is nonspecial of degree 12). So we conclude that

$$\dim H^0(\mathbf{P},\mathscr{I}(2)) \geqslant 1.$$

An element of that space is a form of degree 2, whose zero-set will be a surface $Q \subseteq \mathbf{P}^3$ of degree 2 containing X. It must be irreducible (and reduced), because X is not contained in any $\mathbf{P}^2$. The curve X could not be contained in two distinct irreducible quadric surfaces Q,Q', because then it would be contained in their intersection $Q \cap Q'$ which is a curve of degree 4, and that is impossible because $\deg X = 6$. So we see that X is contained in a unique irreducible quadric surface Q.

Twisting the same sequence by 3 and taking cohomology, a similar calculation shows that

$$\dim H^0(\mathbf{P},\mathscr{I}(3)) \geqslant 5.$$

The cubic forms in here consisting of the quadratic form above times a linear form, form a subspace of dimension 4. Hence there is an irreducible cubic form in that space, so X is contained in an irreducible cubic surface F. Then X must be contained in the complete intersection $Q \cap F$, and since both have degree 6, X is equal to that complete intersection.

Proposition 5.3. *Let X be a hyperelliptic curve of genus $g \geqslant 2$. Then X has a unique g_2^1. If $f_0 : X \to \mathbf{P}^1$ is the corresponding morphism of degree 2, then the canonical morphism $f : X \to \mathbf{P}^{g-1}$ consists of f_0 followed by the*

$(g - 1)$-uple *embedding of* $\mathbf{P}^1$ *in* $\mathbf{P}^{g-1}$. *In particular, the image* $X' = f(X)$ *is a rational normal curve of degree* $g - 1$ *(Ex. 3.4), and* f *is a morphism of degree 2 onto* X'. *Finally, every effective canonical divisor on* X *is a sum of* $g - 1$ *divisors in the unique* g_2^1, *so we write* $|K| = \sum_1^{g-1} g_2^1$.

PROOF. We begin by considering the canonical morphism $f: X \to \mathbf{P}^{g-1}$, and let X' be its image. Since X is hyperelliptic, it has a g_2^1, by definition. We don't yet know that it is unique, so fix one for the moment. For any divisor $P + Q \in g_2^1$, the proof of (5.2) shows that Q is a base point of $|K - P|$, so $f(P) = f(Q)$. Since the g_2^1 has infinitely many divisors in it, we see that f cannot be birational. So let the degree of the map $f: X \to X'$ be $\mu \geqslant 2$, and let $d = \deg X'$. Then since $\deg K = 2g - 2$, we have $d\mu = 2g - 2$, hence $d \leqslant g - 1$.

Next, let $\tilde{X}'$ be the normalization of X', and let $\mathfrak{d}$ be the linear system on $\tilde{X}'$ corresponding to the morphism $\tilde{X}' \to X' \subseteq \mathbf{P}^{g-1}$. Then $\mathfrak{d}$ is a linear system of degree d and dimension $g - 1$. Since $d \leqslant g - 1$, we conclude (Ex. 1.5) that $d = g - 1$, the genus of $\tilde{X}'$ is 0, so $\tilde{X}' \cong \mathbf{P}^1$, and the linear system $\mathfrak{d}$ is the unique complete linear system on $\mathbf{P}^1$ of degree $g - 1$, namely $|(g - 1) \cdot P|$. Therefore, X' is the $(g - 1)$-uple embedding of $\mathbf{P}^1$. In particular, it is nonsingular, and it is a rational normal curve in the sense of (Ex. 3.4).

Next, from $d\mu = 2g - 2$, we conclude that $\mu = 2$. Since f already collapses the pairs of the g_2^1 we chose above, it must be equal to the composition of the map $f_0: X \to \mathbf{P}^1$ determined by our g_2^1 with the $(g - 1)$-uple embedding of $\mathbf{P}^1$. Thus the g_2^1 is determined by f, and so is uniquely determined.

Finally, any effective canonical divisor K on X is f^{-1} of a hyperplane section of X'. Hence it is a sum of $g - 1$ divisors in the unique g_2^1. Conversely, any set of $g - 1$ points of X' is a hyperplane section, so we can identify the canonical linear system $|K|$ with the set of sums of $g - 1$ divisors of the g_2^1. Hence we write

$$|K| = \sum_1^{g-1} g_2^1.$$

Now we come to Clifford's theorem. The idea is this. For a nonspecial divisor D on a curve X, we can compute $\dim|D|$ exactly as a function of $\deg D$ by the Riemann–Roch theorem. However, for a special divisor, $\dim|D|$ does not depend only on the degree. Hence it is useful to have some bound on $\dim|D|$, and this is provided by Clifford's theorem.

Theorem 5.4 (Clifford). *Let* D *be an effective special divisor on the curve* X. *Then*

$$\dim|D| \leqslant \frac{1}{2} \deg D.$$

Furthermore, equality occurs if and only if either $D = 0$ *or* $D = K$ *or* X *is hyperelliptic and* D *is a multiple of the unique* g_2^1 *on* X.

Lemma 5.5. *Let D,E be effective divisors on a curve X. Then*

$$\dim|D| + \dim|E| \leqslant \dim|D + E|.$$

PROOF. We define a map of sets

$$\varphi : |D| \times |E| \rightarrow |D + E|$$

by sending $\langle D', E' \rangle$ to $D' + E'$ for any $D' \in |D|$ and $E' \in |E|$. The map φ is finite-to-one, because a given effective divisor can be written in only finitely many ways as a sum of two other effective divisors. On the other hand, since φ corresponds to the natural bilinear map of vector spaces

$$H^0(X, \mathscr{L}(D)) \times H^0(X, \mathscr{L}(E)) \rightarrow H^0(X, \mathscr{L}(D + E)),$$

we see that φ is a morphism when we endow $|D|,|E|$ and $|D + E|$ with their structure of projective spaces. Therefore, since φ is finite-to-one, the dimension of its image is exactly $\dim|D| + \dim|E|$, and from this the result follows.

PROOF OF (5.4) (following Saint-Donat [1, §1]). If D is effective and special, then $K - D$ is also effective, so we can apply the lemma, and obtain

$$\dim|D| + \dim|K - D| \leqslant \dim|K| = g - 1.$$

On the other hand, by Riemann–Roch we have

$$\dim|D| - \dim|K - D| = \deg D + 1 - g.$$

Adding these two expressions, we have

$$2 \dim|D| \leqslant \deg D,$$

or in other words

$$\dim|D| \leqslant \frac{1}{2} \deg D.$$

This gives the first statement of the theorem. Also, it is clear that we have equality in case $D = 0$ or $D = K$.

For the second statement, suppose that $D \neq 0,K$, and that $\dim|D| = \frac{1}{2} \deg D$. Then we must show that X is hyperelliptic, and that D is a multiple of the g_2^1. We proceed by induction on $\deg D$ (which must be even). If $\deg D = 2$, then $|D|$ itself is a g_2^1, so X is hyperelliptic and there is nothing more to prove.

So suppose now that $\deg D \geqslant 4$, hence $\dim|D| \geqslant 2$. Fix a divisor $E \in |K - D|$, and fix two points $P,Q \in X$ such that $P \in \operatorname{Supp} E$ and $Q \notin \operatorname{Supp} E$. Since $\dim|D| \geqslant 2$, we can find a divisor $D \in |D|$ such that $P,Q \in \operatorname{Supp} D$. Now let $D' = D \cap E$, by which we mean the largest divisor dominated by both D and E. This D' will accomplish our induction.

First note that since $Q \in \operatorname{Supp} D$ but $Q \notin \operatorname{Supp} E$, we must have $Q \notin \operatorname{Supp} D'$, so $\deg D' < \deg D$. On the other hand, $\deg D' > 0$ since $P \in \operatorname{Supp} D'$.

Next, by construction of D', we have an exact sequence

$$0 \to \mathcal{L}(D') \to \mathcal{L}(D) \oplus \mathcal{L}(E) \to \mathcal{L}(D + E - D') \to 0,$$

where we consider these as subspaces of the constant sheaf $\mathcal{K}$ on X, and the first map is addition, the second subtraction. (Think of $\mathcal{L}(D) = \{f \in K(X) | (f) \geqslant -D\}$, cf. (II, 7.7).) Therefore, considering global sections of this sequence, we have

$$\dim|D| + \dim|E| \leqslant \dim|D'| + \dim|D + E - D'|.$$

But $E \sim K - D$ and $D + E - D' \sim K - D'$, so the left-hand side is equal to

$$\dim|D| + \dim|K - D|,$$

which must be $= \dim|K| = g - 1$ since we have $\dim|D| = \frac{1}{2} \deg D$ by hypothesis. On the other hand, the right-hand side is $\leqslant g - 1$ by (5.5) applied to D', so we must have equality everywhere. We conclude, then, that $\dim|D'| = \frac{1}{2} \deg D'$, as above. Now by the induction hypothesis, this implies that X is hyperelliptic.

Now suppose again that $D \neq 0, K$, and $\dim|D| = \frac{1}{2} \deg D$. Let $r = \dim|D|$. Consider the linear system $|D| + (g - 1 - r)g_2^1$. It has degree $2g - 2$, and dimension $\geqslant g - 1$, by (5.5) again, so it must be equal to the canonical system $|K|$. But we have already seen (5.3) that $|K| = (g - 1)g_2^1$. So we conclude that $|D| = rg_2^1$, which completes the proof.

Classification of Curves

To classify curves, we first specify the genus, which as we have seen (1.1.1) can be any nonnegative integer $g \geqslant 0$. If $g = 0$, X is isomorphic to $\mathbf{P}^1$ (1.3.5), so there is nothing further to say. If $g = 1$, then X is classified up to isomorphism by its j-invariant (4.1), so here again we have a good answer to the classification problem. For $g \geqslant 2$, the problem becomes much more difficult, and except for a few special cases (e.g., (Ex. 2.2)), one cannot give an explicit answer.

For $g \geqslant 3$ we can subdivide the set $\mathfrak{M}_g$ of all curves of genus g according to whether the curve admits linear systems of certain degrees and dimensions. For example, we have defined X to be hyperelliptic if it has a g_2^1, and we have seen that there are hyperelliptic curves of every genus $g \geqslant 2$ (Ex. 1.7), and at least for $g = 3$ and 4, that there exist nonhyperelliptic curves (5.2.1) and (5.2.2).

More generally, we can subdivide curves according to whether they have a g_d^1 for various d. If X has a g_3^1 it is called *trigonal*.

Remark 5.5.1. The facts here are as follows. For any $d \geqslant \frac{1}{2}g + 1$, any curve of genus g has a g_d^1; for $d < \frac{1}{2}g + 1$, there exist curves of genus g having no g_d^1. See Kleiman and Laksov [1] for proofs and discussion. Note in particular this implies that there exist nonhyperelliptic curves of every genus $g \geqslant 3$ (V, Ex. 2.10). We give some examples of this result.

Example 5.5.2. For $g = 3,4$ this result states that there exist nonhyperelliptic curves (which we have seen) and that every such curve has a g_3^1. Of course if X is hyperelliptic, this is trivial, by adding a point to the g_2^1. If X is non-hyperelliptic of genus 3, then its canonical embedding is a plane quartic curve (5.2.1). Projecting from any point of X to $\mathbf{P}^1$, we get a g_3^1. Thus X has infinitely many g_3^1's.

If X is nonhyperelliptic of genus 4, then its canonical embedding in $\mathbf{P}^3$ lies on a unique irreducible quadric surface Q (5.2.2). If Q is nonsingular, then X has type $(3,3)$ on Q (II, 6.6.1), and each of the two families of lines on Q cuts out a g_3^1 on X. So in this case X has two g_3^1's (to see that these are the only ones, copy the argument of (5.5.3) below). If Q is singular, it is a quadric cone, and the one family of lines on Q cuts out a unique g_3^1 on X.

Example 5.5.3. Let $g = 5$. Then (5.5.1) says that every curve of genus 5 has a g_4^1, and that there exist such curves with no g_3^1. Let X be a nonhyperelliptic curve of genus 5, in its canonical embedding as a curve of degree 8 in $\mathbf{P}^4$. First we show that X has a g_3^1 if and only if it has a trisecant in this embedding. Let $P,Q,R \in X$. Then by Riemann–Roch, we have

$$\dim|P + Q + R| = \dim|K - P - Q - R| - 1.$$

On the other hand, since X is in its canonical embedding, $\dim|K - P - Q - R|$ is the dimension of the linear system of hyperplanes in $\mathbf{P}^4$ which contain P,Q,R. Hence $\dim|P + Q + R| = 1$ if and only if P,Q,R are contained in a 2-dimensional family of hyperplanes, which is equivalent to saying that P,Q and R are collinear. Thus X has a g_3^1 if and only if it has a trisecant (and in that case it will have a 1-parameter family of trisecants).

Now let X be a nonsingular complete intersection of three quadric hypersurfaces in $\mathbf{P}^4$. Then $\deg X = 8$, and $\omega_X \cong \mathcal{O}_X(1)$, so X is a canonical curve of genus 5. If X had a trisecant L, then L would meet each of the quadric hypersurfaces in three points, so it would have to be contained in these hypersurfaces, and so $L \subseteq X$, which is impossible. So we see that there exist curves of genus 5 containing no g_3^1.

Now projecting this X from one of its own points P to $\mathbf{P}^3$, we obtain a curve $X' \subseteq \mathbf{P}^3$ of degree 7, which is nonsingular (because X had no trisecants). This new curve X' must have trisecants, because otherwise a projection from one of its points would give a nonsingular curve of degree 6 in $\mathbf{P}^2$, which has the wrong genus. So let Q,R,S lie on a trisecant of X'. Then their inverse images on X, together with P, form four points which lie in a plane of $\mathbf{P}^4$. Then the same argument as above shows these points give a g_4^1.

Coming back to the general classification question, for fixed g one would like to endow the set $\mathfrak{M}_g$ of all curves of genus g up to isomorphism with an algebraic structure, in which case we call $\mathfrak{M}_g$ the *variety of moduli* of curves of genus g. Such is the case for $g = 1$, where the j-invariants form an affine line.

The best way to specify the algebraic structure on $\mathfrak{M}_g$ would be to require it to be a universal parameter variety for families of curves of genus g, in the following sense: we require that there be a flat family $\mathfrak{X} \to \mathfrak{M}_g$ of curves of genus g such that for any other flat family $X \to T$ of curves of genus g, there is a unique morphism $T \to \mathfrak{M}_g$ such that X is the pullback of $\mathfrak{X}$. In this case we call $\mathfrak{M}_g$ a *fine moduli variety*. Unfortunately, there are several reasons why such a universal family cannot exist. One is that there are nontrivial families of curves, all of whose fibres are isomorphic to each other (III, Ex. 9.10).

However, Mumford has shown that for $g \geqslant 2$ there is a *coarse moduli variety* $\mathfrak{M}_g$, which has the following properties (Mumford [1, Th. 5.11]):

(1) the set of closed points of $\mathfrak{M}_g$ is in one-to-one correspondence with the set of isomorphism classes of curves of genus g;
(2) if $f : X \to T$ is any flat family of curves of genus g, then there is a morphism $h : T \to \mathfrak{M}_g$ such that for each closed point $t \in T$, X_t is in the isomorphism class of curves determined by the point $h(t) \in \mathfrak{M}_g$.

In case $g = 1$, the affine j-line is a coarse variety of moduli for families of elliptic curves with a section. One verifies condition (2) using the fact that j is a rational function of the coefficients of a plane embedding of the curve (Ex. 4.4).

Remark 5.5.4. In fact, Deligne and Mumford [1] have shown that $\mathfrak{M}_g$ for $g \geqslant 2$ is an irreducible quasi-projective variety of dimension $3g - 3$ over any fixed algebraically closed field.

Example 5.5.5. Assuming that $\mathfrak{M}_g$ exists, we can discover some of its properties. For example, using the method of (Ex. 2.2), one can show that hyperelliptic curves of genus g are determined as two-fold coverings of $\mathbf{P}^1$, ramified at $0, 1, \infty$, and $2g - 1$ additional points, up to the action of a certain finite group. Thus we see that the hyperelliptic curves correspond to an irreducible subvariety of dimension $2g - 1$ of $\mathfrak{M}_g$. If $g = 2$, this is the whole space, which confirms that $\mathfrak{M}_2$ is irreducible of dimension 3.

Example 5.5.6. Let $g = 3$. Then the hyperelliptic curves form an irreducible subvariety of dimension 5 of $\mathfrak{M}_3$. The nonhyperelliptic curves of genus 3 are the nonsingular plane quartic curves. Since the embedding is canonical, two of them are isomorphic as abstract curves if and only if they differ by an automorphism of $\mathbf{P}^2$. The family of all these curves is parametrized by an open set $U \subseteq \mathbf{P}^N$ with $N = 14$, because a form of degree 4 has 15 coefficients. So there is a morphism $U \to \mathfrak{M}_3$, whose fibres are images of the group PGL(2) which has dimension 8. Since any individual curve has only finitely many automorphisms (Ex. 5.2), the fibres have dimension $= 8$, and so the image of U has dimension $14 - 8 = 6$. So we confirm that $\mathfrak{M}_3$ has dimension 6.

EXERCISES

5.1. Show that a hyperelliptic curve can never be a complete intersection in any projective space. Cf. (Ex. 3.3).

5.2. If X is a curve of genus $\geqslant 2$ over a field of characteristic 0, show that the group Aut X of automorphisms of X is finite. [*Hint*: If X is hyperelliptic, use the unique g_2^1 and show that Aut X permutes the ramification points of the 2-fold covering $X \to \mathbf{P}^1$. If X is not hyperelliptic, show that Aut X permutes the hyperosculation points (Ex. 4.6) of the canonical embedding. Cf. (Ex. 2.5).]

5.3. *Moduli of Curves of Genus* 4. The hyperelliptic curves of genus 4 form an irreducible family of dimension 7. The nonhyperelliptic ones form an irreducible family of dimension 9. The subset of those having only one g_3^1 is an irreducible family of dimension 8. [*Hint*: Use (5.2.2) to count how many complete intersections $Q \cap F_3$ there are.]

5.4. Another way of distinguishing curves of genus g is to ask, what is the least degree of a birational plane model with only nodes as singularities (3.11)? Let X be nonhyperelliptic of genus 4. Then:
(a) if X has two g_3^1's, it can be represented as a plane quintic with two nodes, and conversely;
(b) if X has one g_3^1, then it can be represented as a plane quintic with a tacnode (I, Ex. 5.14d), but the least degree of a plane representation with only nodes is 6.

5.5. *Curves of Genus* 5. Assume X is not hyperelliptic.
(a) The curves of genus 5 whose canonical model in $\mathbf{P}^4$ is a complete intersection $F_2.F_2.F_2$ form a family of dimension 12.
(b) X has a g_4^1 if and only if it can be represented as a plane quintic with one node. These form an irreducible family of dimension 11. [*Hint*: If $D \in g_4^1$, use $K - D$ to map $X \to \mathbf{P}^2$.]
*(c) In that case, the conics through the node cut out the canonical system (not counting the fixed points at the node). Mapping $\mathbf{P}^2 \to \mathbf{P}^4$ by this linear system of conics, show that the canonical curve X is contained in a cubic surface $V \subseteq \mathbf{P}^4$, with V isomorphic to $\mathbf{P}^2$ with one point blown up (II, Ex. 7.7). Furthermore, V is the union of all the trisecants of X corresponding to the g_4^1 (5.5.3), so V is contained in the intersection of all the quadric hypersurfaces containing X. Thus V and the g_4^1 are unique.
Note. Conversely, if X does not have a g_4^1, then its canonical embedding is a complete intersection, as in (a). More generally, a classical theorem of Enriques and Petri shows that for any nonhyperelliptic curve of genus $g \geqslant 3$, the canonical model is projectively normal, and it is an intersection of quadric hypersurfaces unless X has a g_3^1 or $g = 6$ and X has a g_5^2. See Saint-Donat [1].

5.6. Show that a nonsingular plane curve of degree 5 has no g_3^1. Show that there are nonhyperelliptic curves of genus 6 which cannot be represented as a nonsingular plane quintic curve.

5.7. (a) Any automorphism of a curve of genus 3 is induced by an automorphism of $\mathbf{P}^2$ via the canonical embedding.
*(b) Assume char $k \neq 3$. If X is the curve given by

$$x^3 y + y^3 z + z^3 x = 0,$$

the group Aut X is the simple group of order 168, whose order is the maximum $84(g - 1)$ allowed by (Ex. 2.5). See Burnside [1, §232] or Klein [1].

*(c) Most curves of genus 3 have no automorphisms except the identity. [*Hint*: For each n, count the dimension of the family of curves with an automorphism T of order n. For example, if $n = 2$, then for suitable choice of coordinates, T can be written as $x \to -x$, $y \to y$, $z \to z$. Then there is an 8-dimensional family of curves fixed by T; changing coordinates there is a 4-dimensional family of such T, so the curves having an automorphism of degree 2 form a family of dimensional 12 inside the 14-dimensional family of all plane curves of degree 4.]

Note: More generally it is true (at least over $\mathbf{C}$) that for any $g \geqslant 3$, a "sufficiently general" curve of genus g has no automorphisms except the identity—see Baily [1].

6 Classification of Curves in $\mathbf{P}^3$

In 1882, a cash prize (the Steiner prize) was offered for the best work on the classification of space curves. It was shared by Max Noether and G. Halphen, each of whom wrote a 200-page treatise on the subject (Noether [1], Halphen [1]). They each proved a number of general results, and then to illustrate their theory, constructed exhaustive tables of curves of low degree (up to about degree 20).

Nowadays the theoretical aspect of this problem is well understood. Using either the Chow variety or the Hilbert scheme, one can show that the nonsingular curves of given degree d and genus g in $\mathbf{P}^3$ are parametrized by a finite union of quasi-projective varieties, in a very natural way. However, the more specific task of determining the number and dimensions of these parameter varieties for each d,g is not solved. It is not even clear exactly for which pairs of integers d,g there exists a curve of degree d and genus g in $\mathbf{P}^3$. Halphen stated the result, but a correct proof was given only recently by Gruson and Peskine.

In this section we will give a few basic results concerning curves in $\mathbf{P}^3$, and then illustrate them by classifying all curves of degree $\leqslant 7$ in $\mathbf{P}^3$.

We begin by investigating when a curve has a nonspecial very ample divisor of a given degree. In the case of $g = 0,1$, this is answered by (3.3.1) and (3.3.3), so we will consider the case $g \geqslant 2$.

Proposition 6.1 (Halphen). *A curve X of genus $g \geqslant 2$ has a nonspecial very ample divisor D of degree d if and only if $d \geqslant g + 3$.*

PROOF. First we show the necessity of the condition. If D is nonspecial and very ample of degree d, then by Riemann–Roch, we have $\dim|D| = d - g$ and $|D|$ gives an embedding of X in $\mathbf{P}^{d-g}$. Since $X \ncong \mathbf{P}^1$, we must have $d - g \geqslant 2$, i.e., $d \geqslant g + 2$. But if $d = g + 2$, then X is a plane curve of degree d. In this case $\omega_X \cong \mathcal{O}_X(d - 3)$, so in order for D to be nonspecial we must have $d \leqslant 3$. But then $g = 0$ or 1, contrary to hypothesis. So we conclude that $d \geqslant g + 3$.

So now we fix $d \geqslant g + 3$, and we search for a nonspecial very ample divisor D of degree d. In order for D to be very ample, by (3.1) it is necessary and sufficient that for all $P, Q \in X$, we have

$$\dim|D - P - Q| = \dim|D| - 2.$$

Since D is nonspecial, this is equivalent, by Riemann–Roch, to saying that $D - P - Q$ is also nonspecial. Replacing D by a linearly equivalent divisor D', we may always assume that $D' - P - Q$ is effective.

Now consider X^d, the product of X with itself d times. We associate an element $\langle P_1, \ldots, P_d \rangle \in X^d$ and all its permutations, with the effective divisor $D = P_1 + \ldots + P_d$, and then by abuse of notation write $D \in X^d$. We will show that the set S of divisors $D \in X^d$ such that there exists $D' \sim D$ and there exist points $P, Q \in X$ with $E = D' - P - Q$ an effective special divisor, has dimension $\leqslant g + 2$. Since $d \geqslant g + 3$, this shows that $S \neq X^d$. Then any $D \notin S$ will be a nonspecial very ample divisor of degree d.

Let E be an effective special divisor of degree $d - 2$. Since $\dim|K| = g - 1$, and since an effective special divisor is a subset of an effective canonical divisor, we see that the set of all such E, as a subset of X^{d-2}, has dimension $\leqslant g - 1$. Thus the set of divisors of the form $E + P + Q$ in X^d has dimension $\leqslant g + 1$. Since the special divisors in X^d form a subset of dimension $\leqslant g - 1$, for the same reason, we may ignore them, so we may assume $E + P + Q$ is nonspecial.

Since E is special, we have $\dim|E| \geqslant d - 1 - g$, by Riemann–Roch. On the other hand, since $E + P + Q$ is nonspecial, we have $\dim|E + P + Q| = d - g$. The difference between these two is 1, so we see that the set of $D \in X^d$ which are linearly equivalent to some divisor of the form $E + P + Q$ has dimension $\leqslant g + 2$, as required.

Corollary 6.2. *There exists a curve X of degree d and genus g in $\mathbf{P}^3$, whose hyperplane section D is nonspecial, if and only if either*

(1) *$g = 0$ and $d \geqslant 1$,*
(2) *$g = 1$ and $d \geqslant 3$, or*
(3) *$g \geqslant 2$ and $d \geqslant g + 3$.*

PROOF. This follows immediately from (3.3.1), (3.3.3), and the proposition. Given D very ample on X, the complete linear system $|D|$ gives an embedding of X into $\mathbf{P}^n$ for some n, and if $n > 3$ we project down to $\mathbf{P}^3$, using (3.5).

Proposition 6.3. *If X is a curve in $\mathbf{P}^3$, not lying in any plane, for which the hyperplane section D is special, then $d \geqslant 6$ and $g \geqslant \frac{1}{2}d + 1$. Furthermore, the only such curve with $d = 6$ is the canonical curve of genus 4 (5.2.2).*

PROOF. If D is special, then by Clifford's theorem (5.4) we have $\dim|D| \leqslant \frac{1}{2}d$. Since X is not in any plane, $\dim|D| \geqslant 3$, so $d \geqslant 6$. And since D is special, $d \leqslant 2g - 2$, hence $g \geqslant \frac{1}{2}d + 1$. Now if $d = 6$, then we have equality in Clifford's theorem, so either $D = 0$ (which is absurd) or $D = K$, in which case X is the canonical curve of genus 4, or X is hyperelliptic and $|D|$ is a multiple of the unique g_2^1. But this last case is impossible, because then $|D|$ would not separate points, so could not be very ample.

Next, we have a result which bounds the genus of a space curve of given degree.

Theorem 6.4 (Castelnuovo [1]). *Let X be a curve of degree d and genus g in* $\mathbf{P}^3$, *which is not contained in any plane. Then $d \geqslant 3$, and*

$$
g \leqslant \begin{cases} \dfrac{1}{4} d^2 - d + 1 & \text{if } d \text{ is even} \\[2mm] \dfrac{1}{4} (d^2 - 1) - d + 1 & \text{if } d \text{ is odd.} \end{cases}
$$

Furthermore, the equality is attained for every $d \geqslant 3$, and any curve for which equality holds lies on a quadric surface.

PROOF. Given X, let D be its hyperplane section. The idea of the proof is to estimate $\dim|nD| - \dim|(n - 1)D|$ for any n, and then add. First of all, we choose the hyperplane section $D = P_1 + \ldots + P_d$ in such a way that no three of the points P_i are collinear. This is possible because not every secant of X is a multisecant (3.8), (3.9), (Ex. 3.9).

Now I claim for each $i = 1, 2, \ldots, \min(d, 2n + 1)$, that P_i is not a base point of the linear system $|nD - P_1 - \ldots - P_{i-1}|$. To show this, it is sufficient to find a surface of degree n in $\mathbf{P}^3$ containing $P_1, \ldots, P_{i-1}$, but not P_i. In fact, a union of n planes will do. We take the first plane to contain P_1 and P_2, but no other P_j, which is possible, since no three P_j are collinear. We take the second plane to contain P_3 and P_4, and so on, until our planes contain $P_1, \ldots, P_{i-1}$, and take the remaining planes to miss all the P_j. This is possible for any i such that $i - 1 \leqslant 2n$, and of course $i \leqslant d$ since there are only d points.

It follows that for any $n \geqslant 1$, we have

$$
\dim|nD| - \dim|(n - 1)D| \geqslant \min(d, 2n + 1),
$$

because $(n - 1)D = nD - P_1 - \ldots - P_d$, and each time we remove a non-base point from a linear system, the dimension drops by 1.

Now we take $n \gg 0$, and add these expressions together, starting with $n = 1$, up to the given n, using the fact that $\dim|0 \cdot D| = 0$. If we let $r = [\frac{1}{2}(d - 1)]$, then we can write the answer as

$$
\dim|nD| \geqslant 3 + 5 + \ldots + (2r + 1) + (n - r)d
$$

or

$$
\dim|nD| \geqslant r(r + 2) + (n - r)d.
$$

On the other hand, for large n, the divisor nD will be nonspecial, so by Riemann–Roch we have

$$\dim|nD| = nd - g.$$

Combining, we find that

$$g \leqslant rd - r(r + 2).$$

To interpret this, we consider two cases. If d is even, then $r = \frac{1}{2}d - 1$, and we get

$$g \leqslant \frac{1}{4}d^2 - d + 1.$$

If d is odd, then $r = \frac{1}{2}(d - 1)$, and we get

$$g \leqslant \frac{1}{4}(d^2 - 1) - d + 1,$$

which is the bound of the theorem.

If X is a curve for which equality holds, then we must have had equality at every step of the way. In particular, we see that $\dim|2D| = 8$ (or even less if $d < 5$), from which it follows that X is contained in a quadric surface. Indeed, from the exact sequence

$$0 \to \mathscr{I}_X \to \mathcal{O}_\mathbf{P} \to \mathcal{O}_X \to 0$$

we obtain

$$0 \to H^0(\mathscr{I}_X(2)) \to H^0(\mathcal{O}_\mathbf{P}(2)) \to H^0(\mathcal{O}_X(2)) \to \ldots.$$

Since $\dim H^0(\mathcal{O}_\mathbf{P}(2)) = 10$ and $\dim H^0(\mathcal{O}_X(2)) = 9$ (or less, by the above), we conclude that $H^0(\mathscr{I}_X(2)) \neq 0$, hence X is contained in a quadric surface.

Finally, to show that equality is achieved, we look at certain curves on a nonsingular quadric surface Q. If d is even, $d = 2s$, we take a curve of type (s,s), which has degree d and genus $s^2 - 2s + 1 = \frac{1}{4}d^2 - d + 1$, by (III, Ex. 5.6). This curve is a complete intersection of Q with a surface of degree s. If d is odd, $d = 2s + 1$, we take a curve of type $(s, s + 1)$ on Q, which has degree d and genus $s^2 - s = \frac{1}{4}(d^2 - 1) - d + 1$.

Remark 6.4.1. Let us gather together everything we know about curves in $\mathbf{P}^3$. First, we recall various classes of curves which we know to exist.

(a) For every $d \geqslant 1$, there are nonsingular plane curves of degree d, and they have $g = \frac{1}{2}(d - 1)(d - 2)$. See (II, 8.20.2) and (II, Ex. 8.4).

(b) For every $a,b \geqslant 1$, there are complete intersections of surfaces of degrees a,b in $\mathbf{P}^3$ which are nonsingular curves. They have degree $d = ab$ and genus $g = \frac{1}{2}ab(a + b - 4) + 1$ (II, Ex. 8.4).

(c) For every $a,b \geqslant 1$, there are nonsingular curves of type (a,b) on a nonsingular quadric surface. They have degree $d = a + b$ and genus $g = ab - a - b + 1$ (III, Ex. 5.6).

(d) We will see later (V, Ex. 2.9) that if X is a curve on a quadric cone Q, there are two cases. If d is even, $d = 2a$, then X is a complete intersection of Q with a surface of degree a, so $g = a^2 - 2a + 1$. If d is odd, $d = 2a + 1$, then X has genus $g = a^2 - a$. Comparing with (c) above, we note that these values of d and g are among those possible on a nonsingular quadric surface.

Example 6.4.2. Now we can classify curves of degree $d \leqslant 7$ in $\mathbf{P}^3$.

$d = 1$. The only curve with $d = 1$ is $\mathbf{P}^1$ (I, Ex. 7.6).

$d = 2$. The only curve of degree 2 is the conic in $\mathbf{P}^2$ (I, Ex. 7.8).

$d = 3$. Here we have the plane cubic with $g = 1$, and the twisted cubic curve in $\mathbf{P}^3$ with $g = 0$. That is all, by (Ex. 3.4).

$d = 4$. The plane quartic has $g = 3$; in $\mathbf{P}^3$ there are rational quartic curves, and elliptic quartic curves, the latter being the complete intersection of two quadric surfaces (Ex. 3.6).

$d = 5$. The plane quintic has $g = 6$. In $\mathbf{P}^3$, there are curves with $\mathcal{O}(1)$ nonspecial of genus 0,1,2, by (6.2), and these are all by (6.3).

$d = 6$. The plane sextic has $g = 10$. In $\mathbf{P}^3$ there are curves with $\mathcal{O}(1)$ nonspecial of genus 0,1,2,3 by (6.2), and a curve of genus 4 which is the canonical curve of genus 4, a complete intersection of a quadric and a cubic surface (6.3).

$d = 7$. The plane septic has $g = 15$. There are curves in $\mathbf{P}^3$ with $\mathcal{O}(1)$ nonspecial of genus 0,1,2,3,4, and any curve of $g \leqslant 4$ must be nonspecial. On the other hand, there is a curve of type (3,4) on a nonsingular quadric surface, which has $g = 6$. This is the maximum genus for this degree, by (6.4), so any curve of $g = 6$ must lie on a quadric.

We are left with the question, does there exist a curve of degree 7 with $g = 5$? By (6.4.1) there is no such curve on a quadric surface. So we approach the question from a different angle. Given an abstract curve X of genus 5, can we embed it as a curve of degree 7 in $\mathbf{P}^3$? We need a very ample divisor D of degree 7, with $\dim|D| \geqslant 3$. By Riemann–Roch, such a divisor must be special. Since $\deg K = 8$, we can write $D = K - P$, and so we see $\dim|D| = 3$. In order for D to be very ample, we must have

$$\dim|K - P - Q - R| = \dim|K - P| - 2$$

for all Q,R in X. Using Riemann–Roch again, this says that

$$\dim|P + Q + R| = 0$$

for all Q,R. But this is possible if and only if X does not have a g_3^1. Indeed, if X has no g_3^1, then $\dim|P + Q + R| = 0$ for all P,Q,R. On the other hand, if X does have a g_3^1, then for any given P, there exist Q,R such that $\dim|P + Q + R| = 1$.

Summing up, we see that the abstract curve X of genus 5 admits an embedding of degree 7 in $\mathbf{P}^3$ if and only if X has no g_3^1. Now from (5.5.3) we know there are such curves, so we see that curves of degree 7 and genus 5 do exist in $\mathbf{P}^3$. This example should give some idea of the complexities which compound themselves when trying to classify curves of higher degree and genus in $\mathbf{P}^3$.

Figure 18 summarizes what we know about the existence of curves of degree d and genus g in $\mathbf{P}^3$, for $d \leqslant 10$ and $g \leqslant 12$, using the results of this section. See (V, 4.13.1) and (V, Ex. 4.14) for further information.

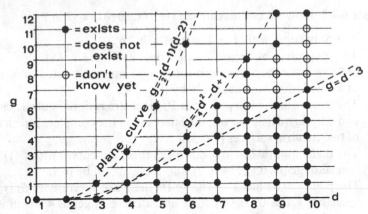

Figure 18.   Curves of degree d and genus g in $\mathbf{P}^3$.

Example 6.4.3. As another example, we consider curves X of degree 9 and genus 10 in $\mathbf{P}^3$. This is the first case where there are two distinct families of curves of the same degree and genus, neither being a special case of the other.

Type 1 is a complete intersection of two cubic surfaces. In this case, $\omega_X \cong \mathcal{O}_X(2)$ (II, Ex. 8.4), so $\mathcal{O}_X(2)$ is special, and dim $H^0(\mathcal{O}_X(2)) = 10$. Furthermore, X is projectively normal (II, Ex. 8.4) so the natural map $H^0(\mathcal{O}_{\mathbf{P}}(2)) \to H^0(\mathcal{O}_X(2))$ is surjective. Since dim $H^0(\mathcal{O}_{\mathbf{P}}(2)) = 10$, we conclude that $H^0(\mathcal{I}_X(2)) = 0$, so X is not contained in any quadric surface.

Type 2 is a curve of type $(3,6)$ on a nonsingular quadric surface Q. In this case, using the calculations of cohomology of (III, Ex. 5.6), from the exact sequence

$$0 \to \mathcal{O}_Q(-3, -6) \to \mathcal{O}_Q \to \mathcal{O}_X \to 0,$$

twisting by 2, and taking cohomology, we find that dim $H^0(\mathcal{O}_X(2)) = 9$. Thus $\mathcal{O}(2)$ is nonspecial. On the other hand, since X cannot be contained in two distinct quadric surfaces, we have dim $H^0(\mathcal{I}_X(2)) = 1$.

Since the dimension of cohomology groups can only increase under specialization, by semicontinuity (III, 12.8), we see that neither of these types can be a specialization of the other. Indeed, dim $H^0(\mathcal{I}_X(2))$ increases from type 1 to type 2, whereas dim $H^0(\mathcal{O}_X(2))$ decreases.

To complete the picture, we show that any curve of degree 9 and genus 10 is one of the two types above. If $\mathcal{O}(2)$ is nonspecial, then dim $H^0(\mathcal{O}(2)) = 9$, so X must be contained in a quadric surface Q. Checking the possibilities for d and g (6.4.1), we see that Q must be nonsingular, and X must be of type $(3,6)$ on Q. On the other hand, if $\mathcal{O}(2)$ is special, then X cannot lie on a quadric surface (because if it did, then it would have to be type 2, in which case $\mathcal{O}(2)$ is nonspecial). Since $\mathcal{O}(3)$ is nonspecial (its degree is $> 2g - 2$), we have dim $H^0(\mathcal{O}_X(3)) = 18$, so we find dim $H^0(\mathscr{I}_X(3)) \geqslant 2$. The corresponding cubic surfaces must be irreducible, so X is contained in the intersection of two cubic surfaces; then by reason of its degree, X is equal to the complete intersection of those two cubic surfaces, so X is type 1.

EXERCISES

6.1. A rational curve of degree 4 in $\mathbf{P}^3$ is contained in a unique quadric surface Q, and Q is necessarily nonsingular.

6.2. A rational curve of degree 5 in $\mathbf{P}^3$ is always contained in a cubic surface, but there are such curves which are not contained in any quadric surface.

6.3. A curve of degree 5 and genus 2 in $\mathbf{P}^3$ is contained in a unique quadric surface Q. Show that for any abstract curve X of genus 2, there exist embeddings of degree 5 in $\mathbf{P}^3$ for which Q is nonsingular, and there exist other embeddings of degree 5 for which Q is singular.

6.4. There is no curve of degree 9 and genus 11 in $\mathbf{P}^3$. [*Hint:* Show that it would have to lie on a quadric surface, then use (6.4.1).]

6.5. If X is a complete intersection of surfaces of degrees a,b in $\mathbf{P}^3$, then X does not lie on any surface of degree $< \min(a,b)$.

6.6. Let X be a projectively normal curve in $\mathbf{P}^3$, not contained in any plane. If $d = 6$, then $g = 3$ or 4. If $d = 7$, then $g = 5$ or 6. Cf. (II, Ex. 8.4) and (III, Ex. 5.6).

6.7. The line, the conic, the twisted cubic curve and the elliptic quartic curve in $\mathbf{P}^3$ have no multisecants. Every other curve in $\mathbf{P}^3$ has infinitely many multisecants. [*Hint:* Consider a projection from a point of the curve to $\mathbf{P}^2$.]

6.8. A curve X of genus g has a nonspecial divisor D of degree d such that $|D|$ has no base points if and only if $d \geqslant g + 1$.

*6.9. Let X be an irreducible nonsingular curve in $\mathbf{P}^3$. Then for each $m \gg 0$, there is a nonsingular surface F of degree m containing X. [*Hint:* Let $\pi : \tilde{\mathbf{P}} \to \mathbf{P}^3$ be the blowing-up of X and let $Y = \pi^{-1}(X)$. Apply Bertini's theorem to the projective embedding of $\tilde{\mathbf{P}}$ corresponding to $\mathscr{I}_Y \otimes \pi^* \mathcal{O}_{\mathbf{P}^3}(m)$.]

CHAPTER V

Surfaces

In this chapter we give an introduction to the study of algebraic surfaces. This includes the basic facts about the geometry on a surface, and about birational transformations of surfaces. Also we treat two special classes of surfaces, the ruled surfaces, and the nonsingular cubic surfaces in $\mathbf{P}^3$, both to illustrate the general theory, and as a first step in the more detailed study of various types of surfaces.

This chapter should be adequate preparation for reading some more advanced works, such as Mumford [2], Zariski [5], Shafarevich [1], Bombieri and Husemoller [1]. We have mentioned the classification of surfaces only very briefly in §6, since it is adequately treated elsewhere.

Sections 1, 3 and 5 are general. Here we develop intersection theory on a surface and prove the Riemann–Roch theorem. As applications we give the Hodge index theorem and the Nakai–Moishezon criterion for an ample divisor. In §3 we study the behavior of a surface and the curves on it under a single monoidal transformation, which is blowing up a point. Then in §5 we prove the theorem of factorization of a birational morphism into monoidal transformations, and prove Castelnuovo's criterion for contracting an exceptional curve of the first kind.

In §2 we discuss ruled surfaces. Here the theory of curves gives a good handle on the ruled surfaces, because many properties of the surface are closely related to the study of certain linear systems on the base curve. Also there is a close connection between ruled surfaces over a curve C and locally free sheaves of rank 2 on C, so as a byproduct, we get some information about the classification of these locally free sheaves on a curve.

In §4, we study the nonsingular cubic surfaces in $\mathbf{P}^3$, and the famous 27 lines which lie on those surfaces. By representing the surface as a $\mathbf{P}^2$ with 6 points blown up, the study of linear systems on the cubic surface is reduced to the study of certain linear system of plane curves with assigned base points. This is a very classical subject, about which whole books have been written, and which we rewrite here in modern language.

1   Geometry on a Surface

We begin our study of surfaces with the internal geometry of a surface. A divisor on a surface is a sum of curves, so (in the absence of a projective embedding) it does not make sense to talk about the degree of a divisor, as in the case of curves. However, we can talk about the intersection of two divisors on a surface, and this gives rise to intersection theory. The Riemann–Roch theorem for surfaces gives a connection between the dimension of a complete linear system $|D|$, which is essentially a cohomological invariant, and certain intersection numbers on the surface. As in the case of curves, the Riemann–Roch theorem is basic to all further work with surfaces, especially questions of classification.

Throughout this chapter, a *surface* will mean a nonsingular projective surface over an algebraically closed field k. It is true that any complete nonsingular surface is projective (cf. II, 4.10.2), but since we will not prove that, we assume that our surfaces are projective. A *curve* on a surface will mean any effective divisor on the surface. In particular, it may be singular, reducible or even have multiple components. A *point* will mean a closed point, unless otherwise specified.

Let X be a surface. We wish to define the intersection number $C.D$ for any two divisors C,D on X in such a way as to generalize the intersection multiplicity defined in (I, Ex. 5.4) and (I, §7). If C and D are curves on X, and if $P \in C \cap D$ is a point of intersection of C and D, we say that C and D meet *transversally* at P if the local equations f,g of C,D at P generate the maximal ideal m_P of $\mathcal{O}_{P,X}$. This implies, by the way, that C and D are each nonsingular at P, because f will generate the maximal ideal of P in $\mathcal{O}_{P,D} = \mathcal{O}_{P,X}/(g)$, and vice versa.

If C and D are two nonsingular curves, which meet transversally at a finite number of points $P_1, \dots ,P_r$, then it is clear that the intersection number $C.D$ should be r. So we take this as our starting point, together with some natural properties the intersection pairing should have, to define our intersection theory. We denote by $\operatorname{Div} X$ the group of all divisors on X, and by $\operatorname{Pic} X$ the group of invertible sheaves up to isomorphism, which is isomorphic to the group of divisors modulo linear equivalence (II, §6).

Theorem 1.1. *There is a unique pairing* $\operatorname{Div} X \times \operatorname{Div} X \to \mathbf{Z}$, *denoted by $C.D$ for any two divisors C,D, such that*

(1) *if C and D are nonsingular curves meeting transversally, then $C.D = \#(C \cap D)$, the number of points of $C \cap D$,*

357

(2) *it is symmetric*: $C.D = D.C$,

(3) *it is additive*: $(C_1 + C_2).D = C_1.D + C_2.D$, *and*

(4) *it depends only on the linear equivalence classes: if* $C_1 \sim C_2$ *then* $C_1.D = C_2.D$.

Before giving the proof, we need some auxiliary results. Our main tool is Bertini's theorem, which we will use to express any divisor as a difference of nonsingular curves, up to linear equivalence.

Lemma 1.2. *Let* $C_1, \ldots, C_r$ *be irreducible curves on the surface* X, *and let* D *be a very ample divisor. Then almost all curves* D' *in the complete linear system* $|D|$ *are irreducible, nonsingular, and meet each of the* C_i *transversally.*

PROOF. We embed X in a projective space $\mathbf{P}^n$ using the very ample divisor D. Then we apply Bertini's theorem (II, 8.18) and (III, 7.9.1) simultaneously to X and to the curves $C_1, \ldots, C_r$. We conclude that most $D' \in |D|$ are irreducible nonsingular curves in X, and that the intersections $C_i \cap D$ are nonsingular, i.e., points with multiplicity one, which means that the C_i and D' meet transversally. Since we did not assume the C_i were nonsingular, we need to use (II, 8.18.1).

Lemma 1.3. *Let* C *be an irreducible nonsingular curve on* X, *and let* D *be any curve meeting* C *transversally. Then*

$$\#(C \cap D) = \deg_C(\mathscr{L}(D) \otimes \mathcal{O}_C).$$

PROOF. Here, of course, $\mathscr{L}(D)$ is the invertible sheaf on X corresponding to D (II, §7), and $\deg_C$ denotes the degree of the invertible sheaf $\mathscr{L}(D) \otimes \mathcal{O}_C$ on C (IV, §1). We use the fact (II, 6.18) that $\mathscr{L}(-D)$ is the ideal sheaf of D on X. Therefore, tensoring with $\mathcal{O}_C$, we have an exact sequence

$$0 \rightarrow \mathscr{L}(-D) \otimes \mathcal{O}_C \rightarrow \mathcal{O}_C \rightarrow \mathcal{O}_{C \cap D} \rightarrow 0$$

where now $C \cap D$ denotes the scheme-theoretic intersection. Thus $\mathscr{L}(D) \otimes \mathcal{O}_C$ is the invertible sheaf on C corresponding to the divisor $C \cap D$. Since the intersection is transversal, the degree of the divisor $C \cap D$ is just the number of points $\#(C \cap D)$.

PROOF OF (1.1). First we show the uniqueness. Fix an ample divisor H on X. Given any two divisors C,D on X, we can find an integer $n > 0$ such that $C + nH, D + nH$, and nH are all very ample. Indeed, we first choose $k > 0$ such that $\mathscr{L}(C + kH), \mathscr{L}(D + kH)$ and $\mathscr{L}(kH)$ are all generated by global sections. This is possible by definition of ampleness (II, §7). Then we choose $l > 0$ so that lH is very ample (II, 7.6). Taking $n = k + l$, it follows that $C + nH, D + nH$, and nH are all very ample (II, Ex. 7.5).

Now using (1.2), choose nonsingular curves

$$C' \in |C + nH|$$
$$D' \in |D + nH|, \quad \text{transversal to } C'$$
$$E' \in |nH|, \quad \text{transversal to } D'$$
$$F' \in |nH|, \quad \text{transversal to } C' \text{ and } E'.$$

Then $C \sim C' - E'$ and $D \sim D' - F'$, so by the properties (1)–(4) of the theorem, we have

$$C.D = \#(C' \cap D') - \#(C' \cap F') - \#(E' \cap D') + \#(E' \cap F').$$

This shows that the intersection number of any two divisors is determined by (1)–(4), so the intersection pairing is unique.

For the existence, we use the same method, and check that everything is well-defined. To simplify matters, we proceed in two steps. Let $\mathfrak{P} \subseteq \text{Div } X$ be the set of very ample divisors. Then $\mathfrak{P}$ is a cone, in the sense that the sum of two very ample divisors is again very ample. For $C, D \in \mathfrak{P}$, we define the intersection number $C.D$ as follows: by (1.2) choose $C' \in |C|$ nonsingular, and choose $D' \in |D|$ nonsingular and transversal to C'. Define $C.D = \#(C' \cap D')$.

To show this is well-defined, first fix C', and let $D'' \in |D|$ be another nonsingular curve, transversal to C'. Then by (1.3), we have

$$\#(C' \cap D') = \deg \mathscr{L}(D') \otimes \mathscr{O}_{C'},$$

and ditto for D''. But $D' \sim D''$, so $\mathscr{L}(D') \cong \mathscr{L}(D'')$, so these two numbers are the same. Thus our definition is independent of D'. Now suppose $C'' \in |C|$ is another nonsingular curve. By the previous step, we may assume D' is transversal to both C' and C''. Then by the same argument, restricting to the curve D', we see that $\#(C' \cap D') = \#(C'' \cap D')$.

So now we have a well-defined pairing $\mathfrak{P} \times \mathfrak{P} \to \mathbf{Z}$, which is clearly symmetric, and by definition it depends only on the linear equivalence classes of the divisors. It also follows from (1.3) that it is additive, since $\mathscr{L}(D_1 + D_2) \cong \mathscr{L}(D_1) \otimes \mathscr{L}(D_2)$, and the degree is additive on a curve. Finally, this pairing on $\mathfrak{P} \times \mathfrak{P}$ satisfies condition (1) by construction.

To define the intersection pairing on all of Div X, let C and D be any two divisors. Then, as above, we can write $C \sim C' - E'$ and $D \sim D' - F'$ where C', D', E', F' are all in $\mathfrak{P}$. So we define

$$C.D = C'.D' - C'.F' - E'.D' + E'.F'.$$

If, for example, we used another expression $C \sim C'' - E''$ with C'', E'' also very ample, then

$$C' + E'' \sim C'' + E',$$

so by what we have shown for the pairing in $\mathfrak{P}$, we have

$$C'.D' + E''.D' = C''.D' + E'.D'$$

and ditto for F' in place of D'. Thus the resulting two expressions for $C.D$ are the same. This shows that the intersection pairing $C.D$ is well-defined on all of Div X.

It satisfies (2), (3), (4) by construction and by the corresponding properties on $\mathfrak{P}$. The condition (1) follows using (1.3) once more.   q.e.d.

Now that we have defined the intersection pairing, it is useful to have a way of calculating it without having to move the curves. If C and D are curves with no common irreducible component, and if $P \in C \cap D$, then we define the *intersection multiplicity* $(C.D)_P$ of C and D at P to be the length of $\mathcal{O}_{P,X}/(f,g)$, where f,g are local equations of C,D at P (I, Ex 5.4). Here *length* is the same as the dimension of a k-vector space.

Proposition 1.4. *If C and D are curves on X having no common irreducible component, then*
$$C.D = \sum_{P \in C \cap D} (C.D)_P.$$

PROOF. As in the proof of (1.3), let $\mathcal{L}(D)$ be the invertible sheaf corresponding to D. Then we have an exact sequence
$$0 \to \mathcal{L}(-D) \otimes \mathcal{O}_C \to \mathcal{O}_C \to \mathcal{O}_{C \cap D} \to 0$$
where we consider $C \cap D$ as a scheme. Now the scheme $C \cap D$ has support at the points of $C \cap D$, and for any such P, its structure sheaf is the k-algebra $\mathcal{O}_{P,X}/(f,g)$. Therefore
$$\dim_k H^0(X, \mathcal{O}_{C \cap D}) = \sum_{P \in C \cap D} (C.D)_P.$$

On the other hand, we can calculate this H^0 from the cohomology sequence of the exact sequence above. We obtain
$$\dim H^0(X, \mathcal{O}_{C \cap D}) = \chi(\mathcal{O}_C) - \chi(\mathcal{L}(-D) \otimes \mathcal{O}_C),$$
where, as usual, for any coherent sheaf $\mathcal{F}$,
$$\chi(\mathcal{F}) = \sum (-1)^i \dim_k H^i(X, \mathcal{F})$$
is the Euler characteristic (III, Ex. 5.1).

This shows that the expression $\sum (C.D)_P$ depends only on the linear equivalence class of D. By symmetry, it also depends only on the linear equivalence class of C. Replacing C and D by differences of nonsingular curves, all transversal to each other as in the proof of (1.1), we see that this quantity is equal to the intersection number $C.D$ defined in (1.1).

Example 1.4.1. If D is any divisor on the surface X, we can define the *self-intersection* number $D.D$, usually denoted by D^2. Even if C is a nonsingular curve on X, the self-intersection C^2 cannot be calculated by the direct method of (1.4). We must use linear equivalence. However, by (1.3), we see that $C^2 = \deg_C(\mathscr{L}(C) \otimes \mathcal{O}_C)$. To reinterpret this, note that since the ideal sheaf $\mathscr{I}$ of C on X is $\mathscr{L}(-C)$ (II, 6.18), we have $\mathscr{I}/\mathscr{I}^2 \cong \mathscr{L}(-C) \otimes \mathcal{O}_C$. Therefore its dual $\mathscr{L}(C) \otimes \mathcal{O}_C$ is isomorphic to the *normal sheaf* $\mathcal{N}_{C/X}$, which is defined as $\mathcal{H}om(\mathscr{I}/\mathscr{I}^2, \mathcal{O}_C)$ (II, §8). So we have $C^2 = \deg_C \mathcal{N}_{C/X}$.

Example 1.4.2. Let $X = \mathbf{P}^2$. Then Pic $X \cong \mathbf{Z}$, and we can take the class h of a line as generator. Since any two lines are linearly equivalent, and since two distinct lines meet in one point, we have $h^2 = 1$. This determines the intersection pairing on $\mathbf{P}^2$, by linearity. Thus if C,D are curves of degrees n,m respectively, we have $C \sim nh$, $D \sim mh$ and so $C.D = nm$. If C and D have no component in common this can be interpreted in terms of the local intersection multiplicities of (1.4), and we get a new proof of Bézout's theorem (I, 7.8).

Example 1.4.3. Let X be the nonsingular quadric surface in $\mathbf{P}^3$. Then Pic $X \cong \mathbf{Z} \oplus \mathbf{Z}$ (II, 6.6.1) and we can take as generators lines l of type $(1,0)$ and m of type $(0,1)$, one from each family. Then $l^2 = 0$, $m^2 = 0$, $l.m = 1$, because two lines in the same family are skew, and two lines of opposite families meet in a point. This determines the intersection pairing on X. So for example if C has type (a,b) and D has type (a',b'), then $C.D = ab' + a'b$.

Example 1.4.4. Using the self-intersection, we can define a new numerical invariant of a surface. Let $\Omega_{X/k}$ be the sheaf of differentials of X/k, and let $\omega_X = \bigwedge^2 \Omega_{X/k}$ be the *canonical sheaf*, as defined in (II, §8). Any divisor K in the linear equivalence class corresponding to ω_X is called a *canonical divisor*. Then K^2, the self-intersection of the canonical divisor, is a number depending only on X. For example, if $X = \mathbf{P}^2$, $K = -3h$, so $K^2 = 9$. If X is the quadric surface (1.4.3), then K has type $(-2,-2)$ (II, Ex. 8.4), so $K^2 = 8$.

Proposition 1.5 (Adjunction Formula). *If C is a nonsingular curve of genus g on the surface X, and if K is the canonical divisor on X, then*

$$2g - 2 = C.(C + K).$$

PROOF. According to (II, 8.20) we have $\omega_C \cong \omega_X \otimes \mathscr{L}(C) \otimes \mathcal{O}_C$. The degree of ω_C is $2g - 2$ (IV, 1.3.3). On the other hand, by (1.3) we have

$$\deg_C(\omega_X \otimes \mathscr{L}(C) \otimes \mathcal{O}_C) = C.(C + K).$$

Example 1.5.1. This gives a quick method of computing the genus of a curve on a surface. For example, if C is a curve of degree d in $\mathbf{P}^2$, then

$$2g - 2 = d(d - 3)$$

so $g = \frac{1}{2}(d - 1)(d - 2)$. Cf. (II, Ex. 8.4).

Example 1.5.2. If C is a curve of type (a,b) on the quadric surface, then $C + K$ has type $(a - 2, b - 2)$, so

$$2g - 2 = a(b - 2) + (a - 2)b,$$

so $g = ab - a - b + 1$. Cf. (III, Ex. 5.6).

Now we come to the Riemann–Roch theorem. For any divisor D on the surface X, we let $l(D) = \dim_k H^0(X, \mathscr{L}(D))$. Thus $l(D) = \dim |D| + 1$, where $|D|$ is the complete linear system of D. We define the *superabundance* $s(D)$ to be $\dim H^1(X, \mathscr{L}(D))$. The reason for this terminology is that before the invention of cohomology, the Riemann–Roch formula was written only with $l(D)$ and $l(K - D)$, and the superabundance was the amount by which it failed to hold. Recall also that the *arithmetic genus* p_a of X is defined by $p_a = \chi(\mathscr{O}_X) - 1$ (III, Ex. 5.3).

Theorem 1.6 (Riemann–Roch). *If D is any divisor on the surface X, then*

$$l(D) - s(D) + l(K - D) = \frac{1}{2} D.(D - K) + 1 + p_a.$$

PROOF. By Serre duality (III, 7.7) we have

$$l(K - D) = \dim H^0(X, \mathscr{L}(D)^\vee \otimes \omega_X) = \dim H^2(X, \mathscr{L}(D)).$$

Thus the left-hand side is just the Euler characteristic, so we have to show for any D that

$$\chi(\mathscr{L}(D)) = \frac{1}{2} D.(D - K) + 1 + p_a.$$

Since both sides depend only on the linear equivalence class of D, as in (1.1) we can write D as the difference $C - E$ of two nonsingular curves. Now let us calculate. Since the ideal sheaves of C, E are $\mathscr{L}(-C), \mathscr{L}(-E)$ respectively, we obtain exact sequences, tensoring with $\mathscr{L}(C)$,

$$0 \to \mathscr{L}(C - E) \to \mathscr{L}(C) \to \mathscr{L}(C) \otimes \mathscr{O}_E \to 0$$

and

$$0 \to \mathscr{O}_X \to \mathscr{L}(C) \to \mathscr{L}(C) \otimes \mathscr{O}_C \to 0.$$

Since χ is additive on short exact sequences (III, Ex. 5.1), we have

$$\chi(\mathscr{L}(C - E)) = \chi(\mathscr{O}_X) + \chi(\mathscr{L}(C) \otimes \mathscr{O}_C) - \chi(\mathscr{L}(C) \otimes \mathscr{O}_E).$$

Now $\chi(\mathscr{O}_X) = 1 + p_a$ by definition of p_a. Using the Riemann–Roch theorem for the curves C and E (IV, 1.3), and using (1.3) to find the degree, we have

$$\chi(\mathscr{L}(C) \otimes \mathscr{O}_C) = C^2 + 1 - g_C$$

and

$$\chi(\mathscr{L}(C) \otimes \mathscr{O}_E) = C.E + 1 - g_E.$$

Finally, we use (1.5) to compute the genus of C and E:

$$g_C = \frac{1}{2} C.(C + K) + 1$$

and

$$g_E = \frac{1}{2} E.(E + K) + 1.$$

Combining all these, we obtain

$$\chi(\mathcal{L}(C - E)) = \frac{1}{2}(C - E).(C - E - K) + 1 + p_a$$

as required.

Remark 1.6.1. There is another formula, which is sometimes considered to be part of the Riemann–Roch theorem, namely

$$12(1 + p_a) = K^2 + c_2,$$

where c_2 is the second Chern class of the tangent sheaf of X. This is a consequence of the generalized Grothendieck–Hirzebruch Riemann–Roch theorem (App. A, 4.1.2).

As applications of the Riemann–Roch theorem, we will prove the Hodge index theorem and Nakai's criterion for an ample divisor.

Remark 1.6.2. In the following, note that if we fix a very ample divisor H on a surface X, then for any curve C on X, the intersection number $C.H$ is just equal to the degree of C in the projective embedding determined by H (Ex. 1.2). In particular, it is positive. More generally, having fixed an ample divisor H on X, the number $C.H$ plays a role similar to the degree of a divisor on a curve.

Lemma 1.7. *Let H be an ample divisor on the surface X. Then there is an integer n_0 such that for any divisor D, if $D.H > n_0$, then $H^2(X, \mathcal{L}(D)) = 0$.*

PROOF. By Serre duality on X, for any divisor D we have $\dim H^2(X, \mathcal{L}(D)) = l(K - D)$. If $l(K - D) > 0$, then the divisor $K - D$ is effective, so $(K - D).H > 0$. In other words, $D.H < K.H$. So we have only to take $n_0 = K.H$ to get the result.

Remark 1.7.1. This result can be regarded as the analogue for surfaces of the result that says on a curve X, there is an integer n_0 (namely $2g_X - 2$) such that if $\deg D > n_0$, then $H^1(X, \mathcal{L}(D)) = 0$ (IV, 1.3.4).

Corollary 1.8. *Let H be an ample divisor on X, and let D be a divisor such that $D.H > 0$ and $D^2 > 0$. Then for all $n \gg 0$, nD is linearly equivalent to an effective divisor.*

PROOF. We apply the Riemann–Roch theorem to nD. Since $D.H > 0$, for $n \gg 0$ we will have $nD.H > n_0$, so by (1.7), $l(K - nD) = 0$. Since $s(nD) \geqslant 0$, the Riemann–Roch theorem gives

$$l(nD) \geqslant \frac{1}{2} n^2 D^2 - \frac{1}{2} nD.K + 1 + p_a.$$

Now since $D^2 > 0$, the right-hand side becomes large for $n \gg 0$, so we see that $l(nD) \to \infty$ as $n \to \infty$. In particular, nD is effective for all $n \gg 0$.

Definition. A divisor D on a surface X is *numerically equivalent to zero*, written $D \equiv 0$, if $D.E = 0$ for all divisors E. We say D and E are numerically equivalent, written $D \equiv E$, if $D - E \equiv 0$.

Theorem 1.9 (Hodge Index Theorem). *Let H be an ample divisor on the surface X, and suppose that D is a divisor, $D \not\equiv 0$, with $D.H = 0$. Then $D^2 < 0$.*

PROOF. Suppose to the contrary that $D^2 \geqslant 0$. We consider two cases. If $D^2 > 0$, let $H' = D + nH$. For $n \gg 0$, H' is ample, as in the proof of (1.1). Furthermore, $D.H' = D^2 > 0$, so by (1.8), we have mD is effective for all $m \gg 0$. But then $mD.H > 0$ (think of the projective embedding defined by a multiple of H), hence $D.H > 0$, which is a contradiction.

If $D^2 = 0$, we use the hypothesis $D \not\equiv 0$ to conclude that there is a divisor E with $D.E \neq 0$. Replacing E by $E' = (H^2)E - (E.H)H$, we may assume furthermore that $E.H = 0$. Now let $D' = nD + E$. Then $D'.H = 0$, and $D'^2 = 2nD.E + E^2$. Since $D.E \neq 0$, by suitable choice of $n \in \mathbf{Z}$ we can make $D'^2 > 0$. But then the previous argument applies to D', and again we have a contradiction.

Remark 1.9.1. We explain the title of this theorem as follows. Let $\mathrm{Pic}^n X$ be the subgroup of $\mathrm{Pic}\, X$ of divisor classes numerically equivalent to zero, and let $\mathrm{Num}\, X = \mathrm{Pic}\, X / \mathrm{Pic}^n X$. Then clearly the intersection pairing induces a nondegenerate bilinear pairing $\mathrm{Num}\, X \times \mathrm{Num}\, X \to \mathbf{Z}$. It is a consequence of the Néron–Severi theorem (Ex. 1.7) that $\mathrm{Num}\, X$ is a free finitely generated abelian group (see also (Ex. 1.8)). So we can consider the vector space $\mathrm{Num}\, X \otimes_{\mathbf{Z}} \mathbf{R}$ over $\mathbf{R}$, and the induced bilinear form. A theorem of Sylvester (Lang [2, XIV, §7, p. 365]) shows that such a bilinear form can be diagonalized with ± 1's on the diagonal, and that the number of $+1$'s and the number of -1's are invariant. The difference of these two numbers is the *signature* or *index* of the bilinear form. In this context, (1.9) says that the diagonalized intersection pairing has one $+1$, corresponding to a (real) multiple of H, and all the rest -1's.

Example 1.9.2. On the quadric surface X (1.4.3) we can take H of type $(1,1)$ and D of type $(1,-1)$. Then $H^2 = 2$, $H.D = 0$, $D^2 = -2$, and D,H form a basis of $\mathrm{Pic}\, X$. In this case the only divisor numerically equivalent to 0 is

0, so Pic X = Num X. The pairing on Num $X \otimes_Z \mathbf{R}$ is diagonalized by taking the basis $(1/\sqrt{2})H$, $(1/\sqrt{2})D$.

Theorem 1.10 (Nakai–Moishezon Criterion). *A divisor D on the surface X is ample if and only if $D^2 > 0$ and $D.C > 0$ for all irreducible curves C in X.*

PROOF. The condition is clearly necessary, because if D is ample, then mD is very ample for some $m > 0$, in which case $m^2 D^2$ is the degree of X in the corresponding embedding, and $mD.C$ is the degree of C, both of which must be positive (Ex. 1.2).

Conversely, suppose $D^2 > 0$ and $D.C > 0$ for all irreducible curves C. If H is a very ample divisor on X, then H is represented by an irreducible curve, so $D.H > 0$ by hypothesis. Therefore by (1.8) some multiple mD for $m > 0$ is effective. Replacing D by mD, we may assume that D is effective. So we think of D as a curve in X, possibly singular, reducible, and nonreduced.

Next, let $\mathscr{L} = \mathscr{L}(D)$. We will show that the sheaf $\mathscr{L} \otimes \mathcal{O}_D$ is ample on the scheme D. For this it is sufficient to show that $\mathscr{L} \otimes \mathcal{O}_{D_{\mathrm{red}}}$ is ample on the reduced scheme D_{red} (III, Ex. 5.7). And if D_{red} is a union of irreducible curves $C_1, \ldots, C_r$, it is enough to show that $\mathscr{L} \otimes \mathcal{O}_{C_i}$ is ample on each C_i (*loc. cit.*). Finally, if $f: \tilde{C}_i \to C_i$ is the normalization of C_i, it is enough to show that $f^*(\mathscr{L} \otimes \mathcal{O}_{C_i})$ is ample on $\tilde{C}_i$, since f is a finite surjective morphism (*loc. cit.*). But $\deg f^*(\mathscr{L} \otimes \mathcal{O}_{C_i})$ is just $D.C_i > 0$, because we can represent $\mathscr{L}$ as a difference of nonsingular curves meeting C_i transversally, so this degree is preserved by f^*. Since the degree is positive, this sheaf is ample on the nonsingular curve $\tilde{C}_i$ (IV, 3.3). Therefore $\mathscr{L} \otimes \mathcal{O}_D$ is ample on D.

Next we will show that $\mathscr{L}^n$ is generated by global sections for $n \gg 0$. We use the exact sequence

$$0 \to \mathscr{L}^{-1} \to \mathcal{O}_X \to \mathcal{O}_D \to 0$$

tensored with $\mathscr{L}^n$, and the resulting cohomology sequence

$$0 \to H^0(X, \mathscr{L}^{n-1}) \to H^0(X, \mathscr{L}^n) \to H^0(D, \mathscr{L}^n \otimes \mathcal{O}_D) \to$$
$$\to H^1(X, \mathscr{L}^{n-1}) \to H^1(X, \mathscr{L}^n) \to H^1(D, \mathscr{L}^n \otimes \mathcal{O}_D) \to \ldots .$$

Since $\mathscr{L} \otimes \mathcal{O}_D$ is ample on D, we have $H^1(D, \mathscr{L}^n \otimes \mathcal{O}_D) = 0$ for $n \gg 0$ (III, 5.3). So we see that for each n,

$$\dim H^1(X, \mathscr{L}^n) \leqslant \dim H^1(X, \mathscr{L}^{n-1}).$$

Since these are finite-dimensional vector spaces, these dimensions must eventually be all equal. Therefore the map

$$H^0(X, \mathscr{L}^n) \to H^0(D, \mathscr{L}^n \otimes \mathcal{O}_D)$$

is surjective for all $n \gg 0$. Again since $\mathscr{L} \otimes \mathcal{O}_D$ is ample on D, the sheaf $\mathscr{L}^n \otimes \mathcal{O}_D$ will be generated by global sections for all $n \gg 0$. These sections lift to global sections of $\mathscr{L}^n$ on X, as we have just seen, so by Nakayama's lemma, the global sections of $\mathscr{L}^n$ generate the stalks at every point of D. But since $\mathscr{L} = \mathscr{L}(D)$, it has a section vanishing only along D, so in fact $\mathscr{L}^n$ is generated by global sections everywhere.

Fixing an n such that $\mathscr{L}^n$ is generated by global sections, we obtain a morphism $\varphi: X \to \mathbf{P}^N$ defined by $\mathscr{L}^n$ (II, 7.1). Next we show that the morphism φ has finite fibres. If not, there would be an irreducible curve C in X with $\varphi(C) = $ a point. In this case, taking a hyperplane in $\mathbf{P}^N$ which misses that point, we would have an effective divisor $E \sim nD$ with $E \cap C = \varnothing$. Therefore $E.C = 0$, which contradicts the hypothesis $D.C > 0$ for all C. So we see that φ has finite fibres.

Then it is a consequence of the Stein factorization theorem (III, 11.5) that φ is actually a finite morphism (III, Ex. 11.2). So $\varphi^*(\mathcal{O}(1)) = \mathscr{L}^n$ is ample on X by (III, Ex. 5.7) and we conclude that D is ample.   q.e.d.

Example 1.10.1. On the quadric surface X (1.4.3), the effective divisors are those of type (a,b) with $a,b \geqslant 0$. So a divisor D of type (a,b) is ample if and only if $a = D.(1,0) > 0$ and $b = D.(0,1) > 0$ (II, 7.6.2). In this case the condition $D.C > 0$ for all irreducible curves C implies $D^2 > 0$. However, there is an example of Mumford of a divisor D on a surface X, with $D.C > 0$ for every irreducible curve, but $D^2 = 0$, hence D not ample. See Hartshorne [5, I, 10.6].

References for § 1. For another approach to intersection theory on a surface, see Mumford [2]. The proof of the Riemann–Roch theorem follows Serre [7, Ch. IV no. 8]. The proof of the Hodge index theorem is due to Grothendieck [2]. The criterion for an ample divisor is due to Nakai [1] and independently Moishezon [1]. See Appendix A for intersection theory and the Riemann–Roch theorem in higher dimensions.

EXERCISES

1.1. Let C, D be any two divisors on a surface X, and let the corresponding invertible sheaves be $\mathscr{L}, \mathscr{M}$. Show that

$$C.D = \chi(\mathcal{O}_X) - \chi(\mathscr{L}^{-1}) - \chi(\mathscr{M}^{-1}) + \chi(\mathscr{L}^{-1} \otimes \mathscr{M}^{-1}).$$

1.2. Let H be a very ample divisor on the surface X, corresponding to a projective embedding $X \subseteq \mathbf{P}^N$. If we write the Hilbert polynomial of X (III, Ex. 5.2) as

$$F(z) = \frac{1}{2} az^2 + bz + c,$$

show that $a = H^2$, $b = \frac{1}{2}H^2 + 1 - \pi$, where π is the genus of a nonsingular curve representing H, and $c = 1 + p_a$. Thus the *degree* of X in $\mathbf{P}^N$, as defined in (I, §7), is just H^2. Show also that if C is any curve in X, then the degree of C in $\mathbf{P}^N$ is just $C.H$.

1.3. Recall that the *arithmetic genus* of a projective scheme D of dimension 1 is defined as $p_a = 1 - \chi(\mathcal{O}_D)$ (III, Ex. 5.3).
  (a) If D is an effective divisor on the surface X, use (1.6) to show that $2p_a - 2 = D.(D + K)$.
  (b) $p_a(D)$ depends only on the linear equivalence class of D on X.

(c) More generally, for *any* divisor D on X, we define the *virtual arithmetic genus* (which is equal to the ordinary arithmetic genus if D is effective) by the same formula: $2p_a - 2 = D.(D + K)$. Show that for any two divisors C,D we have

$$p_a(-D) = D^2 - p_a(D) + 2$$

and

$$p_a(C + D) = p_a(C) + p_a(D) + C.D - 1.$$

1.4. (a) If a surface X of degree d in $\mathbf{P}^3$ contains a straight line $C = \mathbf{P}^1$, show that $C^2 = 2 - d$.

(b) Assume char $k = 0$, and show for every $d \geq 1$, there exists a nonsingular surface X of degree d in $\mathbf{P}^3$ containing the line $x = y = 0$.

1.5. (a) If X is a surface of degree d in $\mathbf{P}^3$, then $K^2 = d(d - 4)^2$.

(b) If X is a product of two nonsingular curves C,C', of genus g,g' respectively, then $K^2 = 8(g - 1)(g' - 1)$. Cf. (II, Ex. 8.3).

1.6. (a) If C is a curve of genus g, show that the diagonal $\Delta \subseteq C \times C$ has self-intersection $\Delta^2 = 2 - 2g$. (Use the definition of $\Omega_{C/k}$ in (II, §8).)

(b) Let $l = C \times$ pt and $m =$ pt $\times C$. If $g \geq 1$, show that $l,m,$ and Δ are linearly independent in $\text{Num}(C \times C)$. Thus $\text{Num}(C \times C)$ has rank ≥ 3, and in particular, $\text{Pic}(C \times C) \neq p_1^* \text{Pic } C \oplus p_2^* \text{Pic } C$. Cf. (III, Ex. 12.6), (IV, Ex. 4.10).

1.7. *Algebraic Equivalence of Divisors.* Let X be a surface. Recall that we have defined an algebraic family of effective divisors on X, parametrized by a nonsingular curve T, to be an effective Cartier divisor D on $X \times T$, flat over T (III, 9.8.5). In this case, for any two closed points $0,1 \in T$, we say the corresponding divisors D_0,D_1 on X are prealgebraically equivalent. Two arbitrary divisors are prealgebraically equivalent if they are differences of prealgebraically equivalent effective divisors. Two divisors D,D' are *algebraically equivalent* if there is a finite sequence $D = D_0,D_1,\dots,D_n = D'$ with D_i and D_{i+1} prealgebraically equivalent for each i.

(a) Show that the divisors algebraically equivalent to 0 form a subgroup of Div X.

(b) Show that linearly equivalent divisors are algebraically equivalent. [*Hint*: If (f) is a principal divisor on X, consider the principal divisor $(tf - u)$ on $X \times \mathbf{P}^1$, where t,u are the homogeneous coordinates on $\mathbf{P}^1$.]

(c) Show that algebraically equivalent divisors are numerically equivalent. [*Hint*: Use (III, 9.9) to show that for any very ample H, if D and D' are algebraically equivalent, then $D.H = D'.H$.]

Note. The theorem of Néron and Severi states that the group of divisors modulo algebraic equivalence, called the *Néron–Severi group*, is a finitely generated abelian group. Over $\mathbf{C}$ this can be proved easily by transcendental methods (App. B, §5) or as in (Ex. 1.8) below. Over a field of arbitrary characteristic, see Lang and Néron [1] for a proof, and Hartshorne [6] for further discussion. Since Num X is a quotient of the Néron–Severi group, it is also finitely generated, and hence free, since it is torsion-free by construction.

1.8. *Cohomology Class of a Divisor.* For any divisor D on the surface X, we define its cohomology class $c(D) \in H^1(X,\Omega_X)$ by using the isomorphism $\text{Pic } X \cong H^1(X,\mathcal{O}_X^*)$ of (III, Ex. 4.5) and the sheaf homomorphism $d\log: \mathcal{O}^* \to \Omega_X$ (III, Ex. 7.4c). Thus we obtain a group homomorphism $c: \text{Pic } X \to H^1(X,\Omega_X)$. On the other hand, $H^1(X,\Omega)$ is dual to itself by Serre duality (III, 7.13), so we have a nondegenerate bilinear map

$$\langle \ , \ \rangle : H^1(X,\Omega) \times H^1(X,\Omega) \to k.$$

(a) Prove that this is compatible with the intersection pairing, in the following sense: for any two divisors D,E on X, we have

$$\langle c(D),c(E) \rangle = (D.E) \cdot 1$$

in k. [*Hint*: Reduce to the case where D and E are nonsingular curves meeting transversally. Then consider the analogous map $c : \text{Pic } D \to H^1(D,\Omega_D)$, and the fact (III, Ex. 7.4) that $c(\text{point})$ goes to 1 under the natural isomorphism of $H^1(D,\Omega_D)$ with k.]

(b) If char $k = 0$, use the fact that $H^1(X,\Omega_X)$ is a finite-dimensional vector space to show that Num X is a finitely generated free abelian group.

1.9. (a) If H is an ample divisor on the surface X, and if D is any divisor, show that

$$(D^2)(H^2) \leqslant (D.H)^2.$$

(b) Now let X be a product of two curves $X = C \times C'$. Let $l = C \times \text{pt}$, and $m = \text{pt} \times C'$. For any divisor D on X, let $a = D.l, b = D.m$. Then we say D has *type* (a,b). If D has type (a,b), with $a,b \in \mathbf{Z}$, show that

$$D^2 \leqslant 2ab,$$

and equality holds if and only if $D \equiv bl + am$. [*Hint*: Show that $H = l + m$ is ample, let $E = l - m$, let $D' = (H^2)(E^2)D - (E^2)(D.H)H - (H^2)(D.E)E$, and apply (1.9). This inequality is due to Castelnuovo and Severi. See Grothendieck [2].]

1.10. *Weil's Proof* [2] *of the Analogue of the Riemann Hypothesis for Curves.* Let C be a curve of genus g defined over the finite field $\mathbf{F}_q$, and let N be the number of points of C rational over $\mathbf{F}_q$. Then $N = 1 - a + q$, with $|a| \leqslant 2g\sqrt{q}$. To prove this, we consider C as a curve over the algebraic closure k of $\mathbf{F}_q$. Let $f : C \to C$ be the k-linear Frobenius morphism obtained by taking qth powers, which makes sense since C is defined over $\mathbf{F}_q$, so $X_q \cong X$ (IV, 2.4.1). Let $\Gamma \subseteq C \times C$ be the graph of f, and let $\Delta \subseteq C \times C$ be the diagonal. Show that $\Gamma^2 = q(2 - 2g)$, and $\Gamma.\Delta = N$. Then apply (Ex. 1.9) to $D = r\Gamma + s\Delta$ for all r and s to obtain the result. See (App. C, Ex. 5.7) for another interpretation of this result.

1.11. In this problem, we assume that X is a surface for which Num X is finitely generated (i.e., any surface, if you accept the Néron–Severi theorem (Ex. 1.7)).

(a) If H is an ample divisor on X, and $d \in \mathbf{Z}$, show that the set of effective divisors D with $D.H = d$, modulo numerical equivalence, is a finite set. [*Hint*: Use the adjunction formula, the fact that p_a of an irreducible curve is $\geqslant 0$, and the fact that the intersection pairing is negative definite on $H^\perp$ in Num X.]

(b) Now let C be a curve of genus $g \geqslant 2$, and use (a) to show that the group of automorphisms of C is finite, as follows. Given an automorphism σ of C, let $\Gamma \subseteq X = C \times C$ be its graph. First show that if $\Gamma \equiv \Delta$, then $\Gamma = \Delta$, using the fact that $\Delta^2 < 0$, since $g \geqslant 2$ (Ex. 1.6). Then use (a). Cf. (IV, Ex. 2.5).

1.12. If D is an ample divisor on the surface X, and $D' \equiv D$, then D' is also ample. Give an example to show, however, that if D is very ample, D' need not be very ample.

2 Ruled Surfaces

In this section we will illustrate some of the general concepts discussed in §1 by studying a particular class of surfaces, the ruled surfaces. By using some results from the theory of curves, we get a good hold on these surfaces, and can describe them and the curves lying on them quite explicitly.

We begin by establishing some general properties of ruled surfaces. Then we will define an invariant e, and give some examples. After that we give a classification of elliptic ruled surfaces, a detailed description of the rational ruled surfaces, and we determine the ample divisors on a ruled surface of any genus.

Definition. A *geometrically ruled surface*, or simply *ruled surface*, is a surface X, together with a surjective morphism $\pi: X \to C$ to a (nonsingular) curve C, such that the fibre X_y is isomorphic to $\mathbf{P}^1$ for every point $y \in C$, and such that π admits a section (i.e., a morphism $\sigma: C \to X$ such that $\pi \circ \sigma = \mathrm{id}_C$).

Note: In fact, one can show using Tsen's theorem that the existence of a section is a consequence of the other provisions of the definition—see, e.g., Shafarevich [1, p. 24].

Example 2.0.1. If C is a curve, then $C \times \mathbf{P}^1$ with its first projection is a ruled surface. In particular, the quadric surface in $\mathbf{P}^3$ is a ruled surface in two different ways. We consider the data π, C as given when we speak of a ruled surface.

Lemma 2.1. *Let $\pi: X \to C$ be a ruled surface, let D be a divisor on X, and suppose that $D.f = n \geqslant 0$, where f is a fibre of π. Then $\pi_* \mathcal{L}(D)$ is a locally free sheaf of rank $n + 1$ on C. In particular, $\pi_* \mathcal{O}_X = \mathcal{O}_C$.*

PROOF. First note that any two fibres of π are algebraically equivalent divisors on X, since they all are parametrized by the curve C. Therefore they are numerically equivalent (Ex. 1.7), so that $D.f$ is independent of the choice of the fibre.

Now for any $y \in C$, we consider the sheaf $\mathcal{L}(D)_y$ on the fibre X_y. This is an invertible sheaf of degree n on $X_y \cong \mathbf{P}^1$, so $H^0(\mathcal{L}(D)_y)$ has dimension $n + 1$. This is independent of y, so by Grauert's theorem (III, 12.9), $\pi_* \mathcal{L}(D)$ is locally free of rank $n + 1$.

In case $D = 0$, $\pi_* \mathcal{O}_X$ is locally free of rank 1. But (III, 12.9) tells us furthermore that the natural map

$$\pi_* \mathcal{O}_X \otimes k(y) \to H^0(X_y, \mathcal{O}_{X_y})$$

is an isomorphism for each y. The right-hand side is canonically isomorphic to k. Therefore the image of the global section 1 of $\mathcal{O}_C$ via the structural map $\mathcal{O}_C \to \pi_*\mathcal{O}_X$ generates the stalk at every point, showing that $\pi_*\mathcal{O}_X \cong \mathcal{O}_C$.

Proposition 2.2. *If $\pi: X \to C$ is a ruled surface, then there exists a locally free sheaf $\mathscr{E}$ of rank 2 on C such that $X \cong \mathbf{P}(\mathscr{E})$ over C. (See (II, §7) for the definition of $\mathbf{P}(\mathscr{E})$.) Conversely, every such $\mathbf{P}(\mathscr{E})$ is a ruled surface over C. If $\mathscr{E}$ and $\mathscr{E}'$ are two locally free sheaves of rank 2 on C, then $\mathbf{P}(\mathscr{E})$ and $\mathbf{P}(\mathscr{E}')$ are isomorphic as ruled surfaces over C if and only if there is an invertible sheaf $\mathscr{L}$ on C such that $\mathscr{E}' \cong \mathscr{E} \otimes \mathscr{L}$.*

PROOF. Given a ruled surface $\pi: X \to C$, then by definition π has a section σ. Let $D = \sigma(C)$. Then D is a divisor on X, and $D.f = 1$ for any fibre. By the lemma, $\mathscr{E} = \pi_*\mathscr{L}(D)$ is a locally free sheaf of rank 2 on C. Furthermore, there is a natural map $\pi^*\mathscr{E} = \pi^*\pi_*\mathscr{L}(D) \to \mathscr{L}(D)$ on X. This map is surjective. Indeed, by Nakayama's lemma, it is enough to check this on any fibre X_y. But $X_y \cong \mathbf{P}^1$, and $\mathscr{L}(D)_y$ is an invertible sheaf of degree 1, which is generated by its global sections, and $\mathscr{E} \otimes k(y) \to H^0(\mathscr{L}(D)_y)$ is surjective by (III, 12.9).

Now we apply (II, 7.12) which shows that the surjection $\pi^*\mathscr{E} \to \mathscr{L}(D) \to 0$ determines a morphism $g: X \to \mathbf{P}(\mathscr{E})$ over C, with the property that $\mathscr{L}(D) \cong g^*\mathcal{O}_{\mathbf{P}(\mathscr{E})}(1)$. Since $\mathscr{L}(D)$ is very ample on each fibre, g is an isomorphism on each fibre, and so g is an isomorphism.

Conversely, let $\mathscr{E}$ be a locally free sheaf of rank 2 on C, let $X = \mathbf{P}(\mathscr{E})$ and let $\pi: X \to C$ be the projection. Then X is a nonsingular projective surface over k, and each fibre of π is isomorphic to $\mathbf{P}^1$. To show the existence of a section, let $U \subseteq C$ be an open subset on which $\mathscr{E}$ is free. Then $\pi^{-1}(U) \cong U \times \mathbf{P}^1$, so we can define a section $\sigma: U \to \pi^{-1}(U)$ by $y \mapsto y \times$ pt. Then, since X is a projective variety, by (I, 6.8) there is a unique extension of σ to a map of C to X, which is necessarily a section.

For the last statement, see (II, Ex. 7.9).

Remark 2.2.1. A surface X is called a *birationally ruled surface* if it is birationally equivalent to $C \times \mathbf{P}^1$ for some curve C. (This includes the rational surfaces, because $\mathbf{P}^2$ is birational to $\mathbf{P}^1 \times \mathbf{P}^1$.) We see from (2.2) that every ruled surface is birationally ruled.

Proposition 2.3. *Let $\pi: X \to C$ be a ruled surface, let $C_0 \subseteq X$ be a section, and let f be a fibre. Then*

$$\operatorname{Pic} X \cong \mathbf{Z} \oplus \pi^* \operatorname{Pic} C,$$

where $\mathbf{Z}$ is generated by C_0. Also

$$\operatorname{Num} X \cong \mathbf{Z} \oplus \mathbf{Z},$$

generated by C_0, f, and satisfying $C_0.f = 1$, $f^2 = 0$.

PROOF. Clearly $C_0.f = 1$, because C_0 and f meet at only one point, and are transversal there. We have $f^2 = 0$ because two distinct fibres don't meet.

Now if $D \in \text{Pic } X$, let $n = D.f$, and let $D' = D - nC_0$. Then $D'.f = 0$. Therefore by (2.1), $\pi_*(\mathscr{L}(D'))$ is an invertible sheaf on C, and clearly $\mathscr{L}(D') \cong \pi^*\pi_*(\mathscr{L}(D'))$. Since $\pi^*:\text{Pic } C \to \text{Pic } X$ is clearly injective, we see that $\text{Pic } X \cong \mathbf{Z} \oplus \pi^* \text{Pic } C$. Then, since any two fibres are numerically equivalent, $\text{Num } X \cong \mathbf{Z} \oplus \mathbf{Z}$, generated by C_0 and f. See also (II, Ex. 7.9) and (III, Ex. 12.5).

Lemma 2.4. *Let D be a divisor on the ruled surface X, and assume that $D.f \geqslant 0$. Then $R^i\pi_*\mathscr{L}(D) = 0$ for $i > 0$; and for all i,*

$$H^i(X,\mathscr{L}(D)) \cong H^i(C,\pi_*\mathscr{L}(D)).$$

PROOF. Since $\mathscr{L}(D)_y$ is an invertible sheaf of degree $D.f \geqslant 0$ on $X_y \cong \mathbf{P}^1$, we have $H^i(X_y,\mathscr{L}(D)_y) = 0$ for all $i > 0$. Therefore $R^i\pi_*\mathscr{L}(D) = 0$ for $i > 0$ (III, Ex. 11.8) or (III, 12.9). The second statement follows from (III, Ex. 8.1).

Corollary 2.5. *If the genus of C is g, then $p_a(X) = -g$, $p_g(X) = 0$, $q(X) = g$.*

PROOF. The arithmetic genus p_a is defined by $1 + p_a = \chi(\mathcal{O}_X)$. Since $\pi_*\mathcal{O}_X = \mathcal{O}_C$ by (2.1), we have $\dim H^0(X,\mathcal{O}_X) = 1$, $\dim H^1(X,\mathcal{O}_X) = g$, $\dim H^2(X,\mathcal{O}_X) = 0$ using (2.4). So $p_a = -g$. By (III, 7.12.3), the geometric genus $p_g = \dim H^2(X,\mathcal{O}_X) = 0$. The irregularity $q = \dim H^1(X,\mathcal{O}_X) = g$. See also (III, Ex. 8.4).

Proposition 2.6. *Let $\mathscr{E}$ be a locally free sheaf of rank 2 on the curve C, and let X be the ruled surface $\mathbf{P}(\mathscr{E})$. Let $\mathcal{O}_X(1)$ be the invertible sheaf $\mathcal{O}_{\mathbf{P}(\mathscr{E})}(1)$ (II, §7). Then there is a one-to-one correspondence between sections $\sigma:C \to X$ and surjections $\mathscr{E} \to \mathscr{L} \to 0$, where $\mathscr{L}$ is an invertible sheaf on C, given by $\mathscr{L} = \sigma^*\mathcal{O}_X(1)$. Under this correspondence, if $\mathscr{N} = \ker(\mathscr{E} \to \mathscr{L})$, then $\mathscr{N}$ is an invertible sheaf on C, and $\mathscr{N} \cong \pi_*(\mathcal{O}_X(1) \otimes \mathscr{L}(-D))$, where $D = \sigma(C)$, and $\pi^*\mathscr{N} \cong \mathcal{O}_X(1) \otimes \mathscr{L}(-D)$.*

PROOF. The correspondence between sections σ and surjections $\mathscr{E} \to \mathscr{L} \to 0$ is given by (II, 7.12). (See also (II, Ex. 7.8).) Given σ, with $\sigma(C) = D$, we consider the exact sequence

$$0 \to \mathcal{O}_X(1) \otimes \mathscr{L}(-D) \to \mathcal{O}_X(1) \to \mathcal{O}_X(1) \otimes \mathcal{O}_D \to 0.$$

Taking π_*, we have

$$0 \to \pi_*(\mathcal{O}_X(1) \otimes \mathscr{L}(-D)) \to \mathscr{E} \to \mathscr{L} \to 0,$$

with 0 on the right because $R^1\pi_*(\mathcal{O}_X(1) \otimes \mathscr{L}(-D)) = 0$ by (2.4). The middle term is $\mathscr{E}$ by (II, 7.11), and the right-hand term is $\mathscr{L}$, because $\mathcal{O}_X(1) \otimes \mathcal{O}_D$ is a sheaf on $D \cong C$, so σ^* and π_* have the same effect. We conclude that

371

$\mathcal{N} \cong \pi_*(\mathcal{O}_X(1) \otimes \mathcal{L}(-D))$. Since the sheaf $\mathcal{O}_X(1) \otimes \mathcal{L}(-D)$ has degree 0 along the fibres, we see that it is isomorphic to $\pi^*\mathcal{N}$ by (2.3) and $\mathcal{N}$ is invertible (2.1).

Corollary 2.7. *Any locally free sheaf $\mathcal{E}$ of rank 2 on a curve C is an extension of invertible sheaves.*

PROOF. Since $\mathbf{P}(\mathcal{E})$ has a section (2.2), we get an exact sequence $0 \to \mathcal{N} \to \mathcal{E} \to \mathcal{L} \to 0$ where $\mathcal{N}$ and $\mathcal{L}$ are invertible sheaves. This also follows from (II, Ex. 8.2).

Remark 2.7.1. The same result holds for locally free sheaves of arbitrary rank (Ex. 2.3).

Proposition 2.8. *If $\pi:X \to C$ is a ruled surface, it is possible to write $X \cong \mathbf{P}(\mathcal{E})$ where $\mathcal{E}$ is a locally free sheaf on C with the property that $H^0(\mathcal{E}) \neq 0$ but for all invertible sheaves $\mathcal{L}$ on C with $\deg \mathcal{L} < 0$, we have $H^0(\mathcal{E} \otimes \mathcal{L}) = 0$. In this case the integer $e = -\deg \mathcal{E}$ is an invariant of X. Furthermore in this case there is a section $\sigma_0:C \to X$ with image C_0, such that $\mathcal{L}(C_0) \cong \mathcal{O}_X(1)$.*

PROOF. First write $X \cong \mathbf{P}(\mathcal{E}')$ for some locally free sheaf $\mathcal{E}'$ on C (2.2). Then we will replace $\mathcal{E}'$ by $\mathcal{E} = \mathcal{E}' \otimes \mathcal{M}$ for a suitable invertible sheaf $\mathcal{M}$ on C so as to have $H^0(\mathcal{E}) \neq 0$ but $H^0(\mathcal{E} \otimes \mathcal{L}) = 0$ for all $\mathcal{L}$ with $\deg \mathcal{L} < 0$. An invertible sheaf of positive degree on C is ample (IV, 3.3), so it is possible to make $H^0(\mathcal{E}) \neq 0$ by taking $\deg \mathcal{M}$ large enough. On the other hand, since $\mathcal{E}'$ is an extension of invertible sheaves (2.7), and since an invertible sheaf of negative degree can have no global sections, we see that $H^0(\mathcal{E}) = 0$ for $\deg \mathcal{M}$ sufficiently negative. So we achieve our result by taking an $\mathcal{M}$ of least degree such that $H^0(\mathcal{E}' \otimes \mathcal{M}) \neq 0$.

Since all possible representations of X as a $\mathbf{P}(\mathcal{E})$ are given by the sheaves $\mathcal{E} = \mathcal{E}' \otimes \mathcal{M}$ (2.2) we see that the integer $e = -\deg \mathcal{E}$ depends only on X. (The *degree* of $\mathcal{E}$ is defined as the degree of the invertible sheaf $\bigwedge^2\mathcal{E}$ (II, Ex. 6.12).)

Finally, let $s \in H^0(\mathcal{E})$ be a nonzero section. It determines an injective map $0 \to \mathcal{O}_C \to \mathcal{E}$. I claim the quotient $\mathcal{L} = \mathcal{E}/\mathcal{O}_C$ is an invertible sheaf on C. Since C is a nonsingular curve, and $\mathcal{L}$ has rank 1 in any case, it is enough to show that $\mathcal{L}$ is torsion-free. If not, let $\mathcal{F} \subseteq \mathcal{E}$ be the inverse image of the torsion subsheaf of $\mathcal{L}$ by the map $\mathcal{E} \to \mathcal{L} \to 0$. In that case $\mathcal{F}$ is torsion-free of rank 1 on C, hence invertible. Furthermore, $\mathcal{O}_C \subsetneqq \mathcal{F}$, so $\deg \mathcal{F} > 0$. But then, since $\mathcal{F} \subseteq \mathcal{E}$, we have $H^0(\mathcal{E} \otimes \mathcal{F}^{\vee}) \neq 0$, and $\deg \mathcal{F}^{\vee} < 0$, so this contradicts the choice of $\mathcal{E}$.

Now, since $\mathcal{L}$ is invertible, it gives a section $\sigma_0:C \to X$ by (2.6). Let C_0 be its image. Then $\mathcal{N} = \mathcal{O}_C$ in the notation of (2.6), so $\mathcal{O}_X(1) \otimes \mathcal{L}(-C_0) \cong \mathcal{O}_X$, which shows that $\mathcal{L}(C_0) \cong \mathcal{O}_X(1)$.

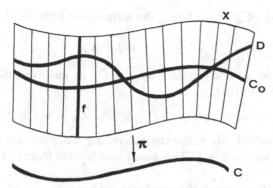

Figure 19. A ruled surface.

Notation 2.8.1. For the rest of this section, we fix the following notation (Fig. 19). Let C be a curve of genus g, and let $\pi: X \to C$ be a ruled surface over C. We write $X \cong \mathbf{P}(\mathscr{E})$, where $\mathscr{E}$ satisfies the conditions of (2.8), in which case we say $\mathscr{E}$ is *normalized*. This does not necessarily determine $\mathscr{E}$ uniquely, but it does determine $\deg \mathscr{E}$. We let $\mathfrak{e}$ be the divisor on C corresponding to the invertible sheaf $\bigwedge^2 \mathscr{E}$, so that $e = -\deg \mathfrak{e}$. (This sign is put in for historical reasons.) We fix a section C_0 of X with $\mathscr{L}(C_0) \cong \mathscr{O}_{\mathbf{P}(\mathscr{E})}(1)$. If $\mathfrak{b}$ is any divisor on C, then we denote the divisor $\pi^*\mathfrak{b}$ on X by $\mathfrak{b}f$, by abuse of notation. Thus any element of Pic X can be written $aC_0 + \mathfrak{b}f$ with $a \in \mathbf{Z}$ and $\mathfrak{b} \in$ Pic C. Any element of Num X can be written $aC_0 + bf$ with $a,b \in \mathbf{Z}$.

Proposition 2.9. *If D is any section of X, corresponding to a surjection $\mathscr{E} \to \mathscr{L} \to 0$, and if $\mathscr{L} = \mathscr{L}(\mathfrak{b})$ for some divisor $\mathfrak{b}$ on C, then $\deg \mathfrak{b} = C_0.D$, and*

$$D \sim C_0 + (\mathfrak{b} - \mathfrak{e})f.$$

In particular, we have $C_0^2 = \deg \mathfrak{e} = -e$.

PROOF. Since $\mathscr{L} = \sigma^*(\mathscr{L}(C_0) \otimes \mathscr{O}_D)$, we have $\deg \mathscr{L} = C_0.D$ by (1.1) and (1.3). Writing

$$0 \to \mathscr{N} \to \mathscr{E} \to \mathscr{L} \to 0,$$

we have $\mathscr{L}(C_0 - D) \cong \pi^*\mathscr{N}$ by (2.6) and the choice of C_0 (2.8.1). But $\mathscr{N} = \mathscr{L}(\mathfrak{e} - \mathfrak{b})$, so we have $D \sim C_0 + (\mathfrak{b} - \mathfrak{e})f$ in Pic X. Finally, in the case $D = C_0$, $\mathscr{N} = \mathscr{O}_C$, so $\mathfrak{b} = \mathfrak{e}$ and we have $C_0^2 = \deg \mathfrak{e} = -e$.

Lemma 2.10. *The canonical divisor K on X is given by*

$$K \sim -2C_0 + (\mathfrak{k} + \mathfrak{e})f$$

where $\mathfrak{k}$ is the canonical divisor on C.

PROOF. Let $K \sim aC_0 + bf$. Using the adjunction formula (1.5) for a fibre f, we have

$$-2 = f.(f + K) = a.$$

Now we use the adjunction formula for C_0 in its invertible sheaf form (II, 8.20), which says that

$$\omega_{C_0} \cong \omega_X \otimes \mathscr{L}(C_0) \otimes \mathcal{O}_{C_0} \cong \mathscr{L}(-C_0 + bf) \otimes \mathcal{O}_{C_0}.$$

Identifying C_0 with C via π, the corresponding statement for divisors on C is $\mathfrak{f} = -\mathfrak{e} + \mathfrak{b}$, so $\mathfrak{b} = \mathfrak{e} + \mathfrak{f}$. This result also follows from (III, Ex. 8.4).

Corollary 2.11. *For numerical equivalence, we have*

$$K \equiv -2C_0 + (2g - 2 - e)f$$

and therefore

$$K^2 = 8(1 - g).$$

PROOF. We have $\deg \mathfrak{f} = 2g - 2$ (IV, 1.3.3) and $\deg \mathfrak{e} = -e$. Then we compute K^2 using (2.3) and (2.9).

Example 2.11.1. For any curve C, the ruled surface $X = C \times \mathbf{P}^1$ corresponds to the (normalized) locally free sheaf $\mathscr{E} = \mathcal{O}_C \oplus \mathcal{O}_C$ on C. In this case $e = 0$, and C_0 is any fibre of the second projection.

Example 2.11.2. If C is a curve of genus $\geqslant 1$, and $\mathscr{E} = \mathcal{O}_C \oplus \mathscr{L}$ where $\deg \mathscr{L} = 0$ but $\mathscr{L} \not\cong \mathcal{O}_C$, then there are two choices of normalized $\mathscr{E}$, namely $\mathscr{E}$ and $\mathscr{E} \otimes \mathscr{L}^{-1}$. We have $e = 0$, $\deg \mathfrak{e} = 0$, but $\mathfrak{e}$ is determined only up to sign. There are exactly two choices of C_0, both with $C_0^2 = 0$.

Example 2.11.3. On any curve C, let $\mathscr{E} = \mathcal{O}_C \oplus \mathscr{L}$ with $\deg \mathscr{L} < 0$. Then the normalized $\mathscr{E}$ is unique, $\mathscr{L} = \mathscr{L}(\mathfrak{e})$ and $\mathfrak{e}$ is unique. The section C_0 is unique, with $C_0^2 = -e < 0$. In this case $e = -\deg \mathscr{L} > 0$.

Example 2.11.4. Let C be any curve embedded in $\mathbf{P}^n$, of degree d. Let X_0 be the cone over C in $\mathbf{P}^{n+1}$, with vertex P_0 (I, Ex. 2.10). If we blow up the point P_0, we will show that we obtain a ruled surface X over C, of the kind (2.11.3) above, with $\mathscr{L} \cong \mathcal{O}_C(-1)$. In particular, $e = d$, and the inverse image of P_0 in X is the section C_0 with $C_0^2 = -d$.

First of all, we show that $\mathbf{P}^{n+1}$ with one point blown up is isomorphic to $\mathbf{P}(\mathcal{O} \oplus \mathcal{O}(1))$ over $\mathbf{P}^n$. Indeed, let $\mathbf{P}^{n+1}$ have coordinates $x_0, \ldots, x_{n+1}$. If we blow up the point $P_0 = (1, 0, \ldots, 0)$, then we get the variety $V \subseteq \mathbf{P}^n \times \mathbf{P}^{n+1}$ defined by the equations $x_i y_j = x_j y_i$ for $i, j = 1, 2, \ldots, n + 1$, where $y_1, \ldots, y_{n+1}$ are the coordinates for $\mathbf{P}^n$ (II, 7.12.1). On the other hand, if $\mathscr{E} = \mathcal{O} \oplus \mathcal{O}(1)$ on $\mathbf{P}^n$, then $\mathbf{P}(\mathscr{E})$ is defined as **Proj** $S(\mathscr{E})$, where $S(\mathscr{E})$ is the symmetric algebra of $\mathscr{E}$ (II, §7). Now $\mathscr{E}$ is generated by the global sections 1 of $\mathcal{O}$ and $y_1, \ldots, y_{n+1}$ of $\mathcal{O}(1)$. Therefore $S(\mathscr{E})$ is a quotient of the polynomial algebra $\mathcal{O}[x_0,\ldots,x_{n+1}]$ by the mapping $x_0 \mapsto 1$, $x_i \mapsto y_i$ for $i = 1,\ldots,n+1$. The kernel of this map is the ideal generated by all $x_i y_j - x_j y_i$, $i = 1,\ldots,n+1$. Therefore $\mathbf{P}(\mathcal{E})$ is isomorphic to the subscheme of $\mathbf{P}^n \times \mathbf{P}^{n+1}$ defined by these equations, which is the same as the variety $V \subseteq \mathbf{P}^n \times \mathbf{P}^{n+1}$ defined above. The first projection makes V look like $\mathbf{P}(\mathcal{E})$, the second projection makes V look like blowing up a point.

Now let Y be any subvariety of $\mathbf{P}^n$, and X_0 its cone in $\mathbf{P}^{n+1}$, with vertex P_0. If we blow up P_0 on X_0, we get a variety X which is the strict transform of X_0 in V (II, 7.15.1). On the other hand, this variety X is clearly the inverse image of Y under the projection $\pi: V \cong \mathbf{P}(\mathcal{E}) \to \mathbf{P}^n$. So we see that $X \cong \mathbf{P}(\mathcal{O}_Y \oplus \mathcal{O}_Y(1))$. Twisting by $\mathcal{O}_Y(-1)$, we still have the same variety, so $X \cong \mathbf{P}(\mathcal{O}_Y \oplus \mathcal{O}_Y(-1))$.

In particular, if Y is a nonsingular curve C of degree d in $\mathbf{P}^n$, then $\mathcal{L} = \mathcal{O}_C(-1)$ has degree $-d$.

Example 2.11.5. As a special case of (2.11.4), we see that $\mathbf{P}^2$ with one point blown up is isomorphic to the rational ruled surface over $\mathbf{P}^1$ defined by $\mathcal{E} = \mathcal{O} \oplus \mathcal{O}(-1)$, having $e = 1$.

Example 2.11.6. For an example of a ruled surface with $e < 0$, let C be an elliptic curve, let $P \in C$ be a point, and construct a locally free sheaf $\mathcal{E}$ of rank 2 as an extension

$$0 \to \mathcal{O} \to \mathcal{E} \to \mathcal{L}(P) \to 0$$

defined by a nonzero element $\xi \in \mathrm{Ext}^1(\mathcal{L}(P),\mathcal{O})$ (III, Ex. 6.1). In this case $\mathrm{Ext}^1(\mathcal{L}(P),\mathcal{O}) \cong H^1(C,\mathcal{L}(-P))$ (III, 6.3) and (III, 6.7). This is dual to $H^0(C,\mathcal{L}(P))$ which has dimension 1. Thus ξ is unique up to a scalar multiple, and so $\mathcal{E}$ is uniquely determined up to isomorphism.

I claim this $\mathcal{E}$ is normalized. Clearly $H^0(\mathcal{E}) \neq 0$ by construction. If $\mathcal{M}$ is any invertible sheaf, then we have an exact sequence

$$0 \to \mathcal{M} \to \mathcal{E} \otimes \mathcal{M} \to \mathcal{L}(P) \otimes \mathcal{M} \to 0.$$

If $\deg \mathcal{M} < 0$, then we have $H^0(\mathcal{M}) = 0$, and $H^0(\mathcal{L}(P) \otimes \mathcal{M}) = 0$, and therefore also $H^0(\mathcal{E} \otimes \mathcal{M}) = 0$, except for the case $\mathcal{M} = \mathcal{L}(-P)$. In that case we look at the cohomology sequence

$$0 \to H^0(\mathcal{M}) \to H^0(\mathcal{E} \otimes \mathcal{M}) \to H^0(\mathcal{L}(P) \otimes \mathcal{M}) \xrightarrow{\delta} H^1(\mathcal{M}) \to \ldots .$$

The image of $1 \in H^0(\mathcal{L}(P) \otimes \mathcal{M}) = H^0(\mathcal{O}_C)$ by δ is just the element ξ defining $\mathcal{E}$ (III, Ex. 6.1), which is nonzero. Therefore δ is injective, and again $H^0(\mathcal{E} \otimes \mathcal{M}) = 0$. Thus $\mathcal{E}$ is normalized.

Now taking $X = \mathbf{P}(\mathcal{E})$, we have an elliptic ruled surface with $e = -1$.

Now that we have established some general properties of ruled surfaces and have given some examples, we can look more closely at some special cases. We begin by discussing the possible values of the invariant e.

Theorem 2.12. *Let X be a ruled surface over the curve C of genus g, determined by a normalized locally free sheaf $\mathscr{E}$.*

(a) *If $\mathscr{E}$ is decomposable (i.e., a direct sum of two invertible sheaves) then $\mathscr{E} \cong \mathcal{O}_C \oplus \mathscr{L}$ for some $\mathscr{L}$ with $\deg \mathscr{L} \leqslant 0$. Therefore $e \geqslant 0$. All values of $e \geqslant 0$ are possible.*

(b) *If $\mathscr{E}$ is indecomposable, then $-2g \leqslant e \leqslant 2g - 2$. (In fact, there are even stronger restrictions on e (Ex. 2.5).)*

PROOF. If $\mathscr{E}$ is decomposable, then $\mathscr{E} \cong \mathscr{L}_1 \oplus \mathscr{L}_2$ for two invertible sheaves $\mathscr{L}_1$ and $\mathscr{L}_2$ on C. We must have $\deg \mathscr{L}_i \leqslant 0$ because of the normalization (2.8) and furthermore $H^0(\mathscr{L}_i) \neq 0$ for at least one of them. Thus one of them is $\mathcal{O}_C$, so we have $\mathscr{E} \cong \mathcal{O}_C \oplus \mathscr{L}$ with $\deg \mathscr{L} \leqslant 0$. From (2.11.1), (2.11.2), and (2.11.3), we see that all values of $e \geqslant 0$ are possible.

Now suppose $\mathscr{E}$ is indecomposable. Then, corresponding to the section C_0, we have an exact sequence

$$0 \to \mathcal{O}_C \to \mathscr{E} \to \mathscr{L} \to 0$$

for some $\mathscr{L}$ (2.8). This must be a nontrivial extension, so it corresponds to a nonzero element $\xi \in \mathrm{Ext}^1(\mathscr{L}, \mathcal{O}_C) \cong H^1(C, \mathscr{L}^{\vee})$ (III, Ex. 6.1). In particular, $H^1(\mathscr{L}^{\vee}) \neq 0$, so we must have $\deg \mathscr{L}^{\vee} \leqslant 2g - 2$ (IV, 1.3.4). Since $e = -\deg \mathscr{L}$, we have $e \leqslant 2g - 2$.

On the other hand, we have $H^0(\mathscr{E} \otimes \mathscr{M}) = 0$ for all $\deg \mathscr{M} < 0$ by the normalization. In particular, taking $\deg \mathscr{M} = -1$, we have

$$0 = H^0(\mathscr{E} \otimes \mathscr{M}) \to H^0(\mathscr{L} \otimes \mathscr{M}) \to H^1(\mathscr{M}) \to \cdots ,$$

so we must have

$$\dim H^0(\mathscr{L} \otimes \mathscr{M}) \leqslant \dim H^1(\mathscr{M}).$$

Since $\deg \mathscr{M} < 0$, $H^0(\mathscr{M}) = 0$, so by Riemann–Roch, we have $\dim H^1(\mathscr{M}) = g$. On the other hand, also by Riemann–Roch, we have

$$\dim H^0(\mathscr{L} \otimes \mathscr{M}) \geqslant \deg \mathscr{L} - 1 + 1 - g.$$

Combining, we get $\deg \mathscr{L} \leqslant 2g$, hence $e \geqslant -2g$.

Corollary 2.13. *If $g = 0$, then $e \geqslant 0$, and for each $e \geqslant 0$ there is exactly one rational ruled surface with invariant e, given by $\mathscr{E} = \mathcal{O} \oplus \mathcal{O}(-e)$ over $C = \mathbf{P}^1$.*

PROOF. If $g = 0$, case (b) of (2.12) cannot occur. Hence $\mathscr{E} \cong \mathcal{O}_C \oplus \mathscr{L}$. But the only invertible sheaves on $\mathbf{P}^1$ are $\mathcal{O}(n)$ for $n \in \mathbf{Z}$ (II, 6.4). So for each $e \geqslant 0$ there is just one possibility.

Corollary 2.14. *Every locally free sheaf $\mathscr{E}$ of rank 2 on $\mathbf{P}^1$ is decomposable.*

PROOF. After tensoring with a suitable invertible sheaf, it becomes normalized, in which case it is isomorphic to $\mathcal{O} \oplus \mathcal{O}(-e)$ by (2.13). See (Ex. 2.6) for a generalization.

Theorem 2.15. *If X is a ruled surface over an elliptic curve C, corresponding to an indecomposable $\mathscr{E}$, then $e = 0$ or -1, and there is exactly one such ruled surface over C for each of these two values of e.*

PROOF. According to (2.12) we must have $e = 0, -1, -2$. If $e = 0$, then we have an exact sequence

$$0 \to \mathcal{O}_C \to \mathscr{E} \to \mathscr{L} \to 0$$

with deg $\mathscr{L} = 0$. This extension corresponds to a nonzero element $\xi \in H^1(\mathscr{L}^\vee)$. In particular, $H^1(\mathscr{L}^\vee) \neq 0$. It is dual to $H^0(\mathscr{L})$, so we must have $\mathscr{L} \cong \mathcal{O}_C$. Conversely, taking $\mathscr{L} = \mathcal{O}_C$, we have dim $H^1(\mathcal{O}_C) = 1$, so there is just one choice of nonzero $\xi \in H^1(\mathcal{O}_C)$, up to isomorphism, which is a nontrivial extension

$$0 \to \mathcal{O}_C \to \mathscr{E} \to \mathcal{O}_C \to 0.$$

Clearly this $\mathscr{E}$ is normalized. Furthermore, this $\mathscr{E}$ is indecomposable, because if $\mathscr{E}$ were decomposable, being normalized, it would be isomorphic to $\mathcal{O}_C \oplus \mathscr{L}$ for some $\mathscr{L}$, by (2.12). But $\bigwedge^2 \mathscr{E} \cong \mathcal{O}_C$, so $\mathscr{L} \cong \mathcal{O}_C$, so in fact this extension would have to split, which it doesn't. Thus we get exactly one elliptic ruled surface X with $e = 0$ and $\mathscr{E}$ indecomposable.

If $e = -1$, then we have an exact sequence

$$0 \to \mathcal{O}_C \to \mathscr{E} \to \mathscr{L}(P) \to 0$$

for some point $P \in C$, because every invertible sheaf of degree 1 on C is of the form $\mathscr{L}(P)$ (IV, 1.3.7). Furthermore, for each P there exists such a normalized bundle, unique up to isomorphism, by (2.11.6). To show that there is just one elliptic ruled surface with $e = -1$, it will be sufficient to show that if $\mathscr{E}$ is defined by P as above, and $\mathscr{E}'$ is similarly defined by $Q \neq P$, then there exists an invertible sheaf $\mathscr{M}$ on C such that $\mathscr{E}' \cong \mathscr{E} \otimes \mathscr{M}$.

Take a point $R \in C$ such that $2R \sim P + Q$. This is possible, because the linear system $|P + Q|$ defines a two-to-one map of C to $\mathbf{P}^1$, ramified at four points (assume char $k \neq 2$), and we can take R to be one of them (IV, §4). We will show that $\mathscr{E}' \cong \mathscr{E} \otimes \mathscr{L}(R - P)$. In any case, we have an exact sequence

$$0 \to \mathscr{L}(R - P) \to \mathscr{E} \otimes \mathscr{L}(R - P) \to \mathscr{L}(R) \to 0.$$

Since $H^0(\mathscr{L}(R)) \neq 0$ and $H^1(\mathscr{L}(R-P)) = 0$, we see that $H^0(\mathscr{E} \otimes \mathscr{L}(R-P)) \neq 0$. So we get an exact sequence

$$0 \to \mathcal{O}_C \to \mathscr{E} \otimes \mathscr{L}(R - P) \to \mathscr{N} \to 0,$$

and the quotient $\mathscr{N}$ must be invertible, as in the proof of (2.8). So we have

$$\mathscr{N} \cong \bigwedge^2(\mathscr{E} \otimes \mathscr{L}(R - P)) \cong (\bigwedge^2 \mathscr{E}) \otimes \mathscr{L}(2R - 2P).$$

Since $\bigwedge^2 \mathscr{E} \cong \mathscr{L}(P)$, we have $\mathscr{N} \cong \mathscr{L}(2R - P) \cong \mathscr{L}(Q)$. Therefore $\mathscr{E} \otimes \mathscr{L}(R - P) \cong \mathscr{E}'$ as required. This proves the uniqueness of the elliptic ruled surface with $e = -1$.

Finally, we will show that the case $e = -2$ does not occur. If it did, we would have a normalized bundle $\mathscr{E}$ with an exact sequence

$$0 \to \mathcal{O}_C \to \mathscr{E} \to \mathscr{L}(P + Q) \to 0$$

for some $P, Q \in C$, since every invertible sheaf of degree 2 is of the form $\mathscr{L}(P + Q)$. Now take any pair of points $R, S \in C$ with $R + S \sim P + Q$, and let $\mathscr{M} = \mathscr{L}(-R)$. Then, since $\mathscr{E}$ is normalized, $H^0(\mathscr{E} \otimes \mathscr{M}) = 0$, so the map $\gamma : H^0(\mathscr{L}(P + Q - R)) \to H^1(\mathscr{L}(-R))$ must be injective. On the other hand, let $\xi \in H^1(\mathscr{L}(-P - Q))$ be the element defining the extension $\mathscr{E}$. Then we have a commutative diagram, writing $\mathscr{L}(P + Q - R)$ as $\mathscr{L}(S)$,

$$
\begin{array}{ccc}
H^0(\mathcal{O}_C) & \xrightarrow{\ \delta\ } & H^1(\mathscr{L}(-P - Q)) \\
\alpha \downarrow & & \downarrow \beta \\
H^0(\mathscr{L}(S)) & \xrightarrow{\ \gamma\ } & H^1(\mathscr{L}(-R))
\end{array}
$$

where $\delta(1) = \xi$, $\alpha(1) = t$, a nonzero section defining the divisor S, and β is induced from the map $\mathcal{O}_C \to \mathscr{L}(S)$ corresponding to t. Now β is dual to the map

$$\beta' : H^0(\mathscr{L}(R)) \to H^0(\mathscr{L}(P + Q))$$

also induced by t. The image of any nonzero element of $H^0(\mathscr{L}(R))$ by β' is a section of $H^0(\mathscr{L}(P + Q))$ corresponding to the effective divisor $R + S \in |P + Q|$.

By varying R and S, we get every divisor in the linear system $|P + Q|$. Therefore the image of β' as R varies fills up the whole 2-dimensional vector space $H^0(\mathscr{L}(P + Q))$. In particular, we can choose R so that the image of β' lands in the kernel of ξ, considered as a linear functional on $H^0(\mathscr{L}(P + Q))$. In that case, $\beta(\xi) = 0$, which contradicts the injectivity of γ. Thus the case $e = -2$ is impossible.

Caution 2.15.1. One point which came up in the first part of this proof should be noted. It is possible for a locally free sheaf of rank 2 to be a nontrivial extension of two invertible sheaves, and yet be decomposable. For example, the sheaf of differentials on $\mathbf{P}^1$ is isomorphic to $\mathcal{O}(-2)$, so we have an exact sequence (II, 8.13)

$$0 \to \mathcal{O}(-2) \to \mathcal{O}(-1) \oplus \mathcal{O}(-1) \to \mathcal{O} \to 0.$$

The sequence cannot be split (because for example $H^0(\mathcal{O}(-1) \oplus \mathcal{O}(-1)) = 0$), but the sheaf in the middle is decomposable.

Corollary 2.16 (Atiyah). *For each integer n, there is a natural one-to-one correspondence (described explicitly in the proof below) between the set of isomorphism classes of indecomposable locally free sheaves of rank 2 and degree n on the elliptic curve C, and the set of points of C.*

PROOF. Fix a point $P_0 \in C$. Let $\mathscr{E}'$ be an indecomposable locally free sheaf of rank 2 and degree n on C. Tensoring with $\mathscr{L}(mP_0)$ for some m, we may assume $n = 0$ or 1. If $n = 0$, then by (2.2), (2.8), and (2.15) there is a (unique) invertible sheaf $\mathscr{L}$ of degree 0 on C such that $\mathscr{E}' \otimes \mathscr{L}$ is isomorphic to the unique nontrivial extension of $\mathscr{O}_C$ by $\mathscr{O}_C$. Since the invertible sheaves of degree 0 are in one-to-one correspondence with the closed points of C (IV, §4), we have the result. If $n = 1$, then as in the proof of (2.15) we find that $\mathscr{E}'$ is an extension of $\mathscr{L}(P)$ by $\mathscr{O}_C$ for some uniquely determined point $P \in C$, whence the result.

Remark 2.16.1. More generally, for any curve C, of genus g, one can consider the problem of classifying all locally free sheaves $\mathscr{E}$ on C up to isomorphism. The rank r and the degree d (which is deg $\bigwedge^r\mathscr{E}$) are numerical invariants. For fixed r and d, one expects some kind of continuous family. For $g = 0$, all locally free sheaves are direct sums of invertible sheaves (Ex. 2.6). For $g = 1$ the general classification, which is similar to the rank 2 case we have just done, has been accomplished by Atiyah [1]. For $g \geq 2$, the situation becomes more complicated. Among the indecomposable locally free sheaves, one has to distinguish between the *stable* ones (Ex. 2.8) in the sense of Mumford [1], and the rest. The stable ones form nice algebraic families, whereas the others do not. See for example, Narasimhan and Seshadri [1]. Similarly, for the ruled surfaces themselves, the ones with $e < 0$ are *stable*, and form nice algebraic families, but the others do not.

Next we will study the rational ruled surfaces, which were classified in (2.13).

Theorem 2.17. Let X_e, for any $e \geq 0$, be the rational ruled surface defined by $\mathscr{E} = \mathscr{O} \oplus \mathscr{O}(-e)$ on $C = \mathbf{P}^1$ (2.13). Then:
  (a) *there is a section* $D \sim C_0 + nf$ *if and only if* $n = 0$ *or* $n \geq e$. *In particular, there is a section* $C_1 \sim C_0 + ef$ *with* $C_0 \cap C_1 = \varnothing$ *and* $C_1^2 = e$;
  (b) *the linear system* $|C_0 + nf|$ *is base-point-free if and only if* $n \geq e$;
  (c) *the linear system* $|C_0 + nf|$ *is very ample if and only if* $n > e$.

PROOF.
  (a) According to (2.6) and (2.9), giving a section $D \sim C_0 + nf$ is equivalent to giving a surjective map $\mathscr{E} \to \mathscr{L} \to 0$ with deg $\mathscr{L} = C_0.D = n - e$. Since we are on $\mathbf{P}^1$, this means a surjective map

$$\mathscr{O} \oplus \mathscr{O}(-e) \to \mathscr{O}(n - e) \to 0.$$

If $n < e$, there are no nonzero maps of $\mathscr{O}$ to $\mathscr{O}(n - e)$, so the map $\mathscr{O}(-e) \to \mathscr{O}(n - e)$ must be an isomorphism, and therefore $n = 0$. This corresponds to the section C_0, which is unique if $e > 0$. Otherwise we have $n \geq e$, and any such n is possible. We have only to take maps $\mathscr{O} \to \mathscr{O}(n - e)$ and $\mathscr{O}(-e) \to \mathscr{O}(n - e)$ corresponding to effective divisors of degrees $n - e$ and n on C which do not meet. Then the corresponding map $\mathscr{O} \oplus \mathscr{O}(-e) \to \mathscr{O}(n - e)$ will be surjective.

In particular, if we take $n = e$, there is a section $C_1 \sim C_0 + ef$. Then $C_1^2 = e$, and $C_0.C_1 = 0$, so $C_0 \cap C_1 = \varnothing$.

(b) If $|C_0 + nf|$ is base-point-free, then $C_0.(C_0 + nf) \geq 0$ so $n \geq e$. Conversely, if $n \geq e$, then $C_0 + nf \sim C_1 + (n - e)f$, and since $C_0 \cap C_1 = \varnothing$, and any f is linearly equivalent to any other, we can find a divisor of the form $C_0 + nf$ or $C_1 + (n - e)f$ which misses any given point.

(c) If $D = C_0 + nf$ is very ample, then we must have $D.C_0 > 0$, so $n > e$. Conversely, suppose $n > e$. Then we will show that D is very ample by showing that the linear system $|D|$ separates points and tangent vectors (II, 7.8.2).

Case 1. Let $P \neq Q$ be two points not both in C_0, and not both in any fibre. Then a divisor of the form $C_0 + nf$ for suitable f will separate them.

Case 2. Let P be a point and t a tangent vector at P, such that P,t are not both in C_0 and not both in any fibre. Then a divisor of the form $C_0 + \sum_{i=1}^{n} f_i$, for suitable fibres f_i, will contain P but not t.

Case 3. Suppose P,Q or P,t are both in C_0. Then a divisor of the form $C_1 + \sum_{i=1}^{n-e} f_i$ will separate them.

Case 4. Suppose P,Q, or P,t are both in the same fibre f. Since $D.f = 1$, the invertible sheaf $\mathscr{L}(D) \otimes \mathcal{O}_f$ is very ample on $f \cong \mathbf{P}^1$. Thus to separate P,Q or P,t, it will be sufficient to show that the natural map $H^0(X,\mathscr{L}(D)) \to H^0(f,\mathscr{L}(D) \otimes \mathcal{O}_f)$ is surjective. The cokernel of this map lands in $H^1(X,\mathscr{L}(D - f))$, which by (2.4) is isomorphic to $H^1(C,\pi_*\mathscr{L}(D - f))$. On the other hand, $D - f \sim C_0 + (n - 1)f$, so

$$\pi_*(\mathscr{L}(D - f)) \cong \pi_*(\mathscr{L}(C_0)) \otimes \mathcal{O}_C(n - 1)$$

by the projection formula (II, Ex. 5.1). Now $\pi_*(\mathscr{L}(C_0)) \cong \mathscr{E}$ by (2.8) and (II, 7.11), so we have

$$\pi_*(\mathscr{L}(D - f)) \cong \mathcal{O}(n - 1) \oplus \mathcal{O}(n - e - 1).$$

Since $n > e \geq 0$, both $n - 1 \geq 0$ and $n - e - 1 \geq 0$, so $H^1 = 0$ and the above map is surjective.   q.e.d.

Corollary 2.18. *Let D be the divisor $aC_0 + bf$ on the rational ruled surface X_e, $e \geq 0$. Then:*

(a) *D is very ample $\Leftrightarrow$ D is ample $\Leftrightarrow$ $a > 0$ and $b > ae$;*

(b) *the linear system $|D|$ contains an irreducible nonsingular curve $\Leftrightarrow$ it contains an irreducible curve $\Leftrightarrow$ $a = 0, b = 1$ (namely f); or $a = 1, b = 0$ (namely C_0); or $a > 0, b > ae$; or $e > 0, a > 0, b = ae$.*

PROOF.

(a) If D is very ample, it is certainly ample (II, 7.4.3). If D is ample, then $D.f > 0$, so $a > 0$, and $D.C_0 > 0$, so $b > ae$ (1.6.2). Now suppose that $a > 0$ and $b > ae$. Then we can write $D = (a - 1)(C_0 + ef) + (C_0 + (b - ae + e)f)$. Since $|C_0 + ef|$ has no base points, and $C_0 + (b - ae + e)f$ is very ample (2.17), we conclude that D is also very ample (II, Ex. 7.5).

(b) If $|D|$ contains an irreducible nonsingular curve, then in particular it contains an irreducible curve. If D is an irreducible curve, then D could be f (in which case $a = 0$, $b = 1$) or C_0 (in which case $a = 1$, $b = 0$). Otherwise, π maps D surjectively to C, so $D.f = a > 0$, and $D.C_0 \geqslant 0$, so $b \geqslant ae$. If $e = 0$ and $b = ae$, then $b = 0$ so $D = aC_0$. But in this case X_0 is $\mathbf{P}^1 \times \mathbf{P}^1$, and C_0 is one of the rulings, so for D to be irreducible, we must have $a = 1$. Thus the restrictions on a,b are necessary. To complete the proof, we must show that if $a > 0$, $b > ae$, or $e > 0$, $a > 0$, $b = ae$, then $|D|$ contains an irreducible nonsingular curve. In the first case, D is very ample by (a), so the result follows from Bertini's theorem (II, 8.18) applied to X. In the second case, we use the fact (2.11.4) that X_e can be obtained from the cone Y over a nonsingular rational curve C of degree $e > 0$ in some $\mathbf{P}^n$, by blowing up the vertex. In this case, the curve C_1 on X_e is the strict transform of the hyperplane section H of Y. By Bertini's theorem applied to the very ample divisor aH on Y (II, 8.18.1), we can find an irreducible nonsingular curve in the linear system $|aH|$, not containing the vertex of Y. Its strict transform on X_e is then an irreducible nonsingular curve in the linear system $|aC_1| = |D|$.

Remark 2.18.1. In case $e = 0$, we get some new proofs of earlier results about curves on the nonsingular quadric surface, which is isomorphic to X_0 (II, 7.6.2), (III, Ex. 5.6), (1.10.1).

Corollary 2.19. *For every $n > e \geqslant 0$, there is an embedding of the rational ruled surface X_e as a rational scroll of degree $d = 2n - e$ in $\mathbf{P}^{d+1}$. (A scroll is a ruled surface embedded in $\mathbf{P}^N$ in such a way that all the fibres f have degree 1.)*

PROOF. Use the very ample divisor $D = C_0 + nf$. Then $D.f = 1$, so the image of X_e in $\mathbf{P}^N$ is a scroll, and $D^2 = 2n - e$, so the image has degree $d = 2n - e$. To find N, we compute $H^0(X, \mathscr{L}(D))$. As in the proof of (2.17), we find that

$$H^0(X, \mathscr{L}(D)) = H^0(C, \pi_* \mathscr{L}(D)) = H^0(C, \mathscr{E} \otimes \mathcal{O}(n)) = H^0(\mathcal{O}(n) \oplus \mathcal{O}(n - e)).$$

This has dimension $2n + 2 - e$, so $N = 2n + 1 - e = d + 1$.

Example 2.19.1. For $e = 0$, $n = 1$, we recover the nonsingular quadric surface in $\mathbf{P}^3$.

For $e = 1$, $n = 2$, we get a rational scroll of degree 3 in $\mathbf{P}^4$, which is isomorphic to $\mathbf{P}^2$ with one point blown up (II, Ex. 7.7).

In $\mathbf{P}^5$, there are two different kinds of rational scrolls of degree 4, corresponding to $e = 0$, $n = 2$, and $e = 2$, $n = 3$.

Remark 2.19.2. In fact, it is known that every nonsingular surface of degree d in $\mathbf{P}^{d+1}$, not contained in any hyperplane, is either one of these rational scrolls (2.19), or $\mathbf{P}^2 \subseteq \mathbf{P}^2$ (if $d = 1$), or the Veronese surface in $\mathbf{P}^5$ (I, Ex. 2.13). See, for example, Nagata [5, I, Theorem 7, p. 365].

Now we will try to determine the ample divisors on a ruled surface over a curve of any genus, as an application of Nakai's criterion (1.10). In order to apply Nakai's criterion, we need to know which numerical equivalence classes of divisors on the surface contain an irreducible curve. On a general ruled surface, we cannot expect to get nearly as precise an answer to this question as in the case of the rational ruled surfaces (2.18), but at least we can get some estimates which allow us to apply Nakai's criterion successfully.

Proposition 2.20. *Let X be a ruled surface over a curve C, with invariant $e \geqslant 0$.*
  (a) *If $Y \equiv aC_0 + bf$ is an irreducible curve $\neq C_0, f$, then $a > 0, b \geqslant ae$.*
  (b) *A divisor $D \equiv aC_0 + bf$ is ample if and only if $a > 0, b > ae$.*

PROOF.
  (a) Since $Y \neq f$, $\pi: Y \to C$ is surjective, so $Y.f = a > 0$. Also since $Y \neq C_0$, $Y.C_0 = b - ae \geqslant 0$.
  (b) If D is ample, then $D.f = a > 0$, and $D.C_0 = b - ae > 0$. Conversely, if $a > 0, b - ae > 0$, then $D.f > 0, D.C_0 > 0, D^2 = 2ab - a^2e > 0$ and if $Y = a'C_0 + b'f$ is any irreducible curve $\neq C_0, f$, then

$$D.Y = ab' + a'b - aa'e > aa'e + aa'e - aa'e = aa'e \geqslant 0.$$

Therefore by (1.10) D is ample.

Proposition 2.21. *Let X be a ruled surface over a curve C of genus g, with invariant $e < 0$, and assume furthermore either $\operatorname{char} k = 0$ or $g \leqslant 1$.*
  (a) *If $Y \equiv aC_0 + bf$ is an irreducible curve $\neq C_0, f$, then either $a = 1$, $b \geqslant 0$ or $a \geqslant 2, b \geqslant \frac{1}{2}ae$.*
  (b) *A divisor $D \equiv aC_0 + bf$ is ample if and only if $a > 0, b > \frac{1}{2}ae$.*

PROOF.
  (a) We will use Hurwitz's theorem (IV, 2.4) to get some information about Y. Let $\tilde{Y}$ be the normalization of Y, and consider the composition of the natural map $\tilde{Y} \to Y$ with the projection $\pi: Y \to C$. If $\operatorname{char} k = 0$, this map is a finite, separable map of degree a, so by (IV, 2.4) we have

$$2g(\tilde{Y}) - 2 = a(2g - 2) + \deg R,$$

where R is the (effective) ramification divisor. On the other hand, $p_a(Y) \geqslant g(\tilde{Y})$ by (IV, Ex. 1.8), so we find that

$$2p_a(Y) - 2 \geqslant a(2g - 2).$$

Furthermore, this last inequality is true in any characteristic if $g = 0, 1$, since in any case $p_a(Y) \geqslant g$ (IV, 2.5.4).
  By the adjunction formula (1.5), we have

$$2p_a(Y) - 2 = Y.(Y + K).$$

Substituting $Y \equiv aC_0 + bf$ and $K \equiv -2C_0 + (2g - 2 - e)f$ from (2.11), and combining with the inequality above, we find that

$$b(a - 1) \geqslant \tfrac{1}{2}ae(a - 1).$$

Therefore if $a \geqslant 2$, we have $b \geqslant \tfrac{1}{2}ae$ as required. Now $Y.f = a > 0$ in any case, so it remains to show that if $a = 1$, then $b \geqslant 0$. In the case $a = 1$, Y is a section, corresponding to a surjective map $\mathscr{E} \to \mathscr{L} \to 0$. Because of the normalization of $\mathscr{E}$, we must have deg $\mathscr{L} \geqslant$ deg $\mathscr{E}$. But deg $\mathscr{L} = C_0.Y$ (2.9), so we have $b - e \geqslant -e$, hence $b \geqslant 0$.

(b) If D is ample, then $D.f = a > 0$, and $D^2 = 2ab - a^2e > 0$, so $b > \tfrac{1}{2}ae$. Conversely, if $a > 0$, $b > \tfrac{1}{2}ae$, then $D.f > 0$, $D^2 > 0$, $D.C_0 = b - ae > -\tfrac{1}{2}ae > 0$, and if $Y \equiv a'C_0 + b'f$ is any irreducible curve $\neq C_0, f$, then

$$D.Y = ab' + a'b - aa'e.$$

Now if $a' = 1$, then $b' \geqslant 0$, so $D.Y > \tfrac{1}{2}ae - ae = -\tfrac{1}{2}ae > 0$. If $a' \geqslant 2$, then $b' \geqslant \tfrac{1}{2}a'e$, so $D.Y > \tfrac{1}{2}aa'e + \tfrac{1}{2}aa'e - aa'e = 0$. Therefore by (1.10), D is ample.

Remark 2.21.1. In the remaining case $e < 0$, char $k = p > 0$, $g \geqslant 2$, we cannot get necessary and sufficient conditions for D to be ample, but it is possible to get some partial results (Ex. 2.14) and (Ex. 2.15).

Remark 2.22.2. The determination of the very ample divisors on a ruled surface with $g \geqslant 1$ is more subtle than in the rational case (2.18), because it does not depend only on the numerical equivalence class of the divisor (Ex. 2.11) and (Ex. 2.12).

References for §2. Since the theory of ruled surfaces is very old, I cannot trace the origins of the results given here. Instead, let me simply list a few recent references: Atiyah [1], Hartshorne [4], Maruyama [1], Nagata [5], Shafarevich [1, Ch. IV, V], Tjurin [1], [2].

EXERCISES

2.1. If X is a birationally ruled surface, show that the curve C, such that X is birationally equivalent to $C \times \mathbf{P}^1$, is unique (up to isomorphism).

2.2. Let X be the ruled surface $\mathbf{P}(\mathscr{E})$ over a curve C. Show that $\mathscr{E}$ is decomposable if and only if there exist two sections C', C'' of X such that $C' \cap C'' = \varnothing$.

2.3. (a) If $\mathscr{E}$ is a locally free sheaf of rank r on a (nonsingular) curve C, then there is a sequence

$$0 = \mathscr{E}_0 \subseteq \mathscr{E}_1 \subseteq \ldots \subseteq \mathscr{E}_r = \mathscr{E}$$

of subsheaves such that $\mathscr{E}_i / \mathscr{E}_{i-1}$ is an invertible sheaf for each $i = 1, \ldots, r$. We say that $\mathscr{E}$ is a *successive extension* of invertible sheaves. [*Hint:* Use (II, Ex. 8.2).]

(b) Show that this is false for varieties of dimension $\geqslant 2$. In particular, the sheaf of differentials Ω on $\mathbf{P}^2$ is not an extension of invertible sheaves.

2.4. Let C be a curve of genus g, and let X be the ruled surface $C \times \mathbf{P}^1$. We consider the question, for what integers $s \in \mathbf{Z}$ does there exist a section D of X with $D^2 = s$? First show that s is always an even integer, say $s = 2r$.
   (a) Show that $r = 0$ and any $r \geqslant g + 1$ are always possible. Cf. (IV, Ex. 6.8).
   (b) If $g = 3$, show that $r = 1$ is not possible, and just one of the two values $r = 2, 3$ is possible, depending on whether C is hyperelliptic or not.

2.5. *Values of e.* Let C be a curve of genus $g \geqslant 1$.
   (a) Show that for each $0 \leqslant e \leqslant 2g - 2$ there is a ruled surface X over C with invariant e, corresponding to an indecomposable $\mathscr{E}$. Cf. (2.12).
   (b) Let $e < 0$, let D be any divisor of degree $d = -e$, and let $\xi \in H^1(\mathscr{L}(-D))$ be a nonzero element defining an extension

$$0 \to \mathcal{O}_C \to \mathscr{E} \to \mathscr{L}(D) \to 0.$$

Let $H \subseteq |D + K|$ be the sublinear system of codimension 1 defined by ker ξ, where ξ is considered as a linear functional on $H^0(\mathscr{L}(D + K))$. For any effective divisor E of degree $d - 1$, let $L_E \subseteq |D + K|$ be the sublinear system $|D + K - E| + E$. Show that $\mathscr{E}$ is normalized if and only if for each E as above, $L_E \not\subseteq H$. Cf. proof of (2.15).
   (c) Now show that if $-g \leqslant e < 0$, there exists a ruled surface X over C with invariant e. [*Hint:* For any given D in (b), show that a suitable ξ exists, using an argument similar to the proof of (II, 8.18).]
   (d) For $g = 2$, show that $e \geqslant -2$ is also necessary for the existence of X.
   Note. It has been shown that $e \geqslant -g$ for any ruled surface (Nagata [8]).

2.6. Show that every locally free sheaf of finite rank on $\mathbf{P}^1$ is isomorphic to a direct sum of invertible sheaves. [*Hint:* Choose a subinvertible sheaf of maximal degree, and use induction on the rank.]

2.7. On the elliptic ruled surface X of (2.11.6), show that the sections C_0 with $C_0^2 = 1$ form a one-dimensional algebraic family, parametrized by the points of the base curve C, and that no two are linearly equivalent.

2.8. A locally free sheaf $\mathscr{E}$ on a curve C is said to be *stable* if for every quotient locally free sheaf $\mathscr{E} \to \mathscr{F} \to 0, \mathscr{F} \neq \mathscr{E}, \mathscr{F} \neq 0$, we have

$$(\deg \mathscr{F})/\text{rank } \mathscr{F} > (\deg \mathscr{E})/\text{rank } \mathscr{E}.$$

Replacing $>$ by $\geqslant$ defines *semistable*.
   (a) A decomposable $\mathscr{E}$ is never stable.
   (b) If $\mathscr{E}$ has rank 2 and is normalized, then $\mathscr{E}$ is stable (respectively, semistable) if and only if deg $\mathscr{E} > 0$ (respectively, $\geqslant 0$).
   (c) Show that the indecomposable locally free sheaves $\mathscr{E}$ of rank 2 that are not semistable are classified, up to isomorphism, by giving (1) an integer $0 < e \leqslant 2g - 2$, (2) an element $\mathscr{L} \in \text{Pic } C$ of degree $-e$, and (3) a nonzero $\xi \in H^1(\mathscr{L}^\vee)$, determined up to a nonzero scalar multiple.

2.9. Let Y be a nonsingular curve on a quadric cone X_0 in $\mathbf{P}^3$. Show that either Y is a complete intersection of X_0 with a surface of degree $a \geqslant 1$, in which case deg $Y = 2a$, $g(Y) = (a - 1)^2$, *or*, deg Y is odd, say $2a + 1$, and $g(Y) = a^2 - a$. Cf. (IV, 6.4.1). [*Hint:* Use (2.11.4).]

2.10. For any $n > e \geqslant 0$, let X be the rational scroll of degree $d = 2n - e$ in $\mathbf{P}^{d+1}$ given by (2.19). If $n \geqslant 2e - 2$, show that X contains a nonsingular curve Y of genus $g = d + 2$ which is a canonical curve in this embedding. Conclude that for every $g \geqslant 4$, there exists a nonhyperelliptic curve of genus g which has a g_3^1. Cf. (IV, §5).

2.11. Let X be a ruled surface over the curve C, defined by a normalized bundle $\mathscr{E}$, and let e be the divisor on C for which $\mathscr{L}(e) \cong \bigwedge^2 \mathscr{E}$ (2.8.1). Let b be any divisor on C.
  (a) If $|b|$ and $|b + e|$ have no base points, and if b is nonspecial, then there is a section $D \sim C_0 + bf$, and $|D|$ has no base points.
  (b) If b and $b + e$ are very ample on C, and for every point $P \in C$, we have $b - P$ and $b + e - P$ nonspecial, then $C_0 + bf$ is very ample.

2.12. Let X be a ruled surface with invariant e over an elliptic curve C, and let b be a divisor on C.
  (a) If $\deg b \geqslant e + 2$, then there is a section $D \sim C_0 + bf$ such that $|D|$ has no base points.
  (b) The linear system $|C_0 + bf|$ is very ample if and only if $\deg b \geqslant e + 3$.
  Note. The case $e = -1$ will require special attention.

2.13. For every $e \geqslant -1$ and $n \geqslant e + 3$, there is an elliptic scroll of degree $d = 2n - e$ in $\mathbf{P}^{d-1}$. In particular, there is an elliptic scroll of degree 5 in $\mathbf{P}^4$.

2.14. Let X be a ruled surface over a curve C of genus g, with invariant $e < 0$, and assume that char $k = p > 0$ and $g \geqslant 2$.
  (a) If $Y \equiv aC_0 + bf$ is an irreducible curve $\neq C_0, f$, then either $a = 1, b \geqslant 0$, or $2 \leqslant a \leqslant p - 1, b \geqslant \frac{1}{2}ae$, or $a \geqslant p, b \geqslant \frac{1}{2}ae + 1 - g$.
  (b) If $a > 0$ and $b > a(\frac{1}{2}e + (1/p)(g - 1))$, then any divisor $D \equiv aC_0 + bf$ is ample. On the other hand, if D is ample, then $a > 0$ and $b > \frac{1}{2}ae$.

2.15. *Funny behavior in characteristic p.* Let C be the plane curve $x^3y + y^3z + z^3x = 0$ over a field k of characteristic 3 (IV, Ex. 2.4).
  (a) Show that the action of the k-linear Frobenius morphism f on $H^1(C, \mathcal{O}_C)$ is identically 0 (Cf. (IV, 4.21)).
  (b) Fix a point $P \in C$, and show that there is a nonzero $\xi \in H^1(\mathscr{L}(-P))$ such that $f^*\xi = 0$ in $H^1(\mathscr{L}(-3P))$.
  (c) Now let $\mathscr{E}$ be defined by ξ as an extension

$$0 \to \mathcal{O}_C \to \mathscr{E} \to \mathscr{L}(P) \to 0,$$

and let X be the corresponding ruled surface over C. Show that X contains a nonsingular curve $Y \equiv 3C_0 - 3f$, such that $\pi: Y \to C$ is purely inseparable. Show that the divisor $D = 2C_0$ satisfies the hypotheses of (2.21b), but is not ample.

2.16. Let C be a nonsingular affine curve. Show that two locally free sheaves $\mathscr{E}, \mathscr{E}'$ of the same rank are isomorphic if and only if their classes in the Grothendieck group $K(X)$ (II, Ex. 6.10) and (II, Ex. 6.11) are the same. This is false for a projective curve.

*2.17. (a) Let $\varphi: \mathbf{P}_k^1 \to \mathbf{P}_k^3$ be the 3-uple embedding (I, Ex. 2.12). Let $\mathscr{I}$ be the sheaf of ideals of the twisted cubic curve C which is the image of φ. Then $\mathscr{I}/\mathscr{I}^2$ is a locally free sheaf of rank 2 on C, so $\varphi^*(\mathscr{I}/\mathscr{I}^2)$ is a locally free sheaf of rank 2 on

$\mathbf{P}^1$. By (2.14), therefore, $\varphi^*(\mathscr{I}/\mathscr{I}^2) \cong \mathcal{O}(l) \oplus \mathcal{O}(m)$ for some $l, m \in \mathbf{Z}$. Determine l and m.

(b) Repeat part (a) for the embedding $\varphi: \mathbf{P}^1 \to \mathbf{P}^3$ given by $x_0 = t^4$, $x_1 = t^3 u$, $x_2 = tu^3$, $x_3 = u^4$, whose image is a nonsingular rational quartic curve. [*Answer*: If char $k \neq 2$, then $l = m = -7$; if char $k = 2$, then $l, m = -6, -8$.]

3 Monoidal Transformations

We define a *monoidal transformation* of a surface X to be the operation of blowing up a single point P. This new terminology is to distinguish it from the more general process of blowing up an arbitrary closed subscheme (II, §7). It also goes by many other names in the literature: locally quadratic transformation, dilatation, σ-process, Hopf map, to mention a few.

We will see later (5.5) that any birational transformation of surfaces can be factored into monoidal transformations and their inverses. Thus the monoidal transformation is basic to the birational study of surfaces.

In this section we will study what happens under a single monoidal transformation. As an application, we will show how to resolve the singularities of a curve on a surface by monoidal transformations, and begin a study of the different types of curve singularities.

First we fix our notation. Let X be a surface, and let P be a point of X. We denote the monoidal transformation with center P by $\pi: \tilde{X} \to X$. Then we know (I, §4) or (II, §7) that π induces an isomorphism of $\tilde{X} - \pi^{-1}(P)$ onto $X - P$. The inverse image of P is a curve E, which we call the *exceptional curve* (I, 4.9.1).

Proposition 3.1. *The new variety $\tilde{X}$ is a nonsingular projective surface. The curve E is isomorphic to $\mathbf{P}^1$. The self-intersection of E on $\tilde{X}$ is $E^2 = -1$.*

PROOF. Since a single point is nonsingular, we can apply (II, 8.24). This tells us that $\tilde{X}$ is nonsingular, and we know already from (II, 7.16) that $\tilde{X}$ is projective, of dimension 2, and birational to X. We also conclude from (II, 8.24) that $E \cong \mathbf{P}^1$, since it is the projective space bundle over the point P corresponding to the two-dimensional vector space $\mathfrak{m}_P/\mathfrak{m}_P^2$. Finally, the normal sheaf $\mathscr{N}_{E/\tilde{X}}$ is just $\mathcal{O}_E(-1)$, so by (1.4.1) we have $E^2 = -1$.

Remark 3.1.1. There is a converse to this result, which we will prove later (5.7), namely, any curve $E \cong \mathbf{P}^1$ in a surface X', with $E^2 = -1$, is obtained as the exceptional curve by a monoidal transformation from some other surface X.

Proposition 3.2. *The natural maps $\pi^*: \operatorname{Pic} X \to \operatorname{Pic} \tilde{X}$ and $\mathbf{Z} \to \operatorname{Pic} \tilde{X}$ defined by $1 \mapsto 1 \cdot E$ give rise to an isomorphism $\operatorname{Pic} \tilde{X} \cong \operatorname{Pic} X \oplus \mathbf{Z}$. The inter-*

section theory on $\tilde{X}$ is determined by the rules:
(a) if $C, D \in \text{Pic } X$, then $(\pi^*C).(\pi^*D) = C.D$;
(b) if $C \in \text{Pic } X$, then $(\pi^*C).E = 0$;
(c) $E^2 = -1$.
Finally, if $\pi_*: \text{Pic } \tilde{X} \to \text{Pic } X$ denotes the projection on the first factor, then:
(d) if $C \in \text{Pic } X$ and $D \in \text{Pic } \tilde{X}$, then $(\pi^*C).D = C.(\pi_*D)$.

PROOF. (See also (II, Ex. 8.5).) From (II, 6.5) we see that $\text{Pic } X \cong \text{Pic}(X - P)$. But $X - P \cong \tilde{X} - E$, so also from (II, 6.5) we have an exact sequence

$$\mathbf{Z} \to \text{Pic } \tilde{X} \to \text{Pic } X \to 0,$$

where the first map sends 1 to $1 \cdot E$. Since for any $n \neq 0$ we have $(nE)^2 = -n^2 \neq 0$, this map is injective. On the other hand, π^* splits this sequence, so we have $\text{Pic } \tilde{X} \cong \text{Pic } X \oplus \mathbf{Z}$.

We have already seen that $E^2 = -1$. To prove (a) and (b), we use the fact (§1) that C and D are linearly equivalent to differences of nonsingular curves, meeting everywhere transversally, and not containing P. For in the proof of (1.2) we can require also that D' misses any given finite set of points. Then π^* does not affect their intersection, which proves (a). Also clearly π^*C does not meet E, So $(\pi^*C).E = 0$. The same argument also proves (d), because we may assume that C is a difference of curves not containing P.

Proposition 3.3. *The canonical divisor of $\tilde{X}$ is given by $K_{\tilde{X}} = \pi^*K_X + E$. Therefore $K_{\tilde{X}}^2 = K_X^2 - 1$.*

PROOF. (See also (II, Ex. 8.5).) Since the canonical sheaf on $\tilde{X} - E$ and $X - P$ is the same, clearly $K_{\tilde{X}} = \pi^*K_X + nE$ for some $n \in \mathbf{Z}$. To determine n, we use the adjunction formula (1.5) for E. It says $-2 = E.(E + K_{\tilde{X}})$, so using (3.2) we find $n = 1$. The formula for K^2 follows directly from (3.2).

Remark 3.3.1. Thus the invariant K^2 of a surface is not a birational invariant. For a specific example, we have K^2 of $\mathbf{P}^2$ is 9 (1.4.4), K^2 of the rational ruled surface X_1 is 8 (2.11), and X_1 is isomorphic to a monoidal transformation of $\mathbf{P}^2$ (2.11.5).

Next we want to show that the arithmetic genus p_a is preserved by a monoidal transformation. For that, we must compare the cohomology of the structure sheaves on X and $\tilde{X}$. We will use the theorem on formal functions (III, 11.1) to compute $R^i\pi_*\mathcal{O}_{\tilde{X}}$.

Proposition 3.4. *We have $\pi_*\mathcal{O}_{\tilde{X}} = \mathcal{O}_X$, and $R^i\pi_*\mathcal{O}_{\tilde{X}} = 0$ for $i > 0$. Therefore $H^i(X, \mathcal{O}_X) \cong H^i(\tilde{X}, \mathcal{O}_{\tilde{X}})$ for all $i \geq 0$.*

PROOF. Since π is an isomorphism of $\tilde{X} - E$ onto $X - P$, it is clear that the natural map $\mathcal{O}_X \to \pi_*\mathcal{O}_{\tilde{X}}$ is an isomorphism except possibly at P, and that the sheaves $\mathscr{F}^i = R^i\pi_*\mathcal{O}_{\tilde{X}}$ for $i > 0$ have support at P. We use the theorem on formal functions (III, 11.1) to compute these $\mathscr{F}^i$. It says (taking completions of the stalks at P) that

$$\hat{\mathscr{F}}^i \cong \varprojlim H^i(E_n, \mathcal{O}_{E_n})$$

where E_n is the closed subscheme of $\tilde{X}$ defined by $\mathscr{J}^n$, where $\mathscr{J}$ is the ideal of E. There are natural exact sequences

$$0 \to \mathscr{J}^n/\mathscr{J}^{n+1} \to \mathcal{O}_{E_{n+1}} \to \mathcal{O}_{E_n} \to 0$$

for each n. Furthermore, by (II, 8.24) we have $\mathscr{J}/\mathscr{J}^2 = \mathcal{O}_E(1)$, and by (II, 8.21Ae), $\mathscr{J}^n/\mathscr{J}^{n+1} \cong S^n(\mathscr{J}/\mathscr{J}^2) \cong \mathcal{O}_E(n)$. Now $E \cong \mathbf{P}^1$, so $H^i(E, \mathcal{O}_E(n)) = 0$ for $i > 0$ and all $n > 0$. Since $E_1 = E$, we conclude from the long exact sequence of cohomology, using induction on n, that $H^i(\mathcal{O}_{E_n}) = 0$ for all $i > 0$, all $n \geq 1$. It follows that $\hat{\mathscr{F}}^i = 0$ for $i > 0$. Since $\mathscr{F}^i$ is a coherent sheaf with support at P, $\mathscr{F}^i = \hat{\mathscr{F}}^i$, so $\mathscr{F}^i = 0$.

The fact that $\mathcal{O}_X \cong \pi_* \mathcal{O}_{\tilde{X}}$ follows simply from the fact that X is normal and π is birational. Cf. proof of (III, 11.4).

Now from (III, Ex. 8.1) we conclude that $H^i(X, \mathcal{O}_X) \cong H^i(\tilde{X}, \mathcal{O}_{\tilde{X}})$ for all $i \geq 0$.

Corollary 3.5. *Let $\pi: \tilde{X} \to X$ be a monoidal transformation. Then $p_a(X) = p_a(\tilde{X})$.*

PROOF. From (III, Ex. 5.3) we have $p_a(X) = \dim H^2(X, \mathcal{O}_X) - \dim H^1(X, \mathcal{O}_X)$ and similarly for $p_a(\tilde{X})$.

Remark 3.5.1. It follows also from (3.4) that X and $\tilde{X}$ have the same irregularity $q(X) = \dim H^1(X, \mathcal{O}_X)$ and the same geometric genus $p_g(X) = \dim H^2(X, \mathcal{O}_X)$ (III, 7.12.3). The invariance of p_g is also of course a consequence of the fact that p_g is a birational invariant in general (II, 8.19).

Next we will investigate what happens to a curve under a monoidal transformation. Let C be an effective divisor on X, and let $\pi: \tilde{X} \to X$ be the monoidal transformation with center P. Recall that the *strict transform* $\tilde{C}$ of C is defined as the closed subscheme of $\tilde{X}$ obtained by blowing up P on C (II, 7.15). It is also the closure in $\tilde{X}$ of $\pi^{-1}(C \cap (X - P))$ (II, 7.15.1). So it is clear that $\tilde{C}$ can be obtained from $\pi^* C$ by throwing away E (with whatever multiplicity it has in $\pi^* C$).

The multiplicity of E in $\pi^* C$ will depend on the behavior of C at the point P. So we make the following definition of the multiplicity, which generalizes the definition for plane curves given in (I, Ex. 5.3).

Definition. Let C be an effective Cartier divisor on the surface X, and let f be a local equation for C at the point P. Then we define the *multiplicity* of C at P, denoted by $\mu_P(C)$, to be the largest integer r such that $f \in \mathfrak{m}_P^r$, where $\mathfrak{m}_P \subseteq \mathcal{O}_{P,X}$ is the maximal ideal.

Remark 3.5.2. We always have $\mu_P(C) \geqslant 0$, since $f \in \mathcal{O}_{P,X}$. Furthermore $\mu_P(C) \geqslant 1$ if and only if $P \in C$, and equality holds if and only if C is non-singular at P, because in that case $\mathfrak{m}_{P,C}$ will be a principal ideal, so $f \notin \mathfrak{m}_{P,X}^2$.

Proposition 3.6. *Let C be an effective divisor on X, let P be a point of multiplicity r on C, and let $\pi: \tilde{X} \to X$ be the monoidal transformation with center P. Then*

$$\pi^*C = \tilde{C} + rE.$$

PROOF. We will go back to the definition of blowing up, and compute explicitly what happens in a neighborhood of P, so that we can trace the local equation of C on X and π^*C on $\tilde{X}$.

Let $\mathfrak{m}$ be the sheaf of ideals of P on X. Then $\tilde{X}$ is defined as **Proj** $\mathcal{S}$, where $\mathcal{S}$ is the graded sheaf of algebras $\mathcal{S} = \bigoplus_{d \geqslant 0} \mathfrak{m}^d$ (II, §7). Let x, y be local parameters at P. Then x, y generate $\mathfrak{m}$ in some neighborhood U of P, which we may assume to be affine, say $U = \operatorname{Spec} A$. The Koszul complex (III, 7.10A) gives a resolution of $\mathfrak{m}$ over U:

$$0 \to \mathcal{O}_U \to \mathcal{O}_U^2 \to \mathfrak{m} \to 0,$$

where we denote the two generators of $\mathcal{O}_U^2$ by t, u, and send t to x, u to y. Then the kernel is generated by $ty - ux$. Therefore $\mathcal{S}$ over U is the sheaf associated to the A-algebra $A[t,u]/(ty - ux)$, so $\tilde{X}$ is the closed subscheme of $\mathbf{P}_U^1$ defined by $ty - ux$, where t, u are the homogeneous coordinates of $\mathbf{P}^1$. (Note how this construction generalizes the example (I, 4.9.1).)

Now let f be a local equation for C on U (shrinking U if necessary). Then by definition of the multiplicity, we can write

$$f = f_r(x, y) + g$$

where f_r is a nonzero homogeneous polynomial of degree r with coefficients in k, and $g \in \mathfrak{m}^{r+1}$. Indeed, $f \in \mathfrak{m}^r$, $f \notin \mathfrak{m}^{r+1}$, and $\mathfrak{m}^r/\mathfrak{m}^{r+1}$ is the k-vector space with basis $x^r, x^{r-1}y, \ldots, y^r$.

Consider the open affine subset V of $\mathbf{P}_U^1$ defined by $t = 1$. Then on $\tilde{X} \cap V$ we have $y = ux$, so we can write

$$\pi^*f = x^r(f_r(1, u) + xh)$$

for some $h \in A[u]$. Indeed, $\mathfrak{m}^{r+1}A[u]$ is generated by $x^{r+1}, x^{r+1}u, \ldots, x^{r+1}u^{r+1}$, so π^*g is divisible by x^{r+1}.

Now x is a local equation for E, and $f_r(1, u)$ is zero at only finitely many points of E, so we see that E occurs with multiplicity exactly r in π^*C, which is locally defined by π^*f.

Corollary 3.7. *With the same hypotheses, we have $\tilde{C}.E = r$, and $p_a(\tilde{C}) = p_a(C) - \frac{1}{2}r(r-1)$.*

PROOF. Since $\tilde{C} = \pi^*C - rE$, we have $\tilde{C}.E = r$ by (3.2). We compute $p_a(\tilde{C})$ by the adjunction formula (1.5) and (Ex. 1.3)

$$2p_a(\tilde{C}) - 2 = \tilde{C}.(\tilde{C} + K_{\tilde{X}})$$
$$= (\pi^*C - rE)(\pi^*C - rE + \pi^*K_X + E)$$
$$= 2p_a(C) - 2 - r(r - 1),$$

so

$$p_a(\tilde{C}) = p_a(C) - \frac{1}{2}r(r - 1).$$

Proposition 3.8. *Let C be an irreducible curve in the surface X. Then there exists a finite sequence of monoidal transformations (with suitable centers) $X_n \to X_{n-1} \to \ldots \to X_1 \to X_0 = X$ such that the strict transform C_n of C on X_n is nonsingular.*

PROOF. If C is already nonsingular, take $n = 0$. Otherwise, let $P \in C$ be a singular point, and let $r \geq 2$ be its multiplicity. Let $X_1 \to X$ be the monoidal transformation with center P, and let C_1 be the strict transform of C. Then from (3.7) we see that $p_a(C_1) < p_a(C)$. If C_1 is nonsingular, stop. Otherwise choose a singular point of C_1 and continue. In this way we obtain a sequence of monoidal transformations

$$\ldots \to X_2 \to X_1 \to X_0 = X,$$

such that the strict transform C_i of C on X_i satisfies $p_a(C_i) < p_a(C_{i-1})$ for each i. Since the arithmetic genus of any irreducible curve is nonnegative $(p_a(C_i) = \dim H^1(\mathcal{O}_{C_i})$ (III, Ex. 5.3)), this process must terminate. Thus for some n, C_n is nonsingular.

Remark 3.8.1. The general problem of *resolution of singularities* is, given a variety V, to find a proper birational morphism $f: V' \to V$ with V' non-singular. If V is a curve, we know this is possible, because each birational equivalence class of curves contains a unique nonsingular projective curve (I, 6.11). In fact, in this case it is sufficient to take V' to be the normalization of V. But in higher dimensions, this method does not work.

So we approach the general problem as follows. Whenever V is singular, blow up some subvariety contained in the singular locus, to get a morphism $f_1: V_1 \to V$. Then (and this is the hard part) find some quantitative way of showing that the singularities of V_1 are less severe than those of V, so that as we repeat this process, we must eventually obtain a nonsingular variety.

Two things become clear quite soon: first, to maintain reasonable control of the singularities, one should only blow up subvarieties which are themselves nonsingular (such as a point); second, to set up an induction on the dimension of V, one should also consider the problem of *embedded resolution*. This problem is, given a variety V, contained in a nonsingular variety W, to find a proper birational morphism $g: W' \to W$ with W' nonsingular, such that not only is the strict transform $\tilde{V}$ of V in W' nonsingular, but the entire inverse image $g^{-1}(V)$ is a *divisor with normal crossings*, which means that each irreducible component of $g^{-1}(V)$ is nonsingular, and whenever r irreducible components $Y_1, \ldots, Y_r$ of $g^{-1}(V)$ meet at a point P, then the local equations $f_1, \ldots, f_r$ of the Y_i form part of a regular system of parameters at P (i.e., $f_1, \ldots, f_r$ are linearly independent $(\bmod\ \mathfrak{m}_P^2)$).

The result just proved (3.8) shows that if C is a curve contained in a nonsingular surface, then one can resolve the singularities of C by successive monoidal transformations. We will prove the stronger theorem of embedded resolution for curves in surfaces below (3.9).

The status of the general resolution problem is as follows. The resolution of curves was known in the late 19th century. The resolution of surfaces (over $\mathbf{C}$) was known to the Italians, but the first "rigorous" proof was given by Walker in 1935. Zariski gave the first purely algebraic proof of resolution for surfaces (char $k = 0$) in 1939. Then in 1944 he proved embedded resolution for surfaces and resolution for threefolds (char $k = 0$). Abhyankar proved resolution for surfaces in characteristic $p > 0$ in 1956, and in 1966 he proved resolution for threefolds in characteristic $p > 5$. Meanwhile in 1964 Hironaka proved resolution and embedded resolution in all dimensions in characteristic 0. For more details and precise references on the resolution problem, see Lipman [1], Hironaka [4] and Hironaka's introduction to Zariski's collected papers on resolution in Zariski [8].

Theorem 3.9 (Embedded Resolution of Curves in Surfaces). *Let Y be any curve in the surface X. Then there exists a finite sequence of monoidal transformations $X' = X_n \to X_{n-1} \to \ldots \to X_0 = X$, such that if $f : X' \to X$ is their composition, then the total inverse image $f^{-1}(Y)$ is a divisor with normal crossings (3.8.1).*

PROOF. Clearly we may assume that Y is connected. Furthermore, since the multiplicities of the irreducible components do not enter into the definition of normal crossings, we may assume that Y is reduced, i.e., each irreducible component has multiplicity 1. Now for any birational morphism $f : X' \to X$, let us denote by $f^{-1}(Y)$ the *reduced inverse image divisor* $f^*(Y)_{\mathrm{red}}$. In other words, $f^{-1}(Y)$ is the sum of all the irreducible components of $f^*(Y)$, with multiplicity 1. If f is a composition of monoidal transformations, then $f^{-1}(Y)$ will also be reduced and connected, so $H^0(\mathcal{O}_{f^{-1}(Y)}) = k$, and $p_a(f^{-1}(Y)) = \dim H^1(\mathcal{O}_{f^{-1}(Y)}) \geq 0$.

Let $\pi : \tilde{X} \to X$ be the monoidal transformation at a point P, and let $\mu_P(Y) = r$. Then the divisor $\pi^{-1}(Y)$ is just $\tilde{Y} + E = \pi^*(Y) - (r-1)E$, by (3.6), so we can easily compute the arithmetic genus, using the adjunction formula as in (3.7). We find

$$p_a(\pi^{-1}(Y)) = p_a(Y) - \frac{1}{2}(r-1)(r-2).$$

To prove our result, we proceed as follows. First we apply (3.8) to each irreducible component of Y. Thus we reduce to the case where each irreducible component of Y is nonsingular, because all the new exceptional curves we add are already nonsingular. Then, if the total curve Y has a singular point P other than a node, we blow it up.

If $\mu_P(Y) \geqslant 3$, then $p_a(\pi^{-1}(Y)) < p_a(Y)$, so there can be only finitely many steps of this kind. If $\mu_P(Y) = 2$, then $p_a(\pi^{-1}(Y)) = p_a(Y)$, and we must look more closely. In that case we have $Y.E = 2$, by (3.7). There are three possibilities. One is that $\tilde{Y}$ meets E transversally in two distinct points, in which case we can stop. The second is that $\tilde{Y}$ meets E in one point Q, $\tilde{Y}$ is nonsingular there, but $\tilde{Y}$ and E have intersection multiplicity 2 at Q. In this case, blowing up Q produces a triple point (check!), so one further blowing-up makes $p_a(Y)$ drop again. The third possibility is that $\tilde{Y}$ has a singular point Q of multiplicity 2 where it meets E. In this case $\tilde{Y} + E$ has multiplicity 3 at Q, so blowing up Q makes p_a drop again.

So we see that any kind of singularity except a node gives rise to a monoidal transformation, or a finite sequence of such, which forces p_a to drop. Therefore the process must terminate. When it does, $f^{-1}(Y)$ will be a divisor with normal crossings, because each irreducible component is nonsingular, and the only singularities of the total curve $f^{-1}(Y)$ are nodes.

Example 3.9.1. If Y is the plane cuspidal curve $y^2 = x^3$, then the singularity of Y is resolved by one monoidal transformation. However, to get $f^{-1}(Y)$ to have normal crossings, we need three monoidal transformations (Fig. 20).

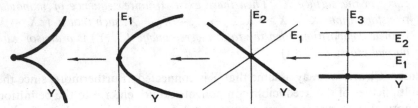

Figure 20. Embedded resolution of a cusp.

In the context of successive monoidal transformations, it is convenient to introduce the language of infinitely near points.

Definition. Let X be a surface. Then any point on any surface X', obtained from X by a finite succession of monoidal transformations, is called an *infinitely near point* of X. If $g : X'' \to X'$ is a further succession of monoidal transformations, and if $Q'' \in X''$ is a point in the open set where g is an isomorphism, then we identify Q'' with $g(Q'')$ as infinitely near points of X. In particular, all the ordinary points of X are included among the infinitely near points. We say "Q is infinitely near P" if P lies on some X' and Q lies on the exceptional curve E obtained by blowing up P. If C is a curve in X, and $Q' \in X'$ is an infinitely near point of X, we say Q' is an *infinitely near point* of C if Q' lies on the strict transform of C on X'.

Example 3.9.2. Let C be an irreducible curve on a surface X, with normalization $\tilde{C}$. Then we have

$$g(\tilde{C}) = p_a(C) - \sum_P \frac{1}{2} r_P(r_P - 1),$$

where r_P is the multiplicity, and the sum is taken over all singular points P of C, including infinitely near singular points. Indeed, by (3.8) we pass from C to $\tilde{C}$ by blowing up the singular points in succession, until there are none left. Each time, by (3.7), the arithmetic genus drops by $\frac{1}{2}r(r - 1)$.

Example 3.9.3. In particular, working with the infinitesimal neighborhood of one point at a time, we see that the integer δ_P of (IV, Ex. 1.8) can be computed as $\sum \frac{1}{2} r_Q(r_Q - 1)$ taken over all infinitely near singular points Q lying over P, including P.

Remark 3.9.4 (Classification of Curve Singularities). With the ideas of this section we can begin a new classification of the possible singularities of a (reduced) curve lying on a surface, which is weaker than the classification by analytic isomorphism introduced in (I, 5.6.1) and (I, Ex. 5.14)—see (Ex. 3.6). For references, see Walker [1, Ch. III, §7] and Zariski [10, Ch. I].

As a first invariant of a singular point P on a curve C (lying always on a surface X) we have its multiplicity. Next we have the multiplicities of the infinitely near singular points of C, and their configuration around P. This data already suffices to determine δ_P (3.9.3).

We define a slightly more complex, but still discrete, invariant of a singular point (or set of singular points), to be its equivalence class for the following equivalence relation. A (reduced) curve C in an open set U of a surface X is *equivalent* to another $C' \subseteq U' \subseteq X'$ if there is a sequence of monoidal transformations $U_n \to U_{n-1} \to \ldots \to U_0 = U$ and another $U'_n \to U'_{n-1} \to \ldots \to U'_0 = U'$, which give embedded resolutions for C and C' respectively, and if there is a one-to-one correspondence between the irreducible components of the reduced total transforms and their singular points at each step, preserving multiplicities, incidence, and compatible with the maps $U_i \to U_{i-1}$ and $U'_i \to U'_{i-1}$ for each i. One checks easily that this is in fact an equivalence relation. To define equivalence of a single singular point $P \in C$, just take U so small that $C \cap U$ has no other singular points.

Example 3.9.5. To illustrate this concept, let us classify all double points up to equivalence. Let $P \in C$ be a double point, and let $\pi: \tilde{X} \to X$ be the monoidal transformation with center P. Then as in the proof of (3.9), there are three possibilities: (a) $\tilde{C}$ meets E transversally in two points, in which

393

case P is a node; (b) $\tilde{C}$ meets E in one point, and is tangent to it there, in which case P is a cusp; (c) $\tilde{C}$ is singular, with a double point Q. In this case E must pass through Q in a direction not equal to any tangent direction of Q, since $\tilde{C}.E = 2$. Therefore the equivalence class of Q determines the equivalence class of P.

Thus we can classify double points according to the number of times n we must blow up to get $\tilde{C}$ nonsingular, and the behavior (a) or (b) at the last step.

In this case it happens that the classification for equivalence coincides with the classification for analytic isomorphism, although that is not true in general (Ex. 3.6). Indeed, up to analytic isomorphism, any double point is given by $y^2 = x^r$ for some $r \geq 2$ (I, Ex. 5.14d). To blow up, set $y = ux$. Then we get $u^2 = x^{r-2}$. So we see inductively that the equivalence class is given by $n = [r/2]$ and the type is (a) if r is even, (b) if r is odd.

EXERCISES

3.1. Let X be a nonsingular projective variety of any dimension, let Y be a nonsingular subvariety, and let $\pi:\tilde{X} \to X$ be obtained by blowing up Y. Show that $p_a(\tilde{X}) = p_a(X)$.

3.2. Let C and D be curves on a surface X, meeting at a point P. Let $\pi:\tilde{X} \to X$ be the monoidal transformation with center P. Show that $\tilde{C}.\tilde{D} = C.D - \mu_P(C) \cdot \mu_P(D)$. Conclude that $C.D = \sum \mu_P(C) \cdot \mu_P(D)$, where the sum is taken over all intersection points of C and D, including infinitely near intersection points.

3.3. Let $\pi:\tilde{X} \to X$ be a monoidal transformation, and let D be a very ample divisor on X. Show that $2\pi^*D - E$ is ample on $\tilde{X}$. [*Hint:* Use a suitable generalization of (I, Ex. 7.5) to curves in $\mathbf{P}^n$.]

3.4. *Multiplicity of a Local Ring.* (See Nagata [7, Ch III, §23] or Zariski–Samuel [1, vol 2, Ch VIII, §10].) Let A be a noetherian local ring with maximal ideal $\mathfrak{m}$. For any $l > 0$, let $\psi(l) = \text{length}(A/\mathfrak{m}^l)$. We call ψ the *Hilbert–Samuel function* of A.
  (a) Show that there is a polynomial $P_A(z) \in \mathbf{Q}[z]$ such that $P_A(l) = \psi(l)$ for all $l \gg 0$. This is the *Hilbert–Samuel polynomial* of A. [*Hint:* Consider the graded ring $\text{gr}_{\mathfrak{m}} A = \bigoplus_{d \geq 0} \mathfrak{m}^d/\mathfrak{m}^{d+1}$, and apply (I, 7.5).]
  (b) Show that $\deg P_A = \dim A$.
  (c) Let $n = \dim A$. Then we define the *multiplicity* of A, denoted $\mu(A)$, to be $(n!) \cdot$ (leading coefficient of P_A). If P is a point on a noetherian scheme X, we define the *multiplicity* of P on X, $\mu_P(X)$, to be $\mu(\mathcal{O}_{P,X})$.
  (d) Show that for a point P on a curve C on a surface X, this definition of $\mu_P(C)$ coincides with the one in the text just before (3.5.2).
  (e) If Y is a variety of degree d in $\mathbf{P}^n$, show that the vertex of the cone over Y is a point of multiplicity d.

3.5. Let $a_1, \dots, a_r$, $r \geq 5$, be distinct elements of k, and let C be the curve in $\mathbf{P}^2$ given by the (affine) equation $y^2 = \prod_{i=1}^r (x - a_i)$. Show that the point P at infinity on the y-axis is a singular point. Compute δ_P and $g(\tilde{Y})$, where $\tilde{Y}$ is the normalization of Y. Show in this way that one obtains hyperelliptic curves of every genus $g \geq 2$.

3.6. Show that analytically isomorphic curve singularities (I, 5.6.1) are equivalent in the sense of (3.9.4), but not conversely.

3.7. For each of the following singularities at (0,0) in the plane, give an embedded resolution, compute δ_P, and decide which ones are equivalent.
   (a) $x^3 + y^5 = 0$.
   (b) $x^3 + x^4 + y^5 = 0$.
   (c) $x^3 + y^4 + y^5 = 0$.
   (d) $x^3 + y^5 + y^6 = 0$.
   (e) $x^3 + xy^3 + y^5 = 0$.

3.8. Show that the following two singularities have the same multiplicity, and the same configuration of infinitely near singular points with the same multiplicities, hence the same δ_P, but are not equivalent.
   (a) $x^4 - xy^4 = 0$.
   (b) $x^4 - x^2y^3 - x^2y^5 + y^8 = 0$.

4   The Cubic Surface in $\mathbf{P}^3$

In this section, as in §2, we consider a very special class of surfaces, to illustrate some general principles. Our main result is that the projective plane with six points blown up is isomorphic to a nonsingular cubic surface in $\mathbf{P}^3$. We use this isomorphism to study the geometry of curves on the cubic surface. The isomorphism is accomplished using the linear system of plane cubic curves with six base points, so we begin with some general remarks about linear systems with base points.

Let X be a surface, let $|D|$ be a complete linear system of curves on X, and let $P_1, \ldots, P_r$ be points of X. Then we will consider the sublinear system $\mathfrak{d}$ consisting of divisors $D \in |D|$ which pass through the points $P_1, \ldots, P_r$, and we denote it by $|D - P_1 - \ldots - P_r|$. We say that $P_1, \ldots, P_r$ are the *assigned base points* of $\mathfrak{d}$.

Let $\pi: X' \to X$ be the morphism obtained by blowing up $P_1, \ldots, P_r$, and let $E_1, \ldots, E_r$ be the exceptional curves. Then there is a natural one-to-one correspondence between the elements of $\mathfrak{d}$ on X and the elements of the complete linear system $\mathfrak{d}' = |\pi^*D - E_1 - \ldots - E_r|$ on X' given by $D \mapsto \pi^*D - E_1 - \ldots - E_r$, because the latter divisor is effective on X' if and only if D passes through $P_1, \ldots, P_r$.

The new linear system $\mathfrak{d}'$ on X' may or may not have base points. We call any base point of $\mathfrak{d}'$, considered as an infinitely near point of X, an *unassigned base point* of $\mathfrak{d}$.

These definitions also make sense if some of the P_i themselves are infinitely near points of X, or if they are given with multiplicities greater than 1. For example, if P_2 is infinitely near P_1, then for $D \in \mathfrak{d}$ we require that D contain P_1, and that $\pi_1^*D - E_1$ contain P_2, where π_1 is the blowing-up of P_1. On the other hand, if P_1 is given with multiplicity $r \geqslant 1$, then we require that D have at least an r-fold point at P_1, and in the definition of $\mathfrak{d}'$, we take $\pi^*D - rE_1$.

(Note that if we assign base points $Q_1, \ldots, Q_s$ infinitely near a point P, then every divisor containing $Q_1, \ldots, Q_s$ will automatically have at least an s-fold point at P. So we make the convention that every point P must be assigned with a multiplicity at least equal to the sum of the multiplicities of the assigned base points infinitely near P.)

The usefulness of this language is that it gives us a way of talking about linear systems on various blown-up models of X, in terms of suitable linear systems with assigned base points on X.

Remark 4.0.1. Using this language, we can rephrase the condition (II, 7.8.2) for a complete linear system $|D|$ to be very ample as follows: $|D|$ is very ample if and only if (a) $|D|$ has no base points, and (b) for every $P \in X$, $|D - P|$ has no unassigned base points. Indeed, $|D|$ separates the points P and Q if and only if Q is not a base point of $|D - P|$, and $|D|$ separates tangent vectors at P if and only if $|D - P|$ has no unassigned base points infinitely near to P.

Remark 4.0.2. If we observe that the dimension drops by exactly one when we assign a base point which was not already an unassigned base point of a linear system, then we can rephrase this condition in a form reminiscent of (IV, 3.1) as follows: $|D|$ is very ample if and only if for any two points $P, Q \in X$, including the case Q infinitely near P,

$$\dim|D - P - Q| = \dim|D| - 2.$$

Remark 4.0.3. Applying (4.0.1) to a blown-up model of X, we see that if $\mathfrak{d} = |D - P_1 - \ldots - P_r|$ is a linear system with assigned base points on X, then the associated linear system $\mathfrak{d}'$ on X' is very ample on X' if and only if (a) $\mathfrak{d}$ has no unassigned base points, and (b) for every $P \in X$, including infinitely near points on X', $\mathfrak{d} - P$ has no unassigned base points.

Now we turn our attention to the particular situation of this section, which is linear systems of plane curves of fixed degree with assigned base points. We ask whether they have unassigned base points, and if not, we study the corresponding morphism of the blown-up model to a projective space. To get the cubic surface in $\mathbf{P}^3$ we will use the linear system of plane cubic curves with six base points. But first we need to consider linear systems of conics with base points. Here we use the word *conic* (respectively, *cubic*) to mean any effective divisor in the plane of degree 2 (respectively, 3).

Proposition 4.1. *Let $\mathfrak{d}$ be the linear system of conics in $\mathbf{P}^2$ with assigned base points $P_1, \ldots, P_r$, and assume that no three of the P_i are collinear. If $r \leqslant 4$, then $\mathfrak{d}$ has no unassigned base points. This result remains true if P_2 is infinitely near P_1.*

PROOF. Clearly it is sufficient to consider the case $r = 4$. First suppose P_1, P_2, P_3, P_4 are all ordinary points. Let L_{ij} denote the line containing P_i and P_j. Then $\mathfrak{d}$ contains $L_{12} + L_{34}$ and $L_{13} + L_{24}$. Since no three of the P_i are collinear, the intersection of these two divisors consists of the points P_1, P_2, P_3, P_4 with multiplicity 1 each, so there are no unassigned base points.

Now suppose P_2 is infinitely near P_1. In this case $\mathfrak{d}$ contains $L_{12} + L_{34}$ and $L_{13} + L_{14}$. (Here, of course, L_{12} denotes the line through P_1 with the tangent direction given by P_2.) This intersection again is just $\{P_1, P_2, P_3, P_4\}$, so there are no further base points.

Corollary 4.2. *With the same hypotheses, we have*:
  (a) *if* $r \leqslant 5$, *then* $\dim \mathfrak{d} = 5 - r$;
  (b) *if* $r = 5$, *then there exists a unique conic containing* $P_1, \ldots, P_5$, *which is necessarily irreducible*.
  Furthermore, these results remain true if P_5 *is infinitely near any one of* $P_1, \ldots, P_4$.

PROOF.
  (a) Every time we prescribe a new base point on a linear system without unassigned base points, the dimension drops by one. Since the linear system of all conics in $\mathbf{P}^2$ has dimension 5, this follows from (4.1).
  (b) For $r = 5$, $\dim \mathfrak{d} = 0$, so there is a unique conic containing $P_1, \ldots, P_5$. It must be irreducible since no three of the P_i are collinear.

Remark 4.2.1. This last statement is the classical result that a conic is uniquely determined by giving 5 points, or 4 points and a tangent direction at one of them, or 3 points with tangent directions at two of them, or even 3 points with a tangent direction and a second order tangent direction at one of them (when P_5 is infinitely near P_2 which is infinitely near P_1).

Example 4.2.2. If $r = 1$, then $\mathfrak{d}$ has no unassigned base points, and for any point P, $\mathfrak{d} - P$ has no unassigned base points, so by (4.0.3), $\mathfrak{d}'$ is very ample on X'. Since $\dim \mathfrak{d}' = 4$, it gives an embedding of X' in $\mathbf{P}^4$, as a surface of degree 3, which is the number of unassigned intersection points of two divisors in $\mathfrak{d}$. In fact, X' is just the rational ruled surface with $e = 1$, and this embedding is the rational cubic scroll (2.19.1).

Example 4.2.3. If $r = 3$, then X' is $\mathbf{P}^2$ with three points blown up, and $\dim \mathfrak{d} = 2$. Since $\mathfrak{d}'$ has no base points, it determines a morphism ψ of X' to $\mathbf{P}^2$. We may take the three points to be $P_1 = (1,0,0)$, $P_2 = (0,1,0)$ and $P_3 = (0,0,1)$. Then the vector space $V \subseteq H^0(\mathcal{O}_{\mathbf{P}^2}(2))$ corresponding to $\mathfrak{d}$ is spanned by $x_1 x_2$, $x_0 x_2$, and $x_0 x_1$, so ψ can be defined by $y_0 = x_1 x_2$, $y_1 = x_0 x_2$, $y_2 = x_0 x_1$, where y_i are the homogeneous coordinates of the new $\mathbf{P}^2$. Considered as a rational map from $\mathbf{P}^2$ to $\mathbf{P}^2$, this is none other than the *quadratic transformation* φ of (I, Ex. 4.6).

Now we will show that ψ identifies X' with the second $\mathbf{P}^2$, blown up at the points $Q_1 = (1,0,0)$, $Q_2 = (0,1,0)$, and $Q_3 = (0,0,1)$, in such a way that the exceptional curve $\psi^{-1}(Q_i)$ is the strict transform of the line L_{jk} joining P_j, P_k on the first $\mathbf{P}^2$, for each $(i,j,k) = (1,2,3)$ in some order. Furthermore $\psi(E_i)$ is the line M_{jk} joining Q_j and Q_k, for each $(i,j,k) = (1,2,3)$. Thus we can say that the quadratic transformation φ is just "blowing up the points P_1, P_2, P_3 and blowing down the lines $\tilde{L}_{12}, \tilde{L}_{13}, \tilde{L}_{23}$" (Fig. 21).

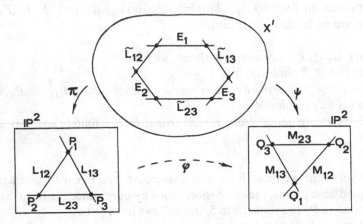

Figure 21. The quadratic transformation of $\mathbf{P}^2$.

To prove this, we consider the variety V in $\mathbf{P}^2 \times \mathbf{P}^2$ defined by the bihomogeneous equations $x_0 y_0 = x_1 y_1 = x_2 y_2$. I claim the first projection $p_1: V \to \mathbf{P}^2$ identifies V with X'. This is a local question, since blowing up a point depends only on a neighborhood of the point, so we consider the open set $U \subseteq \mathbf{P}^2$ defined by $x_0 = 1$. Then $U = \operatorname{Spec} A$ with $A = k[x_1, x_2]$, and $p_1^{-1}(U)$ can be written as

$$p_1^{-1}(U) = \operatorname{Proj} A[y_0, y_1, y_2]/(y_0 - x_1 y_1, x_1 y_1 - x_2 y_2).$$

We can eliminate y_0 from the graded ring, so

$$p_1^{-1}(U) \cong \operatorname{Proj} A[y_1, y_2]/(x_1 y_1 - x_2 y_2).$$

But this shows that $p_1^{-1}(U)$ is isomorphic to U with the point $(x_1, x_2) = (0,0)$ blown up, as in the proof of (3.6).

Doing the same with the open sets $x_1 = 1$ and $x_2 = 1$ of $\mathbf{P}^2$, we see that V, via the first projection, is just $\mathbf{P}^2$ with the points P_1, P_2, P_3 blown up, so $V \cong X'$. By symmetry, V with the second projection is the second $\mathbf{P}^2$ with the points Q_1, Q_2, Q_3 blown up. So $p_2 \circ p_1^{-1}$ gives a birational transformation of $\mathbf{P}^2$ to itself. Solving the equations $x_0 y_0 = x_1 y_1 = x_2 y_2$ in the case $x_0, x_1, x_2 \neq 0$, we get $y_0 = x_1 x_2$, $y_1 = x_0 x_2$, $y_2 = x_0 x_1$, so that this transformation is again the quadratic transformation φ above.

We conclude that $\psi: X' \to \mathbf{P}^2$ is the same as $p_2: V \to \mathbf{P}^2$, so that φ is just blowing up three points and blowing down three lines. Finally, it is clear from the equations that $p_2(\tilde{L}_{ij}) = Q_k$ and $p_2(E_i) = M_{jk}$ for each i,j,k.

Proposition 4.3. *Let $\mathfrak{d}$ be the linear system of plane cubic curves with assigned base points $P_1,\ldots,P_r$, and assume that no 4 of the P_i are collinear, and no 7 of them lie on a conic. If $r \leqslant 7$, then $\mathfrak{d}$ has no unassigned base points. This result remains true if P_2 is infinitely near P_1.*

PROOF. It is sufficient to consider the case $r = 7$. We will show first that if $P_1,\ldots,P_7$ are all ordinary points, then $\mathfrak{d}$ has no unassigned ordinary base points. For this it is sufficient to exhibit, for each point Q not equal to any P_i, a cubic curve containing $P_1,\ldots,P_7$ but not Q.

Case 1. Suppose there exist some three points P_1,P_2,P_3 lying on a line L^* with Q. The points P_4,P_5,P_6,P_7 are not all collinear, so we may assume that P_4,P_5,P_6 are not collinear. Then the conic Γ_{12456} through P_1,P_2,P_4,P_5,P_6, together with the line L_{37} through P_3 and P_7 forms a cubic curve containing $P_1,\ldots,P_7$ but not Q. Indeed, if $Q \in \Gamma_{12456}$, then this conic contains the line L^*, so it is reducible, in which case P_4,P_5,P_6 must be collinear, which is a contradiction. If $Q \in L_{37}$, then $P_7 \in L^*$, so P_1,P_2,P_3,P_7 are collinear, which is a contradiction.

Case 2. Suppose that Q is not collinear with any set of three of the points P_i, but that Q lies on a conic Γ^* (necessarily irreducible) containing 6 of them, say $P_1,\ldots,P_6$. Then $\Gamma_{12347} + L_{56}$ is a cubic not containing Q. Indeed, if $Q \in \Gamma_{12347}$, then P_1,P_2,P_3,P_4,Q are in this conic and also in Γ^*, so by (4.2), $\Gamma_{12347} = \Gamma^*$. But then $P_1,\ldots,P_7$ are all in Γ^*, a contradiction. If $Q \in L_{56}$, then Γ^* is reducible, a contradiction.

Case 3. Q is not collinear with any 3 of the P_i, and not on a conic with any 6 of them. Then consider the three cubic curves $C_i = \Gamma_{1234i} + L_{jk}$, where $(i,j,k) = (5,6,7)$ in some order. We will show that one of these does not contain Q. If $Q \in L_{56}$, then $Q \notin L_{57}$ and $Q \notin L_{67}$, because in either case P_5,P_6,P_7,Q would be collinear. So, ruling out C_7, we may assume $Q \notin L_{57}$ and $Q \notin L_{67}$. Then, if $Q \in C_5$ and $Q \in C_6$, we have $Q \in \Gamma_{12345}$ and $Q \in \Gamma_{12346}$. Consider the conic $\Gamma' = \Gamma_{1234Q}$. If Γ' is irreducible, then from (4.2) we have all three conics equal, so $Q \in \Gamma_{123456}$, which is a contradiction. If Γ' is reducible, then for a suitable relabeling, we have either (a) $\Gamma' = L_{123} + L_{4Q}$ or (b) $\Gamma' = L_{12Q} + L_{34}$. In case (a), $\Gamma_{12345} = L_{123} + L_{45}$ and $\Gamma_{12346} = L_{123} + L_{46}$, so P_4,P_5,P_6 and Q are collinear, a contradiction. In case (b), $\Gamma_{12345} = L_{12} + L_{345}$ and $\Gamma_{12346} = L_{12} + L_{346}$, so P_3,P_4,P_5,P_6 are collinear, a contradiction.

This completes the proof in the case $P_1,\ldots,P_7$ and Q are all ordinary points. The same proof also works in case P_2 is infinitely near P_1, or Q is infinitely near one of $P_1,\ldots,P_7$, or both. One has to relabel the P_i occasionally so that the constructions make sense, and one has to use (4.2) in the case of infinitely near points also. (Details left to reader.)

Corollary 4.4. *With the same hypotheses, we have*
  (a) *if $r \leqslant 8$, then* dim $\mathfrak{d} = 9 - r$, *and*
  (b) *if $r = 8$,* dim $\mathfrak{d} = 1$ *and almost every curve in $\mathfrak{d}$ is irreducible.*

PROOF. For $r \leqslant 7$ there are no unassigned base points, so at each step, the dimension drops by one. The cubics with no base points form a linear system of dimension 9. This proves (a). To prove (b), we observe that with no 4 points collinear and no 7 on a conic, there are only finitely many ways of passing three lines, or one line and one irreducible conic through the 8 points.

Corollary 4.5. *Given 8 points $P_1, \ldots, P_8$ in the plane, no 4 collinear, and no 7 lying on a conic, there is a uniquely determined point P_9 (possibly an infinitely near point) such that every cubic through $P_1, \ldots, P_8$ also passes through P_9. This is still true if P_2 is infinitely near P_1, and P_8 is infinitely near any one of $P_1, \ldots, P_7$.*

PROOF. By (4.4) the linear system $\mathfrak{d}$ of all cubics through $P_1, \ldots, P_8$ has dimension one, and we can choose two distinct irreducible ones $C, C' \in \mathfrak{d}$. Then by Bézout's theorem (1.4.2) (cf. Ex. 3.2), C and C' meet in 9 points, 8 of which are $P_1, \ldots, P_8$. So this determines a ninth point P_9, possibly an infinitely near point. Now since dim $\mathfrak{d} = 1$, any other curve $C'' \in \mathfrak{d}$, irreducible or not, is a linear combination of C and C', so it must also pass through P_9. Thus P_9 is an unassigned base point of $\mathfrak{d}$.

Remark 4.5.1. This classical result has a number of interesting geometrical consequences. See (Ex. 4.4), (Ex. 4.5).

Theorem 4.6. *Let $\mathfrak{d}$ be the linear system of plane cubic curves with assigned (ordinary) base points $P_1, \ldots, P_r$, and assume that no 3 of the P_i are collinear, and no 6 of them lie on a conic. If $r \leqslant 6$, then the corresponding linear system $\mathfrak{d}'$ on the surface X' obtained from $\mathbf{P}^2$ by blowing up $P_1, \ldots, P_r$, is very ample.*

PROOF. According to (4.0.3) we must verify that $\mathfrak{d}$ has no unassigned base points, and that for every point P, possibly infinitely near, $\mathfrak{d} - P$ has no unassigned base points. The first statement is an immediate consequence of (4.3). For the second, we note that since no 3 of the P_i are collinear, and no 6 of them lie on a conic, the $r + 1$ points $P_1, \ldots, P_r, P$ satisfy the hypotheses of (4.3). So this case also follows from (4.3).

Corollary 4.7. *With the same hypotheses, for each $r = 0, 1, \ldots, 6$, we obtain an embedding of X' in $\mathbf{P}^{9-r}$ as a surface of degree $9 - r$, whose canonical sheaf $\omega_{X'}$ is isomorphic to $\mathcal{O}_{X'}(-1)$. In particular, for $r = 6$, we obtain a nonsingular cubic surface in $\mathbf{P}^3$.*

PROOF. We embed X' in $\mathbf{P}^N$ via the very ample linear system $\mathfrak{d}'$. Since dim $\mathfrak{d} = $ dim $\mathfrak{d}' = 9 - r$ by (4.4), we have $N = 9 - r$. If L is a line in $\mathbf{P}^2$, then $\mathfrak{d}' = |\pi^*3L - E_1 - \ldots - E_r|$, so for any $D' \in \mathfrak{d}'$, we have $D'^2 = 9 - r$. Therefore the degree of X' in $\mathbf{P}^N$ is $9 - r$. Finally, since the canonical divisor on $\mathbf{P}^2$ is $-3L$, we see from (3.3) that $K_{X'} = -\pi^*3L + E_1 + \ldots +$

E_r, which is just $-D'$. Therefore $\omega_{X'} \cong \mathcal{O}_{X'}(-1)$ in the given projective embedding.

Remark 4.7.1. A *Del Pezzo surface* is defined to be a surface X of degree d in $\mathbf{P}^d$ such that $\omega_X \cong \mathcal{O}_X(-1)$. So (4.7) gives a construction of Del Pezzo surfaces of degrees $d = 3,4,\ldots,9$. A classical result states that every Del Pezzo surface is either one given by (4.7) for a suitable choice of points $P_i \in \mathbf{P}^2$, or the 2-uple embedding of a quadric surface in $\mathbf{P}^3$, which is a Del Pezzo surface of degree 8 in $\mathbf{P}^8$. In particular, every nonsingular cubic surface in $\mathbf{P}^3$ can be obtained by blowing up 6 points in the plane. Indeed, for a cubic surface in $\mathbf{P}^3$, the condition $\omega_X \cong \mathcal{O}_X(-1)$ is automatic (II, Ex. 8.4). For proofs see, e.g., Manin [3, §24] or Nagata [5, I, Thm. 8, p. 366].

Remark 4.7.2. In the case of cubic surfaces in $\mathbf{P}^3$, we can prove a slightly weaker result by counting constants. The choice of 6 points in the plane requires 12 parameters. Subtract off the automorphisms of $\mathbf{P}^2$ (8 parameters) and add automorphisms of $\mathbf{P}^3$ (15 parameters). Thus we see that the cubic surfaces in $\mathbf{P}^3$ given by (4.7) form a 19-dimensional family. But the family of all cubic surfaces in $\mathbf{P}^3$ has dimension equal to $\dim H^0(\mathcal{O}_{\mathbf{P}^3}(3)) - 1$, which is also 19. Thus we see at least that almost all nonsingular cubic surfaces arise by (4.7).

Notation 4.7.3. For the rest of this section, we specialize to the case of the cubic surface in $\mathbf{P}^3$, and fix our notation. Let $P_1,\ldots,P_6$ be six points of the plane, no three collinear, and not all six lying on a conic. Let $\mathfrak{d}$ be the linear system of plane cubic curves through $P_1,\ldots,P_6$, and let X be the nonsingular cubic surface in $\mathbf{P}^3$ obtained by (4.7). Thus X is isomorphic to $\mathbf{P}^2$ with the six points $P_1,\ldots,P_6$ blown up. Let $\pi:X \to \mathbf{P}^2$ be the projection. Let $E_1,\ldots,E_6 \subseteq X$ be the exceptional curves, and let $e_1,\ldots,e_6 \in \operatorname{Pic} X$ be their linear equivalence classes. Let $l \in \operatorname{Pic} X$ be the class of π^* of a line in $\mathbf{P}^2$.

Proposition 4.8. *Let X be the cubic surface in $\mathbf{P}^3$ (4.7.3). Then:*

(a) $\operatorname{Pic} X \cong \mathbf{Z}^7$, *generated by* $l, e_1,\ldots,e_6$;

(b) *the intersection pairing on X is given by* $l^2 = 1$, $e_i^2 = -1$, $l.e_i = 0$, $e_i.e_j = 0$ *for* $i \neq j$;

(c) *the hyperplane section h is* $3l - \sum e_i$;

(d) *the canonical class is* $K = -h = -3l + \sum e_i$;

(e) *if D is any effective divisor on $X, D \sim al - \sum b_i e_i$, then the degree of D, as a curve in $\mathbf{P}^3$, is*

$$d = 3a - \sum b_i;$$

(f) *the self-intersection of D is* $D^2 = a^2 - \sum b_i^2$;

(g) *the arithmetic genus of D is*

$$p_a(D) = \frac{1}{2}(D^2 - d) + 1 = \frac{1}{2}(a-1)(a-2) - \frac{1}{2}\sum b_i(b_i - 1).$$

PROOF. All of this follows from earlier results. (a) and (b) follow from (3.2). (c) comes from the definition of the embedding in $\mathbf{P}^3$. (d) comes from (3.3). For (e), we note that the degree of D is just $D.h$. (f) is immediate from (b), and (g) follows from the adjunction formula $2p_a(D) - 2 = D.(D + K)$ (Ex. 1.3) and the fact that $D.K = -D.h = -d$ by (d).

Remark 4.8.1. If C is any irreducible curve in X, other than $E_1, \ldots, E_6$, then $\pi(C)$ is an irreducible plane curve C_0, and C in turn is the strict transform of C_0. Let C_0 have degree a, and suppose that C_0 has a point of multiplicity b_i at each P_i. Then $\pi^*C_0 = C + \sum b_i E_i$, by (3.6). Since $C_0 \sim a \cdot \text{line}$, we conclude that $C \sim al - \sum b_i e_i$. Thus for any $a, b_1, \ldots, b_6 \geq 0$, we can interpret an irreducible curve C on X in the class $al - \sum b_i e_i$ as the strict transform of a plane curve of degree a with a b_i-fold point at each P_i. So the study of curves on X is reduced to the study of certain plane curves.

Theorem 4.9 (Twenty-Seven Lines). *The cubic surface X contains exactly 27 lines. Each one has self-intersection -1, and they are the only irreducible curves with negative self-intersection on X. They are*
(a) *the exceptional curves E_i, $i = 1, \ldots, 6$ (six of these),*
(b) *the strict transform F_{ij} of the line in $\mathbf{P}^2$ containing P_i and P_j, $1 \leq i < j \leq 6$ (fifteen of these), and*
(c) *the strict transform G_j of the conic in $\mathbf{P}^2$ containing the five P_i for $i \neq j, j = 1, \ldots, 6$ (six of these).*

PROOF. First of all, if L is any line in X, then $\deg L = 1$ and $p_a(L) = 0$, so by (4.8) we have $L^2 = -1$. (See also (Ex. 1.4).) Conversely, if C is an irreducible curve on X with $C^2 < 0$, then since $p_a(C) \geq 0$, we must have $C^2 = -1$, $p_a(C) = 0$, $\deg C = 1$ again by (4.8), so C is a line.

Next, from (4.8.1) we see that $E_i \sim e_i$, $F_{ij} \sim l - e_i - e_j$, and $G_j \sim 2l - \sum_{i \neq j} e_i$, and we see immediately from (4.8) that each of these has degree 1, i.e., is a line.

It remains to show that if C is any irreducible curve on X with $\deg C = 1$ and $C^2 = -1$, then C is one of those 27 lines listed. Assuming C is not one of the E_i, we can write $C \sim al - \sum b_i e_i$, and by (4.8.1) we must have $a > 0$, $b_i \geq 0$. Furthermore,

$$\deg C = 3a - \sum b_i = 1$$
$$C^2 = a^2 - \sum b_i^2 = -1.$$

We will show that the only integers $a, b_1, \ldots, b_6$ satisfying all these conditions are those corresponding to the F_{ij} and G_j above.

Recall Schwarz's inequality, which says that if $x_1, x_2, \ldots, y_1, y_2, \ldots$ are two sequences of real numbers, then

$$|\sum x_i y_i|^2 \leq |\sum x_i^2| \cdot |\sum y_i^2|.$$

Taking $x_i = 1$, $y_i = b_i$, $i = 1, \ldots, 6$, we find

$$(\sum b_i)^2 \leq 6(\sum b_i^2).$$

Substituting $\sum b_i = 3a - 1$ and $\sum b_i^2 = a^2 + 1$ from above, we obtain
$$3a^2 - 6a - 5 \leqslant 0.$$
Solving the quadratic equation, this implies $a \leqslant 1 + (2/3)\sqrt{6} < 3$. Therefore $a = 1$ or 2. Now one quickly finds all possible values of the b_i by trial: if $a = 1$, then $b_i = b_j = 1$ for some i,j, the rest 0. This gives F_{ij}. If $a = 2$, then all $b_i = 1$ except for one $b_j = 0$. This gives G_j.

Remark 4.9.1. There is lots of classical projective geometry associated with the 27 lines. For example, the configuration of 12 lines $E_1,\ldots,E_6, G_1,\ldots,G_6$ in $\mathbf{P}^3$ with the property that the E_i are mutually skew, the G_j are mutually skew, and E_i meets G_j if and only if $i \neq j$, is called *Schläfli's double-six*. One can show that given a line E_1, and five lines $G_2,\ldots,G_6$ meeting it, but otherwise in sufficiently general position, then other lines $E_2,\ldots,E_6$ and G_1 are uniquely determined so as to form a double-six. Furthermore, each double-six is contained in a unique nonsingular cubic surface, and thus forms part of a set of 27 lines on a cubic surface. See Hilbert and Cohn-Vossen [1, §25]. The 27 lines have a high degree of symmetry, as we see in the next result.

Proposition 4.10. *Let X be a cubic surface as above, and let $E'_1,\ldots,E'_6$ be any subset of six mutually skew lines chosen from among the 27 lines on X. Then there is another morphism $\pi': X \to \mathbf{P}^2$, making X isomorphic to that $\mathbf{P}^2$ with six points $P'_1,\ldots,P'_6$ blown up (no 3 collinear and not all 6 on a conic), such that $E'_1,\ldots,E'_6$ are the exceptional curves for π'.*

PROOF. We proceed stepwise, working with one line at a time. We will show first that it is possible to find π' such that E'_1 is the inverse image of P'_1.

Case 1. If E'_1 is one of the E_i, we take $\pi' = \pi$, but relabel the P_i so that P_i becomes P'_1.

Case 2. If E'_1 is one of the F_{ij}, say $E'_1 = F_{12}$, then we apply the quadratic transformation with centers P_1,P_2,P_3 (4.2.3) as follows. Let X_0 be $\mathbf{P}^2$ with P_1,P_2,P_3 blown up, let $\pi_0: X_0 \to \mathbf{P}^2$ be the projection, and let $\psi: X_0 \to \mathbf{P}^2$ be the other map to $\mathbf{P}^2$ of (4.2.3), so that X_0 via ψ is $\mathbf{P}^2$ with Q_1,Q_2,Q_3 blown up. Since $\pi: X \to \mathbf{P}^2$ expresses X as $\mathbf{P}^2$ with $P_1,\ldots,P_6$ blown up, π factors through π_0, say $\pi = \pi_0 \circ \theta$, where $\theta: X \to X_0$. Now we define π' as $\psi \circ \theta$.

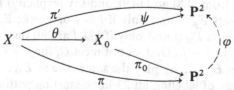

Then, using the notation of (4.2.3), $\theta(F_{12}) = \tilde{L}_{12}$, so $\pi'(F_{12}) = Q_3$. Furthermore, π' expresses X as $\mathbf{P}^2$ with $Q_1,Q_2,Q_3,P'_4,P'_5,P'_6$ blown up, where P'_4,P'_5,P'_6 are the images of P_4,P_5,P_6 under $\psi \circ \pi_0^{-1}$. Now taking $P'_1 = Q_3$, and P'_2,P'_3 to be Q_1,Q_2, we have $E'_1 = \pi'^{-1}(P'_1)$.

We still have to verify that no 3 of $Q_1, Q_2, Q_3, P'_4, P'_5, P'_6$ lie on a line, and no 6 on a conic. Q_1, Q_2, Q_3 are noncollinear by construction. If Q_1, Q_2, P'_4 were collinear, then $\psi^{-1}(P'_4) \in E_3$, so P_4 would be infinitely near P_3. If Q_1, P'_4, P'_5 were collinear, let L' be the line containing them. Then the strict transform of L' by φ^{-1} will be a line L containing P_1, P_4, P_5. Indeed, φ^{-1} is the rational map determined by the linear system of conics through Q_1, Q_2, Q_3. Such a conic has one free intersection with L', so the strict transform of L' is a line L. Furthermore, L' meets M_{23}, so L passes through P_1. Finally, suppose P'_4, P'_5, P'_6 were collinear. Since φ is determined by the conics through P_1, P_2, P_3, the strict transform of the line L' containing P'_4, P'_5, P'_6 would be a conic Γ containing $P_1, \ldots, P_6$, which is impossible. For the same reason, if $Q_1, Q_2, Q_3, P'_4, P'_5, P'_6$ lay on a conic, then P_4, P_5, P_6 would be collinear. This completes Case 2.

Case 3. If E'_1 is one of the G_j, say $E'_1 = G_6$, we again apply the quadratic transformation of (4.2.3) with centers P_1, P_2, P_3. Since $\pi(G_6)$ is the conic through $P_1, \ldots, P_5$, we see that $\pi'(G_6)$ is the line through P'_4, P'_5. Thus E'_1 is the curve F'_{45} for π', which reduces us to Case 2.

Now that we have moved E'_1 to the position of E_1, we may assume $E'_1 = E_1$, and we consider E'_2. Since E'_2 does not meet E_1, the possible values of E'_2 are $E_2, \ldots, E_6, F_{ij}$ with $1 < i, j$, or G_1. We apply the same method as in Cases 1, 2, and 3 above, and find that we can move E'_2 to the role of E_2 without touching P_1. That is to say, we allow ourselves only to relabel $P_2, \ldots, P_5$, or use quadratic transformations based at three points among $P_2, \ldots, P_5$.

Continuing in this manner, we eventually have $E'_1, \ldots, E'_6$ in the position of $E_1, \ldots, E_6$, which proves the proposition. For example, the last step is this. Assuming that $E'_i = E_i$ for $i = 1, 2, 3, 4$, there are only 3 lines left which do not meet $E_1, \ldots, E_4$. They are E_5, E_6, F_{56}. Since F_{56} meets E_5 and E_6, the lines E'_5 and E'_6 must be E_5 and E_6 in some order. So for the last step we have only to permute 5 and 6 if necessary.

Remark 4.10.1. The proposition says that any six mutually skew lines among the 27 lines play the role of $E_1, \ldots, E_6$. Another way of expressing this is to consider the *configuration* of the 27 lines (forgetting the surface X). That means we consider simply the set of 27 elements named E_i, F_{ij}, G_j (the lines) together with the incidence relations they satisfy. These incidence relations are easily deduced from (4.8) and (4.9), and say (explicity) E_i does not meet E_j for $i \neq j$; E_i meets F_{jk} if and only if $i = j$ or $i = k$; E_i meets G_j if and only if $i \neq j$; F_{ij} meets F_{kl} if and only if i, j, k, l are all distinct; F_{ij} meets G_k if and only if $i = k$ or $j = k$; G_j does not meet G_k for $j \neq k$.

Now to say that $E'_1, \ldots, E'_6$ play the same role as $E_1, \ldots, E_6$ means that there is another way of labeling all 27 lines, starting with $E'_1, \ldots, E'_6$, so as to satisfy the same incidence relations. In other words, there is an *automorphism of the configuration* (meaning a permutation of the set of 27 elements, preserving the incidence relations) which sends $E_1, \ldots, E_6$ to $E'_1, \ldots, E'_6$. Notice furthermore that naming $E_1, \ldots, E_6$ uniquely determines the names of the remaining 21 lines: F_{ij} is the unique line which meets E_i, E_j but not other E_k; G_j is the unique line which meets all E_i except E_j.

So (4.10) tells us that for every (ordered) set of six mutually skew lines among the 27 lines, there is a unique automorphism of the configuration taking $E_1, \ldots, E_6$ to those six. Since any automorphism must send skew lines to skew lines, we get all elements of the group G of automorphisms of the configuration this way. From the incidence relations it is easy to count the ways of choosing six mutually skew lines: there are 27 choices for E_1, 16 for E_2, 10 for E_3, 6 for E_4, 2 for E_5 and 1 for E_6. So the order of the group G is $27 \cdot 16 \cdot 10 \cdot 6 \cdot 2 = 51{,}840$.

One can show that G is isomorphic to the *Weyl group* $\mathbf{E}_6$, and that it contains a normal subgroup of index 2 which is a simple group of order 25,920. See (Ex. 4.11) and Manin [3, §25, 26].

We will use this symmetry of the 27 lines to determine the ample and very ample divisor classes on the cubic surface.

Theorem 4.11. *The following conditions are equivalent, for a divisor D on the cubic surface X:*

  (i) *D is very ample;*
  (ii) *D is ample;*
  (iii) *$D^2 > 0$, and for every line $L \subseteq X$, $D.L > 0$;*
  (iv) *for every line $L \subseteq X$, $D.L > 0$.*

PROOF. Of course (i) $\Rightarrow$ (ii) $\Rightarrow$ (iii) $\Rightarrow$ (iv), using the easy direction of Nakai's criterion (1.10). For (iv) $\Rightarrow$ (i) we will first prove a lemma.

Lemma 4.12. *Let $D \sim al - \sum b_i e_i$ be a divisor class on the cubic surface X, and suppose that $b_1 \geqslant b_2 \geqslant \ldots \geqslant b_6 > 0$ and $a \geqslant b_1 + b_2 + b_5$. Then D is very ample.*

PROOF. We use the general fact that a very ample divisor plus a divisor moving in a linear system without base points is very ample (II, Ex. 7.5). Let us consider the divisor classes

$$D_0 = l$$
$$D_1 = l - e_1$$
$$D_2 = 2l - e_1 - e_2$$
$$D_3 = 2l - e_1 - e_2 - e_3$$
$$D_4 = 2l - e_1 - e_2 - e_3 - e_4$$
$$D_5 = 3l - e_1 - e_2 - e_3 - e_4 - e_5$$
$$D_6 = 3l - e_1 - e_2 - e_3 - e_4 - e_5 - e_6.$$

Then $|D_0|, |D_1|$ correspond to the linear systems of lines in $\mathbf{P}^2$ with 0 or 1 assigned base points, which have no unassigned base points. $|D_2|, |D_3|, |D_4|$

have no base points by (4.1), $|D_5|$ has no base points by (4.3), and D_6 is very ample by (4.6). Therefore any linear combination of these, $D = \sum c_i D_i$, with $c_i \geqslant 0$ and $c_6 > 0$, will be very ample.

Clearly $D_0, \ldots, D_6$ form a free basis for Pic $X \cong \mathbf{Z}^7$. Writing $D \sim al - \sum b_i e_i$, we have $b_6 = c_6$, $b_5 = c_5 + c_6, \ldots, b_1 = c_1 + \ldots + c_6$, $a = c_1 + 2(c_2 + c_3 + c_4) + 3(c_5 + c_6)$. Then one checks easily that the conditions $c_i \geqslant 0$, $c_6 > 0$ are equivalent to the conditions $b_1 \geqslant \ldots \geqslant b_6 > 0$ and $a \geqslant b_1 + b_2 + b_5$, so all divisors satisfying these conditions are very ample.

PROOF OF (4.11), CONTINUED. Suppose D is a divisor satisfying $D.L > 0$ for every line $L \subseteq X$. Choose six mutually skew lines $E_1', \ldots, E_6'$ as follows: choose E_6' so that $D.E_6'$ is equal to the minimum value of $D.L$ for any line L; choose E_5' so that $D.E_5'$ is equal to the minimum value of $D.L$ among those lines L which do not meet E_6'; and choose E_4', E_3' similarly. There will be just three remaining lines which do not meet E_6', E_5', E_4', E_3', one of them meeting the other two. Choose E_1', E_2' so that $D.E_1' \geqslant D.E_2'$.

Now according to (4.10), we may assume that $E_i' = E_i$ for each i. Writing $D \sim al - \sum b_i e_i$, we have $D.E_i = b_i$, so by construction we have $b_1 \geqslant b_2 \geqslant \ldots \geqslant b_6 > 0$. On the other hand, F_{12} was available as a candidate at the time we chose E_3, so we have $D.F_{12} \geqslant D.E_3$. This translates as $a - b_1 - b_2 \geqslant b_3$, i.e., $a \geqslant b_1 + b_2 + b_3$. Since $b_3 \geqslant b_5$, these conditions imply the conditions of the lemma, so D is very ample.   q.e.d.

Corollary 4.13. *Let $D \sim al - \sum b_i e_i$ be a divisor class on X. Then:*
   (a) *D is ample $\Leftrightarrow$ very ample $\Leftrightarrow$ $b_i > 0$ for each i, and $a > b_i + b_j$ for each i,j, and $2a > \sum_{i \neq j} b_i$ for each j;*
   (b) *in any divisor class satisfying the conditions of (a), there is an irreducible nonsingular curve.*

PROOF. (a) is just a translation of (4.11), using the enumeration of the 27 lines in (4.9), and (b) is a consequence of Bertini's theorem (II, 8.18) and (III, 7.9.1).

Example 4.13.1. Taking $a = 7$, $b_1 = b_2 = 3$, $b_3 = b_4 = b_5 = b_6 = 2$ we obtain an irreducible nonsingular curve $C \sim al - \sum b_i e_i$, which according to (4.8) has degree 7 and genus 5. This gives another proof of the existence of a curve of degree 7 and genus 5 in $\mathbf{P}^3$ (IV, 6.4.2).

EXERCISES

4.1. The linear system of conics in $\mathbf{P}^2$ with two assigned base points P_1 and P_2 (4.1) determines a morphism ψ of X' (which is $\mathbf{P}^2$ with P_1 and P_2 blown up) to a nonsingular quadric surface Y in $\mathbf{P}^3$, and furthermore X' via ψ is isomorphic to Y with one point blown up.

4.2. Let φ be the quadratic transformation of (4.2.3), centered at P_1, P_2, P_3. If C is an irreducible curve of degree d in $\mathbf{P}^2$, with points of multiplicity r_1, r_2, r_3 at P_1, P_2, P_3, then the strict transform C' of C by φ has degree $d' = 2d - r_1 - r_2 - r_3$, and has points of multiplicity $d - r_2 - r_3$ at Q_1, $d - r_1 - r_3$ at Q_2 and $d - r_1 - r_2$ at Q_3. The curve C may have arbitrary singularities. [*Hint*: Use (Ex. 3.2).]

4.3. Let C be an irreducible curve in $\mathbf{P}^2$. Then there exists a finite sequence of quadratic transformations, centered at suitable triples of points, so that the strict transform C' of C has only *ordinary* singularities, i.e., multiple points with all distinct tangent directions (I, Ex. 5.14). Use (3.8).

4.4. (a) Use (4.5) to prove the following lemma on cubics: If C is an irreducible plane cubic curve, if L is a line meeting C in points P,Q,R, and L' is a line meeting C in points P',Q',R', let P'' be the third intersection of the line PP' with C, and define Q'',R'' similarly. Then P'',Q'',R'' are collinear.

   (b) Let P_0 be an inflection point of C, and *define* the group operation on the set of regular points of C by the geometric recipe "let the line PQ meet C at R, and let P_0R meet C at T, then $P + Q = T$" as in (II, 6.10.2) and (II, 6.11.4). Use (a) to show that this operation is associative.

4.5. Prove Pascal's theorem: if A,B,C,A',B',C' are any six points on a conic, then the points $P = AB'.A'B$, $Q = AC'.A'C$, and $R = BC'.B'C$ are collinear (Fig. 22).

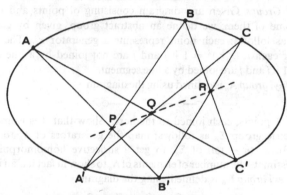

Figure 22. Pascal's theorem.

4.6. Generalize (4.5) as follows: given 13 points $P_1,\ldots,P_{13}$ in the plane, there are three additional determined points P_{14},P_{15},P_{16}, such that all quartic curves through $P_1,\ldots,P_{13}$ also pass through P_{14},P_{15},P_{16}. What hypotheses are necessary on $P_1,\ldots,P_{13}$ for this to be true?

4.7. If D is any divisor of degree d on the cubic surface (4.7.3), show that

$$p_a(D) \leq \begin{cases} \dfrac{1}{6}(d-1)(d-2) & \text{if } d \equiv 1,2 \,(\text{mod } 3) \\[2mm] \dfrac{1}{6}(d-1)(d-2) + \dfrac{2}{3} & \text{if } d \equiv 0 \,(\text{mod } 3). \end{cases}$$

Show furthermore that for every $d > 0$, this maximum is achieved by some irreducible nonsingular curve.

4.8. Show that a divisor class D on the cubic surface contains an irreducible curve $\Leftrightarrow$ it contains an irreducible nonsingular curve $\Leftrightarrow$ it is either (a) one of the 27 lines, or (b) a conic (meaning a curve of degree 2) with $D^2 = 0$, or (c) $D.L \geq 0$ for every

line L, and $D^2 > 0$. [*Hint*: Generalize (4.11) to the surfaces obtained by blowing up 2, 3, 4, or 5 points of $\mathbf{P}^2$, and combine with our earlier results about curves on $\mathbf{P}^1 \times \mathbf{P}^1$ and the rational ruled surface X_1, (2.18).]

4.9. If C is an irreducible non-singular curve of degree d on the cubic surface, and if the genus $g > 0$, then

$$g \geqslant \begin{cases} \dfrac{1}{2}(d-6) & \text{if } d \text{ is even, } d \geqslant 8, \\[2mm] \dfrac{1}{2}(d-5) & \text{if } d \text{ is odd, } d \geqslant 13, \end{cases}$$

and this minimum value of $g > 0$ is achieved for each d in the range given.

4.10. A curious consequence of the implication (iv) $\Rightarrow$ (iii) of (4.11) is the following numerical fact: Given integers $a, b_1, \ldots, b_6$ such that $b_i > 0$ for each i, $a - b_i - b_j > 0$ for each i, j and $2a - \sum_{i \neq j} b_i > 0$ for each j, we must necessarily have $a^2 - \sum b_i^2 > 0$. Prove this directly (for $a, b_1, \ldots, b_6 \in \mathbf{R}$) using methods of freshman calculus.

4.11. *The Weyl Groups.* Given any diagram consisting of points and line segments joining some of them, we define an abstract group, given by generators and relations, as follows: each point represents a generator x_i. The relations are $x_i^2 = 1$ for each i; $(x_i x_j)^2 = 1$ if i and j are not joined by a line segment, and $(x_i x_j)^3 = 1$ if i and j are joined by a line segment.

(a) The *Weyl group* $\mathbf{A}_n$ is defined using the diagram

$$\circ\!\!-\!\!\circ\!\!-\!\!\circ \cdots \,-\!\!\circ$$

of $n - 1$ points, each joined to the next. Show that it is isomorphic to the symmetric group Σ_n as follows: map the generators of $\mathbf{A}_n$ to the elements $(12), (23), \ldots, (n-1, n)$ of Σ_n, to get a surjective homomorphism $\mathbf{A}_n \to \Sigma_n$. Then estimate the number of elements of $\mathbf{A}_n$ to show in fact it is an isomorphism.

(b) The *Weyl group* $\mathbf{E}_6$ is defined using the diagram

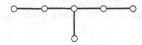

Call the generators $x_1, \ldots, x_5$ and y. Show that one obtains a surjective homomorphism $\mathbf{E}_6 \to G$, the group of automorphisms of the configuration of 27 lines (4.10.1), by sending $x_1, \ldots, x_5$ to the permutations $(12), (23), \ldots, (56)$ of the E_i, respectively, and y to the element associated with the quadratic transformation based at P_1, P_2, P_3.

*(c) Estimate the number of elements in $\mathbf{E}_6$, and thus conclude that $\mathbf{E}_6 \cong G$. *Note*: See Manin [3, §25,26] for more about Weyl groups, root systems, and exceptional curves.

4.12. Use (4.11) to show that if D is any ample divisor on the cubic surface X, then $H^1(X, \mathcal{O}_X(-D)) = 0$. This is Kodaira's vanishing theorem for the cubic surface (III, 7.15).

4.13. Let X be the Del Pezzo surface of degree 4 in $\mathbf{P}^4$ obtained by blowing up 5 points of $\mathbf{P}^2$ (4.7).

(a) Show that X contains 16 lines.

(b) Show that X is a complete intersection of two quadric hypersurfaces in $\mathbf{P}^4$ (the converse follows from (4.7.1)).

4.14. Using the method of (4.13.1), verify that there are nonsingular curves in $\mathbf{P}^3$ with $d = 8$, $g = 6,7$; $d = 9$, $g = 7,8,9$; $d = 10$, $g = 8,9,10,11$. Combining with (IV, §6), this completes the determination of all posible g for curves of degree $d \leqslant 10$ in $\mathbf{P}^3$.

4.15. Let $P_1, \ldots, P_r$ be a finite set of (ordinary) points of $\mathbf{P}^2$, no 3 collinear. We define an *admissible transformation* to be a quadratic transformation (4.2.3) centered at some three of the P_i (call them P_1, P_2, P_3). This gives a new $\mathbf{P}^2$, and a new set of r points, namely Q_1, Q_2, Q_3, and the images of $P_4, \ldots, P_r$. We say that $P_1, \ldots, P_r$ are in *general position* if no three are collinear, and furthermore after any finite sequence of admissible transformations, the new set of r points also has no three collinear.

(a) A set of 6 points is in general position if and only if no three are collinear and not all six lie on a conic.

(b) If $P_1, \ldots, P_r$ are in general position, then the r points obtained by any finite sequence of admissible transformations are also in general position.

(c) Assume the ground field k is uncountable. Then given $P_1, \ldots, P_r$ in general position, there is a dense subset $V \subseteq \mathbf{P}^2$ such that for any $P_{r+1} \in V, P_1, \ldots, P_{r+1}$ will be in general position. [*Hint*: Prove a lemma that when k is uncountable, a variety cannot be equal to the union of a countable family of proper closed subsets.]

(d) Now take $P_1, \ldots, P_r \in \mathbf{P}^2$ in general position, and let X be the surface obtained by blowing up $P_1, \ldots, P_r$. If $r = 7$, show that X has exactly 56 irreducible nonsingular curves C with $g = 0$, $C^2 = -1$, and that these are the only irreducible curves with negative self-intersection. Ditto for $r = 8$, the number being 240.

*(e) For $r = 9$, show that the surface X defined in (d) has infinitely many irreducible nonsingular curves C with $g = 0$ and $C^2 = -1$. [*Hint*: Let L be the line joining P_1 and P_2. Show that there exist finite sequences of admissible transformations such that the strict transform of L becomes a plane curve of arbitrarily high degree.] This example is apparently due to Kodaira—see Nagata [5, II, p. 283].

4.16. For the *Fermat cubic surface* $x_0^3 + x_1^3 + x_2^3 + x_3^3 = 0$, find the equations of the 27 lines explicitly, and verify their incidence relations. What is the group of automorphisms of this surface?

5 Birational Transformations

Up to now we have dealt with one surface at a time, or a surface and its monoidal transforms. Now we will show that in fact any birational transformation of (nonsingular projective) surfaces can be factored into a finite sequence of monoidal transformations and their inverses. This confirms the central role played by the monoidal transformations in the study of surfaces. A consequence of this result is the fact that the arithmetic genus of a surface is a *birational* invariant.

In this section we will also prove Castelnuovo's criterion for contracting exceptional curves of the first kind, and deduce the existence of relatively minimal models for surfaces. See Shafarevich [1, Ch. I, II], Zariski [5], and Zariski [10, Ch. IV] as general references.

We begin by recalling some general facts about birational maps between varieties of any dimension, including Zariski's Main Theorem.

Let X and Y be projective varieties of any dimension. Recall (I, §4) that to give a birational transformation T from X to Y is to give an open subset $U \subseteq X$ and a morphism $\varphi: U \to Y$ which induces an isomorphism of function fields $K(Y) \cong K(X)$. If we have another open set $V \subseteq X$ and another morphism $\psi: V \to Y$ representing T, then φ and ψ agree where both are defined, so we can glue them to obtain a morphism defined on $U \cup V$ (I, Ex. 4.2). So there is a largest open set $U \subseteq X$ on which T is represented by a morphism $\varphi: U \to Y$. We say that T is defined at the points of U, and we call the points of $X - U$ fundamental points of T.

Now let $T: X \to Y$ be a birational transformation, represented by the morphism $\varphi: U \to Y$. Let $\Gamma_0 \subseteq U \times Y$ be the graph of φ, and let $\Gamma \subseteq X \times Y$ be the closure of Γ_0. We call Γ the *graph* of T. For any subset $Z \subseteq X$, we define $T(Z)$ to be $p_2(p_1^{-1}(Z))$, where p_1 and p_2 are the projections of Γ to X and Y. We call $T(Z)$ the *total transform* of Z. If T is defined at a point P, then $T(P)$ will be the point $\varphi(P)$. However, if P is a fundamental point of T, then in general $T(P)$ will consist of more than one point.

Lemma 5.1. *If $T: X \to Y$ is a birational transformation of projective varieties, and if X is normal, then the fundamental points of T form a closed subset of codimension ≥ 2.*

PROOF. If $P \in X$ is a point of codimension 1, then $\mathcal{O}_{P,X}$ is a discrete valuation ring. Since T is defined at the generic point of X, and Y is projective, hence proper, it follows from the valuative criterion of properness (II, 4.7) that T is also defined at P. (We have already used this argument in the proof of (II, 8.19), to show that the geometric genus is a birational invariant.)

Example 5.1.1. Let X be a surface, and let $\pi: \tilde{X} \to X$ be the monoidal transformation with center P. Then π is defined everywhere. Its inverse $\pi^{-1}: X \to \tilde{X}$ is a birational transformation having P as a fundamental point.

Theorem 5.2 (Zariski's Main Theorem). *Let $T: X \to Y$ be a birational transformation of projective varieties, and assume that X is normal. If P is a fundamental point of T, then the total transform $T(P)$ is connected and of dimension ≥ 1.*

PROOF. This is just a variant of the earlier form of Zariski's Main Theorem (III, 11.4). Let Γ be the graph of T, and consider the morphism $p_1: \Gamma \to X$. This is a birational projective morphism, so by (III, 11.4), $p_1^{-1}(P)$ is connected.

If it has dimension 0, then the same is true in a neighborhood V of P (II, Ex. 3.22). In that case $p_1^{-1}(V) \to V$ is a projective, birational morphism, with finite fibres, so it is a finite morphism (III, Ex. 11.2). But V is normal, so it must be an isomorphism. This says that T is defined at P, which is a contradiction. We conclude that $p_1^{-1}(P)$ is connected of dimension $\geqslant 1$. Since p_2 maps this set isomorphically onto $T(P)$, we have the result.

Now we come to the key result which enables us to factor birational transformations of surfaces.

Proposition 5.3. *Let* $f: X' \to X$ *be a birational morphism of (nonsingular, projective) surfaces. Let P be a fundamental point of f^{-1}. Then f factors through the monoidal transformation* $\pi: \tilde{X} \to X$ *with center P.*

PROOF. Let T be the birational transformation of X' to $\tilde{X}$ defined as $\pi^{-1} \circ f$. Our object is to show that T is a morphism. If not, then it has a fundamental (closed) point P'. Clearly $f(P') = P$. Furthermore $T(P')$ has dimension $\geqslant 1$ in $\tilde{X}$, by (5.2). Thus $T(P')$ must be the exceptional curve E of π.

On the other hand, by (5.1), T^{-1} is defined at all except finitely many points of $\tilde{X}$, so we can find a closed point $Q \in E$ where T^{-1} is defined, and hence $T^{-1}(Q) = P'$. We will show that this situation leads to a contradiction.

Choose local coordinates x, y at P on X. Then as in the proof of (3.6), there is an open neighborhood V of P such that $\pi^{-1}(V)$ is defined by the equation $ty - ux$ in $\mathbf{P}_V^1$. By a linear change of variables in x, y and t, u, we may assume that Q is the point $t = 0$, $u = 1$ in E. Then t, y form local coordinates at Q on $\tilde{X}$; the local equation of E is $y = 0$, and $x = ty$.

Since P is a fundamental point of f^{-1}, by (5.2) we have $f^{-1}(P)$ connected of dimension $\geqslant 1$, so there is an irreducible curve C in $f^{-1}(P)$ containing P'. Let $z = 0$ be a local equation for C at P'.

Since $f^{-1}(P)$ is defined by $x = y = 0$, the images of x, y in $\mathcal{O}_{P'}$ are in the ideal generated by z, so we can write $x = az$, $y = bz$, $a, b \in \mathcal{O}_{P'}$. On the other hand, $\mathcal{O}_Q$ dominates $\mathcal{O}_{P'}$. (We consider $\mathcal{O}_P, \mathcal{O}_{P'}, \mathcal{O}_Q$ all as subrings of the common function field K of $X, X', \tilde{X}$.) Since t, y are local coordinates at Q, $y \notin \mathfrak{m}_Q^2$, so we conclude that $y \notin \mathfrak{m}_{P'}^2$ in $\mathcal{O}_{P'}$. Therefore b is a unit in $\mathcal{O}_{P'}$, and so $t = x/y = a/b$ is in the local ring $\mathcal{O}_{P'}$. Since $t \in \mathfrak{m}_Q$, we must have $t \in \mathfrak{m}_{P'}$.

Now we use the fact that $T(P') = E$. This implies that for any $w \in \mathfrak{m}_{P'}$, the image of w in $\mathcal{O}_Q$ must be contained in the ideal generated by y, since y is the local equation for E. In particular, taking $w = t$, we find $t \in (y)$, which is a contradiction, since t and y are local coordinates at Q.

Corollary 5.4. *Let* $f: X' \to X$ *be a birational morphism of surfaces. Let $n(f)$ be the number of irreducible curves $C' \subseteq X'$ such that $f(C')$ is a point. Then $n(f)$ is finite and f can be factored into a composition of exactly $n(f)$ monoidal transformations.*

PROOF. If $f(C')$ is a point P, then P is a fundamental point of f^{-1}. By (5.1) the fundamental points of f^{-1} form a finite set, and for each one, its inverse image $f^{-1}(P)$ is a closed subset of X', having only finitely many irreducible components, so the set of curves C' which are mapped to a point is finite.

Now let P be a fundamental point of f^{-1}. Then by (5.3) f factors through the monoidal transformation $\pi: \tilde{X} \to X$ with center P, i.e., $f = \pi \circ f_1$ for some morphism $f_1: X' \to \tilde{X}$. We will show that $n(f_1) = n(f) - 1$. Indeed, if $f_1(C')$ is a point, then certainly $f(C')$ is a point. Conversely, if $f(C')$ is a point, then either $f_1(C')$ is a point, or $f_1(C') = E$, the exceptional curve of π. Furthermore, since f_1^{-1} is a morphism except at finitely many points, there is a unique irreducible curve E' in X' with $f_1(E') = E$. Thus $n(f_1) = n(f) - 1$.

Continuing in this fashion, after factoring through $n(f)$ monoidal transformations, we reduce to a morphism with $n(f) = 0$. But by (5.2), such a morphism has no fundamental points, so it is an isomorphism. Thus f is factored into $n(f)$ monoidal transformations.

Remark 5.4.1. It is interesting to compare the factorization of (5.3) with the universal property of blowing up proved in (II, 7.14). While the new result implies the old one in the special case of blowing up a point (since $f^{-1} m_P \cdot \mathcal{O}_{X'}$ being invertible implies $f^{-1}(P)$ has dimension 1, so P is a fundamental point), it is actually stronger, since it uses Zariski's Main Theorem. We cannot deduce (5.3) from (II, 7.14), because the hypothesis that $f^{-1} m_P \cdot \mathcal{O}_{X'}$ is invertible is impossible to verify in our case.

Remark 5.4.2. Comparing (5.4) to the earlier theorem (II, 7.17), we see that the new result is more precise, because it uses only monoidal transformations, rather than the more general concept of blowing up an arbitrary sheaf of ideals.

Remark 5.4.3. It is easy to see that (5.3) is false for nonsingular projective varieties of dimension $\geqslant 3$. For example, let $f: X' \to X$ be the blowing-up of a nonsingular curve C in a nonsingular projective 3-fold X. Then any point $P \in C$ is a fundamental point of f^{-1}, but f cannot factor through the monoidal transformation $\pi: \tilde{X} \to X$ with center P, because $f^{-1}(P)$ has dimension 1, while $\pi^{-1}(P)$ has dimension 2.

Remark 5.4.4. The example (5.4.3) suggests posing the following modified problem: given a birational morphism $f: X' \to X$ of nonsingular projective varieties, is it possible to factor f into a finite succession of monoidal transformations along nonsingular subvarieties? This is also false in dimension $\geqslant 3$: see Sally [1] and Shannon [1].

Theorem 5.5. *Let $T: X \to X'$ be a birational transformation of surfaces. Then it is possible to factor T into a finite sequence of monoidal transformations and their inverses.*

PROOF. Using (5.4), it will be sufficient to show that there is a surface X'', and birational morphisms $f:X'' \to X$ and $g:X'' \to X'$ such that $T = g \circ f^{-1}$. To construct X'', we proceed as follows.

Let H' be a very ample divisor on X', and let C' be an irreducible non-singular curve in the linear system $|2H'|$, which does not pass through any of the fundamental points of T^{-1}. In other words, C' is entirely contained in the largest open set $U' \subseteq X'$ on which T^{-1} is represented by a morphism $\varphi:U' \to X$. Let $C = \varphi(C')$ be the image of C' in X. We define an integer m by $m = p_a(C) - p_a(C')$. Since we have a finite birational morphism of C' to C, we see that $m \geqslant 0$, and $m = 0$ if and only if C' is isomorphic to C (IV, Ex. 1.8). Note also that if we replace C' by a linearly equivalent curve C_1', also missing the fundamental points of T^{-1}, then $C_1 = \varphi(C_1')$ is linearly equivalent to C. In fact, if $C' - C_1' = (f)$ for some rational function f on X', then $C - C_1 = (f)$ on X. Since the arithmetic genus of a curve depends only on its linear equivalence class (Ex. 1.3), we see that the integer m depends only on T and H', and not on the particular curve $C' \in |2H'|$ chosen.

Now fix C' temporarily. If $m > 0$, then C must be singular. Let P be a singular point of C, let $\pi:\tilde{X} \to X$ be the monoidal transformation with center P, and let $\tilde{C}$ be the strict transform of C. Then by (3.7), $p_a(\tilde{C}) < p_a(C)$. Thus if $\tilde{T} = T \circ \pi$, then $m(\tilde{T}) < m(T)$.

Continuing in this fashion, as in the proof of (3.8), we see that there is a morphism $f:X'' \to X$, obtained by a finite number of monoidal transformations, such that if $T' = T \circ f$, then $m(T') = 0$.

We will show that in fact T' is a morphism. If not, then T' will have a fundamental point P. By (5.2), $T'(P)$ contains an irreducible curve $E' \subseteq X'$. Since H' is very ample, $E'.H' > 0$, and so $C'.E' \geqslant 2$ for any $C' \in |2H'|$. Let us choose C', not containing any fundamental point of T'^{-1}, such that C' meets E' transversally (1.2). Then C' meets E' in at least two distinct points, so the corresponding curve C in X'' has at least a double point at P. But this contradicts $m(T') = 0$.

We conclude that T' is a morphism of X'' to X', so that applying (5.4) completes the proof, as mentioned above.

Corollary 5.6. *The arithmetic genus of a nonsingular projective surface is a birational invariant.*

PROOF. Indeed, p_a is unchanged by a monoidal transformation (3.5), so this follows directly from the theorem.

Remark 5.6.1. Even though the factorization theorem (5.5) in a form analogous to (5.4.4) is false in dimension $\geqslant 3$, Hironaka [3] is able to deduce the birational invariance of p_a, for nonsingular projective varieties over a field of characteristic 0, from the following statement, which is a consequence of his resolution of singularities: If $T:X \to X'$ is any birational transformation of nonsingular projective varieties over a field of characteristic 0, then there is a morphism $f:X'' \to X$, obtained by a finite succession of monoidal transformations along nonsingular subvarieties, such that the birational map $T' = T \circ f$ is a morphism. There is another proof of the birational invariance of p_a, for varieties over **C**, by Kodaira and Spencer [1], using the equalities $h^{0q} = h^{q0}$ from Hodge theory, and the birational invariance of h^{q0} (II, Ex. 8.8).

Now we come to Castelnuovo's criterion for contracting a curve on a surface. We have seen that if E is the exceptional curve of a monoidal transformation, then $E \cong \mathbf{P}^1$, and $E^2 = -1$ (3.1). In general, any curve Y on a surface X, with $Y \cong \mathbf{P}^1$ and $Y^2 = -1$ is classically called an *exceptional curve of the first kind*. The following theorem tells us that any exceptional curve of the first kind is the exceptional curve of some monoidal transformation.

Theorem 5.7 (Castelnuovo). *If Y is a curve on a surface X, with $Y \cong \mathbf{P}^1$ and $Y^2 = -1$, then there exists a morphism $f: X \to X_0$ to a (nonsingular projective) surface X_0, and a point $P \in X_0$, such that X is isomorphic via f to the monoidal transformation of X_0 with center P, and Y is the exceptional curve.*

PROOF. We will construct X_0 using the image of X under a suitable morphism to a projective space. Choose a very ample divisor H on X such that $H^1(X, \mathcal{L}(H)) = 0$: for example, a sufficiently high multiple of any given very ample divisor (III, 5.2). Let $k = H.Y$, and let us assume $k \geqslant 2$. Then we will use the invertible sheaf $\mathcal{M} = \mathcal{L}(H + kY)$ to define a morphism of X to $\mathbf{P}^N$.

Step 1. First we prove that $H^1(X, \mathcal{L}(H + (k-1)Y)) = 0$. In fact, we will prove more generally that for every $i = 0, 1, \ldots, k$, we have $H^1(X, \mathcal{L}(H + iY)) = 0$. For $i = 0$, this is true by hypothesis, so we proceed by induction on i. Suppose it is true for $i - 1$. We consider the exact sequence of sheaves

$$0 \to \mathcal{L}(H + (i-1)Y) \to \mathcal{L}(H + iY) \to \mathcal{O}_Y \otimes \mathcal{L}(H + iY) \to 0.$$

Now $Y \cong \mathbf{P}^1$, and $(H + iY).Y = k - i$, so

$$\mathcal{O}_Y \otimes \mathcal{L}(H + iY) \cong \mathcal{O}_{\mathbf{P}^1}(k - i).$$

We get an exact cohomology sequence

$$\ldots \to H^1(X, \mathcal{L}(H + (i-1)Y)) \to H^1(X, \mathcal{L}(H + iY)) \to$$
$$\to H^1(\mathbf{P}^1, \mathcal{O}_{\mathbf{P}^1}(k - i)) \to \ldots .$$

So from the induction hypothesis, and the known cohomology of $\mathbf{P}^1$, we conclude that $H^1(X, \mathcal{L}(H + iY)) = 0$ for any $i \leqslant k$.

Step 2. Next we show that $\mathcal{M}$ is generated by global sections. Since H is very ample, the corresponding linear system $|H + kY|$ has no base points away from Y, so $\mathcal{M}$ is generated by global sections off Y. On the other hand, the natural map

$$H^0(X, \mathcal{M}) \to H^0(Y, \mathcal{M} \otimes \mathcal{O}_Y)$$

is surjective, because $\mathcal{M} \otimes \mathcal{I}_Y \cong \mathcal{L}(H + (k - 1)Y)$, and

$$H^1(X, \mathcal{L}(H + (k - 1)Y)) = 0$$

by Step 1. Next observe that $(H + kY).Y = 0$, so $\mathcal{M} \otimes \mathcal{O}_Y \cong \mathcal{O}_{\mathbf{P}^1}$, which is generated by the global section 1. Lifting this section to $H^0(X, \mathcal{M})$, and using Nakayama's lemma, we see that $\mathcal{M}$ is generated by global sections also at every point of Y.

Step 3. Therefore $\mathcal{M}$ determines a morphism $f_1 : X \to \mathbf{P}^N$ (II, 7.1). Let X_1 be its image. Since $f_1^* \mathcal{O}(1) \cong \mathcal{M}$, and since the degree of $\mathcal{M} \otimes \mathcal{O}_Y$ is 0, f_1 must map Y to a point P_1. On the other hand, since H is very ample, the linear system $|H + kY|$ separates points and tangent vectors away from Y, and also separates points of Y from points not on Y, so f_1 is an isomorphism of $X - Y$ onto $X_1 - P_1$ (II, 7.8.2).

Step 4. Let X_0 be the normalization of X_1 (II, Ex. 3.8). Since X is nonsingular, hence normal, the map f_1 factors to give a morphism $f : X \to X_0$. Since Y is irreducible, $f(Y)$ is a point P, and since $X_1 - P_1$ was nonsingular, we still have $f : X - Y \to X_0 - P$ an isomorphism.

Step 5. Now we will show that X_0 is nonsingular at the point P. Since in any case X_0 is normal, and f is birational, we have $f_* \mathcal{O}_X \cong \mathcal{O}_{X_0}$ (see proof of (III, 11.4)). So we can apply the theorem on formal functions (III, 11.1) to conclude that

$$\hat{\mathcal{O}}_P \cong \varprojlim H^0(Y_n, \mathcal{O}_{Y_n}),$$

where Y_n is the closed subscheme of X defined by $\mathfrak{m}_P^n \cdot \mathcal{O}_X$. But since $f^{-1}(P) = Y$, the sequence of ideals $\mathfrak{m}_P^n \mathcal{O}_X$ is cofinal with the sequence of ideals $\mathcal{I}_Y^n$, so we may use these instead in the definition of Y_n (II, 9.3.1).

We will show for each n that $H^0(Y_n, \mathcal{O}_{Y_n})$ is isomorphic to a truncated power series ring $A_n = k[[x, y]]/(x, y)^n$. It will follow that $\hat{\mathcal{O}}_P \cong \varprojlim A_n \cong k[[x, y]]$, which is a regular local ring. This in turn implies that $\mathcal{O}_P$ is regular (I, 5.4A), hence P is a nonsingular point.

For $n = 1$, we have $H^0(Y, \mathcal{O}_Y) = k$. For $n > 1$, we use the exact sequences

$$0 \to \mathcal{I}_Y^n / \mathcal{I}_Y^{n+1} \to \mathcal{O}_{Y_{n+1}} \to \mathcal{O}_{Y_n} \to 0.$$

Since $Y \cong \mathbf{P}^1$, and $Y^2 = -1$, we have $\mathcal{I}/\mathcal{I}^2 \cong \mathcal{O}_{\mathbf{P}^1}(1)$ by (1.4.1), and $\mathcal{I}^n/\mathcal{I}^{n+1} \cong \mathcal{O}_{\mathbf{P}^1}(n)$ for each n, as in the proof of (3.4). Taking cohomology, we have

$$0 \to H^0(\mathcal{O}_{\mathbf{P}^1}(n)) \to H^0(\mathcal{O}_{Y_{n+1}}) \to H^0(\mathcal{O}_{Y_n}) \to 0.$$

For $n = 1$, $H^0(\mathcal{O}_{\mathbf{P}^1}(1))$ is a 2-dimensional vector space. Take a basis x, y. Then $H^0(\mathcal{O}_{Y_2})$, which in any case contains k, is seen to be isomorphic to A_2.

Now inductively, if $H^0(\mathcal{O}_{Y_n})$ is isomorphic to A_n, lift the elements x, y to $H^0(\mathcal{O}_{Y_{n+1}})$. Since $H^0(\mathcal{O}_{\mathbf{P}^1}(n))$ is the vector space with basis $x^n, x^{n-1}y, \ldots, y^n$, we see that $H^0(\mathcal{O}_{Y_{n+1}}) \cong A_{n+1}$. Now as above we see that P is a nonsingular point.

Step 6. We complete the proof using the factorization theorem (5.4). Since X_0 is nonsingular, we can apply (5.4) to $f: X \to X_0$. We have $n(f) = 1$ by construction, so f must be the monoidal transformation with center P.

Step 7. As an addendum, we show in fact that $X_0 = X_1$, so the normalization was unnecessary. The natural map

$$H^0(X, \mathcal{M} \otimes \mathcal{I}_Y) \to H^0(Y, \mathcal{M} \otimes \mathcal{I}_Y / \mathcal{I}_Y^2)$$

is surjective, because the next term of the cohomology sequence is $H^1(X, \mathcal{L}(H + (k-2)Y))$, which is 0 by Step 1. Since $\mathcal{M} \otimes \mathcal{O}_Y \cong \mathcal{O}_Y$, this shows that there are global sections $s, t \in H^0(X, \mathcal{M} \otimes \mathcal{I}_Y) \subseteq H^0(X, \mathcal{M})$, which map to the parameters $x, y \in H^0(\mathcal{O}_{Y_2}) \cong A_2$. On the other hand, these sections s, t become sections of $\mathcal{O}(1)$ on $\mathbf{P}^N$, defining hyperplanes containing P_1. So they give elements $\bar{s}, \bar{t} \in \mathfrak{m}_{P_1}$, whose images in $\mathcal{O}_P$ generate the maximal ideal $\mathfrak{m}_P$. Since in any case $\mathcal{O}_P$ is a finitely generated $\mathcal{O}_{P_1}$-module, we conclude that $\mathcal{O}_P \cong \mathcal{O}_{P_1}$ (II, 7.4), and so $X_0 \cong X_1$.

Example 5.7.1. Let $\pi: X \to C$ be a geometrically ruled surface (§2), let P be a point of X, and let L be the fibre of π containing P. Let $f: \tilde{X} \to X$ be the monoidal transformation with center P. Then the strict transform $\tilde{L}$ of L on $\tilde{X}$ is isomorphic to $\mathbf{P}^1$, and has $\tilde{L}^2 = -1$. Indeed, $L^2 = 0$ by (2.3), and P is a nonsingular point of L, so $\tilde{L} \sim f^*L - E$ by (3.6). It follows that $\tilde{L}^2 = -1$. Therefore by the theorem, we can blow down $\tilde{L}$. In other words, there is a morphism $g: \tilde{X} \to X'$ sending $\tilde{L}$ to a point Q, and such that g is the monoidal transformation with center Q. If $M = g(E)$, then $M \cong \mathbf{P}^1$ and $M^2 = 0$, for a similar reason to the above. Note also that the rational map $\pi': X' \to C$ obtained from π on $X - L \cong X' - M$ is in fact a morphism. Therefore $\pi': X' \to C$ is another geometrically ruled surface. Indeed, the fibres of π' are all isomorphic to $\mathbf{P}^1$, and since π has a section, its strict transform on X' will be a section of π'. This new ruled surface is called the *elementary transform of X with center P*, and is denoted by $\mathrm{elm}_P X$ (Fig. 23). See (Ex. 5.5) for some applications.

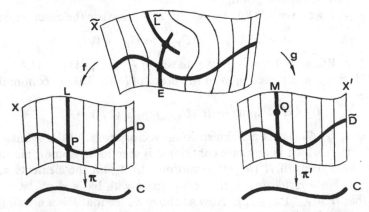

Figure 23. An elementary transformation of a ruled surface.

Remark 5.7.2 (The General Contraction Problem). With the theorem of Castelnuovo (5.7) in mind, one can pose the following general problem: Given a variety X, and a closed subset $Y \subseteq X$, find necessary and sufficient conditions for the existence of a birational morphism $f: X \to X_0$ such that $f(Y)$ is a single point P, and $f: X - Y \to X_0 - P$ is an isomorphism. If such a morphism exists, we say that Y is *contractible*. A number of special cases of this problem have been treated, but a general solution is unknown. See Artin [3], [4], Grauert [2], and Mumford [6].

In case Y is an irreducible curve on a surface X, here is what is known. If we require X_0 to be nonsingular, then by (5.7) the necessary and sufficient conditions are that $Y \cong \mathbf{P}^1$ and $Y^2 = -1$. If we allow X_0 to be singular, then a necessary condition is that $Y^2 < 0$ (Ex. 5.7). If $Y \cong \mathbf{P}^1$, this condition is also sufficient (Ex. 5.2). If Y is arbitrary, with $Y^2 < 0$, and if the base field is $\mathbf{C}$, then a theorem of Grauert [2] shows that X_0 exists as a complex analytic space. However, Y may not be contractible to an algebraic variety, as we show in the following example.

Example 5.7.3 (Hironaka). Let Y_0 be a nonsingular cubic curve in $\mathbf{P}^2$ over an uncountable algebraically closed field k (e.g., $k = \mathbf{C}$). Fix an inflection point $P_0 \in Y_0$ to be the origin of the group law on Y_0. Since the abelian group Y_0 is uncountable, and since the torsion points are countable (IV, 4.8.1), the torsion-free part must have infinite rank. Therefore we can choose 10 points $P_1, \ldots, P_{10} \in Y_0$ which are linearly independent over $\mathbf{Z}$ in the group law.

Now blow up $P_1, \ldots, P_{10}$ in $\mathbf{P}^2$, let X be the resulting surface, and let Y be the strict transform of Y_0. Since $Y_0^2 = 9$, and we have blown up 10 points on Y, we have $Y^2 = -1$, using (3.6). So by Grauert's theorem (5.7.2), if $k = \mathbf{C}$, then Y in X would be contractible to a complex analytic space. We will show, however, that Y is not contractible to a point P in an algebraic variety X_0. If it were, let $P \in U \subseteq X_0$ be an open affine neighborhood of P. Let $C_0' \subseteq U$ be a curve not containing P. Let $C_0 \subseteq X_0$ be its closure, which still does not contain P. Then its inverse image in X will be a curve $C \subseteq X$ which does not meet Y. The image of C in $\mathbf{P}^2$ will be a curve C^* which does not meet Y_0 except at the points $P_1, \ldots, P_{10}$.

But this is impossible. Let $d = \deg C^*$. Then by Bézout's theorem (1.4.2), $C^*.Y_0 = 3d > 0$. So we can write

$$C^*.Y_0 = \sum_{i=1}^{10} n_i P_i$$

on Y_0, with $n_i \geqslant 0$, $\sum n_i = 3d$. But $C^* \sim dL$, where L is a line in $\mathbf{P}^2$, and $L.Y_0 \sim 3P_0$, so we have

$$\sum_{i=1}^{10} n_i P_i = 0$$

in the group law on Y_0 (IV, 1.3.7). This contradicts the fact that $P_1, \ldots, P_{10}$ were chosen to be linearly independent over $\mathbf{Z}$.

417

To conclude this section, we will prove the existence of relatively minimal models of surfaces. The idea is to find, within each birational equivalence class of surfaces, one which is as canonical as possible. Since one can always blow up a point, there is never a unique nonsingular projective model of a function field, as in the case of curves. However, we can look for one which is minimal for the relation of domination. So we say that a (nonsingular projective) surface X is a *relatively minimal model* of its function field, if every birational morphism $f: X \rightarrow X'$ to another (nonsingular projective) surface X' is necessarily an isomorphism. If X is the unique relatively minimal model in its birational equivalence class, then we say that X is a *minimal model*. (This somewhat irregular use of the word "minimal" is retained for historical reasons.)

Theorem 5.8. *Every surface admits a birational morphism to a relatively minimal model.*

PROOF. Combining (5.4) and (5.7), it is clear that a surface is a relatively minimal model if and only if it contains no exceptional curves of the first kind. So given a surface X, if it is already a relatively minimal model, stop. If not, let Y be an exceptional curve of the first kind. By (5.7) there is a morphism $X \rightarrow X_1$ contracting Y.

We continue in this manner, contracting exceptional curves of the first kind whenever one exists, and so we obtain a sequence of birational morphisms $X \rightarrow X_1 \rightarrow X_2 \rightarrow \ldots$. We must show that this process eventually stops.

The following proof is due to Matsumura [1]. Suppose that we have a sequence of n contractions

$$X = X_0 \rightarrow X_1 \ldots \rightarrow X_n$$

as above. For each $i = 1, \ldots, n$, let $E_i' \subseteq X_{i-1}$ be the exceptional curve of the contraction $X_{i-1} \rightarrow X_i$, and let E_i be its total transform on X. Then by (3.2) we have $E_i^2 = -1$ for each i, and $E_i.E_j = 0$ for $i \neq j$.

Now for each i, let $e_i = c(E_i)$ be the cohomology class of E_i in $H^1(X, \Omega)$ (Ex. 1.8). Then we have $\langle e_i, e_i \rangle = -1$ and $\langle e_i, e_j \rangle = 0$ in the intersection pairing on $H^1(X, \Omega)$, by (Ex. 1.8). It follows that $e_1, \ldots, e_n$ are linearly independent elements of the vector space $H^1(X, \Omega)$ over k.

We conclude that $n \leqslant \dim_k H^1(X, \Omega)$. Since this is a finite-dimensional vector space, n is bounded, so the contraction process must terminate.

Note: One can give another proof of this result by showing that the rank of the Néron–Severi group drops by 1 with each contraction, so that $n \leqslant$ rank $NS(X)$, which is finite—Cf. (Ex. 1.7).

Remark 5.8.1. In spite of this result, it is *not* true that a surface necessarily has only finitely many exceptional curves of the first kind. For example, if we blow up r points in general position in $\mathbf{P}^2$, with $r \geqslant 9$, then the resulting surface has infinitely many exceptional curves of the first kind (Ex. 4.15).

Example 5.8.2. In the birational equivalence class of rational surfaces, $\mathbf{P}^2$ is a relatively minimal model, and so is the rational ruled surface X_e for each $e \geqslant 0$, $e \neq 1$. This follows easily from the determination of all irreducible curves on X_e (2.18). On the contrary, X_1 is not relatively minimal (2.11.5).

Example 5.8.3. In the class of surfaces birational to $\mathbf{P}^1 \times C$, where C is a curve of genus $g > 0$, every geometrically ruled surface $\pi: X \to C$ is relatively minimal. Indeed, if Y is any rational curve in X, then $\pi(Y)$ is a point because of (IV, 2.5.4). Thus Y is a fibre of π, $Y^2 = 0$, and so we see that X has no exceptional curves of the first kind.

Remark 5.8.4. A classical theorem, proved in all characteristics by Zariski [5], [6], [9] states that except for the rational and ruled surfaces, every surface is birational to a (unique) minimal model. One can also show that in the case of rational and ruled surfaces, every relatively minimal model is one of those listed in (5.8.2) and (5.8.3). See Nagata [5] or Hartshorne [4].

EXERCISES

5.1. Let f be a rational function on the surface X. Show that it is possible to "resolve the singularities of f" in the following sense: there is a birational morphism $g: X' \to X$ so that f induces a morphism of X' to $\mathbf{P}^1$. [*Hints:* Write the divisor of f as $(f) = \sum n_i C_i$. Then apply embedded resolution (3.9) to the curve $Y = \bigcup C_i$. Then blow up further as necessary whenever a curve of zeros meets a curve of poles until the zeros and poles of f are disjoint.]

5.2. Let $Y \cong \mathbf{P}^1$ be a curve in a surface X, with $Y^2 < 0$. Show that Y is contractible (5.7.2) to a point on a projective variety X_0 (in general singular).

5.3. If $\pi: \tilde{X} \to X$ is a monoidal transformation with center P, show that $H^1(\tilde{X}, \Omega_{\tilde{X}}) \cong H^1(X, \Omega_X) \oplus k$. This gives another proof of (5.8). [*Hints:* Use the projection formula (III, Ex. 8.3) and (III, Ex. 8.1) to show that $H^i(X, \Omega_X) \cong H^i(\tilde{X}, \pi^*\Omega_X)$ for each i. Next use the exact sequence

$$0 \to \pi^*\Omega_X \to \Omega_{\tilde{X}} \to \Omega_{\tilde{X}/X} \to 0$$

and a local calculation with coordinates to show that there is a natural isomorphism $\Omega_{\tilde{X}/X} \cong \Omega_E$, where E is the exceptional curve. Now use the cohomology sequence of the above sequence (you will need every term) and Serre duality to get the result.]

5.4. Let $f: X \to X'$ be a birational morphism of nonsingular surfaces.
(a) If $Y \subseteq X$ is an irreducible curve such that $f(Y)$ is a point, then $Y \cong \mathbf{P}^1$ and $Y^2 < 0$.
(b) (Mumford [6].) Let $P' \in X'$ be a fundamental point of f^{-1}, and let $Y_1, \ldots, Y_r$ be the irreducible components of $f^{-1}(P')$. Show that the matrix $\| Y_i . Y_j \|$ is negative definite.

5.5. Let C be a curve, and let $\pi: X \to C$ and $\pi': X' \to C$ be two geometrically ruled surfaces over C. Show that there is a finite sequence of elementary transformations (5.7.1) which transform X into X'. [*Hints:* First show if $D \subseteq X$ is a section of π containing a point P, and if $\tilde{D}$ is the strict transform of D by elm_P, then $\tilde{D}^2 = D^2 - 1$

419

(Fig. 23). Next show that X can be transformed into a geometrically ruled surface X'' with invariant $e \gg 0$. Then use (2.12), and study how the ruled surface $\mathbf{P}(\mathcal{E})$ with $\mathcal{E}$ decomposable behaves under elm_P.]

5.6. Let X be a surface with function field K. Show that every valuation ring R of K/k is one of the three kinds described in (II, Ex. 4.12). [Hint: In case (3), let $f \in R$. Use (Ex. 5.1) to show that for all $i \gg 0$, $f \in \mathcal{O}_{X_i}$, so in fact $f \in R_0$.]

5.7. Let Y be an irreducible curve on a surface X, and suppose there is a morphism $f : X \to X_0$ to a projective variety X_0 of dimension 2, such that $f(Y)$ is a point P and $f^{-1}(P) = Y$. Then show that $Y^2 < 0$. [Hint: Let $|H|$ be a very ample (Cartier) divisor class on X_0, let $H_0 \in |H|$ be a divisor containing P, and let $H_1 \in |H|$ be a divisor not containing P. Then consider f^*H_0, f^*H_1 and $\tilde{H}_0 = f^*(H_0 - P)^-$.]

5.8. *A surface singularity.* Let k be an algebraically closed field, and let X be the surface in $\mathbf{A}_k^3$ defined by the equation $x^2 + y^3 + z^5 = 0$. It has an isolated singularity at the origin $P = (0,0,0)$.

 (a) Show that the affine ring $A = k[x,y,z]/(x^2 + y^3 + z^5)$ of X is a unique factorization domain, as follows. Let $t = z^{-1}$; $u = t^3x$, and $v = t^2y$. Show that z is irreducible in A; $t \in k[u,v]$, and $A[z^{-1}] = k[u,v,t^{-1}]$. Conclude that A is a UFD.

 (b) Show that the singularity at P can be resolved by eight successive blowings-up. If $\tilde{X}$ is the resulting nonsingular surface, then the inverse image of P is a union of eight projective lines, which intersect each other according to the Dynkin diagram $\mathbf{E}_8$:

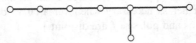

Here each circle denotes a line, and two circles are joined by a line segment whenever the corresponding lines intersect.

Note. This singularity has interesting connections with local algebra, invariant theory, and topology.

In case $k = \mathbf{C}$, Mumford [6] showed that the completion $\hat{A}$ of the ring A at the maximal ideal $\mathfrak{m} = (x, y, z)$ is also a UFD. This is remarkable, because in general the completion of a local UFD need not be UFD, although the converse is true (theorem of Mori)—see Samuel [3]. Brieskorn [2] showed that the corresponding analytic local ring $\mathbf{C}\{x,y,z\}/(x^2 + y^3 + z^5)$ is the *only* nonregular normal 2-dimensional analytic local ring which is a UFD. Lipman [2] generalized this as follows: over any algebraically closed field k of characteristic $\neq 2,3,5$, the only nonregular normal complete 2-dimensional local ring which is a UFD is $k[[x,y,z]]/(x^2 + y^3 + z^5)$. See also Lipman [3] for a report on recent work connected with UFD's.

This singularity arose classically out of Klein's work on the icosahedron. The group I of rotations of the icosahedron, which is isomorphic to the simple group of order 60, acts naturally on the 2-sphere. Identifying the 2-sphere with $\mathbf{P}_{\mathbf{C}}^1$ by stereographic projection, the group I appears as a finite subgroup of Aut $\mathbf{P}_{\mathbf{C}}^1$. This action lifts to give an action of the binary icosahedral group $\bar{I}$ on $\mathbf{C}^2$ by linear transformations of the complex variables t_1 and t_2. Klein [2, I, 2, §13, p.62] found three invariant polynomials x, y, z in t_1 and t_2, related by the equation $x^2 + y^3 + z^5 = 0$. Thus the surface X appears as the quotient of $\mathbf{A}_{\mathbf{C}}^2$ by the action of the group $\bar{I}$. In particular, the local fundamental group of X at P is just $\bar{I}$.

With regard to the topology of algebraic varieties over **C**, Mumford [6] showed that a normal algebraic surface over **C**, whose underlying topological space (in its "usual" topology) is a topological manifold, must be nonsingular. Brieskorn showed that this is not so in higher dimensions. For example, the underlying topological space of the hypersurface in $\mathbf{C}^4$ defined by $x_1^2 + x_2^2 + x_3^2 + x_4^3 = 0$ is a manifold. Later Brieskorn [1] showed that if one intersects such a singularity with a small sphere around the singular point, then one may get a topological sphere whose differentiable structure is not the standard one. Thus for example, by intersecting the singularity

$$x_1^2 + x_2^2 + x_3^2 + x_4^3 + x_5^{6k-1} = 0$$

in $\mathbf{C}^5$ with a small sphere around the origin, for $k = 1,2,\ldots,28$, one obtains all 28 possible differentiable structures on the 7-sphere. See Hirzebruch and Mayer [1] for an account of this work.

6 Classification of Surfaces

In the case of curves, we could achieve a classification as follows. Each birational equivalence class has a unique nonsingular projective model. There is a numerical invariant, the genus g, which can take on every value $g \geqslant 0$. For fixed g, the curves of genus g are parametrized by the points of the variety of moduli $\mathfrak{M}_g$ (IV, §5).

For surfaces, the situation is much more complicated. First of all, the nonsingular projective model is not unique. However, we can standardize by always considering a relatively minimal model. For rational and ruled surfaces, these are known, and for other birational classes there is a unique minimal model (5.8.4).

Next, we have the birational invariants p_a (5.6) and p_g (II, 8.19), and K^2, which is well-defined if we specify the minimal model. However, it is not known exactly which triples of integers can occur as p_a, p_g, K^2 of a surface. As to the existence of varieties of moduli, this question is wide open except in some special cases. So we must settle for less complete information than in the case of curves.

In this section we will mention very briefly a few basic results, and refer to Bombieri and Husemoller [1] and Shafarevich [1] for more details and further references.

To begin with, for any projective variety X over k, we define the *Kodaira dimension* $\kappa(X)$ to be the transcendence degree over k of the ring

$$R = \bigoplus_{n \geqslant 0} H^0(X, \mathscr{L}(nK)),$$

minus 1, where K is the canonical divisor. One sees, as in the proof of (II, 8.19), that R and hence κ are birational invariants. Another way of expressing this is that κ is the largest dimension of the image of X in $\mathbf{P}^N$ by the rational map determined by the linear system $|nK|$, for some $n \geqslant 1$, or

$\kappa = -1$ if $|nK| = \emptyset$ for all $n \geqslant 1$. It is known, for varieties of dimension n, that κ can take on every value from -1 to n. For example, for curves, we have $\kappa = -1 \Leftrightarrow g = 0; \kappa = 0 \Leftrightarrow g = 1; \kappa = 1 \Leftrightarrow g \geqslant 2$.

We will classify surfaces according to $\kappa = -1,0,1,2$. Some more specific information about each group is provided by the following theorems.

Theorem 6.1. $\kappa = -1 \Leftrightarrow |12K| = \emptyset \Leftrightarrow X$ *is either rational or ruled.*

Theorem 6.2 (Castelnuovo). *X is rational* $\Leftrightarrow p_a = P_2 = 0$, *where* $P_2 = \dim H^0(X, \mathcal{L}(2K))$ *is the* second plurigenus.

PROOF. A modern proof over **C**, due to Kodaira, is given in Serre [13]. In characteristic $p > 0$, the proof is due to Zariski [5], [6], [9].

Remark 6.2.1. As a consequence of (6.2), one can prove the analogue of Lüroth's theorem (IV, 2.5.5) in dimension 2: let k be an algebraically closed field, let L be a subfield of a pure transcendental extension $k(t,u)$ of k, containing k, such that $k(t,u)$ is a finite *separable* extension of L. Then L is also a pure transcendental extension of k. This is Castelnuovo's theorem "on the rationality of plane involutions."

For the proof, let X' be a nonsingular projective model of L, and let X be a nonsingular projective model of $k(t,u)$. Then as in (II, 8.19) or (II, Ex. 8.8), using separability, one shows that $p_g(X') \leqslant p_g(X)$ and $P_2(X') \leqslant P_2(X)$, hence $p_g(X') = P_2(X') = 0$ since the same is true of X. One must also show that $q(X') \leqslant q(X)$ to conclude that $p_a(X') = p_g(X') - q(X') = 0$. Then the rationality of X follows from (6.2). See Serre [13] and Zariski [9].

This result is false if one does not assume $k(t,u)$ separable over L—see Zariski [9] or Shioda [1].

Theorem 6.3. $\kappa = 0 \Leftrightarrow 12K = 0$. *A surface in this class must be one of the following (assume* char $k \neq 2,3$):

(1) *a K3 surface, which is defined as a surface with* $K = 0$ *and irregularity* $q = 0$. *These have* $p_a = p_g = 1$;
(2) *an Enriques surface, which has* $p_a = p_g = 0$ *and* $2K = 0$;
(3) *a two-dimensional* abelian variety, *which has* $p_a = -1$, $p_g = 1$; *or*
(4) *a hyperelliptic surface, which is a surface fibred over* $\mathbf{P}^1$ *by a pencil of elliptic curves.*

Theorem 6.4. *A surface with* $\kappa = 1$ *is an* elliptic surface, *which is a surface X with a morphism* $\pi: X \to C$ *to a curve C, such that almost all fibres of* π *are nonsingular elliptic curves (assume* char $k \neq 2,3$).

Theorem 6.5. $\kappa = 2$ *if and only if for some* $n > 0$, $|nK|$ *determines a birational morphism of X onto its image in* $\mathbf{P}^N$. *These are called* surfaces of general type.

EXERCISES

6.1. Let X be a surface in $\mathbf{P}^n$, $n \geqslant 3$, defined as the complete intersection of hypersurfaces of degrees $d_1, \ldots, d_{n-2}$, with each $d_i \geqslant 2$. Show that for all but finitely many choices of $(n, d_1, \ldots, d_{n-2})$, the surface X is of general type. List the exceptional cases, and where they fit into the classification picture.

6.2. Prove the following theorem of Chern and Griffiths. Let X be a nonsingular surface of degree d in $\mathbf{P}_\mathbf{C}^{n+1}$, which is not contained in any hyperplane. If $d < 2n$, then $p_g(X) = 0$. If $d = 2n$, then either $p_g(X) = 0$, or $p_g(X) = 1$ and X is a K3 surface. [*Hint*: Cut X with a hyperplane and use Clifford's theorem (IV, 5.4). For the last statement, use the Riemann–Roch theorem on X and the Kodaira vanishing theorem (III, 7.15).]

Intersection Theory

In this appendix we will outline the generalization of intersection theory and the Riemann–Roch theorem to nonsingular projective varieties of any dimension. To motivate the discussion, let us look at the case of curves and surfaces, and then see what needs to be generalized. For a divisor D on a curve X, leaving out the contribution of Serre duality, we can write the Riemann–Roch theorem (IV, 1.3) as

$$\chi(\mathscr{L}(D)) = \deg D + 1 - g,$$

where χ is the Euler characteristic (III, Ex. 5.1). On a surface, we can write the Riemann–Roch theorem (V, 1.6) as

$$\chi(\mathscr{L}(D)) = \frac{1}{2} D.(D - K) + 1 + p_a.$$

In each case, on the left-hand side we have something involving cohomology groups of the sheaf $\mathscr{L}(D)$, while on the right-hand side we have some numerical data involving the divisor D, the canonical divisor K, and some invariants of the variety X. Of course the ultimate aim of a Riemann–Roch type theorem is to compute the dimension of the linear system $|D|$ or of $|nD|$ for large n (II, Ex. 7.6). This is achieved by combining a formula for $\chi(\mathscr{L}(D))$ with some vanishing theorems for $H^i(X, \mathscr{L}(D))$ for $i > 0$, such as the theorems of Serre (III, 5.2) or Kodaira (III, 7.15).

We will now generalize these results so as to give an expression for $\chi(\mathscr{L}(D))$ on a nonsingular projective variety X of any dimension. And while we are at it, with no extra effort we get a formula for $\chi(\mathscr{E})$, where $\mathscr{E}$ is any coherent locally free sheaf.

To generalize the right-hand side, we need an intersection theory on X. The intersection of two divisors, for example, will not be a number, but a *cycle* of codimension 2, which is a linear combination of subvarieties of codimension 2. So we will introduce the language of cycles and rational equivalence (which generalizes the linear equivalence of divisors), in order to set up our intersection theory.

We also need to generalize the correspondence between the invertible sheaf $\mathscr{L}(D)$ and the divisor D. This is accomplished by the theory of Chern classes: to each locally free sheaf $\mathscr{E}$ of rank r, we associate Chern classes $c_1(\mathscr{E}), \ldots, c_r(\mathscr{E})$, where $c_i(\mathscr{E})$ is a cycle of codimension i, defined up to rational equivalence.

As for invariants of the variety X, the canonical class K and the arithmetic genus p_a are not enough in general, so we use all the Chern classes of the tangent sheaf of X as well.

Then the generalized Riemann–Roch theorem will give a formula for $\chi(\mathscr{E})$ in terms of certain intersection numbers of the Chern classes of $\mathscr{E}$ and of the tangent sheaf of X.

1 Intersection Theory

The intersection theory on a surface (V, 1.1) can be summarized by saying that there is a unique symmetric bilinear pairing $\operatorname{Pic} X \times \operatorname{Pic} X \to \mathbf{Z}$, which is *normalized* by requiring that for any two irreducible nonsingular curves C, D meeting transversally, $C.D$ is just the number of intersection points of C and D. Our main tool in proving this theorem was Bertini's theorem, which allowed us to move any two divisors in their linear equivalence class, so that they became differences of irreducible nonsingular curves meeting transversally.

In higher dimensions, the situation is considerably more complicated. The corresponding moving lemma is weaker, so we need a stronger normalization requirement. It turns out that the most convenient way to develop intersection theory is to do it for all varieties at once, and include some functorial mappings f_* and f^* associated to a morphism $f : X \to X'$ as part of the structure.

Let X be any variety over k. A *cycle of codimension r* on X is an element of the free abelian group generated by the closed irreducible subvarieties of X of codimension r. So we write a cycle as $Y = \sum n_i Y_i$ where the Y_i are subvarieties, and $n_i \in \mathbf{Z}$. Sometimes it is useful to speak of the cycle associated to a closed subscheme. If Z is a closed subscheme of codimension r, let $Y_1, \ldots, Y_t$ be those irreducible components of Z which have codimension r, and define the *cycle associated* to Z to be $\sum n_i Y_i$, where n_i is the length of the local ring $\mathcal{O}_{y_i, Z}$ of the generic point y_i of Y_i on Z.

Let $f : X \to X'$ be a morphism of varieties, and let Y be a subvariety of X. If $\dim f(Y) < \dim Y$, we set $f_*(Y) = 0$. If $\dim f(Y) = \dim Y$, then the function field $K(Y)$ is a finite extension field of $K(f(Y))$, and we set

$$f_*(Y) = [K(Y) : K(f(Y))] \cdot \overline{f(Y)}.$$

Extending by linearity defines a homomorphism f_* of the group of cycles on X to the group of cycles on X'.

Now we come to the definition of rational equivalence. For any subvariety V of X, let $f : \tilde{V} \to V$ be the normalization of V. Then $\tilde{V}$ satisfies the condition (∗) of (II, §6), so we can talk about Weil divisors and linear equivalence on $\tilde{V}$. Whenever D and D' are linearly equivalent Weil divisors on $\tilde{V}$, we say that f_*D and f_*D' are *rationally equivalent* as cycles on X. Then we define *rational equivalence* of cycles on X in general by dividing out by the group generated by all such $f_*D \sim f_*D'$ for all subvarieties V, and all linearly equivalent Weil divisors D,D' on $\tilde{V}$. In particular, if X itself is normal, then rational equivalence for cycles of codimension 1 coincides with linear equivalence of Weil divisors.

For each r we let $A^r(X)$ be the group of cycles of codimension r on X modulo rational equivalence. We denote by $A(X)$ the graded group $\bigoplus_{r=0}^{n} A^r(X)$, where $n = \dim X$. Note that $A^0(X) = \mathbf{Z}$, and that $A^r(X) = 0$ for $r > \dim X$. Note also that if X is complete there is a natural group homomorphism, the *degree*, from $A^n(X)$ to $\mathbf{Z}$, defined by $\deg(\sum n_i P_i) = \sum n_i$, where the P_i are points. This is well-defined on rational equivalence classes because of (II, 6.10).

An intersection theory on a given class of varieties $\mathfrak{B}$ consists of giving a pairing $A^r(X) \times A^s(X) \to A^{r+s}(X)$ for each r,s, and for each $X \in \mathfrak{B}$, satisfying the axioms listed below. If $Y \in A^r(X)$ and $Z \in A^s(X)$ we denote the intersection cycle class by $Y.Z$.

Before stating the axioms, for any morphism $f : X \to X'$ of varieties in $\mathfrak{B}$, we assume that $X \times X'$ is also in $\mathfrak{B}$, and we define a homomorphism $f^* : A(X') \to A(X)$ as follows. For a subvariety $Y' \subseteq X'$ we define

$$f^*(Y') = p_{1*}(\Gamma_f . p_2^{-1}(Y')),$$

where p_1 and p_2 are the projections of $X \times X'$ to X and X', and Γ_f is the graph of f, considered as a cycle on $X \times X'$.

This data is now subject to the following requirements.

A1. The intersection pairing makes $A(X)$ into a commutative associative graded ring with identity, for every $X \in \mathfrak{B}$. It is called the *Chow ring* of X.

A2. For any morphism $f : X \to X'$ of varieties in $\mathfrak{B}$, $f^* : A(X') \to A(X)$ is a ring homomorphism. If $g : X' \to X''$ is another morphism, then $f^* \circ g^* = (g \circ f)^*$.

A3. For any proper morphism $f : X \to X'$ of varieties in $\mathfrak{B}$, $f_* : A(X) \to A(X')$ is a homomorphism of graded groups (which shifts degrees). If $g : X' \to X''$ is another morphism, then $g_* \circ f_* = (g \circ f)_*$.

A4. *Projection formula.* If $f : X \to X'$ is a proper morphism, if $x \in A(X)$ and $y \in A(X')$, then

$$f_*(x.f^*y) = f_*(x).y.$$

A5. *Reduction to the diagonal.* If Y and Z are cycles on X, and if $\Delta: X \to X \times X$ is the diagonal morphism, then

$$Y.Z = \Delta^*(Y \times Z).$$

A6. *Local nature.* If Y and Z are subvarieties of X which *intersect properly* (meaning that every irreducible component of $Y \cap Z$ has codimension equal to codim Y + codim Z), then we can write

$$Y.Z = \sum i(Y,Z; W_j)W_j,$$

where the sum runs over the irreducible components W_j of $Y \cap Z$, and where the integer $i(Y,Z; W_j)$ depends only on a neighborhood of the generic point of W_j on X. We call $i(Y,Z; W_j)$ the *local intersection multiplicity* of Y and Z along W_j.

A7. *Normalization.* If Y is a subvariety of X, and Z is an effective Cartier divisor meeting Y properly, then $Y.Z$ is just the cycle associated to the Cartier divisor $Y \cap Z$ on Y, which is defined by restricting the local equation of Z to Y. (This implies in particular that transversal intersections of nonsingular subvarieties have multiplicity 1.)

Theorem 1.1. *Let $\mathfrak{B}$ be the class of nonsingular quasi-projective varieties over a fixed algebraically closed field k. Then there is a unique intersection theory for cycles modulo rational equivalence on the varieties $X \in \mathfrak{B}$ which satisfies the axioms A1–A7 above.*

There are two main ingredients in the proof of this theorem. One is the correct definition of the local intersection multiplicities; the other is Chow's moving lemma. There are several ways of defining intersection multiplicity. We just mention Serre's definition, which is historically most recent, but has the advantage of being compact. If Y and Z intersect properly, and if W is an irreducible component of $Y \cap Z$, we define

$$i(Y,Z; W) = \sum (-1)^i \text{ length Tor}_i^A(A/\mathfrak{a}, A/\mathfrak{b})$$

where A is the local ring $\mathcal{O}_{w,X}$ of the generic point of W on X, and $\mathfrak{a}$ and $\mathfrak{b}$ are the ideals of Y and Z in A. Serre [11] shows that this is a nonnegative integer, and that it has the required properties. Note in particular that the naive definition, taking the length of $A/(\mathfrak{a} + \mathfrak{b}) = A/\mathfrak{a} \otimes A/\mathfrak{b}$, modeled after the case of curves on a surface (V, 1.4) does not work (1.1.1).

The other ingredient is *Chow's moving lemma*, which says that if Y, Z are cycles on a nonsingular quasi-projective variety X, then there is a cycle Z', rationally equivalent to Z, such that Y and Z' intersect properly. Furthermore, if Z'' is another such, then $Y.Z'$ and $Y.Z''$ are rationally equivalent. There are proofs of this moving lemma by Chevalley [2] and Roberts [1].

The uniqueness of the intersection theory is proved as follows: given cycles Y, Z on X, by the moving lemma we may assume they intersect properly.

Then using the reduction to the diagonal (A5) we reduce to the case of computing $\varDelta.(Y \times Z)$ on $X \times X$. This has the advantage that $\varDelta$ is a local complete intersection. Since the intersection multiplicity is local (A6) we reduce to the case where one of the cycles is a complete intersection of Cartier divisors, and then repeated application of the normalization (A7) gives the uniqueness.

Some general references for intersection theory are Weil [1], Chevalley [2], Samuel [1], and Serre [11]. For discussion of some other equivalence relations on cycles, and attempts to calculate the groups $A^i(X)$, see Harts-horne [6].

Example 1.1.1. To see why the higher Tor's are necessary, let Y be the union of two planes in $\mathbf{A}^4$ meeting at a point, so the ideal of Y is $(x,y) \cap (z,w) = (xz,xw,yz,yw)$. Let Z be the plane $(x - z, y - w)$. Since Z meets each component of Y in one point P, we have $i(Y,Z; P) = 2$ by linearity. However, if we naively take $A/(\mathfrak{a} + \mathfrak{b})$ where $\mathfrak{a},\mathfrak{b}$ are the ideals of Y and Z, we get

$$k[x,y,z,w]/(xz,xw,yz,yw,x - z,y - w) \cong k[x,y]/(x^2,xy,y^2),$$

which has length 3.

Example 1.1.2. We cannot expect to have an intersection theory like the one of the theorem on singular varieties. For example, suppose there was an intersection theory on the quadric cone Q given by $xy = z^2$ in $\mathbf{P}^3$. Let L be the ruling $x = z = 0$, and M the ruling $y = z = 0$. Then $2M$ is linearly equivalent to a hyperplane section, which could be taken to be a conic C on Q which meets L,M each transversally in one point. So

$$1 = L.C = L.(2M).$$

By linearity we would have to have $L.M = \frac{1}{2}$, which is not an integer.

2 Properties of the Chow Ring

For any nonsingular quasi-projective variety X we now consider the Chow ring $A(X)$, and list some of its properties. See Chevalley [2] for proofs.

A8. Since the cycles in codimension 1 are just Weil divisors, and rational equivalence is the same as linear equivalence for them, and X is nonsingular, we have $A^1(X) \cong \text{Pic } X$.

Thus, for example, if X is a nonsingular projective surface, we recover the intersection theory of (V, 1.1), using the pairing $A^1(X) \times A^1(X) \to A^2(X)$ followed by the degree map.

A9. For any affine space $\mathbf{A}^m$, the projection $p: X \times \mathbf{A}^m \to X$ induces an isomorphism $p^*: A(X) \to A(X \times \mathbf{A}^m)$.

A10 *Exactness.* If Y is a nonsingular closed subvariety of X, and $U = X - Y$, there is an exact sequence

$$A(Y) \overset{i_*}{\to} A(X) \overset{j^*}{\to} A(U) \to 0,$$

where $i: Y \to X$ is one inclusion, and $j: U \to X$ is the other.

The proofs of these two results are similar to the corresponding results for divisors (II, 6.5), (II, 6.6).

Example 2.0.1. $A(\mathbf{P}^n) \cong \mathbf{Z}[h]/h^{n+1}$, where h in degree 1 is the class of a hyperplane. One can prove this inductively from (A9) and (A10), or directly, by showing that any subvariety of degree d in $\mathbf{P}^n$ is rationally equivalent to d times a linear space of the same dimension (Ex. 6.3).

The next property is important for the definition of the Chern classes in the next section.

A11. Let $\mathscr{E}$ be a locally free sheaf of rank r on X, let $\mathbf{P}(\mathscr{E})$ be the associated projective space bundle (II, §7), and let $\xi \in A^1(\mathbf{P}(\mathscr{E}))$ be the class of the divisor corresponding to $\mathcal{O}_{\mathbf{P}(\mathscr{E})}(1)$. Let $\pi: \mathbf{P}(\mathscr{E}) \to X$ be the projection. Then π^* makes $A(\mathbf{P}(\mathscr{E}))$ into a free $A(X)$-module generated by $1, \xi, \xi^2, \ldots, \xi^{r-1}$.

3 Chern Classes

Here we follow the treatment of Grothendieck [3].

Definition. Let $\mathscr{E}$ be a locally free sheaf of rank r on a nonsingular quasi-projective variety X. For each $i = 0, 1, \ldots, r$, we define the ith *Chern class* $c_i(\mathscr{E}) \in A^i(X)$ by the requirement $c_0(\mathscr{E}) = 1$ and

$$\sum_{i=0}^{r} (-1)^i \pi^* c_i(\mathscr{E}) . \xi^{r-i} = 0$$

in $A^r(\mathbf{P}(\mathscr{E}))$, using the notation of (A11).

This makes sense, because by (A11), we can express ξ^r as a unique linear combination of $1, \xi, \ldots, \xi^{r-1}$, with coefficients in $A(X)$, via π^*. Here are some properties of the Chern classes. For convenience we define the *total Chern class*

$$c(\mathscr{E}) = c_0(\mathscr{E}) + c_1(\mathscr{E}) + \ldots + c_r(\mathscr{E})$$

and the *Chern polynomial*

$$c_t(\mathscr{E}) = c_0(\mathscr{E}) + c_1(\mathscr{E})t + \ldots + c_r(\mathscr{E})t^r.$$

C1. If $\mathscr{E} \cong \mathscr{L}(D)$ for a divisor D, $c_t(\mathscr{E}) = 1 + Dt$. Indeed, in this case $\mathbf{P}(\mathscr{E}) = X$, $\mathcal{O}_{\mathbf{P}(\mathscr{E})}(1) = \mathscr{L}(D)$, so $\xi = D$, so by definition $1.\xi - c_1(\mathscr{E}).1 = 0$, so $c_1(\mathscr{E}) = D$.

C2. If $f : X' \to X$ is a morphism, and $\mathscr{E}$ is a locally free sheaf on X, then for each i

$$c_i(f^*\mathscr{E}) = f^*c_i(\mathscr{E}).$$

This follows immediately from the functoriality properties of the $\mathbf{P}(\mathscr{E})$ construction and f^*.

C3. If $0 \to \mathscr{E}' \to \mathscr{E} \to \mathscr{E}'' \to 0$ is an exact sequence of locally free sheaves on X, then

$$c_t(\mathscr{E}) = c_t(\mathscr{E}') \cdot c_t(\mathscr{E}'').$$

In fact, forgetting the definition for a moment, one can show that there is a unique theory of Chern classes, which for each locally free sheaf $\mathscr{E}$ on X assigns $c_i(\mathscr{E}) \in A^i(X)$, satisfying (C1), (C2), and (C3). For the proof of this uniqueness and for the proof of (C3) and the other properties below, one uses the *splitting principle*, which says that given $\mathscr{E}$ on X, there exists a morphism $f : X' \to X$ such that $f^* : A(X) \to A(X')$ is injective, and $\mathscr{E}' = f^*\mathscr{E}$ *splits*, i.e., it has a filtration $\mathscr{E}' = \mathscr{E}'_0 \supseteq \mathscr{E}'_1 \supseteq \ldots \supseteq \mathscr{E}'_r = 0$ whose successive quotients are all invertible sheaves. Then one uses the following property.

C4. If $\mathscr{E}$ splits, and the filtration has the invertible sheaves $\mathscr{L}_1, \ldots, \mathscr{L}_r$ as quotients then

$$c_t(\mathscr{E}) = \prod_{i=1}^{r} c_t(\mathscr{L}_i).$$

(And of course we know each $c_t(\mathscr{L}_i)$ from (C1).)

Using the splitting principle, we can also calculate the Chern classes of tensor products, exterior products, and dual locally free sheaves. Let $\mathscr{E}$ have rank r, and let $\mathscr{F}$ have rank s. Write

$$c_t(\mathscr{E}) = \prod_{i=1}^{r} (1 + a_i t)$$

and

$$c_t(\mathscr{F}) = \prod_{i=1}^{s} (1 + b_i t),$$

where $a_1, \ldots, a_r, b_1, \ldots, b_s$ are just formal symbols. Then we have

C5.

$$c_t(\mathscr{E} \otimes \mathscr{F}) = \prod_{i,j} (1 + (a_i + b_j)t)$$

$$c_t(\wedge^p \mathscr{E}) = \prod_{1 \le i_1 < \ldots < i_p \le r} (1 + (a_{i_1} + \ldots + a_{i_p})t)$$

$$c_t(\mathscr{E}^{\vee}) = c_{-t}(\mathscr{E}).$$

These expressions make sense, because when multiplied out, the coefficients of each power of t are symmetric functions in the a_i and the b_j. Hence by a well-known theorem on symmetric functions, they can be expressed as polynomials in the elementary symmetric functions of the a_i and b_j, which are none other than the Chern classes of $\mathscr{E}$ and $\mathscr{F}$. For a further reference on this formalism, see Hirzebruch [1, Ch. I, §4.4].

C6. Let $s \in \Gamma(X,\mathscr{E})$ be a global section of a locally free sheaf $\mathscr{E}$ of rank r on X. Then s defines a homomorphism $\mathscr{O}_X \to \mathscr{E}$ by sending 1 to s. We define the *scheme of zeros* of s to be the closed subscheme Y of X defined by the exact sequence

$$\mathscr{E}^\vee \xrightarrow{s^\vee} \mathscr{O}_X \to \mathscr{O}_Y \to 0$$

where $s^\vee$ is the dual of the map s. Let Y also denote the associated cycle of Y. Then if Y has codimension r, we have $c_r(\mathscr{E}) = Y$ in $A^r(X)$.

This generalizes the fact that a section of an invertible sheaf gives the corresponding divisor (II, 7.7).

C7. *Self-intersection formula.* Let Y be a nonsingular subvariety of X of codimension r, and let $\mathscr{N}$ be the normal sheaf (II, §8). Let $i: Y \to X$ be the inclusion map. Then

$$i^* i_*(1_Y) = c_r(\mathscr{N}).$$

Therefore, applying the projection formula (A4) we have

$$i_*(c_r(\mathscr{N})) = Y.Y$$

on X.

This result, due to Mumford (see Lascu, Mumford, and Scott [1]), generalizes the self-intersection formula (V, 1.4.1) for a curve on a surface.

4 The Riemann–Roch Theorem

Let $\mathscr{E}$ be a locally free sheaf of rank r on a nonsingular projective variety X of dimension n, and let $\mathscr{T} = \mathscr{T}_X$ be the tangent sheaf of X (II, §8). We want to give an expression for $\chi(\mathscr{E})$ in terms of the Chern classes of $\mathscr{E}$ and $\mathscr{T}$. For this purpose we introduce two elements of $A(X) \otimes \mathbf{Q}$, which are defined as certain universal polynomials in the Chern classes of a sheaf $\mathscr{E}$. Let

$$c_t(\mathscr{E}) = \prod_{i=1}^{r} (1 + a_i t)$$

as above, where the a_i are formal symbols. Then we define the *exponential Chern character*

$$\mathrm{ch}(\mathscr{E}) = \sum_{i=1}^{r} e^{a_i},$$

where

$$e^x = 1 + x + \frac{1}{2}x^2 + \cdots,$$

and the *Todd class* of $\mathscr{E}$,

$$\text{td}(\mathscr{E}) = \prod_{i=1}^{r} \frac{a_i}{1 - e^{-a_i}}$$

where

$$\frac{x}{1 - e^{-x}} = 1 + \frac{1}{2}x + \frac{1}{12}x^2 - \frac{1}{720}x^4 + \cdots.$$

As before, these are symmetric expressions in the a_i, so can be expressed as polynomials in the $c_i(\mathscr{E})$, with rational coefficients. By elementary but tedious calculation, one can show using these definitions that

$$\text{ch}(\mathscr{E}) = r + c_1 + \frac{1}{2}(c_1^2 - 2c_2) + \frac{1}{6}(c_1^3 - 3c_1 c_2 + 3c_3)$$

$$+ \frac{1}{24}(c_1^4 - 4c_1^2 c_2 + 4c_1 c_3 + 2c_2^2 - 4c_4) + \cdots$$

and

$$\text{td}(\mathscr{E}) = 1 + \frac{1}{2}c_1 + \frac{1}{12}(c_1^2 + c_2) + \frac{1}{24}c_1 c_2$$

$$- \frac{1}{720}(c_1^4 - 4c_1^2 c_2 - 3c_2^2 - c_1 c_3 + c_4) + \cdots$$

where we set $c_i = c_i(\mathscr{E})$, $c_i = 0$ if $i > r$.

Theorem 4.1 (Hirzebruch–Riemann–Roch). *For a locally free sheaf $\mathscr{E}$ of rank r on a nonsingular projective variety X of dimension n,*

$$\chi(\mathscr{E}) = \deg(\text{ch}(\mathscr{E}).\text{td}(\mathscr{T}))_n,$$

where $(\ \ )_n$ denotes the component of degree n in $A(X) \otimes \mathbf{Q}$.

This theorem was proved by Hirzebruch [1] over $\mathbf{C}$, and by Grothendieck in a generalized form (5.3) over any algebraically closed field k (see Borel and Serre [1]).

Example 4.1.1. If X is a curve, and $\mathscr{E} = \mathscr{L}(D)$, we have $\text{ch}(\mathscr{E}) = 1 + D$. The tangent sheaf $\mathscr{T}_X$ is the dual of Ω_X. Therefore $\mathscr{T}_X \cong \mathscr{L}(-K)$, where K is the canonical divisor, and so $\text{td}(\mathscr{T}_X) = 1 - \frac{1}{2}K$. Thus (4.1) tells us that

$$\chi(\mathscr{L}(D)) = \deg\left((1 + D)\left(1 - \frac{1}{2}K\right)\right)_1$$

$$= \deg\left(D - \frac{1}{2}K\right).$$

For $D = 0$, this says that $1 - g = -\frac{1}{2}\deg K$, so we can write the theorem as

$$\chi(\mathscr{L}(D)) = \deg D + 1 - g,$$

which is the Riemann–Roch theorem for curves proved earlier (IV, 1.3).

Example 4.1.2. Now let X be a surface, and again let $\mathscr{E} = \mathscr{L}(D)$. Then $\mathrm{ch}(\mathscr{E}) = 1 + D + \frac{1}{2}D^2$. We denote by c_1 and c_2 the Chern classes of the tangent sheaf $\mathscr{T}_X$. These depend only on X, so they are sometimes called the *Chern classes of* X. Since $\mathscr{T}_X$ is the dual of Ω_X, and since $c_1(\Omega_X) = c_1(\wedge^2\Omega_X)$ by (C5), and since $\wedge^2\Omega_X = \omega_X$ is just $\mathscr{L}(K)$, where K is the canonical divisor, we see that $c_1(\mathscr{T}_X) = -K$. But c_2 (or rather its degree) is a new numerical invariant of a surface which we have not met before.

Using $c_1 = -K$ and c_2, we have

$$\mathrm{td}(\mathscr{T}_X) = 1 - \frac{1}{2}K + \frac{1}{12}(K^2 + c_2).$$

Multiplying, and taking degrees (by abuse of notation we let D^2 denote both the class in $A^2(X)$, and its degree), we can write (4.1) as

$$\chi(\mathscr{L}(D)) = \frac{1}{2}D.(D - K) + \frac{1}{12}(K^2 + c_2).$$

In particular, for $D = 0$, we find that

$$\chi(\mathscr{O}_X) = \frac{1}{12}(K^2 + c_2).$$

By definition of the arithmetic genus (III, Ex. 5.3), this says

$$1 + p_a = \frac{1}{12}(K^2 + c_2).$$

So the new Riemann–Roch theorem for surfaces gives us the earlier one (V, 1.6), together with the additional information that c_2 can be expressed in terms of the invariants p_a, K^2 by this last formula.

Example 4.1.3. As an application, we derive a formula relating the numerical invariants of a surface in $\mathbf{P}^4$. For perspective, note that if X is a surface of degree d in $\mathbf{P}^3$, then the numerical invariants p_a, K^2, and hence c_2, are uniquely determined by d (I, Ex. 7.2) and (V, Ex. 1.5). On the other hand, any projective surface can be embedded in $\mathbf{P}^5$ (IV, Ex. 3.11), so for surfaces in $\mathbf{P}^5$ we do not expect any particular connection between these invariants. However, a surface cannot in general be embedded in $\mathbf{P}^4$, so for those which can, we expect some condition to be satisfied.

So let X be a nonsingular surface of degree d in $\mathbf{P}^4$. In the Chow ring of $\mathbf{P}^4$, X is equivalent to d times a plane, so $X.X = d^2$. On the other hand, we can compute $X.X$, using the self-intersection formula (C7), as $\deg c_2(\mathscr{N})$ where $\mathscr{N}$ is the normal bundle of X in $\mathbf{P}^4$. There is an exact sequence

$$0 \to \mathscr{T}_X \to i^*\mathscr{T}_{\mathbf{P}^4} \to \mathscr{N} \to 0,$$

where $i: X \to \mathbf{P}^4$ is the inclusion. We use this sequence to compute $c_2(\mathscr{N})$, and thus get our condition.

First we use the exact sequence

$$0 \to \mathscr{O}_{\mathbf{P}^4} \to \mathscr{O}_{\mathbf{P}^4}(1)^5 \to \mathscr{T}_{\mathbf{P}^4} \to 0.$$

Letting $h \in A^1(\mathbf{P}^4)$ be the class of a hyperplane, we see that

$$c_t(\mathscr{T}_{\mathbf{P}^4}) = (1 + ht)^5 = 1 + 5ht + 10h^2t^2 + \cdots .$$

On the other hand,

$$c_t(\mathscr{T}_X) = 1 - Kt + c_2t^2$$

as in (4.1.2). Therefore, denoting by $H \in A^1(X)$ the class of a hyperplane section of X, we have from the exact sequence above, using (C3), that

$$(1 - Kt + c_2t^2)(1 + c_1(\mathscr{N})t + c_2(\mathscr{N})t^2) = 1 + 5Ht + 10H^2t^2.$$

Comparing coefficients of t and t^2, we find that

$$c_1(\mathscr{N}) = 5H + K$$
$$c_2(\mathscr{N}) = 10H^2 - c_2 + 5H.K + K^2.$$

Now take degrees and combine with $\deg c_2(\mathscr{N}) = d^2$. Also note that $\deg H^2 = d$, and use the expression for c_2 in (4.1.2). The final result is

$$d^2 - 10d - 5H.K - 2K^2 + 12 + 12p_a = 0.$$

This holds for any nonsingular surface of degree d in $\mathbf{P}^4$. See (Ex. 6.9) for some applications.

5 Complements and Generalizations

Having developed an intersection theory for n-dimensional varieties, we can ask whether some of the other theorems we proved for surfaces in Ch. V also extend. They do.

Theorem 5.1 (Nakai–Moishezon Criterion). *Let D be a Cartier divisor on a scheme X which is proper over an algebraically closed field k. Then D is ample on X if and only if for every closed integral subscheme $Y \subseteq X$ (including the case $Y = X$ if X is integral), we have $D^r.Y > 0$ where $r = \dim Y$.*

This theorem was proved by Nakai [1] for X projective over k, and independently by Moishezon [1] for X an abstract complete variety. The proof was clarified and simplified by Kleiman [1]. Strictly speaking, this theorem uses a slightly different intersection theory than the one we have developed. We do not assume X nonsingular projective, so we do not have Chow's moving lemma. On the other hand, the only intersections we need to consider are those of a number of Cartier divisors with a single closed subscheme. And this intersection theory is in fact more elementary to develop than the one we outlined in §1. See Kleiman [1] for details. Notice that this theorem extends the one given earlier for a surface (V, 1.10), because taking $Y = X$ gives $D^2 > 0$, and for Y a curve we have $D.Y > 0$.

We can also generalize the Hodge index theorem (V, 1.9) to a nonsingular projective variety X over $\mathbf{C}$. We consider the associated complex manifold X_h (App. B) and its complex cohomology $H^i(X_h, \mathbf{C})$. For any cycle Y of codimension r on X, one can define its cohomology class $\eta(Y) \in H^{2r}(X_h, \mathbf{C})$. We say that Y is *homologically equivalent to zero*, written $Y \sim_{\text{hom}} 0$, if $\eta(Y) = 0$.

Theorem 5.2 (Hodge Index Theorem). *Let X be a nonsingular projective variety over $\mathbf{C}$, of even dimension $n = 2k$. Let H be an ample divisor on X, let Y be a cycle of codimension k, and assume that $Y.H \sim_{\text{hom}} 0$, and $Y \not\sim_{\text{hom}} 0$. Then $(-1)^k Y^2 > 0$.*

This theorem is proved using Hodge's theory of harmonic integrals— see Weil [5, Th. 8, p. 78]. It generalizes the earlier result for surfaces (V, 1.9), because for divisors, one can show that homological and numerical equivalence coincide. It is conjectured by Grothendieck [9] to be true over an arbitrary algebraically closed field k, using l-adic cohomology for the definition of homological equivalence. He suggests that it might also be true using numerical equivalence of cycles, but that isn't known even over $\mathbf{C}$. See Kleiman [2] for a discussion of these and Grothendieck's other "standard conjectures."

Now let us turn to further generalizations of the Riemann–Roch theorem following Borel and Serre [1]. The first step is to extend the definition of the Chern classes to the Grothendieck group $K(X)$ (II, Ex. 6.10). For X nonsingular, we can compute $K(X)$ using only locally free sheaves (III, Ex. 6.9). Then, because of the additivity property (C3) of Chern classes, it is clear that the Chern polynomial c_t extends to give a map

$$c_t : K(X) \rightarrow A(X)[t].$$

Thus we have Chern classes defined on $K(X)$. The exponential Chern character ch extends to give a mapping

$$\text{ch} : K(X) \rightarrow A(X) \otimes \mathbf{Q}.$$

One shows that $K(X)$ has a natural ring structure (defined by $\mathscr{E} \otimes \mathscr{F}$ for locally free sheaves $\mathscr{E}$ and $\mathscr{F}$), and that ch is a ring homomorphism. If $f : X' \rightarrow X$ is a morphism of nonsingular varieties, then there is a ring homomorphism

$$f^! : K(X) \rightarrow K(X')$$

defined by $\mathscr{E} \mapsto f^*\mathscr{E}$ for locally free $\mathscr{E}$. The exponential Chern character ch commutes with $f^!$.

If $f: X \to Y$ is a proper morphism, one defines an additive map $f_!: K(X) \to K(Y)$ by

$$f_!(\mathscr{F}) = \sum (-1)^i R^i f_*(\mathscr{F})$$

for $\mathscr{F}$ coherent.

This map $f_!$ does not commute with ch. The extent to which it fails to commute is the generalized Riemann–Roch theorem of Grothendieck.

Theorem 5.3 (Grothendieck–Riemann–Roch). *Let $f: X \to Y$ be a smooth projective morphism of nonsingular quasi-projective varieties. Then for any $x \in K(X)$ we have*

$$\mathrm{ch}(f_!(x)) = f_*(\mathrm{ch}(x).\mathrm{td}(\mathscr{T}_f))$$

in $A(Y) \otimes \mathbf{Q}$, where $\mathscr{T}_f$ is the relative tangent sheaf of f.

If Y is a point, this reduces to the earlier form (4.1).

After wrestling with formidable technical obstacles, this theorem has been further generalized in the Paris seminar of Grothendieck [SGA 6], to the case where Y is a noetherian scheme admitting an ample invertible sheaf, and f is a projective locally complete intersection morphism. See Manin [1] for a readable account of this work.

We should also mention another Riemann–Roch formula, for the case of a closed immersion $f: X \hookrightarrow Y$ of nonsingular varieties, due to Jouanolou [1]. In the case of a closed immersion, the formula of (5.3) gives a way of computing the Chern classes $c_i(f_*\mathscr{F})$, for any coherent sheaf $\mathscr{F}$ on X, in terms of $f_*(c_i(\mathscr{F}))$ and $f_*(c_i(\mathscr{N}))$, where $\mathscr{N}$ is the normal sheaf. It turns out this can be done using polynomials with *integer* coefficients, but the proof of (5.3) gives the result only in $A(Y) \otimes \mathbf{Q}$, i.e., mod torsion. Jouanolou's result is that the result actually holds in $A(Y)$ itself.

Recently another kind of generalization of the Riemann–Roch theorem to singular varieties has been developed by Baum, Fulton, and MacPherson. See Fulton [2].

EXERCISES

6.1. Show that the definition of rational equivalence in §1 is equivalent to the equivalence relation generated by the following relation: two cycles Y, Z of codimension r on X are equivalent if there exists a cycle W of codimension r on $X \times \mathbf{A}^1$, which intersects $X \times \{0\}$ and $X \times \{1\}$ properly, and such that $Y = W.(X \times \{0\})$, $Z = W.(X \times \{1\})$.

6.2. Prove the following result about Weil divisors, which generalizes (IV, Ex. 2.6), and which is needed to show that f_* is well-defined modulo rational equivalence (A3): Let $f: X \to X'$ be a proper, generically finite map of normal varieties, and let D_1 and D_2 be linearly equivalent Weil divisors on X. Then $f_* D_1$ and $f_* D_2$ are linearly equivalent Weil divisors on X'. [*Hint*: Remove a subset of codimension ≥ 2 from X' so that f becomes a finite flat morphism, then generalize (IV, Ex. 2.6).]

6.3. Show directly that any subvariety of degree d in $\mathbf{P}^n$ is rationally equivalent to d times a linear space of the same dimension, by using a projection argument similar to (III, 9.8.3).

6.4. Let $\pi:X \to C$ be a ruled surface (V, §2) over a nonsingular curve C. Show that the group $A^2(X)$ of zero-cycles modulo rational equivalence is isomorphic to Pic C.

6.5. Let X be a surface, let $P \in X$ be a point, and let $\pi:\tilde{X} \to X$ be the monoidal transformation with center P (V, §3). Show that $A(\tilde{X}) \cong \pi^*A(X) \oplus \mathbf{Z}$, where $\mathbf{Z}$ is generated by the exceptional curve $E \in A^1(X)$, and the intersection theory is determined by $E^2 = -\pi^*(P)$.

6.6. Let X be a nonsingular projective variety of dimension n, and let $\Delta \subseteq X \times X$ be the diagonal. Show that $c_n(\mathcal{T}_X) = \Delta^2$ in $A^n(X)$, under the natural isomorphism of X with Δ.

6.7. Let X be a nonsingular projective 3-fold, with Chern classes c_1, c_2, c_3. Show that

$$1 - p_a = \frac{1}{24} c_1 c_2,$$

and for any divisor D,

$$\chi(\mathscr{L}(D)) = \frac{1}{12} D.(D - K).(2D - K) + \frac{1}{12} D.c_2 + 1 - p_a.$$

6.8. Let $\mathscr{E}$ be a locally free sheaf of rank 2 on $\mathbf{P}^3$, with Chern classes c_1, c_2. Since $A(\mathbf{P}^3) = \mathbf{Z}[h]/h^4$, we can think of c_1 and c_2 as integers. Show that $c_1 c_2 \equiv 0$ (mod 2). [*Hint:* In the Riemann–Roch theorem for $\mathscr{E}$, the left-hand side is automatically an integer, while the right-hand side is *a priori* only a rational number.]

6.9. *Surfaces in* $\mathbf{P}^4$.
  (a) Verify the formula of (4.1.3) for the rational cubic scroll in $\mathbf{P}^4$ (V, 2.19.1).
  (b) If X is a K3 surface in $\mathbf{P}^4$, show that its degree must be 4 or 6. (Examples of such are the quartic surface in $\mathbf{P}^3$, and the complete intersection of a quadric and a cubic hypersurface in $\mathbf{P}^4$.)
  (c) If X is an abelian surface in $\mathbf{P}^4$, show that its degree must be 10. (Horrocks and Mumford [1] have shown that such abelian surfaces exist.)
  *(d) Determine which of the rational ruled surfaces X_e, $e \geqslant 0$ (V, §2), admit an embedding in $\mathbf{P}^4$.

6.10. Use the fact that the tangent sheaf on an abelian variety is free, to show that it is impossible to embed an abelian 3-fold in $\mathbf{P}^5$.

APPENDIX B

Transcendental Methods

If X is a nonsingular variety over $\mathbf{C}$, then we can also consider X as a complex manifold. All the methods of complex analysis and differential geometry can be used to study this complex manifold. And, given an adequate dictionary between the language of abstract algebraic geometry and complex manifolds, these results can be translated back into results about the original variety X.

This is an extremely powerful method, which has produced and is still producing many important results, proved by these so-called "transcendental methods," for which no purely algebraic proofs are known.

On the other hand, one can ask where do the algebraic varieties fit into the general theory of complex manifolds, and what special properties characterize them among all complex manifolds?

In this appendix we will give a very brief report of this vast and important area of research.

1 The Associated Complex Analytic Space

A *complex analytic space* (in the sense of Grauert) is a topological space $\mathfrak{X}$, together with a sheaf of rings $\mathcal{O}_x$, which can be covered by open sets, each of which is isomorphic, as a locally ringed space, to one of the following kind Y: let $U \subseteq \mathbf{C}^n$ be the polydisc $\{|z_i| < 1 | i = 1, \ldots, n\}$, let $f_1, \ldots, f_q$ be holomorphic functions on U, let $Y \subseteq U$ be the closed subset (for the "usual" topology) consisting of the common zeros of $f_1, \ldots, f_q$, and take $\mathcal{O}_Y$ to be the sheaf $\mathcal{O}_U/(f_1, \ldots, f_q)$, where $\mathcal{O}_U$ is the sheaf of germs of holomorphic functions on U. Note that the structure sheaf may have nilpotent elements. See Gunning and Rossi [1] for a development of the general theory

438

of complex analytic spaces, coherent analytic sheaves, and cohomology. See also Bănică and Stănăşilă [1] for a survey of recent techniques in the cohomology of complex analytic spaces, parallel to the algebraic techniques in Chapter III.

Now if X is a scheme of finite type over $\mathbf{C}$, we define the *associated complex analytic space* X_h as follows. Cover X with open affine subsets $Y_i = \operatorname{Spec} A_i$. Each A_i is an algebra of finite type over $\mathbf{C}$, so we can write it as $A_i \cong \mathbf{C}[x_1, \ldots, x_n]/(f_1, \ldots, f_q)$. Here $f_1, \ldots, f_q$ are polynomials in $x_1, \ldots, x_n$. We can regard them as holomorphic functions on $\mathbf{C}^n$, so that their set of common zeros is a complex analytic subspace $(Y_i)_h \subseteq \mathbf{C}^n$. The scheme X is obtained by glueing the open sets Y_i, so we can use the same glueing data to glue the analytic spaces $(Y_i)_h$ into an analytic space X_h. This is the associated complex analytic space of X.

The construction is clearly functorial, so we obtain a functor h from the category of schemes of finite type over $\mathbf{C}$ to the category of complex analytic spaces. In a similar way, if $\mathscr{F}$ is a coherent sheaf on X, one can define the associated coherent analytic sheaf $\mathscr{F}_h$ as follows. The sheaf $\mathscr{F}$ is locally (for the Zariski topology) a cokernel

$$\mathscr{O}_U^m \xrightarrow{\varphi} \mathscr{O}_U^n \to \mathscr{F} \to 0$$

of a morphism φ of free sheaves. Since the usual topology is finer than the Zariski topology, U_h is open in X_h. Furthermore, since φ is defined by a matrix of local sections of $\mathscr{O}_U$, these give local sections of $\mathscr{O}_{U_h}$, so we can define $\mathscr{F}_h$ as the cokernel of the corresponding map φ_h of free coherent analytic sheaves locally.

One can prove easily some basic facts about the relationship between a scheme X and its associated analytic space X_h (see Serre [4]). For example, X is separated over $\mathbf{C}$ if and only if X_h is Hausdorff. X is connected in the Zariski topology if and only if X_h is connected in the usual topology. X is reduced if and only if X_h is reduced. X is smooth over $\mathbf{C}$ if and only if X_h is a complex manifold. A morphism $f : X \to Y$ is proper if and only if $f_h : X_h \to Y_h$ is proper in the usual sense, i.e., the inverse image of a compact set is compact. In particular, X is proper over $\mathbf{C}$ if and only if X_h is compact.

One can also compare the cohomology of coherent sheaves on X and X_h. There is a continuous map $\varphi : X_h \to X$ of the underlying topological spaces, which sends X_h bijectively onto the set of closed points of X, but of course the topology is different. There is also a natural map of the structure sheaves $\varphi^{-1}\mathscr{O}_X \to \mathscr{O}_{X_h}$, which makes φ into a morphism of locally ringed spaces. It follows from our definitions that for any coherent sheaf $\mathscr{F}$ of $\mathscr{O}_X$-modules, $\mathscr{F}_h \cong \varphi^* \mathscr{F}$. From this one can show easily that there are natural maps of cohomology groups

$$\alpha_i : H^i(X, \mathscr{F}) \to H^i(X_h, \mathscr{F}_h)$$

for each i. Here we always take cohomology in the sense of derived functors (III, §2), but one can show that on the analytic space X_h, this coincides with

439

the other cohomology theories in use in the literature—see the historical note at the end of (III, §2).

2 Comparison of the Algebraic and Analytic Categories

To stimulate our thinking about the comparison between schemes of finite type over **C** and their associated complex analytic spaces, let us consider five questions which arise naturally from contemplating the functor h.

Q1. Given a complex analytic space $\mathfrak{X}$, does there exist a scheme X such that $X_h \cong \mathfrak{X}$?

Q2. If X and X' are two schemes such that $X_h \cong X'_h$, then is $X \cong X'$?

Q3. Given a scheme X, and a coherent analytic sheaf $\mathfrak{F}$ on X_h, does there exist a coherent sheaf $\mathscr{F}$ on X such that $\mathscr{F}_h \cong \mathfrak{F}$?

Q4. Given a scheme X, and two coherent sheaves $\mathscr{E}$ and $\mathscr{F}$ on X such that $\mathscr{E}_h \cong \mathscr{F}_h$ on X_h, then is $\mathscr{E} \cong \mathscr{F}$?

Q5. Given a scheme X, and a coherent sheaf $\mathscr{F}$, are the maps α_i on cohomology isomorphisms?

As one might expect, when phrased in this generality, the answer to all five questions is NO. It is fairly easy to give counterexamples to Q1, Q3, and Q5—see (Ex. 6.1), (Ex. 6.3), (Ex. 6.4). Q2 and Q4 are more difficult, so we mention the following example.

Example 2.0.1 (Serre). Let C be an elliptic curve, let X be the unique nontrivial ruled surface over C with invariant $e = 0$ (V, 2.15), and let C_0 be the section with $C_0^2 = 0$ (V, 2.8.1). Let $U = X - C_0$. On the other hand, let $U' = (\mathbf{A}^1 - \{0\}) \times (\mathbf{A}^1 - \{0\})$. Then one can show that $U_h \cong U'_h$, but $U \not\cong U'$, because U is not affine. In particular, U_h is Stein although U is not affine. Furthermore, one can show that Pic $U \cong$ Pic C, whereas Pic $U_h \cong$ **Z**. In particular, there are nonisomorphic invertible sheaves $\mathscr{L}$ and $\mathscr{L}'$ on U such that $\mathscr{L}_h \cong \mathscr{L}'_h$. For details see Hartshorne [5, p. 232].

In contrast, if one restricts one's attention to *projective* schemes, then the answer to all five questions is YES. These results were proved by Serre in his beautiful paper GAGA (Serre [4]). The main theorem is this.

Theorem 2.1 (Serre). *Let X be a projective scheme over* **C**. *Then the functor h induces an equivalence of categories from the category of coherent sheaves on X to the category of coherent analytic sheaves on X_h. Further-*

more, for every coherent sheaf $\mathcal{F}$ on X, the natural maps

$$\alpha_i : H^i(X, \mathcal{F}) \to H^i(X_h, \mathcal{F}_h)$$

are isomorphisms, for all i.

This answers questions Q3, Q4, and Q5. The proof requires knowing the analytic cohomology groups $H^i(\mathbf{P}_h^n, \mathcal{O}(q))$ for all i, n, q, which can be computed using Cartan's Theorems A and B. The answer is the same as in the algebraic case (III, 5.1). Then the result follows from the standard technique of embedding X in $\mathbf{P}^n$, and resolving $\mathcal{F}$ by sheaves of the form $\sum \mathcal{O}(q_i)$, as in (III, §5). This theorem has also been generalized by Grothendieck [SGA 1, XII] to the case when X is proper over $\mathbf{C}$.

As a corollary, Serre obtains a new proof of a theorem of Chow [1]:

Theorem 2.2 (Chow). *If $\mathfrak{X}$ is a compact analytic subspace of the complex manifold $\mathbf{P}_{\mathbf{C}}^n$, then there is a subscheme $X \subseteq \mathbf{P}^n$ with $X_h = \mathfrak{X}$.*

This answers Q1 in the projective case. We leave Q2 as an exercise (Ex. 6.6).

3   When is a Compact Complex Manifold Algebraic?

If $\mathfrak{X}$ is a compact complex manifold, then one can show that a scheme X such that $X_h \cong \mathfrak{X}$, if it exists, is unique. So if such an X exists, we will simply say $\mathfrak{X}$ *is algebraic.* Let us consider the modified form of question 1:

Q1′. Can one give reasonable necessary and sufficient conditions for a compact complex manifold $\mathfrak{X}$ to be algebraic?

The first result in this direction is

Theorem 3.1 (Riemann). *Every compact complex manifold of dimension 1 (i.e., a compact Riemann surface) is projective algebraic.*

This is a deep result. To understand why, recall that the notion of complex manifold is very local. It is defined by glueing small discs, with holomorphic transition functions. We make one global hypothesis, namely that it is compact, and our conclusion is that it can be embedded globally in some projective space. In particular, by considering a projection to $\mathbf{P}^1$, we see that it has nonconstant meromorphic functions, which is not at all obvious a priori. One proves the theorem in two steps:

(a) One shows that $\mathfrak{X}$ admits a global nonconstant meromorphic function. This requires some hard analysis. One proof, given by Weyl [1], following Hilbert, uses Dirichlet's minimum principle to prove the existence of harmonic functions, and hence of meromorphic functions. Another proof, given by Gunning [1], uses distributions to first prove the finite-dimensionality of cohomology of coherent analytic sheaves, and then deduce the existence of meromorphic functions.

(b) The second step is to take a nonconstant meromorphic function f on $\mathfrak{X}$, and to regard it as giving a finite morphism of $\mathfrak{X}$ to $\mathbf{P}^1$. Then one shows that $\mathfrak{X}$ is a nonsingular algebraic curve, hence projective.

Part (b), which is often called the "Riemann existence theorem," is elementary by comparison with part (a). It has been generalized to higher dimensions by Grauert and Remmert [1]. Recently Grothendieck [SGA 1, XII] has given an elegant proof of their generalization, using Hironaka's resolution of singularities. The result is this:

Theorem 3.2 (Generalized Riemann Existence Theorem). *Let X be a normal scheme of finite type over $\mathbf{C}$. Let $\mathfrak{X}'$ be a normal complex analytic space, together with a finite morphism $\mathfrak{f}:\mathfrak{X}' \to X_h$. (We define a* finite *morphism of analytic spaces to be a proper morphism with finite fibres.) Then there is a unique normal scheme X' and a finite morphism $g:X' \to X$ such that $X'_h \cong \mathfrak{X}'$ and $g_h = \mathfrak{f}$.*

One corollary of this theorem is that the algebraic fundamental group of X, $\pi_1^{\text{alg}}(X)$, defined as the inverse limit of the Galois groups of finite étale covers of X (IV, Ex. 4.8), is isomorphic to the completion $\pi_1^{\text{top}}(X_h)\hat{\ }$ of the usual fundamental group of X_h with respect to subgroups of finite index. Indeed, if $\mathfrak{Y}$ is any finite unramified *topological* covering space of X_h, then $\mathfrak{Y}$ has a natural structure of normal complex analytic space, so by the theorem it is algebraic (and étale) over X.

In dimensions greater than 1, it is no longer true that every compact complex manifold is algebraic. But we have the following result which gives a necessary condition.

Proposition 3.3 (Siegel [1]). *Let $\mathfrak{X}$ be a compact complex manifold of dimension n. Then the field $K(\mathfrak{X})$ of meromorphic functions on $\mathfrak{X}$ has transcendence degree $\leqslant n$ over $\mathbf{C}$, and (at least in the case tr.d. $K(\mathfrak{X}) = n$) it is a finitely generated extension field of $\mathbf{C}$.*

If $\mathfrak{X}$ is algebraic, say $\mathfrak{X} \cong X_h$, then one can show that $K(\mathfrak{X}) \cong K(X)$, the field of rational functions on X, so in this case we must have tr.d. $K(\mathfrak{X}) = n$. Compact complex manifolds $\mathfrak{X}$ with tr.d. $K(\mathfrak{X}) = \dim \mathfrak{X}$ were studied by Moishezon [2], so we call them *Moishezon manifolds*.

In dimension $n \geqslant 2$, there are compact complex manifolds with no nonconstant meromorphic functions at all, so these cannot be algebraic. For example, a complex torus $\mathbf{C}^n/\Lambda$, where $\Lambda \cong \mathbf{Z}^{2n}$ is a sufficiently general lattice, for $n \geqslant 2$, will have this property. See for example Morrow and Kodaira [1].

Restricting our attention to Moishezon manifolds, we have the following theorem in dimension 2.

Theorem 3.4 (Chow and Kodaira [1]). *A compact complex manifold of dimension 2, with two algebraically independent meromorphic functions, is projective algebraic.*

In dimensions ≥ 3, both Hironaka [2] and Moishezon [2] have given examples of Moishezon manifolds which are not algebraic. They exist in *every* birational equivalence class of algebraic varieties of dimension ≥ 3 over **C**. However, Moishezon shows that any Moishezon manifold becomes projective algebraic after a finite number of monoidal transformations with nonsingular centers, so they are not too far from being algebraic.

Example 3.4.1 (Hironaka [2]). We describe two examples with a similar construction. The first is a nonsingular complete algebraic three-fold over **C** which is not projective. The second is a Moishezon manifold of dimension three which is not algebraic.

For the first example, let X be any nonsingular projective algebraic three-fold. Take two nonsingular curves $c,d \subseteq X$ which meet transversally at two points P,Q, and nowhere else (Fig. 24). On $X - Q$, first blow up the curve c, then blow up the strict transform of the curve d. On $X - P$, first

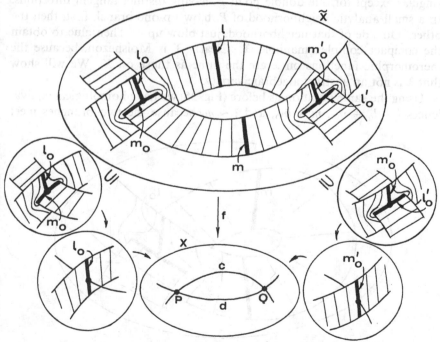

Figure 24. A complete nonprojective variety.

blow up the curve d, then blow up the strict transform of the curve c. On $X - P - Q$ it doesn't matter in which order we blow up the curves c and d, so we can glue our blown-up varieties along the inverse images of $X - P - Q$. The result is a nonsingular complete algebraic variety $\tilde{X}$. We will show that $\tilde{X}$ is not projective. It follows, incidentally, that the birational morphism $f: \tilde{X} \to X$ cannot be factored into any sequence of monoidal transformations, because it is not a projective morphism.

To do this, we must examine what happens in a neighborhood of P (Fig. 24). Let l be the inverse image in $\tilde{X}$ of a general point of c. Let m be the inverse image of a general point of d. Note that l and m are projective lines. Then the inverse image of P consists of two lines l_0 and m_0, and we have algebraic equivalence of cycles $l \sim l_0 + m_0$ and $m \sim m_0$. Note the asymmetry resulting from the order in which we blew up the two curves. Now in the neighborhood of Q the opposite happens. So $f^{-1}(Q)$ is the union of two lines l'_0 and m'_0, and we have algebraic equivalence $l \sim l_0$ and $m \sim l'_0 + m'_0$. Combining these equivalences, we find that $l_0 + m'_0 \sim 0$. This would be impossible on a projective variety, because a curve has a degree, which is a positive integer, and degrees are additive and are preserved by algebraic equivalence. So $\tilde{X}$ is not projective.

Example 3.4.2. For the second example, we start with any nonsingular projective algebraic threefold, as before. Let c be a curve in X which is nonsingular except for one double point P, having distinct tangent directions. In a small analytic neighborhood of P, blow up one branch first, then the other. Outside of that neighborhood, just blow up c. Then glue to obtain the compact complex manifold $\mathfrak{X}$. Clearly $\mathfrak{X}$ is Moishezon, because the meromorphic functions on $\mathfrak{X}$ are the same as those on X. We will show that $\mathfrak{X}$ is not an abstract algebraic variety.

Using the same notation as before (Fig. 25) we have homological equivalences $l \sim l_0 + m_0$, $m \sim m_0$, and $l \sim m$, because the two branches meet

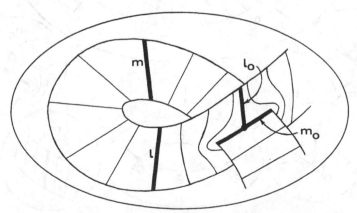

Figure 25.   A nonalgebraic Moishezon manifold.

away from P. So we find that $l_0 \sim 0$. But this is impossible if $\mathfrak{X}$ is algebraic. Indeed, let T be a point of l_0. Then T has an affine neighborhood U in $\mathfrak{X}$. Let Y be an irreducible surface in U which passes through T but does not contain l_0. Extend Y by closure to a surface Y in $\mathfrak{X}$. Now Y meets l_0 in a finite nonzero number of points, so the intersection number of Y with l_0 is defined and is $\neq 0$. But the intersection number is defined on homology classes, so we cannot have $l_0 \sim 0$. Hence $\mathfrak{X}$ is not algebraic.

Remark 3.4.3. At this point we should also mention the algebraic spaces of Artin [2] and Knutson [1]. Over any field k, they define an *algebraic space* to be something which is locally a quotient of a scheme by an étale equivalence relation. The category of algebraic spaces contains the category of schemes. If X is an algebraic space of finite type over $\mathbf{C}$, one can define its associated complex analytic space X_h. Artin shows that the category of smooth proper algebraic spaces over $\mathbf{C}$ is equivalent, via the functor h, to the category of Moishezon manifolds. Thus every Moishezon manifold is "algebraic" in the sense of algebraic spaces. In particular, Hironaka's example (3.4.2) gives an example of an algebraic space over $\mathbf{C}$ which is not a scheme.

4 Kähler Manifolds

The methods of differential geometry provide a powerful tool for the study of compact complex manifolds, and hence of algebraic varieties over the complex numbers. Notable among such applications of differential geometry are Hodge's [1] theory of harmonic integrals, and the resulting decomposition of the complex cohomology into its (p,q)-components (see also Weil [5]); the vanishing theorems of Kodaira [1] and Nakano [1], recently generalized by Grauert and Riemenschneider [1]; and the work of Griffiths on the intermediate Jacobians and the period mapping. Here we will only mention the definition of a Kähler manifold, and how that notion helps to characterize algebraic complex manifolds.

Any complex manifold admits a Hermitian metric (in many ways). A Hermitian metric is said to be *Kähler* if the associated differential 2-form of type $(1,1)$ is closed. A complex manifold with a Kähler metric is called a *Kähler manifold*. One can show easily that complex projective space has a natural Kähler metric on it, and hence that every projective algebraic manifold is a Kähler manifold with the induced metric. A compact Kähler manifold $\mathfrak{X}$ is called a *Hodge manifold* if the cohomology class in $H^2(\mathfrak{X},\mathbf{C})$ of the 2-form mentioned above is in the image of the integral cohomology $H^2(\mathfrak{X},\mathbf{Z})$. Now a fundamental result is

Theorem 4.1 (Kodaira [2]). *Every Hodge manifold is projective algebraic.*

445

This can be thought of as a generalization of the theorem of Riemann (3.1) quoted above, because every compact complex manifold of dimension one is trivially seen to be a Hodge manifold. We also have the following

Theorem 4.2 (Moishezon [2]). *Every Moishezon manifold which is Kähler is projective algebraic.*

Summing up, we have the following implications among properties of compact complex manifolds, and there are examples to show that no further implications are possible.

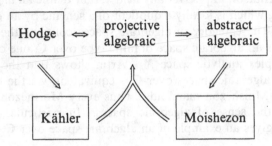

5 The Exponential Sequence

Let us give one simple example of the use of transcendental methods, by looking at the exponential sequence. The exponential function $f(x) = e^{2\pi i x}$ gives an exact sequence of abelian groups

$$0 \to \mathbf{Z} \to \mathbf{C} \xrightarrow{f} \mathbf{C}^* \to 0,$$

where $\mathbf{C}$ has its additive structure, and $\mathbf{C}^* = \mathbf{C} - \{0\}$ has its multiplicative structure. If $\mathfrak{X}$ is any reduced complex analytic space, by considering holomorphic functions with values in the above sequence, we get an exact sequence of sheaves

$$0 \to \mathbf{Z} \to \mathcal{O}_{\mathfrak{X}} \xrightarrow{f} \mathcal{O}_{\mathfrak{X}}^* \to 0,$$

where $\mathbf{Z}$ is the constant sheaf, $\mathcal{O}_{\mathfrak{X}}$ is the structure sheaf, and $\mathcal{O}_{\mathfrak{X}}^*$ is the sheaf of invertible elements of $\mathcal{O}_{\mathfrak{X}}$ under multiplication.

The cohomology sequence of this short exact sequence of sheaves is very interesting. Let us apply it to X_h, where X is a projective variety over $\mathbf{C}$. At the H^0 level, we recover the original exact sequence of groups $0 \to \mathbf{Z} \to \mathbf{C} \to \mathbf{C}^* \to 0$, because the global holomorphic functions are constant. Then starting with H^1, we have an exact sequence

$$0 \to H^1(X_h, \mathbf{Z}) \to H^1(X_h, \mathcal{O}_{X_h}) \to H^1(X_h, \mathcal{O}_{X_h}^*) \to H^2(X_h, \mathbf{Z}) \to H^2(X_h, \mathcal{O}_{X_h}) \to \dots .$$

By Serre's theorem (2.1), we have $H^i(X_h, \mathcal{O}_{X_h}) \cong H^i(X, \mathcal{O}_X)$. On the other hand

$$H^1(X_h, \mathcal{O}_{X_h}^*) \cong \operatorname{Pic} X_h$$

by (III, Ex. 4.5), which is valid for any ringed space. But Serre's theorem (2.1) also gives an equivalence of categories of coherent sheaves, so in particular, Pic $X_h \cong$ Pic X. So we can rewrite our sequence as

$$0 \to H^1(X_h, \mathbf{Z}) \to H^1(X, \mathcal{O}_X) \to \text{Pic } X \to H^2(X_h, \mathbf{Z}) \to H^2(X, \mathcal{O}_X) \to \ldots .$$

The only nonalgebraic part is the integral cohomology of X_h. Since any algebraic variety is triangulable (see for example, Hironaka [5]), the cohomology groups $H^i(X_h, \mathbf{Z})$ are finitely generated abelian groups. From this sequence we can deduce some information about the Picard group of X.

First of all, one sees easily that algebraically equivalent Cartier divisors give the same element in $H^2(X_h, \mathbf{Z})$. Therefore the Néron–Severi group of X is a subgroup of $H^2(X_h, \mathbf{Z})$, and hence is finitely generated (V, Ex. 1.7). On the other hand, the group Pic$^\circ X$ of divisors algebraically equivalent to zero modulo linear equivalence is isomorphic to $H^1(X, \mathcal{O}_X)/H^1(X_h, \mathbf{Z})$. One shows that this is a complex torus, and in fact it is an abelian variety, the *Picard variety* of X.

If X is a nonsingular curve of genus g, we can see even more clearly what is happening. In that case X_h is a compact Riemann surface of genus g. As a topological space, it is a compact oriented real 2-manifold which is homeomorphic to a sphere with g handles. So we have

$$H^0(X_h, \mathbf{Z}) = H^2(X_h, \mathbf{Z}) = \mathbf{Z}, \qquad \text{and} \qquad H^1(X_h, \mathbf{Z}) \cong \mathbf{Z}^{2g}.$$

On the other hand, $H^1(X, \mathcal{O}_X) \cong \mathbf{C}^g$, so

$$\text{Pic}^\circ X \cong \mathbf{C}^g / \mathbf{Z}^{2g}.$$

This is the Jacobian variety of X (IV, §4), which is an abelian variety of dimension g. Of course $\text{NS}(X) \cong \mathbf{Z}$ in this case, the isomorphism being given by the degree function.

EXERCISES

6.1. Show that the unit disc in $\mathbf{C}$ is not isomorphic to X_h for any scheme X.

6.2. Let $z_1, z_2, \ldots$ be an infinite sequence of complex numbers with $|z_n| \to \infty$ as $n \to \infty$. Let $\mathfrak{I} \subseteq \mathcal{O}_\mathbf{C}$ be the sheaf of ideals of holomorphic functions vanishing at all of the z_n. Show that there is no coherent algebraic sheaf of ideals $\mathcal{I} \subseteq \mathcal{O}_X$, where $X = \mathbf{A}^1_\mathbf{C}$ is the affine line, such that $\mathfrak{I} = \mathcal{I}_h$ as an ideal in $\mathcal{O}_\mathbf{C}$. Show on the other hand that there is a coherent sheaf $\mathcal{F}$ on X such that $\mathcal{F}_h \cong \mathfrak{I}$ as coherent sheaves.

6.3. (Serre [12].) On $\mathbf{C}^2 - \{0, 0\}$, we define an invertible analytic sheaf $\mathfrak{L}$ as follows: $\mathfrak{L} \cong \mathcal{O}$ when $z \neq 0$; $\mathfrak{L} \cong \mathcal{O}$ when $w \neq 0$, and when both $z, w \neq 0$, the two copies of $\mathcal{O}$ are glued by multiplication by $e^{-1/zw}$ in the local ring at the point (z, w). Show that there is no invertible algebraic sheaf $\mathcal{L}$ on $\mathbf{A}^2 - \{0, 0\}$ with $\mathcal{L}_h \cong \mathfrak{L}$.

6.4. Show directly that if X is a scheme which is reduced and proper over $\mathbf{C}$, then $H^0(X, \mathcal{O}_X) \cong H^0(X_h, \mathcal{O}_{X_h})$. Conversely, show that if X is not proper over $\mathbf{C}$, then there is a coherent sheaf $\mathcal{F}$ on X with $H^0(X, \mathcal{F}) \neq H^0(X_h, \mathcal{F}_h)$.

6.5. If X,X' are nonsingular affine algebraic curves, with $X_h \cong X'_h$, show that $X \cong X'$.

6.6. Show that if X and Y are projective schemes over $\mathbf{C}$, and $\mathfrak{f}: X_h \to Y_h$ is a morphism of analytic spaces, then there exists a (unique) morphism $f: X \to Y$ with $f_h = \mathfrak{f}$. [*Hint*: First reduce to the case $Y = \mathbf{P}^n$. Then consider the invertible analytic sheaf $\mathfrak{L} = \mathfrak{f}^* \mathcal{O}(1)$ on X_h, use (2.1), and the techniques of (II, §7).]

APPENDIX C
The Weil Conjectures

In 1949, André Weil [4] stated his now famous conjectures concerning the number of solutions of polynomial equations over finite fields. These conjectures suggested a deep connection between the arithmetic of algebraic varieties defined over finite fields and the topology of algebraic varieties defined over the complex numbers. Weil also pointed out that if one had a suitable cohomology theory for abstract varieties, analogous to the ordinary cohomology of varieties defined over $\mathbf{C}$, then one could deduce his conjectures from various standard properties of the cohomology theory. This observation has been one of the principal motivations for the introduction of various cohomology theories into abstract algebraic geometry. In 1963, Grothendieck was able to show that his l-adic cohomology had sufficient properties to imply part of the Weil conjectures (the rationality of the zeta function). Deligne's [3] proof in 1973 of the remainder of the Weil conjectures (specifically the analogue of the "Riemann hypothesis") may be regarded as the culmination of the study of l-adic cohomology begun by Grothendieck, M. Artin, and others in the Paris seminars [SGA 4], [SGA 5], and [SGA 7].

1 The Zeta Function and the Weil Conjectures

Let $k = \mathbf{F}_q$ be a finite field with q elements. Let X be a scheme of finite type over k. For example, X could be the set of solutions in affine or projective space over k of a finite number of polynomial equations with coefficients in k. Let $\bar{k}$ be an algebraic closure of k, and let $\bar{X} = X \times_k \bar{k}$ be the corresponding scheme over $\bar{k}$. For each integer $r \geq 1$, let N_r be the number of points of $\bar{X}$ which are rational over the field $k_r = \mathbf{F}_{q^r}$ of q^r elements. In other words, N_r is the number of points of $\bar{X}$ whose coordinates lie in k_r.

The numbers $N_1, N_2, N_3, \ldots$ are clearly of great importance in studying arithmetical properties of the scheme X. To study them, we form the *zeta function* of X (following Weil), which is defined as

$$Z(t) = Z(X; t) = \exp\left(\sum_{r=1}^{\infty} N_r \frac{t^r}{r}\right).$$

Note that by definition, it is a power series with rational coefficients: $Z(t) \in \mathbf{Q}[[t]]$.

For example, let $X = \mathbf{P}^1$. Over any field, $\mathbf{P}^1$ has one more point than the number of elements of the field. Hence $N_r = q^r + 1$. Thus

$$Z(\mathbf{P}^1, t) = \exp\left(\sum_{r=1}^{\infty} (q^r + 1) \frac{t^r}{r}\right).$$

It is easy to sum this series, and we find that

$$Z(\mathbf{P}^1, t) = \frac{1}{(1 - t)(1 - qt)}.$$

In particular, it is a rational function of t.

Now we can state the Weil conjectures. Let X be a smooth projective variety of dimension n defined over $k = \mathbf{F}_q$. Let $Z(t)$ be the zeta function of X. Then

1.1. *Rationality.* $Z(t)$ is a rational function of t, i.e., a quotient of polynomials with rational coefficients.

1.2. *Functional equation.* Let E be the self-intersection number of the diagonal Δ of $X \times X$ (which is also the top Chern class of the tangent bundle of X (App. A, Ex. 6.6)). Then $Z(t)$ satisfies a functional equation, namely

$$Z\left(\frac{1}{q^n t}\right) = \pm q^{nE/2} t^E Z(t).$$

1.3. *Analogue of the Riemann hypothesis.* It is possible to write

$$Z(t) = \frac{P_1(t) P_3(t) \cdots P_{2n-1}(t)}{P_0(t) P_2(t) \cdots P_{2n}(t)}$$

where $P_0(t) = 1 - t$; $P_{2n}(t) = 1 - q^n t$; and for each $1 \leqslant i \leqslant 2n - 1$, $P_i(t)$ is a polynomial with integer coefficients, which can be written

$$P_i(t) = \prod (1 - \alpha_{ij} t),$$

where the α_{ij} are algebraic integers with $|\alpha_{ij}| = q^{i/2}$. (Note that these conditions uniquely determine the polynomials $P_i(t)$, if they exist.)

1.4. *Betti numbers.* Assuming (1.3), we can define the *i*th Betti number $B_i = B_i(X)$ to be the degree of the polynomial $P_i(t)$. Then we have $E = \sum (-1)^i B_i$. Furthermore, suppose that X is obtained from a variety Y defined over an algebraic number ring R, by reduction modulo a prime ideal $\mathfrak{p}$ of R. Then $B_i(X)$ is equal to the *i*th Betti number of the topological space $Y_h = (Y \times_R \mathbf{C})_h$ (App. B), i.e., $B_i(X)$ is the rank of the ordinary cohomology group $H^i(Y_h, \mathbf{Z})$.

Let us verify the conjectures for the case $X = \mathbf{P}^1$. We have already seen that $Z(t)$ is rational. The invariant E of $\mathbf{P}^1$ is 2, and one verifies immediately the functional equation which says in this case

$$Z\left(\frac{1}{qt}\right) = qt^2 Z(t).$$

The analogue of the Riemann hypothesis is immediate, with $P_1(t) = 1$. Hence $B_0 = B_2 = 1$ and $B_1 = 0$. These are indeed the usual Betti numbers of $\mathbf{P}^1_{\mathbf{C}}$, which is a sphere, and finally we have $E = \sum (-1)^i B_i$.

2 History of Work on the Weil Conjectures

Weil was led to his conjectures by consideration of the zeta functions of some special varieties. See his article Weil [4] for number-theoretic background, and calculations for the "Fermat hypersurfaces" $\sum a_i x_i^n = 0$. One of Weil's major pieces of work was the proof that his conjectures hold for curves. This is done in his book Weil [2]. The rationality and the functional equation follow from the Riemann–Roch theorem on the curve. The analogue of the Riemann hypothesis is deeper (V, Ex. 1.10). He deduces it from an inequality of Castelnuovo and Severi about correspondences on a curve (V, Ex. 1.9). This proof was later simplified by Mattuck and Tate [1] and Grothendieck [2]. Weil [3] also gave another proof using the *l*-adic representation of Frobenius on abelian varieties, which inspired the later cohomological approaches. Recently a completely independent elementary proof of the Riemann hypothesis for curves has been discovered by Stepanov, Schmidt and Bombieri (see Bombieri [1]).

For higher-dimensional varieties, the rationality of the zeta function and the functional equation were first proved by Dwork [1], using methods of *p*-adic analysis. See also Serre [8] for an account of this proof.

Most other work on the Weil conjectures has centered around the search for a good cohomology theory for varieties defined over fields of characteristic *p*, which would give the "right" Betti numbers as defined in (1.4) above. Furthermore, the cohomology theory should have its coefficients in a field of characteristic zero, so that one can count the fixed points of a morphism as a sum of traces on cohomology groups, à la Lefschetz.

The first cohomology introduced into abstract algebraic geometry was that of Serre [3] using coherent sheaves (Ch. III). Although it could not satisfy the present need, because of its coefficients being in the field over which the variety is defined, it served as a basis for the development of later cohomology theories. Serre [6] proposed a cohomology with coefficients in the Witt vectors, but was unable to prove much about it. Grothendieck, inspired by some of Serre's ideas, saw that one could obtain a good theory by considering the variety together with all its unramified covers. This was the beginning of his theory of étale topology, developed jointly with M. Artin, which he used to define the l-adic cohomology, and thus to obtain another proof of the rationality and functional equation of the zeta function. See Grothendieck [4] for a brief announcement; Artin [1] and Grothendieck [SGA 4] for the foundations of étale cohomology; Grothendieck [6] for the proof of rationality of the zeta function, modulo general facts about l-adic cohomology which are supposed to appear in [SGA 5] (as yet unpublished). Lubkin [1] more or less independently developed a p-adic cohomology theory which led also to a proof of rationality and the functional equation, for varieties which could be lifted to characteristic zero. The crystalline cohomology of Grothendieck [8] and Berthelot [1] gives another similar cohomological interpretation of the Weil conjectures.

The analogue of the Riemann hypothesis has proved more difficult to handle. Lang and Weil [1] established an inequality for n-dimensional varieties, which is equivalent to the analogue of the Riemann hypothesis if $n = 1$, but falls far short of it if $n \geqslant 2$. Serre [9] established another analogue of the Riemann hypothesis for the eigenvalues of certain operators on the cohomology of a Kähler manifold, using the powerful results of Hodge theory. This suggests that one should try to establish in abstract algebraic geometry some results known for varieties over $\mathbf{C}$ via Hodge theory, in particular the "strong Lefschetz theorem" and the "generalized Hodge index theorem." Grothendieck [9] optimistically calls these the "standard conjectures," and notes that they immediately imply the analogue of the Riemann hypothesis. See also Kleiman [2] for a more detailed account of these conjectures and their interrelations.

Until Deligne's proof [3] of the general analogue of the Riemann hypothesis, only a few special cases were known: curves (above), rational threefolds by Manin [2]—see also Demazure [1], K3 surfaces by Deligne [2], and certain complete intersections by Deligne [5].

In addition to the references given above, I would like to mention Serre's survey article [10] and Tate's companion article [1] suggesting further (as yet untouched) conjectures about cycles on varieties over fields of characteristic p. Also, for the number-theorists, Deligne's article [1] which shows that the analogue of the Riemann hypothesis implies the conjecture of Ramanujan about the τ-function.

3 The *l*-adic Cohomology

In this and the following section we will describe the cohomological interpretation of the Weil conjectures in terms of the *l*-adic cohomology of Grothendieck. Similar results would hold in any cohomology theory with similar formal properties. See Kleiman [2] for an axiomatic treatment of a "Weil cohomology theory."

Let X be a scheme of finite type over an algebraically closed field k of characteristic $p \geqslant 0$. Let l be a prime number $l \neq p$. Let $\mathbf{Z}_l = \varprojlim \mathbf{Z}/l^r\mathbf{Z}$ be the ring of *l*-adic integers, and $\mathbf{Q}_l$ its quotient field. We consider the étale topology of X (see Artin [1] or [SGA 4]), and then using étale cohomology, we define the *l*-adic cohomology of X by

$$H^i(X,\mathbf{Q}_l) = (\varprojlim H^i_{\text{ét}}(X,\mathbf{Z}/l^r\mathbf{Z})) \otimes_{\mathbf{Z}_l} \mathbf{Q}_l.$$

We will not go into a detailed explanation of this definition here (see [SGA $4\frac{1}{2}$]). Rather, we will content ourselves with listing some of the main properties of the *l*-adic cohomology.

3.1. The groups $H^i(X,\mathbf{Q}_l)$ are vector spaces over $\mathbf{Q}_l$. They are zero except in the range $0 \leqslant i \leqslant 2n$, where $n = \dim X$. They are known to be finite-dimensional if X is proper over k. (They are expected to be finite-dimensional in general, but there is no proof yet because of the problem of resolution of singularities in characteristic $p > 0$.)

3.2. $H^i(X,\mathbf{Q}_l)$ is a contravariant functor in X.

3.3. There is a cup-product structure

$$H^i(X,\mathbf{Q}_l) \times H^j(X,\mathbf{Q}_l) \to H^{i+j}(X,\mathbf{Q}_l)$$

defined for all i,j.

3.4. *Poincaré duality.* If X is smooth and proper over k, of dimension n, then $H^{2n}(X,\mathbf{Q}_l)$ is 1-dimensional, and the cup-product pairing

$$H^i(X,\mathbf{Q}_l) \times H^{2n-i}(X,\mathbf{Q}_l) \to H^{2n}(X,\mathbf{Q}_l)$$

is a perfect pairing for each i, $0 \leqslant i \leqslant 2n$.

3.5. *Lefschetz fixed-point formula.* Let X be smooth and proper over k. Let $f:X \to X$ be a morphism with isolated fixed points, and for each fixed point $x \in X$, assume that the action of $1 - df$ on Ω^1_X is injective. This last condition says that the fixed point has "multiplicity 1." Let $L(f,X)$ be the number of fixed points of f. Then

$$L(f,X) = \sum (-1)^i \operatorname{Tr}(f^*; H^i(X,\mathbf{Q}_l))$$

where f^* is the induced map on the cohomology of X.

3.6. If $f : X \to Y$ is a smooth proper morphism, with Y connected, then $\dim H^i(X_y, \mathbf{Q}_l)$ is constant for $y \in Y$. In particular, $\dim H^i(X, \mathbf{Q}_l)$ is constant under base field extension.

3.7. *Comparison theorem.* If X is smooth and proper over $\mathbf{C}$, then

$$H^i(X, \mathbf{Q}_l) \otimes_{\mathbf{Q}_l} \mathbf{C} \cong H^i(X_h, \mathbf{C})$$

where X_h is the associated complex manifold in its classical topology (App. B).

3.8. *Cohomology class of a cycle.* If X is smooth and proper over k, and if Z is a subvariety of codimension q, then there is associated to Z a cohomology class $\eta(Z) \in H^{2q}(X, \mathbf{Q}_l)$. This map extends by linearity to cycles. Rationally equivalent cycles have the same cohomology class. Intersection of cycles becomes cup-product of cohomology classes. In other words, η is a homomorphism from the Chow ring $A(X)$ to the cohomology ring $H^*(X, \mathbf{Q}_l)$. Finally, it is non-trivial: if $P \in X$ is a closed point, then $\eta(P) \in H^{2n}(X, \mathbf{Q}_l)$ is nonzero.

This list of properties has no pretensions to completeness. In particular, we have not mentioned sheaves of twisted coefficients, higher direct images, Leray spectral sequence, and so forth. For further properties, as well as for the proofs of the properties given above, we refer to [SGA 4] for the corresponding statement with torsion coefficients, and to [SGA 5] for the passage to the limit of $\mathbf{Z}_l$ or $\mathbf{Q}_l$ coefficients.

4 Cohomological Interpretation of the Weil Conjectures

Using the l-adic cohomology described above, we can give a cohomological interpretation of the Weil conjectures. The main idea, which goes back to Weil, is very simple. Let X be a projective variety defined over the finite field $k = \mathbf{F}_q$, and let $\bar{X} = X \times_k \bar{k}$ be the corresponding variety over the algebraic closure $\bar{k}$ of k. We define the *Frobenius morphism* $f : \bar{X} \to \bar{X}$ by sending the point P with coordinates (a_i), $a_i \in \bar{k}$, to the point $f(P)$ with coordinates (a_i^q). This is the $\bar{k}$-linear Frobenius morphism, where $\bar{X}_q$ is identified with $\bar{X}$ (IV, 2.4.1). Since $\bar{X}$ is defined by equations with coefficients in k, $f(P)$ is also a point of $\bar{X}$. Furthermore, P is a fixed point of f if and only if its coordinates lie in k. More generally, P is a fixed point of the iterate f^r if and only if it has coordinates in the field $k_r = \mathbf{F}_{q^r}$. Thus in the notation of §1, we have

$$N_r = \# \ \{\text{fixed points of } f^r\} = L(f^r, \bar{X}).$$

If X is smooth, we can calculate this number by the Lefschetz fixed-point formula (3.5). We find

$$N_r = \sum_{i=0}^{2n} (-1)^i \, \mathrm{Tr}(f^{r*}; H^i(\bar{X}, \mathbf{Q}_l)).$$

Substituting in the definition of the zeta function, we have

$$Z(X,t) = \prod_{i=0}^{2n} \left[\exp \left(\sum_{r=1}^{\infty} \text{Tr}(f^{r*}; H^i(\bar{X}, \mathbf{Q}_l)) \frac{t^r}{r} \right) \right]^{(-1)^i}$$

To simplify this expression, we need an elementary lemma.

Lemma 4.1. *Let φ be an endomorphism of a finite-dimensional vector space V over a field K. Then we have an identity of formal power series in t, with coefficients in K,*

$$\exp \left(\sum_{r=1}^{\infty} \text{Tr}(\varphi^r; V) \frac{t^r}{r} \right) = \det(1 - \varphi t; V)^{-1}.$$

PROOF. If $\dim V = 1$, then φ is multiplication by a scalar $\lambda \in K$, and it says

$$\exp \left(\sum_{r=1}^{\infty} \lambda^r \frac{t^r}{r} \right) = \frac{1}{1 - \lambda t}.$$

This is an elementary calculation, which we already did in computing the zeta function of $\mathbf{P}^1$. For the general case, we use induction on $\dim V$. Furthermore, we may clearly assume that K is algebraically closed. Hence φ has an eigenvector, so we have an invariant subspace $V' \subseteq V$. We use the exact sequence

$$0 \to V' \to V \to V/V' \to 0$$

and the fact that both sides of the above equation are multiplicative for short exact sequences of vector spaces. By induction, this gives the result.

Using the lemma, we immediately obtain the following result.

Theorem 4.2. *Let X be projective and smooth over $k = \mathbf{F}_q$, of dimension n. Then*

$$Z(X,t) = \frac{P_1(t) \cdots P_{2n-1}(t)}{P_0(t) \cdots P_{2n}(t)},$$

where

$$P_i(t) = \det(1 - f^*t; H^i(\bar{X}, \mathbf{Q}_l))$$

and f^* is the map on cohomology induced by the Frobenius morphism f: $\bar{X} \to \bar{X}$.

This theorem shows immediately that $Z(t)$ is a quotient of polynomials with $\mathbf{Q}_l$ coefficients. One can show by an elementary argument on power series (Bourbaki [2, Ch. IV §5, Ex. 3, p. 66]) that $\mathbf{Q}[[t]] \cap \mathbf{Q}_l(t) = \mathbf{Q}(t)$. Since we know that $Z(t)$ is a power series with rational coefficients, we deduce that $Z(t)$ is a rational function, which proves (1.1). Notice, however, that we do not know yet whether the $P_i(t)$ have rational coefficients, and we do not know whether they are the polynomials referred to in (1.3) above.

We can extract a bit more information from this theorem. Since f^* acts on $H^0(\bar{X},\mathbf{Q}_l)$ as the identity, $P_0(t) = 1 - t$. Furthermore, we can determine $P_{2n}(t)$. The Frobenius morphism is a finite morphism of degree q^n. Hence it must act as multiplication by q^n on a generator of $H^{2n}(X,\mathbf{Q}_l)$. So $P_{2n}(t) = 1 - q^n t$. If we provisionally define the ith Betti number B_i as dim $H^i(\bar{X},\mathbf{Q}_l)$, then $B_i = $ degree $P_i(t)$, and one can show easily that the invariant E of X is given by

$$E = \sum(-1)^i B_i.$$

So we call E the "topological Euler–Poincaré characteristic" of X. We do not yet know that this definition of the Betti numbers agrees with the one in (1.4) above. However, once we do know this, the statement (1.4) will follow from the general properties (3.6) and (3.7) of the l-adic cohomology.

Next, we will show that the functional equation follows from Poincaré duality. Again we need a lemma from linear algebra.

Lemma 4.3. *Let $V \times W \to K$ be a perfect pairing of vector spaces V,W of dimension r over K. Let $\lambda \in K$, and let $\varphi : V \to V$ and $\psi : W \to W$ be endomorphisms such that*

$$\langle \varphi v, \psi w \rangle = \lambda \langle v, w \rangle$$

for all $v \in V$, $w \in W$. Then

$$\det(1 - \psi t, W) = \frac{(-1)^r \lambda^r t^r}{\det(\varphi; V)} \det\left(1 - \frac{\varphi}{\lambda t}; V\right)$$

and

$$\det(\psi; W) = \frac{\lambda^r}{(\det(\varphi; V)}.$$

Theorem 4.4. *With the hypotheses of (4.2), the zeta function $Z(X,t)$ satisfies the functional equation (1.2).*

PROOF. One applies the lemma (whose proof is elementary) to the pairings $H^i(X,\mathbf{Q}_l) \times H^{2n-i}(X,\mathbf{Q}_l) \to H^{2n}(X,\mathbf{Q}_l)$ given us by Poincaré duality (3.4). Using the fact that f^* is compatible with cup-product, and that it acts by multiplication by q^n on $H^{2n}(X,\mathbf{Q}_l)$, we get an expression for P_{2n-i} in terms of P_i, namely

$$P_{2n-i}(t) = (-1)^{B_i} \frac{q^{nB_i} t^{B_i}}{\det(f^*; H^i)} P_i\left(\frac{1}{q^n t}\right).$$

Furthermore, we have

$$\det(f^*; H^{2n-i}) = \frac{q^{nB_i}}{\det(f^*; H^i)}.$$

Substituting these in the formula of (4.2) and using $E = \sum(-1)^i B_i$, we obtain the functional equation.

So we see that the conjectures (1.1), (1.2), and (1.4) follow from the formal properties of l-adic cohomology once we have interpreted the zeta function as in (4.2). The analogue of the Riemann hypothesis is much deeper.

Theorem 4.5 (Deligne [3]). *With the hypotheses of* (4.2), *the polynomials* $P_i(t)$ *have integer coefficients, independent of l, and they can be written*

$$P_i(t) = \prod(1 - \alpha_{ij}t)$$

where the α_{ij} are algebraic integers with $|\alpha_{ij}| = q^{i/2}$.

This result completes the solution of the Weil conjectures. Note that it implies that the polynomials $P_i(t)$ of (4.2) are the same as those of (1.3), and hence the two definitions of the Betti numbers agree.

We cannot describe the proof of Deligne's theorem here, except to say that it relies on the deeper properties of l-adic cohomology developed in [SGA 4], [SGA 5] and [SGA 7]. In particular it makes use of Lefschetz's technique of fibering a variety by a "Lefschetz pencil," and studying the monodromy action on the cohomology near a singular fibre.

EXERCISES

5.1. Let X be a disjoint union of locally closed subschemes X_i. Then show that

$$Z(X,t) = \prod Z(X_i,t).$$

5.2. Let $X = \mathbf{P}_k^n$, where $k = \mathbf{F}_q$, and show from the definition of the zeta function that

$$Z(\mathbf{P}^n,t) = \frac{1}{(1 - t)(1 - qt) \cdots (1 - q^n t)}.$$

Verify the Weil conjectures for $\mathbf{P}^n$.

5.3. Let X be a scheme of finite type over $\mathbf{F}_q$, and let $\mathbf{A}^1$ be the affine line. Show that

$$Z(X \times \mathbf{A}^1,t) = Z(X,qt).$$

5.4. The *Riemann zeta function* is defined as

$$\zeta(s) = \prod \frac{1}{1 - p^{-s}},$$

for $s \in \mathbf{C}$, the product being taken over all prime integers p. If we regard this function as being associated with the scheme Spec $\mathbf{Z}$, it is natural to define, for any scheme X of finite type over Spec $\mathbf{Z}$,

$$\zeta_X(s) = \prod(1 - N(x)^{-s})^{-1}$$

where the product is taken over all closed points $x \in X$, and $N(x)$ denotes the number of elements in the residue field $k(x)$. Show that if X is of finite type over $\mathbf{F}_q$, then this function is connected to $Z(X,t)$ by the formula

$$\zeta_X(s) = Z(X,q^{-s}).$$

[*Hint*: Take $d$log of both sides, replace q^{-s} by t, and compare.]

5.5. Let X be a curve of genus g over k. Assuming the statements (1.1) to (1.4) of the Weil conjectures, show that $N_1, N_2, \ldots, N_g$ determine N_r for all $r \geqslant 1$.

5.6. Use (IV, Ex. 4.16) to prove the Weil conjectures for elliptic curves. First note that for any r,

$$N_r = q^r - (f^r + \hat{f}^r) + 1,$$

where $f = F'$. Then calculate $Z(t)$ formally and conclude that

$$Z(t) = \frac{(1 - ft)(1 - \hat{f}t)}{(1 - t)(1 - qt)}$$

and hence

$$Z(t) = \frac{1 - at + qt^2}{(1 - t)(1 - qt)},$$

where $f + \hat{f} = a_X$. This proves rationality immediately. Verify the functional equation. Finally, if we write

$$1 - at + qt^2 = (1 - \alpha t)(1 - \beta t),$$

show that $|a| \leqslant 2\sqrt{q}$ if and only if $|\alpha| = |\beta| = \sqrt{q}$. Thus the analogue of the Riemann hypothesis is just (IV, Ex. 4.16d).

5.7. Use (V, Ex. 1.10) to prove the analogue of the Riemann hypothesis (1.3) for any curve C of genus g defined over $\mathbf{F}_q$. Write $N_r = 1 - a_r + q^r$. Then according to (V, Ex. 1.10),

$$|a_r| \leqslant 2g\sqrt{q^r}.$$

On the other hand, by (4.2) the zeta function of C can be written

$$Z(t) = \frac{P_1(t)}{(1 - t)(1 - qt)}$$

where

$$P_1(t) = \prod_{i=1}^{2g} (1 - \alpha_i t)$$

is a polynomial of degree $2g = \dim H^1(C, \mathbf{Q}_l)$.

(a) Using the definition of the zeta function and taking logs, show that

$$a_r = \sum_{i=1}^{2g} (\alpha_i)^r$$

for each r.

(b) Next show that

$$|a_r| \leqslant 2g\sqrt{q^r} \quad \text{for all } r \quad \Leftrightarrow \quad |\alpha_i| \leqslant \sqrt{q} \quad \text{for all } i.$$

[*Hint*: One direction is easy. For the other, use the power series expansion

$$\sum_{i=1}^{2g} \frac{\alpha_i t}{1 - \alpha_i t} = \sum_{r=1}^{\infty} a_r t^r$$

for suitable $t \in \mathbf{C}$.]

(c) Finally, use the functional equation (4.4) to show that $|\alpha_i| \leqslant \sqrt{q}$ for all i implies that $|\alpha_i| = \sqrt{q}$ for all i.

Bibliography

Altman, A. and Kleiman, S.
1. *Introduction to Grothendieck Duality Theory*, Lecture Notes in Math. *146* Springer-Verlag, Heidelberg (1970), 185 pp.

Artin, M.
1. *Grothendieck topologies*, Harvard Math. Dept. Lecture Notes (1962).
2. The implicit function theorem in algebraic geometry, in *Algebraic Geometry, Bombay 1968*, Oxford Univ. Press, Oxford (1969), 13–34.
3. Some numerical criteria for contractibility of curves on algebraic surfaces, *Amer. J. Math. 84* (1962) 485–496.
4. Algebraization of formal moduli II: Existence of modifications, *Annals of Math. 91* (1970) 88–135.

Atiyah, M. F.
1. Vector bundles over an elliptic curve, *Proc. Lond. Math. Soc.* (3) VII *27* (1957), 414–452.

Atiyah, M. F. and Macdonald, I. G.
1. *Introduction to Commutative Algebra.* Addison-Wesley, Reading, Mass. (1969), ix + 128 pp.

Baily, W. L., Jr.
1. On the automorphism group of a generic curve of genus >2, *J. Math. Kyoto Univ. 1* (1961/2) 101–108; correction p. 325.

Bănică, C. and Stănăşilă, O.
1. *Algebraic Methods in the Global Theory of Complex Spaces*, John Wiley, New York (1976) 296 pp.

Berthelot, P.
1. *Cohomologie Cristalline des Schémas de Caractéristique p > 0*, Lecture Notes in Math. *407*, Springer-Verlag, Heidelberg (1974), 604 pp.

Birch, B. J. and Kuyk, W., ed.
1. *Modular Functions of One Variable, IV (Antwerp)*, Lecture Notes in Math. *476*, Springer-Verlag, Heidelberg (1975).

459

Bombieri, E.
   1. Counting points on curves over finite fields (d'après S. A. Stepanov), *Séminaire Bourbaki 430* (1972/73).

Bombieri, E. and Husemoller, D.
   1. Classification and embeddings of surfaces, in *Algebraic Geometry, Arcata 1974*, Amer. Math. Soc. Proc. Symp. Pure Math. *29* (1975), 329–420.

Borel, A. and Serre, J.-P.
   1. Le théorème de Riemann-Roch, *Bull. Soc. Math. de France 86* (1958), 97–136.

Borelli, M.
   1. Divisorial varieties, *Pacific J. Math. 13* (1963), 375–388.

Bourbaki, N.
   1. *Algèbre Commutative*, Eléments de Math. *27, 28, 30, 31* Hermann, Paris (1961–1965).
   2. *Algèbre*, Eléments de Math. *4, 6, 7, 11, 14, 23, 24*, Hermann, Paris (1947–59).

Brieskorn, E.
   1. Beispiele zur Differentialtopologie von Singularitäten, *Invent. Math. 2* (1966) 1–14.
   2. Rationale Singularitäten Komplexer Flächen, *Invent. Math. 4* (1968) 336–358.

Burnside, W.
   1. *Theory of Groups of Finite Order*, Cambridge Univ. Press, Cambridge (1911); reprinted by Dover, New York.

Cartan, H. and Chevalley, C.
   1. *Géométrie Algébrique*, Séminaire Cartan–Chevalley, Secrétariat Math., Paris (1955/56).

Cartan, H. and Eilenberg, S.
   1. *Homological Algebra*, Princeton Univ. Press, Princeton (1956), xv + 390 pp.

Cassels, J. W. S.
   1. Diophantine equations with special reference to elliptic curves, *J. Lond. Math. Soc. 41* (1966), 193–291. Corr. *42* (1967) 183.

Cassels, J. W. S. and Fröhlich, A., ed.
   1. *Algebraic Number Theory*, Thompson Book Co, Washington D.C. (1967).

Castelnuovo, G.
   1. Sui multipli di una serie lineare di gruppi di punti appartenente ad una curva algebrica, *Rend. Circ. Mat. Palermo 7* (1893). Also in *Memorie Scelte*, pp. 95–113.

Chevalley, C.
   1. Intersections of algebraic and algebroid varieties, *Trans. Amer. Math. Soc. 57* (1945), 1–85.
   2. *Anneaux de Chow et Applications*, Séminaire Chevalley, Secrétariat Math., Paris (1958).

Chow, W. L.
   1. On compact complex analytic varieties, *Amer. J. of Math. 71* (1949), 893–914; errata *72*, p. 624.
   2. On Picard varieties, *Amer. J. Math. 74* (1952), 895–909.
   3. The Jacobian variety of an algebraic curve, *Amer. J. of Math. 76* (1954), 453–476.

Chow, W. L. and Kodaira, K.
   1. On analytic surfaces with two independent meromorphic functions, *Proc. Nat. Acad. Sci. USA 38* (1952), 319–325.

Clemens, C. H. and Griffiths, P. A.
1. The intermediate Jacobian of the cubic threefold, *Annals of Math. 95* (1972), 281–356.

Deligne, P.
1. Formes modulaires et représentations *l*-adiques, *Séminaire Bourbaki 355*, Lecture Notes in Math. *179*, Springer-Verlag, Heidelberg (1971), 139–172.
2. La conjecture de Weil pour les surfaces K3, *Invent. Math. 15* (1972), 206–226.
3. La conjecture de Weil, I, *Publ. Math. IHES 43* (1974), 273–307.
4. Théorèmes de Lefschetz et critères de dégénérescence de suites spectrales, *Publ. Math. IHES 35* (1968) 107–126.
5. Les intersections complètes de niveau de Hodge un, *Invent. Math. 15* (1972) 237–250.

Deligne, P. and Mumford, D.
1. The irreducibility of the space of curves of given genus, *Publ. Math. IHES 36* (1969), 75–110.

Demazure, M.
1. Motifs des variétés algébriques, *Séminaire Bourbaki 365*, Lecture Notes in Math. *180*, Springer-Verlag, Heidelberg (1971), 19–38.

Deuring, M.
1. Die Typen der Multiplikatorenringe elliptischer Funktionenkörper, *Abh. Math. Sem. Univ. Hamburg 14* (1941), 197–272.
2. *Die Klassenkörper der Komplexen Multiplikation*, Enz. der Math. Wiss., 2nd ed., I_2, Heft 10, II, §23 (1958) 60 pp.

Dieudonné, J.
1. *Cours de Géométrie Algébrique, I. Aperçu Historique sur le Développement de la Géométrie Algébrique*, Presses Univ. France, Collection Sup. (1974), 234 pp.

Dwork, B.
1. On the rationality of the zeta function of an algebraic variety, *Amer. J. Math. 82* (1960), 631–648.

Freyd, P.
1. *Abelian Categories, an Introduction to the Theory of Functors*, Harper & Row, New York (1964), 164 pp.

Fulton, W.
1. *Algebraic Curves*, W. A. Benjamin, New York (1969), xii + 226 pp.
2. Riemann–Roch for singular varieties, in *Algebraic Geometry, Arcata 1974*, Amer. Math. Soc. Proc. Symp. Pure Math *29* (1975), 449–457.

Godement, R.
1. *Topologie Algébrique et Théorie des Faisceaux*, Hermann, Paris (1958).

Grauert, H.
1. Ein Theorem der analytischen Garbentheorie und die Modulräume komplexer Strukturen, *Pub. Math. IHES 5* (1960), 233–292.
2. Über Modifikationen und exzeptionelle analytische Mengen, *Math. Ann. 146* (1962), 331–368.

Grauert, H. and Remmert, R.
1. Komplex Räume, *Math. Ann. 136* (1958), 245–318.

Grauert, H. and Riemenschneider, O.
1. Verschwindungssätze fur analytische Kohomologiegruppen auf komplexen Räumen, *Inv. Math. 11* (1970), 263–292.

Grothendieck, A.
1. Sur quelques points d'algèbre homologique, *Tôhoku Math. J. 9* (1957), 119–221.
2. Sur une note de Mattuck-Tate, *J. Reine u. Angew. Math. 200* (1958), 208–215.
3. La théorie des classes de Chern, *Bull. Soc. Math. de France 86* (1958), 137–154.
4. The cohomology theory of abstract algebraic varieties, *Proc. Int. Cong. Math., Edinburgh* (1958), 103–118.
5. *Fondements de la Géométrie Algébrique*, Séminaire Bourbaki 1957–62, Secrétariat Math., Paris (1962).
6. Formule de Lefschetz et rationalité des fonctions *L*, *Séminaire Bourbaki 279* (1965).
7. *Local Cohomology* (notes by R. Hartshorne), Lecture Notes in Math. *41*, Springer-Verlag, Heidelberg (1967), 106 pp.
8. Crystals and the De Rham cohomology of schemes (notes by I. Coates and O. Jussila), in *Dix Exposés sur la Cohomologie des Schémas*, North-Holland, Amsterdam (1968), 306–358.
9. Standard conjectures on algebraic cycles, in *Algebraic Geometry, Bombay 1968*, Oxford University Press, Oxford (1969), 193–199.

Grothendieck, A. and Dieudonné, J.
Eléments de Géométrie Algébrique.
EGA I.   Le langage des schémas, *Publ. Math. IHES 4* (1960).
EGA II. Étude globale élémentaire de quelques classes de morphismes, *Ibid. 8* (1961).
EGA III. Étude cohomologique des faisceaux cohérents, *Ibid. 11* (1961), and *17* (1963).
EGA IV. Étude locale des schémas et des morphismes de schémas, *Ibid. 20* (1964), *24* (1965), *28* (1966), *32* (1967).
EGA I.   *Eléments de Géométrie Algébrique, I*, Grundlehren *166*, Springer-Verlag, Heidelberg (new ed., 1971), ix + 466 pp.

Grothendieck, A. et al.
Séminaire de Géométrie Algébrique.
SGA 1. *Revêtements étales et Groupe Fondemental*, Lecture Notes in Math. *224*, Springer-Verlag, Heidelberg (1971).
SGA 2. *Cohomologie Locale des Faisceaux Cohérents et Théorèmes de Lefschetz Locaux et Globaux*, North-Holland, Amsterdam (1968).
SGA 3. (with Demazure, M.) *Schémas en Groupes I, II, III*, Lecture Notes in Math. *151, 152, 153*, Springer-Verlag, Heidelberg (1970).
SGA 4. (with Artin, M. and Verdier, J. L.) *Théorie des Topos et Cohomologie Étale des Schémas*, Lecture Notes in Math. *269, 270, 305*, Springer-Verlag, Heidelberg (1972–1973).
SGA 4½. (by Deligne, P., with Boutot, J. F., Illusie, L., and Verdier, J. L.) *Cohomologie Etale*, Lecture Notes in Math. *569*, Springer-Verlag, Heidelberg (1977).
SGA 5. *Cohomologie l-adique et fonctions L* (unpublished).
SGA 6. (with Berthelot, P. and Illusie, L.) *Théorie des Intersections et Théorème de Riemann–Roch*, Lecture Notes in Math. *225*, Springer-Verlag, Heidelberg (1971).
SGA 7. (with Raynaud, M. and Rim, D. S.) *Groupes de Monodromie en Géométrie Algébrique*, Lecture Notes in Math. *288*, Springer-Verlag, Heidelberg (1972). Part II (by Deligne, P. and Katz, N.) *340* (1973).

Gunning, R. C.
1. *Lectures on Riemann Surfaces*, Princeton Math. Notes, Princeton U. Press, Princeton (1966), 256 pp.

462

Gunning, R. C. and Rossi, H.
1. *Analytic Functions of Several Complex Variables*, Prentice-Hall (1965), xii + 317 pp.

Halphen, G.
1. Mémoire sur la classification des courbes gauches algébriques, *J. Éc. Polyt. 52* (1882), 1–200.

Hartshorne, R.
1. Complete intersections and connectedness, *Amer. J. of Math. 84* (1962), 497–508.
2. *Residues and Duality*, Lecture Notes in Math. *20*, Springer-Verlag, Heidelberg (1966).
3. Cohomological dimension of algebraic varieties, *Annals of Math. 88* (1968), 403–450.
4. Curves with high self-intersection on algebraic surfaces, *Publ. Math. IHES 36* (1969), 111–125.
5. *Ample Subvarieties of Algebraic Varieties*, Lecture Notes in Math. *156*, Springer-Verlag, Heidelberg (1970), xiii + 256 pp.
6. Equivalence relations on algebraic cycles and subvarieties of small codimension, in *Algebraic Geometry, Arcata 1974*, Amer. Math. Soc. Proc. Symp. Pure Math. *29* (1975), 129–164.
7. On the De Rham cohomology of algebraic varieties, *Publ. Math. IHES 45* (1976), 5–99.

Hartshorne, R., ed.
1. *Algebraic Geometry, Arcata 1974*, Amer. Math. Soc. Proc. Symp. Pure Math. *29* (1975).

Hilbert, D. and Cohn-Vossen, S.
1. *Geometry and the Imagination*, Chelsea Pub. Co., New York (1952), ix + 357 pp. (translated from German *Anschauliche Geometrie* (1932)).

Hilton, P. J. and Stammbach, U.
1. *A Course in Homological Algebra*, Graduate Texts in Mathematics *4*, Springer-Verlag, Heidelberg (1970), ix + 338 pp.

Hironaka, H.
1. A note on algebraic geometry over ground rings. The invariance of Hilbert characteristic functions under the specialization process, *Ill. J. Math. 2* (1958), 355–366.
2. On the theory of birational blowing-up, Thesis, Harvard (1960) (unpublished).
3. On resolution of singularities (characteristic zero), *Proc. Int. Cong. Math.* (1962), 507–521.
4. Resolution of singularities of an algebraic variety over a field of characteristic zero, *Annals of Math. 79* (1964). I: 109–203; II: 205–326.
5. Triangulations of algebraic sets, in *Algebraic Geometry, Arcata 1974*, Amer. Math. Soc. Proc. Symp. Pure Math. *29* (1975), 165–184.

Hironaka, H. and Matsumura, H.
1. Formal functions and formal embeddings, *J. of Math. Soc. of Japan 20* (1968), 52–82.

Hirzebruch, F.
1. *Topological Methods in Algebraic Geometry*, Grundlehren *131*, Springer-Verlag, Heidelberg (3rd ed., 1966), ix + 232 pp.

Hirzebruch, F. and Mayer, K. H.
1. *O(n)-Mannigfaltigkeiten, Exotische Sphären und Singularitäten*, Lecture Notes in Math. *57*, Springer-Verlag, Heidelberg (1968).

Hodge, W. V. D.
1. *The Theory and Applications of Harmonic Integrals*, Cambridge Univ. Press, Cambridge (2nd ed., 1952), 282 pp.

Horrocks, G. and Mumford, D.
1. A rank 2 vector bundle on P^4 with 15,000 symmetries, *Topology 12* (1973), 63–81.

Hurwitz, A.
1. Über algebraische Gebilde mit eindeutigen Transformationen in sich, *Math. Ann. 41* (1893), 403–442.

Hurwitz, A. and Courant, R.
1. *Allgemeine Funktionentheorie und elliptische Funktionen*; *Geometrische Funktionentheorie*, Grundlehren *3*, Springer-Verlag, Heidelberg (1922), xi + 399 pp.

Igusa, J.-I.
1. Arithmetic genera of normal varieties in an algebraic family, *Proc. Nat. Acad. Sci. USA 41* (1955) 34–37.
2. Class number of a definite quaternion with prime discriminant, *Proc. Nat. Acad. Sci. USA 44* (1958) 312–314.

Iskovskih, V. A. and Manin, Ju. I.
1. Three-dimensional quartics and counterexamples to the Lüroth problem, *Math. USSR—Sbornik 15* (1971) 141–166.

Jouanolou, J. P.
1. Riemann–Roch sans dénominateurs, *Invent. Math. 11* (1970), 15–26.

Kleiman, S. L.
1. Toward a numerical theory of ampleness, *Annals of Math. 84* (1966), 293–344.
2. Algebraic cycles and the Weil conjectures, in *Dix Exposés sur la Cohomologie des Schémas*, North-Holland, Amsterdam (1968), 359–386.
3. The transversality of a general translate, *Compos. Math. 28* (1974), 287–297.

Kleiman, S. L. and Laksov, D.
1. Another proof of the existence of special divisors, *Acta Math. 132* (1974), 163–176.

Klein, F.
1. Ueber die Transformationen siebenter Ordnung der elliptischen Funktionen, *Math. Ann. 14* (1879). Also in Klein, *Ges. Math. Abh. 3* (1923), 90–136.
2. *Lectures on the Icosahedron and the Solution of Equations of the Fifth Degree*, Kegan Paul, Trench, Trübner, London (1913). Dover reprint (1956).

Knutson, D.
1. *Algebraic spaces*, Lecture Notes in Math. *203*, Springer-Verlag, Heidelberg (1971) vi + 261 pp.

Kodaira, K.
1. On a differential-geometric method in the theory of analytic stacks, *Proc. Nat. Acad. Sci. USA 39* (1953), 1268–1273.
2. On Kähler varieties of restricted type. (An intrinsic characterization of algebraic varieties.), *Annals of Math. 60* (1954), 28–48.

Kodaira, K. and Spencer, D. C.
1. On arithmetic genera of algebraic varieties, *Proc. Nat. Acad. Sci. USA 39* (1953), 641–649.

Kunz, E.
1. Holomorphe Differentialformen auf algebraischen Varietäten mit Singularitäten, I, *Manus. Math. 15* (1975) 91–108.

Lang, S.
1. *Abelian Varieties*, Interscience Pub., New York (1959), xii + 256 pp.
2. *Algebra*, Addison-Wesley (1971), xvii + 526 pp.

Lang, S. and Néron, A.
1. Rational points of abelian varieties over function fields, *Amer. J. Math. 81* (1959), 95–118.

Lang, S. and Trotter, H.
1. *Frobenius Distributions in* GL$_2$-*Extensions*, Lecture Notes in Math. *504*, Springer-Verlag, Heidelberg (1976).

Lang, S. and Weil, A.
1. Number of points of varieties in finite fields, *Amer. J. Math. 76* (1954), 819–827.

Lascu, A. T., Mumford, D., and Scott, D. B.
1. The self-intersection formula and the "formule-clef," *Math. Proc. Camb. Phil. Soc. 78* (1975), 117–123.

Lichtenbaum, S. and Schlessinger, M.
1. The cotangent complex of a morphism, *Trans. Amer. Math. Soc. 128* (1967), 41–70.

Lipman, J.
1. Introduction to resolution of singularities, in *Algebraic Geometry, Arcata 1974*, Amer. Math. Soc. Proc. Symp. Pure Math. *29* (1975), 187–230.
2. Rational singularities with applications to algebraic surfaces and unique factorization, *Publ. Math. IHES 36* (1969) 195–279
3. Unique factorization in complete local rings, in *Algebraic Geometry, Arcata 1974*, Amer. Math. Soc. Proc. Symp. Pure Math. *29* (1975) 531–546.

Lubkin, S.
1. A *p*-adic proof of Weil's conjectures, *Annals of Math. 87* (1968), 105–194, and *87* (1968), 195–255.

Macbeath, A. M.
1. On a theorem of Hurwitz, *Proc. Glasgow Math. Assoc. 5* (1961) 90–96.

Manin, Yu. I.
1. *Lectures on the K-functor in Algebraic Geometry*, Russian Mathematical Surveys *24* (5) (1969), 1–89.
2. Correspondences, motifs, and monoidal transformations, *Math USSR—Sbornik 6* (1968) 439–470.
3. *Cubic forms: Algebra, Geometry, Arithmetic*, North-Holland, Amsterdam (1974), vii + 292 pp.

Maruyama, M.
1. *On Classification of Ruled Surfaces*, Kyoto Univ., Lectures in Math. *3*, Kinokuniya, Tokyo (1970).

Matsumura, H.
1. Geometric structure of the cohomology rings in abstract algebraic geometry, *Mem. Coll. Sci. Univ. Kyoto (A) 32* (1959), 33–84.
2. *Commutative Algebra*, W. A. Benjamin Co., New York (1970), xii + 262 pp.

Mattuck, A. and Tate, J.
1. On the inequality of Castelnuovo-Severi, *Abh. Math. Sem. Univ. Hamburg 22* (1958), 295–299.

Moishezon, B. G.
1. A criterion for projectivity of complete algebraic abstract varieties, *Amer. Math. Soc. Translations 63* (1967), 1–50.

465

2. On n-dimensional compact varieties with n algebraically independent meromorphic functions, *Amer. Math. Soc. Translations 63* (1967), 51–177.

Morrow, J. and Kodaira, K.
1. *Complex Manifolds*, Holt, Rinehart & Winston, New York (1971), vii + 192 pp.

Mumford, D.
1. *Geometric Invariant Theory*, Ergebnisse, Springer-Verlag, Heidelberg (1965), vi + 146 pp.
2. *Lectures on Curves on an Algebraic Surface*, Annals of Math. Studies *59*, Princeton U. Press, Princeton (1966).
3. Pathologies, III, *Amer. J. Math. 89* (1967), 94–104.
4. Varieties defined by quadratic equations (with an Appendix by G. Kempf), in *Questions on Algebraic Varieties*, Centro Internationale Matematica Estivo, Cremonese, Rome (1970), 29–100.
5. *Abelian Varieties*, Oxford Univ. Press, Oxford (1970), ix + 242 pp.
6. The topology of normal singularities of an algebraic surface and a criterion for simplicity, *Publ. Math. IHES 9* (1961) 5–22.

Nagata, M.
1. On the embedding problem of abstract varieties in projective varieties, *Mem. Coll. Sci. Kyoto (A) 30* (1956), 71–82.
2. A general theory of algebraic geometry over Dedekind domains, *Amer. J. of Math. 78* (1956), 78–116.
3. On the imbeddings of abstract surfaces in projective varieties, *Mem. Coll. Sci. Kyoto (A) 30* (1957), 231–235.
4. Existence theorems for non-projective complete algebraic varieties, *Ill. J. Math. 2* (1958), 490–498.
5. On rational surfaces I, II, *Mem. Coll. Sci. Kyoto (A) 32* (1960), 351–370, and *33* (1960), 271–293.
6. Imbedding of an abstract variety in a complete variety, *J. Math. Kyoto Univ. 2* (1962), 1–10.
7. *Local Rings*, Interscience Tracts in Pure & Applied Math. *13*, J. Wiley, New York (1962).
8. On self-intersection number of a section on a ruled surface, *Nagoya Math. J. 37* (1970), 191–196.

Nakai, Y.
1. A criterion of an ample sheaf on a projective scheme, *Amer. J. Math. 85* (1963), 14–26.
2. Some fundamental lemmas on projective schemes, *Trans. Amer. Math. Soc. 109* (1963), 296–302.

Nakano, S.
1. On complex analytic vector bundles, *J. Math. Soc. Japan, 7* (1955) 1–12.

Narasimhan, M. S. and Seshadri, C. S.
1. Stable and unitary vector bundles on a compact Riemann surface, *Annals of Math. 82* (1965), 540–567.

Noether, M.
1. *Zur Grundlegung der Theorie der Algebraischen Raumcurven*, Verlag der Königlichen Akademie der Wissenschaften, Berlin (1883).

Olson, L.
1. An elementary proof that elliptic curves are abelian varieties, *Ens. Math. 19* (1973), 173–181.

Ramanujam, C. P.
1. Remarks on the Kodaira vanishing theorem, *J. Indian Math. Soc. (N.S.) 36* (1972), 41–51.

Ramis, J. P. and Ruget, G.
1. Complexes dualisants et théorème de dualité en géométrie analytique complexe, *Pub. Math. IHES 38* (1970), 77–91.

Ramis, J. P., Ruget, G., and Verdier, J. L.
1. Dualité relative en géométrie analytique complexe, *Invent. Math. 13* (1971), 261–283.

Roberts, J.
1. Chow's moving lemma, in *Algebraic Geometry, Oslo 1970* (F. Oort, ed.), Wolters-Noordhoff (1972), 89–96.

Roquette, P.
1. Abschätzung der Automorphismenanzahl von Funktionenkörpern bei Primzahlcharakteristik, *Math. Zeit. 117* (1970) 157–163.

Rotman, J. J.
1. *Notes on Homological Algebra*, Van Nostrand Reinhold Math. Studies *26*, New York (1970).

Saint-Donat, B.
1. On Petri's analysis of the linear system of quadrics through a canonical curve, *Math. Ann. 206* (1973), 157–175.

Sally, J.
1. Regular overrings of regular local rings, *Trans. Amer. Math. Soc. 171* (1972) 291–300.

Samuel, P.
1. *Méthodes d'Algèbre Abstraite en Géométrie Algébrique*, Ergebnisse *4*, Springer-Verlag, Heidelberg (1955).
2. *Lectures on old and new results on algebraic curves* (notes by S. Anantharaman), Tata Inst. Fund. Res. (1966), 127 pp.
3. *Anneaux Factoriels* (rédaction de A. Micali), Soc. Mat. de São Paulo (1963) 97 pp.

Schlessinger, M.
1. Functors of Artin rings, *Trans. Amer. Math. Soc. 130* (1968), 208–222.

Serre, J.-P.
1. Cohomologie et géométrie algébrique, *Proc. ICM* (1954), vol. III, 515–520.
2. Un théorème de dualité, *Comm. Math. Helv. 29* (1955), 9–26.
3. Faisceaux algébriques cohérents, *Ann. of Math. 61* (1955), 197–278.
4. Géométrie algébrique et géométrie analytique, *Ann. Inst. Fourier 6* (1956), 1–42.
5. Sur la cohomologie des variétés algébriques, *J. de Maths. Pures et Appl. 36* (1957), 1–16.
6. Sur la topologie des variétés algébriques en caractéristique *p*, *Symposium Int. de Topologia Algebraica, Mexico* (1958), 24–53.
7. *Groupes Algébriques et Corps de Classes*, Hermann, Paris (1959).
8. Rationalité des fonctions ζ des variétés algébriques (d'après B. Dwork) *Séminaire Bourbaki 198* (1960).
9. Analogues kählériens de certaines conjectures de Weil, *Annals of Math. 71* (1960), 392–394.
10. Zeta and *L* functions, in *Arithmetical Algebraic Geometry* (Schilling, ed.), Harper & Row, New York (1965), 82–92.
11. *Algèbre Locale—Multiplicités* (rédigé par P. Gabriel), Lectures Notes in Math. *11*, Springer-Verlag, Heidelberg (1965).

12. Prolongement de faisceaux analytiques cohérents, *Ann. Inst. Fourier 16* (1966), 363–374.
13. Critère de rationalité pour les surfaces algébriques (d'après K. Kodaira), *Séminaire Bourbaki 146* (1957).
14. *A Course in Arithmetic*, Graduate Texts in Math. 7, Springer-Verlag, Heidelberg (1973) 115 pp.

Severi, F.
1. Intorno ai punti doppi impropri di una superficie generale dello spazio a quattro dimensioni, e a suoi punti tripli apparenti, *Rend. Circ. Matem. Palermo 15* (1901), 33–51.
2. *Vorlesungen über Algebraische Geometrie* (transl. by E. Löffler), Johnson Pub. (rpt., 1968; 1st. ed., Leipzig 1921).
3. Über die Grundlagen der algebraischen Geometrie, *Hamb. Abh. 9* (1933) 335–364.

Shafarevich, I. R.
1. *Algebraic surfaces*, Proc. Steklov Inst. Math. 75 (1965) (trans. by A.M.S. 1967).
2. *Basic Algebraic Geometry*, Grundlehren 213, Springer-Verlag, Heidelberg (1974), xv + 439 pp.

Shannon, D. L.
1. Monoidal transforms of regular local rings, *Amer. J. Math. 95* (1973) 294–320.

Shioda, T.
1. An example of unirational surfaces in characteristic p, *Math. Ann. 211* (1974) 233–236.

Siegel, C. L.
1. Meromorphe Funktionen auf kompakten analytischen Mannigfaltigkeiten, *Nach. Akad. Wiss. Göttingen* (1955), 71–77.

Singh, B.
1. On the group of automorphisms of a function field of genus at least two, *J. Pure Appl. Math. 4* (1974) 205–229.

Spanier, E. H.
1. *Algebraic Topology*, McGraw-Hill, New York (1966).

Stichtenoth, H.
1. Über die Automorphismengruppe eines algebraischen Funktionenkörpers von Primzahlcharakteristik, *Archiv der Math. 24* (1973) 527–544.

Suominen, K.
1. Duality for coherent sheaves on analytic manifolds, *Ann. Acad. Sci. Fenn (A) 424* (1968), 1–19.

Tate, J. T.
1. Algebraic cycles and poles of the zeta function, in *Arithmetical Algebraic Geometry* (Schilling, ed.), Harper & Row, New York (1965), 93–110.
2. Residues of differentials on curves, *Ann. Sci. de l'E.N.S.* (4) *1* (1968), 149–159.
3. The arithmetic of elliptic curves, *Inv. Math. 23* (1974), 179–206.

Tjurin, A. N.
1. On the classification of two-dimensional fibre bundles over an algebraic curve of arbitrary genus (in Russian) *Izv. Akad. Nauk SSSR Ser. Mat. 28* (1964) 21–52; **MR** *29* (1965) # 4762.
2. Classification of vector bundles over an algebraic curve of arbitrary genus, *Amer. Math. Soc. Translations 63* (1967) 245–279.

Verdier, J.-L.
1. Base change for twisted inverse image of coherent sheaves, in *Algebraic Geometry, Bombay 1968*, Oxford Univ. Press, Oxford (1969), 393–408.

468

Vitushkin, A. G.
1. On polynomial transformation of C^n, in *Manifolds, Tokyo 1973*, Tokyo Univ. Press, Tokyo (1975), 415–417.

Walker, R. J.
1. *Algebraic Curves*, Princeton Univ., Princeton (1950), Dover reprint (1962).

van der Waerden, B. L.
1. *Modern Algebra*, Frederick Ungar Pub. Co, New York: I (1953), xii + 264 pp.; II (1950), ix + 222 pp.

Weil, A.
1. *Foundations of Algebraic Geometry*, Amer. Math. Soc., Colloquium Publ. *29* (1946) (revised and enlarged edition 1962), xx + 363 pp.
2. *Sur les Courbes Algébriques et les Variétés qui s'en Déduisent*, Hermann, Paris (1948). This volume and the next have been republished in one volume, *Courbes Algébriques et Variétés Abéliennes*, Hermann, Paris (1971), 249 pp.
3. *Variétés Abéliennes et Courbes Algébriques*, Hermann, Paris (1948).
4. Number of solutions of equations over finite fields, *Bull. Amer. Math. Soc. 55* (1949), 497–508.
5. *Variétés Kählériennes*, Hermann, Paris (1958), 175 pp.
6. On the projective embedding of abelian varieties, in *Algebraic Geometry and Topology (in honor of S. Lefschetz)*, Princeton Univ., Princeton (1957) 177–181.

Wells, R. O., Jr.
1. *Differential Analysis on Complex Manifolds*, Prentice-Hall (1973), x + 252 pp.

Weyl, H.
1. *Die Idee der Riemannschen Fläche*, Teubner (3rd ed., 1955), vii + 162 pp. (1st ed., 1913).

Zariski, O.
1. The concept of a simple point on an abstract algebraic variety, *Trans. Amer. Math. Soc. 62* (1947), 1–52.
2. A simple analytical proof of a fundamental property of birational transformations, *Proc. Nat. Acad. Sci. USA 35* (1949), 62–66.
3. *Theory and Applications of Holomorphic Functions on Algebraic Varieties over Arbitrary Ground Fields*, Memoirs of Amer. Math. Soc. New York (1951).
4. Complete linear systems on normal varieties and a generalization of a lemma of Enriques-Severi, *Ann. of Math. 55* (1952), 552–592.
5. *Introduction to the Problem of Minimal Models in the Theory of Algebraic Surfaces*, Pub. Math. Soc. of Japan *4* (1958), vii + 89 pp.
6. The problem of minimal models in the theory of algebraic surfaces, *Amer. J. Math. 80* (1958), 146–184.
7. The theorem of Riemann-Roch for high multiples of an effective divisor on an algebraic surface, *Ann. Math. 76* (1962), 560–615.
8. *Collected papers. Vol. I. Foundations of Algebraic Geometry and Resolution of Singularities*, ed. H. Hironaka and D. Mumford, M.I.T. Press, Cambridge (1972), *xxi* + 543 pp. Vol. II, *Holomorphic Functions and Linear Systems*, ed. M. Artin and D. Mumford, M.I.T. Press (1973), xxiii + 615 pp.
9. On Castelnuovo's criterion of rationality $p_a = P_2 = 0$ of an algebraic surface, *Ill. J. Math. 2* (1958) 303–315.
10. *Algebraic Surfaces*, 2nd suppl. ed., Ergebnisse *61*, Springer-Verlag, Heidelberg (1971).

Zariski, O. and Samuel, P.
1. *Commutative Algebra (Vol. I, II)*, Van Nostrand, Princeton (1958, 1960).

Results from Algebra

470

Glossary of Notations

Reg Y	set of nonsingular points,	54
$\mathbf{Q}$	the rational numbers,	58
Spec A	spectrum of a ring,	59
$\mathfrak{Top}(X)$	category of open sets of X,	61
$\mathfrak{Ab}$	category of abelian groups,	61
$\Gamma(U,\mathscr{F})$	sections of a sheaf,	61
ker	kernel,	63
coker	cokernel,	63
im	image,	63
$f_*\mathscr{F}$	direct image sheaf,	65
$\mathfrak{Ab}(X)$	category of sheaves of abelian groups on X,	65
$\mathscr{F}\oplus\mathscr{G}$	direct sum of sheaves,	66
$\varinjlim \mathscr{F}_i$	direct limit of sheaves,	66
$\varprojlim \mathscr{F}_i$	inverse limit of sheaves,	67
$\mathrm{Sp\acute{e}}(\mathscr{F})$	espace étalé of a presheaf,	67
Supp s	support of a section of a sheaf,	67
Supp $\mathscr{F}$	support of a sheaf,	67
$\mathscr{H}om(\mathscr{F},\mathscr{G})$	sheaf of local morphisms,	67
$j_!(\mathscr{F})$	extension of a sheaf by zero,	68
$\Gamma_Z(X,\mathscr{F})$	sections with support in Z,	68
$\mathscr{H}_Z^0(\mathscr{F})$	subsheaf with support in Z,	68
$\mathscr{O}_X$	sheaf of regular functions on a variety,	68
$\mathscr{I}_Y$	sheaf of ideals of a subvariety,	69
Spec A	spectrum of a ring,	70
$\mathscr{O}$	sheaf of rings,	70
$V(\mathfrak{a})$	closed subset of an ideal,	70, 76
$D(f)$	open subset of Spec A,	70
$(X,\mathscr{O}_X)$	scheme,	74
sp(X)	space of X,	74
$\mathbf{A}_k^1$	affine line (as a scheme),	74
S_+	ideal of positive elements,	76
Proj S	Proj of a graded ring,	76
$D_+(f)$	open subset of Proj,	76
$\mathbf{P}_A^n$	projective n-space over a ring,	77
$\mathfrak{Sch}(S)$	category of schemes over S,	78
$\mathfrak{Var}(k)$	category of varieties over k,	78
$t(V)$	scheme associated to a variety,	78
$(\mathscr{O}_X)_{\mathrm{red}}$	associated reduced sheaf of rings,	79
A_{red}	reduced ring,	79
X_{red}	reduced scheme,	79
$\mathscr{O}_x$	local ring of a point on a scheme,	80
$\mathfrak{m}_x$	maximal ideal of local ring at x,	80
$k(x)$	residue field at x,	80
T_x	Zariski tangent space at x,	80
$\mathbf{C}$	complex numbers,	80
$\mathbf{F}_p$	finite field of p elements,	80
X_f	open set defined by f,	81
nil A	nilradical of a ring,	82
Y_n	nth infinitesimal neighborhood,	85

$\dim X$	dimension of a scheme,	86
$\mathrm{codim}(Z,X)$	codimension of a subscheme,	86
$X \times_S Y$	fibred product of schemes,	87
$X \times Y$	product of schemes,	87
$K(X)$	function field of an integral scheme,	91
k_s	separable closure of a field,	93
k_p	perfect closure of a field,	93
$x_1 \rightsquigarrow x_2$	specialization,	93
Δ	diagonal morphism,	96
Γ_f	graph morphism,	106
$\mathrm{Hom}_{\mathcal{O}_X}(\mathscr{F},\mathscr{G})$	group of morphisms of sheaves of $\mathcal{O}_X$-modules,	109
$\mathscr{H}om_{\mathcal{O}_X}(\mathscr{F},\mathscr{G})$	sheaf Hom,	109
$\mathscr{F} \otimes_{\mathcal{O}_X} \mathscr{G}$	tensor product,	109
$f^*\mathscr{G}$	inverse image sheaf,	110
$\tilde{M}$	sheaf associated to an A-module,	110
$\mathscr{I}_Y$	ideal sheaf of a closed subscheme,	115
$\tilde{M}$	sheaf associated to a graded S-module,	116
$M_{(\mathfrak{p})}$	degree zero localization,	116
$M_{(f)}$	degree zero localization,	117
$\mathcal{O}_X(1)$	twisting sheaf,	117
$\mathscr{F}(n)$	twisted sheaf,	117
$\Gamma_*(\mathscr{F})$	graded module associated to a sheaf,	118
X_f	open set defined by a section of an invertible sheaf,	118
$\check{\mathscr{E}}$	dual of a locally free sheaf,	123
$\mathrm{Ann}\,m$	annihilator of an element of a module,	124
$\Gamma_{\mathfrak{a}}(M)$	submodule with supports in $\mathfrak{a}$,	124
$S^{(d)}$	the graded ring $\bigoplus_{n \geqslant 0} S_{nd}$,	126
$T(M)$	tensor algebra of M,	127
$S(M)$	symmetric algebra of M,	127
$\wedge(M)$	exterior algebra of M,	127
Spec $\mathscr{A}$	spectrum of a sheaf of algebras,	128
$\mathbf{V}(\mathscr{E})$	vector bundle associated to a locally free sheaf,	128
$\mathrm{Div}(X)$	group of divisors,	130
v_Y	valuation of a prime divisor,	130
(f)	divisor of a rational function,	131
K^*	multiplicative group of a field,	131
$D \sim D'$	linear equivalence of divisors,	131
$\mathrm{Cl}\,X$	divisor class group,	131
$\deg D$	degree of a divisor,	132
$[K(X):K(Y)]$	degree of a finite field extension,	137
f^*	inverse image for divisors,	137
$\mathrm{Cl}^{\circ}X$	divisor class group of degree 0,	139
$\mathscr{K}^*$	sheaf of invertible elements,	141
$\mathrm{CaCl}\,X$	Cartier divisor class group,	142
$\mathrm{Pic}\,X$	Picard group of X,	143
$\mathscr{L}(D)$	sheaf associated to a Cartier divisor,	144
$K(X)$	Grothendieck group,	148
$\gamma(\mathscr{F})$	image of a sheaf in the Grothendieck group,	148
$\mathrm{PGL}(n,k)$	projective general linear group,	151
$\mathrm{GL}(n,k)$	general linear group,	151

$l(D)$	dimension of $H^0(X, \mathscr{L}(D))$,	295
K	canonical divisor on a curve,	295
X_{reg}	set of regular points of a curve,	298
δ_P	measure of a curve singularity,	298
e_P	ramification index,	299
dt/du	quotient of differentials,	300
R	ramification divisor,	301
X_p	modified scheme in characteristic p,	302
$\text{Sec}\,X$	secant variety,	310
$\text{Tan}\,X$	tangent variety,	310
j	the j-invariant of an elliptic curve,	317
$\text{Aut}\,X$	group of automorphisms of an elliptic curve,	318
Σ_n	symmetric group on n letters,	318
ω	a cube root of 1,	320
$\text{Aut}(X, P_0)$	group of automorphisms leaving P_0 fixed,	320
n_X	multiplication by n,	322
$\text{End}(X, P_0)$	ring of endomorphisms,	323
$\text{Pic}^\circ(X/T)$	relative Picard group,	323
e	identity section of a group scheme,	324
ρ	inverse morphism of a group scheme,	324
μ	multiplication of a group scheme,	324
$\wp(z)$	Weierstrass $\wp$-function,	327
$J(\tau)$	J-invariant of a lattice,	328
$\mathbf{Z}[i]$	ring of Gaussian integers,	331
$\pi_1(X)$	fundamental group,	338
$\mathbf{Z}_l$	l-adic integers,	338
g_d^r	a linear system of dimension r and degree d,	341
$\mathfrak{M}_g$	set of all curves of genus g,	345
$C.D$	intersection number,	357
$\deg_C$	degree of an invertible sheaf on a curve,	358
$(C.D)_P$	intersection multiplicity at P,	360
D^2	self-intersection number,	360
K^2	self-intersection of the canonical divisor,	361
$l(D)$	dimension of $H^0(X, \mathscr{L}(D))$,	362
$s(D)$	superabundance,	362
c_2	second Chern class,	363
$D \equiv E$	numerical equivalence of divisors,	364
$\text{Pic}^n X$	divisor classes numerically equivalent to 0,	364
$\text{Num}\,X$	group of divisors modulo numerical equivalence,	364
$c(D)$	cohomology class of a divisor,	367
$\pi: X \to C$	a ruled surface,	369
f	a fibre of a ruled surface,	369
e	invariant of a ruled surface,	372
$\mathfrak{e}$	divisor of $\wedge^2 \mathscr{E}$,	373
$\pi: \tilde{X} \to X$	a monoidal transformation,	386
E	the exceptional curve,	386
$\mu_P(C)$	multiplicity of a curve C at a point P,	388
$f^{-1}(Y)$	reduced inverse image divisor,	391
$\text{gr}_{\mathfrak{m}}(A)$	associated graded ring of a local ring,	394

$\mu(A)$ multiplicity of a local ring,   394
$\pi : X \to \mathbf{P}^2$ projection of a cubic surface,   401
$E_1, \ldots, E_6$ the exceptional curves,   401
$e_1, \ldots, e_6$ their linear equivalence classes,   401
l the class of a line,   401
$\mathbf{E}_6$ a Weyl group,   405
$\mathbf{A}_n$ a Weyl group,   408
$T(Z)$ total transform of Z by T,   410
$\mathrm{elm}_P X$ elementary transformation of a ruled surface,   416
$\kappa(X)$ Kodaira dimension,   421
P_2 second plurigenus,   422
$f_*(Y)$ direct image cycle,   425
$A^r(X)$ cycles modulo rational equivalence,   426
$A(X)$ Chow ring,   426
$Y.Z$ intersection cycle class,   426
$i(Y, Z; W_j)$ local intersection multiplicity,   427
$c_i(\mathscr{E})$ ith Chern class,   429
$c(\mathscr{E})$ total Chern class,   429
$c_t(\mathscr{E})$ Chern polynomial,   429
$\mathrm{ch}(\mathscr{E})$ exponential Chern character,   431
$\mathrm{td}(\mathscr{E})$ Todd class,   432
$\eta(Y)$ cohomology class of Y,   435
$Y \sim_{\mathrm{hom}} 0$ homological equivalence,   435
$f^!$ inverse image in $K(X)$,   435
$f_!$ direct image in $K(X)$,   436
X_h associated complex analytic space,   439
N_r number of points of $\bar{X}$ rational over $\mathbf{F}_{q^r}$,   449
$Z(t)$ the zeta function,   450
E self-intersection of the diagonal,   450
B_i ith Betti number,   451
$\mathbf{Z}_l$ l-adic integers,   453
$\mathbf{Q}_l$ l-adic numbers,   453
$H^i(X, \mathbf{Q}_l)$ l-adic cohomology,   453
$L(f, X)$ number of fixed points of a morphism,   453
$\zeta(s)$ Riemann zeta function,   457

Index

Index

Index

Index

Index

Valuation, 39, 43, 299
  center of, 106
  of a prime divisor, 130, 135
Valuation ring, 40, 97, 101, 105. *See also*
  Discrete valuation ring
  examples of, 108, 420
  nondiscrete, 108
Valuative criterion of properness, 101, 107,
  181, 259, 325, 410
Valuative criterion of separatedness, 97, 107
Vanishing theorem
  of Grothendieck, 208
  of Kodaira, 248, 249, 408, 423, 424, 445
  of Nakano, 445
  of Serre, 228, 424
Variety
  abelian. *See* Abelian variety
  abstract, 58, 105
  affine. *See* Affine variety
  algebraic family of. *See* Family
  complete. *See* Complete variety
  isomorphisms of, 16, 18
  morphisms of, 14–23
  nonprojective, 171, 443 (Fig)
  normal. *See* Normal
  of moduli, 56, 58, 266, 317, 346, 421
  over k, 15, 78
  projective. *See* Projective variety
  projectively normal. *See* Projectively
    normal
  quasi-affine, 3, 21, 223
  rational. *See* Rational variety
  scheme associated to, 78, 84, 104
Vector bundle, 128, 170. *See also* Locally
  free sheaf
  sheaf of sections of, $\mathscr{S}(X/Y)$, 129
Verdier, Jean-Louis, 249
Veronese surface, 13, 170, 316, 381
Vertex of a cone, 37, 394. *See also* Cone
Very ample divisors, 307, 308, 358
  form a cone, 359
Very ample invertible sheaf, 120, 126,
  153–156, 228, 236
  on a curve, 307
  on **Proj** $\mathscr{S}$, 161

Very flat family, 266
Virtual arithmetic genus, 367
Vitushkin, A. G., 23

Walker, R. J., 391, 393
Weierstrass $\wp$-function, 327
Weil, André, 48, 58, 105, 248, 368, 428, 435,
  445, 449–452, 454
Weil cohomology, 453
Weil conjectures, 449–458
  cohomological interpretation of, 454–457
Weil divisor, 130–136, 294, 426, 428, 436.
  See also Divisor
Weyl group, 405, 408
Weyl, Hermann, 441
Wild ramification, 299
Witt vectors, 452

Zariski, Oscar, 32, 57, 58, 60, 105, 170, 190,
  244, 276, 356, 391, 393, 394, 410, 419,
  422
Zariski's Main Theorem, 57, 276, 280, 410,
  412
Zariski space, 93, 94, 213
Zariski tangent space, 37, 80, 158, 265, 270,
  324
Zariski topology, 2, 5, 7, 10, 14, 70
  base of open affine subsets, 25
  is not Hausdorff, 2, 95
  weaker than usual topology, 439
Zeros
  common, 35
  of a polynomial, 2, 9
  of a rational function, 130, 131
  of a section of a locally free sheaf, 157,
  431
Zeta function, 449–452, 455–458
  functional equation of, 450
  of a curve, 458
  of an elliptic curve, 458
  of P^1, 450, 451
  of P^n, 457
  of Riemann, 457
  rationality of, 449, 450, 452

Printed in the United States
By Bookmasters